Earth Systems Data and Models

Volume 9

Series Editor

Avelino Núñez-Delgado, Soil Science and Agricultural Chemistry, University of Santiago de Compostela, Lugo, Spain

The book series Earth Systems Data & Models publishes state-of-the-art research and technologies aimed at understanding processes and interactions in the earth system. A special emphasis is given to theory, methods, and tools used in earth, planetary and environmental sciences for: modeling, observation and analysis; data generation, assimilation and visualization; forecasting and simulation; and optimization. Topics in the series include but are not limited to: numerical, data-driven and agent-based modeling of the earth system; uncertainty analysis of models; geodynamic simulations, climate change, weather forecasting, hydroinformatics, and complex ecological models; model evaluation for decision-making processes and other earth science applications; and remote sensing and GIS technology. The series publishes monographs, edited volumes and selected conference proceedings addressing an interdisciplinary audience, which not only includes geologists, hydrologists, meteorologists, chemists, biologists and ecologists but also physicists, engineers and applied mathematicians, as well as policy makers who use model outputs as the basis of decision-making processes.

Dulal Chandra Naskar

Advances in Engineering Geophysics: Concepts and Applications

Innovative Methods and Multidisciplinary Applications for Subsurface Investigation and Infrastructure Resilience

Springer

Dulal Chandra Naskar
Formerly Geological Survey of India
Kolkata, West Bengal, India

DMT Pvt. Solution Ltd.,
Kolkata
West Bengal, India

National Centre for Seismology
New Delhi, Delhi
India

ISSN 2364-5830 ISSN 2364-5849 (electronic)
Earth Systems Data and Models
ISBN 978-3-032-27999-6 ISBN 978-3-032-28000-8 (eBook)
https://doi.org/10.1007/978-3-032-28000-8

This Springer imprint is published by the registered company Springer Nature Switzerland AG
The registered company address is: Gewerbestrasse 11, 6330 Cham, Switzerland

Dedicated to
My beloved elder brother,
Late Shri Gopal Chandra Naskar

Foreword

In an era defined by the confluence of environmental uncertainty, infrastructure vulnerability, and digital transformation, the relevance of engineering geophysics has never been more profound. From non-invasive diagnostics of buried infrastructure to advanced sensing for defense installations and climate resilience, geophysics is evolving into a vital pillar of sustainable development and strategic planning.

This book, *Advances in Engineering Geophysics: Concepts and Applications*, authored by Dr. Dulal Chandra Naskar, is both timely and visionary. With a career that spans more than 35 years across the Geological Survey of India and National Centre for Seismology, Dr. Naskar brings a rare combination of field expertise, policy engagement, and academic depth to the subject. His insights are grounded in decades of geophysical exploration, research, and leadership—ranging from foundational studies in seismology and geoelectrics to cutting-edge work in infrastructure geophysics and hazard mitigation.

The book sets itself apart by offering a comprehensive synthesis of classical geophysical techniques alongside emerging trends such as quantum sensing, AI-enabled interpretation, autonomous survey platforms, and digital twin technology. Its structure is designed to serve not only as a technical reference, but also as a roadmap for interdisciplinary integration, particularly in areas like environmental science, archaeology, defense, urban development, and energy systems.

A defining strength of this work is its commitment to bridging theory with real-world application. Through richly detailed chapters, supported by case studies from India and beyond, the author successfully demonstrates how geophysics can inform decision-making in a variety of sectors—from planning resilient metro systems to mapping cultural heritage without excavation.

Another notable contribution of this volume is its alignment with the Sustainable Development Goals (SDGs), where geophysical science is shown to be instrumental in achieving cleaner water, safer cities, and sustainable energy systems. Dr. Naskar's framing of geophysics within a sustainability and ethical context will be particularly valuable to students, practitioners, and policy architects navigating the increasingly interconnected demands of modern science and society.

This book is an essential resource for students, researchers, policymakers, and practitioners involved in geoscience, engineering, disaster management, and environmental studies. It not only illuminates the vast potential of subsurface exploration technologies but also reaffirms the critical role of scientific expertise in shaping a resilient and informed future.

Dr. Santanu Bhattacharya
Deputy Director General
Geological Survey of India
Ministry of Mines (MoM)
Government of India
Kolkata, India

Preface

The ever-increasing complexity of global challenges—ranging from natural disasters, infrastructure vulnerability, energy security, heritage preservation, to climate change adaptation—demands advanced tools and integrated approaches to better understand and manage the Earth's subsurface environment. Engineering geophysics has emerged as a pivotal discipline that bridges scientific theory with practical applications, enabling the non-invasive investigation, monitoring, and management of subsurface conditions critical to society, industry, and national security.

This book, *Advances in Engineering Geophysics: Concepts and Applications*, is the culmination of my lifelong engagement with geophysical sciences and their diverse applications in national and international contexts. Having served as the Director (Geophysics) of the Geological Survey of India (GSI) and presently working as a Senior Consultant at the National Centre for Seismology (NCS) under the Ministry of Earth Sciences, Government of India, I have had the privilege to witness and contribute to the evolution of geophysical technologies, policies, and practices spanning more than three decades. This book embodies that accumulated knowledge and experience, distilled into a comprehensive resource aimed at researchers, professionals, policymakers, and students alike.

The motivation for this book arises from the urgent need to expand the frontiers of engineering geophysics beyond traditional boundaries. While classical geophysical methods have been extensively applied to resource exploration, this volume emphasizes their critical roles in emerging domains such as infrastructure resilience, defense and security, renewable energy integration, environmental remediation, and cultural heritage conservation. In addition, the rapid technological advances in data acquisition, machine learning, robotics, quantum sensing, and digital transformation have opened new horizons for subsurface intelligence. This work systematically explores these innovations while addressing their operational challenges and ethical considerations.

The book is structured into 18 chapters, progressing from fundamental principles to specialized applications, interdisciplinary collaborations, and emerging trends. Key highlights include: Comprehensive coverage of geophysical techniques tailored to engineering and environmental contexts; Case studies that demonstrate practical applications in diverse geological settings and sectors, both in India and globally; In-depth discussions on the integration of geophysics with remote sensing, artificial intelligence, and big data platforms; Examination of policy frameworks, ethical issues, and international collaboration essential for responsible geoscience; Future-oriented chapters addressing quantum sensing, autonomous platforms, climate resilience, and smart cities.

This volume is designed for a broad spectrum of readers engaged in geosciences and allied disciplines: Graduate and postgraduate students seeking a rigorous yet accessible introduction to engineering geophysics and its modern applications; Early-career and experienced geophysicists aiming to deepen their understanding of interdisciplinary methods and sector-specific challenges; Engineers, environmental scientists, urban planners, and policymakers who interface with subsurface data and need practical insights into geophysical tools and interpretation; Researchers and innovators exploring new frontiers in geophysical instrumentation, data analytics, and sustainable development; By fostering an integrated understanding of geophysical science and technology, I hope this book will contribute to training the next generation of scientists and professionals equipped to address complex, real-world problems through sound subsurface intelligence.

Throughout my career, the Geological Survey of India has been a source of inspiration and professional rigor, providing opportunities to lead cutting-edge projects across varied terrains and geological challenges. This institutional backdrop, combined with my ongoing association with the National Centre for Seismology, has enriched my perspectives on the evolving requirements of geophysical applications in national security, disaster risk reduction, and scientific research.

Writing this book has been both a reflective journey and a forward-looking endeavour. It allowed me to revisit foundational principles while embracing the transformative technologies reshaping our discipline. My sincere hope is that the knowledge encapsulated here will serve as a valuable resource for those dedicated to applying geophysical science for societal benefit and environmental stewardship.

Finally, I would like to acknowledge the numerous colleagues, field experts, academic mentors, and institutional partners who have supported and contributed to my work over the years. Their collaboration and encouragement have been invaluable in shaping this volume. Most importantly, I am grateful to my family for their unwavering patience and support throughout this demanding process.

I look forward to continued dialogue and innovation within the geophysical community, and to the many exciting developments that lie ahead for engineering geophysics as a discipline committed to global well-being.

August 2025

Dr. Dulal Chandra Naskar
M.Sc. (Physics), Ph.D. (Science)
Former Director (Geophysics)
Geological Survey of India (GSI)
Kolkata, India

Chief Consultant in Geophysics
DMT Pvt. Solution Ltd.
Kolkata, India

Senior Consultant, National Centre for Seismology (NCS)
New Delhi, India

Ministry of Earth Sciences (MoES)
Government of India
Kolkata, India

Acknowledgments The completion of this volume, *Advances in Engineering Geophysics: Concepts and Applications*, marks a significant milestone in my professional journey spanning over three decades in the geosciences. As a single-authored work, it carries not just the scientific knowledge I have gathered, but also the deep professional experiences and collaborative insights accumulated through years of working with some of the finest minds in geophysics and allied disciplines.

At the outset, I express my heartfelt gratitude to the Geological Survey of India (GSI)—an institution that has been instrumental in shaping my scientific and leadership capacities. Serving in various capacities and culminating as Director (Geophysics) has given me unparalleled opportunities to work across India's diverse geological terrains, engage with national and international geophysical missions, and lead multidisciplinary teams dedicated to solving some of the country's most pressing geotechnical and exploration challenges.

My ongoing association with the National Centre for Seismology (NCS), under the Ministry of Earth Sciences, Government of India, has been equally enriching. As a Senior Consultant, I have been fortunate to contribute to critical national-level initiatives involving seismic hazard assessments, infrastructure resilience strategies, and advanced geophysical studies related to earthquake-prone zones. This position has provided valuable perspective on the strategic applications of engineering geophysics for national safety and policy development.

I would like to acknowledge the contributions of numerous colleagues, scientists, engineers, and field personnel with whom I have had the privilege of working during various geophysical surveys, technical evaluations, and collaborative research programs. Their professional integrity and scientific acumen have deeply influenced my thinking and execution of geophysical methodologies.

A special mention must be made of those who supported me directly during the preparation of this manuscript. Sayani Naskar, my daughter and a Computer Science Engineer, provided consistent moral support and valuable digital and formatting assistance. Pranav Muktibodh, Geophysicist at the Geological Survey of India, offered technical insight and professional encouragement. Their help—each in their own way—played an important part in seeing this work through to its final shape.

I am also grateful to the academic and research community, both in India and abroad, whose publications, workshops, and dialogues have helped shape this book's interdisciplinary framework. The blending of geophysics with fields such as environmental science, civil engineering, artificial intelligence, and heritage conservation owes much to the insights drawn from ongoing global research.

A special note of thanks is due to Springer and its editorial team, whose support and publishing guidance have helped shape this manuscript into a cohesive academic volume. Their professional standards and commitment to scientific communication resonate with my own aspirations for this book.

I am indebted to my students, mentees, and young professionals, whose curiosity and questions have often challenged me to revisit foundational ideas with fresh perspectives. Their enthusiasm is a reminder of the vitality and future of the geophysical sciences.

Lastly, I express my deepest appreciation to my family, especially to my wife, Mrs. Arati Naskar, for their steadfast support and patience throughout this demanding endeavor. Their understanding and encouragement have been my bedrock during the long hours of writing and research.

To all who have walked this path with me in various capacities—thank you.

August 2025

Dr. Dulal Chandra Naskar
M.Sc. (Physics), Ph.D. (Science)
Former Director (Geophysics),
Geological Survey of India (GSI)
Kolkata, India

Chief Consultant in Geophysics
DMT Pvt. Solution Ltd., Kolkata, India

Senior Consultant, National Centre for
Seismology (NCS)
New Delhi, India

Ministry of Earth Sciences (MoES)
Government of India, Kolkata, India

Competing Interests The author has no competing interests to declare that are relevant to the content of this manuscript.

Contents

1 Introduction to Engineering Geology 1
1.1 Introduction 1
1.2 Rocks and Geological Structures 2
1.2.1 Rock Types 2
1.2.2 Rock Structures 2
1.2.3 Plate Tectonic 3
1.2.4 Engineering Geological Maps 3
1.3 Geohazards 3
1.3.1 Boundary Hazards 3
1.3.2 Ground Subsidence 4
1.3.3 Landslide 4
1.3.4 Slope Failure 5
1.3.5 Rock Slope Stability Under Temperature Fluctuations 5
1.3.6 Subsidence Risk Assessment Using Microgravity: A Case Study from Candado Hill, Malaga, Southern Spain 8
1.4 Geological Masses 9
1.4.1 Mass Movement 10
1.4.2 Mass Fabric 10
1.4.3 Ground Mass Description 10
1.4.4 Weathering 11
1.4.5 Rocks Discontinuities 11
1.5 Field Tests and Measurements 12
1.5.1 Tests in Boreholes and Excavations 12
1.5.2 Engineering Geophysics 13
1.5.3 Seismic Methods 14
1.5.4 Magnetic and Electrical Methods 14
1.5.5 Global Perspective 15
1.5.6 Geophysical Investigations for Offshore Renewables 15

1.6 Ground Improvement ... 17
1.6.1 Shallow and Deep Impaction ... 18
1.6.2 Grout Treatment ... 18
1.6.3 Bentonite Suspension ... 18
1.6.4 Ground Anchor ... 19
1.7 Water Reservoirs and Dams ... 19
1.7.1 Dam Design Parameters ... 20
1.7.2 Geological Influences Upon the Selection of Reservoir Sites ... 20
1.7.3 Dam Foundations ... 21
1.7.4 Seismic Considerations for Dams ... 21
1.7.5 Structural Vibration Analysis and Engineering Standards ... 22
1.7.6 Embankment Dams Response to Actual PGA During Earthquakes ... 26
1.7.7 Concrete Dams Response to Observed PGAs During Earthquakes ... 29
1.8 Integration with Modern Infrastructure Development ... 30
1.8.1 Role of Engineering Geology in Smart Cities ... 30
1.8.2 Subsurface Data for Urban Planning and Resilience ... 30
1.8.3 Case Links to Digital Twins and Geospatial Infrastructure ... 31
1.9 Environmental and Climatic Considerations in Engineering Geology ... 31
1.9.1 Climate-Responsive Site Selection ... 31
1.9.2 Geological Impacts of Extreme Weather Events ... 32
1.9.3 Geo-Environmental Risk Assessments for Infrastructure Projects ... 32
1.10 Ethical, Regulatory, and Societal Aspects ... 33
1.10.1 Environmental Ethics in Engineering Geology ... 33
1.10.2 Regulatory Frameworks Governing Geological Investigations ... 33
1.10.3 Community Engagement and Geo-Social Impacts ... 33
1.11 Trends and Future Challenges in Engineering Geology ... 34
1.11.1 Advances in Remote Sensing and Data Analytics ... 34
1.11.2 Emerging Challenges: Urbanization, Deep Excavation, and Subsurface Congestion ... 34
1.11.3 Future Skillsets for Engineering Geologists ... 35
References ... 35

Part I Foundations and Theoretical Framework

2 Introduction to the Engineering Geophysics 39
2.1 Introduction to Engineering Geophysics 39
2.2 Exploring the Frontiers of Engineering Geophysics 40
2.2.1 Overview of the Evolution of Engineering Geophysics 41
2.2.2 Importance of Advances in Geophysical Techniques and Technologies 44
2.2.3 The Relevance of Engineering Geophysics in Modern Infrastructure and Environmental Challenges 50
2.2.4 The Changing Role of Geophysics in Environmental Evaluation, Monitoring, and Management of Contaminant Plumes from Landfill Leachate: A Comprehensive Overview 54
2.2.5 The Future of Geophysics in Environmental Engineering: Innovations and Applications 67
2.3 Conclusions 79
References 79

3 Fundamental Concepts in Engineering Geophysics 83
3.1 Introduction to Fundamental Concepts 83
3.2 Introduction to Engineering Geophysics 84
3.2.1 Overview of Geophysics in Civil, Construction, and Environmental Engineering 84
3.2.2 Importance of Geophysical Methods in Subsurface Exploration and Risk Assessment 85
3.2.3 Challenges in Subsurface Exploration and the Role of Geophysics 85
3.3 Basic Geophysical Principles 87
3.4 Major Geophysical Methods in Engineering 89
3.4.1 Seismic Methods 89
3.4.2 Electrical and Electromagnetic Methods 96
3.4.3 Magnetic and Gravitational Methods in Engineering Geophysics 112
3.4.4 Acoustic and Ultrasonic Methods in Engineering Geophysics 118
3.5 Conclusions 121
References 122

Part II Technological Advances and Core Applications

4 Recent Advances in Geophysical Surveying Techniques 129
4.1 Introduction to Recent Advances in Geophysical Surveying 129
4.2 Emerging Tools for Data Collection and Analysis 130
4.2.1 New Technologies in Geophysical Surveying (e.g., UAVs, Drones, Remote Sensing) 130
4.2.2 Advances in Instrumentation for Higher Accuracy and Efficiency 132
4.2.3 Integration of Geophysical Techniques with Other Technologies like AI and Machine Learning 134
4.2.4 Types of Drones 138
4.2.5 Overview of Drone Technology in Engineering Geophysics 139
4.2.6 Integration of Drones with Geophysical Surveying Techniques 142
4.3 Conclusions 171
References 172

5 Geophysics for Subsurface Exploration and Site Characterization 177
5.1 Introduction to Subsurface Exploration and Site Characterization 177
5.2 Innovative Approaches in Subsurface Imaging, Characterization, and Mapping 178
5.2.1 Advances in Subsurface Imaging Technologies 178
5.2.2 Characterization of Subsurface Properties 210
5.2.3 Subsurface Mapping Applications 211
5.2.4 Challenges and Future Directions 212
5.3 Applications in Civil Engineering, Construction, and Environmental Studies 214
5.4 Geophysical Exploration for Foundations, Tunnelling, and Infrastructure 215
5.4.1 Case Study on Urban Geophysical Exploration in Chengdu International Bio-City 216
5.4.2 Case Study on Pre-foundation Geophysical Investigation in Oka-Akoko, Nigeria 219
5.5 Conclusions 225
References 228

6 Geophysical Methods in Environmental Monitoring 231
6.1 Introduction to Geophysical Methods in Environmental Monitoring 231
6.2 Addressing Environmental Challenges Through Geophysical Monitoring of Groundwater, Contamination, and Land Use 232
6.2.1 Groundwater Monitoring Using Geophysical Techniques 232
6.2.2 Contamination Monitoring in Environmental Settings 234
6.2.3 Land Use and Urban Planning 239
6.2.4 Challenges and Future Directions 243
6.3 Advancements in Environmental Geophysics for Sustainable Development 244
6.3.1 Key Advancements in Environmental Geophysics 244
6.4 Application of Geophysics in Managing Natural Resources and Protecting Ecosystems 251
6.4.1 Water Resources Management 251
6.4.2 Forest and Land Management 253
6.4.3 Agricultural Applications 255
6.4.4 Energy Resources and Conservation 256
6.4.5 Ecosystem Protection and Biodiversity Conservation 260
6.4.6 Pollution Control and Remediation 262
6.4.7 Disaster Risk Reduction 264
6.4.8 Mineral Exploration (or Coal Exploration) 266
6.5 Master Table with the Geophysical Methods, Their Properties, and Key Parameters 268
6.5.1 Explanation of Key Parameters 268
6.6 Conclusions 270
References 271

7 Geophysics for Disaster Risk Management and Hazard Mitigation 275
7.1 Introduction 275
7.1.1 The Role of Geophysics in Disaster Risk Management and Hazard Mitigation 276
7.2 Innovations in Geophysical Approaches for Natural Hazard Assessment and Risk Mitigation 276
7.2.1 Seismology 277
7.2.2 Geodesy 280
7.2.3 Geophysical Surveys 286

7.3 Mitigation Strategies ... 289
7.3.1 Identifying Vulnerabilities Through Geophysics ... 289
7.3.2 Targeted Structural Reinforcement ... 289
7.3.3 Monitoring and Ongoing Assessment ... 290
7.3.4 Case Studies and Practical Applications ... 290
7.3.5 Land Use Planning for Hazard Reduction ... 291
7.3.6 Early Warning Systems for Earthquakes ... 292
7.3.7 Early Warning Systems for Floods ... 293
7.3.8 Early Warning Systems for Landslides ... 293
7.3.9 Integrated Early Warning Systems ... 294
7.4 Recent Advances in Seismic, Geodetic, and Remote Sensing Technologies ... 295
7.4.1 Advances in Seismic Network Technologies ... 295
7.4.2 Seismic Networks for Earthquake Forecasting and Risk Assessment ... 296
7.4.3 Seismic Networks for Post-earthquake Damage Assessment ... 296
7.4.4 Case Studies and Recent Research ... 297
7.4.5 Overview of Microseismic Monitoring ... 297
7.4.6 Advances in Microseismic Monitoring Technologies ... 298
7.4.7 Applications in Disaster Risk Management ... 299
7.4.8 Recent Case Studies and Applications ... 300
7.4.9 Challenges in Microseismic Monitoring ... 300
7.4.10 Geodetic Technologies ... 301
7.5 Remote Sensing Technologies ... 303
7.5.1 LiDAR (Light Detection and Ranging) ... 304
7.5.2 Satellites and UAVs (Unmanned Aerial Vehicles) ... 305
7.6 Case Studies of Geophysical Applications ... 307
7.6.1 Sentinel-1 InSAR and GNSS on the San Andreas Fault, California ... 307
7.7 Results ... 309
7.7.1 Seismic Characterization of SAF Using Fault-Zone Trapped Waves, California ... 311
7.7.2 Landslide Monitoring in the Swiss Alps ... 314
7.7.3 Landslide Prediction with Geosensors and Wireless Networks ... 316
7.7.4 Gokseong Landslide—South Korea, August 2020 ... 319
7.7.5 Flood Prediction Using Machine Learning and Remote Sensing ... 322
7.7.6 Flood Extent Monitoring in the Lower Amazon with ALOS/PALSAR ... 325

7.7.7 Flood Risk in Upper Solo River, Central Java, Indonesia ... 326
7.7.8 Earth Observation and ML Integration for Flood Risk Management ... 330
7.8 Conclusions ... 333
References ... 334

8 Geophysical Applications in Civil Engineering and Infrastructure ... 341
8.1 Introduction: The Role of Geophysics in Modern Civil Infrastructure ... 341
8.2 From Site Investigation to Structural Health Monitoring ... 342
8.2.1 Advanced Geophysical Methods in Civil Engineering Projects ... 342
8.2.2 Integrating Geophysical Methods in Infrastructure Monitoring ... 349
8.3 Innovations in Geophysical Monitoring of Structural Health and Integrity ... 354
8.3.1 Overview of Geophysical Techniques in Structural Health Monitoring (SHM) ... 355
8.3.2 Advances in Monitoring Technologies ... 355
8.3.3 Applications in Key Infrastructure ... 356
8.3.4 Case Studies ... 357
8.3.5 Emerging Trends in Geophysical Monitoring ... 360
8.4 Examples of Geophysical Tools Used for Ongoing Infrastructure Monitoring ... 361
8.4.1 Ground Penetrating Radar (GPR) ... 361
8.4.2 Seismic Tomography ... 362
8.4.3 Electrical Resistivity Imaging (ERI) ... 362
8.4.4 Acoustic Emission (AE) Monitoring ... 363
8.4.5 Vibration Monitoring (Accelerometers and Seismic Sensors) ... 363
8.4.6 Satellite-Based Remote Sensing and InSAR (Interferometric Synthetic Aperture Radar) ... 364
8.4.7 Distributed Fiber Optic Sensing (DFOS) ... 364
8.4.8 Multi-channel Analysis of Surface Waves (MASW) ... 365
8.4.9 Magnetometry ... 365
8.4.10 Electromagnetic (EM) Induction ... 366
8.4.11 Electrical Resistivity Tomography (ERT) ... 366
8.4.12 Gravity Surveying ... 366
8.4.13 Refraction Seismology ... 367
8.4.14 Borehole Geophysics ... 367

8.5 Case Studies: Recent Applications of Geophysical Tools in Infrastructure Monitoring ... 368
8.5.1 GPR for Bridge Stability—United States ... 368
8.5.2 Tunnel Infrastructure Monitoring ... 369
8.5.3 Electrical Resistivity Imaging (ERI) for Foundation Settlement—United Kingdom ... 373
8.5.4 Interferometric Synthetic Aperture Radar (InSAR) for Urban Subsidence—Italy ... 374
8.5.5 DFOS for Pipeline Integrity—Australia ... 374
8.5.6 GPR for Bridge Inspection—Mumbai, India ... 375
8.5.7 Seismic Tomography for Tunnel Safety—Delhi Metro Project ... 375
8.5.8 ERI for Foundation Monitoring—Bengaluru, India ... 376
8.5.9 InSAR (Interferometric Synthetic Aperture Radar) for Subsidence—Chennai, Tamil Nadu ... 376
8.5.10 DFOS for Pipeline Integrity—Gujarat, India ... 376
8.5.11 GPR Inspection of Dam Spillway—Alabama, USA ... 377
8.5.12 ERI for Hazard Investigations—Taiwan ... 381
8.5.13 Urban Subsidence via Small Baseline Subset-Interferometric Synthetic Aperture Radar (SBAS-InSAR)—Beijing Plain, China ... 383
8.5.14 Distributed Fiber Optic Sensing (DFOS) for Pipeline Structural Health Monitoring (SHM)—Case Study ... 388
8.6 Conclusions ... 391
References ... 392

9 Geophysics in Mining and Resource Exploration ... 397
9.1 Overview of Advancements and Sustainability in Geophysical Exploration ... 397
9.2 Optimizing Resource Extraction with Geophysical Approaches Sectorwise ... 398
9.2.1 Mineral Exploration ... 398
9.2.2 Oil and Gas Exploration ... 407
9.2.3 Geothermal Energy Exploration ... 408
9.3 Emerging Techniques and Technologies ... 409
9.3.1 Novel Methods ... 409
9.3.2 Machine Learning and Big Data in Geophysics ... 411
9.4 Sustainable and Environmental Applications ... 413
9.4.1 Environmental Monitoring ... 413
9.4.2 Sustainable Practices ... 414
9.4.3 Integrated Approaches ... 415

9.5 Tools and Techniques for Improved Exploration 416
9.5.1 New Geophysical Tools 416
9.5.2 Environmental Impact Reduction 419
9.5.3 Machine Learning Integration 420
9.6 Case Studies of Geophysical Innovations in Mining 428
9.6.1 Airborne EM Surveys—Australia 430
9.6.2 3D Seismic Imaging—Russia (Khokhlovka Gold Deposit) 431
9.6.3 Gravity and Magnetic Surveys—Iran (Chah Gaz Iron Ore) 431
9.6.4 GPR for Ore Mapping—Canada 432
9.6.5 Superconducting Gravimeter—South America (Waste Monitoring) 433
9.6.6 AEM Surveys—Rajasthan, India 433
9.6.7 3D Seismic Imaging—Jharkhand (Coal), India 434
9.6.8 Gravity and Magnetic—Odisha (Iron Ore), India 434
9.6.9 GPR—Chhattisgarh (Shallow Ore Bodies), India 435
9.6.10 Gravimeter—Goa (Environmental Monitoring), India 435
9.7 Conclusions 436
References 437

Part III Strategic and Specialized Applications

10 Engineering Geophysics and the Sustainable Development Goals 443
10.1 Introduction 444
10.1.1 Background on the Sustainable Development Goals (SDGs) 444
10.1.2 The Role of Engineering Geophysics in Sustainable Development 444
10.1.3 Chapter Objectives and Structure 445
10.2 Engineering Geophysics in the Context of the SDGs 445
10.2.1 Overview of Engineering Geophysics Applications 445
10.2.2 Alignment of Geophysical Techniques with Selected SDGs 446
10.3 Geophysical Contributions to Key SDGs 447
10.3.1 Water Resources Management (SDG 6) 447
10.3.2 Urban Resilience and Infrastructure (SDG 11) 447
10.3.3 Climate Monitoring and Mitigation (SDG 13) 448
10.3.4 Renewable and Clean Energy Exploration (SDG 7) 449

10.3.5 Innovation and Frontier Exploration (SDG 9, 13, 14) 450
10.3.6 Geohazards and Risk Reduction (SDG 1, 9, 11, 13, 14, 15) 451
10.3.7 Near-Surface Geosciences for Sustainable Development (SDG 1, 8, 9, 11, 13, 14, 15) 454
10.3.8 Minerals and Rock Materials for Sustainable Development (SDGs 1, 2, 7, 8, 11, 12, 13) 457
10.3.9 Energy: Extraction and Storage for Sustainable Development (SDGs 1, 7, 8, 9, 11, 12, 13) 459
10.3.10 Hydrogeology, Waste, and Contamination 463
10.3.11 Climate Science and Earth System Change 465
10.3.12 Geoheritage, Public Engagement, and Education for the SDGs 466
10.3.13 Higher Education and Research 468
10.3.14 Data and Skill Sets 469
10.3.15 Social Geology and Human–Environment Interaction 470
10.4 Engineering Geophysics for Resilience and Environmental Justice 473
10.4.1 Geophysics for Community Resilience 473
10.4.2 Inclusive and Participatory Approaches 474
10.4.3 Case Examples in Environmental Justice 474
10.5 Geoethics and Global Equity in Geophysical Technology 474
10.5.1 Defining Geoethics in the Geosciences 474
10.5.2 Technology Access and Capacity Building 475
10.5.3 Ethical Dilemmas in Practice and Research 475
10.6 Case Studies and Practical Applications 476
10.6.1 Climate-Resilient Infrastructure in South Asia 476
10.6.2 Groundwater Mapping in Water-Scarce Regions 478
10.6.3 Subsurface Data for Sustainable Urban Development 482
10.7 Challenges, Gaps, and Opportunities 483
10.7.1 Technical and Operational Limitations 485
10.7.2 Integration with Emerging Technologies (AI, Remote Sensing) 485
10.7.3 Future Directions for SDG-Aligned Geophysics 486
10.8 Supporting Resources and Practical Tools 487
10.8.1 SDG-Geophysics Alignment Checklist 487
10.9 Glossary of Terms 487
10.9.1 Supplementary Case Studies 488
10.10 Conclusion: Engineering Geophysics and the SDGs—A Roadmap for the Future 488
References 489

11 Geophysics for Risk-Resilient and Critical Infrastructure 499
11.1 Introduction 499
11.1.1 Assessing Earthquake Resilience in Critical Infrastructure 501
11.2 Infrastructure Resilience Planning 508
11.2.1 Understanding Risk and Resilience 508
11.2.2 Integration of Geophysical Techniques in Planning 508
11.2.3 Decision-Making Tools and Data Integration 509
11.2.4 Resilience Assessment Frameworks of Critical Infrastructures: State-of-the-Art Review 509
11.2.5 Assessing Infrastructure Resilience: Theoretical Models, Scenario-Based Approaches, and Case Applications 511
11.3 Applications in Utilities, Transportation, and Energy Networks 516
11.3.1 Utilities (Water, Gas, Sewerage) 516
11.3.2 Transportation Systems (Roads, Railways, Airports) 516
11.3.3 Energy Networks (Pipelines, Substations, Renewables) 518
11.3.4 Building Disaster-Resilient Seaports: Exploring Adaptation in Practice 518
11.4 Pre-failure Detection and Maintenance Strategies 520
11.4.1 Subsurface Change Detection and Monitoring 520
11.4.2 Predictive Maintenance Using Geophysics 520
11.4.3 Case Studies and Best Practices 521
11.5 Challenges and Opportunities 521
11.5.1 Technical and Environmental Constraints 521
11.5.2 Standardization and Quality Control Needs 522
11.5.3 Opportunities from AI, IOT, and Smart Infrastructure 522
11.6 Future Directions 524
11.6.1 Digital Twins and Geophysical Integration 524
11.6.2 Role of Geophysics in Climate-Resilient Infrastructure Planning 524
11.6.3 Policy Recommendations and Funding Avenues 525
11.7 Case Studies 525
11.7.1 Case Study: Climate Disaster Vulnerability and Resilience of Railway Infrastructure in Northeast India 525
11.7.2 Lessons from Case Studies of Flood Resilience: Institutions and Built Systems 528

11.7.3 Social Sensing and Community Response to Flood Impacts: A Case Study of the 2018 Kerala Floods, India ... 530
11.7.4 Resilience of Urban Underground Infrastructure Under Multi-hazard Impacts: From Structural Integrity to Network-Level Functionality ... 532
11.7.5 Risk Assessment and Management of Underground Metro Construction (Bengaluru Scenario) ... 536
11.7.6 Case Study: Improving Climate Resilience at Kansai International Airport, Osaka Bay, Japan ... 538
11.7.7 An Integrated Infrastructure Resilience Approach: From the Geotechnical Asset to the Transport Network (Case Study: Greece) ... 540
11.8 Conclusions and Recommendations ... 543
References ... 545

12 Defense and Strategic Installations ... 549
12.1 Securing National Assets with Subsurface Intelligence ... 549
12.2 Detection of Subsurface Threats: Tunnels, Bunkers, and Concealed Infrastructure ... 550
12.2.1 Nature of Subsurface Threats ... 550
12.2.2 Geophysical Detection Techniques ... 550
12.2.3 Operational Challenges ... 552
12.2.4 Case Studies and Applications ... 553
12.3 Military Engineering Geophysics: Field Applications and Tactical Advantage ... 554
12.3.1 Introduction to Military Geophysics ... 554
12.3.2 Engineering Support ... 554
12.3.3 Rapid Assessment and Deployment Tools ... 557
12.3.4 Training and Multidisciplinary Teams ... 558
12.4 National Security and Dual-Use Technologies ... 559
12.4.1 Dual-Use Potential of Geophysical Technologies ... 559
12.4.2 Homeland and Critical Infrastructure Security ... 560
12.4.3 Technology Integration and Policy Frameworks ... 566
12.4.4 Future Directions ... 567
12.5 Environmental and Ethical Considerations in Defense Geophysics ... 568
12.5.1 Environmental Impact of Subsurface Surveys ... 568
12.5.2 Ethical Issues in Military Geophysical Applications ... 569

12.5.3 Compliance with International Norms and Treaties 569
12.5.4 Toward Responsible Defense Geophysics 569
12.6 Interoperability, Standards, and Data Sharing 570
12.6.1 Geophysical Data Standards for Defense Applications 570
12.6.2 Interagency and International Collaboration on Data Sharing 571
12.6.3 Challenges and Solutions in Multinational Operations 571
12.7 Limitations and Challenges of Current Geophysical Methods in Defense 572
12.7.1 Technical and Environmental Limitations 572
12.7.2 False Positives and Ambiguities in Data Interpretation 572
12.7.3 Resource and Deployment Constraints 573
12.8 Conclusions and Recommendations 573
References 575

13 Energy Resilience in Defense Installations: Geophysical and Engineering Perspectives 577
13.1 Introduction: Strategic Importance of Energy Resilience 577
13.2 Subsurface Investigations for Energy Infrastructure 579
13.3 Natural Hazard Risk Mitigation for Critical Energy Systems 581
13.4 Renewable Integration and Subsurface Considerations 582
13.5 Design and Protection of Hardened Energy Systems 584
13.6 Operational Independence: Microgrids and Remote Energy Solutions 585
13.7 Policy, Collaboration, and Future Pathways 588
13.8 Conclusions and Recommendations 591
References 593

Part IV Synthesis, Future Directions, and Interdisciplinarity

14 Archaeological, Heritage, and Hidden Manmade Structures 599
14.1 Introduction 600
14.2 Principles of Geophysical Methods for Subsurface Exploration 601
14.2.1 Ground Penetrating Radar (GPR) 601
14.2.2 Electrical Resistivity Tomography (ERT) 602
14.2.3 Magnetometry 602
14.2.4 Seismic Methods 602

14.2.5 Electromagnetic (EM) Induction 603
14.2.6 Comparative Suitability and Integration of Methods 603
14.3 Non-invasive Archaeological Mapping 604
14.3.1 Strategies for Detecting Ancient Structures Without Excavation 604
14.3.2 GPR and Magnetometry in Site Prospection 605
14.3.3 Multimethod Surveys and Data Integration 605
14.3.4 Case Study: Mapping an Ancient Burial Site—Qubbet el-Hawa (Egypt) 606
14.4 Heritage Conservation with Geophysical Tools 607
14.4.1 Case Study: Geophysical Monitoring of a UNESCO World Heritage Site—The Leaning Tower of Pisa (Italy) 608
14.4.2 Case Study: Foundation Features Investigation of the Sun Temple, Konark (India), Using Coupled Non-invasive and Invasive Techniques 609
14.4.3 Archaeological Excavation, Protection, and Display Engineering Design Practice: A Case Study in the Ruins of the Imperial City of the Minyue Kingdom 611
14.4.4 Archaeological Analysis of the Newly Discovered Tomb with a Relief of a Couple at the Funerary Area of Porta Sarno in Pompeii 614
14.5 Identification of Buried and Clandestine Structures 618
14.5.1 Case Study: Discovery of Clandestine Tunnels Beneath Urban Infrastructure—Mexico–U.S. Border Zone 621
14.6 Data Processing and Interpretation Techniques 622
14.6.1 Geospatial Technologies for Heritage: GIS and Remote Sensing in the DBAR-Heritage Mission 624
14.6.2 GPR Data Processing and Visualization for Cultural Heritage 625
14.7 Ethical, Legal, and Conservation Considerations 627
14.7.1 Heritage Conservation in India: Challenges and New Paradigms 629
14.7.2 Case Study: Archaeological Site Management and Tourism at Bibracte, France 630
14.8 Future Trends and Innovations 632
14.9 Conclusions and Recommendations 635
References 636

15 Interdisciplinary Approaches in Engineering Geophysics 643
15.1 Integrating Geophysics with Geology, Hydrology, and Remote Sensing 643
15.1.1 The Role of Interdisciplinary Collaborations in Solving Complex Geophysical Problems 644
15.1.2 Interdisciplinary Approaches to Geophysical Challenges: Case Studies 649
15.2 Integration of Geophysics with Other Scientific Fields to Address Environmental and Engineering Challenges 654
15.2.1 Geophysics and Environmental Science Integration 655
15.2.2 Integration of Geophysics with Geology and Hydrology 656
15.2.3 Geophysics and Remote Sensing: Synergistic Collaboration 657
15.2.4 Case Studies and Applications in Engineering Projects 658
15.2.5 Environmental Remediation Projects 660
15.2.6 Geophysical Monitoring for Disaster Mitigation 667
15.2.7 Challenges and Opportunities 668
15.3 The Synergy Between Geophysical Methods and Other Technologies for Enhanced Solutions 669
15.3.1 Introduction to Synergy Between Geophysics and Other Technologies 669
15.3.2 Geophysics and Remote Sensing Technologies 669
15.3.3 Geophysics and Geospatial Technologies 670
15.3.4 Geophysics and Machine Learning/Artificial Intelligence 670
15.3.5 Geophysics and Environmental Monitoring Technologies 671
15.3.6 Geophysics and Biotechnology 671
15.3.7 Geophysics and Civil Engineering Technologies 672
15.3.8 The Role of Data Fusion and Multidisciplinary Collaboration 672
15.3.9 Future Directions in Geophysics and Technology Synergy 673
15.4 Conclusions 674
References 674

16 The Next Frontier: Emerging Trends and Technologies 681
16.1 Introduction 681
16.2 Quantum Sensing in Geophysics 682
16.2.1 Principles of Quantum Sensing 682
16.2.2 Real-World Applications 683

16.2.3 Advantages Over Classical Instruments 683
16.2.4 Challenges and Future Directions 684
16.3 Robotics and Autonomous Geophysical Platforms 685
16.3.1 Development of UAVs, ROVs, and Autonomous Land Rovers 685
16.3.2 Integration of Geophysical Instruments with Mobile Platforms 685
16.3.3 Remote and Hazardous Site Surveying 686
16.3.4 Swarm Robotics and Synchronized Sensor Arrays 686
16.4 Artificial Intelligence and Machine Learning 688
16.4.1 AI for Data Processing and Pattern Recognition 688
16.4.2 Machine Learning Models for Anomaly Detection and Classification 688
16.4.3 Predictive Modeling and Inversion Optimization 689
16.4.4 Case Studies in Smart Geophysical Interpretation 689
16.5 Geophysics in Climate Adaptation and Resilience 689
16.5.1 Subsurface Monitoring for Flood, Drought, and Permafrost Change 689
16.5.2 Role in Early Warning Systems and Risk Modeling 690
16.5.3 Ground Stability Assessments for Climate-Impacted Regions 690
16.6 Smart Cities and Urban Geophysics 692
16.6.1 Non-invasive Assessment of Underground Infrastructure 692
16.6.2 Utility Mapping and Sinkhole Prediction 694
16.6.3 Real-Time Geophysical Monitoring in Intelligent Urban Systems 694
16.6.4 Integrating Geophysics with IoT and Digital Twins 694
16.7 Deep Subsurface Mapping and Exploration 695
16.7.1 Next-Gen Seismic, EM, and Gravity Technologies 695
16.7.2 High-Resolution 3D/4D Subsurface Modeling 695
16.7.3 Role in Geothermal Energy, Carbon Capture, and Hydrogen Storage 696
16.8 Digital Transformation of Infrastructure Geophysics 696
16.8.1 Sensor Fusion and Edge Computing in Field Systems 696
16.8.2 Digital Twins of Civil Structures Informed by Geophysical Data 697

16.8.3 Big Data Platforms and Real-Time Interpretation ... 698
16.8.4 Integration with BIM and GIS Technologies ... 699
16.9 Ethical and Policy Considerations ... 699
16.9.1 Data Privacy and Proprietary Models ... 699
16.9.2 AI Transparency and Decision-Making Accountability ... 700
16.9.3 Equitable Access to Emerging Geophysical Technologies ... 700
16.10 The Road Ahead: Opportunities and Barriers ... 700
16.10.1 Translating Research into Practical Field Applications ... 701
16.10.2 Talent Development and Interdisciplinary Collaboration ... 701
16.10.3 Funding, Infrastructure, and Regulatory Needs ... 701
16.11 Summary and Outlook ... 702
References ... 703

Part V Reference Materials

17 Conclusions: Bridging Innovation and Practical Application in Geophysics ... 707
17.1 Summary of Advances ... 707
17.2 Strategic Importance for Global Challenges ... 708
17.3 Reflections on Future Impact Across Sectors ... 708
17.4 Challenges and Limitations ... 708
17.5 Recommendations for Practice and Policy ... 709
17.6 Role in Education and Workforce Development ... 710
17.7 Ethical and Social Considerations ... 710
17.8 Vision for the Discipline ... 711
17.9 Final Thoughts ... 711

18 Technical Reference and Glossary: A Comprehensive Companion for Terminology, Symbols, and Equations ... 713
18.1 Glossary of Terms ... 713
18.2 List of Acronyms and Abbreviations in Engineering Geophysics ... 719
18.3 Mathematical Symbols and Notation ... 721
18.4 Key Mathematical Equations and Models ... 723
18.5 Data and Measurement Units Reference ... 729
18.5.1 SI and CGS Unit Conversions (Select Parameters) ... 729
18.5.2 Common Measurement Scales and Units ... 730
18.5.3 Temperature and Heat Flow ... 730
18.5.4 Seismic and Wave Units ... 731

18.6 Further Reading and Online Resources (Optional) 731
18.6.1 Core Textbooks on Engineering Geophysics 731
18.6.2 Journals for Advanced Study and Research 731
18.6.3 Online Resources and Tools 732
18.6.4 Open Datasets and Field Examples 732
18.6.5 Suggested Reading by Chapter 733

Book Synopsis: Advances in Engineering Geophysics: Concepts and Applications 735

About the Author

Dr. Dulal Chandra Naskar is a distinguished geophysicist with over 35 years of experience in geophysical mapping, research, and applied earth sciences. He currently serves as a Senior Consultant at the National Centre for Seismology (NCS), Ministry of Earth Sciences, Government of India, New Delhi. Over the course of his illustrious career, he has held a wide range of professional roles, including Director (Geophysics) at the Geological Survey of India (GSI), Chief Consultant at DMT Pvt. Ltd., and earlier positions such as Senior Technical Assistant (STA) at the West Bengal State Electricity Board (WBSEB), Chemical and Metallurgical Assistant (CMA) with South Eastern Railway, and Assistant Teacher in Physics at a government-aided school in West Bengal.

Dr. Naskar holds an M.Sc. in Physics from Calcutta University and earned his Ph.D. in Science from Jadavpur University. His doctoral research, titled "*Geophysical Studies of Crust and Upper Mantle Structures in Parts of Madhya Pradesh and Maharashtra, Central India*," employed Audio-Magnetotellurics (AMT) and Magnetotellurics (MT) techniques—marking a pioneering contribution to the application of deep geophysical methods beneath the Deccan Traps.

Throughout his tenure at GSI, Dr. Naskar has spearheaded and participated in numerous high-impact geophysical investigations across India. His work has incorporated gravity, magnetic, resistivity, GPS, seismic, and MT surveys, leading to vital insights into crustal structures and tectonic processes. Notable achievements include the detection of Moho upwarp near Jabalpur, the identification of active fault systems in key seismic zones, and multidisciplinary investigations in geothermal, tectonic, and neotectonic environments. His leadership in studies of the Central Indian Shield (CIS), Narmada-Son Lineament (NSL), 1998 Pandhana earthquake swarm, and post-tsunami surveys in the Andaman-Nicobar Islands has made significant contributions to both regional and national hazard mapping.

Dr. Naskar is widely recognized for introducing and advancing AMT and MT techniques in India, particularly in the deep imaging of subsurface geological formations beneath complex terrains like the Deccan Traps. His research has also spanned the Singhbhum–Orissa Iron Ore Craton, the Narmada Valley, and the Shillong Plateau, with a special focus on fault detection, rift structures, seismic hazard evaluation, and geothermal anomalies.

An accomplished author and editor, Dr. Naskar has published over 140 research papers in reputed national and international journals. He has served as a guest editor for special issues on geophysics, Earth sciences, and disaster risk reduction for Elsevier, and is the author of several scholarly books, including:

- *Geophysical Methods: Fundamentals, Applications, and Case Studies* (CRC Press, Taylor & Francis, USA)
- *A Field Manual of Magnetotelluric (MT) Surveys with Case Studies for Earth Scientists and Engineers* (CRC Press, Taylor & Francis, USA)
- *Geothermal Resource Exploration and Development: Geophysical Techniques for Sustainable Energy* (CRC Press, Taylor & Francis, USA)
- *Baseline Geoscience Data Generation for Sustainable Exploration of Natural Resources* (Elsevier, in production and will be published shortly)

A dedicated mentor, Dr. Naskar has guided more than 300 students, researchers, and professionals, contributing significantly to academic development and capacity-building in geophysics and Earth sciences. His work has earned him several prestigious accolades, including:

- H. S. Pareek Award for best paper in coal science (2008)
- ONGC–AEG Award for best Ph.D. thesis
- Recognition for outstanding contributions to national and international geophysical research initiatives

Dr. Naskar is an active life member of many prominent professional organizations, such as the American Geophysical Union (AGU), Asia Oceania Geosciences Society (AOGS), International Union of Geodesy and Geophysics (IUGG), and the Indian Journal of Earth Science and Technology (IJOEST), among others—highlighting his enduring engagement with the global scientific community.

His areas of expertise include geophysical mapping and exploration, seismic hazard detection, geothermal and tectonic studies, groundwater investigations, and urban infrastructure planning for disaster risk reduction. At GSI, he led the National Geophysical Mapping Programme, overseeing extensive surveys and contributing to the identification of mineralized zones, hidden faults, and vulnerable seismic areas.

In his current role at NCS, Dr. Naskar continues to provide strategic guidance on seismological studies, including the establishment of seismic observatories, earthquake hazard analysis, and the design of resilient infrastructure systems. His career stands as a testament to scientific excellence, institutional leadership, and a lifelong commitment to advancing geophysics for national and societal benefit.

List of Figures

Fig. 1.1 Observation from a thermal monitoring station at the eastern cliff of Mount Masada, Israel. **a** Location map showing Masada along the Dead Sea Transform Fault. **b** The Snake Path cliff with the old cable car station located below the monitored zone. **c** The unstable rock mass ("Block 1") situated near the newly constructed station and tourist path. **d** Seasonal monitoring results from 1998 showing a clear correlation between temperature fluctuations and joint displacement, demonstrating the cumulative effect of thermal cycling on slope stability. *Source* Bakun-Mazor (2022)—Open Access 6

Fig. 1.2 Photos of blocks from the field that illustrate the ratchet mechanism (The ratchet mechanism refers to the process where repeated expansion and contraction from temperature changes cause incremental, one-directional movement of rock blocks, eventually leading to permanent displacement or failure). **a** Masada Mountain, Israel, **b** Arugot Valley, Israel, **c** Yellow Mountain, China, **d** Zohar Valley, Israel and **e** Machtesh Ramon, Israel. Each image shows displaced or partially detached blocks affected by thermal cycling. Scale bars represent 1 m. *Source* Bakun-Mazor (2022)—Open Access 7

Fig. 1.3 Laboratory simulation of thermal-induced block displacement. **a** Concrete block system representing a stratified rock slope, showing the placement of displacement transducers for monitoring block movement (dimensions in cm). **b** Experimental setup of the block system placed on an inclined steel platform inside a temperature-controlled chamber to simulate daily thermal cycles and observe cumulative displacement effects. *Source* Bakun-Mazor (2022)—Open Access 7
Fig. 1.4 Visualization of typical offshore geophysical survey vessel using sonar and echo sounders. *Source* Wikipedia—Open Access 16
Fig. 1.5 Timeframe for a typical offshore wind farm development. *Source* Wikipedia—Open Access 17
Fig. 1.6 Rapid Impact Compaction on variable geological strata. This figure illustrates the application of Rapid Impact Compaction (RIC) on a site with heterogeneous subsurface conditions. The compaction mechanism and resulting ground response are shown in relation to soil type, density, and groundwater conditions, emphasizing the importance of geological assessment in designing and evaluating ground improvement strategies. *Source* Baca (2025)—Open Access 23
Fig. 1.7 Building under vibration monitoring in a geologically sensitive area. This illustration depicts vibration monitoring of a structure situated on varying subsurface geology, highlighting the influence of local lithology, soil stratigraphy, and groundwater conditions on wave propagation and structural response. *Source* Baca (2025)—Open Access 24
Fig. 1.8 Monitoring of vibrations in warehouse facilities on geologically variable ground. This figure illustrates the monitoring of structural vibrations in a warehouse setting, emphasizing the role of underlying geological conditions such as soil type, compaction, and groundwater presence. Vibration monitoring devices are deployed to assess how subsurface materials influence wave transmission and structural response, aiding in geotechnical risk assessment and mitigation. *Source* Baca (2025)—Open Access 25

Fig. 1.9 Svantek SVAN 958A monitoring station deployed in a geotechnical setting. The figure shows the SVAN 958A vibration and noise monitoring station used in an engineering geophysics application. The setup is positioned on a site with variable subsurface geology, highlighting the need to understand local soil conditions, stratigraphy, and groundwater influence to interpret recorded vibration data effectively. *Source* Baca (2025)—Open Access ... 25
Fig. 2.1 Overview of geophysical methods in engineering geophysics ... 43
Fig. 2.2 Graphical representation of electrical resistivity survey results showing the impact of landfill leachate plume contaminants on soils and groundwater. *Source* Alao (2025)—Open Access ... 56
Fig. 2.3 Seismic refraction tomography results for mapping landfill leachate plume and subsurface characteristics, with the dashed white line indicating the base of the loess layer. *Source* Hakimov et al. (2021)—Open Access ... 57
Fig. 2.4 Ground magnetic survey for mapping contaminant plumes in landfills. *Source* Alao (2025)—Open Access ... 58
Fig. 2.5 GPR radargram showing the areas where leachate plumes (LLPs) have accumulated. *Source* Alao (2025)—Open Access ... 60
Fig. 2.6 GPR radargram results of buried targets at the USP shallow GTS-II: **a** spatial distribution across the site, showing targets (target 1 is an empty PVC pipe, target 2 is an empty ceramic tube, and targets 3 and 4 are ceramic tubes filled with half water and half air). **b** GPR radargram results at 900 MHz. **c** GPR radargram results at 270 MHz for the profile. **d** GPR depth slice at 0.62 m for 270 MHz frequency showing the targets. *Source* Alao (2025)—Open Access ... 61
Fig. 2.7 Satellite image of Weidenpesch landfill location with geophysical survey points. *Source* Ibraheem et al. (2024)—Open Access ... 63
Fig. 2.8 The RTP magnetic map of the Weidenpesch landfill area after applying an upward continuation filter at an altitude of 5 m. *Source* Ibraheem et al. (2024)—Open Access ... 64
Fig. 2.9 A 3D fence diagram of subsurface resistivity distribution from the 2D inversion of ERT data at the Weidenpesch landfill. Blue anomalies at shallow depths indicate the waste body, with the dashed black line marking the southern boundary. *Source* Ibraheem et al. (2024)—Open Access ... 64

Fig. 2.10 2D resistivity cross-section from 1D Occam inversion results for TEM data along profile ERT4 at the Weidenpesch landfill. The lower panel shows χ values for Occam (R1 and R2) and MQ inversions. Borehole W4 log is shown between TEM stations T10 and T11. *Source* Ibraheem et al. (2024)—Open Access 65
Fig. 2.11 Subsurface lithology logs from boreholes around the Weidenpesch landfill. Groundwater tables for 2005 and 2006 are marked by dashed blue lines. *Source* Ibraheem et al. (2024)—Open Access 65
Fig. 2.12 Geophysical data acquisition using the SuperSting R8/IP 8 resistivity (**a**) and the EM38 MK2 system electrical conductivities (**b**) at Al-Ahsa Oasis, Saudi Arabia. *Source* Mohammed et al. (2022)—Open Access 72
Fig. 2.13 **a** Two-dimensional resistivity section for location P-6 at Al-Ahsa Oasis. The upper section shows the measured resistivity (**A**), the middle section displays the calculated resistivity (**B**), and the lower section represents the inverted resistivity (**C**). **b** Electrical conductivities along EM38 at four tested locations (P-4, P-7, P-8, and P-10) across Al-Ahsa Oasis. Line 4 at P-4 (**a**), Line 7 at P-7 (**b**), Line 8 at P-8 (**c**), and Line 10 at P-10 (**d**). *Source* Mohammed et al. (2022)—Open Access 73
Fig. 2.14 Magnetic anomaly map at Sungai Batu, Lembah Bujang archaeological site, Kedah, Malaysia. *Source* Sapiai et al. (2010)—Open Access 74
Fig. 2.15 2-D resistivity pseudosection of inversion model for survey line LB006 at Sungai Batu, Lembah Bujang, Kedah, Malaysia. *Source* Sapiai et al. (2010)—Open Access 75
Fig. 2.16 Ground penetrating radar (GPR) image for survey line M3 at Sungai Batu, Lembah Bujang, Kedah, Malaysia. *Source* Sapiai et al. (2010)—Open Access 75
Fig. 2.17 Small-scale field testbed for sinkhole monitoring with optical fiber acoustic sensor in Gauteng Province, South Africa. *Source* Jena et al. (2024)—Open Access 77
Fig. 2.18 Sinkhole damage in South Africa's dolomite land (Gauteng Province). *Source* Jena et al. (2024)—Open Access 77
Fig. 2.19 Optical fiber sensor monitoring of sinkhole trigger events during cavity growth in Gauteng, South Africa. **a** Steady state ambient noise signals, **b** void establishment and **c** cavity enlargement. *Source* Jena et al. (2024)—Open Access 78

Fig. 3.1 Seismic survey sketch: layer model, seismic rays (green: direct wave, blue: refracted/head wave, red: reflected waves), and resulting seismogram with seismic signals. *Source* Venkiteswaran and Naskar (2000) 90
Fig. 3.2 Seismic waves: body waves (P-waves and S-waves) and surface waves (Rayleigh and Love waves). *Source* Wikipedia 91
Fig. 3.3 Cross-section of current paths and equipotential surfaces (surfaces of equal potential) between a source and sink, showing current distribution. *Source* Robinson and Coruh (1988) 97
Fig. 3.4 General four-electrode configuration for resistivity measurement, showing current electrodes (**a**, **b**) and potential electrodes (**c**, **d**). *Source* Naskar and Singh (2002) 98
Fig. 3.5 Geometries of current and potential electrodes in the **a** Wenner and **b** Schlumberger configurations. *Source* Naskar et al. (2011) 99
Fig. 3.6 Construction of a pseudo-section for a dipole–dipole resistivity survey: Plotting measured values at the intersection of 45° lines from the transmitter and receiver mid-points. *Source* GSI (1992) 100
Fig. 3.7 IP-related decay of potential: **a** illustration of potential decay after primary current interruption, **b** effect of IP decay time on potential waveform with square-wave input current, **c** overvoltage (temporary excess voltage after current interruption), and **d** chargeability. *Source* Lowrie (2007) 102
Fig. 3.8 Reflection and refraction of an incoming radio wave in ground penetrating radar (GPR). *Source* Wikipedia 105
Fig. 4.1 Types of drones. *Source* Wikipedia 138
Fig. 4.2 Location of study area, highlighting Quaternary volcanic units and faults with Quaternary activity. *Source* Silva-Fragoso et al. (2024)—Open Access 143
Fig. 4.3 Volcano-tectonic map of the Casamicciola Holocene Graben area with shaded relief derived from 20 cm-resolution DEM. *Source* Silva-Fragoso et al. (2024)—Open Access 143
Fig. 4.4 Location of the study area on the Sicilian Coast, Italy, highlighting key geological features for engineering geophysics analysis. *Source* Wierzbicki et al. (2019)—Open Access 145
Fig. 4.5 System components: SR10 receiver, M10 radiosondes, GNSS-based ground control kit, and antennas (400 MHz). *Source* Wierzbicki et al. (2019)—Open Access 146

Fig. 4.6 Atmospheric scattering model in a humid atmosphere for UAV-acquired imagery. *Source* Wierzbicki et al. (2019)—Open Access 146
Fig. 4.7 Entropy values before and after dehazing for different methods (September–November 2018). *Source* Wierzbicki et al. (2019)—Open Access 147
Fig. 4.8 Maps and geological context of the study area. **a** Location of Israel, **b** Beer Sheva city map with site marked (red triangle), **c** orthophoto of the Children's Park construction site (red line), and **d** geological map fragment showing site location (red triangle). *Source* Frid and Frid (2024)—Open Access 147
Fig. 4.9 DJI M300 drone with GPR system for subsurface surveying. *Source* Frid and Frid (2024)—Open Access 148
Fig. 4.10 Processed radargram showing subsurface boundaries. *Source* Frid and Frid (2024)—Open Access 149
Fig. 4.11 Triangular gradiometer system for Ranbogen survey. Visualization of a downward-pointed triangular gradiometer with 2 m baseline magnetometers at its corners. Arrows depict across-sensor and same-sensor differences, flight direction, and ideal orthogonal differences. *Source* Petersen and Dossing (2024)—Open Access 152
Fig. 4.12 Comparison of equivalent source maps and simulated gradiometer data. *Source* Petersen and Dossing (2024)—Open Access 153
Fig. 4.13 Equivalent source-based processing results for Ranbogen gradiometer data. *Source* Petersen and Dossing (2024)—Open Access 154
Fig. 4.14 Flight mission planning for Chaoyang University of Technology (CYUT) situated in Taichung, Taiwan solar photovoltaic experimental field. *Source* Liao et al. (2022)—Open Access 157
Fig. 4.15 Flowchart of the image filtering process for enhanced analysis. *Source* Liao et al. (2022)—Open Access 159
Fig. 4.16 Verification experiment with irradiance at 515 W/m^2: **a** original IR image, **b** grayscale treatment with improved box filter, **c** 3D image with improved box filter. *Source* Liao et al. (2022)—Open Access 160
Fig. 4.17 **a** Image captured with multispectral camera for geophysical evaluation. *Source* Ioja et al. (2022)—Open Access. **b** Multispectral map of agricultural crop for geophysical analysis. *Source* Ioja et al. (2022)—Open Access 162

Fig. 4.18 3D terrain map generated using LiDAR technology for geophysical analysis. *Source* Ioja et al. (2022)—Open Access ... 162
Fig. 4.19 Operation technique of LiDAR device for geophysical surveys. *Source* Ioja et al. (2022)—Open Access ... 163
Fig. 4.20 Pre-processing normalization of infrared images of building facade and roof. *Source* Ioja et al. (2022)—Open Access ... 164
Fig. 4.21 Photogrammetric reconstruction of 3D models (polygon and strip paths) and 2D orthophotos (a geometrically corrected image, typically derived from aerial or satellite imagery) of roof and facade surfaces (front or exterior surface of a building). *Source* Ioja et al. (2022)—Open Access ... 165
Fig. 4.22 In-flight EM sensor measurements from a drone: setup with GPS, ultrasonic sensor, Wi-Fi transceiver, vertical tail fin, and mounting jigs (field test in Kumamoto, Japan). *Source* Mitsuhata et al. (2022)—Open Access ... 167
Fig. 4.23 Variations in EM data with different drone-sensor spacings: standard deviations of **a** quadrature and **b** in-phase components. *Source* Mitsuhata et al. (2022)—Open Access ... 167
Fig. 4.24 Resistivity distributions and measurement points: **a** resistivity distributions at 0–1 m depth, **b** 2.8–3.2 m depth, and **c** 4.5–5 m depth derived from 1-D inversion of EM data and in-situ electrical conductivity (EC) measurements. **d** Resistivity measured using a soil EC meter, with white circles marking measurement points. White lines in (**a**) indicate profiles L2, L6, and L9 used for constructing resistivity cross-sections in Fig. 4.25a. *Source* Mitsuhata et al. (2022)—Open Access ... 169
Fig. 4.25 Resistivity cross sections and depth profiles: **a** cross sections along profiles L2, L6, and L9. **b** Depth profiles at stations L6-1 to L6-4, shown on the L6 profile cross section. *Source* Mitsuhata et al. (2022)—Open Access ... 170
Fig. 5.1 **a** Detailed schematic of the Full Waveform Inversion (FWI) process workflow. *Source* Georgia Institute of Technology. **b** Velocity profiles along line 2 from 2-D and 3-D Full Waveform Inversion (FWI) models. *Source* Boddupalli et al. (2021)—Open Access ... 179
Fig. 5.2 Geospatial overview of the North Sea Grane Oil Field. *Source* Wikipedia ... 180

Fig. 5.3 Upper panel: Map showing the location of the Galicia 3-D (G3D) experiment in the DGM (black box), with surrounding features including the Galicia Bank, Galicia Interior Basin (GIB), Porto Basin (PB), and Lusitania Basin (LB), and the Southern Iberia Abyssal Plain to the south. Bathymetry and topography are indicated by color. Lower panel: Time-migrated seismic profile across the G3D area, highlighting the Peridotite Ridge (PR), pre- and syn-rift sediments (below the dashed blue line), crystalline crust (below the dashed red line), and the S-reflector (dashed black line). *Source* Boddupalli et al. (2021)—Open Access ... 182
Fig. 5.4 Map showing ocean bottom seismometers (OBS) locations, multichannel seismic shooting, and 3-D FWI study area. *Source* Boddupalli et al. (2021)—Open Access ... 183
Fig. 5.5 Difference plots comparing the 2-D and 3-D models to the starting model, with red and black borders highlighting significant anomalies. **a** 2DB model difference plot, **b** recovered 2DB anomaly difference, **c** 3DA model difference plot, **d** recovered 3DA anomaly difference, **e** 3DB model difference plot, and **f** recovered 3DB anomaly difference. Black arrows highlight a blue anomaly that extends below the S-reflector from 2-D to 3-D features. *Source* Boddupalli et al. (2021)—Open Access ... 184
Fig. 5.6 On-vehicle Ground Penetrating Radar system for subsurface utility mapping. *Source* Mizutani et al. (2024)—Open Access ... 186
Fig. 5.7 Flowchart of the proposed methodology for subsurface utility mapping. *Source* Mizutani et al. (2024)—Open Access ... 187
Fig. 5.8 Reflected hyperbola from the target depth z_0 and the dielectric constant of soil ε. *Source* Mizutani et al. (2024)—Open Access ... 188
Fig. 5.9 Pipes under operating road in the (dataset from the road in operation collected in Japan) field site. *Source* Mizutani et al. (2024)—Open Access ... 189
Fig. 5.10 **a** Side view of all segments showing pipes on the operating road in Japan. *Source* Mizutani et al. (2024)—Open Access. **b** Ground-penetrating radar (GPR) scan of a subsurface utility line in a city infrastructure project, illustrating its effectiveness during urban redevelopment in Japan. Clear detection of 5 pipes in the dataset from the actual road. *Source* Mizutani et al. (2024)—Open Access ... 190

Fig. 5.11 **a** Migrated vertical profile at 0.1 m in the cross-survey direction. **b** Migrated profile at 1 m in the cross-survey direction. **c** Migrated profile at 2 m in the cross-survey direction. **d** Migrated profile at 2.9 m along the cross-survey direction. *Source* Zou et al. (2024)—Open Access 191
Fig. 5.12 **a** GPR peak frequency division profile at 0.1 m in the cross-survey direction. **b** GPR peak frequency distribution profile at 1 m along the cross-survey direction. **c** GPR peak frequency division profile at 2 m; along the cross-survey direction. **d** GPR peak frequency division profile at 2.9 m; along the cross-survey direction. *Source* Zou et al. (2024)—Open Access 192
Fig. 5.13 **a** Migrated horizontal slice at a depth of 0.21 m. **b** Migrated horizontal slice at a depth of 0.36 m. *Source* Zou et al. (2024)—Open Access 193
Fig. 5.14 Geographic map of the South-Link Line Railway (SLRT) in Taiwan. The red pentagon marks the location of the tunnel along this line, while the red-dashed line, blue line, and orange line represent the Neishih Fault, the Shihwen Anticline, and the Peihulushan Syncline, respectively. *Source* Puntu et al. (2021)—Open Access 194
Fig. 5.15 **a** Filtered radargram with a grayscale palette from GPR scanning in the Fangdian I tunnel, covering a total distance of 40 m. **b** Interpretation of the tunnel layers in the B-scan. The blue color represents the air layer, peach represents the second lining, green represents the first lining, and brown represents the wall rock. The contrast between these layers is due to variations in relative permittivity, which is used for identifying the lining layers. *Source* Puntu et al. (2021)—Open Access ... 194
Fig. 5.16 Clustered radargram showing rebar identification using Hierarchical Agglomerative Clustering (HAC). This figure demonstrates the identification of rebars (marked in blue) at a depth of approximately 0.2 m (around 4 ns). The clustered radargram illustrates how the rebar response remains distinct despite varying numbers of clusters (k = 2–10), confirming the robustness of the rebar detection process in the radar data. *Source* Puntu et al. (2021)—Open Access 195

Fig. 5.17 GPR instruments used at the site: **a** DJI M-300 drone with a 500 MHz shielded dipole antenna and altimeter for flight control. **b** Intelligent cart with a 500 MHz shielded dipole antenna, both capable of sensing up to 5 m below the surface. *Source* Frid and Frid (2024)—Open Access 197
Fig. 5.18 Radargram excerpts from drone-based (**a**) and ground-based (**b**) GPR surveys showing a geological transition at 1 m depth. The upper layer (yellow dashed lines) exhibits anomalies at 0.3 m, indicating irregular terrain, while the lower layer (light blue dashed lines) suggests potential archaeological walls made of limestone boulders, with key features highlighted by circles and arrows. *Source* Frid and Frid (2024)—Open Access 198
Fig. 5.19 Numerically mapped locations of overlapping ground-based and drone-based GPR findings, indicating potential archaeological objects, such as walls or concentrations of pottery. The irregular scattering pattern suggests a link to Early Bronze Age construction practices in the Hulata region, characterized by structures made from broken limestone and sandstone boulders. *Source* Frid and Frid (2024)—Open Access 199
Fig. 5.20 Resistivity tomography results for recirculation pipe H2 at the Denton landfill, showing moisture distribution before and after leachate recirculation. *Source* Alam et al. (2024)—Open Access 201
Fig. 5.21 ERT experiment on pavement slope and schematic of the survey location: **a** field setup of ERT experiment on pavement slope; **b** location map of survey line. *Source* Alam et al. (2024)—Open Access 201
Fig. 5.22 Change in resistivity with rainfall at different depths: **a** 0.91 m (2 ft.); **b** 3.05 m (10 ft.). *Source* Alam et al. (2024)—Open Access 202
Fig. 5.23 Cracks observed on the shoulder of the slope (**a**) and the failure at the crest of the slope (**b**). *Source* Alam et al. (2024)—Open Access 203
Fig. 5.24 **a** Failure condition of SH 183 slope; **b** resistivity lines along the slope. *Source* Alam et al. (2024)—Open Access 204
Fig. 5.25 **a** Test results of the resistivity line RL-1 at the crest; **b** test results of the resistivity line RL-2 at the middle of the slope. *Source* Alam et al. (2024)—Open Access 204
Fig. 5.26 **a** Field setup with resistivity meter; **b** conducting the ERT test. *Source* Alam et al. (2024)—Open Access 205

Fig. 5.27 Resistivity tomography results for lysimeter-2. *Source* Alam et al. (2024)—Open Access 206
Fig. 5.28 Number of case studies by method. *Source* Vallee et al. (2024)—Open Access 207
Fig. 5.29 Number of case studies by continent and method. *Source* Vallee et al. (2024)—Open Access 207
Fig. 5.30 Distribution of case studies by continent and method. *Source* Vallee et al. (2024)—Open Access 208
Fig. 5.31 Number of case studies by types of mineralization. *IOCG* Iron-Oxide Copper-Gold, *IONC* Iron-Oxide Nickel-Copper, *MS* Massive Sulfides, *MVLZ* Mississippi Valley Lead-Zinc, *SEDEX* Sedimentary Exhalative Lead-Zinc. *Source* Vallee et al. (2024)—Open Access 209
Fig. 5.32 The rose diagram of strike estimates (N5°E) for the magnetotelluric (MT) dataset collected from the Al-Mubazzarah Geothermal Area (AMGA). The strike analysis was conducted over the entire frequency range of the MT data to better understand the regional geological structures and fault orientations. The rose diagram reveals the dominant strike directions and provides insights into the extensional tectonic features of the study area, which play a significant role in the geothermal system by influencing fluid pathways and subsurface fluid migration. *Source* Saibi et al. (2021)—Open Access 211
Fig. 5.33 Resistivity cross-section derived from the 2D inversion of MT data. The model reveals shallow sediments with moderate resistivity, overlying a low-resistivity layer, which is followed by a progressively increasing resistivity zone at greater depths. This cross-section provides a detailed visualization of the subsurface structure, helping to delineate the geological layers and potential geothermal reservoir within the Al-Mubazzarah Geothermal Area. *Source* Saibi et al. (2021)—Open Access 212

Fig. 5.34 **a** Profile location used for dissecting inverted gravity and magnetic results from 3D potential field inversion, with black dots indicating the locations of MT stations. **b** Magnetic inversion (magnetic susceptibility in SI units), and **c** gravity inversion (density contrasts in g/cm^3). The gravity inversion cross-section reveals a potential normal fault dipping to the north at approximately 25°, or alternatively, it may represent a buried erosional surface (non-conformity). If interpreted as a fault, it could link the Al-Mubazzarah Geothermal Area (AMGA) to the Ain Faidha hot springs, as fault orientations in this region are predominantly E–W. A probable fault is also observed in the resistivity cross-section derived from the MT data. *Source* Saibi et al. (2021)—Open Access 213

Fig. 5.35 Resistivity and S-wave velocity characteristics of different lithologies in Chengdu's International Bio-City reveal key geological insights. The resistivity varies across layers, with lower values for clay and mudstone (10–40 Ωm) and higher values for sandy mudstone and gypsum-rich layers (above 100 Ωm). S-wave velocities also differ, showing lower values in soft sediments like clay and mudstone (180–300 m/s) and higher values in consolidated layers like weathered sandstone (1300–2100 m/s), indicating variations in rock stiffness and lithological composition. These characteristics were crucial in constructing a detailed 3D geological model for the area. *Source* Wang et al. (2023)—Open Access 217

Fig. 5.36 A comprehensive view of the resistivity data and geological interpretation for the study area. **a** The measured apparent resistivity pseudosection shows the raw resistivity data along the profile. **b** The calculated apparent resistivity pseudosection represents the resistivity model based on theoretical assumptions. **c** The inverted resistivity section illustrates the result of resistivity inversion, highlighting subsurface variations, though it shows discrepancies due to urban infrastructure interference. **d** The illustrated geological section combines the resistivity data with borehole information, providing a final interpretation of the geological layers in the area. *Source* Wang et al. (2023)—Open Access 218

Fig. 5.37 **a** 3D geo-electric structure of the study area, showing the resistivity distribution at various depths. **b** 3D geological model of the study area, integrating geophysical and borehole data to depict key subsurface layers, including clay, sandy pebbles, weathered mudstone, and gypsum-rich zones. *Source* Wang et al. (2023)—Open Access 219
Fig. 5.38 Geophysical Data Acquisition Map for Pre-Foundation Investigation in Oka-Akoko, Nigeria. *Source* Ademila (2022)—Open Access 220
Fig. 5.39 **a** VLF-EM plot of real components and **b** corresponding 2-D inverted section of Traverse 1 in Oka-Akoko, Nigeria. *Source* Ademila (2022)—Open Access 221
Fig. 5.40 **a** VLF-EM plot of real components and **b** corresponding 2-D inverted section of Traverse 4 in Oka-Akoko, Nigeria. *Source* Ademila (2022)—Open Access 222
Fig. 5.41 **a** Two-dimensional (2D) subsurface structure of Oka-Akoko, Nigeria and **b** geoelectrical section obtained from VES curves 1 to 5 of Traverse 1 in Oka-Akoko, Nigeria. *Source* Ademila (2022)—Open Access 223
Fig. 5.42 **a** Two-dimensional (2D) subsurface structure and **b** geoelectrical section obtained from interpretation of VES curves 15 to 19 along Traverse 4 in in Oka-Akoko, Nigeria. *Source* Ademila (2022)—Open Access 224
Fig. 5.43 The geoelectrical section of Traverses 5 and 6 in Oka, Nigeria, obtained from the interpretation of VES data, showing subsurface resistivity variations. *Source* Ademila (2022)—Open Access 225
Fig. 5.44 **a** Fresh bedrock resistivity map of the engineering site in Oka, Nigeria. **b** 3D perspective resistivity view of the fresh bedrock of the site in Oka, Nigeria. *Source* Ademila (2022)—Open Access 226
Fig. 5.45 Subsurface characterisation map of the engineering site in Oka, Nigeria, showing the variations in subsurface geological features and suitability for construction. *Source* Ademila (2022)—Open Access 227
Fig. 6.1 Location and geology of the study area. **a** Study area situated in Wuwei city, Gansu Province, China; **b** Location of the Leitai cultural heritage site, east of the Hexi Corridor; **c** Satellite image of the Leitai platform, generated using Locaspace (Version 4.3.9), available at http://www.locaspace.cn/ **d** Site photograph of the Leitai platform; **c** Geological details of the Leitai platform. *Source* Lu et al. (2025)—Open Access 237

Fig. 6.2 SP results. **a** Distribution of spontaneous potential on the loess platform; **b** Measured SP curve and fitted curve; **c** SP inversion results; **d** Comparison between SP inversion results and the inverted resistivity cross-section. *Source* Lu et al. (2025)—Open Access 238
Fig. 6.3 Inferred channel locations and subsequent excavation verification results. **a** SP plane with integrated inferred seepage channels; **b** ERT slice at 2 m depth with integrated inferred seepage channels; **c** ERT slice at 5 m depth with integrated inferred seepage channels. *Source* Lu et al. (2025)—Open Access 239
Fig. 6.4 The model obtained using robust inversion (L1-norm) of the ERT data from profile ERT4 after 10 iterations (absolute error = 1.4%) is presented. The borehole log from W4 is superimposed on the 2D cross section for comparison. *Source* Ibraheem et al. (2024)—Open Access 240
Fig. 6.5 Shows the 1D inversion results for station TEM27. The Marquardt–Levenberg (MQ) model is shown in red, while the Occam models with roughness values R1 and R2 are in green and blue, respectively. The alongside equivalent (EQ) models are in light grey. The data fit (χ) is shown in the legend next to the model. *Source* Ibraheem et al. (2024)—Open Access 241
Fig. 6.6 **a** The RTP magnetic map after applying an upward continuation filter at a height of 5 m. *Source* Ibraheem et al. (2024)—Open Access 241
Fig. 6.7 Shows depth slices at 2, 5, 10, 12, 15, 20, 25, and 30 m, derived from the 1D Occam inversion results of the TEM data. The waste body is visible up to a depth of 10 m. Below this depth, a contaminated gravelly sand layer extends from 20 to 23 m. The underlying layer, made up of saturated coal and clay, is also likely contaminated due to the high porosity of the gravelly sand layer above it. There is a noticeable difference in resistivity values at similar depths inside and outside the landfill, especially between the gravelly sand and the coal/clay layers. This could suggest the presence of contamination leachate. The average χ value shows a good fit. *Source* Ibraheem et al. (2024)—Open Access 242

Fig. 6.8 Presents a 2D resistivity cross-section derived from the 1D Occam inversion results with the roughness parameter R1 for the TEM data along profile ERT4. The lower panel shows the χ values for both the Occam (R1 and R2) and MQ inversions. The log of borehole W4 is included between TEM stations T10 and T11. *Source* Ibraheem et al. (2024)—Open Access 243
Fig. 6.9 1D Geo-electric Section for the TE Mode along the Barcuri-Bakreswar-Tantloi Traverse. The vertical scale represents depth (in meters), while resistivity values are expressed in Ohm-m. *Source* Naskar (2017) 257
Fig. 6.10 **a** Collinear dipole–dipole resistivity profiling along the Bakreswar-Asanshuli traverse. **b** Collinear dipole–dipole profiling along the Chandrapur-Tantipara traverse. *Source* Naskar (2017) 258
Fig. 6.11 1D inversion along the Bonhat-Sherpur profile (Trav-1), eastern part of the Birbhum coalfield, West Bengal. *Source* Naskar and Lahiri (2014) 267
Fig. 7.1 Bouguer gravity anomaly map of Tonk and Dausa districts, Rajasthan (Toposheets 45 N/13, 14 and 54 B/1, 2, 5, 6) for risk and hazard mitigation analysis. *Source* Geological Survey of India (2013)—Naskar and Lashkari 284
Fig. 7.2 Free air bouguer gravity anomaly map of toposheets 45 N/13, 14 and 54 B/1, 2, 5, 6 in Tonk and Dausa districts, Rajasthan, for risk and hazard mitigation analysis. *Source* Geological Survey of India (2013)—Naskar and Lashkari 285
Fig. 7.3 Map of California highlighting the seismic hazard hotspots, indicating areas with the highest risk of earthquake activity. *Source* Li (2021)—Open Access 308
Fig. 7.4 Surface velocity maps for the SAFS in the ITRF2014 system, including decomposed velocity components and GNSS model comparisons. *Source* Xu et al. (2021)—Open Access 309
Fig. 7.5 Strain rate maps for the San Andreas Fault system showing GNSS and InSAR/GNSS-derived components and their differences. *Source* Xu et al. (2021)—Open Access 310
Fig. 7.6 Map showing the locations of explosions (SP01, SP02, SP03) and seismic arrays at Parkfield and Cienega Valley, with borehole seismic network stations and key fault features. Updated from Li et al. (1997). *Source* Li (2021)—Open Access 312

Fig. 7.7 Shows seismic data from the San Andreas Fault at Parkfield. Panel (**a**) compares vertical-component seismograms from repeated events in 2002 and 2004, with P-wave arrivals aligned. Cross-correlation of the seismograms at stations ST0 and W8 within and outside the fault zone is shown. Panel (**b**) presents changes in S-wave velocity within the rupture zone from the M6 Parkfield earthquake. A 1.3% decrease in S-wave velocity occurred between the two events, likely due to co-seismic damage. Aftershock measurements show a 1.2% velocity increase, indicating fault healing. These results suggest that fault strength recovers quickly post-mainshock, but healing slows over time. *Source* Li (2021)—Open Access 313

Fig. 7.8 Illustrates the flow chart of the configuration process for the landslide monitoring system. It outlines the step-by-step procedure for setting up the system, beginning with user login and sensor configuration, followed by the creation of nodes and assignment of specific sensor parameters. Each sensor's alert stage is set (Green, Yellow, Orange, or Red) based on predefined threshold values. The process also includes setting user details and ensuring the proper display of node positions and real-time data visualization on the monitoring interface. This configuration ensures seamless integration of sensors, nodes, and the central server for continuous monitoring and early warning generation. *Source* Chaulya et al. (2024)—Open Access 317

Fig. 7.9 Shows a view of the Karshingsa landslide-prone area, highlighting the toe erosion caused by the Dikrong River. This erosion, primarily driven by continuous rainfall and the river's water flow, exacerbates slope instability in the region, making it highly susceptible to landslides. The image emphasizes the significant role of riverbank erosion in triggering landslide events, a critical factor monitored by the landslide monitoring system deployed in this area. *Source* Chaulya et al. (2024)—Open Access 318

Fig. 7.10 **a** Optical and Geospatial analysis of the Gokseong Landslide Site **a** Sentinel-2 image showing the landslide-affected areas (Sites 1 and 2) after the event, **b** pre-event image of Site 1 captured by the Korea National Geographic Information Institute (KNGII) in 2019, **c** location map of Gokseong County in South Korea, and **d** cross-sectional profiles illustrating the topographic changes before and after the landslide. The red polygons highlight the landslide-impacted areas. **b** Digital photographs depicting key features of the Gokseong landslide. **a** Overview of the landslide at Site 1, **b** initiation zone of Site 1, **c** deposition zone of Site 1, and **d** overview of the landslide at Site 2. *Source* Choi et al. (2024)—Open Access 320
Fig. 7.11 Digital elevation models of the Gokseong landslide region. **a** Pre-event elevation model, **b** Post-event elevation model, and **c** Elevation difference map highlighting the source and deposition areas. *Source* Choi et al. (2024)—Open Access 321
Fig. 7.12 Classification of remote sensing methods for flood prediction. *Source* Munawar et al. (2022)—Open Access 322
Fig. 7.13 Year-wise distribution of articles from each category of remote sensing methods for flood prediction. *Source* Munawar et al. (2022)—Open Access 323
Fig. 7.14 **a** Methods adopted in studies for flood prediction using remote sensing and machine learning techniques. **b** Schematic diagram of the proposed deep learning-based system for real-time flood detection and extent mapping. *Source* Munawar et al. (2022)—Open Access 324
Fig. 7.15 Study area: upper Bengawan Solo Watershed, Java, Indonesia (16,100 km^2). *Source* Jumadi et al. (2024)—Open Access 326
Fig. 7.16 Reclassified parameters: **a** Slope (Sl), **b** Rainfall (RF), **c** Digital Density (DD), **d** Soil Moisture (SM), **e** Land Use/Land Cover (LULC), **f** Elevation (El), **g** Drainage Ratio (DR), **h** Normalized Difference Vegetation Index (NDVI), **i** Curvature (Cu), **j** Forest Area (FA), **k** Topographic Wetness Index (TWI). *Source* Jumadi et al. (2024)—Open Access 328
Fig. 7.17 Vulnerability class in the Upper Bengawan Solo Watershed, Java, Indonesia. *Source* Jumadi et al. (2024)—Open Access 328
Fig. 7.18 Flood risk map of the Upper Bengawan Solo Watershed, Java, Indonesia. *Source* Jumadi et al. (2024)—Open Access 329

Fig. 7.19 European summer floods, 2021: red indicates flooded areas, with blue showing the mean annual flood extent. Dark gray represents previously mapped floods, and light gray shows city lights. Flood data were collected using various remote sensing sources, including MODIS and Sentinel-1 SAR. *Source* Schumann et al. (2023)—Open Access ... 331

Fig. 7.20 Comparison of wet season flooding—a comparison between a 2-D flood model simulation and a Landsat satellite image of the Niger Inland Delta. *Source* Schumann et al. (2023)—Open Access ... 332

Fig. 8.1 Photograph of the project site: hydroelectric dam spillway in Alabama, USA. *Source* Bigman and Day (2022)—open access ... 377

Fig. 8.2 Example of GPR response from PVC drainage pipes in the hydroelectric dam spillway, Alabama, USA—**a** exposed PVC drainage pipe embedded in the spillway, **b** A-scan (1D, Amplitude Scan) trace rendering of the response from the PVC pipe, indicating the polarity of the response, **c** GPR B-scan (2D, bridge scan or cross-sectional scan) profile of data collected over the PVC drainage pipes onsite, approximating the expected response from a concrete/air boundary in the spillway. *Source* Bigman and Day (2022)—open access ... 378

Fig. 8.3 Photograph of data collection with Proceq GP8000 at hydroelectric dam spillway, Alabama, USA. *Source* Bigman and Day (2022)—open access ... 379

Fig. 8.4 Examples of two shallow responses indicative of possible delamination in Alabama, USA: **a** 10 cm deep and **b** almost 15 cm below the surface. Both responses extend for more than 100 cm. *Source* Bigman and Day (2022)—open access ... 379

Fig. 8.5 Examples of two deeper responses indicative of possible voids, which turned out to be weathered rock, in Alabama, USA: **a** 38 cm and **b** almost 29 cm below the surface. Both responses extend for more than 200 cm. *Source* Bigman and Day (2022)—open access ... 380

Fig. 8.6 Photograph of the spillway in Alabama, USA showing large (right) and small (left) concrete patches from repair and resurfacing following the GPR investigation. *Source* Bigman and Day (2022)—open access ... 380

Fig. 8.7 ERI data distribution for hydrogeological and land subsidence studies in Yilan, Yunlin, and Taichung basins. *Source* Chang et al. (2023)—open access ... 382

Fig. 8.8 ERI data distribution and inversion models for the Yunlin Chousui River Alluvial Fan: **a** apparent resistivity data, **b** calculated apparent resistivity data, and **c** an inverted resistivity model. *Source* Chang et al. (2023)—open access ... 383

Fig. 8.9 Flow chart of resistivity imaging data acquisition, processing, and inversion for the Yilan Alluvial Basin. *Source* Chang et al. (2023)—open access ... 384

Fig. 8.10 Mean vertical deformation velocities measured from the two data stacks: **a** Mean vertical deformation acquired from ASAR data during 2003–2010, highlighting significant subsidence in Chaoyang, Shunyi, and parts of Tongzhou districts. **b** Mean vertical deformation acquired from Radarsat-2 data during 2010–2013, showing a similar subsidence pattern, with maximum deformation occurring in Chaoyang, Shunyi, and Changping districts. **c** Disparities of mean deformation velocities between the two time periods, showing notable differences in Changping, the eastern part of Chaoyang, and the western part of Tongzhou district, indicating rapid deformation in these regions during the second period (2010–2013). *Source* Liu et al. (2020)—open access ... 386

Fig. 8.11 The corresponding relationship between the dynamic variation of groundwater and mean subsidence velocity: **a** the subsidence result derived from 2003 to 2010, with the contour representing the groundwater level of 2007. Significant subsidence is observed in Chaoyang, Shunyi, and parts of Tongzhou districts, correlating with lower groundwater levels. **b** The variation of groundwater during 2003–2010, showing a marked decline in groundwater levels, particularly in the northeastern part of the Beijing Plain, which corresponds to the observed subsidence areas. **c** The subsidence result derived from 2010 to 2013, where the contour represents the groundwater level of 2011. Similar subsidence patterns are observed in Chaoyang, Shunyi, and Tongzhou, with a clear relationship to groundwater level changes. **d** The variation of groundwater during 2010–2013, indicating continued decline in groundwater levels, reinforcing the correlation with land subsidence in these key districts. *Source* Liu et al. (2020)—open access ... 387

Fig. 8.12 **a** Schematic diagram of a distributed fiber optic sensor system, and **b** application of DFOS technology in civil engineering for structural health monitoring of a pile. *Source* Soga et al. (2018)—open access 389

Fig. 8.13 Example of the operation of a distributed fiber optic sensor, illustrating spatial resolution, readout, measurement resolution, and sensing distance where T represents temperature, σ represents stress, and ε represents strain. *Source* Soga et al. (2018)—open access 390

Fig. 9.1 **a** Location of the study area in Northeast China and **b** geological setting of the study area. The inset in **a** shows the location of the study area within China. Key geological features include the North China Craton (NCC), Central Asian Orogenic Belt (CAOB), and several major faults such as the Chifeng-Kaiyuan Fault (CKF), Yilan-Yitong Fault (YYF), and Dunhua-Mishan Fault (DMF). Sections of the Huangchagou (Pt1h3, Pt1h2, Pt1h1) and Mayihe (Pt1m3, Pt1m2, Pt1m1) Formations are labeled and Qh: Quaternary sediment. Magnetic investigation areas and fault lines are indicated, along with measurement sites for magnetic method investigations (black triangles). *Source* Liu et al. (2025)—open access 400

Fig. 9.2 Magnetic anomaly data and geological mapping of profile AB in the study area of Northeast China. *Source* Liu et al. (2025)—open access 400

Fig. 9.3 Magnetic anomaly in the study area of Northeast China. The blue dashed lines and labels indicate the three magnetic anomalous zones. Magenta dots and labels represent the self-potential sites and the survey line names, respectively (Liu et al. 2025—open access) 401

Fig. 9.4 Self-potential anomaly data and geological mapping of profile CD in the study area of Northeast China (Fig. 9.1b) (Liu et al. 2025—open access) 402

Fig. 9.5 Self-potential anomaly in the study area of Northeast China. Magenta lines represent the anomaly zones (Z1), black dotted lines indicate the peak anomaly values, and orange lines represent the designed trenches (Liu et al. 2025—open access) 402

Fig. 9.6 Borehole histogram of drillings ZK4-5, ZK2-5, and ZK1-1 in the study area of Northeast China. *Source* Liu et al. (2025)—open access 403

Fig. 9.7 Graphite ore determined based on drillings and trenches in the study area of Northeast China. The purple polygons represent the graphite ore. The yellow dashed line indicates profile 4. The black lines on the ore bodies represent the trenches. *Source* Liu et al. (2025)—open access . . . 404
Fig. 9.8 Case studies of geophysical methods by continent and technique: a regional analysis of magnetic and electromagnetic surveys in mineral exploration. *Source* Vallee et al. (2024)—open access . . . 405
Fig. 9.9 Number of case studies by types of mineralization: a regional overview of mineral targets (IOCG: iron-oxide copper–gold, IONC: iron-oxide nickel-copper, MS: massive sulfides, MVLZ: mississippi valley lead–zinc, SEDEX: sedimentary exhalative lead–zinc). *Source* Vallee et al. (2024)—open access . . . 406
Fig. 9.10 First vertical derivative of RTP data from red lake gold camp, overlain with calculated magnetic ridges, inferred faults, and locations of past and present gold mines (total gold production indicated). *Source* Kwan and Reford (2025)—open access . . . 424
Fig. 9.11 Contact occurrence density (COD) map of the red lake gold camp, overlain with calculated magnetic ridges, inferred faults, and locations of Campbell-red lake (1, 2) and Cochenour-Willans (3) gold mines. *Source* Kwan and Reford (2025)—open access . . . 425
Fig. 9.12 Orientation entropy (OE) map of the red lake gold camp, overlain with calculated magnetic ridges, inferred faults, and locations of Campbell-Red Lake (1, 2) and Cochenour-Willans (3) gold mines. *Source* Kwan and Reford (2025)—open access . . . 426
Fig. 9.13 RTP magnetic data of the eastern Athabasca basin, overlain with known faults and uranium deposits. *Source* Kwan and Reford (2025)—open access . . . 427
Fig. 9.14 Structural vectors over the first vertical derivative of the RTP data, red lake gold camp. *Source* Kwan and Reford (2025)—open access . . . 428
Fig. 9.15 Contact occurrence density (COD) map of the eastern Athabasca Basin. *Source* Kwan and Reford (2025)—open access . . . 429
Fig. 9.16 Orientation entropy (OE) map of the eastern Athabasca Basin. *Source* Kwan and Reford (2025)—open access . . . 430

Fig. 10.1 Soil sample resonant column test setup used for evaluating the dynamic properties of soils in seismic and infrastructure risk assessments (*Source* Gabrys 2025—Open Access) 448
Fig. 10.2 Coal power plant flow diagram illustrating major pollution control components: FGD—Flue-gas desulfurization; ESP—Electrostatic precipitators; SCR—Selective catalytic reduction (*Source* Gabrys 2025—Open Access) 449
Fig. 10.3 Global reported natural disasters from 1970 to 2024 (to April), sorted by hazard type. Both weather- and non-weather-related disasters are included (*Source* Senger 2024—Open Access) 451
Fig. 10.4 Timeline of global disaster-related deaths from 1900 to 2020, with bubble sizes indicating the estimated annual death toll. This visual illustrates long-term mortality trends from natural disasters and emphasizes the need for proactive risk reduction strategies (*Source* Senger 2024—Open Access) 452
Fig. 10.5 Conceptual diagram illustrating the alignment between engineering geology and the United Nations Sustainable Development Goals (SDGs). The figure highlights areas where geological science contributes to hazard mitigation, infrastructure development, environmental protection, and sustainable resource use (*Source* Senger 2024—Open Access) 455
Fig. 10.6 Changes in society's use of minerals over time and global mineral production trends. **a** Increase in mineral use by society over time, highlighting the growing reliance on diverse mineral resources in modern times. **b** Inflation-adjusted global value (in 1998 US dollars) of metal and mineral production by major sector, from 1900 to 2015 (*Source* Senger 2024—Open Access) 457
Fig. 10.7 Minerals required in the transport and power generation sectors, including selected green energy technologies, nuclear energy, and fossil fuels. The figure illustrates the dependency of emerging energy systems on a wide range of critical minerals, emphasizing the role of geophysical methods in their sustainable discovery and management (*Source* Senger 2024—Open Access) 459
Fig. 10.8 Global energy mix development. **a** Global energy consumption from 1850, with projection to 2150. **b** Timeline illustrating the major phases of the energy transition from 1800 to 2016, alongside the global rise of democracy (*Source* Senger 2024—Open Access) 462

Fig. 10.9 Key factors influencing global water crises and hotspot locations. The figure shows an overview of the social, technical, and hydrological factors linked to six major types of water crises. The inset map highlights global regions most affected by these crises (*Source* Senger 2024—Open Access) . . . 464
Fig. 10.10 Visual framework of the science communication landscape (*Source* Senger 2024—Open Access) . . . 467
Fig. 10.11 **a** Data science at the intersection of mathematics, computer technology, and geosciences. **b** Integration of various data types, data analysis contents, and data products (*Source* Senger 2024—Open Access) . . . 471
Fig. 10.12 Essential climate variables and their interactions (modified from NASA) (*Source* Senger 2024—Open Access) . . . 472
Fig. 10.13 Blue-Green Infrastructure supports climate resilience and provides multiple benefits in Chinese cities. It combines natural water management systems (blue) and green spaces to improve both urban and ecological functions (*Source* Kadic et al. 2025—Open Access; Siehr et al. 2022; Ramboll 2016; Catalano and Baumann 2017) . . . 477
Fig. 10.14 Number of selected peer-reviewed articles on the role of geophysical information (GI) in climate change adaptation, categorized by year of publication (2010–2024) (*Source* Kadic et al. 2025—Open Access) . . . 478
Fig. 10.15 Illustrates a typical layout of green infrastructure elements within a city, showing how they contribute to stormwater management, cooling, and improved quality of life (*Source* Kurniawan et al. 2024) . . . 481
Fig. 10.16 Climate-resilient water management framework highlighting sustainable groundwater use, recharge enhancement, and policy-integration approaches under climate stress (*Source* Kadic et al. 2025—Open Access) . . . 482
Fig. 10.17 Simplified version of a global groundwater resources map, highlighting major regional aquifer systems (*Source* Taylor et al. 2013) . . . 483
Fig. 10.18 Overview of the 17 Sustainable Development Goals, highlighting the geoscience links based on the number of geoscientific categories addressing each goal. SDG icons courtesy of the United Nations (United Nations 2022). The full list of goals and related targets is available from United Nations (2022) (*Source* Senger 2024—Open Access) . . . 484

Fig. 11.1 Performance-based resilience components of a Critical Infrastructure System (CIS). The framework emphasizes system functionality over time, integrating aspects of robustness, redundancy, resourcefulness, and rapidity (*Source* Sathurshan et al. 2022—Open Access) 501
Fig. 11.2 The Preferred Reporting Items for Systematic Reviews and Meta-Analyses (PRISMA) flowchart used to identify and select studies relevant to infrastructure resilience frameworks (*Source* Sathurshan et al. 2022—Open Access) 502
Fig. 11.3 The summary of studies evaluated by the year of publication (*Source* Sathurshan et al. 2022—Open Access) 503
Fig. 11.4 Summary of infrastructure resilience framework development approaches (*Source* Sathurshan et al. 2022—Open Access) 504
Fig. 11.5 The geographic scope of the infrastructure resilience frameworks evaluated (*Source* Sathurshan et al. 2022—Open Access) 505
Fig. 11.6 Summary of studies categorized by infrastructure type (*Source* Sathurshan et al. 2022—Open Access) 505
Fig. 11.7 Attributes of the damage-analysis-based framework from past studies in terms of social, functional, and physical metrics, considering the indicators and methodology used (*Source* Sathurshan et al. 2022—Open Access) 506
Fig. 11.8 Classification of resilience factors along with road performance curve and disaster phase (*Source* Irankunda et al. 2024—Open Access) 512
Fig. 11.9 Affected part of road infrastructure at mileage points MP15 (**a**) and MP28 (**b**). The visible damage in these segments highlights the importance of sustainable recovery efforts that not only restore functionality but also account for long-term resilience against similar events (*Source* Irankunda et al. 2024—Open Access) 513
Fig. 11.10 NR9 recovery management roadmap (*Source* Irankunda et al. 2024—Open Access) 514
Fig. 11.11 NR9 after recovery at MP15 (**a**) and MP28 (**b**), respectively (*Source* Irankunda et al. 2024—Open Access) ... 515
Fig. 11.12 Major hydro-meteorological hazards and their impact on railway infrastructure in Assam, India, 2022 (*Source* Roy et al. 2023—Open Access) 526
Fig. 11.13 Railway tunnels (**a, b**) and bridges (**c, d**) along the railway network in Manipur (*Source* Roy et al. 2023—Open Access) 526

Fig. 11.14 Railway infrastructure and earthquake sites in Manipur, Northeast India (*Source* Roy et al. 2023—Open Access) 527

Fig. 11.15 Flood risk exposure of major highways in New York city: overlay with 100 year floodplain map (*Source* Singh et al. 2021—Open Access) 529

Fig. 11.16 Roads in Kerala, India destroyed by landslides following a flood in August 2018 (*Source* World Bank) 531

Fig. 11.17 Damaged roads and infrastructure in the Pamba Region of Kerala, India following a series of deadly floods and landslides in August 2018 (*Source* World Bank) 532

Fig. 11.18 Topographical map of the road tunnel network in Suzhou: Showing the spatial layout and elevation context of key underground routes (*Source* Huang et al. 2022—Open Access) 533

Fig. 11.19 Network disruption in the Suzhou road tunnel system under varying flood levels: Illustrating the impact of rising water on tunnel connectivity and efficiency (*Source* Huang et al. 2022—Open Access) 534

Fig. 11.20 Location of Japan on the world map—context for the Kansai International Airport case study (*Source* Coalition for Disaster Resilient Infrastructure) 538

Fig. 11.21 Osaka maps and orientation: Osaka, Kansai, Japan (*Source* World Guides) 539

Fig. 11.22 Framing infrastructure as social, ecological, and technological (SET) systems (*Source* Belokas et al. 2024—Open Access) 541

Fig. 11.23 SET environment interaction due to ecological component; case of Tsakona area (Greece) landslide: **a** prior to onset of major activation, **b** After major activation (*Source* Belokas et al. 2024—Open Access) 542

Fig. 11.24 The Tsakona landslide: **a** the network scale—showing disruption across regional connectivity, **b** the asset scale—localized failure of the highway section (*Source* Belokas et al. 2024—Open Access) 543

Fig. 11.25 The Tempi rockfall (Greece): **a** the network scale—regional transport impact, **b** the asset scale—rockfall site adjacent to roadway infrastructure (*Source* Belokas et al. 2024—Open Access) 543

Fig. 12.1 Fictitious military base map (*Source* Mallery et al. 2022) 556

Fig. 12.2 System resilience curve (*Source* Mallery et al. 2022) 560

Fig. 12.3 Recommended mission priorities for critical load determination (*Source* Mallery et al. 2022) 561

Fig. 12.4 Initial microgrid composition (*Source* Mallery et al. 2022) 562

Fig. 12.5 Final microgrid composition (*Source* Mallery et al. 2022) 562

Fig. 12.6 Energy storage system (ESS) and generator fuel storage in 72-H generator failure scenario (*Source* Peterson et al. 2021—Open access) 563
Fig. 12.7 ESS storage during 72-H loss of the PV array and the connection between the BUS1-BUS2 feeders. BUS1 and BUS2 typically refer to redundant or separate power distribution lines (*Source* Peterson et al. 2021—Open access) 564
Fig. 13.1 Resilience Venn diagram. Defines installation energy resilience by showing the overlap between energy resilience and installation resilience, focusing on mission-critical energy use (*Source* Varley et al. 2022—Open Access) 579
Fig. 13.2 Standard resilience curve—installation versus mobile microgrid recovery. A resilience performance comparison curve between an installation microgrid and a mobile microgrid post-disruption (*Source* Varley et al. 2022—Open Access) 586
Fig. 13.3 CONOPS diagram—mobile microgrid deployment to contingency operation. Concept of operations (CONOPS) showing the logistical deployment and setup sequence of a mobile microgrid in a contingency environment (*Source* Varley et al. 2022—Open Access) 587
Fig. 13.4 System behavior snapshot during stress test conditions. Operational simulation of a mobile microgrid during a 14-day period under high load and solar input, illustrating system performance under stress (*Source* Varley et al. 2022—Open Access) 591
Fig. 14.1 Location analysis of the ruins of the imperial city of the Minyue Kingdom in Chengcun village, Wuyishan City, Fujian Province, China; the objective is to assess the site's geographical setting for archaeological excavation, structural preservation, and display planning (*Source* Zhou et al. 2024—Open Access) 611
Fig. 14.2 Typical erosion of the soil remains (image source: provided by the excavation team for the Ruins of the Imperial City of the Minyue Kingdom) (*Source* Zhou et al. 2024—Open Access) 612
Fig. 14.3 Typical collapse of soil remains (image source: provided by the excavation team for the Ruins of the Imperial City of the Minyue Kingdom) (*Source* Zhou et al. 2024—Open Access) 612

Fig. 14.4 Typical growth of trees and shrubs in the soil remains (image source: provided by the excavation team for the Ruins of the imperial city of the Minyue Kingdom) (*Source* Zhou et al. 2024—Open Access) 613

Fig. 14.5 Aerial photo of the remains of the east city wall (image source: provided by the excavation team for the Ruins of the imperial city of the Minyue Kingdom) (*Source* Zhou et al. 2024—Open Access) 613

Fig. 14.6 Rendering of the design scheme for the east city gate relic exhibition project (image source: provided by the excavation team for the Ruins of the imperial city of the Minyue Kingdom) (*Source* Zhou et al. 2024—Open Access) 614

Fig. 14.7 Map of Pompeii. Location of the funerary areas outside Porta Sarno (map: Pompeii archaeological park) (*Source* Alapont et al. 2025—Open Access) 615

Fig. 14.8 General plan of the excavated trenches (map: Pilar Mas and Joaquin Alfonso) (*Source* Alapont et al. 2025—Open Access) 615

Fig. 14.9 Funerary monument located at a distance of 0.94 m from the boundary wall situated behind, suggesting that the monument is within a bounded enclosure (photo: Joaquin Alfonso) (*Source* Alapont et al. 2025—Open Access) 616

Fig. 14.10 The aspergillum of laurel leaves preserves elements of green paint (photo: Alfio Giannotti) (*Source* Alapont et al. 2025—Open Access) 618

Fig. 14.11 Ring on the proximal phalanx of the ring finger (photo: Alfio Giannotti) (*Source* Alapont et al. 2025—Open Access) 619

Fig. 14.12 Sarcophagus with battle scenes from the graveyard in the via Portuense, in Rome, uncovered in the Viale Gianicolense (1226) (photo: Llorenç Alapont) (*Source* Alapont et al. 2025—Open Access) 620

Fig. 14.13 Satellite imagery of the Myra Valley. **a** Location and scope of the Myra Valley Corridor within Tunisia, **b** Landsat 8 OLI data of the corridor. (RGB: bands 7, 4, and 3), and **c** Elevation of the corridor derived from ASTER GDEM data (*Source* Yao et al. 2023—Open Access) 625

Fig. 14.14 Presents examples of GPR applications in archaeological and heritage investigations: **a** A 3D overlaid time-slice produced at the Roman site of Aquis Querquennis (Bande, Spain), revealing part of a buried barrack structure; **b** A 500 MHz 3D cube and volumetric reconstruction of an ancient underground tunnel with a central corridor and lateral galleries; **c** High-frequency radargrams (2.3 GHz and 1.2 GHz) from an anthropomorphic grave in the rock necropolis of San Vítor de Barxacova (Ourense, Spain), illustrating sub-vertical fractures marked in red (*Source* Lombardi et al. 2022—Open Access) ... 627

Fig. 14.15 Location of the Bibracte archaeological site on Mont Beuvray, Burgundy, France. This map illustrates the site's strategic setting within a rural landscape, highlighting its cultural and environmental context essential for heritage management planning (*Source* Ababneh 2021—Open Access) ... 631

Fig. 14.16 Combined interpretive techniques including panels, tactile screens, 3D reconstruction, dioramas, and maquette (*Source* Ababneh, 2021—Open Access) ... 632

Fig. 14.17 Management problems faced by Bibracte using three circles and levels (*Source* Ababneh 2021—Open Access) ... 633

Fig. 15.1 Geophysical methods and applications project flowchart: interdisciplinary approaches in environmental and engineering challenges ... 649

Fig. 15.2 Location map of coastal districts in Gujarat State, India (*Source* Patel 2022—Open Access) ... 652

Fig. 15.3 Methodology flow-chart of Shoreline and erosion monitoring in Gujarat state (*Source* Patel 2022—Open Access) ... 652

Fig. 15.4 Normalized sea surface temperature (SST) data covering Gujarat coastline from 1860 to 2020: analysis of Gulf of Kachchh, Khambhat, and Saurashtra coast (*Source* Patel 2022—Open Access) ... 653

Fig. 15.5 Sea surface temperature (SST) and annual mean sunspot activity data from 1960 to 2020: analysis of Gujarat coastline, including Gulf of Kachchh, Khambhat, and Saurashtra coast (*Source* Patel 2022—Open Access) ... 653

Fig. 15.6 Estimated coastal erosion under different categories during the last 42 years (1978–2020) in Gujarat state (*Source* Patel 2022—Open Access) ... 654

Fig. 15.7 **a** Relationship between sea surface temperature (SST) and CO_2 emissions: analysis of Gujarat coastline, including Gulf of Kachchh, Khambhat, and Saurashtra coast (*Source* Patel 2022—Open Access). **b** Regression analysis of sea surface temperature (SST) and CO_2 emissions: analysis of Gujarat coastline, including Gulf of Kachchh, Khambhat, and Saurashtra coast (*Source* Patel, 2022—Open Access) ... 655
Fig. 15.8 Resistivity profiles illustrating leachate migration in uncontrolled landfill sites in Portugal (*Source* Martinho 2023—Open Access, after Casado et al. 2015a, b) ... 662
Fig. 15.9 Interpretation of 2D electrical resistivity tomography (ERT) and induced polarization (IP) inversion models for Profile 1, obtained in the Digha area (West Bengal, India). The aim of the study was to delineate possible zones of shallow aquifers contaminated by seawater. The resistivity values (ρ) are indicated as 10^x, where x represents the values corresponding to the color scale shown in the figure (for a to c) (*Source* Martinho 2023—Open Access, after Kumar et al. 2021) ... 665
Fig. 15.10 Crosshole imaging results from a hydrocarbon-contaminated site in Strasbourg-Entzheim, France. **a** Real component, **b** Imaginary component, and **c** Phase of complex conductivity (*Source* Martinho 2023—Open Access, after Kemna et al. 2004) ... 666
Fig. 16.1 Ground penetrating radar (GPR) applications in forensic and demining scenarios, demonstrating the integration of GPR with autonomous platforms for hazardous site investigations: **a** a 500 MHz radargram revealing a potential shallow grave (green shadow), **b** a 900 MHz radargram detecting a subsurface cavity with a test subject, **c** a 2 GHz 3D GPR dataset mapping inert landmines and mortar shells in a controlled minefield, and **d** high-resolution radar slices showing the internal structure of a simulated antipersonnel landmine. These examples highlight the capability of GPR-equipped unmanned systems to conduct high-risk surveys safely and effectively. *Source* Lombardi et al. (2022)—Open Access ... 687

Fig. 16.2 GPR applications in environmental monitoring: **a** 100 MHz 3D data delineating alluvial aquifer geometry, **b** 25 MHz profile revealing glacier bed and reflection classifications, **c** 270 MHz data showing sinkhole anomalies, and **d** 30 MHz radar profile of a kimberlite pipe. These applications support climate resilience planning. *Source* Lombardi et al. (2022)—Open Access 691

Fig. 16.3 Ground penetrating radar (GPR) applications for subsurface utility detection and infrastructure assessment in urban environments: **a** 1 GHz GPR data revealing underground pipes made of various materials—metal (1), cement (2), PVC (3), polypropylene (4), and polyethylene (5); **b** 500 MHz GPR scan of a rigid pavement structure; **c** 2.3 GHz data used for non-invasive inspection of a reinforced concrete (RC) building slab. These examples demonstrate the integration potential of GPR into smart city systems for underground asset management, infrastructure maintenance, and digital twin modeling. *Source* Lombardi et al. (2022)—Open Access 693

Fig. 16.4 Subsurface structural imaging from the Chang'E-4 mission on the far side of the Moon using a 500 MHz GPR system. Data reveal stratified regolith and underlying structures, showcasing how GPR extends beyond terrestrial geophysics and supports space-based exploration. *Source* Lombardi et al. (2022)—Open Access 697

List of Tables

Table 2.1	Evolution of engineering geophysics	44
Table 2.2	Overview of geophysical techniques and their applications	45
Table 2.3	Applications of engineering geophysics in modern infrastructure, environmental challenges, and disaster risk reduction	53
Table 3.1	Common geophysical methods used for site characterization	86
Table 3.2	Interaction of waves with subsurface materials	88
Table 3.3	Summary of seismic methods and their applications	93
Table 3.4	Seismic wave velocities (Pelton 2005)	95
Table 3.5	Typical electrical resistivity values for some Earth materials (Loke 2002; Gibson and George 2003; Society of Exploration Geophysicist of Japan 2004; Advanced Geosciences Incorporated 2018; Naskar and Das 2009)	98
Table 3.6	Electrical and electromagnetic methods in engineering geophysics	109
Table 3.7	Interpretation of Self Potential (SP) measurements and their subsurface implications	112
Table 3.8	Gravity surveying in engineering geophysics	114
Table 3.9	Magnetic surveying in engineering geophysics	116
Table 3.10	Comparative table of magnetic and gravitational methods	118
Table 3.11	Wave propagation characteristics for different materials (Telford et al. 1990; Stokoe and Santamarina 2000; Dobrin and Savit 1988)	121
Table 3.12	Seismic refraction and reflection methods (Sheriff and Geldart 1995; Kearey et al. 2002; Woods and Kurrle 2007)	122
Table 3.13	Ultrasonic pulse velocity testing applications (RILEM (International Union of Laboratories and Experts in Construction Materials and Structures) 1984; Krautblatter and Haider 2010; Malhotra and Carino 2004)	122

Table 4.1 Recent advancements in geophysical surveying techniques ... 135
Table 4.2 AI-driven geophysical data processing workflow ... 136
Table 4.3 Overview of the stabilization method for gradiometer data ... 151
Table 6.1 Geophysical methods, properties, and parameters ... 269
Table 7.1 Flood risk parameter data sources for the Upper Bengawan Solo Watershed, Java, Indonesia (Jumadi et al. 2024) ... 327
Table 8.1 Classification of tunnel defects with concrete lining—revised in light of Bien (2019) ... 370
Table 8.2 Degradation mechanisms of tunnel structures ... 370
Table 9.1 Lists the selected critical metals, their associated mineral deposit types, and major producing countries ... 422
Table 10.1 Basic metadata of selected articles, including year, region, focus area, and method applied (after Kadic et al. 2025) ... 478
Table 10.2 Top journals by number of published papers on GI and climate change adaptation (after Kadic et al. 2025) ... 479
Table 10.3 Key literature sources with corresponding geophysical strategies and application techniques (after Kadic et al. 2025) ... 480
Table 11.1 Observed differences in flood management policies and suggested contributing institutional and cultural factors (after Singh et al. 2021) ... 530
Table 11.2 Post-mitigated schedule risk analysis results for iteration 1000 (after Faruq et al. 2022) ... 537
Table 12.1 Mission capabilities of operational stakeholders at the military base camp (after Mallery et al. 2022) ... 556
Table 12.2 Mission impact assessment of power loss scenarios ... 566
Table 13.1 Geophysical tools and applications for defense energy resilience ... 590
Table 14.1 Comparative summary of geophysical methods for archaeological applications ... 603

Chapter 1
Introduction to Engineering Geology

Abstract This chapter introduces the basic principles and applications of engineering geology, a vital field in infrastructure planning and safety. It begins with an overview of rock types, geological structures, and plate tectonics. Readers learn how geological maps support engineering decisions. The chapter also discusses natural hazards such as landslides, ground subsidence, and slope failures, including a case study from Southern Spain. Geological masses, weathering, and discontinuities are explained, highlighting their importance in construction. Field tests, borehole studies, and geophysical methods like seismic and magnetic surveys are explored. The chapter covers ground improvement techniques and examines the role of geology in dam design and performance during earthquakes. It also addresses the integration of geology with modern infrastructure, such as smart cities and digital twins. Environmental, societal, and regulatory aspects are discussed, along with ethical considerations. Finally, it explores future trends, including remote sensing and the skills needed for tomorrow's engineering geologists.

1.1 Introduction

Engineering geology is the application of geological science to engineering practice, aiming to ensure that geological factors affecting construction, infrastructure, and land use are properly understood and considered (Lwisa 2021). It supports safe, sustainable, and cost-effective engineering by analyzing how earth materials and processes interact with human-made structures (Jackson 1981). Engineering geologists assess geological risks, advise on site suitability, and recommend design adaptations based on subsurface conditions. Their expertise is especially vital in addressing natural hazards like landslides, earthquakes, and ground subsidence (Kiersh 1991).

Key areas of engineering geology include:

(i) Rock types and structures (e.g., faults, folds, tectonics)

This chapter is intended to provide foundational context for readers by exploring essential geological principles that underpin engineering geophysics.

D. C. Naskar, *Advances in Engineering Geophysics: Concepts and Applications*, Earth Systems Data and Models 9, https://doi.org/10.1007/978-3-032-28000-8_1

(ii) Geohazards (e.g., slope failure, subsidence, earthquakes)
(iii) Ground characterization (e.g., weathering, discontinuities)
(iv) Field and geophysical testing (e.g., seismic, electrical methods)
(v) Ground improvement techniques (e.g., grouting, compaction)
(vi) Dams and reservoirs (e.g., foundation geology, seismic safety).

Understanding these aspects is crucial for interpreting geophysical data effectively and forms the groundwork for the methods and applications discussed in the chapters that follow.

1.2 Rocks and Geological Structures

Geological structures and rock types form the foundation of engineering geology. They result from dynamic Earth processes, especially tectonic forces, which shape the subsurface environment that engineers must understand for safe and efficient design.

1.2.1 Rock Types

Rocks are classified into three main categories:

1. Igneous rocks form through the cooling and solidification of molten magma
2. Sedimentary rocks develop from the accumulation of mineral or organic particles, often layered, and typically formed in aquatic environments
3. Metamorphic rocks originate from pre-existing rocks transformed by heat, pressure, or chemically active fluids.

Each rock type has distinct physical properties that influence strength, durability, and suitability for engineering applications (American Museum of Natural History 2021).

1.2.2 Rock Structures

Rocks exhibit structures on multiple scales, from microscopic grain textures to large faults and folds. These structures are critical for understanding rock behavior under stress and for identifying potential hazards.

(i) Primary structures form during rock formation (e.g., bedding, flow lines).
(ii) Secondary structures arise after rock formation, often due to tectonic stress (e.g., joints, faults, folds) (United States Department of Agriculture 2012).

Understanding these features helps infer geological history and predict rock performance in the field.

1.2.3 Plate Tectonic

Plate tectonics explains the large-scale movement of Earth's lithospheric plates. Driven by internal heat and mantle convection, these movements shape continents, create mountains, and generate earthquakes and volcanoes.

(i) Convergent (plates collide—mountains, subduction zones)
(ii) Divergent (plates separate—mid-ocean ridges)
(iii) Transform (plates slide past—fault lines).

Tectonic activity directly affects site stability, seismic risk, and long-term ground behavior—key concerns in engineering geophysics (Butler 1992).

1.2.4 Engineering Geological Maps

Engineering geological maps integrate geological information with practical engineering needs. They depict rock types, faults, soil properties, slope stability, and groundwater conditions, serving as essential tools for planning, risk assessment, and design in construction and infrastructure development.

1.3 Geohazards

Geohazards are natural geological or environmental processes that pose risks to human life, infrastructure, and the environment. These hazards, which include both sudden and gradual events, can range from localized effects—such as ground settlement—to large-scale disasters like landslides, tsunamis, and earthquakes. Understanding geohazards is crucial for site selection, hazard mitigation, and the safe design of engineering works (Gutierrez et al. 2019).

1.3.1 Boundary Hazards

Tectonic plate boundaries are key sources of geohazards due to intense geological activity:

(i) Oceanic-Oceanic Boundaries—one oceanic plate subduct beneath another, forming deep ocean trenches and potentially triggering undersea volcanic eruptions.
(ii) Oceanic-Continental Boundaries—the denser oceanic plate subducts beneath a continental plate, often resulting in earthquakes and volcanic arcs.
(iii) Continental-Continental Boundaries—when two continental plates collide, neither subducts, leading to the formation of mountain ranges and large-scale seismic activity.

These plate interactions can generate high-magnitude earthquakes and initiate related hazards such as landslides and tsunamis.

1.3.2 Ground Subsidence

Ground subsidence is the gradual or sudden sinking of the Earth's surface due to natural or anthropogenic causes. It may result from:

(i) Collapse of underground voids (natural or man-made)
(ii) Consolidation of compressible soils, and withdrawal of groundwater or hydrocarbons.

Subsidence can damage structures, disrupt utilities, and increase flood risk. In severe cases, it may result in sinkholes that pose immediate threats to life and infrastructure.

1.3.3 Landslide

Landslides involve the downslope movement of rock, soil, or debris under gravity—a type of mass wasting. They can take several forms:

(i) Falls (ii) Slides (iii) Spreads (iv) Flows (v) Topples.

Landslides may be triggered by:

(i) Heavy rainfall
(ii) Rapid snowmelt
(iii) Earthquakes and volcanic activity
(iv) Erosion or undercutting of slopes
(v) Human activities such as excavation or deforestation.

Underwater (submarine) landslides can also occur, often leading to *tsunamis* that impact coastal regions.

1.3.4 Slope Failure

Slope failure refers to the sudden collapse of a slope due to the loss of shear strength in soil or rock. It is often initiated by:

(i) Heavy rainfall infiltration
(ii) Seismic loading
(iii) Poor slope geometry or construction practices.

In inhabited areas, slope failures can cause significant casualties and property loss. Preventive engineering and early warning systems are essential for risk mitigation.

1.3.5 Rock Slope Stability Under Temperature Fluctuations

Monitoring studies from various locations around the world have shown that air temperature fluctuations can cause intermittent shrinkage and expansion of rock blocks near the surface. While temperature variations are typically cyclical and predictable, the resulting deformations in the rock masses are often non-reversible, leading to plastic behavior that can accumulate over time. This cumulative, irreversible displacement can cause progressive detachment of rock blocks down slopes, creating hazardous conditions in both natural and built environments.

This phenomenon was notably documented at Mount Masada, Israel, during a 1998 monitoring campaign. Joint meters installed on a block known as "Block 1" revealed clear correlations between seasonal temperature changes and joint expansions in unstable rock blocks located along the eastern cliff face (Fig. 1.1). The site, located in a dry desert climate along the Dead Sea transform, experiences high daily and seasonal temperature gradients that contribute to these movements. As a result of the recorded deformations, the block was later anchored to the main cliff to mitigate the risk of detachment (Bakun-Mazor 2022).

A subsequent research station installed in 2009 on the western cliff of Masada, near the Roman Ramp, provided additional confirmation. Equipped with joint meters and environmental sensors, the station monitored cumulative joint opening over two years, clearly linked to seasonal temperature variations. These findings illustrate that thermal cycling alone, even without seismic or hydrological influences, can lead to the progressive failure of rock slopes. Figure 1.2 presents field photographs from various locations illustrating this ratchet mechanism, where repeated expansion and contraction leads to the step-by-step movement of blocks.

In addition to field observations, a laboratory experiment was conducted to simulate the thermal behavior of a stratified and cracked rock slope. A prismatic block, inserted into a tensile crack, was subjected to controlled temperature cycles. The results showed that this wedge-shaped block accelerated the rate of movement when exposed to repeated heating and cooling, supporting the theory of thermal

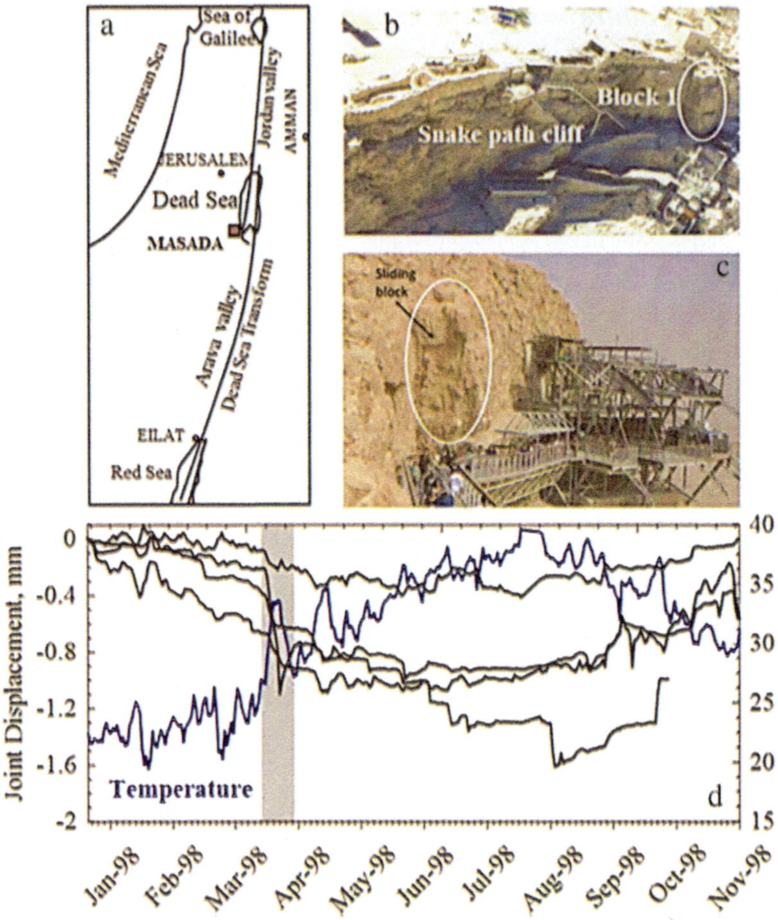

Fig. 1.1 Observation from a thermal monitoring station at the eastern cliff of Mount Masada, Israel. **a** Location map showing Masada along the Dead Sea Transform Fault. **b** The Snake Path cliff with the old cable car station located below the monitored zone. **c** The unstable rock mass ("Block 1") situated near the newly constructed station and tourist path. **d** Seasonal monitoring results from 1998 showing a clear correlation between temperature fluctuations and joint displacement, demonstrating the cumulative effect of thermal cycling on slope stability. *Source* Bakun-Mazor (2022)—Open Access

ratcheting (Fig. 1.3). The block's geometric configuration, particularly in the presence of directional discontinuities, was a critical factor influencing the degree of displacement.

These experimental findings were then used to calibrate analytical and 3D numerical models, which accurately predicted the observed displacements. When applied

Fig. 1.2 Photos of blocks from the field that illustrate the ratchet mechanism (The ratchet mechanism refers to the process where repeated expansion and contraction from temperature changes cause incremental, one-directional movement of rock blocks, eventually leading to permanent displacement or failure). **a** Masada Mountain, Israel, **b** Arugot Valley, Israel, **c** Yellow Mountain, China, **d** Zohar Valley, Israel and **e** Machtesh Ramon, Israel. Each image shows displaced or partially detached blocks affected by thermal cycling. Scale bars represent 1 m. *Source* Bakun-Mazor (2022)—Open Access

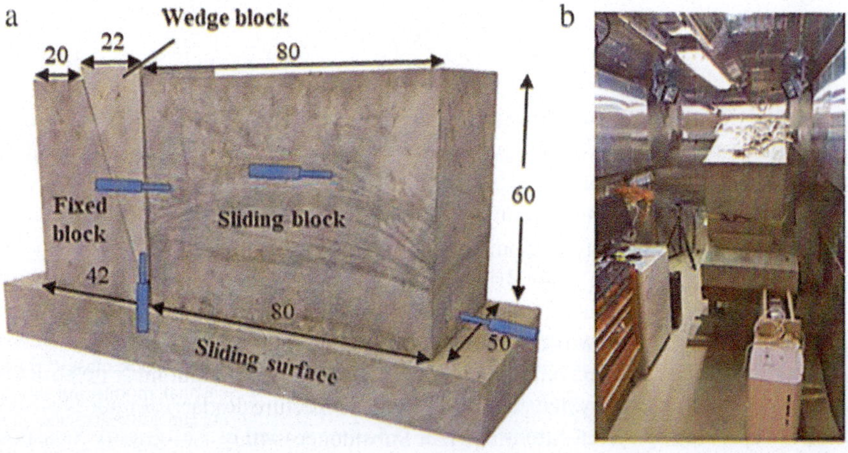

Fig. 1.3 Laboratory simulation of thermal-induced block displacement. **a** Concrete block system representing a stratified rock slope, showing the placement of displacement transducers for monitoring block movement (dimensions in cm). **b** Experimental setup of the block system placed on an inclined steel platform inside a temperature-controlled chamber to simulate daily thermal cycles and observe cumulative displacement effects. *Source* Bakun-Mazor (2022)—Open Access

to the real conditions at Masada, the models demonstrated that temperature fluctuations alone were sufficient to explain the displacements and potential detachment of blocks, even in the absence of rainfall, seismic activity, or anthropogenic disturbance.

This evidence underscores the importance of incorporating thermal stress effects in the stability assessment of rock slopes, especially in arid or high-exposure environments. Sites with cultural heritage significance, such as Masada, and those with critical infrastructure, like mountainous highways or cliffside developments, require close monitoring and geotechnical evaluation that considers temperature-driven deformation processes.

1.3.6 Subsidence Risk Assessment Using Microgravity: A Case Study from Candado Hill, Malaga, Southern Spain

This study presents the application of microgravity surveying to generate a collapse susceptibility map in an abandoned mining area at Candado Hill, located in the Malaguide Complex of the Internal Zones of the Betic Cordillera in southern Spain. The goal was to identify subsurface mine galleries whose extent and exact location were previously unknown (Martinez-Moreno et al. 2016).

Geological Context

Candado Hill spans 700 × 400 m, rising from 65 to 125 m, and features a stratified geology of Permo-Triassic clays, gypsum, and sandy conglomerates hosting the mine, overlain by SE-dipping Jurassic–Cretaceous limestone and dolostone, and capped by Quaternary sands and conglomerates (Chalouan and Michard 1990; Martin-Algarra et al. 2009).

Microgravity Survey Design

The microgravity survey was conducted using a Scintrex Autograv CG-5 gravimeter with a high precision of ± 0.001 mGal, supported by differential GPS positioning through a Leica 1200+ system connected to the Andalusian Positioning Network (RAP), achieving spatial accuracy up to ± 0.5 mm. A total of 317 gravity measurements were collected across a 10 × 10 m grid, providing uniform spatial coverage, including within urbanized zones. To ensure data accuracy, standard gravimetric corrections were applied, including adjustments for tidal effects and instrument drift, calculation of the Bouguer anomaly using a reference density of 2.67 g/cm^3, and terrain corrections performed using the Hammer method in combination with high-resolution digital terrain models (DTMs).

Residual Gravity and Forward Modeling

Residual gravity anomalies were derived by removing regional trends through first-order polynomial regression, allowing the identification of localized subsurface density variations. The most prominent residual minima were concentrated near the center of Candado Hill, indicating potential underground voids attributed to abandoned mine galleries. To further interpret these anomalies, two 2.5D forward gravity models were developed using GRAVMAG v1.7 software. The models revealed major voids measuring up to 90 $\times$ 80 m and reaching depths of approximately 50 m. Additionally, smaller, shallow cavities ranging from 5 to 20 m in width and 10 to 20 m in depth were identified, corresponding to secondary localized gravity lows. Across the modeled profiles, residual gravity anomalies varied between − 0.3 and − 0.55 mGal.

Engineering Implications: How the Map Informs Land-Use Planning and Risk Mitigation

The collapse susceptibility map derived from the microgravity survey provides essential guidance for engineering decision-making and land-use planning in the Candado Hill area. By accurately delineating the location, size, and depth of underground voids, the map identifies zones at risk of ground subsidence, enabling planners to avoid or reinforce these areas during development. This geophysical insight is particularly critical in urban or peri-urban settings where uncharted mine galleries pose hidden hazards beneath roads, buildings, or infrastructure corridors. The results support informed zoning, the design of mitigation measures such as ground improvement or structural reinforcement, and the prioritization of monitoring efforts. Ultimately, the integration of microgravity data into geotechnical assessments enhances public safety, reduces construction risks, and promotes more resilient and sustainable land-use strategies in areas with a legacy of underground mining.

Conclusion

Microgravity proved effective in detecting hidden voids, enabling the creation of a collapse susceptibility map that can guide urban planning, construction safety, and hazard mitigation in post-mining environments. This case study exemplifies the value of high-resolution geophysical methods in geohazard assessment.

1.4 Geological Masses

Understanding geological masses is essential in engineering geophysics, as they represent the physical framework upon which infrastructure is designed and constructed. The stability, composition, and internal structure of rock and soil masses influence site safety, engineering design, and the selection of appropriate geophysical techniques.

1.4.1 Mass Movement

Mass movement—also referred to as mass wasting—involves the downslope movement of soil, rock, or debris under the influence of gravity. These processes may be rapid (e.g., landslides) or gradual (e.g., soil creep), and they often result in significant changes to the Earth's surface. In some contexts, the term *mass movement* has evolved to encompass both traditional gravity-driven displacement and localized subsidence or collapse events. These movements are frequently influenced or facilitated by water, whether through saturation, pore pressure changes, or erosion. Mass movements play a critical role in shaping terrestrial landscapes and pose risks to engineered structures. Understanding the triggers, mechanisms, and progression of such phenomena is crucial in geotechnical assessments and hazard mitigation strategies.

1.4.2 Mass Fabric

In geology, fabric refers to the spatial arrangement and geometric relationships of components within a rock or soil mass. The analysis of fabric provides insights into both the formation processes and subsequent deformation history. In sedimentary rocks, fabric is typically controlled by the depositional environment and can offer clues about paleocurrents, sediment transport mechanisms, and post-depositional modifications. In metamorphic and structural geology, fabric may reveal the direction and magnitude of tectonic stresses or strain patterns that shaped a particular rock body.

Types of geologic fabric include:

1. Primary fabric—Formed during original deposition or crystallization.
2. Shape fabric—Defined by the alignment of grains or clasts.
3. Crystallographic preferred orientation—Common in metamorphic rocks, indicating aligned crystal lattices.
4. S-fabric and L-fabric—Represent planar and linear structures, respectively, related to deformation.
5. Penetrative fabric—A fabric that extends continuously throughout the rock mass.
6. Magnetic fabric—Reflects the alignment of magnetic minerals and can indicate deformation or flow processes.

1.4.3 Ground Mass Description

The ground mass, also known as the matrix, is the fine-grained material in which larger mineral grains, crystals, or voids are embedded. In igneous rocks, the ground mass comprises microscopic crystals surrounding larger phenocrysts, often forming a porphyritic texture indicative of complex cooling histories. In sedimentary rocks, the matrix consists of fine materials such as clay or silt that fill the spaces between larger clasts. It often determines rock porosity and permeability, impacting fluid flow. In metamorphic rocks, the ground mass refers to the finer components between porphyroblasts or mineral assemblages, especially in low-grade metamorphic settings. The characterization of the matrix is vital for interpreting rock genesis and assessing its mechanical behavior in engineering contexts.

1.4.4 Weathering

Weathering is the process of breaking down rocks and minerals at the Earth's surface due to environmental factors such as water, temperature changes, and biological activity. Unlike erosion, weathering does not involve the physical removal of material. The three main types of weathering are:

1. Physical weathering—Mechanical breakdown through freeze–thaw cycles, abrasion, or thermal expansion.
2. Chemical weathering—Alteration of minerals through reactions with water, oxygen, acids, and other chemicals.
3. Biological weathering—Involvement of organisms, such as plant roots or microbial activity, in the degradation of rock.

Understanding weathering processes is fundamental for evaluating rock durability, slope stability, and the long-term integrity of engineering structures.

1.4.5 Rocks Discontinuities

In geotechnical engineering, discontinuities (commonly referred to as joints) are surfaces where there is a distinct change in the physical or chemical properties within a rock or soil mass. These may be natural features such as:

(i) Bedding planes
(ii) Schistosity
(iii) Foliation
(iv) Joints
(v) Fractures
(vi) Cracks
(vii) Fault planes.

Discontinuities can be categorized as:

(i) Mechanical discontinuities—Features that interrupt the mechanical continuity of the mass.

(ii) Integral discontinuities—Features that influence, but do not fully interrupt, mechanical integrity.

Discontinuities may occur individually or in groups and significantly impact the anisotropy and stability of geological masses. Their presence alters stress distribution and fluid flow paths, making their identification and characterization critical in engineering geophysics.

1.5 Field Tests and Measurements

Engineering geophysics plays a pivotal role in understanding subsurface conditions critical for modern engineering projects. It supports the evaluation of ground properties, offering vital data that supplements or enhances traditional geotechnical investigations. This section outlines key in-situ testing methods and geophysical approaches, highlighting their relevance to a global audience engaged in infrastructure development, environmental assessment, and natural hazard mitigation.

1.5.1 Tests in Boreholes and Excavations

Subsurface investigations are essential for assessing soil and rock properties at construction sites. Geotechnical engineers and engineering geologists conduct these evaluations to ensure safe and sustainable design and construction practices. The information gathered helps define foundation conditions, address ground-related construction challenges, and support planning for structures ranging from buildings to pipelines and energy infrastructure.

Common objectives of these tests include:

(i) Evaluating soil strength and composition
(ii) Determining foundation suitability
(iii) Identifying subsurface anomalies or hazards
(iv) Assessing thermal conductivity for buried utilities and energy storage systems.

Standard field tests include:

(i) Standard Penetration Test (SPT)—Measures soil resistance to penetration.
(ii) Dynamic Cone Penetration Test (DCPT)—Provides stratigraphic profiling and compaction levels.
(iii) Cone Penetration Test (CPT)—Offers continuous data on soil behavior and stratigraphy.

These tests often accompany geophysical surveys to provide a more comprehensive view of the subsurface.

1.5.2 *Engineering Geophysics*

Engineering geophysics is the application of geophysical techniques to solve engineering problems. It involves the study of the Earth's surface and shallow subsurface using physical principles to measure and analyze anomalies in soil and rock properties. These techniques are valuable in various domains including civil engineering, environmental assessment, energy infrastructure, and land-use planning.

Typical engineering geophysics applications include:

(i) Locating bedrock and identifying weak zones
(ii) Detecting voids, underground cavities, or buried structures
(iii) Characterizing subsurface layers for foundations
(iv) Investigating groundwater contamination or landfill integrity
(v) Mapping seismic hazards or fault zones
(vi) Locating utilities, pipelines, and archaeological remains.

These methods provide non-intrusive, cost-effective, and scalable solutions, often yielding insights unattainable through drilling alone.

Use of Engineering Geophysics to Investigate a Site for a Building Foundation

In recent years, geophysical methods have become a valuable tool in civil and environmental engineering, particularly for assessing the suitability of building foundations. These methods—such as Electrical Resistivity Tomography (ERT) and Electromagnetic Terrain Conductivity (ETC)—enable early detection of subsurface anomalies like cavities, inhomogeneities, and weak zones that may pose risks during or after construction (Loke and Barker 1996; Loke 1997; Ward 1990). By integrating geophysical surveys with geological and geotechnical data, engineers can map near-surface structures, assess soil conditions, and optimize foundation design. A case study in Hania, Crete, demonstrated how geophysical investigation improved decision-making on footing placement, confirming the value of these methods in reducing geotechnical risk and ensuring structural stability (Soupios et al. 2007).

The combined use of 2D/3D Electrical Resistivity Tomography (ERT) and Electromagnetic (EM) terrain conductivity mapping enabled effective identification of two significant subsurface anomalies (A1 and A2). Excavation confirmed the geophysical findings, revealing a 3 m thick clay-filled discontinuity (A1) interpreted as a former hand-dug well. This matched the predicted location and geometry from the models. The presence of soft Neogenic clay, undetected by boreholes alone, was found to significantly influence foundation stability. The study demonstrated that integrating geophysical and geotechnical methods enhances subsurface characterization, avoids foundation placement on unsuitable soils, and improves site selection and design decisions for safe building construction.

1.5.3 Seismic Methods

Seismic techniques are fundamental in subsurface exploration, leveraging the behavior of elastic waves to infer material properties and structural features.

(i) Seismic Tomography—Uses wave propagation from natural or artificial sources to create 3D models of subsurface conditions. It helps in hazard mapping, tunnel design, and deep-earth studies.
(ii) Seismic Reflection—Analyzes the return of seismic waves from subsurface boundaries to characterize stratigraphy and detect buried anomalies. Widely used in oil exploration and groundwater mapping.
(iii) Seismic Refraction—Based on Snell's Law, this method measures wave velocity changes across layers to assess soil/rock interfaces and mechanical properties. It is commonly used in road design, foundation assessment, and landslide studies.

These methods are applicable across a range of geological settings and adaptable to different project scales.

1.5.4 Magnetic and Electrical Methods

Magnetic Techniques

1. Aeromagnetic Surveys—Aircraft-mounted magnetometers detect variations in Earth's magnetic field, useful for regional geological mapping, fault detection, and locating ferrous objects.
2. Ground Magnetic Surveys—Applied at smaller scales to map buried infrastructure, mineral deposits, or unexploded ordnance.

Electrical and Electromagnetic Techniques

1. Electrical Resistivity Tomography (ERT)—Maps subsurface resistivity variations to identify groundwater, contamination plumes (a concentrated zone of pollutants that spreads through soil or groundwater from a specific source, often moving outward over time), and stratigraphic boundaries (stratigraphic boundaries are the distinct layers or interfaces between different rock or soil strata that represent changes in deposition conditions, composition, or geological time periods) (Grelle and Guadagno 2016).
2. Induced Polarization (IP)—Measures the chargeability of subsurface materials. Useful for mineral exploration and mapping clay-rich layers or waste zones.
3. Magnetotellurics (MT)—A passive technique that records natural electromagnetic signals to explore to depths of 10 km or more. Used in geothermal studies and crustal research.

4. Ground Penetrating Radar (GPR)—High-resolution imaging of shallow subsurface features. Commonly used for utility mapping, archaeology, and forensic investigations.
5. Time-Domain Electromagnetics (TDEM)—Uses pulsed electromagnetic fields to map conductive and resistive features in the subsurface. Particularly effective in groundwater and mineral exploration.

1.5.5 Global Perspective

Field-based testing and geophysical surveying are critical in addressing global engineering challenges, from rapid urban development and disaster resilience to energy transition and sustainable infrastructure. These techniques bridge the gap between theoretical geoscience and real-world application, enabling more informed decision-making for stakeholders worldwide. Whether assessing earthquake risks in Japan, locating groundwater in arid African regions, or mapping ancient foundations in Europe, engineering geophysics delivers universally relevant tools for twenty-first-century challenges.

1.5.6 Geophysical Investigations for Offshore Renewables

Geophysical investigations for offshore renewable energy projects are non-intrusive, vessel-based surveys designed to gather critical information about the seabed and shallow sub-seafloor in areas designated for infrastructure installation, such as offshore wind farms, tidal energy converters, and subsea cabling. These investigations help identify the physical and geological characteristics of the seabed—including topography, sediment type, and subsurface features—as well as detect potential geohazards, archaeological artifacts, abandoned structures, and environmentally sensitive habitats. This data ensures informed, safe, and sustainable decision-making during the design and construction phases.

Typical Equipment and Techniques

Geophysical investigations in offshore environments typically utilize a suite of acoustic and magnetic instruments, including:

(i) Multi-beam echo sounders (MBES)—to generate detailed bathymetric maps of the seafloor.
(ii) Sub-bottom profilers (SBP)—to image shallow geological layers beneath the seabed.
(iii) Side scan sonar—to provide acoustic imagery of the seabed surface.
(iv) Magnetometers—to detect buried ferrous objects, such as shipwrecks or unexploded ordnance.
(v) Underwater cameras—to visually validate acoustic survey findings.

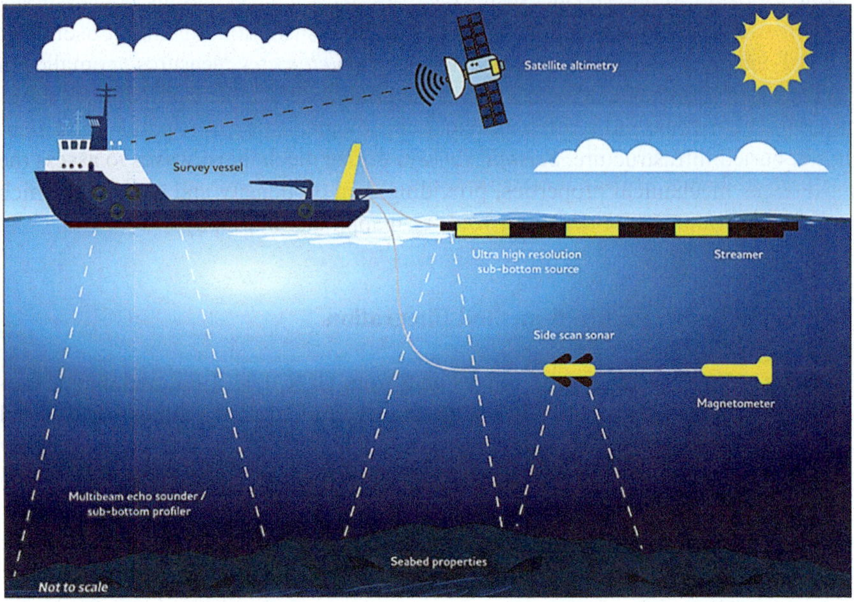

Fig. 1.4 Visualization of typical offshore geophysical survey vessel using sonar and echo sounders. *Source* Wikipedia—Open Access

These instruments are often deployed from various types of vessels, including research ships, commercial survey vessels, fishing boats, or recreational craft, depending on project scale and location (Fig. 1.4).

Objectives of Offshore Renewable Geophysical Surveys

At the early stages of project development, geophysical surveys serve several essential functions:

(i) Determine seabed composition (e.g., soft sediments, rock outcrops, sandbanks).
(ii) Map topographic features such as hills, gullies, reefs, and depressions.
(iii) Inform planners with a comprehensive and reliable geological model, reducing uncertainty and enhancing design confidence.
(iv) Integrate with geotechnical investigations to support foundation design and layout planning for turbines and cable routes.

This process is analogous to examining the subsurface through a keyhole—offering high-resolution insight without excavation.

Comparison with Marine Seismic Surveys for Hydrocarbon Exploration

While both offshore renewable energy and oil and gas sectors utilize geophysical methods, their objectives and environmental footprints differ substantially:

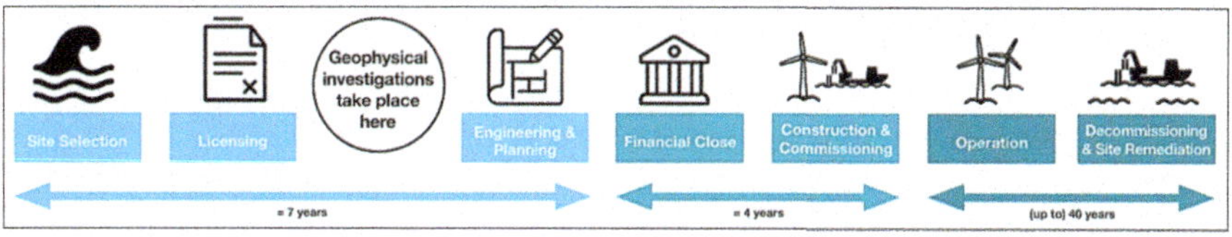

Fig. 1.5 Timeframe for a typical offshore wind farm development. *Source* Wikipedia—Open Access

Aspect	Offshore petroleum	Offshore renewables
Survey goal	Image deep geological structures (kilometers below seabed)	Investigate seabed and shallow sub-surface
Acoustic intensity	High-intensity air guns	Low-intensity sources
Environmental impact	Higher potential impacts; often controversial	Lower impact; mitigation measures common
Use case	Locate hydrocarbon reservoirs	Support engineering and environmental design

Environmental Considerations

Due to the potential sensitivity of marine ecosystems, offshore geophysical surveys for renewables adopt impact-mitigation strategies, such as:

(i) Scheduling surveys outside critical periods for marine life (e.g., breeding or migration seasons).
(ii) Operating at reduced vessel speeds.
(iii) Employing marine fauna observers and passive acoustic monitoring (PAM) to detect and avoid disturbance to marine species.

Development Timeline

Geophysical surveys are typically one of the first steps in the timeline of offshore renewable energy development (Fig. 1.5). The data collected is vital for site selection, engineering design, environmental assessments, and regulatory approvals.

1.6 Ground Improvement

Ground improvement encompasses a set of techniques aimed at enhancing the engineering properties of in-situ soils to make them more suitable for construction and infrastructure development. The primary objectives typically include increasing *shear strength*, improving *stiffness*, and reducing *permeability*. These enhancements allow weak or problematic soils to support structural loads and withstand environmental conditions more effectively.

1.6.1 Shallow and Deep Impaction

Soil compaction involves increasing the density of soil by expelling air from the voids between soil particles, usually through mechanical means. This improves load-bearing capacity and reduces settlement potential (Roshani et al. 2022).

Compaction techniques are categorized based on the mode of energy application:

(i) Static compaction (e.g., rollers applying constant pressure)
(ii) Impact compaction (e.g., dropping heavy weights)
(iii) Vibratory compaction (e.g., using vibrating plates or rollers)
(iv) Gyratory and kneading compaction (commonly used in lab testing or asphalt compaction).

Shallow compaction is typically used for roadbeds, foundations, and embankments, while deep compaction methods like dynamic compaction or vibroflotation are employed for densifying soils at greater depths.

1.6.2 Grout Treatment

Grouting is a ground improvement method in which various materials are injected under pressure into the subsurface to reduce permeability and/or enhance strength. It is widely used in dam foundations, tunnels, retaining structures, and remediation of sinkholes or voids.

Types of grouting include:

1. Permeation grouting (injection into porous soils)
2. Compaction grouting (displacement of soil to increase density)
3. Jet grouting (high-pressure injection that breaks up and mixes soil in-place).

Injectable materials range from cementitious grouts to chemical solutions, selected based on project requirements and subsurface conditions. Grouting has a long history, dating back to the nineteenth century, particularly in hydraulic structures such as canals and dams where grout curtains are used to minimize water seepage.

1.6.3 Bentonite Suspension

Bentonite is a naturally occurring clay formed from the weathering of volcanic ash, widely used in geotechnical and environmental engineering due to its swelling and sealing properties. Its primary mineral is montmorillonite, and it is categorized by its dominant cation:

(i) Sodium bentonite—High swelling capacity, excellent for sealing
(ii) Calcium bentonite—Lower swelling, more stable.

Bentonite suspensions are commonly used in:

(i) Drilling muds for borehole stabilization
(ii) Slurry walls for groundwater control
(iii) Liners and barriers in landfills and containment systems.

In geological applications, weathered volcanic ash layers rich in potassium are often referred to as K-bentonites, and similar clay-rich layers associated with coal seams may be classified as tonsteins (often dominated by kaolinite).

1.6.4 Ground Anchor

Ground anchors are structural elements used to transfer tensile loads into the ground. They serve as critical support systems for both temporary and permanent structures. Anchors are typically installed by drilling a hole, inserting the anchor, and then grouting it into place.

Common applications include:

(i) Stabilizing retaining walls and slopes
(ii) Supporting temporary or permanent structures
(iii) Anchoring marine platforms, pipelines, and floating docks
(iv) Securing utility poles and wind turbine bases
(v) Tethering masts and towers
(vi) Landscape anchoring of trees, especially large transplants
(vii) Security and recreational purposes, such as anchoring aircraft or slacklines.

These systems are designed to resist uplift or lateral forces and can be tensioned to meet specific load requirements.

1.7 Water Reservoirs and Dams

Water reservoirs—typically artificial lakes—are created to store water for various uses, including irrigation, drinking water, hydroelectric power, flood control, and industrial activities. Most reservoirs are formed by constructing dams across rivers, though they can also be developed by modifying natural lakes (e.g., by blocking natural outlets to regulate water levels). Dams control the outflow of water from these reservoirs to meet operational objectives.

1.7.1 Dam Design Parameters

The engineering design of dams, particularly earth dams, requires careful consideration of several parameters to ensure safety, stability, and functional reliability (Mishra 2020). Key criteria include:

(i) Spillway and basin capacity—The design must ensure sufficient capacity for water drainage and retention to avoid overtopping, which could lead to dam failure.
(ii) Seepage control—Seepage through the dam body and its foundation must be minimized to prevent erosion. The seepage line should be confined within the downstream slope, and drainage systems should be provided to manage flow and maintain structural integrity.
(iii) Uplift pressure—Hydraulic uplift forces acting beneath the dam must not exceed tolerable limits. Excessive uplift can reduce stability and should be mitigated through foundation drainage or other methods.
(iv) Slope stability—Both upstream and downstream slopes must remain stable under varying reservoir conditions, including during rapid drawdown or prolonged saturation. Design should account for slope loading and potential seepage forces.
(v) Structural stresses—The stresses imposed by the dam structure on its foundation should not exceed the bearing capacity of the underlying materials.
(vi) Erosion protection—The upstream face must be protected from wave action using materials like riprap or revetments. The downstream face should be shielded from rainfall erosion, often using grass cover or other erosion control methods.

1.7.2 Geological Influences Upon the Selection of Reservoir Sites

The geology of a proposed dam and reservoir site is critical to assessing its feasibility and long-term performance. Important factors include:

(i) Topography—The terrain's physical features—such as valleys, hills, and drainage patterns—play a key role in determining the suitability of the site. A preliminary review using topographic maps and field surveys can help identify viable locations.
(ii) Hydrology and Hydrogeology
    (a) Hydrology examines surface water availability, including rainfall, streamflow, and catchment area characteristics. A viable reservoir must be supplied by an adequate and reliable inflow throughout the year.
    (b) Hydrogeology involves groundwater behavior. The interaction between the reservoir and groundwater systems must be understood—whether

groundwater contributes to the reservoir or whether the reservoir will lose water through seepage into the subsurface.

(iii) Geological Structure and Composition—Understanding the type and condition of bedrock and soil materials at the site is essential for foundation design and reservoir tightness. Aspects to consider include:

(a) Rock types and their mechanical properties.
(b) Depth of weathered or altered zones.
(c) Presence of geological discontinuities such as faults, fractures, and joints.
(d) Potentially permeable layers (e.g., sand, silt) that could lead to leakage.
(e) Ancient or relict soil deposits with variable stability characteristics.

These assessments help ensure the dam structure is both geotechnically sound and capable of minimizing water loss.

1.7.3 Dam Foundations

Before construction, detailed site investigations are required to assess the suitability of foundation materials (Best et al. 1998; UKEssays 2018). A foundation must provide:

(i) Stability under various load and saturation conditions, and
(ii) Sufficient resistance to seepage to prevent excessive water loss or piping.

Foundations are typically categorized into three types:

1. Rock foundations—Often desirable for their strength, but must be assessed for joints, fractures, and susceptibility to uplift pressure or leakage.
2. Coarse-grained material foundations—Such as gravel or sand, which may require treatment to enhance stability and impermeability.
3. Fine-grained material foundations—Including clays and silts, which may exhibit strength under dry conditions but need evaluation under saturated loading.

In cases of poor foundation quality, ground improvement techniques (e.g., grouting, cut-off walls, or drainage systems) may be necessary.

1.7.4 Seismic Considerations for Dams

Induced seismicity refers to earthquakes triggered by human activities such as reservoir impoundment, mining, or fluid injection. Although typically of low magnitude, such events can pose safety concerns for dams (Gosschalk et al. 1994). Seismic hazard assessments for dams exposed to induced or natural earthquakes use methodologies similar to those applied in tectonic seismic zones. These include:

(i) Evaluation of ground motion parameters using empirical models, including those based on databases of natural seismic events.
(ii) Site-specific seismic response analysis to understand how local geology and soil conditions might amplify shaking.
(iii) Risk assessments that account for structural vulnerability and potential impacts on surrounding communities and infrastructure.

Mitigation strategies may involve structural reinforcement, reservoir operation guidelines to manage induced stress changes, and relocation or protection of at-risk populations.

1.7.5 Structural Vibration Analysis and Engineering Standards

Vibration monitoring is a vital tool in engineering geology for assessing the impact of construction activities on nearby structures. Using sensors, engineers can track ground and structural vibrations to ensure safety and compliance with design standards.

(i) Structural vibration monitoring helps calibrate geotechnical methods and safeguard adjacent buildings.
(ii) Vibration limits are guided by international standards (e.g., ISO, DIN, BS), tailored to structure type and sensitivity.
(iii) These standards act as references, but expert judgment remains essential in interpreting results.
(iv) Adjustments to construction techniques, such as changing the frequency of vibratory hammers, can reduce vibration impacts.
(v) Vibration data also improves the accuracy of numerical models, aligning simulations with real ground conditions.
(vi) Alternative vibration control methods can be applied, but must be calibrated for effectiveness.

As geotechnical engineering evolves with more invasive technologies, vibration monitoring is becoming increasingly important.

Case Studies

(i) Urban Construction Monitoring, Warsaw Metro Expansion (Poland)—Sensors monitored building response during deep excavations near heritage sites.
(ii) Sheet Pile Installation, Netherlands—Frequency adjustments of vibratory hammers reduced damage risk to adjacent structures.
(iii) Seismic Retrofitting Project, Japan—Vibration data guided model calibration and validated safety assumptions in dense urban areas.
(iv) Vibration Monitoring of a Two-Storey Brick House—Wrocław, Poland.

Location: Wybrzeze Wyspianskiego 27, 50-370 Wrocław, Poland.

Context: Vibration monitoring was conducted during road construction using the Rapid Impact Compaction (RIC) method (Fig. 1.6) near a two-storey brick building (Fig. 1.7).

Site Conditions

(a) Original soil—Organic, 2–4 m deep.
(b) Replaced with—Medium dry sand, compacted using an 8-tonne hydraulic hammer.
(c) Equipment—Steel footplates (100 and 60 cm in diameter), ~ 40 blows/min from 80 cm drop height.

Main Objective

To evaluate the impact of compaction vibrations on a nearby structure and assess compliance with standards.

Fig. 1.6 Rapid Impact Compaction on variable geological strata. This figure illustrates the application of Rapid Impact Compaction (RIC) on a site with heterogeneous subsurface conditions. The compaction mechanism and resulting ground response are shown in relation to soil type, density, and groundwater conditions, emphasizing the importance of geological assessment in designing and evaluating ground improvement strategies. *Source* Baca (2025)—Open Access

Fig. 1.7 Building under vibration monitoring in a geologically sensitive area. This illustration depicts vibration monitoring of a structure situated on varying subsurface geology, highlighting the influence of local lithology, soil stratigraphy, and groundwater conditions on wave propagation and structural response. *Source* Baca (2025)—Open Access

Key Observations and Results

(a) Vibration velocity increased significantly with the larger footplate (100 cm); peak velocities reached up to 5 mm/s in the 0–15 Hz frequency range.
(b) These values exceeded vibration limits for sensitive buildings based on DIN [Deutsches Institut fur Normung (German Institute for Standardization)] and Circulaire (French National Guideline) standards.
(c) However, vibration levels remained within acceptable limits according to BS (British Standard) standards.
(d) After switching to the smaller footplate (60 cm), vibration velocities dropped below all standard thresholds.
(e) High-frequency vibrations from other sources were observed but were under 1 mm/s and had no structural impact.

Conclusion

Changing compaction equipment size can significantly reduce vibration levels. Monitoring based on multiple standards is essential for accurate structural impact assessment.

(v) Vibration Monitoring During Adjacent Hall Construction—Wrocław, Poland.

Location: Wybrzeze Wyspianskiego 27, 50-370 Wrocław, Poland.

Context: Vibration monitoring was performed inside an existing warehouse hall during impact compaction for a new adjacent hall (Fig. 1.8).

Site Conditions

(a) Soil Type—Anthropogenic, non-cohesive soil with post-mining waste.
(b) Compaction Method—100 cm diameter footplate, same machine specs as case study (iv).
(c) Sensor Placement—On the warehouse floor using Svantek SVAN 958A (Fig. 1.9).

Fig. 1.8 Monitoring of vibrations in warehouse facilities on geologically variable ground. This figure illustrates the monitoring of structural vibrations in a warehouse setting, emphasizing the role of underlying geological conditions such as soil type, compaction, and groundwater presence. Vibration monitoring devices are deployed to assess how subsurface materials influence wave transmission and structural response, aiding in geotechnical risk assessment and mitigation. *Source* Baca (2025)—Open Access

Fig. 1.9 Svantek SVAN 958A monitoring station deployed in a geotechnical setting. The figure shows the SVAN 958A vibration and noise monitoring station used in an engineering geophysics application. The setup is positioned on a site with variable subsurface geology, highlighting the need to understand local soil conditions, stratigraphy, and groundwater influence to interpret recorded vibration data effectively. *Source* Baca (2025)—Open Access

Main Objective

To assess whether impact compaction affected the structural safety of the existing warehouse hall.

Key Observations and Results

(a) Vibration velocity during impulse compaction was in the 0–5 mm/s range, similar to case study (iv).
(b) The frequency range of 0–12 Hz corresponded to the compaction activity, while higher frequencies originated from other sources.
(c) Evaluation against DIN, BS, and Circulaire standards showed that vibration levels remained within acceptable limits for this building type.
(d) Unlike in case study (iv), no changes were required in compaction methods due to the lower sensitivity of the warehouse structure.

Conclusion

Although vibration levels were comparable to those in case study (iv), the structural characteristics and usage of the warehouse meant that the vibrations posed no risk. This highlights the importance of structure-specific evaluation in vibration impact assessments.

1.7.6 Embankment Dams Response to Actual PGA During Earthquakes

This section examines the dynamic behavior of embankment dams under the influence of measured Peak Ground Acceleration (PGA) during actual earthquake events. Through case studies and monitoring data, it demonstrates how geological conditions, foundation response, and structural configurations interact to influence seismic performance. The role of in-situ geophysical testing, structural health monitoring, and post-event back-analysis is also discussed in the context of engineering geology.

Case Study: Matahina Dam, New Zealand—Response to Edgecumbe Earthquake (1987)

The seismic response of embankment dams varies significantly based on earthquake characteristics, geological conditions, and dam design (USCOLD 2000). Understanding this variability is critical for improving seismic resilience. This section summarizes the response of Matahina Dam in New Zealand to the 1987 Edgecumbe Earthquake (Mw 6.3) (Adamo et al. 2021).

Case Summary: Matahina Dam, New Zealand

(i) Dam Type—Rockfill embankment with upstream sloping core
(ii) Earthquake PGA at Base—0.33 g ("g" refers to the acceleration due to gravity, which is approximately 9.81 m/s^2. For example: 0.1 g = 10% of gravity → light shaking; 0.33 g (as in Matahina Dam) → moderate to strong shaking; 1.0 g = acceleration equal to gravity (1.0 g = 9.81 m/s^2 approximately) → very intense shaking)
(iii) Peak Crest Acceleration—0.42–0.48 g
(iv) Crest Settlement—~ 100 mm
(v) Downstream Displacement—~ 250 mm

(vi) Rockfill Shoulder Settlement—up to 800 mm.

Key Observations

(i) Seepage—Post-quake seepage rose sharply but later stabilized. Increased flow from abutments indicated possible internal erosion (de Gali Price and Christianson 2019; USCOLD 1992).
(ii) Settlements and Displacements—Crest and shoulder settlements occurred progressively, consistent with seismic compaction. No signs of underwater slope failure were detected.
(iii) Instrumentation—Five accelerometers recorded strong motion. Crest motions were amplified, with the highest acceleration (0.48 g) at mid-height on the rockfill shoulder.
(iv) Geological Factors—Variability in foundation materials and abutment conditions played a key role in the dam's response. High pore pressures and elevated permeability were observed in the left abutment.

Engineering Implications

This case highlights that embankment dam response is site-specific, shaped by local geology and earthquake parameters. Globally, such data supports more accurate seismic design, risk assessment, and post-earthquake evaluation protocols for critical infrastructure.

Case Study: La Villita Dam, Mexico—Response to the 1985 Michoacan Earthquake

La Villita Dam is a 60 m high zoned earth dam built on 70 m of Balsas River alluvium, featuring a central clay core, rockfill shells, filter zones, and a concrete cut-off wall. On September 19, 1985, the dam experienced strong ground shaking from the Mw 8.1 Michoacan earthquake, located ~ 75 km from the site.

Seismic Records and Instrumentation

(i) Crest PGA—0.45 g (mainshock), 0.16 g (aftershock)
(ii) Bedrock PGA—0.13 g (mainshock), 0.04 g (aftershock)
(iii) Five strong-motion accelerometers, 45 piezometers, inclinometers, extensometers, and survey monuments provided comprehensive seismic data.

Observed Seismic Response

(i) Cracking

Longitudinal cracks developed along the dam crest, up to 1000 ft. in length and 2 in. wide. These were attributed to differential settlement between the core and filters, but did not penetrate the impervious core, except near the right abutment.

(b) Settlements

Crest settlements ranged from 20 to 32 cm upstream and 9 to 22 cm downstream, more pronounced in the central section. Settlement trends suggested cumulative degradation from repeated seismic events over prior years.

(iii) Displacements

Horizontal displacements of up to 16.5 cm upstream and 10 cm downstream were recorded. Deformations decreased with depth and did not affect foundation materials, as confirmed by inclinometers.

Engineering Significance

The La Villita case demonstrates how earth fill dams on thick alluvium can withstand significant seismic shaking with manageable deformation, provided proper design, monitoring, and maintenance are in place. The case highlights:

(i) The importance of dense instrumentation, especially for strong-motion monitoring.
(ii) The role of material stiffness evolution in long-term dam response.
(iii) The benefit of comparative analysis (e.g., El Infiernillo Dam) in regional seismic performance studies. "El Infiernillo" translates to "The Little Hell" in English. It likely refers to the intense heat and rugged terrain in the region where the dam is located—a common naming convention in remote or harsh natural environments.

Globally, this underscores the necessity of performance-based design and site-specific seismic hazard assessment for critical infrastructure in seismic regions.

Case Study: Cogoti Dam, Chile—Response to the 1943 Illapel Earthquake

The Cogoti Dam is an 85 m high concrete face rockfill dam (CFRD) located in a seismic region of north-central Chile, constructed in 1938 for irrigation. The dam impounds 148 million m^3 of water and is founded on andesitic breccia within a narrow gorge. It features an uncompacted rockfill embankment and a segmented concrete face laid on a hand-placed rock bedding layer.

Seismic Events and Dam Response

On April 6, 1943, the dam experienced a Mw 7.9 Illapel Earthquake, with an estimated PGA of 0.19 g at the site. The dam sustained a crest settlement of up to 41 cm, mostly near the center. No cracking of the concrete face occurred. Only minor rockslides on the downstream slope were reported. Leakage, which had existed since first filling, did not significantly increase due to the quake. In subsequent earthquakes (1965, 1971, 1985, and 1997), the dam withstood shaking without major damage. Notably, in 1997, the Punitaqui Earthquake (Mw 6.8) caused a 15 cm crest settlement but again no damage to the face slab.

Key Observations

(i) Seismic performance—Despite older construction methods and lack of compaction, the dam performed well.
(ii) No face cracking occurred even during strong seismic loading (0.19 g), validating the resilience of CFRDs under earthquake stress.
(iii) Long-term leakage increase has been linked to aging, not seismic damage.

(iv) Instrumentation gap—No accelerometers were present at the time of any earthquake, which limits analysis of true dynamic behavior.

Engineering Significance

Cogoti Dam's history highlights the seismic durability of CFRDs even when constructed using early techniques without mechanical compaction. Its case supports the global engineering belief that CFRDs, if properly designed and maintained, can withstand significant seismic motions without experiencing structural failure. The absence of instrumentation remains a missed opportunity for seismic research, underscoring the importance of monitoring infrastructure in seismic zones (Wieland 2003, 2009).

1.7.7 Concrete Dams Response to Observed PGAs During Earthquakes

Concrete dams have historically performed better than embankment dams under seismic loading, with relatively few cases of major structural damage despite experiencing strong ground motions. Over the past 50 years, many concrete dams have been subjected to peak horizontal ground accelerations (PHGA) exceeding 0.3 g, including arch, buttress, and roller-compacted concrete (RCC) types (USSD 2017).

Most arch dams have shown excellent seismic performance, with no recorded structural failures, even under high-intensity shaking. However, buttress dams have occasionally developed horizontal cracks at transition zones where stiffness changes, such as the intersection of buttresses and vertical chimney sections. RCC dams, such as Shapai Dam (China) and Miyatoko Dam (Japan), have shown seismic behavior comparable to conventional concrete dams.

The only known failure of a concrete dam during an earthquake is the Shih Kang gravity dam in Taiwan, which was built in 1977 and severely damaged during the 1999 Chi-Chi Earthquake (M7.6). The dam was located directly above the surface fault rupture, leading to extreme displacements and cracking. PGAs recorded just 0.3 miles from the site reached 0.51 g (horizontal) and 0.53 g (vertical). The dam experienced cracking along its ogee crest and separation from the foundation. Despite the failure, this case is considered exceptional due to its direct fault intersection (Hansen et al. 2011).

The widespread deployment of high-quality strong-motion accelerometers in recent decades has greatly improved our understanding of concrete dams seismic response. These instruments have enabled detailed correlations between observed damage and recorded ground motions, informing the seismic design of new dams. A review of 19 concrete dams shaken by earthquakes highlights cases where PGAs exceeded 0.3 g. Most dams remained operational with only minor damage, emphasizing the inherent seismic resilience of well-designed concrete structures (Hansen and Nuss 2001). Nevertheless, the dynamic analysis of concrete dams remains a complex task. In strong earthquakes, nonlinear behavior and tensile cracking—especially at the dam heel and geometric transitions—must be accounted for. Accurate seismic evaluation requires nonlinear dynamic modeling, incorporating material behavior and site-specific spectral input. These insights guide modern dam engineering practice, particularly for regions of high seismicity, where design must accommodate not only recorded PGAs but also the potential for near-fault effects and structural nonlinearity.

1.8 Integration with Modern Infrastructure Development

The rapid growth of urban centers and complex infrastructure demands a new paradigm where engineering geology integrates seamlessly with digital technologies, sustainable planning, and resilience-oriented design. Engineering geology now extends beyond foundational studies to influence smart infrastructure, underground space planning, and long-term resilience frameworks.

1.8.1 Role of Engineering Geology in Smart Cities

Smart cities rely on data-driven systems for managing infrastructure, energy, utilities, and transport. Engineering geology supports this transformation through:

(i) Subsurface characterization for utility placement, tunnel routing, and structural foundation design.
(ii) Hazard zoning for landslides, seismicity, and subsidence integrated into smart planning tools.
(iii) Real-time geotechnical monitoring using IoT-linked (Internet of Things) sensors embedded in critical assets (e.g., slopes, embankments).

In smart cities, geological inputs are no longer static maps but dynamic layers feeding into urban information systems for proactive decision-making.

1.8.2 Subsurface Data for Urban Planning and Resilience

Modern cities require three-dimensional understanding of the subsurface to avoid conflicts and ensure sustainable development. Engineering geological data contributes to:

(i) Subsurface zoning for infrastructure corridors, especially in crowded urban areas.

(ii) Risk assessment of geohazards (e.g., liquefaction potential, soft soil zones) for urban safety and compliance (Stabile et al. 2020).
(iii) Resilience planning by identifying vulnerable zones and supporting adaptive design (e.g., flexible foundations, waterproofing zones).

This data informs zoning laws, land-use plans, and building codes in a spatially explicit and predictive manner.

1.8.3 Case Links to Digital Twins and Geospatial Infrastructure

Digital twins—virtual replicas of physical assets—require accurate geological and geotechnical models. Applications include:

(i) Digital twin of a metro system, integrating borehole logs, fault lines, and groundwater maps to optimize tunneling and maintenance.
(ii) Geospatial infrastructure management systems using GIS-based (Geographic Information System) integration of geological, hydrological, and structural data for lifecycle monitoring.
(iii) Urban sinkhole prediction models, where geophysical imaging and geological profiles are integrated with real-time sensor data.

Engineering geology provides the foundational data that feeds these digital platforms, making it a core discipline in future-ready infrastructure design.

1.9 Environmental and Climatic Considerations in Engineering Geology

Engineering geology increasingly interfaces with climate science and environmental assessments due to the rising influence of climate variability on ground conditions. Infrastructure projects must now consider geological responses to climatic extremes and long-term environmental sustainability.

1.9.1 Climate-Responsive Site Selection

Geological site assessments must account for climatic variables, especially in regions prone to:

(i) Freeze–thaw (Freeze–thaw is a weathering process where water enters cracks in rocks, freezes and expands, causing the rock to fracture and break apart over repeated cycles) cycles affecting slope stability and foundation integrity.

(ii) High precipitation zones where increased infiltration may trigger landslides or reduce bearing capacity.
(iii) Arid regions, where desiccation (process of extreme drying in soils or sediments) and soil erosion can affect surface structures.

Incorporating regional climate models with geological mapping allows planners to select safer, more adaptive sites for infrastructure development.

1.9.2 Geological Impacts of Extreme Weather Events

Extreme weather—such as floods, heatwaves, and cyclones—can have profound impacts on geological conditions:

(i) Flooding may lead to subsurface erosion, piping, or unexpected sediment transport.
(ii) Intense rainfall can cause landslides in mountainous terrains with weak lithology or fractured rocks.
(iii) Droughts may accelerate soil shrinkage and settlement, especially in clay-rich areas.

Engineering geology contributes predictive insights and mitigation strategies (e.g., slope reinforcement, subsurface drainage, flexible foundation design) to minimize damage.

1.9.3 Geo-Environmental Risk Assessments for Infrastructure Projects

Environmental risk assessments now include geo-environmental parameters, evaluating how geology interacts with ecological and anthropogenic factors. Key components include:

(i) Assessment of contaminated land (e.g., brownfields, old industrial sites) through geophysical and geochemical surveys (Matsuda and Tanimoto 2021).
(ii) Hydrogeological impacts of infrastructure on aquifers, including seepage, contamination, and drawdown effects.
(iii) Cumulative risk mapping combining geological sensitivity, land use, and projected climate impacts.

These assessments ensure regulatory compliance, reduce liability, and guide sustainable infrastructure design from pre-construction to long-term monitoring.

1.10 Ethical, Regulatory, and Societal Aspects

Engineering geology does not operate in isolation—it intersects with ethical responsibilities, legal compliance, and social accountability. As infrastructure development increasingly affects communities and ecosystems, integrating ethical and societal dimensions becomes essential.

1.10.1 Environmental Ethics in Engineering Geology

Engineering geologists are ethically obligated to:

(i) Minimize environmental disruption, particularly in sensitive ecosystems.
(ii) Avoid excessive resource extraction and ensure sustainable land use.
(iii) Balance human development with natural preservation, recognizing geological heritage as a non-renewable resource.

Ethical practice also involves transparency in reporting risks and uncertainties in site evaluations, especially in hazard-prone regions.

1.10.2 Regulatory Frameworks Governing Geological Investigations

Legal and policy frameworks regulate geological activities to ensure public safety and environmental protection. These may include:

(i) National building codes and geological survey standards (e.g., IS codes in India, Eurocode 7 in Europe).
(ii) Environmental impact assessment (EIA) regulations, requiring baseline geological and geotechnical investigations.
(iii) Permits and compliance checks for subsurface exploration, drilling, and waste disposal.

Geologists must remain updated on evolving legislation, including those tied to climate adaptation, disaster resilience, and land use zoning.

1.10.3 Community Engagement and Geo-Social Impacts

Large-scale geological investigations and infrastructure projects often affect communities. Responsible practice includes:

(i) Stakeholder consultations, especially in displacement-prone or indigenous areas.
(ii) Clear communication of risks such as ground subsidence, slope instability, or groundwater depletion.
(iii) Integrating local knowledge with scientific assessments to enhance acceptance and effectiveness.

Community engagement fosters trust and leads to more inclusive and resilient infrastructure outcomes.

1.11 Trends and Future Challenges in Engineering Geology

As engineering demands evolve, engineering geology must adapt to complex environments, cutting-edge technologies, and multidisciplinary expectations. This section highlights key future-facing trends and challenges.

1.11.1 Advances in Remote Sensing and Data Analytics

Modern geological investigations increasingly rely on:

(i) High-resolution satellite imagery and UAV (drone) (Unmanned Aerial Vehicle) data for mapping inaccessible terrains.
(ii) LiDAR (Light Detection and Ranging) and InSAR (Interferometric Synthetic Aperture Radar) for detecting surface deformation and subsidence.
(iii) Big data and AI-driven analytics to interpret vast geological datasets for predictive modeling.

These tools enhance precision in site assessment and monitoring, especially in dynamic or urban environments.

1.11.2 Emerging Challenges: Urbanization, Deep Excavation, and Subsurface Congestion

Urban infrastructure growth presents new geological issues:

(i) Limited surface space drives deeper excavations for basements, tunnels, and utilities—requiring accurate subsurface characterization.
(ii) Subsurface congestion with intersecting foundations, pipelines, and transit systems increases geotechnical risk.
(iii) Climate-adaptive design must account for shifting hydrogeological conditions, especially in coastal and flood-prone zones.

Geologists must innovate to manage uncertainty in complex stratigraphic and anthropogenic settings.

1.11.3 Future Skillsets for Engineering Geologists

To stay relevant, future engineering geologists will need:

(i) Interdisciplinary expertise across geology, geophysics, environmental science, and civil engineering.
(ii) Proficiency in digital tools, including GIS, BIM (Building Information Modeling), and geospatial AI.
(iii) Strong communication and policy literacy to engage stakeholders, planners, and regulatory bodies.

The field is moving beyond static investigations to dynamic, systems-oriented thinking aligned with smart cities, resilience, and sustainability goals.

References

Adamo N, Al-Ansari N, Sissakian V, Laue J, Knutsson S (2021) Dam safety: use of seismic monitoring instrumentation in dams. J Earth Sci Geotech Eng 11(1):203–247

American Museum of Natural History (2021) Exhibitions. American Museum of Natural History, 612021. [Online]. Available: https://www.amnh.org/exhibitions/permanent/planet-earth/how-do-we-read-the-rocks/three-types

Baca M (2025) Structural vibration analysis with reference to different standards. Buildings 15:1951

Bakun-Mazor D (2022) Rock slope stability under temperature fluctuations. IntechOpen, 16 p. https://doi.org/10.5772/intechopen.108464

Best E, Herschy RW, Herschy RW, Gâştescu P, Starosolsky Ö (1998) Danube river: development. Dordrecht

Butler RF (1992) Applications to paleogeography. IUPAC

Chalouan A, Michard A (1990) The Ghomarides nappes, Rif coastal range, Morocco: a Variscan chip in the Alpine belt. Tectonics 9:1565–1583

de Gali Price CH, Christianson KJ (2019) The legacy of Edgecumbe the 1987 earthquake. In: Silvesti F, Moraci N (eds) Proceeding of the 7th international conference on earthquake geotechnical engineering for protection and construction, Rome, Italy, 17–20 June 2019, pp 1994–2001

Gosschalk EM, Severn RT, Charles JA, Hinks JL (1994) An engineering guide to seismic risk to dams in the United Kingdom, and its international relevance. Soil Dyn Earthq Eng 1(3):163–179

Grelle G, Guadagno FM (2016) Electrical resistivity tomography for engineering geological applications. Geosciences 6(3):50

Gutierrez F et al (2019) Geophysical methods for landslide monitoring: a review. Geosciences 9(7):314

Hansen KD, Nuss LK (2001) Lessons learned from the earthquake performance of concrete dams. In: Proceedings of 15th European conference on soil mechanics and geotechnical engineering, 12–15 Sept 2011, p 7

Hansen KD, Nuss LK, Roehm L (2011) Lessons learned from the earthquake performance of concrete dams. In: International waterpower and dam construction magazine, Table (1), 27 Sept

2011. https://www.waterpowermagazine.com/features/featurelessons-learned-from-the-earthquake-performance-of-concrete-dams/

Jackson BA (1981) Glossary of geology. AGI

Kiersh (1991) The heritage of engineering geology: the first 100 years. Geological Society of America, p 3

Loke MH (1997) Rapid 2D resistivity inversion using the least-squares method RES2DINV program manual (Penang, Malaysia)

Loke MH, Barker RD (1996) Practical techniques for 3D resistivity surveys and data inversion. Geophys Prospect 44:499–523

Lwisa EG (2021) Introductory chapter: engineering geology. IntechOpen, 13 p. https://doi.org/10.5772/intechopen.95991

Martin-Algarra A, Mazzoli S, Perrone V, Rodriguez-Canero R, Navas-Parejo P (2009) Variscan tectonics in the Malaguide complex (Betic Cordillera, southern Spain): stratigraphic and structural alpine versus pre-alpine constraints from the Ardales area (Province of Malaga). I. Stratigraphy. J Geol 117:241–262

Martinez-Moreno FJ, Galindo-Zaldivar JG, Gonzalez-Castillo L, Azanon JM (2016) Collapse susceptibility map in abandoned mining areas by microgravity survey: a case study in Candado Hill (Málaga, southern Spain). J Appl Geophys 130:101–109

Matsuda H, Tanimoto T (2021) Geophysical surveys in civil engineering: advances and case studies. Geosciences 11(2):58

Mishra G (2020) Criteria for design of earth dams. The Constructor. The Construction Encyclopedia, 12 Dec 2020. [Online]. Available: https://theconstructor.org/water-resources/criteria-for-design-of-earth-dams/2278/

Roshani A et al (2022) Ground improvement techniques for soft soils: a review. Appl Sci 12(4):2083

Soupios PM, Georgakopoulos P, Papadopoulos N, Saltas V, Andreadakis A, Vallianatos F, Sarris A, Makris JP (2007) Use of engineering geophysics to investigate a site for a building foundation. J Geophys Eng 4(2007):94–103

Stabile TA et al (2020) Seismic hazard assessment for critical infrastructures. Appl Sci 10(21):7731

UKEssays (2018) Design and construction fundamentals of earth dams environmental sciences essay. UKEssays, 12122018. [Online]. Available: https://www.ukessays.com/essays/environmental-sciences/design-and-construction-fundamentals-of-earth-dams-environmental-sciences-essay.php

United States Department of Agriculture (2012) National engineering handbook. United States Department of Agriculture

USCOLD (1992) Matahina dam, New Zealand. Observed performance of dams during earthquakes, vol 1, pp 101–109

USCOLD (2000) Observed performance of dams during earthquakes, vol II, Oct 2000, pp 15–20

USSD (2017) Seismic analysis of concrete dams. In: Workshop annual conference, 6–7 Apr 2017. http://www.ussdams.org/wp-co

Ward SH (1990) Resistivity and induced polarization methods. In: Ward S (ed) Geotechnical and environmental geophysics. Society of Exploration Geophysicists, Tulsa, OK, pp 147–189

Wieland M (2003) Seismic aspects of dams. Q83 general report. In: Proceedings of the 21st congress of the international commission on large dams, Montreal, June 2003, pp 1233–1362

Wieland M (2009) CFRD's in highly seismic regions. In: Water power and dam construction magazine, published on 14 Apr 2010. Originally presented at the First international symposium on rockfill dams, Chengdu, China, 18–21 Oct 2009

Part I
Foundations and Theoretical Framework

Chapter 2
Introduction to the Engineering Geophysics

Abstract This chapter introduces engineering geophysics and its growing role in solving complex infrastructure and environmental challenges. It traces the field's evolution—from early academic work to today's use of digital tools, automation, and artificial intelligence. Modern geophysical methods are helping design safer, more sustainable infrastructure while addressing environmental risks like contamination and climate change. The chapter presents real-world case studies using electrical resistivity, seismic imaging, and ground-penetrating radar (GPR) to track pollution, detect underground pipes, and assess landfill sites. Applications range from trenchless pipe installations in India to archaeological studies in Malaysia. The future of engineering geophysics lies in multi-method integration, advanced sensing, and better data interpretation. Despite its power, the field also faces limitations and misconceptions. This chapter highlights the importance of geophysics in environmental monitoring, disaster risk reduction, and smart infrastructure planning, showing how it continues to adapt in an increasingly complex world.

2.1 Introduction to Engineering Geophysics

"Advances in Engineering Geophysics: Concepts and Applications" is a comprehensive volume that delves into the integration of geophysical methods within engineering contexts. The introductory chapter, "Introduction to Engineering Geophysics," sets the stage by defining engineering geophysics and highlighting its significance in various engineering disciplines.

The chapter begins by outlining the fundamental principles of geophysics, emphasizing its role in investigating subsurface conditions without direct excavation. It underscores the importance of non-invasive techniques in assessing soil and rock properties, which are crucial for the design and construction of infrastructure such as buildings, bridges, and dams.

A key focus is placed on the diverse geophysical methods employed in engineering applications. These include seismic refraction and reflection, electrical resistivity

D. C. Naskar, *Advances in Engineering Geophysics: Concepts and Applications*, Earth Systems Data and Models 9, https://doi.org/10.1007/978-3-032-28000-8_2

tomography, ground-penetrating radar, and electromagnetic methods. Each technique is discussed in terms of its principles, advantages, limitations, and typical applications in engineering projects. The chapter also addresses the challenges and considerations in applying geophysical methods to engineering problems. It discusses factors such as site-specific conditions, data interpretation complexities, and the integration of geophysical data with geological and engineering information. The importance of a multidisciplinary approach, combining geophysical expertise with engineering judgment, is emphasized to ensure accurate assessments and informed decision-making.

Furthermore, the chapter highlights the evolving nature of engineering geophysics, noting advancements in technology and methodology that have expanded its applications. It points to the increasing reliance on geophysical surveys for site characterization, risk assessment, and monitoring of infrastructure health, reflecting the growing recognition of the value of geophysical techniques in engineering practice. In summary, the introductory chapter of "Advances in Engineering Geophysics: Concepts and Applications" provides a thorough overview of the field, establishing a foundation for the subsequent chapters that delve into specific methods, case studies, and emerging trends in engineering geophysics.

2.2 Exploring the Frontiers of Engineering Geophysics

Advances in Engineering Geophysics: Concepts and Applications delves into the evolving field of engineering geophysics, exploring its foundational principles, technological advancements, and applications in modern infrastructure and environmental challenges. This chapter introduces the historical trajectory of engineering geophysics, beginning with its early academic foundations and gradual integration of advanced digital tools. It highlights the pivotal developments from the twentieth century through the twenty-first century, marked by innovation, automation, and the advent of geophysical techniques crucial to infrastructure development and environmental sustainability (Lowrie 2016).

The chapter emphasizes the transformative role of advances in geophysical techniques and technologies. These advancements have revolutionized infrastructure design and construction, addressing contemporary challenges in sustainability, risk mitigation, and environmental preservation (Telford et al. 1990). The chapter also discusses the growing influence of automation, remote sensing, artificial intelligence (AI), and machine learning in advancing geophysical research and application, further enhancing precision and efficiency in geophysical surveys and data interpretation.

Furthermore, the chapter examines the critical relevance of engineering geophysics in addressing modern challenges, including infrastructure development, disaster risk reduction, and resilience. It outlines how geophysics plays an essential role in mitigating environmental risks and enhancing sustainability efforts, particularly in the context of urbanization, climate change, and disaster preparedness. The chapter concludes by underscoring the importance of continued innovation in engineering geophysics, emphasizing its significant contributions to solving complex, interdisciplinary challenges in infrastructure and environmental protection.

2.2.1 *Overview of the Evolution of Engineering Geophysics*

Engineering geophysics has undergone significant transformation over the last century, evolving from a predominantly academic and theoretical field to a dynamic, applied science that plays a crucial role in modern infrastructure and environmental management (Annan 1997). The history of engineering geophysics is marked by continuous advancements in both techniques and technologies, which have expanded its scope of applications and its relevance in solving practical engineering challenges. This evolution has been driven by the need for more accurate, efficient, and non-invasive methods of subsurface exploration in various engineering domains, such as civil engineering, environmental protection, and resource management.

Early Development and Academic Foundations

The origins of engineering geophysics can be traced back to the late nineteenth and early twentieth centuries when geophysical techniques were first developed primarily for mineral and petroleum exploration. Early geophysical methods focused on identifying and mapping subsurface geological formations for resource extraction purposes (Culshaw et al. 2014). The first significant geophysical technique was seismic refraction, developed in the early 1900s, which enabled geophysicists to study the Earth's subsurface by measuring the travel times of seismic waves through different geological layers. This technique laid the foundation for the broader field of geophysics, providing a method for identifying variations in subsurface properties based on seismic wave velocities (Telford et al. 1990).

However, it was not until the mid-twentieth century that the field of engineering geophysics began to specialize in techniques directly relevant to civil and environmental engineering (Hatch 2024). The post-World War II era marked a turning point, as the demand for infrastructure projects, particularly in rapidly developing urban areas, grew significantly. Engineers needed reliable methods for assessing subsurface conditions without resorting to extensive and disruptive drilling. This spurred the development of specialized geophysical methods tailored to civil engineering, such as ground-penetrating radar (GPR), electrical resistivity tomography (ERT) (Panek et al. 2010; Batayneh and Al-Diabat 2002; Drahor et al. 2006; Hauck and Kneisel 2006), and magnetometry. These tools allowed engineers to obtain detailed information about the subsurface, including the presence of utilities, buried structures, and potential hazards, without the need for invasive exploration (Yardimci and Arslan 2020).

Technological Advancements and the Rise of Digital Tools

The second half of the twentieth century witnessed substantial technological advancements that revolutionized engineering geophysics (Eddleston et al. 1997). The integration of computational power and digital signal processing played a significant role in the refinement of geophysical methods. The introduction of digital processing algorithms in the 1960s and 1970s, which allowed for the more precise interpretation of geophysical data, was a major milestone. Engineers and geophysicists could now analyze data more efficiently and with higher resolution, leading to improved accuracy in subsurface imaging and analysis (Schleicher and Sloane 2019).

By the 1980s and 1990s, the availability of more affordable and portable geophysical equipment, combined with advances in data acquisition and interpretation software, made geophysics more accessible and practical for large-scale infrastructure projects. Techniques such as seismic reflection and electrical resistivity imaging became more widely used in both urban and rural areas, allowing for more detailed subsurface assessments for projects such as foundation design, tunnelling, and environmental remediation (Kovari and Kozak 2016). These advances were accompanied by the development of GIS (Geographic Information Systems) and 3D modeling, which enabled engineers to visualize subsurface data in ways that were previously impossible.

The Twenty-First Century: The Age of Innovation and Automation

The twenty-first century has brought about significant innovations in engineering geophysics, largely driven by the integration of automation, AI (Artificial Intelligence), and machine learning (ML). Advances in sensor technologies, coupled with the rise of autonomous drones, robotic systems, and satellite-based geophysical measurements, have expanded the capabilities of geophysical surveys. Today, data acquisition can be carried out in remote or hazardous environments with minimal human intervention, reducing both risk and cost. Autonomous platforms, such as drones equipped with GPR, magnetometers, or electromagnetic sensors, can cover large areas quickly and provide high-resolution data without the need for traditional ground-based surveys (Chavez and Lopez 2018).

The application of machine learning algorithms to geophysical data interpretation is also revolutionizing the field. ML and AI have improved the ability to process and analyze large volumes of geophysical data, uncovering patterns and anomalies that might be missed by traditional methods. These tools not only enhance the efficiency of subsurface analysis but also contribute to more accurate risk assessments, enabling engineers to make data-driven decisions in real-time (Schleicher and Sloane 2019).

Furthermore, the integration of multi-method geophysics, which combines data from several geophysical techniques such as seismic, electrical, and electromagnetic methods, has allowed for more comprehensive and reliable subsurface imaging (Fig. 2.1). These integrated approaches provide a more holistic understanding of subsurface conditions, crucial for complex engineering projects such

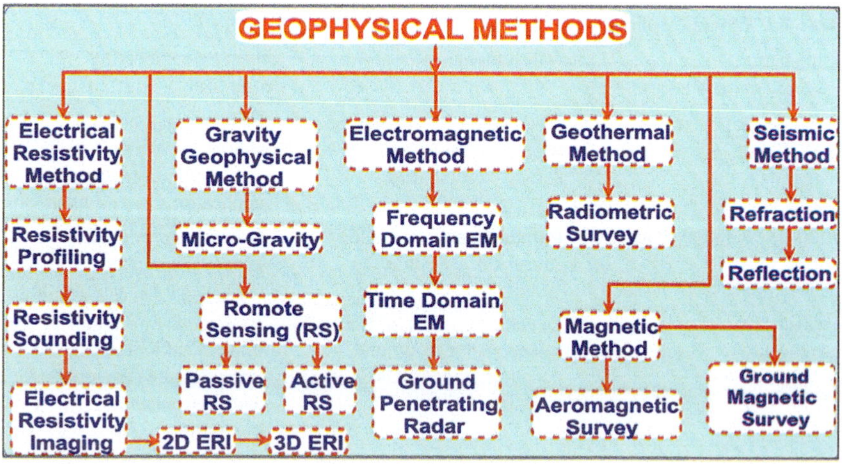

Fig. 2.1 Overview of geophysical methods in engineering geophysics

as tunnel construction, transportation infrastructure, and environmental remediation (McDowell et al. 2002; Yardımcı and Arslan 2020).

Geophysics for Environmental and Sustainable Development

In recent decades, engineering geophysics has seen an increased focus on environmental applications. With the growing awareness of environmental degradation and the need for sustainable development, geophysical techniques have become essential tools for assessing environmental risks and promoting sustainability. Geophysics is now widely used to study and monitor groundwater flow, soil contamination, and the effects of climate change, such as land subsidence and erosion. Techniques such as ERT (Naskar et al. 2002; Loke and Barker 1995) and GPR are used to detect groundwater contamination, monitor subsurface water movement, and assess soil stability—critical factors in sustainable urban planning and infrastructure development (Kovari and Kozak 2016).

The continued development of more efficient, non-invasive geophysical methods has also contributed to the growing emphasis on green technologies in engineering. The ability to perform in situ testing of subsurface conditions without the need for drilling or excavation aligns with the principles of minimizing environmental disruption and preserving ecosystems. As geophysics continues to evolve, it is increasingly incorporated into environmental management strategies, contributing to better resource management, disaster risk reduction, and environmental conservation efforts (Chavez and Lopez 2018).

Table 2.1 summarizes the major milestones in the development of engineering geophysics, from early foundational techniques to the integration of cutting-edge

Table 2.1 Evolution of engineering geophysics

Time period	Key developments	Impact on engineering geophysics
Early development (pre-1900)	– Basic geophysical techniques (e.g., magnetism, gravity surveys) established	– Fundamental tools for subsurface exploration. Initial steps towards engineering applications of geophysics
1900–1950: initial growth	– Introduction of seismology for subsurface studies	– Foundation for modern geophysical methods in exploration and engineering
1950–1980: technological advances	– Development of sophisticated seismic surveys, ground penetrating radar (GPR), and magnetometry	– Expanded applications in mineral exploration, infrastructure development, and environmental monitoring
1980–2000: digital revolution	– Adoption of computer-aided data processing, 3D imaging, and modeling	– Increased accuracy and efficiency in subsurface analysis. Enhanced ability to detect geological features and risks
2000–2010: automation era	– Introduction of automated geophysical sensors and UAVs (Unmanned Aerial Vehicles)	– Enabled large-scale data collection, real-time analysis, and more precise engineering assessments
2010–present: age of innovation	– Rise of AI, machine learning, and big data analytics in processing geophysical data	– Revolutionized predictive modeling, anomaly detection, and data interpretation in engineering projects

technologies in the modern era, with a focus on their impact on engineering practices and environmental sustainability.

2.2.2 *Importance of Advances in Geophysical Techniques and Technologies*

Advances in geophysical techniques and technologies are integral to addressing the multifaceted challenges posed by modern engineering, infrastructure development, and environmental sustainability (Table 2.2). Here's an expanded table that includes various additional geophysical techniques commonly used in engineering geophysics. These methods are critical for applications related to infrastructure, environmental monitoring, and geological investigations.

These methods are essential for a wide range of engineering applications, from infrastructure development to environmental monitoring and resource exploration.

Table 2.2 Overview of geophysical techniques and their applications

Geophysical techniques	Description	Icon/symbol	Applications
Seismic surveys	Uses seismic waves to map subsurface structures, identifying layers and faults	Seismic wave (e.g., waves)	Infrastructure site investigation, mineral exploration, fault detection
Ground penetrating radar (GPR)	Uses radar waves to detect and map subsurface structures	Radar symbol (e.g., wave)	Mapping foundations, detecting buried objects, archaeological surveys
Magnetometry	Measures variations in the Earth's magnetic field to detect mineral deposits or buried structures	Magnetic field (e.g., N/S)	Geological exploration, identifying underground structures or pipelines
Electrical resistivity tomography (ERT)	Measures the electrical resistivity of the ground to identify water, minerals, and soil composition	Electric current (e.g., resistor)	Groundwater exploration, detecting contaminants, mapping soil properties
Electromagnetic methods	Measures electromagnetic fields to identify subsurface conductivity	Magnetic/EM wave (e.g., wave)	Detecting contaminants, mapping geological features, infrastructure planning
Gravimetry	Measures variations in Earth's gravitational field to detect density anomalies underground	Gravity symbol (e.g., weight)	Oil and gas exploration, identifying underground cavities or faults
Geoelectrical imaging	Uses electrical methods to map subsurface structures and materials	Electrical circuit (e.g., wires)	Groundwater assessment, contamination detection, subsurface imaging
Microgravity	A detailed measurement of gravitational variations to detect small-scale subsurface anomalies like cavities, tunnels, and voids	Microgravity symbol	Identifying subsurface voids, infrastructure assessment, tunnel mapping
Tomography (acoustic or electrical)	Uses a cross-sectional imaging technique based on either acoustic or electrical resistivity for subsurface imaging	Tomographic scan (e.g., slices)	Geological exploration, civil engineering, soil contamination analysis

(continued)

Table 2.2 (continued)

Geophysical techniques	Description	Icon/symbol	Applications
Borehole geophysics	A suite of techniques applied in boreholes to directly measure properties of the subsurface such as resistivity, acoustic velocities, and porosity	Borehole symbol (e.g., drilling)	Site characterization, water resource management, contamination detection
Vibro seismic testing	Uses vibrations generated by mechanical sources to explore subsurface conditions by measuring wave propagation velocity	Vibration symbol	Foundation studies, seismic site investigations, geotechnical engineering
Soil resistivity testing	Measures the resistivity of soil for applications in assessing corrosion risk, grounding systems, and understanding soil structure	Electric resistor symbol	Corrosion monitoring, electrical grounding system design, soil assessment
Induced polarization (IP)	Measures chargeability or resistivity properties of the ground, which can help detect materials like clay, ores, and groundwater	Polarization symbol	Mineral exploration, groundwater resource evaluation, environmental studies
Magnetic Resonance Imaging (MRI)	Uses nuclear magnetic resonance to measure the water content and properties of subsurface materials	MRI symbol	Characterization of soil and rock properties, water content measurement
Thermal conductivity measurements	Measures the heat transfer rate of materials to infer subsurface properties like porosity, water content, and soil composition	Heat flow symbol	Site thermal properties mapping, geothermal exploration, construction planning
Acoustic Emission (AE)	Monitors high-frequency elastic waves produced by sudden displacements or cracks in subsurface materials	AE symbol	Structural health monitoring, detecting fractures, underground fault monitoring

(continued)

Table 2.2 (continued)

Geophysical techniques	Description	Icon/symbol	Applications
Downhole seismic	Measures seismic wave velocities at various depths to determine the material properties of the subsurface layers	Seismic wave and borehole symbol	Subsurface characterization, site evaluation for construction
Reflection seismology	A seismic survey method where reflected seismic waves are analyzed to create an image of subsurface structures	Seismic reflection symbol	Subsurface imaging, fault and fracture detection, resource exploration
Well logging	Uses a variety of geophysical techniques in boreholes (e.g., gamma-ray, resistivity, and porosity logs) to analyze subsurface conditions	Logging tool symbol	Mineral exploration, oil and gas reservoir characterization, site investigation

Each technique helps provide crucial subsurface data that supports engineering decisions in construction, hazard assessment, and resource management.

As the global population continues to grow and the demand for robust, sustainable infrastructure intensifies, the need for reliable, accurate, and non-invasive methods for subsurface exploration has never been more critical. Engineering geophysics, which provides a range of non-destructive techniques to investigate subsurface conditions, plays a pivotal role in optimizing the design, construction, and maintenance of various infrastructure projects. These techniques help engineers, geologists, and environmental scientists gather crucial data, mitigate risks, and make informed decisions.

Revolutionizing Infrastructure Design and Construction

The rapid development of geophysical technologies has enabled engineers to move away from traditional, invasive methods like drilling and excavation. Through the application of non-invasive geophysical methods, engineers can now gather vital subsurface data without the need to disturb the ground. This has important economic, environmental, and safety implications. Techniques such as seismic imaging, ground-penetrating radar (GPR), and electrical resistivity tomography (ERT) have allowed engineers to analyze site conditions in unprecedented detail, providing valuable insights into soil composition, groundwater flow, and rock properties.

For instance, seismic imaging, particularly seismic reflection and refraction, has significantly improved the ability to map subsurface structures, detect fault lines, and assess seismic hazards. This technology is particularly important in earthquake-prone regions where understanding fault lines and seismic hazards is essential for designing safe and resilient infrastructure (Yardimci and Arslan 2020). The ability to assess the ground stability and detect potential hazards before construction begins can prevent costly and dangerous structural failures in the future. By accurately identifying subsurface conditions early on, geophysics enables engineers to adjust their designs, ensuring structures are built to withstand the specific conditions of the site.

Additionally, methods such as GPR and ERT are widely used to detect underground utilities, such as pipes, cables, and other infrastructure components, without the need for excavation. This capability is particularly valuable in urban environments, where the risk of damaging existing infrastructure is high. For example, GPR can be used to locate underground utilities and structural weaknesses, facilitating the planning and maintenance of infrastructure while minimizing disruption to the surrounding area (Chavez and Lopez 2018). ERT, on the other hand, is effective in mapping the resistivity of the ground, which can help identify areas with high water saturation, contamination, or other subsurface anomalies that could pose a risk to the stability of structures (Schleicher and Sloane 2019).

Addressing Environmental and Sustainability Challenges

Geophysical techniques have become essential in addressing environmental and sustainability concerns, particularly in the areas of groundwater management, environmental monitoring, and disaster risk assessment. The ability to monitor and assess subsurface conditions using non-invasive techniques is crucial for mitigating the environmental impact of infrastructure projects and ensuring that construction activities do not harm local ecosystems.

One of the most significant advances in geophysics has been the development of techniques for monitoring groundwater and subsurface water movement. Groundwater is a critical resource, and its management is crucial for both environmental sustainability and infrastructure development. Geophysical methods such as ERT and electromagnetic surveys have been used to map aquifers and track groundwater flow patterns, helping engineers assess the availability and movement of water. These methods are especially important in arid regions where groundwater is scarce, as they provide a more accurate and cost-effective way of monitoring these vital resources (Kovari and Kozak 2016).

Furthermore, geophysical surveys are increasingly being used to monitor environmental factors such as soil stability, land subsidence, and contamination. In many regions, industrial activities, urbanization, and climate change have led to concerns about soil degradation, erosion, and the potential for landslides. Geophysical tools such as seismic methods, GPR, and ERT are now commonly used to monitor these phenomena, providing valuable data that can inform risk assessments and help mitigate the environmental impact of infrastructure development (Schleicher and Sloane 2019). These technologies can also detect subsurface contamination from industrial activities, providing early warning signs of pollution and enabling remediation efforts before the problem worsens.

In the context of climate change, geophysical techniques have also been employed to monitor phenomena such as land subsidence, which can occur due to excessive groundwater extraction or the effects of rising sea levels. By using advanced geophysical methods, scientists can track these changes over time, helping policymakers and engineers make informed decisions about mitigation strategies (Chavez and Lopez 2018).

Advancements in Automation and Remote Sensing

A particularly transformative aspect of recent advances in geophysical technologies is the integration of automation and remote sensing. The introduction of autonomous drones, robots, and remotely operated vehicles (ROVs) has greatly enhanced the efficiency and safety of geophysical surveys, particularly in hazardous or hard-to-reach environments. Drones equipped with sensors such as magnetometers, GPR, or LiDAR can be deployed over large areas to gather data quickly and efficiently. These drones provide valuable insights into areas that may otherwise be difficult or dangerous to access, such as steep or unstable terrain, urban areas with complex infrastructure, or areas affected by natural disasters (Schleicher and Sloane 2019).

For example, autonomous drones are increasingly used in large-scale environmental monitoring projects, such as tracking forest health, surveying contaminated sites, or mapping the extent of damage caused by natural disasters. In such cases, drones can quickly cover vast areas, collect high-resolution data, and transmit that data back to operators in real-time. This not only accelerates the data collection process but also reduces the risks associated with human involvement in dangerous environments (Yardimci and Arslan 2020).

The development of robotic systems has also facilitated the collection of data from more challenging environments. Robotic systems, which can be equipped with a variety of geophysical sensors, are used in applications such as monitoring the structural integrity of aging infrastructure, inspecting underground pipelines, and surveying hazardous waste sites. These systems can be controlled remotely, allowing operators to collect data from dangerous environments without putting themselves at risk (Kovari and Kozak 2016). This integration of robotics and geophysics not only improves data accuracy but also opens up new opportunities for geophysical surveys in previously inaccessible areas.

The Role of Artificial Intelligence (AI) and Machine Learning

As the volume of data generated by geophysical surveys continues to grow, the role of artificial intelligence (AI) and machine learning in geophysics is becoming increasingly important. AI algorithms are now being applied to geophysical data to assist with data processing, anomaly detection, and pattern recognition. Machine learning techniques enable the analysis of large datasets, identifying trends and relationships that might be difficult for human analysts to discern.

For example, AI-driven software can automatically analyze seismic data to identify fault lines, predict seismic activity, and assess risk levels. Machine learning models can also be used to predict the location of underground utilities, detect soil contamination, and even forecast subsurface conditions in areas with little prior data. By automating these processes, AI not only speeds up data analysis but also improves the accuracy and reliability of geophysical surveys, enabling engineers and geoscientists to make better-informed decisions (Chavez and Lopez 2018).

Moreover, AI and machine learning are being integrated into geophysical data acquisition systems, improving the efficiency of field surveys. Automated data collection systems, guided by AI algorithms, can optimize survey parameters in real-time, adapting to changing conditions and ensuring that the most relevant data is collected. This has the potential to significantly reduce survey costs and improve the overall effectiveness of geophysical investigations (Yardimci and Arslan 2020).

2.2.3 The Relevance of Engineering Geophysics in Modern Infrastructure and Environmental Challenges

Engineering geophysics has become an indispensable tool in addressing the complexities of modern infrastructure and environmental challenges. With the increasing demand for sustainable development, efficient resource management, and resilient infrastructure, the role of geophysical methods has expanded significantly. As cities grow, and environmental concerns become more pressing, engineers and geoscientists rely on advanced geophysical techniques to obtain crucial subsurface data. These techniques help guide decisions on infrastructure design, environmental protection, and hazard mitigation, ensuring the safety, efficiency, and sustainability of construction projects.

Role in Infrastructure Development and Risk Mitigation

Infrastructure development, particularly in urban and densely populated areas, requires a thorough understanding of subsurface conditions to ensure safety and cost-effectiveness. Geophysical surveys offer a non-invasive method of assessing the physical characteristics of the ground, which is essential for optimizing designs and reducing the likelihood of construction-related failures. For example, before the construction of foundations for buildings, roads, or bridges, geophysical methods, such as seismic refraction, GPR, and electrical resistivity tomography (ERT), are used to evaluate ground stability and detect potential subsurface hazards, such as voids, cavities, and faults (Yardimci and Arslan 2020).

Modern geophysical techniques, such as high-resolution seismic imaging, allow engineers to create detailed profiles of subsurface structures, aiding in the assessment of geohazards such as landslides, soil liquefaction, or fault movement. This is particularly important in areas prone to natural disasters like earthquakes, floods, and landslides. Seismic imaging helps engineers to identify the location and extent of fault lines and assess the risk to infrastructure in earthquake-prone zones. Understanding the structural integrity of the soil and the likelihood of movement or subsidence is critical for designing buildings and bridges that can withstand these forces (Chavez and Lopez 2018).

Furthermore, geophysical techniques like GPR and magnetometry are used to locate and map underground utilities and infrastructure, such as water pipes, cables, and tunnels. Identifying the exact location of these utilities before excavation helps prevent costly damage, service disruptions, and safety hazards during construction (Schleicher and Sloane 2019). These techniques also aid in detecting structural weaknesses in existing infrastructure, such as cracks in roads or bridges, corrosion in pipelines, and deterioration in foundations, enabling early intervention and maintenance that reduces the risk of catastrophic failures (Kovari and Kozak 2016).

Geophysics in Environmental Challenges and Sustainability

In addition to infrastructure, engineering geophysics is a critical tool for addressing environmental challenges. One of the primary concerns today is the degradation of the environment due to urbanization, industrialization, and climate change. Geophysical methods provide a powerful means of monitoring, assessing, and managing environmental impacts, particularly those related to groundwater, soil stability, and contamination.

For example, geophysical methods like ERT and seismic reflection can be used to monitor groundwater flow and detect contamination in aquifers. The ability to map the distribution of contaminants, such as hydrocarbons or heavy metals, in groundwater helps prevent the spread of pollution and provides the data necessary for effective remediation efforts (Yardimci and Arslan 2020). Groundwater is a vital resource, and ensuring its protection is critical for both environmental sustainability and public health. Non-invasive geophysical techniques enable regular monitoring of groundwater resources, providing real-time data to help authorities address contamination issues before they escalate (Chavez and Lopez 2018).

In regions affected by climate change, geophysical methods have become essential for tracking land subsidence and erosion. For example, land subsidence, often caused by excessive groundwater extraction, can lead to significant damage to infrastructure, including roads, buildings, and bridges. Geophysical surveys, such as seismic refraction and geodetic surveys, help monitor these changes over time and provide early warnings of potential problems. By using geophysical data, engineers can better understand the causes of subsidence and implement strategies to mitigate its effects, ensuring the resilience of infrastructure in areas affected by climate change (Kovari and Kozak 2016).

Geophysical tools are also critical in assessing the environmental impacts of large-scale construction projects, such as dams, bridges, and highways. For instance, seismic and geoelectrical methods are used to evaluate soil stability, identify potential hazards such as landslides or fault zones, and monitor soil compaction, all of which are crucial for preventing disasters and ensuring the long-term sustainability of these projects (Schleicher and Sloane 2019).

Moreover, geophysics plays a vital role in assessing the environmental sustainability of resource extraction activities. As society moves toward more sustainable practices, it is essential to assess the environmental impact of extracting resources like oil, gas, minerals, and even geothermal energy. Geophysical methods allow for detailed exploration of the subsurface to locate these resources while minimizing disruption to the surrounding environment. Methods such as seismic surveys, electromagnetic methods, and gravity measurements provide essential data on the location and quantity of these resources, aiding in the responsible management of these valuable assets (Yardimci and Arslan 2020).

Geophysics in Disaster Risk Reduction and Resilience

Another crucial application of engineering geophysics is in disaster risk management and resilience. Geophysical surveys can identify areas of high seismic, flooding, or landslide risk, helping planners design more resilient cities and infrastructure. In earthquake-prone regions, seismic hazard mapping based on geophysical data helps local governments and engineers assess the likelihood of ground shaking, fault movement, and liquefaction, which are key to creating earthquake-resistant buildings (Chavez and Lopez 2018). Moreover, geophysical methods are employed in post-disaster scenarios to assess damage, locate survivors, and guide rescue operations, enhancing the overall disaster response capacity.

In flood-prone areas, geophysical techniques such as GPR and ERT are used to study soil properties and groundwater conditions that influence flooding behavior. These methods allow for the mapping of water table fluctuations and the identification of areas where floodwaters may accumulate, helping engineers design drainage systems and flood barriers that can mitigate flood risks. Geophysical data can also help determine the best locations for dams, levees, and flood protection infrastructure, ensuring their effectiveness in managing floodwaters (Kovari and Kozak 2016).

Additionally, geophysical methods are employed in landslide risk assessments, especially in mountainous or coastal areas. Seismic refraction and GPR can be used to investigate soil and rock properties, identify weak layers, and detect signs of instability in slopes. This data is critical for predicting potential landslides and implementing preventive measures to protect infrastructure and communities at risk (Schleicher and Sloane 2019).

Table 2.3 illustrates the key applications of geophysics in infrastructure, environmental challenges, and disaster risk reduction. This would provide a clear, concise visual summary of how geophysical techniques are being applied in modern engineering and environmental scenarios.

Table 2.3 Applications of engineering geophysics in modern infrastructure, environmental challenges, and disaster risk reduction

Focus area	Geophysical techniques	Key applications	Impact on development and sustainability
Infrastructure development	Seismic surveys, ground penetrating radar (GPR), electrical resistivity tomography (ERT), borehole geophysics	– Site characterization for construction – Foundation design – Underground utilities mapping	– Ensures structural safety by detecting subsurface anomalies – Optimizes resource use
Risk mitigation	Seismic surveys, gravimetry, electromagnetic methods, GPR, microgravity	– Fault zone identification – Detection of soil instability – Mapping of potential landslide areas	– Minimizes risks from seismic activity, landslides, and other geological hazards – Reduces infrastructure damage in vulnerable areas
Environmental challenges	Electrical resistivity tomography (ERT), magnetometry, geoelectrical imaging, ground penetrating radar (GPR)	– Mapping of groundwater resources – Contamination detection – Soil pollution monitoring	– Protects ecosystems by assessing contamination – Enhances sustainable management of natural resources
Sustainability and resource management	Magnetometry, electromagnetic methods, borehole geophysics	– Mineral and water resource exploration – Geothermal energy assessments	– Supports sustainable extraction and efficient resource use – Reduces environmental impact of resource extraction
Disaster risk reduction and resilience	Seismic surveys, microgravity, gravimetry, ground penetrating radar (GPR)	– Earthquake preparedness – Flood risk assessment – Mapping of subsurface voids for sinkhole prevention	– Increases community resilience by assessing risks – Aids in disaster preparedness and early warning systems
Climate change adaptation	Ground penetrating radar (GPR), electrical resistivity tomography (ERT), thermal conductivity measurements	– Coastal erosion monitoring – Monitoring of soil moisture and temperature – Permafrost stability analysis	– Helps in planning for climate-induced risks – Supports adaptation strategies in vulnerable regions

2.2.4 The Changing Role of Geophysics in Environmental Evaluation, Monitoring, and Management of Contaminant Plumes from Landfill Leachate: A Comprehensive Overview

Landfill leachate, a harmful mix of pollutants, is a serious threat to water, ecosystems, and public health. Traditional methods of monitoring struggle to fully understand the spread of this contamination. This review shows how geophysics can help detect and monitor landfill leachate plumes (LLPs) by looking beneath the surface. Geophysical methods like resistivity, seismic, electromagnetic (EM), and magnetic surveys can map and study LLPs in great detail. Electrical resistivity imaging (ERI) and ground penetrating radar (GPR) are especially good at monitoring LLPs, while seismic tomography (SRT) and ground magnetic surveys (GMS) are somewhat helpful. Very low-frequency EM (VLF-EM) is less effective at detecting and mapping contaminated areas underground. Studies show that LLPs often have different geophysical properties, like density or conductivity, compared to the surrounding environment, which helps geophysical tools quickly find and map them with high accuracy and low cost. These methods do not disturb the surface or cause additional contamination. However, while geophysics is useful, it cannot replace direct methods like geological borehole data and water chemical analysis for full environmental monitoring. Factors like soil porosity, water flow, and contamination types can affect the accuracy of geophysical data, sometimes leading to incorrect results. Still, case studies show that geophysics is valuable for understanding LLPs and improving waste management practices (Alao 2025).

Several case studies show the effectiveness of geophysical methods like electromagnetic (EM), electrical resistivity (ER), seismic, and magnetic surveys in assessing, monitoring, and managing landfill leachate plumes (LLPs). For instance, electrical resistivity surveys have been used to map the spread of leachate plumes and identify areas where the flow of contaminants is concentrated in the subsurface (Ekeocha et al. 2012; Okpoli 2013; Frid et al. 2017). Ground penetrating radar (GPR) surveys have been helpful in mapping waste material distribution and detecting possible contamination sources (Junior et al. 2016; Umar et al. 2022; Zhang et al. 2022). Electromagnetic surveys are also used to monitor changes in soil moisture and identify areas with high conductivity linked to leachate migration. These geophysical methods have been applied to a range of landfill management tasks, such as outlining landfill boundaries, mapping zones where leachate has accumulated, monitoring soil moisture content, checking the integrity of landfill liners, and detecting leaks or pathways for leachate (Netto et al. 2023).

Case Study Electrical Resistivity Imaging (ERI) for Mapping Contaminant Pathways

Electrical Resistivity Imaging (ERI) is a commonly used geophysical method for detecting and mapping landfill leachate plumes (LLPs). ERI works by injecting electrical current into the ground to measure the electrical resistivity of subsurface materials. Changes in resistivity can highlight variations in soil properties, moisture content, and the presence of contaminants. This technique is especially useful for mapping the extent of LLPs and identifying their migration pathways, helping to assess the impact on water sources, soil, and public health.

Several studies around the world have demonstrated the application of ERI for monitoring landfill sites. In Poland, ERI detected low resistivity values in areas affected by leachate, which helped to track how contaminants spread into the soil and groundwater. In Northern Nigeria, ERI was used to investigate dumpsites, revealing that leachate had infiltrated the soil and posed a threat to local groundwater. Similarly, in Malaysia, ERI was applied to urban landfills, finding low resistivity values between 1.5 and 4.0 m in depth, which were linked to the presence of chemical contaminants.

ERI has also been applied in countries like Denmark, South Africa, and Poland to identify contaminant pathways and evaluate landfill liners effectiveness. In the United States, ERI was used to measure the degree of saturation in landfill liners, helping to ensure their quality during construction and maintenance.

The results from these studies clearly show that ERI is an effective tool for tracking the spread of leachate, assessing potential environmental risks, and guiding waste management decisions. These applications highlight the value of ERI in providing critical insights for the protection of groundwater and soil, helping decision-makers implement effective landfill management strategies (Fig. 2.2).

Case Study Seismic Refraction Tomography for Plume Delineation

Seismic refraction tomography (SRT) is a valuable geophysical technique that has been widely used to delineate shallower targets (less than 30 m deep) and assess landfill leachate plume (LLP) contaminants (Kirkpatrick et al. 2012; Molua 2022). SRT plays a crucial role in monitoring and understanding the subsurface structure of landfill sites. It maps various subsurface parameters, including LLP contamination zones, and helps assess waste degradation and stability.

In landfills with leachate recirculation, seismic surveys are used to monitor moisture distribution and changes in the mechanical properties of municipal solid waste (MSW) (Carpenter et al. 2012). These surveys provide data on waste stability, including shear modulus and Poisson's ratio, offering insights into seismic stability. Seismic refraction surveys, in particular, provide high-resolution data that complements other geophysical methods for better landfill assessments.

A study in the Al Minya basin, Palestine, used seismic refraction to map the distribution of subsurface rock types and identify pathways for contaminants (El-Kelani and Khader 2019). The study identified three velocity zones, with velocities

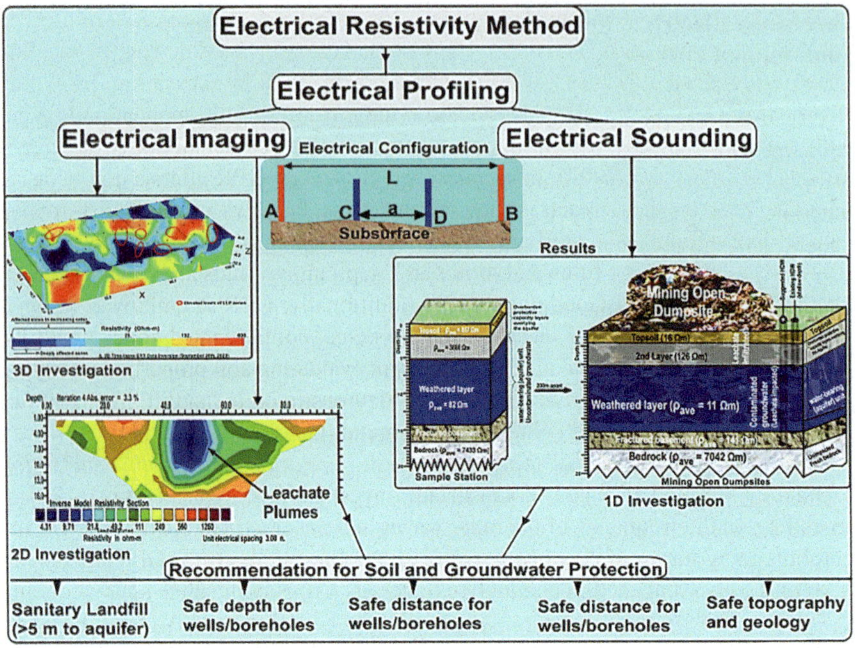

Fig. 2.2 Graphical representation of electrical resistivity survey results showing the impact of landfill leachate plume contaminants on soils and groundwater. *Source* Alao (2025)—Open Access

ranging from 192 to 2400 m/s. Although minor fractures in the bedrock could allow seepage, the subsurface geology was found to be suitable for the proposed landfill, with an aquitard layer reducing leachate infiltration.

In another study, SRT and surface wave analysis (MASW) were used to explore a landfill site's topography, revealing that the seismic wave velocity increased sharply below 5 m depth, indicating compaction of waste material (Carpenter et al. 2012). The study also showed differing S-wave velocities in different parts of the landfill, with higher velocities in older, compacted waste compared to newer, uncompacted sections. The surveys were limited to depths of 8–9 m.

On Crete, an SRT study at the largest landfill on the island identified saturated unconsolidated waste as the second layer with velocities around 1670 m/s, corresponding to conductive leachates (Soupios et al. 2007). This study demonstrated the effectiveness of combining seismic refraction with geoelectrical tomography to map subsurface contamination and delineate the extent of LLP migration.

In conclusion, seismic refraction tomography (SRT) is a powerful tool for detecting and monitoring LLP contaminants in landfills. The technique provides valuable insights into subsurface conditions, helping to track the movement of contaminants and support effective environmental management (Fig. 2.3).

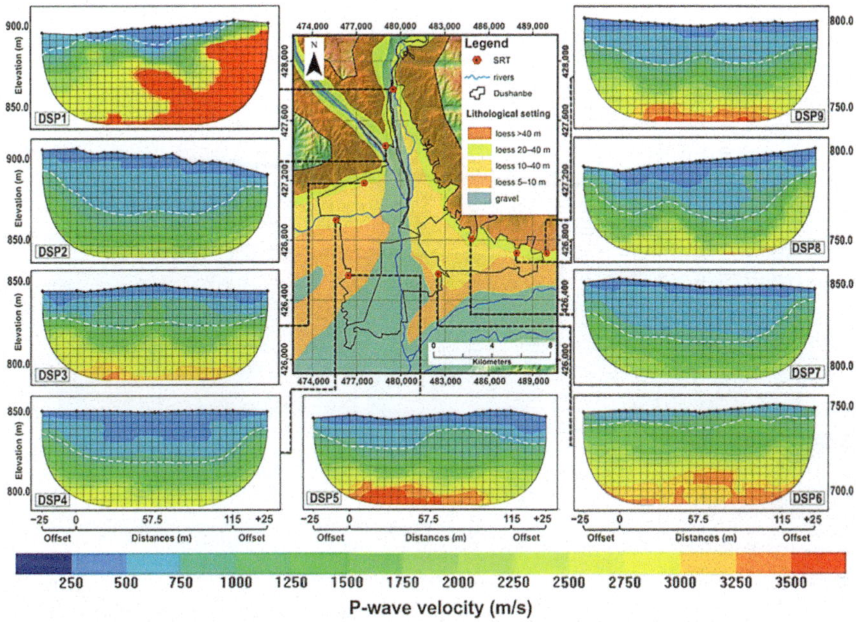

Fig. 2.3 Seismic refraction tomography results for mapping landfill leachate plume and subsurface characteristics, with the dashed white line indicating the base of the loess layer. *Source* Hakimov et al. (2021)—Open Access

Case Study Ground Magnetic Survey for Mapping Contaminant Plumes

Ground magnetic surveys (GMS) are a powerful tool for detecting and mapping leachate-laden plumes (LLPs), as landfill environments often contain ferromagnetic materials with high magnetic susceptibility. GMS works by detecting variations in the Earth's magnetic field caused by subsurface anomalies, such as metallic objects or buried waste, which can indicate contamination from LLPs. These contaminants produce magnetic signatures that differ from surrounding soil, making it possible to identify areas of elevated magnetic response.

GMS can map the spatial distribution of magnetic anomalies linked to LLPs, helping to delineate the extent and boundaries of contaminated zones (Fig. 2.4). It is effective in detecting metallic items like iron and steel and reacts to ferromagnetic materials. Studies show that vertical gradient magnetic measurements are more sensitive to smaller magnetic targets and provide better resolution than total field measurements. This is important for understanding buried items at hazardous waste sites.

GMS is a fast, non-intrusive, and cost-effective method for surveying landfill sites. It can generate magnetic profiles or maps to highlight areas of interest. In one study, magnetic forces ranging from 120 to 360 μT improved leachate quality by reducing contaminants like ammonia–nitrogen and bacteria. The results suggest that magnetic

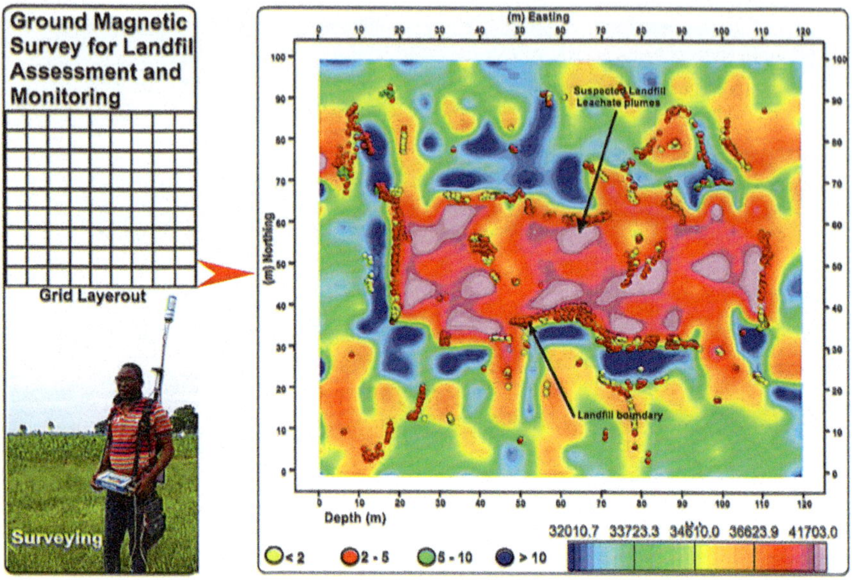

Fig. 2.4 Ground magnetic survey for mapping contaminant plumes in landfills. *Source* Alao (2025)—Open Access

force can be used as an inexpensive pre-treatment method to enhance landfill leachate characteristics.

A comprehensive GMS was also conducted at three dumpsites to examine the intrusion of dumpsite leachate plumes (DLPs), based on changes in the total magnetic intensity (TMI) both within and outside the dumpsite. This investigation revealed that the TMI values within the dumpsites ranged from 37,600 to 44,500 nT, whereas off-site TMI values ranged from 31,095 to 37,300 nT, typically within a depth of 0.0–5.0 m. The differences in TMI values between the dumpsites and the surrounding area suggested that leachate contamination altered the magnetic properties of the soil, resulting in more distinct contamination indicators. Shallow leachate plumes produced stronger magnetic anomalies, making them easier to detect and map. However, the presence of other magnetic sources, such as buried infrastructure or natural geological features, can interfere with the magnetic survey data, making it challenging to accurately delineate leachate plumes, especially at greater depths.

In the U.S., GMS has been used as part of multi-technique geophysical investigations to distinguish between deep and superficial magnetic sources within landfill sites (Roberts et al. 1990; Freidman et al. 2014). For example, at Wilkes Station, a landfill located near the ocean, magnetic anomalies stretched from northwest to southeast, indicating the extent of the landfill and revealing magnetic changes caused by the material orientation and dumping activities. Magnetic susceptibilities greater than 0.02 SI units (2% more magnetically susceptible compared to the reference material, which is typically air which has a magnetic susceptibility of approximately

0) were observed beyond the landfill area, highlighting elevated magnetic characteristics of the basement geology. Furthermore, the study also identified major magnetic anomalies within the landfill, allowing for accurate identification of contamination zones. This survey method proved to be both economical and non-intrusive.

In conclusion, the depth of leachate plumes significantly affects the ability to detect and interpret magnetic anomalies. Shallow plumes are easier to identify, while deeper plumes may be more challenging and subject to interference. GMS is crucial for mapping LLPs, characterizing contaminant pathways, and monitoring changes over time. It is a valuable tool for managing landfill contamination and understanding the potential risks to groundwater and surface water resources.

Case Study Electromagnetic Surveys for Monitoring Plume Dynamics

Electromagnetic (EM) geophysical survey methods use EM fields to analyze the Earth's subsurface properties by exploiting interactions between EM waves and electrical conductivity (EC), magnetic susceptibility, and permittivity (Geology Science 2023). These methods are based on electromagnetic induction (EMI) and wave propagation. Types of EM surveys include transient EM, Very Low Frequency (VLF), Ground-Penetrating Radar (GPR), magnetotelluric (MT), and audio-magnetotelluric (AMT). This review focuses on the use of VLF-EM and GPR for landfill characterization, helping to identify waste distribution and potential leachate pathways, aiding in environmental evaluation, planning, and remediation (Splajt et al. 2003).

VLF-EM and GPR are effective tools for mapping leachate plumes (LLPs) due to their ability to detect variations in material properties associated with contaminants (Umar et al. 2022; Nobes et al. 1994; Zhou and Yang 2011). GPR, for example, sends high-frequency EM pulses into the ground and measures reflections from subsurface layers, identifying contaminants through distinct radar signatures. The high resolution of GPR imaging allows for the visualization of LLP distribution and helps map the depth, extent, and shape of the plume.

Boundary characterization is also crucial, as it helps identify transitions between contaminated and uncontaminated zones. GPR can detect fractures, conduits, and permeable zones that influence the mobility of LLP contaminants, enhancing understanding of how plumes spread (Proulx-McInnis et al. 2013). Combining GPR with other geophysical methods, such as ERT, VLF-EM, or seismic surveys, provides more comprehensive insights. VLF-EM is particularly useful for detecting zones of high electrical conductivity, such as LLPs (Nobes et al. 1994).

A study at a dumpsite used EM and GPR surveys and found that EC and hydraulic properties affect GPR depth penetration (Nobes et al. 1994). Conductive leachate plumes caused radar signal attenuation, appearing as blank spots on profiles. The study also observed that topographic variations influenced EM results, which were clarified by removing topographic effects.

In Brazil, GPR has become an important tool in geotechnical and environmental studies. For example, a study at the Gramacho landfill in Rio de Janeiro (Figs. 2.5 and 2.6) revealed small voids indicating slip surfaces, large voids suggesting gas pools,

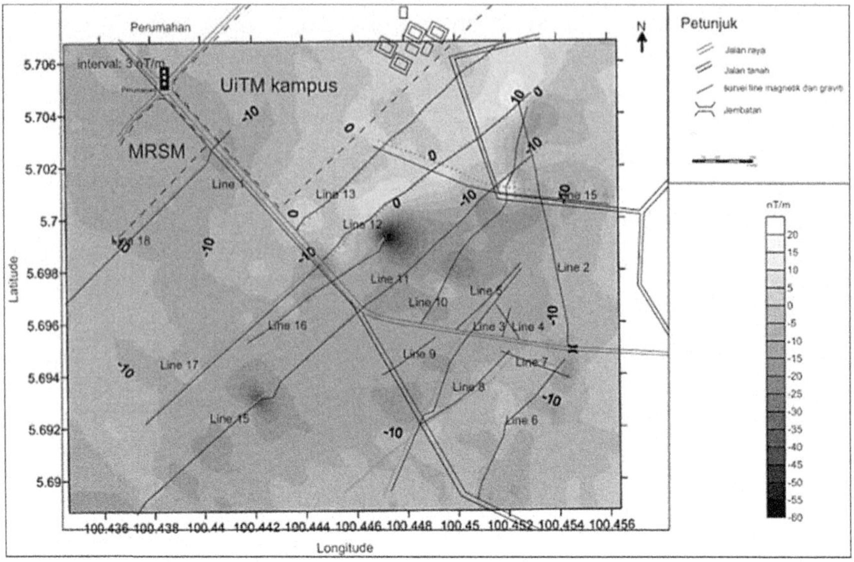

Fig. 2.5 GPR radargram showing the areas where leachate plumes (LLPs) have accumulated. *Source* Alao (2025)—Open Access

and high conductivity zones indicating leachate pools (Junior et al. 2016). Another study at an established landfill found that GPR and EM techniques accurately mapped waste boundaries, depth, and volume, providing a 3D model of the landfill's structure (Zhang et al. 2022).

At the University of Sao Paulo (USP), GPR was used to detect buried objects simulating pipelines, confirming GPR's effectiveness for subsurface utility mapping (Porsani et al. 2017). At the Songshuling landfill in China, GPR identified fracture zones and karst fissures that allowed leachate to seep into groundwater, further confirming GPR's utility in detecting leachate leakage (Zhou and Yang 2011).

In a landfill on Crete Island, VLF-EM surveys identified conductive zones indicative of potential leachate plumes, although the results were less precise compared to ERI (Soupios et al. 2007). VLF-EM also detected and mapped LLP flow paths at a landfill operating since 1998, confirming contamination via fractures in the waste complex (Monteiro Santos et al. 2006).

In conclusion, both GPR and VLF-EM technologies are valuable for mapping subsurface features and managing LLP contaminants in landfills. GPR offers higher precision, particularly in visualizing contaminant flow paths and plume boundaries, while VLF-EM provides complementary data. These technologies are essential for improving landfill management, protecting groundwater resources, and mitigating environmental risks.

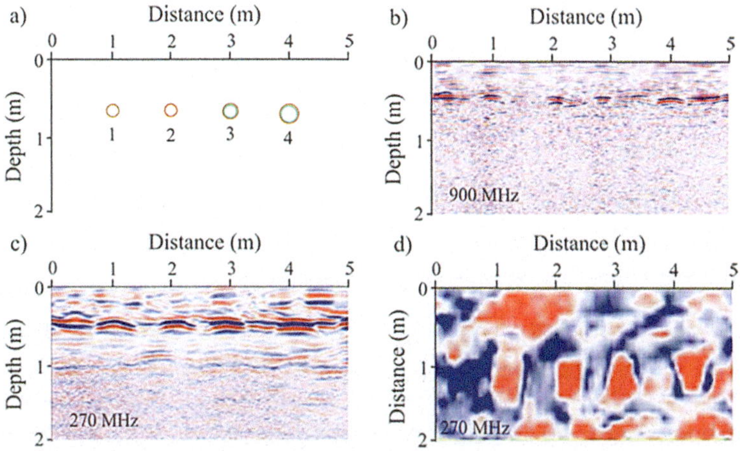

Fig. 2.6 GPR radargram results of buried targets at the USP shallow GTS-II: **a** spatial distribution across the site, showing targets (target 1 is an empty PVC pipe, target 2 is an empty ceramic tube, and targets 3 and 4 are ceramic tubes filled with half water and half air). **b** GPR radargram results at 900 MHz. **c** GPR radargram results at 270 MHz for the profile. **d** GPR depth slice at 0.62 m for 270 MHz frequency showing the targets. *Source* Alao (2025)—Open Access

Case Study Use of Integrated Geophysical Methods in Assessing Contaminant Plumes

Integrated geophysical methods are essential for evaluating landfill contaminant plumes. These methods provide detailed subsurface mapping, which helps in identifying and monitoring contaminant distribution (Eze et al. 2022).

One study in Gaeiras, Central Portugal, used combined radio frequency electromagnetics (RF-EM), Geonics EM34, spontaneous potential (SP), vertical electrical soundings (VES), and magnetic prospecting surveys. The aim was to study geological conditions, waste distribution, and leachate characteristics at a landfill site. The results showed the presence of faults, bedrock depth, and pathways for leachate movement. The integration of geophysical data, hydrogeological information, and chemical analyses allowed the development of a schematic model of the contamination plume. The study concluded that the landfill was poorly sited, and remedial actions are needed to prevent further environmental damage. The researchers recommended using multidisciplinary techniques for future landfill environmental assessments (Ramalho et al. 2012).

In another study, combined geophysical methods (CGMs), including electrical resistivity tomography (ERT) and Very Low Frequency Electromagnetic (VLF-EM) surveys, were used along with water quality analysis to assess the impact of uncontrolled landfills on groundwater in Kaduna, Nigeria. The results showed that landfill leachate plumes had compromised groundwater quality, leading to high levels of heavy metals and dissolved organic substances. Low resistivity anomalies ($< 9\ \Omega$m) within the dumpsite and higher resistivity off-site (111–413 Ωm) indicated the presence of contaminant plumes. However, VLF-EM was less effective in detecting leachate contaminants compared to ERT. This highlights how these integrated techniques complement each other in monitoring and managing leachate contamination (Alao et al. 2023).

A similar study in Ilero, southwestern Nigeria, used electrical resistivity (ER) and frequency domain electromagnetic (FDEM) methods to map subsurface contamination from cassava processing waste. The study produced 1D, 2D, and 3D resistivity and conductivity models, revealing the contamination plume. Low resistivity and high conductivity zones (21–30 Ωm and 11–40 mS/m) were identified as indicating contamination. This combined approach of ERI and FDEM effectively mapped the subsurface leachate contamination (Osinowo et al. 2020).

Another study investigated the impact of landfill leachate on surrounding ecosystems by combining hydrochemical analysis with induced polarization (IP) surveys. The study found high polarization effects near steel sheet piles, indicating leachate distribution. Low chargeability zones (< 100 mV/V) were correlated with high conductivity areas, confirming leachate presence. The study highlighted that IP and hydrochemical data integration is effective for locating contaminated zones and monitoring plume behavior. It also showed that steel sheet piles can cut off contamination sources, which has important implications for environmental monitoring (Xia et al. 2023).

In summary, each geophysical method has its strengths and limitations. Electrical Resistivity Imaging (ERI) excels in delineating plumes and identifying flow paths. Seismic Refraction Tomography (SRT) provides valuable structural data. Ground Magnetic Surveys (GMS) help map spatial distributions of magnetic anomalies linked to leachate. Ground Penetrating Radar (GPR) offers high-resolution imaging of plumes and pathways. Combining these techniques provides a comprehensive understanding of landfill contaminant plumes, making it possible to choose suitable landfill sites, assess leachate migration potential, and design protective barriers.

Case Study Integrated Geophysical Techniques for Landfill Characterization and Contamination Detection at Weidenpesch Landfill, Cologne, Germany

Geophysical techniques have become essential for characterizing landfills and assessing contamination, offering non-invasive methods for subsurface exploration. This study employed an integrated geophysical approach—using magnetic, electrical resistivity tomography (ERT), and transient electromagnetic (TEM) surveys—to characterize the Weidenpesch landfill in Cologne, Germany. The goal was to assess potential groundwater contamination and gain insights into subsurface landfill conditions (Fig. 2.7).

The study used a combination of magnetic surveys to detect variations in the magnetic susceptibility of materials (Fig. 2.8), electrical resistivity tomography (ERT) to measure variations in electrical resistivity indicating the presence of waste

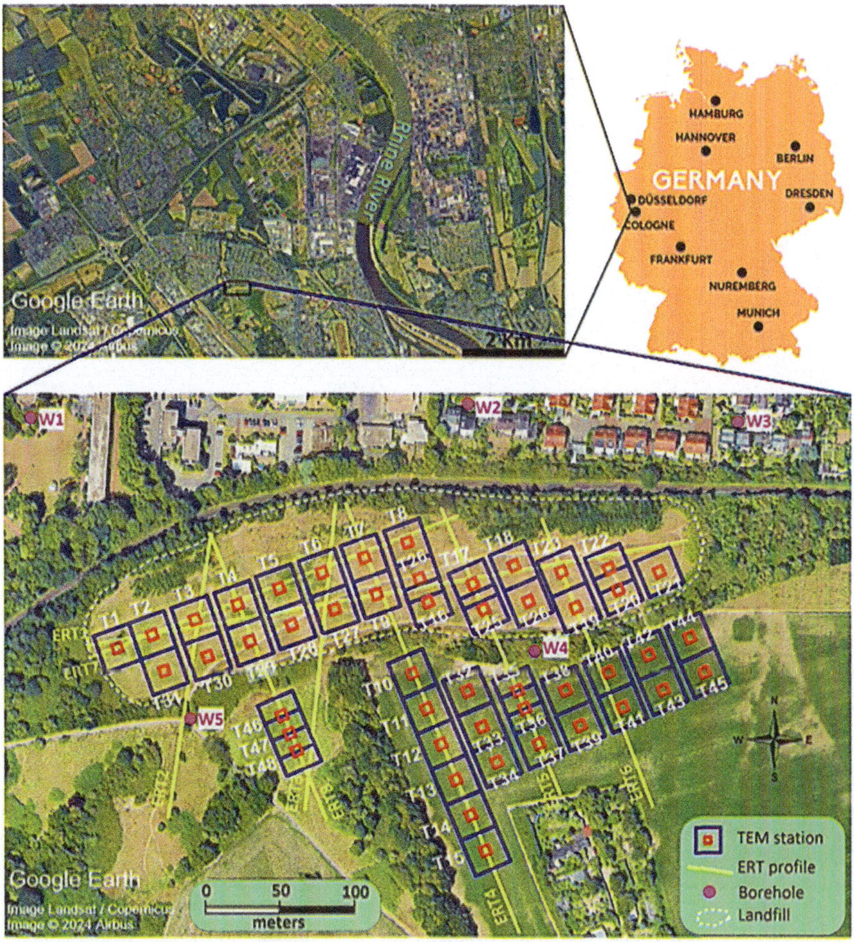

Fig. 2.7 Satellite image of Weidenpesch landfill location with geophysical survey points. *Source* Ibraheem et al. (2024)—Open Access

or contamination (Fig. 2.9), and transient electromagnetic (TEM) surveys to assess subsurface conductivity (Fig. 2.10), which is particularly useful for detecting contamination. Subsurface lithology logs from boreholes around the landfill show that the Holocene layer, made up of silt, clayey silt, or sandy silt, reaches up to 2.8 m thick (Fig. 2.11).

The Pleistocene gravel and sand layers range from 17 to 25 m in thickness, while the Oligocene base layer consists of fine sand, sandy coal, silty coal, or brown coal. These surveys were conducted over multiple areas of the landfill to gather a comprehensive dataset for analysis.

The integrated geophysical approach provided consistent and reliable results, allowing for a clear delineation of landfill boundaries and identification of potential

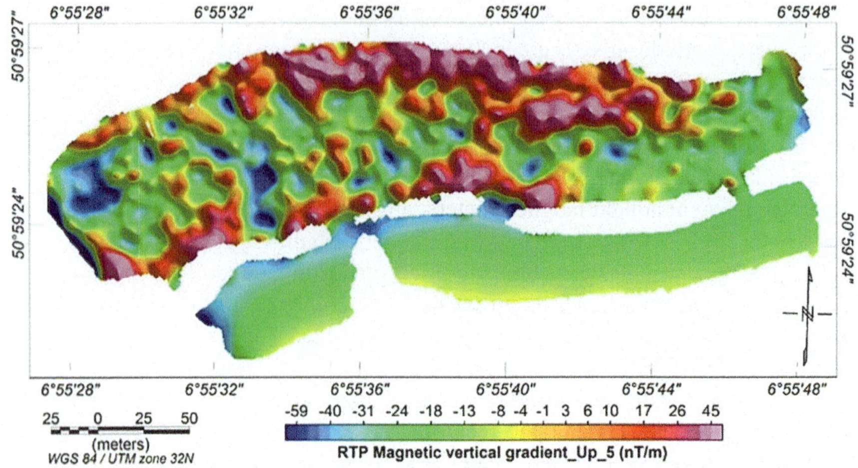

Fig. 2.8 The RTP magnetic map of the Weidenpesch landfill area after applying an upward continuation filter at an altitude of 5 m. *Source* Ibraheem et al. (2024)—Open Access

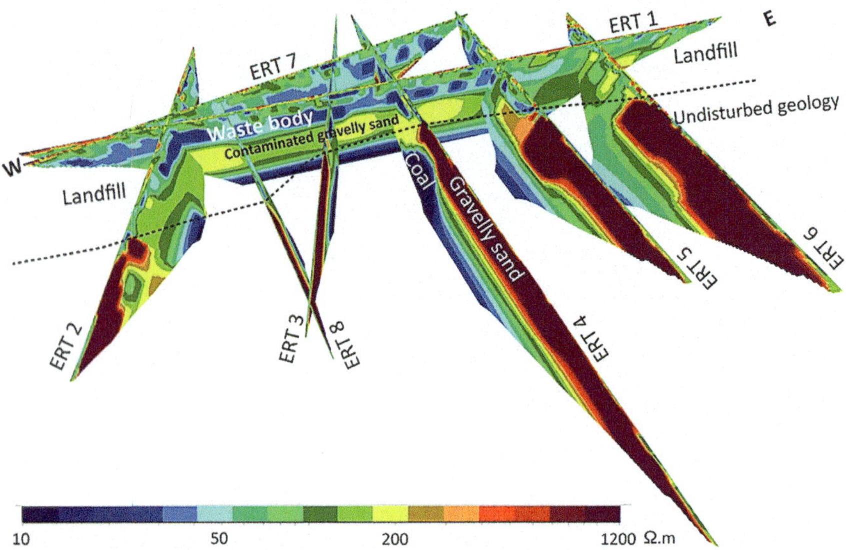

Fig. 2.9 A 3D fence diagram of subsurface resistivity distribution from the 2D inversion of ERT data at the Weidenpesch landfill. Blue anomalies at shallow depths indicate the waste body, with the dashed black line marking the southern boundary. *Source* Ibraheem et al. (2024)—Open Access

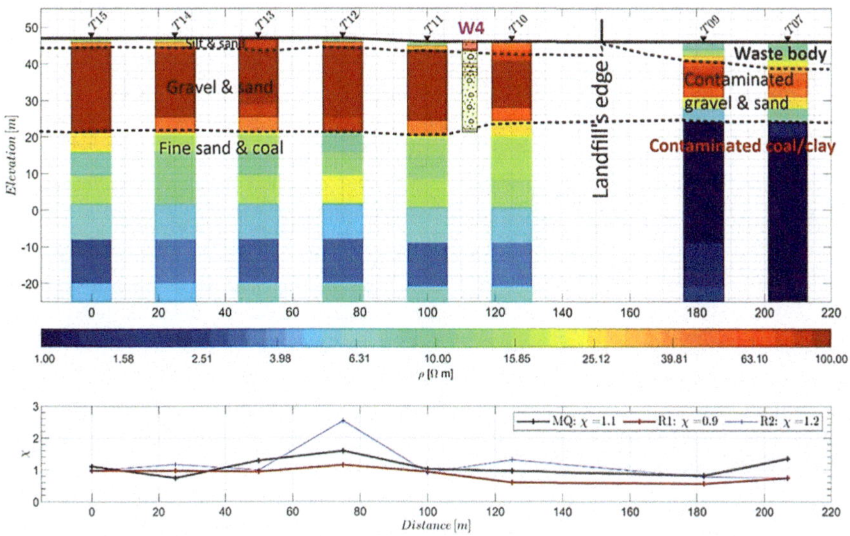

Fig. 2.10 2D resistivity cross-section from 1D Occam inversion results for TEM data along profile ERT4 at the Weidenpesch landfill. The lower panel shows χ values for Occam (R1 and R2) and MQ inversions. Borehole W4 log is shown between TEM stations T10 and T11. *Source* Ibraheem et al. (2024)—Open Access

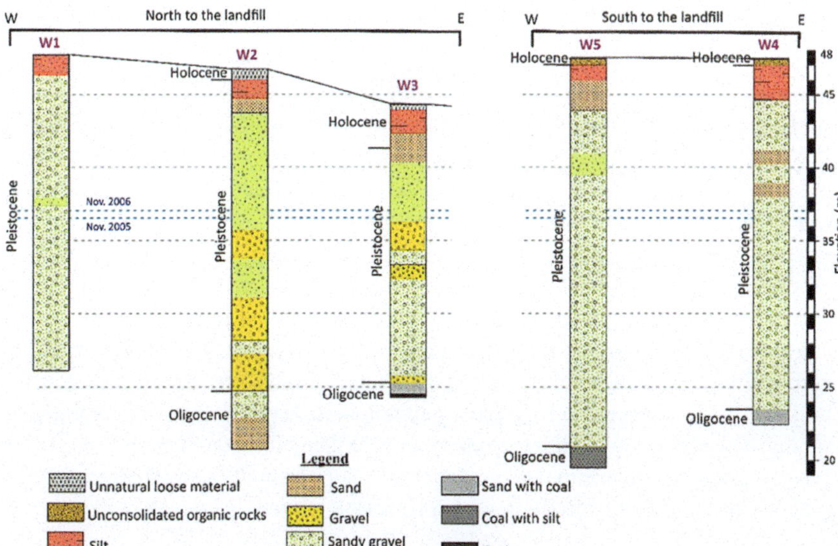

Fig. 2.11 Subsurface lithology logs from boreholes around the Weidenpesch landfill. Groundwater tables for 2005 and 2006 are marked by dashed blue lines. *Source* Ibraheem et al. (2024)—Open Access

contamination zones. The waste body was characterized by low resistivity values in the western and central parts of the landfill, with an average resistivity of 1–10 Ωm. The eastern part of the landfill exhibited a slightly higher resistivity range of 20–50 Ωm, contrasting with the surrounding high-resistivity gravelly sand layer (several hundred Ωm). The depth of the landfill was estimated to be up to 15 m.

Beneath the waste body, a gravelly sand layer was observed at depths ranging from 20 to 23 m, with resistivity values between 80 and 100 Ωm. A coal/clay layer was detected at depths of 20–25 m, exhibiting low resistivity values (less than 5 Ωm). This layer is suspected to be contaminated, likely due to migration from the landfill. The low resistivity observed within the gravelly sand layer beneath the landfill and the highly conductive coal/clay layer suggests potential migration of contaminants. TEM data further revealed increased conductivity in the coal/clay layer over time, indicating a possible contamination plume spreading downward and laterally.

The integrated geophysical methods provided a comprehensive understanding of the subsurface conditions at Weidenpesch landfill. The magnetic, ERT, and TEM data consistently highlighted the presence of a heterogeneous waste body with variable resistivity values, possible contamination migration, especially within the coal/clay layer beneath the landfill, and significant time-dependent changes in conductivity, particularly in the deeper coal/clay layer. These findings are crucial for assessing the landfill's environmental impact, guiding remediation efforts, and preventing further contamination of the surrounding groundwater.

The study demonstrates the effectiveness of an integrated geophysical approach for landfill characterization and contamination detection. The results provide valuable information for the Environmental Office of the City of Cologne and local authorities to make informed decisions about landfill management and remediation. The waste body is present up to 10–15 m deep, with low resistivity values in the central and western areas and higher resistivity in the eastern parts. The coal/clay layer below the landfill exhibits low resistivity, indicating possible contamination, and the TEM data showed variations in conductivity over time, suggesting that continuous monitoring is needed.

Future studies should incorporate additional techniques like induced polarization (IP) for enhanced characterization of subsurface conditions. Long-term monitoring using TEM, ERT, and IP is also recommended to track contamination migration. Continuous borehole measurements of temperature, groundwater levels, and conductivity will be essential for ongoing contamination monitoring and remediation planning. This study underlines the importance of using integrated geophysical techniques for effective environmental management, supporting both waste management practices and sustainable environmental protection strategies.

2.2.5 *The Future of Geophysics in Environmental Engineering: Innovations and Applications*

In the rapidly evolving world of environmental engineering, geophysics plays an essential role in enabling more sustainable and informed decisions. Near-surface geophysics, in particular, has become increasingly crucial for site characterization, utility locating, and mapping both known and unknown subsurface structures.

Geophysics involves the study of the Earth's physical properties and processes, ranging from oil and gas exploration to mineral exploration and seismology. However, near-surface geophysics, which typically focuses on depths up to 50 m, is where environmental engineering and site investigations converge. This discipline uses non-destructive testing (NDT) methods to assess the subsurface without disrupting the environment.

Key Techniques in Near-Surface Geophysics

Key techniques in near-surface geophysics include *Ground Penetrating Radar (GPR)*, which transmits radar waves into the ground and detects changes in subsurface properties based on the reflection of these waves. GPR provides a high-resolution image of electrical property variations but has limitations in clay-rich soils. *Electromagnetic Induction (EM)* uses varying electromagnetic fields to induce eddy currents in the subsurface, measuring electrical conductivity and magnetic susceptibility. It is well-suited for mapping contamination but has lower resolution for deeper investigations. *Electrical Resistivity Tomography (ERT)* involves injecting electrical currents into the ground and measuring the resulting voltage variations along a survey line, creating a 2D cross-section of the resistivity distribution, making it useful for bedrock mapping and groundwater analysis. *Magnetometry* measures anomalies in the Earth's magnetic field to detect buried materials with magnetic properties, such as minerals and metals, and is often used in archaeological site investigations and identifying buried infrastructure. *Seismic surveys* utilize mechanical waves to assess the subsurfaces mechanical properties, providing data on density and elasticity, which is especially valuable for detecting voids or mapping stratigraphy.

GeoScan's Approach: Multi-method Integration

GeoScan's geophysical team combines data from GPR, EM, ERT, and magnetometry to build a more reliable and comprehensive model of the subsurface. Aerial imagery and digital elevation models are also often used for site characterization. By integrating multiple datasets, GeoScan's avoids relying on a single method, which may only provide partial information. For example, a GPR survey might identify a subsurface structure, while EM or magnetometry data can help in understanding material composition. Together, these datasets provide a clearer picture of subsurface conditions, allowing engineers to make more informed decisions.

Applications in Environmental Engineering

Geophysical techniques are indispensable in a variety of environmental engineering applications. For example, Ground Penetrating Radar (GPR) is commonly used in *highway and railway inspections* to assess asphalt thickness and road base conditions, thereby ensuring the safety and integrity of transportation infrastructure.

Case Study GPR for Civil Engineering Applications—Trenchless Technology for Pipe Installation in Tamil Nadu, India

The main objective of this study was to assess the feasibility of continuing the installation of a 1700 cm diameter, 2 cm thick mild steel (MS) pipe using trenchless technology under the road near the Head Works at Project Circle Ramnad, Tiruchirappalli, Tamil Nadu, India (Sonkamble and Chandra 2021). The project was part of the Tamil Nadu Water Supply and Drainage (TWAD) Board's ongoing efforts to improve water infrastructure in the region. The study area covered a 20 m wide and 75 m long section of the road, which was scanned to a depth of 8 m using Ground Penetrating Radar (GPR). The road had sustained damage due to subsidence caused by the trenchless pipe installation. The GPR scans were conducted using seven profiles with a 200 MHz antenna and a DIEL 4 system to cover the required depth range of 7–8 m, scanning both sides of the road.

The GPR survey of the underground strata around the trenchless pipe installation area revealed no anomalies or obstacles in the surveyed region, except for Profile-A. This profile indicated a deformation at approximately 3.0 m depth, suggesting the pipe had lost its circular shape and was no longer aligned horizontally. The top portion of the pipe was clearly visible in the GPR image at a depth of 3 m, between 10 and 12.5 m from the starting point of the profile. A significant reduction in the pipe's diameter was observed both during the field investigation and in the GPR anomaly. This deformation could be attributed to the buckling of welded joints, possibly caused by poor welding practices and excessive hydraulic pressure applied during the trenchless installation. The deformation could also have been triggered by an underlying obstacle, which the pipe encountered during installation.

According to the site engineer's observations, the road filling comprised weathered material with occasional boulders. The pipe likely struck a strong obstacle, possibly a large boulder, which hindered its further installation despite high hydraulic pressure being applied. The collision with the obstacle and the excessive pressure led to the pipe deforming and flattening, as confirmed by both the field observations and the GPR anomaly. Based on the GPR results and site engineer feedback, it was clear that the initial route for the trenchless installation was unsuitable due to the encountered obstacle. To avoid further complications, an alternative location for the trenchless pipe installation was recommended. The GPR scanning was extended westward to locate a potential new route. At approximately 10 m west of the original location, a new site was identified. This site consisted of weathered strata and appeared free of obstacles, extending to a depth of 8 m, which made it a feasible route for the trenchless installation.

It was recommended that the new site be used for the trenchless installation of the pipe. Additionally, it was advised that extra care be taken during the welding of pipe joints to avoid further deformation. The saturated zone, identified at approximately 3 m depth, should be carefully managed by regularly withdrawing water using high-capacity pump sets to facilitate a smoother trenchless pipe installation without encountering obstructions. The shallow water table, which consistently returned a strong reflection signal in the GPR images, clearly showed its distinct anomaly, as previously reported in studies by Mahmoudzadeh et al. (2012). This highlighted the need for effective water management to ensure the trenchless installation process remains unobstructed. The GPR-based investigation successfully identified the cause of the deformed pipe and helped locate a more suitable route for the trenchless technology installation. This approach not only ensured a more efficient solution to the problem but also facilitated better planning for future installations, avoiding obstacles and minimizing potential delays in civil engineering projects.

Case Study Ground Penetrating Radar (GPR) for Detection of Clandestine Pipes in Industrial Areas

Overview

Clandestine pipes, often used to illegally discharge industrial effluents into nearby water bodies, pose significant environmental risks by contaminating groundwater and surface water systems. This study aims to detect such underground pipes using Ground Penetrating Radar (GPR), focusing on locations with suspected clandestine discharges. The following sections detail individual case studies where GPR was successfully employed to identify these underground pipes (Sonkamble and Chandra 2021).

Case Study Detection of Underground Effluent-Carrying Pipes at Periyar River, Kerala, India

Periyar River, the longest river in Kerala, is a crucial water resource for the central region of the state. However, the river faces significant environmental challenges, including ecological degradation due to pollution from domestic, agricultural, and industrial sources. The pollution in the river is particularly severe in the lower reaches, approximately 15 km upstream from Cochin city, where a series of small-scale industries are located. Clandestine discharges of industrial effluents directly into the river during the night through underground pipelines have been reported. This study focused on detecting such effluent-carrying pipes along a 5 km stretch of both banks of the river in the Eloor-Edayar industrial belt.

A total of 128 GPR profiles were conducted along both riverbanks to detect underground pipes, cables, drains, and other structures. The GPR scans used a 200 MHz antenna and the DIEL 16 setting, appropriate for wet sand conditions, in utility scan mode. The survey aimed to identify anomalies indicative of effluent leaks from industrial plants, including underground drains, storage pond leaks, and effluent treatment plant (ETP) discharges. The GPR profiles revealed several key anomalies: A reflection indicating moist material at 1 m depth between 10 and 20 m distances, identified as accumulated leachate leaking from the ETP and the Endosulfan plant. A parabolic-shaped anomaly at 0.56 m depth, corresponding to a clandestine pipe, which was later verified during excavation. These findings helped local pollution control authorities take preventive measures against further pollution from clandestine discharges into the Periyar River.

Case Study Detection of Clandestine PVC Pipes in Perundurai, Tamil Nadu, India

Perundurai, located in Erode district, Tamil Nadu, is home to an industrial cluster supported by the State Industries Promotion Corporation of Tamil Nadu (SIPCOT). The area houses 565 industrial units within a 10.5 km^2 region. These industries collectively discharge 22,500 metric tons per annum of industrial waste. Over time, groundwater, surrounding ponds, and channels in the region have been observed to be contaminated by clandestine industrial effluent discharges. A comprehensive GPR survey was conducted to detect such clandestine discharges around the SIPCOT industrial area. The GPR profiles used a 400 MHz antenna with DIEL 3–5 settings (optimized for PVC and granitic terrain) in utility scan mode.

The GPR profiles identified several underground pipes leading to low-lying areas, channels, and sewerage systems: Parabolic anomalies at 4 m and 12 m surface distances indicated the presence of pipes at depths of 0.5 m and 0.9 m, respectively. A hyperbolic anomaly at 0.1 m depth between 1.5 and 3 m distances overlapped with a parabolic anomaly at 0.5–1.2 m depth, revealing a small cement pipe (80 cm diameter) lying beneath a drain. The presence of PVC pipes was identified through distinct GPR anomalies, which exhibited a dielectric constant (DIEL) of 3.5. A total of three PVC pipes were detected at depths of 10 cm, located at 2.2, 5, and 5.3 m distances. The GPR findings were submitted to the Tamil Nadu Pollution Control Board (TNPCB) for further action, including the implementation of preventive measures to curb unauthorized discharges.

Contamination mapping is another key application of geophysical techniques in environmental engineering. Electromagnetic Induction (EM) and Electrical Resistivity Tomography (ERT) are employed to map the extent of subsurface contamination, such as leachates originating from industrial sites and landfills, enabling more effective site assessment and remediation planning.

Case Study Agricultural Monitoring at Al-Ahsa Oasis Using Geophysical Techniques

The study explores the use of electromagnetic induction (EMI) and electrical resistivity tomography (ERT) for monitoring and assessing soil properties in Al-Ahsa Oasis, Saudi Arabia (Mohammed et al. 2022). EMI and ERT techniques offer non-invasive ways to evaluate soil conditions, including moisture, salinity, and composition. These methods allow for real-time data collection and mapping of subsurface features, which is crucial for agricultural management, water resource optimization, and environmental monitoring in arid regions like Al-Ahsa. The case study highlights their potential to improve land management and irrigation strategies in the oasis.

Al-Ahsa Oasis, one of the largest natural oases in the world, is located in Saudi Arabia and serves as a vital agricultural center. Covering approximately 12,000 ha, it is the largest irrigated agricultural oasis globally. Groundwater has traditionally been the primary water source for irrigation and other agricultural needs. However, due to significant depletion of these groundwater resources, treated wastewater is now used to supplement irrigation in the region. This shift in water sources has become essential to maintain agricultural productivity in the oasis.

The oasis faces several challenges that threaten agricultural productivity, including increasing soil salinity and the presence of a hardpan layer in certain areas. The soil texture plays a crucial role in addressing these issues, as it affects water retention and plant growth. These problems, if not managed effectively, can lead to reduced crop yields and soil degradation. Thus, understanding the soil's characteristics and its salinity distribution is critical for sustainable agricultural practices in Al-Ahsa.

This study aimed to assess the effectiveness of modern geophysical techniques, particularly electromagnetic induction (EMI) and electrical resistivity tomography (ERT), in evaluating the agricultural conditions of Al-Ahsa Oasis (Fig. 2.12). The primary objectives were to investigate the variations in soil salinity, moisture content, and the impact of soil texture on irrigation and crop production. By utilizing these advanced methods, the study sought to provide a clearer understanding of the geophysical properties influencing agricultural productivity in the region.

To conduct this investigation, ten 2-D electrical tomography profiles were implemented using the SuperSting R8/IP 8-channel multielectrode resistivity and induced polarization (IP) imaging system. The system included 112 electrodes spaced one meter apart, which allowed for detailed imaging of the subsurface. In addition, twenty electrical conductivity profiles were measured at 2240 points using the EM38-MK2 device in vertical mode. This device, with transmitter–receiver coils of 0.5 and one-meter spacing, was used to map soil conductivity and provide insights into the salinity variations within the oasis.

The results revealed notable variations in soil salinity, both vertically and laterally, influenced by factors such as soil type, texture, moisture content, and the presence of sabkha (salt flats) (Fig. 2.13a, b). These variations were critical in understanding the spatial distribution of salinity across the oasis. Furthermore, the analysis of the resistivity tomograms from the ten 2-D profiles identified distinct geoelectric units that corresponded to different hydrogeological conditions and soil textures. These

Fig. 2.12 Geophysical data acquisition using the SuperSting R8/IP 8 resistivity (**a**) and the EM38 MK2 system electrical conductivities (**b**) at Al-Ahsa Oasis, Saudi Arabia. *Source* Mohammed et al. (2022)—Open Access

findings underscored the areas where salinity and hardpan layers posed significant challenges to agricultural practices.

The integration of geophysical techniques like EMI and ERT provided valuable insights into the agricultural conditions of Al-Ahsa Oasis. The results of this study can assist farmers, agricultural developers, and research institutions in better managing soil salinity, improving irrigation strategies, and ensuring the sustainability of agriculture in the oasis. By leveraging modern geophysical methods, stakeholders can make informed decisions to address soil-related challenges and optimize agricultural productivity in this historically important region.

Archaeological site mapping is an important application of geophysical techniques, allowing researchers to identify buried structures and features without the need for excavation, thereby preserving the integrity of archaeological sites for future study and investigation.

Case Study Geophysical Techniques for Mapping Subsurface Structures at the Archaeological Site of Lembah Bujang, Kedah, Malaysia

This case study focuses on the use of geophysical methods to map the subsurface structure of an archaeological site located at Sungai Batu, Lembah Bujang in Kedah, Malaysia (Sapiai et al. 2010). Lembah Bujang is historically recognized as one of Southeast Asia's significant ancient trading hubs. Evidence suggests the area is rich in archaeological sites, making it an ideal location for exploration. Geophysics provides non-destructive techniques that assist archaeologists in obtaining valuable preliminary results, aiding in more accurate excavation planning. The primary objective of this study is to identify the location of ancient archaeological monuments in Lembah Bujang before excavation, particularly in the Sungai Batu region. Located in the southern part of Kedah, the area is characterized by its flat terrain, which includes sandy clay soils and is surrounded by palm oil plantations, rubber trees, small rivers,

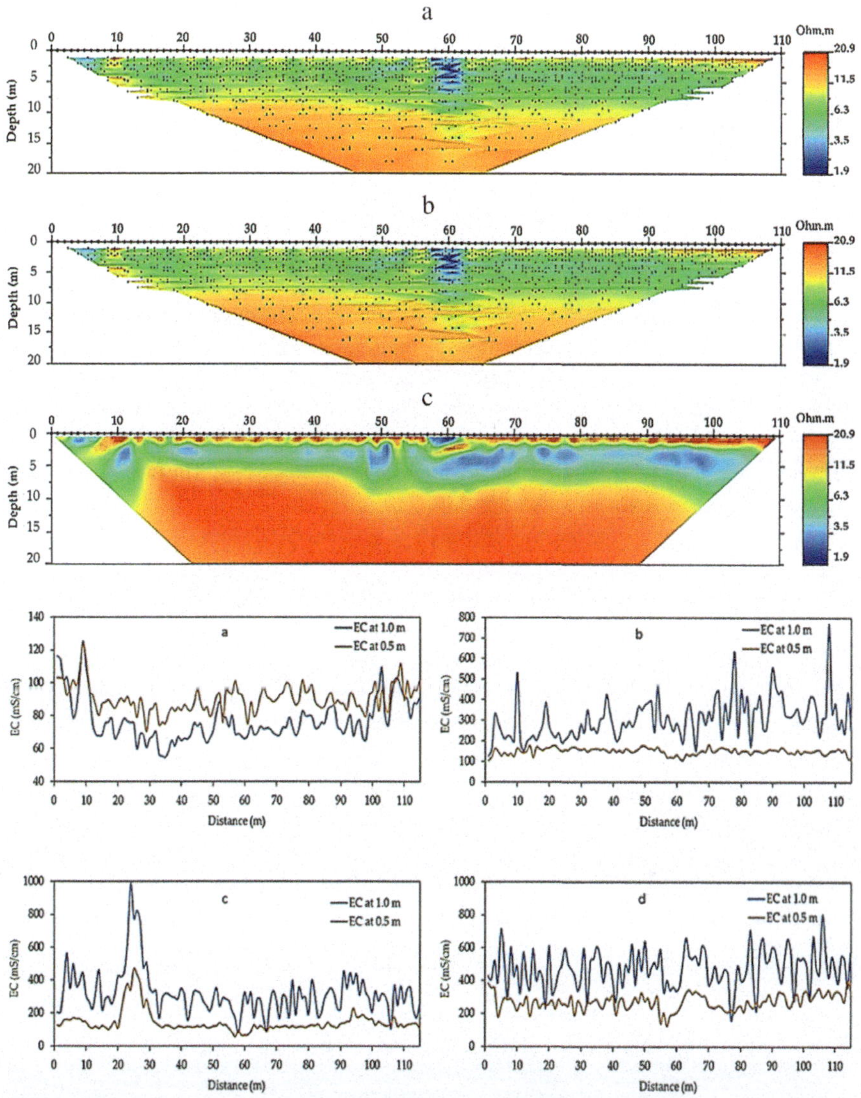

Fig. 2.13 **a** Two-dimensional resistivity section for location P-6 at Al-Ahsa Oasis. The upper section shows the measured resistivity (**A**), the middle section displays the calculated resistivity (**B**), and the lower section represents the inverted resistivity (**C**). **b** Electrical conductivities along EM38 at four tested locations (P-4, P-7, P-8, and P-10) across Al-Ahsa Oasis. Line 4 at P-4 (**a**), Line 7 at P-7 (**b**), Line 8 at P-8 (**c**), and Line 10 at P-10 (**d**). *Source* Mohammed et al. (2022)—Open Access

and swamps. It is important to note that the coastline of Kedah experienced significant progradation during the early historic period, altering the local landscape, and leading to the regression of the ancient Kedah port that was initially situated in an estuarine environment. To achieve accurate subsurface mapping, a combination of geophysical techniques was employed, with magnetic gradiometry being the primary method. This was followed by 2-D resistivity imaging and Ground Penetrating Radar (GPR) surveys to confirm subsurface anomalies.

Magnetic gradiometry is a method used to measure the gradients of the magnetic field in different directions. This technique is advantageous over traditional magnetic field measurements as it helps reduce noise from temporal field changes, enhances shallow anomalies, and improves lateral resolution. A regional magnetic survey covering an area of 5 km^2 was conducted to produce a magnetic anomaly map (Fig. 2.14). The data were processed using Microsoft Excel and Surfer software. The magnetic anomalies map revealed regions with low magnetic susceptibility, which were interpreted as laterite or shale formations with a susceptibility range of 0.01–15 × 10^3 nT/m. This result was aligned with earlier discoveries of brick structures (candi) in the area, suggesting the presence of ancient monuments.

The 2-D resistivity method measures the electrical resistivity of subsurface materials, offering insight into the composition and structure of the soil layers. In this study, a pole-dipole array was used due to its effective horizontal coverage and lower sensitivity to telluric noise compared to other arrays. A roll-along technique with four cable rolls was applied to survey a large area. The data were processed using RES2DINV software to generate a pseudosection (Fig. 2.15). One of the survey lines (LB006) coincided with the current archaeological excavation site. The resistivity results revealed three main subsurface layers: a low-resistivity zone (less than 100 Ωm), a middle-resistivity zone (150–800 Ωm), and a high-resistivity zone (greater than 1200 Ωm). The middle resistivity zone, lying between 0 and 15 m depth, was associated with alluvial deposits, sandy clay, and boulders, which are likely related to the base of the ancient monument made from laterite or granite—materials commonly used in Southeast Asian temple constructions.

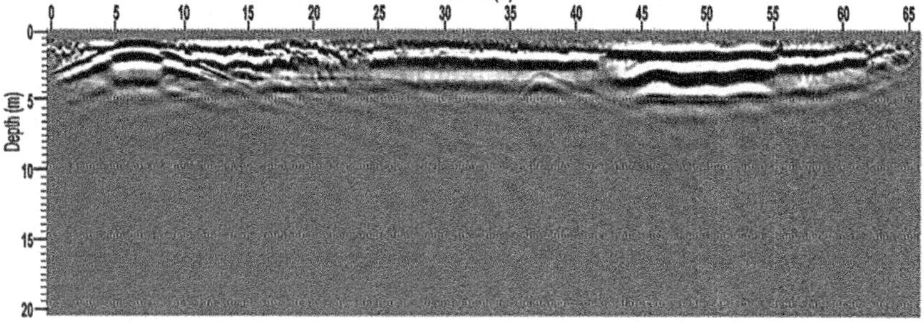

Fig. 2.14 Magnetic anomaly map at Sungai Batu, Lembah Bujang archaeological site, Kedah, Malaysia. *Source* Sapiai et al. (2010)—Open Access

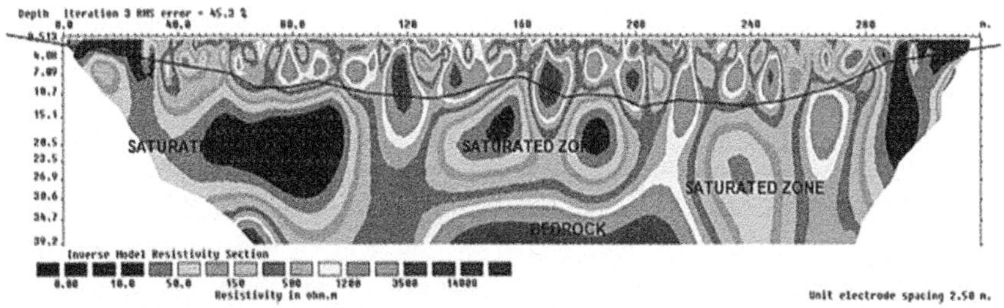

Fig. 2.15 2-D resistivity pseudosection of inversion model for survey line LB006 at Sungai Batu, Lembah Bujang, Kedah, Malaysia. *Source* Sapiai et al. (2010)—Open Access

GPR was employed to acquire data using electromagnetic (EM) energy reflections, transmitted into the ground at high frequencies ranging from 10 to 1000 MHz. The data were collected along a 42 m line (M3), which was oriented perpendicular to the resistivity line. GPR data processing was performed using Ground Vision v1.3.1 software. The results showed several interesting anomalies, particularly in the shape of triangles and circular patterns (Fig. 2.16). The triangular patterns were interpreted as potential pipelines, while the round shapes were suspected to represent iron objects, further supporting the evidence of industrial activity in the area.

The magnetic gradiometry survey revealed areas with low magnetic susceptibility, which were consistent with laterite or shale formations. These low-susceptibility zones were correlated with earlier findings of brick structures in the area. The 2-D resistivity survey revealed that the subsurface materials consisted of three distinct resistivity zones, with the middle zone likely corresponding to the construction materials of the monument. The GPR survey identified anomalies that were interpreted as possible iron artifacts, including pipes and iron-related structures. These findings

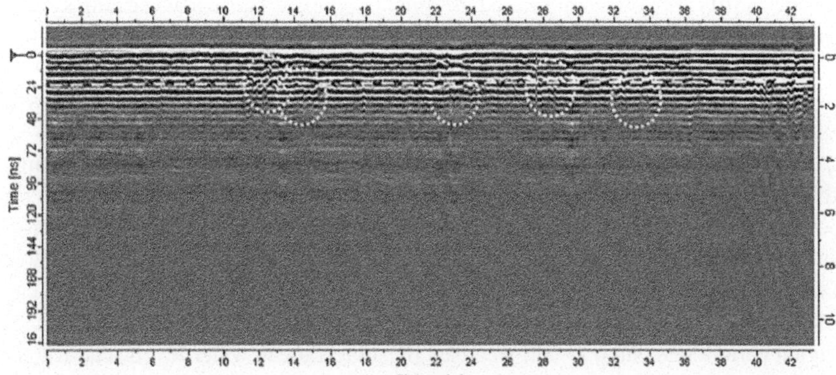

Fig. 2.16 Ground penetrating radar (GPR) image for survey line M3 at Sungai Batu, Lembah Bujang, Kedah, Malaysia. *Source* Sapiai et al. (2010)—Open Access

collectively pointed to the presence of an ancient monument at the location, providing a more accurate target for subsequent excavation.

The geophysical investigations confirmed that Sungai Batu was likely an ancient river system, characterized by slow-flowing waters. This conclusion was supported by the sedimentation layers identified in the resistivity and GPR surveys, which consisted of sandy clay, alluvium, and boulders. Additionally, sandstone and shale were observed on the surface, indicating a shallow-water environment in the past. The use of magnetic, resistivity, and GPR methods provided valuable insights into the subsurface structure, confirming the presence of an ancient monument. Excavations carried out at the site subsequently verified the position of the monument and revealed further evidence of a third-century iron smelting site, making this geophysical survey a valuable tool in archaeological exploration.

Sinkhole detection is a critical application of geophysical methods, as they can identify subsurface voids and anomalies that may indicate potential sinkhole formation. This is especially important in regions prone to such geohazards, helping to ensure public safety and inform land-use planning.

Case Study Early Warning Detection of Sinkhole Collapse and Subsidence Using Optical Fiber Seismic Sensors

This study demonstrates the use of a seismic ambient noise monitoring optical fiber sensor for early detection of sinkhole collapse and subsidence (Fig. 2.17). The sensor detects changes in the subsurface caused by the progressive development of sinkholes, which affects the acoustic modes of the ground. By monitoring variations in the spectral content of ambient noise, the sensor identifies structural weaknesses and instability before collapse. The sensor uses a polarization-based single-mode fiber in a phase-sensitive distributed acoustic sensor setup. Peak frequencies in the microseismic noise band (0.1–1.0 Hz) were observed as the sinkhole developed. Larger frequency shifts above 1 Hz signalled imminent collapse. This early warning system helps to reduce risks and socio-economic impacts from sudden sinkhole collapses. The system can also be applied to monitor earthquakes and landslides. Figure 2.18 shows subsurface subsidence and sinkholes in dolomitic and non-dolomitic areas of South Africa.

An optical fiber sensor has been successfully developed to detect sinkholes using ambient noise signals (Fig. 2.19). This system provides early warnings, allowing for timely responses to prevent damage to infrastructure and protect lives. The sensor monitors subsurface structural health, detecting vibrations that indicate sinkhole development. While it cannot pinpoint the exact location or depth of a sinkhole, it can track its growth and predict collapse. The system operates continuously, and changes in signal parameters, such as shifts caused by subterranean water levels, can trigger early warnings. Additionally, the sensor can monitor seismic events like earthquakes and landslides, offering further protection by detecting potential sinkhole-related triggers.

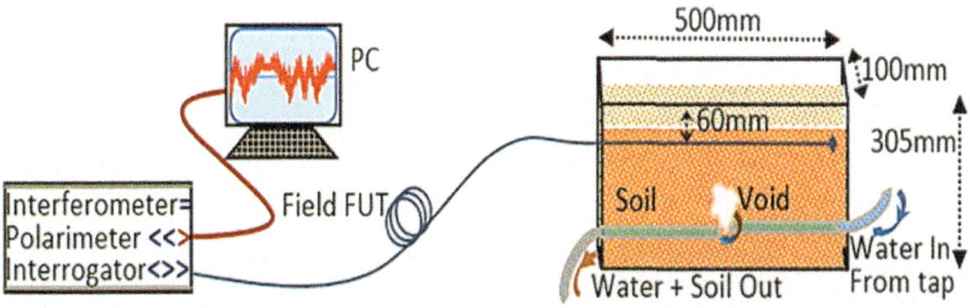

Fig. 2.17 Small-scale field testbed for sinkhole monitoring with optical fiber acoustic sensor in Gauteng Province, South Africa. *Source* Jena et al. (2024)—Open Access

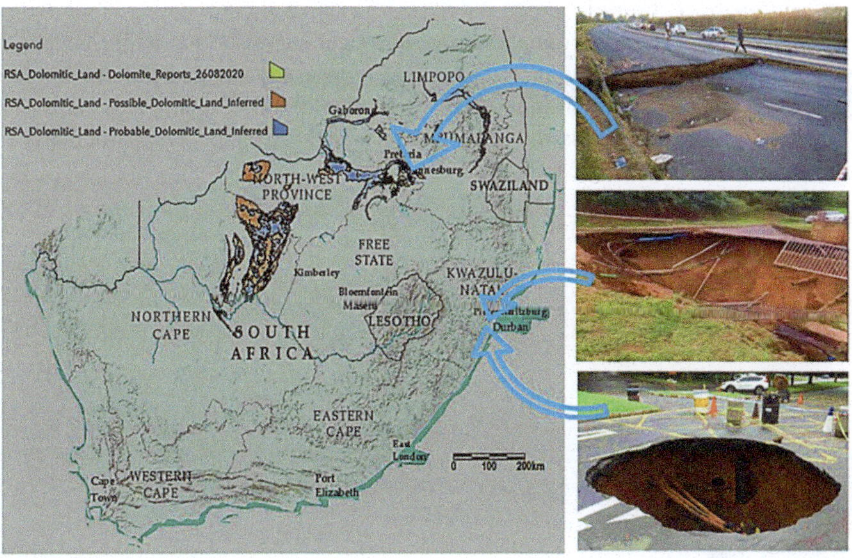

Fig. 2.18 Sinkhole damage in South Africa's dolomite land (Gauteng Province). *Source* Jena et al. (2024)—Open Access

Misconceptions and Limitations

Several misconceptions about geophysics hinder its adoption. One common belief is that geophysics provides a perfect "view" of the subsurface. In reality, these surveys detect changes in physical properties, and interpreting these variations requires expertise and an understanding of the surrounding geology.

Each geophysical method has its limitations, which must be considered when selecting the appropriate technique for a given application. *Ground Penetrating Radar (GPR)* struggles in conductive environments such as clay-rich soils, where signal attenuation reduces effectiveness. *Electromagnetic Induction (EM)* offers lower resolution compared to GPR, making it less suitable for detecting small-scale features. *Electrical Resistivity Tomography (ERT)* is relatively slow and requires

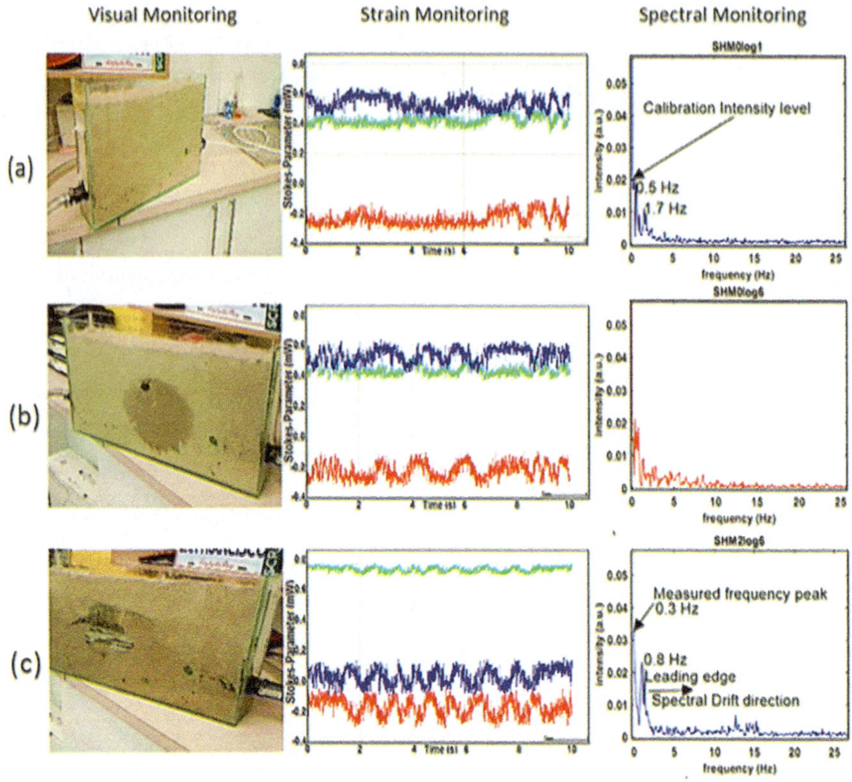

Fig. 2.19 Optical fiber sensor monitoring of sinkhole trigger events during cavity growth in Gauteng, South Africa. **a** Steady state ambient noise signals, **b** void establishment and **c** cavity enlargement. *Source* Jena et al. (2024)—Open Access

substantial equipment and logistical support, which can limit its practicality in large-scale or rapid assessments. *Magnetometry* is highly sensitive to small magnetic disturbances, necessitating careful calibration and control of background noise to ensure accurate results.

Looking to the Future: What's Ahead for Geophysics?

As environmental challenges become more complex, geophysics is set to evolve alongside them. Advances in visualization technologies, machine learning, and artificial intelligence (AI) promise more precise and automated interpretations of geophysical data. Drone-based surveys using radiometric and thermal imaging methods offer new possibilities for rapid site assessments without extensive ground surveys.

While the fundamental physics of these techniques remains the same, the tools and methods for interpreting and visualizing data are constantly improving. AI may one day automate much of the interpretation work, making geophysical insights more accessible to engineers and decision-makers.

Geophysics plays a critical role in the future of environmental engineering, offering non-invasive methods for assessing subsurface conditions. As tools and methodologies evolve, geophysics will continue to provide valuable insights that help engineers and planners make more informed, sustainable decisions. With the potential for faster and more accurate processes, the future of geophysics in environmental engineering looks promising.

2.3 Conclusions

This chapter provides a comprehensive overview of the evolution, significance, and transformative impact of engineering geophysics on modern infrastructure, environmental challenges, and sustainable development. From its early academic foundations to the rise of cutting-edge technologies such as automation, remote sensing, and artificial intelligence, engineering geophysics has become an indispensable tool for tackling complex environmental and infrastructural problems.

The chapter highlights how geophysical techniques, such as electrical resistivity imaging, seismic refraction tomography, and electromagnetic surveys, have revolutionized the assessment and management of contaminants, especially in the context of landfill leachate plumes. Case studies demonstrate the power of integrated geophysical methods in providing accurate, real-time monitoring and risk mitigation.

As we look to the future, the integration of advanced geophysical tools with AI and machine learning promises to further enhance the precision and efficiency of environmental assessments, infrastructure development, and disaster risk reduction. Despite some misconceptions and limitations, the continued innovation in this field holds significant promise for addressing the pressing environmental and infrastructural challenges of the twenty-first century.

In conclusion, engineering geophysics stands at the forefront of creating sustainable solutions for infrastructure and environmental management, with its role only expected to grow in importance as new technologies and methodologies emerge. The future of geophysics is poised to offer even more powerful applications that will contribute to a safer, more sustainable world.

References

Alao JO (2025) The evolving roles of geophysics in environmental assessment, monitoring, and management of landfill leachate contaminant plumes: an overview. Case Stud Chem Environ Eng 11:101124

Alao JO, Abdo HG, Ayejoto DA, Mohammed M, Danladi E, Saqr MA, Almohamad H, Fahad A (2023) Evaluation of groundwater contamination and the health risk due to landfills using integrated geophysical methods and physiochemical water analysis. Case Stud Chem Environ Eng 8:100523

Annan AP (1997) Engineering and environmental geophysics: the future. Geol Soc Lond Eng Geol Spec Publ 12:419–426

Batayneh AT, Al-Diabat AA (2002) Application of a two-dimensional electrical tomography technique for investigating landslides along the Amman-Dead Sea Highway, Jordan. Environ Geol 42:399–403

Carpenter PJ, Reddy KR, Thompson MD (2012) Seismic imaging of a leachate-recirculation landfill: spatial changes in dynamic properties of municipal solid waste. J Hazard Toxic Radioact Waste 17:4

Chavez S, Lopez R (2018) Technological advances in geophysical methods: a review. Geophys Prospect 66(2):335–357

Culshaw MG, Northmore KJ, McCann DM (2014) A short history of engineering geology and geophysics at the British Geological Survey. In: Lollino G, Arattano M, Giardino M, Oliveira R, Peppoloni S (eds) Engineering geology for society and territory, vol 7. Springer, Cham, pp 257–260

Drahor MG, Gokturkler G, Berge MA, Kurtulmus TO (2006) Application of electrical resistivity tomography technique for investigation of landslides a case from Turkey. Environ Geol 50:147–155

Eddleston M, McCann DM, Cripps JC, Johnson P (1997) Modern geophysics in engineering geology: an overview. Geol Soc Lond Eng Geol Spec Publ 12:427–432

Ekeocha NE, Okereke II, Okonkwo S (2012) Electrical resistivity investigation of solid waste dumpsite at Rumuekpolu in Obio Akpor LGA, Rivers State, Nigeria. Int J Sci Technol (IJST) 1(11):631–637

El-Kelani R, Khader A (2019) Refraction seismic study over a proposed landfill site in South West Bank, Palestine. Springer, Cham. https://doi.org/10.1007/978-3-030-01656-2_22

Eze O, Orji O, Onoriode A, Saleh S, Abolarin M (2022) Integrated geoelectrical resistivity method for environmental assessment of landfill leachate pollution and aquifer vulnerability studies. J Geosci Environ Prot 10:1–26

Freidman BL, Camenzuli D, Lackie M (2014) Locating an ice-covered Antarctic landfill using ground magnetometry. Antarct Sci 26(4):361–368

Frid V, Sharabi I, Frid MEA (2017) Leachate detection via statistical analysis of electrical resistivity and induced polarization data at a waste disposal site (Northern Israel). Environ Earth Sci 76:233. https://doi.org/10.1007/s12665-017-655

Geology Science (2023) Electromagnetic (EM) methods. Modified date: 06/12/2023. https://geologyscience.com/geology-branches/geophysics/electromagnetic-em-methods/

Hakimov F, Domej G, Ischuk A, Reicherter K, Cauchie L, Havenith H-B (2021) Site amplification analysis of Dushanbe city area, Tajikistan to support seismic microzonation. Geosciences 11:154

Hatch M (2024) Environmental geophysics. Preview 2024(233):40

Hauck C, Kneisel (2006) Application of capacitively-coupled and DC electrical resistivity imaging for mountain permafrost studies. In: Permafrost and periglacial processes, vol 17, pp 169–177

Ibraheem IM, Yogeshwar P, Bergers R, Tezkan B (2024) Joint interpretation of magnetic, transient electromagnetic, and electric resistivity tomography data for landfill characterization and contamination detection. Sci Rep 14:30616

Junior T, Pires P, Domingues D, de Campos T, Junior TJA, Pires PJM, Domingues DLP, de Campos TMP (2016) The use of ground penetrating radar in municipal solid waste landfill geotechnical investigation. J Civ Eng Archit 10:822–827

Kirkpatrick JD, Fagereng A, Shelly DR (2012) Geological constraints on the mechanisms of slow earthquakes. Nat Rev Earth Environ 2:285–301

Kovari K, Kozak T (2016) Applications of geophysics for engineering and environmental purposes. Springer

Loke MH, Barker RD (1995) Least-squares deconvolution of apparent resistivity pseudo sections. Geophysics 60:1682–1690

Lowrie W (2016) Fundamentals of geophysics, 2nd edn. Cambridge University Press

Mahmoudzadeh MR, Frances AP, Lubczynski M, Lambot S (2012) Using ground penetrating radar to investigate the water table depth in weathered granites—Sardon case study, Spain. J Appl Geophys 79:17–26

McDowell PW, Barker RD, Butcher AP, Culshaw MP, Jackson PD, McCann DM, Skipp BO, Matthews SL, Arthur JCR (2002) Geophysics in engineering investigations. Geol Soc Lond Eng Geol Spec Publ 19

Mohammed M, El Mahmoudi A, Almolhem Y (2022) Applications of electromagnetic induction and electrical resistivity tomography for digital monitoring and assessment of the soil: a case study of Al-Ahsa Oasis, Saudi Arabia. Appl Sci 12:2067

Molua CO (2022) Assessing leachate migration and gas emissions in landfill sites using seismic and electrical resistivity tomography (ERT) methods. J Environ Impact Manag Policy 2:6

Monteiro Santos F, Mateus A, Figueiras J, Goncalves MA, Monteiro Santos F, Mateus A, Figueiras J, Goncalves MA (2006) Mapping groundwater contamination around a landfill facility using the VLF-EM method—a case study. J Appl Geophys 60(2):115–125

Naskar DC, Choudhury K, Roy KK, Singh OP, Majumdar RK (2002) Geophysical approach for delineation of shallow crustal structure along Borgaon-Sanwer transect, Madhya Pradesh. Geol Soc India 60:173–181

Netto LG, Guimaraes CC, Barbosa AEA (2023) Characterization of an inactive dumpsite using electrical resistivity tomography and hydrochemical data: a case study in Brazil. Environ Earth Sci 82:529

Nobes DC, Armstrong MJ, Broadbent M (1994) Delineation of a landfill leachate plume using shallow electromagnetic and ground-penetrating radar surveys. In: SEG technical program expanded abstracts

Okpoli CC (2013) Application of 2D electrical resistivity tomography in landfill site: a case study of Iku, Ikare Akoko, southwestern Nigeria. J Geol Res 8:895160

Osinowo OO, Agbaje MA, Ariyo SO (2020) Integrated geophysical investigation techniques for mapping cassava effluent leachate contamination plume, at a dumpsite in Ilero, southwestern Nigeria. Sci Afr 8:e00374

Panek T, Margieleski W, Taborik P, Urban J, Hradecky J, Szura C (2010) Gravitationally induced caves and other discontinuities detected by 2D electrical resistivity tomography: case studies from the Polish Flysch Carpathians. Geomorphology 123:165–180

Porsani JL, Elis VR, Ferreira AA, Poluha B (2017) IAG/USP test site-II: construction and preliminary GPR results on PVC/metal pipes and ceramic tubes. In: Near surface geoscience, Malmo, Sweden, 3–7 Sept 2017. https://doi.org/10.3997/2214-4609.201702002

Proulx-McInnis S, St-Hilaire A, Rousseau AN, Jutras S (2013) A review of ground-penetrating radar studies related to peatland stratigraphy with a case study on the determination of peat thickness in a northern boreal fen in Quebec, Canada. Prog Phys Geogr. https://doi.org/10.1177/0309133313501106

Ramalho EC, Dill AC, Rocha R (2012) Assessment of the leachate movement in a sealed landfill using geophysical methods. Environ Earth Sci. https://doi.org/10.1007/s12665-012-1742-8

Roberts RL, Hinze WJ, Leap DI (1990) Data enhancement procedures on magnetic data from landfill investigations. https://doi.org/10.1190/1.9781560802785.ch26

Sapiai SM, Saad R, Nawawi MNM, Shyeh SK, Saidin MM (2010) Geophysical applications in mapping the subsurface structure of archaeological site at Lembah Bujang, Kedah, Malaysia. AIP Conf Proc 1250:189

Schleicher J, Sloane A (2019) Geophysics for infrastructure and environment: applications and techniques. Elsevier

Sonkamble S, Chandra S (2021) GPR for earth and environmental applications: case studies from India. J Appl Geophys 193:104422

Soupios P, Papadopoulos N, Papadopoulos IEA (2007) Application of integrated methods in mapping waste disposal areas. Environ Geol 53:661–675

Splajt T, Ferrier G, Frostick LE (2003) Application of ground penetrating radar in mapping and monitoring landfill sites. Environ Geol 44:963–967

Telford WM, Geldart LP, Sheriff RE (1990) Applied geophysics, vol 1, 2nd edn. Cambridge University Press

Umar MS, Ghazali MD, Samad AM, Zainon O, Zainuddin K, Mazurki F (2022) Implementation of ground penetrating radar in assessing leachate in sanitary landfill. In: 2022 IEEE 20th student conference on research and development (SCOReD), Bangi, Malaysia, 8–9 Nov 2022

Xia T, Meng J, Ding B, Chen Z, Liu S, Titov K, Mao D (2023) Integration of hydrochemical and induced polarization analysis for leachate localization in a municipal landfill. Waste Manage 157:130–140

Yardimci E, Arslan M (2020) Recent advances in engineering geophysics: trends and innovations. J Appl Geophys 176:1–10

Zhang T, Zhang D, Zheng D, Guo X, Zhao W (2022) Construction waste landfill volume estimation using ground penetrating radar. Waste Manag Res

Zhou Q, Yang F (2011) Application of ground penetrating radar in songs using landfill site, Huzhou city to investigate the leakage passage of pollutants. In: 2011 international symposium on water resource and environmental protection

Chapter 3
Fundamental Concepts in Engineering Geophysics

Abstract This chapter explains the key concepts and principles behind engineering geophysics, focusing on how geophysical methods help in understanding subsurface conditions. It highlights the importance of these techniques in civil, construction, and environmental engineering projects, especially for subsurface exploration and risk assessment. The chapter discusses the challenges in underground investigations and how geophysics provides safer and cost-effective solutions. It introduces major geophysical methods, including seismic, electrical, electromagnetic, magnetic, gravity, acoustic, and ultrasonic techniques. Each method's basic theory, operational principles, and practical applications—such as soil profiling, fault detection, and structure evaluation—are explained. The chapter also covers mathematical models and wave behaviors like Rayleigh waves in geophysical surveys. These methods are essential for detecting underground features, ensuring construction safety, and improving environmental monitoring. Overall, this chapter gives a strong foundation in the tools and science behind engineering geophysics for practical and research applications.

3.1 Introduction to Fundamental Concepts

The chapter titled "Fundamental Concepts in Engineering Geophysics" in the book *Advances in Engineering Geophysics: Concepts and Applications* provides a comprehensive introduction to the core principles and methods used in engineering geophysics. Geophysics plays a crucial role in modern civil, construction, and environmental engineering by offering non-invasive techniques for subsurface exploration, site characterization, and risk assessment. This chapter outlines the fundamental geophysical principles that underpin various geophysical methods, including seismic, electrical, electromagnetic, magnetic, gravitational, and acoustic techniques.

The chapter emphasizes the importance of geophysical methods in addressing challenges faced during subsurface investigations, such as identifying soil properties, detecting faults, and assessing geological hazards. It also highlights the critical role

D. C. Naskar, *Advances in Engineering Geophysics: Concepts and Applications*, Earth Systems Data and Models 9, https://doi.org/10.1007/978-3-032-28000-8_3

of geophysical techniques in evaluating the feasibility of engineering projects, particularly in terms of foundation design, groundwater management, and environmental monitoring.

In addition to providing theoretical insights, the chapter discusses the applications and limitations of various methods. Seismic methods are explored for soil profiling, fault detection, and seismic site characterization, while electrical and electromagnetic techniques, such as resistivity and ground-penetrating radar, are analyzed for their utility in mapping subsurface features and detecting anomalies. The chapter also covers the use of acoustic and ultrasonic techniques for assessing material properties and structural health.

By presenting both the basic principles and the practical applications of engineering geophysics, this chapter serves as an essential resource for engineers, geophysicists, and researchers looking to deepen their understanding of geophysical methods and their applications in engineering practice.

3.2 Introduction to Engineering Geophysics

3.2.1 Overview of Geophysics in Civil, Construction, and Environmental Engineering

Engineering geophysics is the application of geophysical methods to investigate the subsurface of the Earth in order to inform civil engineering (Orozco and Vidal 2019; Sharma 1997; Akintorinwa and Adesoji 2009), construction, and environmental studies. This discipline employs non-invasive techniques to gather critical information about subsurface conditions, such as the composition, structure, and potential risks of materials below the surface. The insights gained from geophysical surveys support a variety of engineering projects, from site selection and foundation design to hazard identification and environmental monitoring (Sullivan and Johnson 2018; Anbazhagan et al. 2016).

In civil engineering, engineering geophysics plays a crucial role in providing information necessary for designing and constructing structures like buildings, bridges, dams, and tunnels (Attewell and Farmer 2021; Huang and Wang 2020). Geophysical methods help determine the nature of the soil and rock beneath a proposed site, ensuring that the foundations of structures are designed to withstand the local conditions (Zhang and Li 2020). In construction, these methods contribute to assessing the stability of a site and identifying any risks that might complicate the project. Geophysical surveys are commonly used to assess soil strength, detect fault zones, or map the extent of groundwater contamination. In environmental studies, engineering geophysics helps monitor and manage resources such as groundwater and minerals, and assess environmental hazards like soil contamination or seismic risks (Akingboye and Ogunyele 2019; Naskar et al. 2024; Attewell and Farmer 1976; Sun et al. 2020; Telford et al. 1990; Kearey et al. 2002; Lwisa 2021; Sullivan and Johnson 2018; Geophysical Society of America 2019).

3.2.2 Importance of Geophysical Methods in Subsurface Exploration and Risk Assessment

Geophysical methods provide essential information for subsurface exploration, helping engineers and environmental scientists (Kearey et al. 2002; Naskar 2005) gain a deeper understanding of the ground conditions without the need for invasive and expensive techniques like drilling or excavation. These methods are particularly useful in site characterization, where they provide data on soil layers, groundwater levels, rock formations, and the presence of contaminants. This data is critical for determining the suitability of a site for construction and ensuring that the design is based on accurate subsurface conditions.

Geophysics is also invaluable in risk assessment. By identifying subsurface features such as fault lines, areas prone to soil liquefaction, or zones of groundwater contamination (Schoenberg 2015), engineers can better plan for and mitigate potential hazards. For example, seismic surveys help assess the seismic risk of a region (Doucet and Boulanger 2019), while resistivity surveys can detect contamination plumes in groundwater. These methods allow for early identification of risks and provide data that can be used to inform safer, more cost-effective construction practices.

3.2.3 Challenges in Subsurface Exploration and the Role of Geophysics

Subsurface exploration faces several significant challenges, primarily due to the difficulty of directly accessing and observing underground conditions. Traditional methods like drilling are often costly, time-consuming, and provide limited data, making it difficult to fully understand the subsurface environment (Whiteley 1990). Furthermore, the subsurface is highly heterogeneous, with variations in soil, rock, and groundwater conditions that change from one location to another, complicating predictions about the nature of the underground materials. This variability can make it difficult to assess site conditions accurately. Additionally, subsurface exploration in environmentally sensitive areas presents its own set of challenges, as invasive techniques like drilling can cause disruption or contamination, which may be unacceptable in such locations.

Geophysical methods help address these challenges by offering a non-invasive alternative to traditional exploration methods. Techniques like seismic surveys, electrical resistivity (Naskar et al. 2022), and ground-penetrating radar (GPR) enable engineers to gather critical subsurface data without the need for drilling or excavation. These methods also provide a more continuous and comprehensive view of subsurface conditions, allowing for better site characterization. Moreover, geophysics is a cost-effective solution, particularly for large-scale surveys or when working in sensitive environments where conventional methods are not feasible. By overcoming the limitations of traditional techniques, geophysics enables more efficient and less disruptive subsurface exploration (Jiang et al. 2020).

Site characterization is vital for civil engineering projects, providing essential data on subsurface conditions like soil layers, rock formations, and groundwater tables. Geophysical surveys (Table 3.1), such as seismic refraction and electrical resistivity (Naskar and Das 2009), help determine bedrock depth and soil properties, aiding in foundation design and assessing site suitability for construction.

Table 3.1 Common geophysical methods used for site characterization

Geophysical method	Application	Parameters measured	Example use case
Seismic refraction	Depth to bedrock and soil layering	Wave velocity, layer thickness	Foundation design, determining soil properties
Seismic tomography	Detection of voids and fractures in materials	Seismic velocity, attenuation	Inspection of tunnels, dams, and foundations
Ultrasonic testing	Material integrity and quality control	Wave velocity, signal reflection	Detection of cracks in concrete or rock
Ground penetrating radar (GPR)	Mapping shallow subsurface features (utilities, voids)	Electromagnetic wave reflection time	Locating underground utilities, voids, and cracks
Electrical resistivity	Soil type and groundwater conditions	Electrical resistivity	Identifying water table, mapping contaminated areas
Induced polarization	Detection of contaminants and groundwater	Chargeability and resistivity	Monitoring the spread of contaminants in groundwater
Magnetic surveys	Large-scale geological structures (faults, voids)	Magnetic field anomalies	Detecting fault lines and subsurface voids
Gravity surveys	Maps subsurface density variations, mineral deposits	Gravity anomalies, density variations	Mineral exploration, oil and gas exploration, and assess the potential of geothermal reservoirs

3.3 Basic Geophysical Principles

Geophysical methods are based on fundamental principles of physics, particularly those related to wave propagation, the interaction of waves with subsurface materials, and the physical properties of the materials themselves. These principles are used to infer the composition, structure, and behavior of subsurface environments, which is essential for applications in civil engineering, construction, and environmental studies. One of the primary methods for subsurface exploration is seismic waves, which include both P-waves and S-waves. P-waves, also known as primary or compressional waves, travel through a medium by alternately compressing and expanding it in the direction of wave propagation. S-waves, or shear waves, move the material in a perpendicular direction to the wave's motion, and they cannot travel through liquids. These seismic waves are analyzed to measure the velocity at which they propagate through different subsurface materials, providing important insights into the composition and structure of the Earth (Telford et al. 1990; Kearey et al. 2002).

Another key principle in geophysics is the propagation of electromagnetic waves. These waves, which are fundamental to techniques like electrical resistivity and induced polarization, propagate through subsurface materials by inducing electric and magnetic fields. The way these waves interact with different materials—based on their resistivity or conductivity—can reveal a great deal about subsurface conditions. For example, areas with higher resistivity may indicate the presence of dry or non-conductive materials like rock, while lower resistivity areas could point to the presence of water or contaminants (Kearey et al. 2002; Milsom 2003).

The basic concept of wave propagation involves how waves travel through different media, with the speed of wave travel being affected by the material's physical properties. For seismic waves, this speed is governed by factors such as density, elasticity, and the compressibility of the material. In electromagnetic methods, the rate at which waves travel is influenced by the electrical properties of the materials they encounter. Reflection and refraction occur when waves encounter boundaries between different materials, such as layers of soil, rock, or water. Reflection refers to the bouncing back of waves from a surface or interface, while refraction involves the bending of waves as they pass from one material to another. These interactions provide valuable data for identifying subsurface structures such as faults, mineral deposits, or groundwater layers (Naskar and Das 2009; Milsom 2003).

Additionally, absorption and scattering are important phenomena in geophysics. Absorption occurs when waves lose energy as they pass through a material, often being converted to heat. This is particularly relevant for seismic waves passing through highly porous materials or for electromagnetic waves interacting with conductive materials like water. Scattering, on the other hand, occurs when irregularities in the subsurface cause the wave energy to disperse in different directions, which can affect the clarity and resolution of the collected data. Understanding the absorption and scattering characteristics of materials helps improve the interpretation of geophysical survey results and the accuracy of subsurface models (Sullivan and Johnson 2018; Shragge 2016).

In practical applications, geophysics relies on these principles to address various challenges in subsurface exploration (Grant and West 1965). Techniques like seismic reflection and refraction surveys provide high-resolution data on subsurface structures, while electrical resistivity and induced polarization methods are particularly useful for identifying contamination, water tables, or mineral resources. The ability of geophysical techniques to non-invasively investigate the subsurface makes them invaluable for assessing site conditions, ensuring the safety of infrastructure, and providing insights into environmental conditions (Lwisa 2021; Milsom 2003).

Table 3.2 illustrates the different ways waves interact with subsurface materials, including reflection, refraction, absorption, and scattering. These interactions are fundamental to geophysical methods as they determine how waves travel through and are detected by sensors. For example, seismic waves reflect off different geological layers, while electromagnetic waves may be absorbed or refracted depending on the material properties. Understanding these wave behaviors is crucial for interpreting data and accurately assessing subsurface conditions in engineering and environmental studies.

These principles help geophysicists and engineers gather valuable data to make informed decisions about site conditions, environmental hazards, and resource management. As technology advances, the integration of geophysical methods into engineering practices continues to improve the safety, sustainability, and efficiency of civil infrastructure and environmental management projects.

Table 3.2 Interaction of waves with subsurface materials

Interaction type	Effect on waves	Example application
Reflection	Waves bounce back when encountering a boundary	Seismic surveys for structural mapping
Refraction	Waves bend when passing through different media	Seismic and electromagnetic surveys for depth estimation
Absorption	Wave energy is absorbed by the material	Seismic waves in porous soils or electromagnetic waves in conductive materials
Scattering	Waves are dispersed by material irregularities	GPR surveys to detect voids or fractures in concrete

3.4 Major Geophysical Methods in Engineering

3.4.1 *Seismic Methods*

Seismic methods play a crucial role in engineering geophysics by providing essential data for subsurface investigations. These methods rely on the propagation of seismic waves through various subsurface materials, which are generated artificially, typically through small explosions or weight drops, and recorded using geophones or seismometers. The interaction of seismic waves with subsurface materials—whether through reflection or refraction—provides detailed information about the physical properties of these materials, including their density, composition, and depth (Telford et al. 1990).

Principles of Seismic Refraction and Reflection

The principles of seismic refraction and reflection are fundamental in engineering geophysics, providing key insights into subsurface structures. Seismic refraction involves the measurement of seismic wave travel times through various subsurface layers. When seismic waves encounter boundaries between materials with different seismic velocities, they refract, or bend, at specific angles. The critical angle of refraction is determined by the velocity contrast between the materials at the interface (Fig. 3.1).

This critical angle (i_c) is given by the equation (Kearey et al. 2002)

$$\frac{\sin i_c}{\sin 90^\circ} = \frac{V_1}{V_2} \Rightarrow \sin i_c = \frac{V_1}{V_2} \tag{3.1}$$

where V_1 is the wave velocity in the first layer and V_2 is the wave velocity in the second layer.

Seismic reflection, on the other hand, involves seismic waves reflecting off subsurface boundaries such as bedrock or faults (Burdick and Lyons 2017). These reflections are captured by sensors at the surface, providing detailed information about deeper geological structures. The reflection coefficient (R) at an interface between two materials is expressed as

$$R = \frac{\rho_2 V_2 - \rho_1 V_1}{\rho_2 V_2 + \rho_1 V_1} = \frac{Z_2 - Z_1}{Z_2 + Z_1} \tag{3.2}$$

where Z_1 and Z_2 represent the acoustic impedances of the two materials.

The transmission coefficient (T), which quantifies the energy transmitted through the interface, is given by (Sullivan and Johnson 2018)

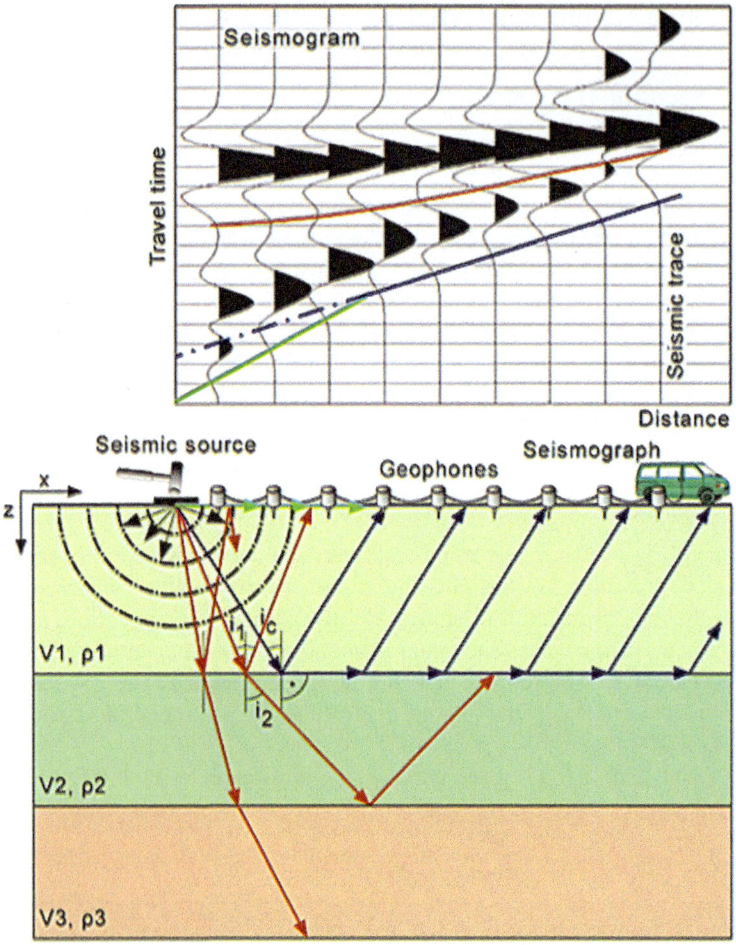

Fig. 3.1 Seismic survey sketch: layer model, seismic rays (green: direct wave, blue: refracted/head wave, red: reflected waves), and resulting seismogram with seismic signals. *Source* Venkiteswaran and Naskar (2000)

$$T = 2\frac{Z_1}{Z_2 + Z_1} \tag{3.3}$$

Seismic waves are classified into body waves and surface waves. Body waves include Primary (P) waves, which travel faster and can propagate through solids, liquids, and gases, and Secondary (S) waves, which are slower and can only move through solids (Milsom and Eriksen 2011). Surface waves include Rayleigh waves, which move along the surface with elliptical particle motion, and Love waves, which cause horizontal shaking and move through the surface horizontally (Fig. 3.2).

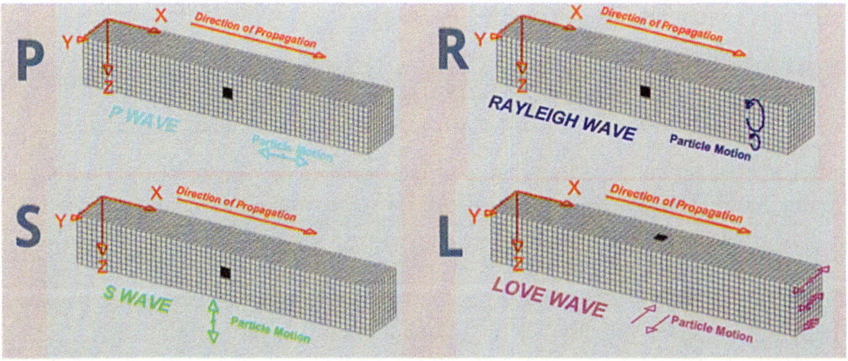

Fig. 3.2 Seismic waves: body waves (P-waves and S-waves) and surface waves (Rayleigh and Love waves). *Source* Wikipedia

The velocities of seismic waves depend on the elastic properties and density of the materials through which they travel. For P-waves, the velocity (V_P) is given by the formula

$$V_P = \sqrt{\frac{k + \frac{4}{3}\mu}{\rho}} \tag{3.4}$$

where k is the bulk modulus, μ is the shear modulus, and ρ is the density of the material. The velocity of S-waves

$$V_S = \sqrt{\frac{\mu}{\rho}} \tag{3.5}$$

Poisson's ratio (σ) is an important material property and is given by

$$\text{Poisson's coefficient } (\sigma) = \frac{\gamma^2 - 2}{\gamma^2 - 1} \tag{3.6}$$

where γ is the velocity ratio between P-waves and S-waves. Other related material properties include Young's modulus (E), which is calculated as

$$\text{Young's modulus } (E) = \rho V_P^2 \frac{(1 - 2\sigma)(1 + \sigma)}{(1 - \sigma)} \tag{3.7}$$

and shear modulus (μ), which is given by

$$\text{Shear modulus } (\mu) = \frac{E}{2(1 + \gamma)} \tag{3.8}$$

The bulk modulus (K) can be calculated using

$$\text{Bulk modulus (K)} = \rho\left(V_P^2 - \frac{4}{3}V_S^2\right) \tag{3.9}$$

Seismic waves carry energy that is proportional to the square of their amplitude (Pelton 2005). The general wave equation for wave propagation is

$$V = f\lambda \tag{3.10}$$

where V is the wave velocity, f is the frequency, and λ is the wavelength.

Seismic waves also interact with material boundaries, where they are partly reflected and refracted. At the interface between two materials, the reflection coefficient at normal incidence is given by

$$R = \frac{\rho_2 V_2 - \rho_1 V_1}{\rho_2 V_2 + \rho_1 V_1} \tag{3.11}$$

and Snell's law of refraction, which describes the refraction of seismic waves at an interface, is expressed as (Rabbel 2006)

$$\frac{\sin i}{V_{P_1}} = \frac{\sin r}{V_{P_2}} = \frac{\sin P_1}{V_{S_1}} = \frac{\sin P_2}{V_{S_2}} \tag{3.12}$$

where i and r are the angles of incidence and refraction, and V_{P_1} and V_{P_2} are the velocities of P-waves in the two materials. Table 3.3 summarizes the principles, applications, advantages, and limitations of seismic refraction and reflection methods, highlighting their key roles in engineering projects.

Table 3.4 presents seismic wave velocities for different subsurface materials, as reported by Pelton (2005). Seismic waves, including P-waves (Primary or compressional waves) and S-waves (Secondary or shear waves), travel through the Earth's subsurface and provide valuable insights into the material properties and structure beneath the surface.

In engineering geophysics, seismic methods have various applications. Seismic refraction is used to identify the depth of soil cover and bedrock, locate faults, fractures, cavities, and groundwater tables, estimate seismic velocities and moduli in rock layers, and map subsurface material properties (Miller and Stewart 2018; Steeples and Miller 1996). Seismic reflection is particularly useful for mapping deep geological structures, such as faults and bedrock, in applications like oil and gas exploration and assessing subsurface features for engineering projects (Dobbins and Voigt 2019; Hanna and Taylor 2021).

Instrumentation for seismic surveys plays a crucial role in data acquisition. Seismic surveys utilize geophones to detect seismic wave arrivals, and seismic pulses are generated by artificial sources such as explosives or vibrators. The time it takes for seismic waves to return to the surface is recorded, which helps create images of

Table 3.3 Summary of seismic methods and their applications

Seismic method	Principle	Applications	Equipment used
Seismic refraction	Measures travel times of seismic waves as they pass through different subsurface layers. Waves refract at interfaces between materials with differing seismic velocities	– Identifying the depth of soil cover and bedrock – Locating faults, fractures, cavities, and groundwater tables – Estimating seismic velocities and moduli of rock layers – Mapping subsurface material properties	– Geophones (for detecting wave arrivals) – Seismic sources (explosives or vibrators)
Seismic reflection	Seismic waves reflect off subsurface boundaries, such as bedrock, faults, or interfaces between different geological materials	– Mapping deep geological structures (e.g., faults, bedrock) – Oil and gas exploration – Assessing subsurface conditions for engineering projects	– Geophones – Seismic vibrators or explosives – Multi-channel recording systems (e.g., 24 or 48 channels)
Seismic tomography	Uses seismic waves passing through the Earth to create images of subsurface structures. This technique involves both reflection and refraction data	– Detailed mapping of subsurface structures – Identifying variations in material properties across large areas	– Arrays of geophones – Seismic sources
Microseismic monitoring	Detects small seismic events caused by human activities or natural phenomena, focusing on the location and magnitude of these events	– Monitoring induced seismicity from mining, reservoir construction, and geothermal projects – Tracking small earthquakes	– High-sensitivity geophones – Seismic sensors

(continued)

Table 3.3 (continued)

Seismic method	Principle	Applications	Equipment used
Surface wave methods	Utilizes surface waves (Rayleigh and Love waves) to study the shallow subsurface. Surface waves are used to evaluate material properties like stiffness	– Geotechnical investigations for foundation design – Estimating shear wave velocity profiles of the shallow subsurface	– Geophones – Seismic sources, often active sources like hammer blows or vibrators
Seismic refraction tomography	Combines seismic refraction with tomographic imaging to produce high-resolution subsurface images	– Identifying heterogeneities in the subsurface – Detailed subsurface mapping, especially in complex geological settings	– Geophones – Seismic sources

subsurface structures. Seismic refraction instruments typically involve multi-channel systems (e.g., 12, 24, or 48 channels) and are used for shallow investigations, while seismic reflection instruments are more complex and suited for deeper subsurface exploration (Malehmir et al. 2012).

In conclusion, seismic refraction and reflection methods are essential tools in engineering geophysics, providing valuable information for various geotechnical and geological applications. These methods, alongside the detailed analysis of seismic wave velocities and material properties, enable engineers and geophysicists to assess subsurface conditions and make informed decisions for construction, exploration, and environmental studies.

Seismic Method in Soil Profiling, Fault Detection, and Seismic Site Characterization

Seismic methods, including seismic refraction, reflection, and surface wave techniques, are essential tools in engineering geophysics for evaluating subsurface conditions. These methods have proven effective in soil profiling, fault detection, and seismic site characterization, providing valuable insights into the structural and material properties of the subsurface.

Soil profiling involves analyzing the properties of soil layers, such as density, elasticity, and thickness, and seismic methods like refraction and surface wave techniques are crucial for this purpose. Seismic refraction measures the travel times of seismic waves through different soil layers, helping geophysicists determine the depth to bedrock and other geological interfaces, which is essential for foundation design and soil stability assessment (Steeples and Miller 1996). Surface wave methods, including Rayleigh and Love waves, assess shear wave velocities in the shallow subsurface,

Table 3.4 Seismic wave velocities (Pelton 2005)

Materials	V_P (m/s)	V_S (m/s)	ρ (g/cm^3)
Unconsolidated materials			
Dry sands	400–1200	100–500	1.5–1.7
Wet sands	1500–4000	400–1200	1.9–2.1
Clay	1100–2500	200–800	2.0–2.4
Glacial till (water saturated)	1500–2500		
Permafrost	3500–4000		
Pore fluids			
Ice	3400–3800	1700–1900	0.9
Oil (petroleum)	1200–1250		0.6–0.9
Water	1400–1500		
Air	300		
Sedimentary rocks			
Marl/shale	2000–3000	750–1500	2.1–2.6
Gypsum	2000–3500		
Coal	2200–2700	1000–1400	1.3–1.8
Sandstone	3000–4500	1200–2800	2.1–2.4
Limestone	3500–6000	2000–3300	2.4–2.7
Chalk	2300–2600	1100–1300	1.8–2.3
Salt	4500–5500	2500–3100	2.1–2.3
Anhydrite	4000–5500	2200–3100	2.9–3.0
Dolomite	3500–6500	1900–3600	2.5–2.9
Igneous/Metamorphic rocks			
Granite	4500–6000	2500–3300	2.5–2.7
Basalt	5000–6000	2800–3400	
Gabbro	6500–7000		
Ultramafic rocks	7500–8500		
Serpentinite	5500–6500		
Other materials			
Steel	6100		
Iron	5800		
Aluminium	6600		
Concrete	3600		

providing insights into soil stiffness and suitability for construction (Milsom and Eriksen 2011).

Seismic methods are essential for *detecting faults and fractures* in subsurface structures, which is crucial for the safety and design of engineering projects. Seismic reflection is commonly used to identify discontinuities such as faults and fractures, as it captures the reflection of seismic waves at material boundaries, providing detailed images of subsurface structures (Sullivan and Johnson 2018). Seismic tomography, which combines refraction and reflection data, offers high-resolution subsurface images, making it particularly useful in complex geological settings where faults are deep or multiple fault zones exist (Grant and West 1965).

Seismic site characterization evaluates a site's seismic risk and suitability for construction by assessing shear wave velocities, which are crucial for determining soil types and how the site will respond to seismic events (Pelton 2005). Seismic reflection techniques also help map deeper geological structures, assessing ground shaking and seismic wave amplification, which are vital for designing stable structures in seismic zones (Kearey et al. 2002; Malehmir et al. 2012). In conclusion, seismic methods are essential for soil profiling, fault detection, and seismic site characterization. These techniques help engineers and geophysicists assess subsurface conditions, reducing risks associated with construction in seismic zones and ensuring the safety and stability of infrastructure.

3.4.2 Electrical and Electromagnetic Methods

Electrical and electromagnetic methods are crucial tools in engineering geophysics, providing valuable insights into subsurface conditions. These methods focus on the electrical properties of materials, such as resistivity, conductivity, and the ability to retain electrical charge. Electrical and electromagnetic methods are employed to detect variations in soil composition, groundwater, fractures, and voids, all of which are essential for engineering applications such as site characterization, foundation design, and environmental assessments. The primary techniques used in electrical and electromagnetic geophysics are resistivity, induced polarization (IP), and ground-penetrating radar (GPR).

Resistivity Method in Engineering Geophysics

The resistivity method is one of the most versatile and widely used geophysical techniques in engineering and environmental geophysics. It is based on the measurement of the ground's resistance to the flow of an electrical current. In the ideal case of a perfectly uniform conducting half-space, the current flows in a dipole pattern, and the resistivity measured with a four-electrode setup gives the true resistivity of the half-space (Fig. 3.3). However, in reality, the resistivity varies because of different types of rocks and geological structures, making the ground inhomogeneous. The

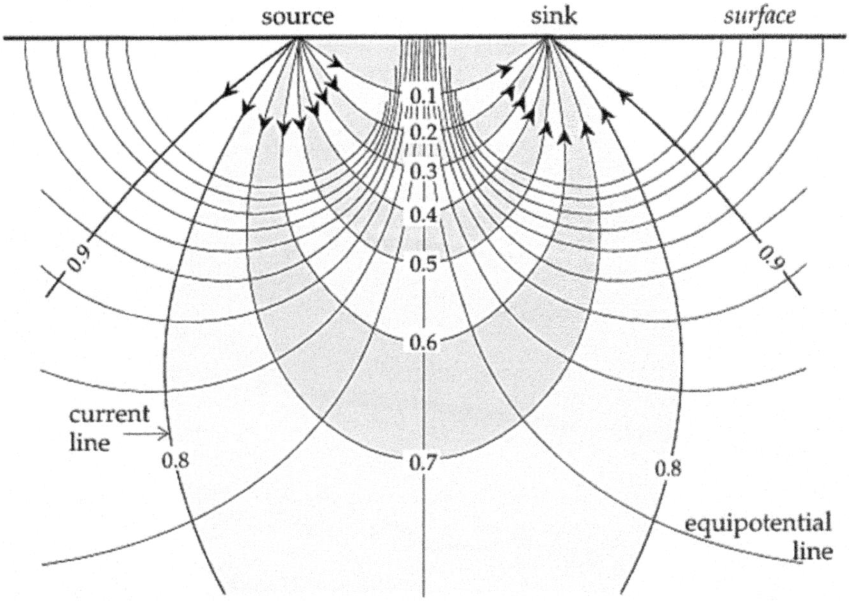

Fig. 3.3 Cross-section of current paths and equipotential surfaces (surfaces of equal potential) between a source and sink, showing current distribution. *Source* Robinson and Coruh (1988)

four-electrode method assumes the ground is uniform, but this isn't true in most cases. As a result, the resistivity measured is the "apparent resistivity" of an imaginary uniform half-space, which usually doesn't reflect the actual resistivity of the ground.

This method provides valuable insights into the distribution and properties of subsurface materials, including soil type, groundwater content, and the presence of contaminants. By analyzing variations in subsurface resistivity, engineers and geophysicists can map the physical properties of soil and rock formations, identify geological boundaries, and assess site conditions for construction projects (Table 3.5).

The *basic principle of resistivity* is that it refers to a material's ability to resist the flow of an electrical current. Different materials, such as rock, soil, and groundwater, have different resistivities depending on factors like composition, moisture content, and porosity. For example, dry rock typically has higher resistivity than saturated soil, and groundwater with high salinity will have lower resistivity compared to freshwater. The electrical resistivity method is based on Ohm's law, which relates voltage, current, and resistance in an electrical circuit:

$$V = IR \tag{3.13}$$

where V is the potential difference (voltage) measured in V, I is the electrical current injected into the ground in A, R is the resistance encountered by the current in Ω. In the arrangement consisting of a pair of current electrodes (a and b) and a pair

Table 3.5 Typical electrical resistivity values for some Earth materials (Loke 2002; Gibson and George 2003; Society of Exploration Geophysicist of Japan 2004; Advanced Geosciences Incorporated 2018; Naskar and Das 2009)

Common materials	Cited resistivity values (Ωm)				
	Loke (2002)	Gibson and George (2003)	SEGJ (2004)	Ewusi (2006)	AGI (2018)
Clay	1–100	1–100	1–300	1–100	10–100
Sand	10–800	50–1050	1–1100	30–1050	$600–1 \times 10^4$
Laterite soil	–	–	–	120–750	–
Gravel	$600–10^4$	100–1400	20–7000	100–1400	$600–1 \times 10^4$
Mudstone	–	–	–	20–120	–
Siltstone	–	–	–	20–150	–
Limestone	80–6000	$50–10^6$	–	–	$100–1 \times 10^6$
Shale	20–2000	–	3–200	$20–2 \times 10^3$	–
Sandstone	10–5000	$1–7.4 \times 10^8$	10–700	200–5000	$100–1 \times 10^3$
Granite	$5000–10^6$	$100–10^6$	300–40,000	$3 \times 10^2–10^6$	–

of potential electrodes (c and d) (as shown in Fig. 3.4), the current electrodes are used to inject a known electrical current into the ground. The potential electrodes, placed at a distance from the current electrodes, measure the resulting voltage. By using the measured voltage and the known current, the resistivity of the ground can be determined. This method is commonly used in geophysical surveys to investigate the electrical properties of the subsurface. The configuration assumes the ground is uniform, but real-world conditions often lead to variations in resistivity.

The apparent resistivity (ρ_a) is the effective resistivity of the ground that is inferred from measurements at the surface. Apparent resistivity is derived from the ratio of the measured voltage (ΔV) to the applied current (I) and the geometric factor of the measurement configuration. For a Wenner array, which is a commonly used electrode

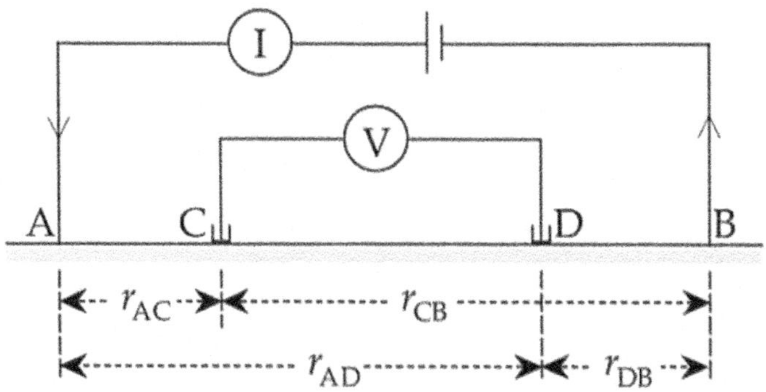

Fig. 3.4 General four-electrode configuration for resistivity measurement, showing current electrodes (**a**, **b**) and potential electrodes (**c**, **d**). *Source* Naskar and Singh (2002)

configuration, the apparent resistivity is given by:

$$\rho_a = 2\pi a \frac{\Delta V}{I} \tag{3.14}$$

where ρ_a is the apparent resistivity (Ωm), a is the distance between the electrodes (in m), ΔV is the measured potential difference (V), I is the injected current (A).

The geometric factor (related to the electrode spacing) plays a crucial role in the final resistivity calculation. For the Wenner array (Fig. 3.5a), the spacing of the electrodes significantly affects the depth of investigation. Other electrode configurations, such as Schlumberger (Fig. 3.5b) or dipole–dipole (Fig. 3.6), provide variations in the depth and resolution of subsurface profiling (Kearey et al. 2002).

The resistivity method often involves *mathematical models* and *inversion techniques* to generate 2D or 3D images of subsurface resistivity. This is done by using numerical models to iteratively refine the apparent resistivity values based on measured data. Inversion allows for the creation of more accurate subsurface resistivity profiles, which can be used for detailed analysis of soil layers, rock formations, and the identification of geological features such as fractures or voids (Sullivan and Johnson 2018).

A common approach to inversion is least-squares inversion, which minimizes the difference between the observed data and the model predictions. The mathematical formulation for inversion typically involves solving a system of equations that represent the relationship between the observed apparent resistivity values and the

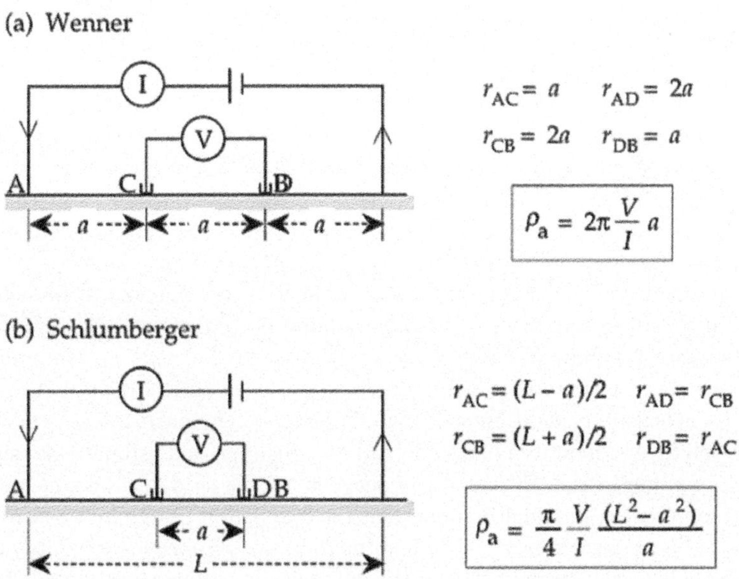

Fig. 3.5 Geometries of current and potential electrodes in the **a** Wenner and **b** Schlumberger configurations. *Source* Naskar et al. (2011)

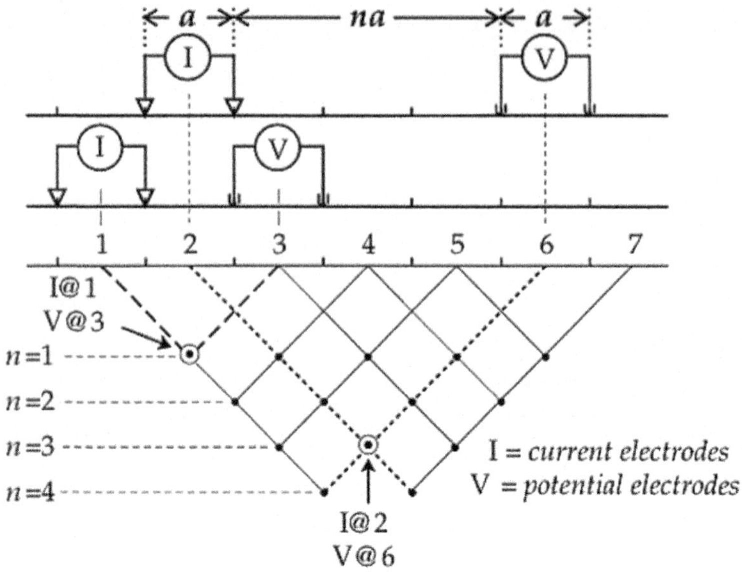

Fig. 3.6 Construction of a pseudo-section for a dipole–dipole resistivity survey: Plotting measured values at the intersection of 45° lines from the transmitter and receiver mid-points. *Source* GSI (1992)

subsurface resistivity distribution:

$$F(r) = d \tag{3.15}$$

where F(r) is the forward modeling operator that predicts the apparent resistivity based on the resistivity distribution r, d is the observed data (apparent resistivity), r is the resistivity model to be estimated. These inversion techniques are essential for creating detailed, high-resolution images of subsurface conditions, which are useful for applications such as environmental contamination detection and mineral exploration.

The resistivity method is employed in a wide range of *engineering and environmental applications* due to its versatility and ability to map subsurface properties non-destructively. Some of the primary applications include *soil profiling and site characterization*, where the resistivity method provides critical data by measuring variations in resistivity at different depths. This enables engineers to assess the thickness of soil layers, identify changes in soil composition, and estimate the depth to bedrock or groundwater. The resistivity data can also be used to assess the mechanical properties of soil, which is important for foundation design and construction stability (Milsom and Eriksen 2011). In *groundwater exploration*, resistivity surveys are commonly used in hydrogeology to locate aquifers, estimate their size, and determine the salinity and quality of groundwater. Freshwater aquifers typically exhibit high resistivity, whereas saline groundwater or brines have lower resistivity. Resistivity data can help map the lateral and vertical extent of groundwater resources, which is essential for water resource management (Steeples and Miller 1996). For *contamination detection and environmental monitoring*, the resistivity method is useful for identifying subsurface contamination from hazardous materials such as heavy metals, chemicals, and petroleum products. Contaminants often alter the resistivity of surrounding soil and water, which can be detected through resistivity anomalies. This method is particularly effective for monitoring the spread of pollutants in landfills, industrial waste sites, or areas affected by spills (Malehmir et al. 2012). In *engineering and construction*, resistivity surveys are used to assess subsurface conditions, determine the depth to bedrock, and evaluate the presence of voids or fractures that may compromise the structural integrity of foundations. For instance, resistivity surveys can help identify zones of weak or soft soil that may require special foundation designs (Kearey et al. 2002). The resistivity method is also applied in *mineral exploration* to detect ore bodies and other valuable minerals [Copper (Cu) ores, Gold (Au) ores, Silver (Ag) ores, Lead (Pb) and Zinc (Zn) ores, Iron (Fe) ore, Nickel (Ni) ores, Manganese (Mn) ores, Cobalt (Co) ores, Uranium (U) ores, Rare Earth Elements (REE)]. Many mineral deposits, such as sulfide ores [copper sulfides (chalcopyrite, bornite), gold is often found in association with other minerals like pyrite (iron sulfide) or arsenopyrite, silver sulfides (such as argentite or stephanite), lead and zinc sulfide minerals (galena, sphalerite), nickel-bearing sulfide minerals (pentlandite), manganese-rich ores (pyrolusite or rhodochrosite), cobalt-bearing sulfide minerals (chalcopyrite or cobaltite), uranium ores (pitchblende or uraninite)] exhibit distinctive resistivity contrasts with the surrounding host rock, allowing for their identification through resistivity anomalies (Sullivan and Johnson 2018).

The *resistivity method* has certain *limitations*. The *depth of penetration* is limited by the spacing of the electrodes and the resistivity of the material. For highly resistive materials like granite, penetration can be shallow, while in more conductive materials like clays or wet soils, the depth of investigation can be greater. The method may also face challenges in resolving *geological heterogeneity*, as it may struggle to identify complex geological structures, such as fractured rock or heterogeneous soil, unless combined with other geophysical techniques like seismic or electromagnetic surveys. Additionally, *signal noise* from external sources, such as power lines or nearby construction, can affect resistivity surveys, potentially impacting the accuracy of the measurements.

Induced Polarization (IP) Method in Engineering Geophysics

The Induced Polarization (IP) method is an essential tool in geophysics, particularly useful for detecting subsurface materials that exhibit chargeability—i.e., the ability to retain an electrical charge after the current has been switched off. This method is particularly sensitive to conductive minerals such as sulfide ores (e.g., pyrite, chalcopyrite), clays, and metallic minerals, which are capable of storing and slowly releasing electrical charge. The IP effect is driven by the electrical properties of the pore spaces in the ground, where charge builds up under an applied current and is released once the current is switched off.

The *basic principle behind the IP (Induced Polarization) method* relies on measuring the time decay of the voltage after an electrical current is turned off (Fig. 3.7). The voltage decay, which occurs over time, reflects the chargeability of the material in the subsurface. Materials that exhibit high chargeability, like sulfides and clays, will show a slow voltage decay, while non-chargeable materials, such as sandstones or granites, will show a faster decay.

The *key parameter in the IP method* is chargeability (M), which refers to the ability of a material to store and release electrical charge. Chargeability is calculated as:

$$M = \frac{V_{off}}{V_{on}} \times 100 \tag{3.16}$$

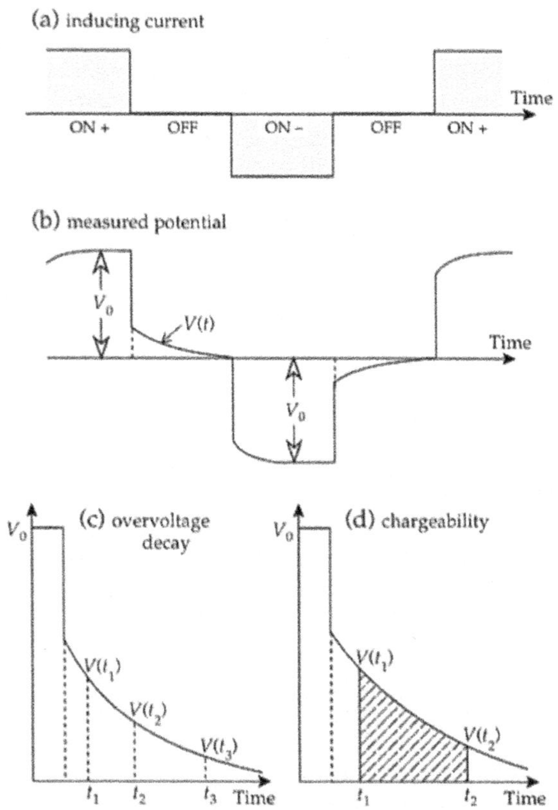

Fig. 3.7 IP-related decay of potential: **a** illustration of potential decay after primary current interruption, **b** effect of IP decay time on potential waveform with square-wave input current, **c** overvoltage (temporary excess voltage after current interruption), and **d** chargeability. *Source* Lowrie (2007)

where V_{on} is the voltage when the current is applied (during the current injection phase), V_{off} is the voltage immediately after the current is switched off (during the voltage decay phase).

Note: The ratio of $\frac{V_{off}}{V_{on}}$ gives the chargeability in percentage, with higher values indicating more chargeable materials.

Another important parameter is apparent chargeability (M_a), which can be derived from the ratio of the voltage decay to the current applied:

$$M_a = \frac{\Delta V_{decay}}{I} \times 100 \tag{3.17}$$

where ΔV_{decay} is the voltage change over time during the decay, and I is the injected current.

This relationship helps to quantify the chargeability observed in the subsurface and can be influenced by the electrical properties of the materials, the geometry of the survey, and the electrode configuration used.

The *mathematical model of IP response* that governs the decay of voltage after current injection is often based on the Cole–Cole equation, which describes the resistivity and chargeability response of materials over time:

$$V(t) = V_0\left(1 - \left(1 + \frac{t}{\tau}\right)^{-n}\right) \tag{3.18}$$

where V(t) is the voltage at time t, V_0 is the initial voltage, τ is the characteristic time constant (related to the chargeability), and n is a parameter related to the heterogeneity of the material.

The decay curve follows a characteristic pattern, where faster decays correspond to materials with low chargeability, and slower decays are typical of more chargeable materials, such as sulfide-rich zones.

Survey techniques in the IP method involve injecting an electrical current into the ground and measuring the resulting voltage decay at various points to analyze the chargeability and resistivity of subsurface materials. Induced polarization surveys are typically performed in conjunction with resistivity surveys to provide a comprehensive view of subsurface properties.

The electrical resistivity and chargeability data can be gathered using a variety of electrode arrays, including the *Dipole–Dipole array*, which is widely used for IP surveys and provides high-resolution data at shallow depths by placing a pair of electrodes for current injection and a separate pair for potential measurement. The *Wenner array*, primarily used for resistivity measurements, can also be applied in IP surveys; it features a uniform arrangement of electrodes with constant spacing and is commonly used for near-surface studies. The *Schlumberger array* is effective for deeper investigations, as it involves varying electrode spacings to obtain data on both shallow and deep subsurface features. The data obtained from the IP method is typically presented as a chargeability map or chargeability profile, highlighting regions of higher chargeability—often indicative of high sulfide or clay content—as anomalies.

The IP method is *applied* in mineral exploration, groundwater studies, and environmental investigations to detect chargeable materials such as sulfide ores, clay-rich zones, and contamination plumes. In *mineral exploration*, the IP method is particularly effective in identifying deposits with chargeable properties, including sulfide minerals like chalcopyrite and pyrite, gold associated with sulfides, and copper in porphyry systems. These sulfide-rich zones often produce strong chargeability anomalies, making IP surveys a valuable tool for guiding exploration drilling in mining operations (Routh et al. 2017). In *hydrogeological surveys*, the method is used to map clay layers and aquifers and detect groundwater contamination. Since clay minerals have high chargeability compared to materials like sand or gravel, IP surveys can help distinguish between different subsurface units; for example, clay-rich zones in Paleocene-Eocene formations have been mapped to assess groundwater contamination (Ward 2011). In *environmental studies*, the IP method is employed to detect subsurface contamination at landfills, industrial sites, and other polluted locations. The presence of contaminants such as heavy metals or hydrocarbons alters the chargeability of surrounding materials, allowing for the identification of pollution sources—IP has been successfully used to detect oil spills and heavy metal contamination (Birks et al. 2010). Within *engineering geophysics*, IP surveys contribute to site characterization and the assessment of construction suitability, particularly in identifying sulfide minerals or subsurface voids that may pose structural risks.

Despite its wide range of applications, the IP method has *limitations*: its *depth penetration* is typically restricted to shallow to moderate depths (up to about 100 m), depending on electrode configuration; *environmental factors* like moisture and temperature can affect data quality; and in *geologically complex settings*, interpreting chargeability anomalies can be difficult due to the influence of overlapping subsurface features.

Ground-Penetrating Radar (GPR) in Engineering Geophysics

Ground-penetrating radar (GPR) is a non-invasive geophysical technique that utilizes high-frequency electromagnetic waves to probe the subsurface. GPR systems consist of a transmitter that emits radar waves and a receiver that detects the reflected waves from different subsurface materials. The primary advantage of GPR is its ability to produce high-resolution images of the shallow subsurface, making it an invaluable tool in engineering geophysics for a wide range of applications, including shallow subsurface mapping, infrastructure inspections (roads, bridges, buildings, and tunnels), utility (water pipes, sewer lines, gas pipelines, electric cables, telecommunication lines, and drainage systems) mapping, and archaeological surveys.

The *basic principle of GPR* involves transmitting electromagnetic waves into the ground and analyzing their reflection and refraction at interfaces between materials with different dielectric properties, and the operation of GPR is based on these interactions (Fig. 3.8), where properties such as permittivity (ε) and conductivity (σ) influence how the radar waves propagate through the material. GPR systems emit electromagnetic pulses through the ground, and when these waves encounter a boundary between materials (e.g., between soil and bedrock), some of the energy is reflected back to the surface, where it is detected by the receiver. The time it takes for the waves to return provides information about the depth of the reflecting interfaces, as well as the material properties of the subsurface.

The *mathematical foundation of GPR* is based on the primary equation that relates to the two-way travel time t and the velocity of the electromagnetic wave v in the subsurface. The depth d to a reflecting interface (e.g., bedrock, water table, or buried objects) can be calculated using the following equation:

$$d = \frac{vt}{2} \tag{3.19}$$

where d is the depth to the reflecting surface (in m), v is the velocity of the wave in the material (in m/s), and t is the two-way travel time of the radar pulse (in s).

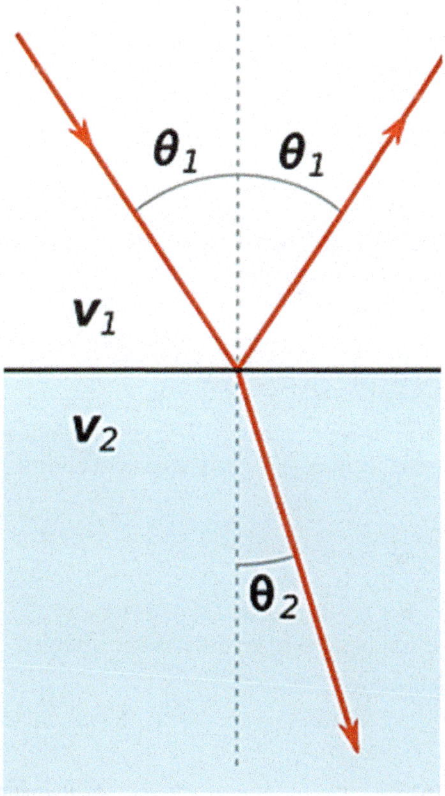

Fig. 3.8 Reflection and refraction of an incoming radio wave in ground penetrating radar (GPR). *Source* Wikipedia

The *velocity v of electromagnetic waves* in the subsurface depends on the dielectric permittivity ε_r of the material and can be calculated using the following equation:

$$v = \frac{c}{\sqrt{\varepsilon_r}} \tag{3.20}$$

where v is the velocity of the wave in the material (in m/s), c is the speed of light in a vacuum (approximately 3×10^8 m/s), and ε_r is the relative dielectric permittivity (also known as the relative permittivity or dielectric constant) of the material. The velocity equation implies that the propagation of radar waves is faster in materials with lower dielectric permittivity and slower in materials with higher permittivity. The velocity is influenced by the composition and moisture content of the subsurface material.

The *resolution and depth penetration of GPR* are critically influenced by the frequency of the radar waves, with higher frequencies providing better resolution for detecting shallow targets but limiting depth penetration, while lower frequencies allow for deeper penetration but offer lower resolution. Higher frequencies, typically ranging from 100 MHz (1 MHz = 10^6 Hz) to 2 GHz (1 GHz = 10^9 Hz), provide better resolution but have a limited depth of penetration. In contrast, lower frequencies (e.g., 20–100 MHz) offer deeper penetration but with lower resolution. The relationship between frequency f, wavelength λ, and the velocity v of electromagnetic waves is given by:

$$\lambda = \frac{v}{f} \tag{3.21}$$

where λ is the wavelength of the radar wave (in m), v is the velocity of the wave (in m/s), and f is the frequency of the radar waves (in Hz). Lower frequencies result in longer wavelengths, which penetrate deeper but provide less detail. Higher frequencies provide shorter wavelengths, improving the resolution but limiting penetration.

The *skin depth (δ) of electromagnetic wave propagation* defines the propagation distance at which the amplitude of an electromagnetic wave is reduced by a factor of $\frac{1}{e}$, or approximately 37% of its original amplitude. By definition, the skin depth is the reciprocal of the attenuation constant (α):

$$\delta = \frac{1}{\alpha} \tag{3.22}$$

For non-magnetic Earth materials ($\mu_r = 1$), the skin depth can be approximated as follows, based on different conditions:

$$\delta \approx \frac{503}{\sqrt{\sigma f}} \text{ for } \omega\varepsilon \ll \sigma \tag{3.23}$$

$$\delta \approx 0.0053\sqrt{\frac{\varepsilon_r}{\sigma}} \text{ for } \sigma \ll \omega\varepsilon \tag{3.24}$$

where σ is the electrical conductivity of the material (in S/m), f is the frequency of the wave (in Hz), ε_r is the relative dielectric permittivity of the material, ω is the angular frequency, μ_r represents relative permeability ($\mu_r = \frac{\mu}{\mu_0}$), μ is the permeability of the material (in Henry's per meter, H/m), and μ_0 is the permeability of free space (approximately $4\pi \times 10^{-7}$ H/m).

Key observations regarding skin depth include that higher frequencies lead to smaller skin depths, making them suitable for shallow investigations. At high frequencies, the skin depth reaches a limit that becomes independent of frequency, defining the boundary of effective wave propagation. Additionally, materials with lower conductivity or higher dielectric permittivity result in larger skin depths, affecting how electromagnetic waves interact with subsurface layers.

In engineering geophysics, skin depth is important for understanding how deep waves can penetrate the ground. It helps choose the right frequency for surveys, ensuring waves reach the right depth. Skin depth also aids in identifying subsurface materials like moisture or metals and improves survey design by reducing signal loss, making it easier to interpret data accurately.

GPR is a versatile tool with broad *applications in engineering geophysics.* One of its primary uses is *shallow subsurface mapping*, where it is highly effective for investigations such as soil profiling, bedrock mapping, and water table detection. GPR can identify geological interfaces like the boundary between soil and bedrock, detect groundwater, and map soil stratigraphy, which is essential for site characterization, foundation design, and civil engineering projects (Porsani et al. 2004). In *infrastructure inspections*, GPR is widely employed for the non-destructive evaluation of structures like concrete, pavements, and roadways. It can detect voids, cracks, corrosion in reinforcement, and delamination of concrete surfaces, which are critical for assessing structural integrity. It has been extensively used to inspect bridge decks, locate hidden voids or deterioration, and evaluate the condition of reinforced steel bars (Sharma 2005). For *utility mapping*, GPR plays a key role in locating buried utilities such as pipes, cables, and conduits without excavation, providing valuable information for construction and urban planning. This application is especially important in urban environments where avoiding accidental utility damage is crucial (Huisman et al. 2003). In *archaeological surveys*, GPR offers a non-invasive means to detect buried artefacts, structures, and ruins. It allows archaeologists to map ancient sites without soil disturbance, preserving site integrity, and has been successfully used in identifying ruins and tombs in Egyptian pyramids and other historical locations (Conyers and Goodman 1997). Additionally, GPR is employed in *environmental and geotechnical investigations* to detect contaminated zones, landfills, and underground storage tanks. It also assists in evaluating soil properties, detecting voids, and mapping underground cavities, as demonstrated in landfill investigations that assess buried waste and contamination spread (Gavazzi et al. 2011).

While GPR is a powerful tool, it has some *limitations*. Its *depth penetration* is generally limited to shallow subsurface investigations, typically up to 30 m, depending on the frequency of the radar waves and the electrical conductivity of the materials being surveyed. *Signal attenuation* is another challenge, as GPR signals can be significantly weakened in conductive materials like clay or saline soils, reducing its effectiveness in such geological settings. Additionally, the *interpretation of GPR data can be complex* and requires expert knowledge, as radar reflections may be difficult to analyze in areas with heterogeneous materials or multiple subsurface interfaces.

Table 3.6 provides a concise overview of key electrical and electromagnetic geophysical methods used in engineering applications. These methods, which primarily involve measuring the ground's electrical properties or response to electromagnetic fields, are essential tools for subsurface investigation. The methods can be categorized into those that rely on direct current (DC) and alternating current (AC) techniques, each with specific applications, advantages, and limitations.

This table gives a concise overview of the key electrical and electromagnetic methods used in engineering geophysics, their principles, applications, advantages, disadvantages, and the typical equipment involved.

Self Potential (SP) Method in Engineering Geophysics

The Self Potential (SP) method is a passive geophysical technique used to measure natural electric potential differences in the Earth. These potential differences arise from subsurface processes such as fluid movement (e.g., groundwater flow) or mineralization (e.g., the presence of metallic minerals). Unlike active geophysical methods, SP does not require an external energy source. Instead, it relies on natural signals produced by the Earth's processes.

The SP method works on the *principle of measuring natural voltage differences* between two points on the Earth's surface. These differences reflect subsurface conditions like groundwater flow, mineral deposits, and electrochemical processes. In simple terms, when water moves through a porous medium (e.g., soil or rock), it generates an electric field due to the interaction between the water and the surrounding materials. This causes a natural electric potential that can be measured at the surface. The SP method can be mathematically described by the equation:

$$V = \int_{a}^{b} E \cdot dl \tag{3.25}$$

where V is the potential difference between two measurement points, E is the electric field strength, and dl is the small element of the measurement path. This equation shows how the electric potential is calculated based on the electric field along the path between two points. It is important for understanding how the subsurface structure affects the measured electric potential.

The *SP method* is widely used in various *engineering geophysics applications*, particularly in environmental and hydrogeological studies (Leprince and Caubet 2001). One common application is *groundwater flow detection*, where the SP method proves highly effective in identifying areas of water movement, especially through porous rocks or fractured zones, as the flow of water generates electric potentials

Table 3.6 Electrical and electromagnetic methods in engineering geophysics

Method	Principle	Applications	Advantages	Disadvantages	Equipment used
Resistivity method	Measures the resistance of the ground to the flow of electrical current	Mapping of subsurface structures, groundwater exploration, environmental studies	Simple, cost-effective, provides high-resolution data in shallow depths	Sensitive to surface conditions (e.g., moisture, salinity). May have low penetration in conductive environments	Electrodes, power supply, resistivity meter
Induced polarization (IP)	Measures chargeability of the subsurface by observing the decay of induced voltage	Mineral exploration, groundwater contamination, engineering site investigations	Can detect both resistivity and polarization effects. Effective for detecting low-conductivity materials	Expensive, requires more complex data analysis, slower than resistivity	Power supply, electrodes, receiver, signal generator
Ground penetrating radar (GPR)	Measures the time it takes for radar waves to return after reflecting off subsurface structures	Detection of buried objects, mapping of groundwater, utility location, and geotechnical investigations	High-resolution, real-time data, non-invasive, able to resolve small-scale features	Limited depth penetration in conductive materials (e.g., clay, saline soils)	GPR system with transmitting and receiving antennas
Electromagnetic (EM) induction	Uses alternating electromagnetic fields to induce currents in the ground. Measures the secondary field induced by the subsurface	Detecting mineral deposits, mapping of subsurface conductivity, environmental studies	Non-contact, rapid data acquisition, good for wide-area surveys	Shallow depth penetration, accuracy affected by surface conditions	Transmitter and receiver coils, electromagnetic sensor

(continued)

Table 3.6 (continued)

Method	Principle	Applications	Advantages	Disadvantages	Equipment used
Self Potential (SP) method	Measures natural voltage differences in the ground. Caused by electrochemical processes or water movement	Hydrogeological studies, mineral exploration, geothermal studies	Non-invasive, easy to use, useful in detecting groundwater flow and ore bodies	Low resolution, data interpretation can be challenging	Potentiometer, electrodes
TDEM (time domain electromagnetic)	Measures the time decay of secondary electromagnetic fields after a transient signal is applied	Mineral exploration, groundwater investigations, mapping of subsurface conductivity	Effective for deep exploration, can detect large-scale features	Requires advanced data interpretation, expensive equipment	Transmitter and receiver coils, control units
Frequency domain electromagnetic (FDEM)	Measures the response of the ground to electromagnetic fields at different frequencies	Mineral exploration, environmental site assessments, geotechnical investigations	Versatile, capable of mapping subsurface conductivity at different depths	Depth penetration limited by soil conditions, complex data interpretation	Frequency transmitter and receiver, coils, control unit

measurable at the surface. In *mineral exploration*, the SP method helps detect mineral deposits; for instance, the movement of groundwater through mineralized zones, such as sulfide ores, produces measurable electrical potentials that can indicate the presence of valuable subsurface minerals. Additionally, the SP method is applied in *detecting environmental contamination*, as the movement of pollutants like oil or chemicals through soil and rock can alter the natural electric field, changes that the SP method can detect and map effectively (Jardin and Selensky 2002; Bierwirth and White 1994; Blanco and Ruiz 2007).

The electric potential at a point in the Earth can be modeled using the *Poisson equation*, which is used to describe how electric fields behave in a medium. The equation is:

$$\nabla^2 \varphi = -\frac{\rho}{\varepsilon} \cdot J \tag{3.26}$$

where φ is the electric potential, ρ is the resistivity of the subsurface material, ε is the permittivity of the medium, and J is the current density. This equation helps model the subsurface electrical properties, and it can be solved numerically to interpret SP measurements in terms of subsurface conditions (Sasaki 2000).

The *SP method* has several *key advantages*. It is *non-invasive*, as it does not require drilling or excavation, making it a safe and simple way to gather data about subsurface conditions. It is also *cost-effective*, since SP measurements rely on natural electrical potentials rather than artificial signals, reducing equipment and operational costs. Additionally, the method involves *passive measurement*, meaning it can operate continuously over time without the need for an external power source, which makes it ideal for long-term monitoring of subsurface conditions. However, the *SP method* also has *limitations*. It offers *low resolution* compared to other geophysical methods like seismic or resistivity surveys, which means it may not detect smaller or deeper subsurface features. *Environmental noise*, such as rainfall or human activity, can interfere with SP measurements and make interpretation more challenging. Moreover, the method has *limited depth penetration* and is generally more effective for shallow subsurface investigations, making it less suitable for detecting deeper geological structures.

Table 3.7 provides a clear illustration of how SP measurements can be used to infer subsurface processes. The *interpretation* of these measurements helps engineers and geophysicists understand the nature of fluid flow, mineralization, and electrochemical activities beneath the Earth's surface. By analyzing the potential values—whether positive, negative, or zero—important inferences can be made about groundwater dynamics (Zhou et al. 2008), the presence of valuable mineral deposits, and potential environmental concerns such as contamination. These insights are crucial for various applications, including resource exploration, environmental monitoring, and site assessments for construction projects.

Detailed understanding of the SP measurements reveals that *positive and negative anomalies* provide valuable insight into subsurface processes. Generally, positive anomalies are indicative of upward fluid flow, mineralization, or electrochemical

Table 3.7 Interpretation of Self Potential (SP) measurements and their subsurface implications

Measurement point	Potential (mV)	Interpretation
Point A	+ 10	Possible upward groundwater flow or mineralization at depth
Point B	– 5	Possible downward flow or electrochemical anomaly
Point C	0	No significant electrokinetic or electrochemical effect

activity, suggesting the presence of geochemical gradients or migrating fluids that generate measurable potentials. In contrast, negative anomalies are often associated with downward fluid movement, contaminant flow (Ramos and Comas 2015), or other electrochemical reactions occurring below the surface. A *reading of 0 mV, or a neutral measurement*, typically implies an absence of significant fluid movement or electrochemical activity. This can be interpreted as a zone with relatively stable subsurface conditions, where processes such as mineralization or active groundwater flow are not currently occurring.

Practical example in an *engineering geophysics survey*, such as for a construction project or environmental study, the SP data could help engineers identify areas where *groundwater* is rising to the surface, which is important for assessing flood risks or managing the water table. The data could also reveal zones with *mineral deposits*, such as sulfide-rich ores, that might require further exploration for resource extraction. Additionally, the SP data can pinpoint *locations of contamination*, such as underground oil or chemical migration, which could affect the safety or environmental integrity of a site.

3.4.3 Magnetic and Gravitational Methods in Engineering Geophysics

Magnetic and gravitational methods are passive geophysical techniques widely used in engineering investigations to detect subsurface variations in density and magnetic properties. These methods aid in mapping geological structures, locating voids, buried objects, faults, and assessing ground stability. Their non-invasive nature makes them valuable tools in site characterization, infrastructure planning, and environmental studies.

Gravity Surveying and Engineering Applications

The *principle of gravitational surveys* measures variations in the Earth's gravitational field caused by subsurface density contrasts. These anomalies arise from differences in density between geological formations and can help detect geological features like faults, voids, or mineral deposits (Nettleton 1976; Blakely 1996). The fundamental law governing gravitational interactions is Newton's Law of Gravitation, expressed mathematically as:

$$F = G\frac{Mm}{r^2} \tag{3.27}$$

where F = Gravitational force between two masses, G = Gravitational constant (6.67×10^{-11} Nm^2/kg^2), M = Mass of the Earth (or another large mass), m = Mass of the object experiencing the force, and r = Distance between the centers of the two masses. Gravitational acceleration, g (= 9.8 m/s^2) on the Earth's surface, but local variations in density (e.g., mountains, voids) cause gravity anomalies, is derived from this force and is given by:

$$g = \frac{GM}{r^2} \tag{3.28}$$

In engineering geophysics, the gravity anomaly is typically the difference between the measured gravitational value and the theoretical or expected value, expressed as:

$$\Delta g = g_{measured} - g_{theoretical} \tag{3.29}$$

Gravity surveys have limited but valuable *applications in engineering geophysics*. They are primarily used to identify *subsurface cavities or voids, fault zones, mineral deposits*, and *disturbed or weak ground*, particularly in areas with significant density contrasts. These applications help in assessing site conditions, detecting potential hazards, and guiding construction and exploration efforts.

Table 3.8 summarizes the primary applications of gravity surveying in engineering geophysics. Gravity surveys are crucial for detecting subsurface anomalies that result from variations in density, such as cavities, fault zones, and mineral deposits. According to various studies, gravity methods provide insight into the structure and composition of the Earth's subsurface by measuring minute variations in gravitational forces caused by density contrasts (Kearey et al. 2002). However, the technique's effectiveness can be limited by factors such as the resolution of the instruments and the presence of external noise (Blakely 1996). Despite these limitations, gravity surveying remains a valuable tool for understanding subsurface conditions, especially in regions where other geophysical methods, such as resistivity or seismic methods, may not perform effectively (Burger et al. 2006).

Instruments used for gravity surveys include the *Autograv Gravity Meter*, which provides high-precision measurements with an accuracy of up to 0.001 milliGal (0.001 mGal), enabling the detection of minute gravitational variations in geophysical surveys. The sensitivity of the instrument refers to its ability to detect extremely small changes in the Earth's gravitational field, typically measured in microGals (10^{-6} Gal) or milliGals (10^{-3} Gal). This sensitivity is vital for accurately identifying subsurface features such as density variations, voids, or mineral deposits, which are crucial in engineering and geological surveys. The sensitivity is primarily determined

Table 3.8 Gravity surveying in engineering geophysics

Application	Description	Limitations
Cavity detection	Identifies underground voids or cavities	Limited resolution for small cavities (Blakely 1996)
Fault zone mapping	Detects subsurface faults and fractures	Low sensitivity to deep or narrow features (Kearey et al. 2002)
Mineral deposit mapping	Detects areas of high-density mineralization	Can be affected by overburden (Kearey et al. 2002)
Subsurface engineering	Used to assess disturbed ground, especially in urban areas	Requires high precision in measurement (Burger et al. 2006)

by the core material of the meter, which responds to gravitational fluctuations by producing measurable shifts. High-density, stable materials are often chosen for their ability to detect even the smallest changes, ensuring that the meter provides reliable and precise data, even under challenging field conditions. This precise sensitivity enables engineers and geophysicists to obtain detailed subsurface information, which is essential for making informed decisions in projects like foundation design, site assessments, and resource exploration.

In engineering geophysics, *Differential GPS (DGPS)* (Trimble R10, Leica GS18 T, Topcon HiPer VR, Javad Triumph-2), is often integrated with gravity meters like the Autograv for high-precision location measurements. DGPS enhances the accuracy of gravity surveys by providing precise geographic coordinates for each measurement, often down to the centimeter level. This is essential for creating accurate gravity maps and correlating gravitational data with exact locations, critical in engineering applications like site investigations and foundation design.

Magnetic surveying, on the other hand, involves measuring variations in the Earth's magnetic field to detect subsurface features like mineral deposits, faults, or voids. In engineering geophysics, it complements gravity surveys by revealing different aspects of subsurface conditions. When used together, gravity and magnetic surveys provide a more comprehensive understanding of subsurface structures, which is vital for resource exploration, site assessments, and environmental studies, enabling engineers to make better-informed decisions for construction and infrastructure projects.

Magnetic Surveying and Engineering Applications

The *principle of magnetic surveys* measure variations in the Earth's magnetic field caused by the presence of magnetic materials such as magnetite-rich rocks, minerals, and man-made metal objects. The interaction between the Earth's magnetic field and the magnetization of subsurface materials is governed by the magnetic force equation:

$$F = \frac{1}{\mu}\frac{P_1 P_2}{r^2} \tag{3.30}$$

where F_m = Magnetic force between two monopoles, μ = Magnetic permeability of the medium, P_1, P_2 = Strength of the magnetic poles, and r = Distance between the poles. Magnetic field strength H is defined by:

$$H = \frac{F}{P_2} = \frac{P_1}{\mu r^2} \tag{3.31}$$

The unit of magnetic field strength H can be expressed as $\frac{\text{Ampere} \times \text{meter}}{\text{meter}^2} = \frac{\text{Am}}{\text{m}^2} = \frac{\text{A}}{\text{m}}$, which is also referred to as amperes per meter (A/m). This unit is equivalent to 1 Tesla (T), a unit named after the Croatian inventor Nikola Tesla. When describing the Earth's magnetic field, it is common to use the unit of nanotesla (nT), where 1 nT = 10^{-9} T. The average strength of the Earth's magnetic field is approximately 50,000 nT, with variations ranging from 20,000 to 70,000 nT depending on geographical location. Additionally, a nanotesla is equivalent to the older unit of gamma, where 1 nT = 1 gamma. When a magnetic material is placed in a magnetic field, it becomes magnetized, and its magnetization J_i is given by:

$$J_i = kT \tag{3.32}$$

where k = Magnetic susceptibility of the material and T = Inducing magnetic field strength.

Magnetic surveys are particularly useful for detecting *magnetic anomalies* caused by underground metal objects or magnetic rocks. They are also valuable in *archaeological and engineering applications*, such as locating buried metal structures, pipelines, or utilities. Additionally, magnetic surveys can help identify *subsurface mineral deposits* with high magnetic susceptibility, such as iron ore.

Table 3.9 outlines the key applications of magnetic surveying in engineering geophysics. Magnetic surveys are instrumental in detecting subsurface features such as mineral deposits, fault zones, and archaeological remnants, as well as identifying the presence of magnetic anomalies. Magnetic methods are particularly effective for mapping areas with ferromagnetic materials and understanding structural geological features like faults and ore bodies (Kearey et al. 2002). The magnetic survey can be conducted using handheld magnetometers, airborne surveys, or ship-towed magnetometers for large-scale regional surveys. The most common instruments used today, such as the cesium magnetometer, offer high precision and can measure variations as small as 0.01 gamma (Al-Khafaji 2020).

However, magnetic surveys have certain limitations. The presence of man-made structures and electrical noise can significantly affect the measurements, particularly in urban or industrial areas (Blakely 1996). The method is also limited by the magnetic susceptibility of the materials being surveyed, which varies widely across different rock types (Kearey et al. 2002). Despite these challenges, magnetic surveying remains a valuable technique for engineering applications, particularly for preliminary investigations in archaeological sites, infrastructure planning, and mineral exploration (Burger et al. 2006).

Table 3.9 Magnetic surveying in engineering geophysics

Application	Description	Limitations
Cavity and fault detection	Identifies subsurface cavities and fault zones using magnetic anomalies	Sensitive to variations in magnetic susceptibility (Kearey et al. 2002)
Mineral exploration	Detects ore bodies or mineral deposits through their magnetic properties	Limited by overburden and geological complexity (Blakely 1996)
Archaeological investigations	Used to find buried archaeological sites and structures by detecting magnetic anomalies	Affected by artificial structures and nearby metallic objects (Al-Khafaji 2020)
Infrastructure planning	Helps identify subsurface metallic objects or materials before construction	Requires careful calibration to avoid interference from local magnetic noise (Burger et al. 2006)

Magnetic instruments are indispensable in engineering geophysics for investigating subsurface structures and identifying geological features like faults, cavities, and mineral deposits. Depending on the specific needs of the project, various types of magnetometers—ranging from fluxgate and proton precession to cesium vapor and gradiometers—can be employed to achieve the desired resolution and accuracy. Despite their strengths, each instrument has its limitations, including susceptibility to environmental noise and the challenge of distinguishing magnetic anomalies from background noise, particularly in urban settings. Nonetheless, with advancements in instrument technology, magnetic surveying continues to play a critical role in both small-scale and large-scale geophysical investigations (Blakely 1996; Kearey et al. 2002; Burger et al. 2006).

The *fluxgate magnetometer* operates based on the principle of detecting variations in the magnetic field through a ferromagnetic core material, typically made of soft iron. An alternating current (AC) is passed through coils wound around the core, causing the core to saturate magnetically in a periodic manner. The external magnetic field, such as the Earth's magnetic field, disturbs this saturation process, altering the magnetization and producing a detectable signal. This change in the signal is directly proportional to the strength and direction of the external magnetic field. In engineering geophysics, fluxgate magnetometers are used for precise applications like mineral exploration, fault zone mapping, and archaeological surveys. The instrument's high sensitivity allows it to detect magnetic field variations as small as 1 nT, enabling detailed subsurface mapping and anomaly detection.

The *proton precession magnetometer* operates on the principle of measuring the precession of protons in a hydrogen-rich fluid, such as water or alcohol, when exposed to an external magnetic field. The hydrogen nuclei (protons) align with the magnetic field, and when a pulse of radio frequency is applied, the protons are knocked out of alignment. As they return to their original orientation, they precess, and the frequency of this precession is directly proportional to the magnetic field strength. The core material used in this instrument is typically a fluid containing hydrogen, such as water or alcohol. In engineering geophysics, proton precession magnetometers are used for high-precision surveys, detecting subsurface magnetic anomalies related to geological features, such as mineral deposits and igneous intrusions. They have a high sensitivity, capable of detecting magnetic field variations as small as 0.1 nT, making them ideal for detailed mapping and anomaly detection over large areas.

The *cesium vapor magnetometer* operates on the principle of detecting magnetic field variations by measuring the precession of cesium atoms in a vapor state. In this instrument, cesium atoms are exposed to a magnetic field, which causes them to align in a specific direction. A laser then excites the cesium atoms, and when a radio frequency pulse is applied, the atoms precess, producing a signal that is proportional to the magnetic field strength. This instrument is highly sensitive and can detect variations as small as 0.01 nT. In engineering geophysics, cesium vapor magnetometers are used for precise magnetic surveys, such as detecting geological features, fault zones, and mineral deposits. Their high sensitivity makes them ideal for detailed anomaly detection and large-area surveys, both on the ground and in airborne applications.

A *gradiometer* (magnetic gradient meter) measures the spatial gradient of the magnetic field by using two or more magnetometers placed at a fixed distance apart. It detects small variations in magnetic anomalies while minimizing the effects of large-scale magnetic fields. Gradiometers typically use the same core materials as fluxgate, proton precession, or cesium vapor magnetometers. Their sensitivity is higher, allowing detection of variations as small as 0.1 nT. In engineering geophysics, gradiometers are used for precise mapping of localized magnetic anomalies, such as fault zones, buried structures, and mineral deposits, making them valuable in infrastructure planning, archaeological surveys, and mineral exploration.

A *total field magnetometer* measures the total strength of the magnetic field at a specific location, including both the Earth's magnetic field and any local anomalies caused by subsurface structures or materials. It typically uses a magnetometer with a core material such as a fluxgate or proton precession sensor. The instrument detects the magnetic field strength in units like nanotesla (nT). In engineering geophysics, total field magnetometers are commonly used for regional surveys, mineral exploration, and archaeological investigations, as they provide a broad overview of magnetic anomalies across large areas. The sensitivity of these instruments can typically detect variations as small as 0.1 nT, making them useful for detecting large-scale geological features.

Table 3.10 provides a comparative overview of magnetic and gravitational methods in engineering geophysics, focusing on principles, applications, sensitivity, and typical uses. Magnetic methods measure variations in the Earth's magnetic field, detecting anomalies caused by ferromagnetic materials, fault zones, and other geological structures. These methods are highly sensitive, with applications in mineral exploration, archaeological surveys, and site investigations for infrastructure planning. Gravitational methods, on the other hand, measure variations in gravity due to subsurface density changes, often used for detecting large-scale geological features such as faults, cavities, and mineral deposits. In engineering geophysics, gravity

Table 3.10 Comparative table of magnetic and gravitational methods

Method	Principle	Primary applications	Limitations
Gravity surveying	Measures variations in Earth's gravitational field due to density contrasts	Cavity detection, fault mapping, disturbed ground identification	Low resolution for small or deep features
Magnetic surveying	Measures variations in Earth's magnetic field due to magnetic materials	Buried metal detection, mineral exploration, archaeological surveys	Limited to areas with significant magnetic contrast

methods are valuable for assessing subsurface structures in environments with high electrical or acoustic noise, offering insights into potential ground disturbances and subsurface cavities. While magnetic methods are generally more sensitive, gravity methods are useful for large-area surveys, especially when precise depth or volume data are required.

3.4.4 Acoustic and Ultrasonic Methods in Engineering Geophysics

Acoustic and ultrasonic methods are essential tools in engineering geophysics for evaluating subsurface materials, assessing rock and soil properties, and ensuring the integrity of construction materials. These methods rely on the propagation of acoustic waves, which interact with the material they travel through, allowing engineers to infer various physical characteristics such as elasticity, porosity, density, and homogeneity. Acoustic waves can be broadly classified into compressional (P-waves) and shear (S-waves), with both types providing valuable information about the subsurface media and material behavior.

Fundamentals of Acoustic Wave Propagation in Subsurface Media

The propagation of acoustic waves through subsurface materials is a fundamental concept in geophysics. When acoustic waves encounter different materials, they propagate at different speeds depending on the material's physical properties, such as density and elasticity. The velocity of acoustic waves, which is a critical parameter in seismic surveying, is influenced by the material's stiffness, elasticity, and density. These waves primarily include compressional waves (P-waves) and shear waves (S-waves).

P-waves (compressional waves) are longitudinal waves that cause particles in the medium to move in the same direction as the wave's propagation. They are the fastest type of seismic waves and travel through both solids and liquids. The velocity of P-waves (V_P) in a material can be expressed by the following equation:

$$V_P = \sqrt{\frac{K + \frac{4}{3}G}{\rho}} \tag{3.33}$$

where V_P = P-wave velocity (m/s), K = Bulk modulus (Pa), G = Shear modulus (Pa), and ρ = Density of the material (kg/m^3). P-waves are crucial for assessing the compressibility and elasticity of the material, and they are often used in seismic refraction and reflection surveys to determine the depth and composition of subsurface layers (Telford et al. 1990).

S-waves (shear waves) are transverse waves, where the particle motion is perpendicular to the wave's direction. These waves travel more slowly than P-waves and cannot propagate through fluids. The velocity of S-waves (V_s) is determined by the shear modulus G and density ρ and is given by:

$$V_s = \sqrt{\frac{G}{\rho}} \tag{3.34}$$

where V_s = S-wave velocity (m/s), G = Shear modulus (Pa), and ρ = Density of the material (kg/m^3). S-waves are valuable for determining the shear strength and rigidity of materials, and they are particularly useful in seismic surveys to characterize subsurface structures such as faults and fractures (Haeni 1990).

Ultrasonic Methods: Pulse Velocity Testing

Ultrasonic methods, particularly ultrasonic pulse velocity (UPV) testing, are widely used in engineering geophysics to evaluate the quality of concrete and other construction materials. This technique involves sending high-frequency ultrasonic pulses through materials and measuring the time it takes for the pulse to travel. The pulse velocity is directly related to the material's properties, such as porosity, elasticity, and compressive strength. The relationship between the pulse velocity (V_{up}) and the material's properties is given by the following equation:

$$V_{up} = \frac{L}{T} \tag{3.35}$$

where V_{up} = Pulse velocity (m/s), L = Distance travelled by the pulse (m), and T = Time taken for the pulse to travel the distance (s). In concrete, a higher pulse velocity typically indicates a denser and less porous material, which is associated with higher strength and durability (Raghavan 2003). Ultrasonic pulse velocity testing is extensively used to assess the quality of concrete and detect internal voids, cracks, or other defects in both fresh and hardened concrete.

Mathematical Models and Principles

The understanding of wave velocity in subsurface media allows engineers and geophysicists to model the properties of the materials more effectively. Several mathematical models are used to estimate material properties from the wave velocity, such as the Rayleigh-wave model for surface waves. This model is particularly useful for evaluating shallow subsurface layers and their elastic properties.

Rayleigh Wave Velocity Model

The velocity of Rayleigh waves (V_R) is an important parameter in surface wave analysis, and it is related to the shear modulus G and density ρ of the material. The Rayleigh wave velocity is given by:

$$V_R = 0.87 \times \sqrt{\frac{G}{\rho}} \tag{3.36}$$

where V_R = Rayleigh wave velocity (m/s), G = Shear modulus (Pa), and ρ = Density of the material (kg/m^3). This model is used in seismic surveys to estimate the stiffness and elastic properties of shallow subsurface materials (Telford et al. 1990).

Applications in Engineering Geophysics

Acoustic and ultrasonic methods have diverse applications in engineering geophysics, particularly in the evaluation of subsurface and construction materials for civil engineering projects. Acoustic waves (seismic waves) are used to map soil and rock layers (*subsurface and profiling mapping*), providing essential data for foundation design in construction projects (Telford et al. 1990). Ultrasonic pulse velocity testing helps assess the integrity of materials (*material integrity assessment*) like concrete, detecting internal defects (such as cracks) that could impact structural strength and safety (Haeni 1990). Acoustic methods can identify anomalies such as cavities or voids in the subsurface (*void detection and soil quality*), which is crucial for assessing ground stability of buildings, roads, and bridges and preventing potential hazards (Raghavan 2003). Ultrasonic methods are used to evaluate the quality of materials in construction (*quality control in construction*), ensuring they meet safety standards and are free from defects before use in infrastructure.

Table 3.11 compares how acoustic and ultrasonic waves propagate through various subsurface materials, such as soils, rocks, and concrete. It includes key velocities like P-wave, S-wave, and ultrasonic pulse velocities, which are essential for evaluating material properties and integrity in engineering geophysics. These velocities are influenced by factors like density, elasticity, and porosity, helping in subsurface profiling and construction applications.

Table 3.11 Wave propagation characteristics for different materials (Telford et al. 1990; Stokoe and Santamarina 2000; Dobrin and Savit 1988)

Material	P-wave velocity (m/s)	S-wave velocity (m/s)	Ultrasonic pulse velocity (m/s)	Remarks
Soil	200–1500	50–800	1500–3000	Varies with soil type and moisture content
Sandstone	2000–5500	1000–3000	2500–5000	Highly dependent on porosity and grain size
Limestone	2000–6000	1000–3500	3000–6000	Dependent on composition and porosity
Granite	3000–6000	1500–3500	4000–6500	Highly dense material with higher wave velocities
Concrete	2500–4500	1000–2500	3000–5500	Dependent on mix and curing
Water	1500–1600	N/A	N/A	Affects wave velocity, used in underwater surveys

Table 3.12 outlines the principles of seismic refraction and reflection methods, showcasing key wave types (P-wave and S-wave), their velocities, and common applications in subsurface mapping. It helps to differentiate between these seismic techniques based on how they measure the velocity of waves through various layers of the Earth and their use in geological surveys and engineering geophysics.

Table 3.13 presents various construction materials (such as concrete and asphalt) along with their expected ultrasonic velocities, as well as the ability of ultrasonic pulse velocity testing to detect defects in these materials.

3.5 Conclusions

The chapter "Fundamental Concepts in Engineering Geophysics" highlights the vital role that geophysical methods play in modern engineering, particularly in subsurface exploration, site characterization, and risk assessment. It provides a clear understanding of the core principles and various geophysical techniques, such as seismic, electrical, electromagnetic, magnetic, and acoustic methods, which are essential for evaluating subsurface conditions and informing engineering decisions. By exploring both the applications and limitations of these methods, the chapter underscores

Table 3.12 Seismic refraction and reflection methods (Sheriff and Geldart 1995; Kearey et al. 2002; Woods and Kurrle 2007)

Method	Wave type	Velocity (m/s)	Application
Seismic refraction	P-wave, S-wave	Varies by material	Used for determining subsurface layers and depth to bedrock
Seismic reflection	P-wave, S-wave	Varies by material	Used to map deeper geological structures and faults
Primary waves (P-wave)	Compressional	Fastest velocity	Used for subsurface profiling and layer identification
Secondary waves (S-wave)	Shear waves	Slower than P-waves	Used for mapping material properties and fault locations

Table 3.13 Ultrasonic pulse velocity testing applications (RILEM (International Union of Laboratories and Experts in Construction Materials and Structures) 1984; Krautblatter and Haider 2010; Malhotra and Carino 2004)

Material	Ultrasonic velocity (m/s)	Defect detection capability
Concrete	4000–5000	Detects cracks, voids, and inconsistencies in density
Asphalt	1500–2000	Identifies voids and material inconsistencies
Brick	2500–3500	Detects cracks, fractures, and voids
Steel	5000–6000	Detects surface and internal defects like cracks

their significance in optimizing engineering designs, ensuring structural integrity, and minimizing risks. Ultimately, this chapter serves as a foundational resource for professionals in engineering geophysics, offering valuable insights for both theoretical knowledge and practical application.

References

Advanced Geosciences Incorporated (AGI) (2018) 1D Geophysical resistivity survey: vertical electrical sounding. Accessed 20 Mar 2018. https://www.agiusa.com/1d-resistivity-survey-vertical-electrical-sounding

Akingboye AS, Ogunyele AC (2019) Insight into seismic refraction and electrical resistivity tomography techniques in subsurface investigations. Min Geol Petrol Eng Bull UDC 550:3. https://doi.org/10.17794/rgn.2019.1.9

Akintorinwa OJ, Adesoji JI (2009) Application of geophysical and geotechnical investigations in engineering site evaluation. Int J Phys Sci 4(8):443–454

Al-Khafaji A (2020) Advanced geophysical methods for engineering applications. Elsevier

Anbazhagan P, Divyesh R, Athul P (2016) Investigations existing dam using integrated geotechnical and geophysical methods. In: Second national dam safety conference, Bengaluru, 12–13 Jan 2016, pp 165–170

Attewell PB, Farmer IW (1976) Principles of engineering geology. Springer, 1045 pp. https://doi.org/10.1007/978-94-009-5707-7

Attewell PB, Farmer IW (2021) Geophysics in civil engineering: site characterization and risk assessment. Elsevier

Bierwirth P, White W (1994) Geophysical self-potential measurements in environmental investigations. Environ Geol 24(4):281–289

Birks HJB et al (2010) Application of induced polarization for environmental site assessment. J Environ Eng

Blakely RJ (1996) Potential theory in gravity and magnetic applications. Cambridge University Press

Blanco J, Ruiz M (2007) Application of the self-potential method to the detection of abandoned mine drainage. Geophysics 72(4):B199–B206

Burdick LJ, Lyons S (2017) Advancements in seismic reflection techniques for geotechnical investigations. Geophys J Int 208(3):1101–1115

Burger HR, Sheehan AF, Jones CH (2006) Introduction to applied geophysics: exploring the shallow subsurface. W.W. Norton & Company

Conyers LB, Goodman D (1997) Ground-penetrating radar in archaeology. Altamira Press

Dobbins R, Voigt S (2019) Exploring the potential of seismic reflection for subsurface fault detection in urban engineering. Eng Geol 249:62–74

Dobrin MB, Savit CH (1988) Introduction to geophysical prospecting, 4th edn. McGraw-Hill

Doucet S, Boulanger RW (2019) Seismic site characterization for earthquake engineering. Soil Dyn Earthq Eng 121:104–117

Ewusi A (2006) Groundwater exploration and management using geophysics: northern region of Ghana. International course of study: environmental and resource management. Ph.D. thesis, Faculty of Environmental Sciences and Process Engineering, Brandenburg Technical University of Cottbus, p 186

Gavazzi E et al (2011) Application of GPR for landfill characterization and site assessment. J Environ Eng Geophys

Geological Survey of India (GSI) (1992) Field season (1989–1990 and 1990–1991): a report on the deep resistivity survey over the Singhbhum craton. Geological Survey of India. Unpublished report

Geophysical Society of America (2019) Geophysics in civil engineering projects. Geophysical Society Press

Gibson PJ, George MD (2003) Environmental applications of geophysical surveying techniques. Nova Science Publishers, Hauppauge, New York, USA

Grant F, West G (1965) Geophysics for the mineral exploration geoscientist. Cambridge University Press

Haeni FP (1990) Geophysical methods in engineering applications. In: SEG geophysical monograph series

Hanna A, Taylor H (2021) Seismic refraction and reflection methods in fault zone mapping: a case study. Geophys Prospect 69(4):1020–1036

Huang Y, Wang L (2020) Seismic site characterization for sustainable construction in earthquake-prone regions. J Earthq Eng 24(8):1550–1565

Huisman JA et al (2003) Utility detection using ground-penetrating radar in urban areas. Geophys Prospect

Jardin E, Selensky J (2002) Using the self-potential method for geochemical prospecting. Geophys Prospect 50(2):155–165

Jiang Z, Li Z, Wang Z (2020) Integrated seismic and geophysical surveys for fault detection in urban areas. J Appl Geophys 178:68–79

Kearey P, Brooks M, Hill I (2002) An introduction to geophysical exploration. Blackwell Science

Krautblatter M, Haider T (2010) Ultrasonic pulse velocity testing for quality control in concrete. Constr Build Mater 24(9):1749–1755
Leprince S, Caubet J (2001) Application of self-potential methods to the study of geological hazards. J Appl Geophys 46(2):103–113
Loke MH (2002) Electrical imaging surveys for environmental and engineering studies: a practical guide to 2-D and 3-D surveys, p 127
Lowrie W (2007) Fundamentals of Geophysics, 2nd edn. Cambridge University Press, New York
Lwisa EG (2021) Introductory chapter: engineering geology. IntechOpen, p 13. https://doi.org/10.5772/intechopen.95991
Malehmir A et al (2012) Seismic methods for environmental and geotechnical investigations. Springer
Malhotra VM, Carino NJ (2004) Handbook on non-destructive testing of concrete, 2nd edn. CRC Press
Miller RD, Stewart RR (2018) Seismic refraction for near-surface geophysics. Springer
Milsom J (2003) Field geophysics, 3rd edn. Wiley-Blackwell
Milsom J, Eriksen A (2011) Field geophysics. Wiley-Blackwell
Naskar DC (2005) Geophysical studies of crust and upper mantle structures in parts of Madhya Pradesh and Maharashtra, Central India. Ph.D. thesis, Jadavpur University, Kolkata, West Bengal, India
Naskar DC, Das LK (2009) Resistivity survey across the Narmada-Son lineament and along the Sanwer-Borgaon traverse, Madhya Pradesh, central India. Boll Geofis Teor Appl 50(4):305–318
Naskar DC, Singh OP (2002) Field season (1998–1999): a report on the integrated geophysical survey along Borgaon-Indore-Ujjain transect. Geological Survey of India (GSI). Unpublished report
Naskar DC, Singh OP, Das LK, Syiem SM (2011) Integrated geophysical survey for delineation of active fault in the foothill of Bhutan Himalaya, Northeast India. Indian J Geosci 65(3):211–222
Naskar DC, Regar RL, Gupta D, Das PC, Chopra M (2022) Integrated geophysical survey for delineating aquifer zones in Sagar area, South 24 Parganas, West Bengal. In: Moitra AK, Bhattacharya J, Kayal JR, Mukerji B, Das AK (eds) Innovative exploration methods for minerals, oil, gas, and groundwater for sustainable development. Elsevier, pp 225–240
Naskar DC, Yadav PK, Ray J, Maurya D (2024) Insight into structural fabric and crustal architecture across Eastern Dharwar Craton, Southern India from combined analysis of gravity and magnetic data. J Indian Geophys Union 28(5):350–370
Nettleton LL (1976) Gravity and magnetics in oil prospecting. McGraw-Hill
Orozco J, Vidal R (2019) Seismic and geophysical techniques for fault detection and characterization in seismic-prone regions. Geophys Res Lett 46(9):4898–4905
Pelton M (2005) Seismic data interpretation. Elsevier
Porsani MJ et al (2004) GPR in soil stratigraphy and groundwater detection. Geophysics
Rabbel W (2006) Seismic tomography and exploration. Springer
Raghavan SR (2003) Acoustic and ultrasonic wave propagation in soils and rocks. Springer
Ramos J, Comas R (2015) Self-potential method applied to the detection of underground seepage in dams. J Appl Geophys 119:81–88
RILEM (International Union of Laboratories and Experts in Construction Materials and Structures) (1984). Recommendation for testing of concrete—ultrasonic pulse velocity measurement
Robinson ES, Coruh C (1988) Basic exploration geophysics. Wiley, New York
Routh PS et al (2017) Geophysical methods in mineral exploration and mine development. Explor Geophys
Sasaki Y (2000) Three-dimensional inversion of resistivity and self-potential data. Geophys J Int 142(1):347–354
Schoenberg M (2015) Geophysical methods in environmental studies. CRC Press
Sharma PV (1997) Environmental and engineering geophysics, chap 1. In: Introductory observations. Cambridge University Press, pp 1–10, https://doi.org/10.1017/CB09781139171168.002. Published online by Cambridge University Press, 05 June 2012

Sharma P (2005) Ground-penetrating radar for infrastructure inspections. NDT & E International
Sheriff RE, Geldart LP (1995) Exploration seismology, 3rd edn. Cambridge University Press
Shragge J (2016) Introduction to seismology. Wiley-Blackwell
Society of Exploration Geophysicist of Japan (SEGJ) (2004) Application of geophysical methods to engineering and environmental problems. Society of Exploration Geophysicist, Tokyo, Japan
Steeples D, Miller RD (1996) Seismic refraction methods for engineering applications. Springer
Stokoe KH, Santamarina JC (2000) Soil and rock characterization by geophysical methods. In: Geotechnical and environmental geophysics, vol 1. Society of Exploration Geophysicists
Sullivan JP, Johnson MA (2018) Introduction to engineering geophysics: methods and applications. Springer
Sun J, Niu Z, Innanen KA, Li J, Trad DO (2020) A theory guided deep learning formulation and optimization seismic waveform inversion. Geophysics 85(2):R87–R99. https://doi.org/10.1190/geo2019-0138.1
Telford WM, Geldart LP, Sheriff RE (1990) Applied geophysics. Cambridge University Press
Venkiteswaran CS, Naskar DC (2000) Field season (1999–2000): report on the geophysical investigation seismic surveys in Nuagaon-Telesahi block, Talchir Gondwana basin, District: Angul, Orissa. Geological Survey of India (GSI). Unpublished report
Ward SH (2011) Principles of geophysical methods in exploration and engineering. Society of Exploration Geophysicists
Whiteley RJ (1990) Engineering geophysics—a geophysicist's view. Explor Geophys 21(1–2):7–16
Woods R, Kurrle (2007) Seismic methods for site characterization and engineering applications. J Appl Geophys 63(4):302–312
Zhang S, Li H (2020) Use of seismic methods in site characterization and soil stability analysis for construction. Geotech Test J 43(1):150–165
Zhou B, Li X, Zhang L (2008) Self-potential method for groundwater flow study in a fractured rock aquifer. J Hydrol 353(1–2):85–94

Part II
Technological Advances and Core Applications

Chapter 4
Recent Advances in Geophysical Surveying Techniques

Abstract This chapter explores the latest developments in geophysical surveying, focusing on how modern technologies are transforming data collection and analysis. It introduces new tools such as drones (UAVs), remote sensing, and AI-based systems, which have greatly improved the speed, safety, and accuracy of surveys. Key advancements in instrumentation and drone-based methods have enhanced subsurface mapping, site monitoring, and terrain analysis. The chapter highlights how drones are used in various applications, including LiDAR, GPR, magnetometry, and thermal imaging. Real-world case studies from Italy and Israel demonstrate these tools in action, such as mapping soil-rock boundaries or conducting coastal investigations. Integration with machine learning and data analytics is also discussed, showing how smart technologies help manage large geophysical datasets. The benefits—like reduced costs, better environmental safety, and improved efficiency—are balanced with a discussion on current limitations and challenges. This chapter presents a forward-looking view of surveying in modern engineering geophysics.

4.1 Introduction to Recent Advances in Geophysical Surveying

This chapter explores recent advancements in geophysical surveying techniques, highlighting emerging tools and methodologies that enhance data collection and analysis. It focuses on the integration of modern technologies such as Unmanned Aerial Vehicles (UAVs), drones, and remote sensing in geophysical surveys, emphasizing their key advantages, applications, and contributions to various engineering and environmental fields. The chapter discusses the latest advancements in geophysical instrumentation for improved accuracy and efficiency, along with the integration of Artificial Intelligence (AI) and Machine Learning (ML) for enhanced data processing and interpretation. It also delves into specific types of drones and their role in geophysical investigations, covering applications in site investigation, monitoring, and cost reduction. Furthermore, the chapter examines case studies of drone-based geophysical techniques, including LiDAR, magnetometry, thermal imaging, and

D. C. Naskar, *Advances in Engineering Geophysics: Concepts and Applications*, Earth Systems Data and Models 9, https://doi.org/10.1007/978-3-032-28000-8_4

electromagnetic surveys, demonstrating their effectiveness in various engineering and environmental applications. The challenges and limitations associated with these technologies are also addressed, providing a comprehensive overview of the evolving landscape in geophysical surveying.

4.2 Emerging Tools for Data Collection and Analysis

The field of geophysical surveying has been significantly impacted by the rapid advancement of technology. These developments have led to a transformation in how geophysical surveys are conducted, bringing about significant improvements in data collection, processing, and interpretation. The growing complexity of subsurface problems, combined with increasing demands for efficiency and accuracy, has driven the development of innovative tools that address these challenges more effectively than traditional methods. In this section, we explore several cutting-edge tools and methodologies reshaping geophysical surveying, focusing on new technologies such as Unmanned Aerial Vehicles (UAVs), drones, remote sensing, and innovations in instrumentation. Additionally, we explore the integration of Artificial Intelligence (AI) and Machine Learning (ML) in geophysical techniques, offering more automated, data-driven approaches to subsurface exploration.

4.2.1 New Technologies in Geophysical Surveying (e.g., UAVs, Drones, Remote Sensing)

The application of UAVs and drones in geophysical surveying has emerged as one of the most transformative advances in recent years. These technologies allow for high-resolution data collection in areas that are difficult to access, such as remote, hazardous, or ecologically sensitive environments. UAVs equipped with geophysical sensors—such as magnetometers, electromagnetic sensors, and thermal imaging equipment—can survey large areas quickly and with minimal disturbance to the environment.

Key Advantages of UAVs/Drones in Geophysical Surveys

UAVs offer several key advantages in geophysical surveys. One major benefit is their ability to access hard-to-reach areas. UAVs can easily navigate difficult terrains such as mountainous regions, dense forests, or even flooded zones. This capability opens up previously inaccessible locations for geophysical surveys, particularly in environmental and geological monitoring.

Another advantage is their cost-effectiveness. The use of UAVs significantly reduces the need for expensive ground crews, helicopters, or larger equipment traditionally employed in geophysical investigations. Their lower deployment costs make them a practical option for both small- and large-scale projects.

UAVs also enhance safety and operational efficiency. They reduce the risks associated with hazardous fieldwork by enabling surveys to be conducted in potentially dangerous environments such as volcanoes, earthquake-prone areas, and landslide zones without putting personnel at risk.

Finally, UAVs support real-time data transmission. When equipped with advanced communication technology, they can transmit collected data to researchers immediately, enabling rapid analysis and decision-making. This is especially valuable in dynamic environments like environmental monitoring, where conditions can change quickly.

Applications

UAVs have become valuable tools in various geophysical survey applications. In magnetic and gravitational surveys, UAVs equipped with magnetometers are used to detect magnetic anomalies, which can indicate the presence of mineral deposits, fault lines, or other subsurface geological features of interest. Similarly, drones fitted with gravimeters can identify density anomalies in the subsurface, providing insights into geological structures.

In electromagnetic surveys, drones carrying electromagnetic sensors are capable of performing ground conductivity mapping. This is especially useful in detecting groundwater resources, locating archaeological sites, and identifying environmental contamination zones.

Recent Studies

Recent studies (Smith et al. 2022) have shown that Unmanned Aerial Vehicles (UAVs) can achieve resolution in data collection that is either comparable to or even surpasses traditional methods when paired with high-resolution sensors. This advancement in sensor technology allows UAVs to capture more precise data from aerial perspectives, improving the accuracy of various measurements. As a result, UAVs are now being increasingly utilized in a broad range of applications, including geological surveys and environmental monitoring, where high precision is critical.

The integration of UAVs with remote sensing techniques has notably expanded their versatility in different fields, particularly in geotechnical investigations. For example, drones equipped with advanced sensors can be used to map topography, identify geological hazards, and monitor terrain changes over time. This capability is especially beneficial in areas that are difficult to access or dangerous for human surveyors, such as active volcanoes or remote mountainous regions. Zhao et al. (2023) emphasize that the combination of UAVs and remote sensing allows for efficient and safe data collection over large areas, making them indispensable tools in modern geotechnical and geological research.

4.2.2 Advances in Instrumentation for Higher Accuracy and Efficiency

The performance and efficiency of geophysical surveys have been significantly improved by recent advancements in instrumentation (McMillan et al. 2023). These innovations have led to a reduction in survey time, increased accuracy, and the ability to gather more detailed subsurface information.

Key Instrumentation Advances

Recent advances in geophysical instrumentation have significantly improved the precision, portability, and efficiency of subsurface investigations. Portable magnetometers and gradiometers, such as new-generation fluxgate magnetometers and proton precession magnetometers, are now far more sensitive and portable than earlier models. These improvements allow for the detection of weaker magnetic anomalies in complex terrains, such as those caused by small faults, mineral deposits, or buried infrastructure.

Multi-electrode resistivity systems have also evolved, with enhanced configurations like time-domain induced polarization (IP) systems enabling high-resolution subsurface imaging. These systems facilitate deeper and more efficient investigations into soil and rock properties, groundwater contamination, and underground storage structures.

In the field of seismic surveys, advanced seismometers featuring improved frequency response, dynamic range, and noise filtering capabilities have greatly increased the accuracy of seismic imaging. These enhancements make it possible to resolve smaller-scale features, including micro faults and subtle subsurface variations that were previously undetectable with older technology.

Ground-penetrating radar (GPR) systems have likewise seen substantial upgrades. Modern GPR units are now equipped with high-frequency antennas and sophisticated signal processing algorithms, which enable more precise imaging of the subsurface at both shallow and moderate depths. These systems are increasingly used for locating and assessing buried structures such as pipelines, cables, and archaeological remains.

Applications

Advanced geophysical instruments are playing an increasingly vital role in various engineering and environmental applications. In infrastructure monitoring, high-resolution geophysical tools are used to assess the integrity of critical structures such as bridges, dams, and underground utilities. These instruments provide early warning signs of structural issues, enabling preventive maintenance and reducing the risk of sudden failures.

In environmental studies, improvements in resistivity and induced polarization (IP) surveys are supporting efforts to assess soil and groundwater contamination. These methods are particularly valuable for detecting pollutants resulting from industrial operations, agricultural runoff, or mining activities.

These technological advancements have significantly improved the ability to gather high-resolution data, even in challenging environmental or geological conditions. Additionally, the development of new sensor technologies has enabled the miniaturization of geophysical instruments. This miniaturization facilitates easier transportation, seamless integration with mobile platforms like UAVs and autonomous vehicles, and greater flexibility in field deployment.

Recent Studies

In 2023, Li et al. introduced a new multi-electrode resistivity system that revolutionized geophysical surveys by significantly reducing the time required for data collection without compromising accuracy (Li et al. 2023). This system, designed to improve efficiency in subsurface investigations, is particularly useful in large urban development projects where rapid site characterization is crucial. Traditionally, resistivity surveys involve the placement of electrodes in the ground to measure the electrical resistivity of the subsurface, a method that can be time-consuming, especially in expansive urban environments. However, Li et al.'s multi-electrode resistivity system integrates advanced technology to automate and streamline the electrode placement and data collection processes, which drastically reduces survey time.

The system's design allows for the simultaneous measurement of multiple locations within the survey area, enhancing spatial coverage and data collection speed. Despite the increased speed, the system maintains high data accuracy by using advanced calibration methods and ensuring precise measurements across multiple electrodes. This capability allows geophysicists to gather comprehensive subsurface data quickly, which is essential for assessing the suitability of sites for construction, infrastructure projects, or urban planning.

In large urban development projects, where time is often a critical factor due to project deadlines and the need to assess various sites efficiently, this new resistivity system provides a powerful tool. Its ability to maintain high data integrity while reducing survey time makes it ideal for rapid site characterization, helping engineers and developers make informed decisions faster and more cost-effectively.

4.2.3 *Integration of Geophysical Techniques with Other Technologies like AI and Machine Learning*

The integration of artificial intelligence (AI) and machine learning (ML) into geophysical surveying has opened new frontiers in data processing, interpretation, and predictive modeling. Machine learning algorithms can be applied to large datasets obtained from geophysical surveys, allowing for the identification of patterns and anomalies that might be missed using traditional methods (Lee et al. 2024).

AI and ML in Geophysical Data Processing

The integration of artificial intelligence (AI) and machine learning (ML) into geophysical data processing has brought significant advancements in efficiency, accuracy, and decision-making. One major development is automated data interpretation, where machine learning algorithms—such as deep learning models—are used to classify and interpret complex data sets from seismic, resistivity, and ground-penetrating radar (GPR) surveys. This automation significantly reduces human error and processing time while generating accurate subsurface models (Lee et al. 2024).

Another important application is predictive modeling, where AI-driven tools are employed to estimate subsurface properties based on geophysical measurements. These models can help predict soil composition, groundwater flow, and potential contamination zones, supporting better environmental and engineering assessments (Patel et al. 2023).

AI and ML also play a critical role in data integration, enabling the combination of information from various geophysical methods—including seismic, magnetic, and resistivity surveys—to create comprehensive 3D models of the subsurface. These integrated models offer a more holistic view of subsurface conditions, enhancing the quality of engineering decisions.

Additionally, real-time data analysis is now possible through AI algorithms that can process incoming field data on-site, providing immediate feedback to field crews and allowing for dynamic adjustments to survey strategies.

The combination of AI with geophysical techniques has shown promising results across multiple domains, including mineral exploration, infrastructure monitoring, and environmental hazard assessment. These tools have not only improved the precision of geophysical models but also expanded their predictive capabilities, empowering engineers to make more informed decisions during project planning and design.

Table 4.1 provides a summary of the key emerging tools and technologies in the field of geophysical surveying, including UAVs, drones, remote sensing, advanced geophysical instruments, and the integration of AI and ML. It highlights their applications, advantages, and impact on improving data collection, processing, and interpretation in subsurface exploration. The table serves as a quick reference for understanding how these tools are reshaping modern geophysical practices.

Table 4.1 Recent advancements in geophysical surveying techniques

Technology	Application	Key advancement	References
UAVs/drones	Remote sensing	High-resolution aerial geophysical surveys, real-time data transmission	Smith et al. (2022)
Portable magnetometers	Magnetic surveys	Increased sensitivity and portability for detecting weak anomalies	McMillan et al. (2023)
Enhanced resistivity meters	Resistivity surveys	Multi-electrode systems for real-time ground profiling	Li et al. (2023)
Seismometers	Seismic surveys	Improved frequency response and noise filtering	Zhao et al. (2023)
AI/ML integration	Data interpretation	Automated analysis and predictive modeling for geophysical surveys	Lee et al. (2024)

Table 4.2 shows that an AI-Driven Geophysical Data Processing Workflow typically involves several stages where artificial intelligence is applied to streamline and enhance geophysical data analysis. Table 4.2 is a breakdown of the general workflow for AI-driven geophysical data processing.

Throughout this workflow, AI plays a crucial role in automating complex processes, improving the efficiency of data processing, enhancing the accuracy of models, and providing actionable insights. The use of AI reduces human error, allows for faster data processing, and facilitates more reliable predictions and decision-making, making it an invaluable tool in geophysical studies and applications.

Applications

AI technologies are increasingly being applied in various geophysical and engineering contexts to enhance efficiency and accuracy. In mineral exploration, AI is used to analyze geophysical data in order to predict the location of mineral deposits. This approach improves the efficiency and success rate of exploration efforts by identifying high-potential zones more effectively than traditional methods.

In geotechnical engineering, AI models assist in predicting soil behavior and supporting foundation design. By analyzing seismic and resistivity data, these models reduce the reliance on extensive and often expensive field tests. This leads to more cost-effective project planning while maintaining engineering reliability and safety.

Table 4.2 AI-driven geophysical data processing workflow

Stage	Description
1. Data acquisition	The collection of raw geophysical data through methods such as seismic surveys, gravity, magnetic, or electrical resistivity. This data forms the foundation for subsequent analysis
2. Data pre-processing	Noise removal: AI algorithms filter out noise from the data Normalization: Standardizing data for consistency Data imputation: Filling missing data using AI techniques Outlier detection: Identifying and removing data points that deviate significantly from expected values
3. Feature extraction	Geophysical feature identification: AI identifies key geophysical features, such as subsurface structures Dimensionality reduction: Techniques like PCA or autoencoders reduce data complexity while preserving essential information
4. Data interpretation and inversion	Modeling subsurface: AI models generate subsurface property models based on geophysical data Inversion: AI inverts data to model earth's interior properties like rock types, mineral content, etc. Automated interpretation: AI identifies and interprets geophysical anomalies, reducing manual analysis
5. Geophysical data fusion	Data integration: AI combines data from multiple geophysical sources (e.g., seismic, gravity, and magnetic) into a unified model for more accurate interpretations Cross-validation: AI ensures consistency and reliability between integrated data sources
6. Machine learning for predictive modeling	Training predictive models: AI models are trained on historical geophysical data to predict future conditions or resources Uncertainty quantification: AI assesses the uncertainty in predictions and generates confidence levels
7. Visualization and decision support	AI-powered visualization: 3D or 4D models are generated to visually represent the processed data Automated reporting: AI creates reports summarizing findings and recommendations to support decision-making

(continued)

Table 4.2 (continued)

Stage	Description
8. Feedback and model optimization	Model improvement: Continuous feedback helps refine AI models to improve accuracy and reduce errors Real-time feedback: AI adapts to incoming data for dynamic analysis and decision-making
9. Final output and application	Resource exploration: AI helps identify potential resources such as oil, gas, or minerals Environmental monitoring: AI enables monitoring of environmental conditions like groundwater or seismic activity Engineering and construction: AI-driven analysis guides decisions in site selection and infrastructure design based on subsurface conditions

Recent Studies

Patel et al. (2023) demonstrated the successful application of deep learning models to interpret seismic data, marking a significant breakthrough in geophysical data processing. Their approach leverages advanced neural networks to analyze seismic data, achieving substantial improvements in both processing speed and interpretation accuracy.

Seismic data interpretation traditionally involves time-consuming manual analysis and complex algorithms that may require extensive computational resources. However, by integrating deep learning techniques, Patel et al. were able to automate much of this process, significantly speeding up the analysis. The deep learning models can quickly identify patterns and anomalies in seismic data, improving the efficiency of data interpretation.

Moreover, these models improve the accuracy of seismic data interpretation by learning from vast amounts of data, recognizing subtle features that may have been overlooked using traditional methods. This increased precision is particularly important in critical phases of engineering design, where accurate subsurface information is essential for safe and efficient construction planning.

By reducing the time needed for seismic data interpretation and enhancing the accuracy of the results, the integration of deep learning allows for faster and more informed decision-making. This is especially valuable in industries such as civil engineering, oil and gas exploration, and environmental monitoring, where timely and reliable seismic data can significantly impact project outcomes.

4.2.4 Types of Drones

There are several types of drones, each designed for specific purposes and applications across various industries. *Multi-rotor drones* are known for their stability and ability to hover, commonly used for aerial photography, surveying, and monitoring. *Fixed-wing drones*, resembling small airplanes, are designed for long-duration flights and are often used in mapping and environmental monitoring. *Hybrid drones* combine the features of multi-rotor and fixed-wing systems, allowing for vertical take-off and landing while covering long distances, making them ideal for large-scale surveying and agricultural monitoring. *Single-rotor drones*, similar to helicopters, are more efficient for long-range operations but harder to control, and they are used in industrial inspections and aerial filming. *Nano drones* are small, lightweight UAVs intended for indoor use or close-range operations, suitable for surveillance and recreational flying. *Underwater drones (ROVs)* are built for underwater exploration, capable of withstanding high-pressure environments while capturing images and videos. *Fixed-wing vertical take-off and landing (VTOL) drones* combine the endurance of fixed-wing systems with the versatility of vertical take-off and landing, commonly used in precision agriculture and infrastructure inspection. Finally, *cargo drones* are designed for transporting goods, with high payload capacities, and are increasingly used in logistics and emergency services. Each type of drone excels in specific tasks based on flight duration, payload capacity, and environmental requirements.

Each of these drone types is designed to excel in particular tasks, depending on factors like flight duration, payload capacity, and the environment in which they operate. Figure 4.1 illustrates the different types of drones, highlighting their unique features and applications. It categorizes drones into various types, such as multi-rotor, fixed-wing, hybrid, and nano drones, each suited for specific tasks like surveying, aerial photography, and logistics.

Fig. 4.1 Types of drones. *Source* Wikipedia

4.2.5 *Overview of Drone Technology in Engineering Geophysics*

Drone technology has made significant advancements in the field of engineering geophysics, where it is now widely used for a variety of applications, improving the efficiency, accuracy, and safety of geophysical surveys and investigations. Drones, also known as unmanned aerial vehicles (UAVs), are increasingly being deployed in a range of geophysical tasks due to their ability to access difficult-to-reach areas, collect high-resolution data, and reduce the need for expensive or hazardous fieldwork.

Data Collection in Geophysical Surveys

Drones equipped with various sensors are revolutionizing data collection for geophysical investigations, allowing for high-quality, efficient surveys over vast or challenging terrains. One key sensor used is *LiDAR (Light Detection and Ranging)*, which is mounted on drones to generate precise topographical maps. LiDAR is especially valuable for creating digital elevation models (DEMs) in areas that are difficult to access on the ground (Shan and Hyyppa 2018; Baltsavias 1999). Another important tool is *multispectral and hyperspectral imaging*, where drones carry advanced cameras and spectral sensors to capture data across various wavelengths of light. This enables the analysis of vegetation, soil properties, and detection of mineral deposits. Drones are also equipped with *Ground Penetrating Radar (GPR)* systems, enabling non-invasive subsurface investigations to detect buried structures, underground utilities, and geological features. *Magnetometers* on UAVs perform airborne magnetic surveys, detecting variations in the Earth's magnetic field, which can reveal the presence of mineral deposits, fault lines, or other geophysical anomalies. Additionally, *thermal imaging* sensors are used to detect temperature variations that can indicate geothermal activity, water table levels, or the presence of underground infrastructure. Finally, *electromagnetic (EM) surveys* are a crucial method in engineering geophysics. Drones equipped with EM systems map subsurface electrical properties and detect conductive materials, such as metals, minerals, and groundwater. The principle behind EM methods involves the interaction of electromagnetic fields with conductive materials, and drone-based EM surveys enhance this by leveraging UAV mobility to collect data over large areas with high spatial resolution.

Remote Sensing

Drones play a pivotal role in remote sensing, allowing for the collection of data from locations that are hard to reach or dangerous to access. For example, drones are widely used in mountainous regions, forests, or areas with dense vegetation where ground-based surveying is impractical or expensive. UAVs can quickly map large areas and gather data that would otherwise require extensive ground crews or heavy equipment.

Site Investigation and Monitoring

Engineering geophysics often requires the monitoring of changes in the physical environment over time, and drones are well-suited for conducting long-term investigations and assessments. For *subsidence monitoring*, drones equipped with photogrammetry and LiDAR systems can detect and track ground deformation, which is especially useful in mining, tunneling, and construction projects. Additionally, drones are valuable for *seismic activity monitoring* by deploying seismic sensors or equipment to monitor ground movements and assess the impacts of construction on surrounding geology. Finally, drones play an essential role in *pipeline and infrastructure inspection*, where they can examine pipelines, dams, and other critical infrastructure for signs of wear, damage, or degradation, such as cracks, corrosion, or subsidence. This capability enhances preventive maintenance programs by enabling early detection and timely intervention.

Efficiency and Cost Reduction

Drone technology greatly reduces the costs associated with geophysical surveys. Traditional geophysical surveying methods, such as seismic surveys or borehole drilling, are expensive and labor-intensive. Drones offer a more cost-effective alternative for gathering preliminary data and performing routine surveys. They also help reduce time in field operations, allowing engineers and geophysicists to collect more data in less time and with fewer personnel.

Enhanced Safety

Drones improve safety during fieldwork by reducing the need for personnel to enter hazardous or hard-to-reach areas. For example, drones can be deployed in areas prone to landslides, earthquakes, or extreme weather conditions, avoiding potential risks to workers.

Applications in Specific Geophysical Areas

Drones have become increasingly valuable in various specific areas of engineering geophysics, offering innovative solutions for data collection and analysis. In *mineral exploration*, drones are employed in remote regions to conduct mineral surveys by collecting spectral data, which helps identify areas rich in minerals. This enables the detection of underground resources without the need for extensive physical exploration. In *geotechnical engineering*, drones are used to monitor and analyze soil conditions, such as detecting moisture content, soil stability, and other key parameters critical for projects involving foundations and slope stability. Additionally, in the field of *hydrogeology*, drones facilitate surface water surveys, gathering data about regional hydrology, including river flow, flood risks, and the identification of aquifers, thus aiding in groundwater studies and management.

Data Integration and Analysis

The data collected by drones can be integrated with other geophysical and geotechnical data sets to enhance decision-making. Geographic Information Systems (GIS) and modeling software can be used to process and analyze drone data, providing detailed insights into the physical characteristics of a site. This integration helps engineers, geophysicists, and decision-makers to better understand subsurface conditions, plan projects effectively, and mitigate risks.

Environmental and Regulatory Benefits

Drones can provide valuable environmental data for projects in sensitive ecological areas. They allow engineers to assess environmental impacts before, during, and after a construction project, helping to ensure compliance with environmental regulations and reducing the carbon footprint of traditional surveying methods.

Challenges and Limitations

Despite the numerous advantages of drone technology in engineering geophysics, there are several challenges and limitations that need to be addressed. *Regulatory issues* are a significant concern, as airspace restrictions and the requirement for licenses to operate drones in certain regions can complicate their deployment. Additionally, drones are highly susceptible to *weather conditions* such as high winds, rain, or extreme temperatures, which can severely limit their operational effectiveness. Another limitation is *battery life*, as the relatively short duration of a drone's flight restricts the area it can cover in a single flight, although advancements in battery technology are gradually improving this aspect. Lastly, *data processing* remains a challenge, as drones generate large volumes of data that require specialized software and expertise for efficient processing and analysis.

4.2.6 *Integration of Drones with Geophysical Surveying Techniques*

Enhancing Digital Terrain Models with Drone-Based LiDAR for Morpho-Structural Analysis: The Case of Ischia Island, Italy

This study explores the use of drone-based LiDAR technology to create high-resolution Digital Terrain Models (DTMs) for improved understanding of fault zones in active tectonic regions. Airborne LiDAR has been widely used in active tectonics research over the last two decades, providing detailed topographic data that reveals fault lines and ground deformation. However, challenges remain in areas with rapid ground movement, dense vegetation, and human impact (Mouginis-Mark and Garbeil 2005; Mazzarini et al. 2005; Dong and Chen 2018; Liu 2008; Conforti et al. 2021; Neri et al. 2008; Kosik et al. 2022; Favalli et al. 2009; Ventura and Vilardo 2008; Davila et al. 2007).

To address these issues, drone-based LiDAR offers a promising alternative, providing higher resolution data and lower costs, both for initial setup and ongoing operation. In this context, the study focuses on the Casamicciola Terme area on Ischia Island, Italy (Fig. 4.2), which is known for its seismic activity, including a destructive earthquake in 2017 (magnitude 4). A drone-based LiDAR survey was conducted to create a DTM, and the results were compared with existing elevation datasets to assess data quality, particularly for fault mapping (Silva-Fragoso et al. 2024).

The analysis of the DTM data revealed fault scarps in the Casamicciola Holocene Graben (CHG), a structural feature formed by two major fault zones (the Maio and Sentinella faults) and smaller internal faults (the Purgatorio and Nizzola faults) (Sbrana et al. 2021; Aucelli et al. 2021; Natale et al. 2022; Chiocci et al. 2023). The high-resolution data obtained through drone-based LiDAR enabled detailed geomorphic analysis and mapping of fault activity in the region (Fig. 4.3).

Figure 4.3 illustrates the structural geometry of the Casamicciola Holocene Graben, showing both major and minor faults, as well as fault scarps, based on data obtained from drone-based LiDAR surveys. It contributes to the understanding of the area's tectonic activity. This approach demonstrates the value of drone-based LiDAR for mapping fault zones, especially in areas where traditional methods may be hindered by dense vegetation or difficult accessibility. The technique provides an efficient and cost-effective solution for studying high slip-rate faults and active deformation, with potential applications in other regions facing similar challenges.

Dehazing UAV Images for Remote Sensing in Engineering Geophysics: Case Study on the Sicilian Coast, Italy

Unmanned aerial vehicles (UAVs), equipped with compact digital cameras and multi-spectral sensors, have become increasingly important in remote sensing, environmental studies, and engineering geophysics (Dalla Mura et al. 2010). Due to the

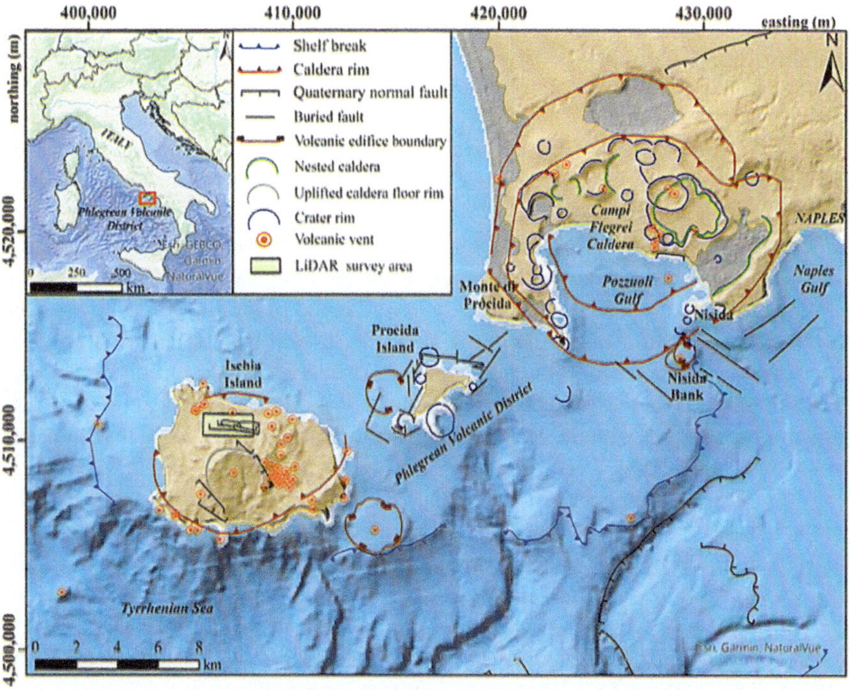

Fig. 4.2 Location of study area, highlighting Quaternary volcanic units and faults with Quaternary activity. *Source* Silva-Fragoso et al. (2024)—Open Access

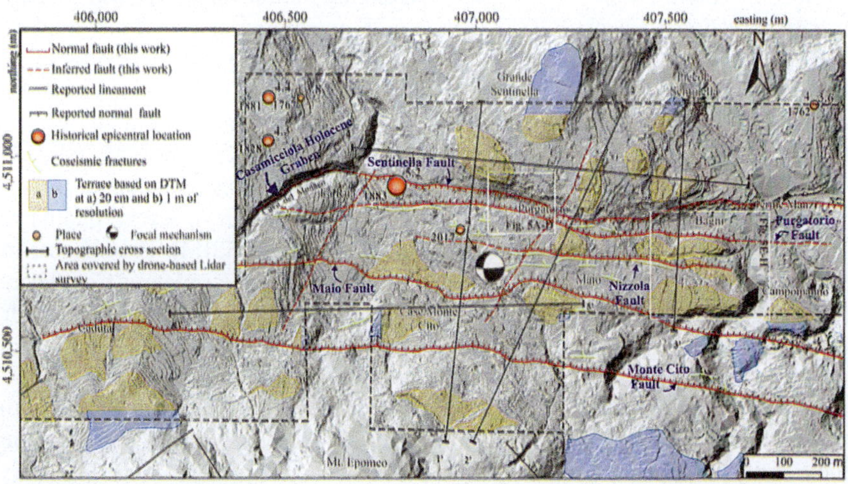

Fig. 4.3 Volcano-tectonic map of the Casamicciola Holocene Graben area with shaded relief derived from 20 cm-resolution DEM. *Source* Silva-Fragoso et al. (2024)—Open Access

decreasing costs of UAV systems, their increased reliability, and the ability to capture images with very high spatial resolution, low-altitude imaging is widely used for both qualitative and quantitative analyses in remote sensing. In recent years, significant advancements in the processing of UAV-acquired imagery have been made. However, much of the research has focused primarily on geometric and radiometric corrections, with little attention given to the effects of low atmospheric conditions, such as haze and fog, on image quality.

In engineering geophysics, particularly for environmental monitoring and structural assessments, the need for clear and accurate imagery is paramount. Images captured at low altitudes can suffer degradation due to scattering caused by fog and other weather conditions (Shao et al. 2019). These atmospheric factors negatively affect contrast and color reproduction, leading to a decrease in the radiometric quality of the images. This study presents a method for dehazing UAV-acquired images, focusing on imagery obtained under low-altitude conditions during high-pressure fronts (Wierzbicki et al. 2019; Woloszyn 2009). The research introduces a methodology for acquiring imagery at low altitudes and applies atmospheric calibration methods based on an atmospheric scattering model. The case study is conducted in the Sicilian Coast region (Fig. 4.4), an area known for its complex geological features and susceptibility to varying weather conditions, making it a valuable location for applying UAV-based imaging in engineering geophysics. The study area includes coastal geophysical investigations, where image quality is crucial for analyzing geological formations, fault zones, and infrastructure stability.

The system used in this study consists of the SR10 receiver, M10 radiosondes (Fig. 4.5), and a ground control kit for radiosonde calibration based on Global Navigation Satellite System (GNSS) (12 channels) initialization. It also includes an omnidirectional antenna (400 MHz), a GPS antenna (400 MHz), and a tripod. The SR10 system operates within a range of at least 350 km and works in the 400–406 MHz frequency range. Additionally, Fig. 4.6 shows the optical model of atmospheric diffusion based on the Lambert reflection model, commonly used in similar studies.

Additionally, a modified dehazing approach using Wiener's adaptive filter is proposed. To evaluate the performance of the dehazing method, several qualitative indices were used, including Structural Similarity Index (SSIM), Peak Signal-to-Noise Ratio (PSNR), Root Mean Square Error (RMSE), Correlation Coefficient, Universal Image Quality Index (Q Index), and Entropy. The experimental results show that the proposed dehazing method significantly improves image quality, with an average PSNR increase of 34% compared to other similar methods.

This study demonstrates the effectiveness of the proposed dehazing technique in mitigating the negative effects of low atmospheric conditions, such as high humidity and radiation fog, on image quality (Fig. 4.7). By enhancing UAV-acquired imagery, this method proves especially useful for engineering geophysics, where accurate visual assessments are essential for geological analysis and infrastructure monitoring. The findings from this research can be applied in other regions with similar environmental challenges, ensuring better image quality for remote sensing applications in engineering geophysics.

Fig. 4.4 Location of the study area on the Sicilian Coast, Italy, highlighting key geological features for engineering geophysics analysis. *Source* Wierzbicki et al. (2019)—Open Access

Drone-Based GPR for Mapping Soil-Rock Boundaries and Groundwater Levels in Foundation Engineering in Beer Sheva, Israel

This case study highlights the application of drone-based ground-penetrating radar (GPR) to map soil-to-rock boundaries and groundwater levels at the construction site of Beer Sheva's Children's Park, Israel (Fig. 4.8a–d) (Frid and Frid 2024). The site presented challenges such as waterlogged clayey soils and urban obstacles, including trees, fences, and streetlights, which often impede traditional methods like borehole drilling and ground-based GPR surveys.

The site is in the northwestern part of Beer Sheva, Israel (Fig. 4.8b). It lies within the old Wadi area, filled with disturbed clayey soil during the construction of the Children's Park (Fig. 4.8c). The filling soil is classified as low-plasticity inorganic clay to silty clay (CL/ML) based on the Unified Soil Classification System. This soil contains carbonate pebble-sized material and overlays calcareous sandstone from the Pliocene period (Fig. 4.8d). Government topographic data suggest natural rock or soil surfaces at depths of 8–15 m but lack precise details.

Combining ground-penetrating radar (GPR) with drones has revolutionized foundation engineering surveys, offering enhanced efficiency and precision. As shown in Fig. 4.9, the GPR unit is mounted beneath the drone, with the dipole antenna positioned for optimal ground distance to ensure accurate data collection. Real-time

Fig. 4.5 System components: SR10 receiver, M10 radiosondes, GNSS-based ground control kit, and antennas (400 MHz). *Source* Wierzbicki et al. (2019)—Open Access

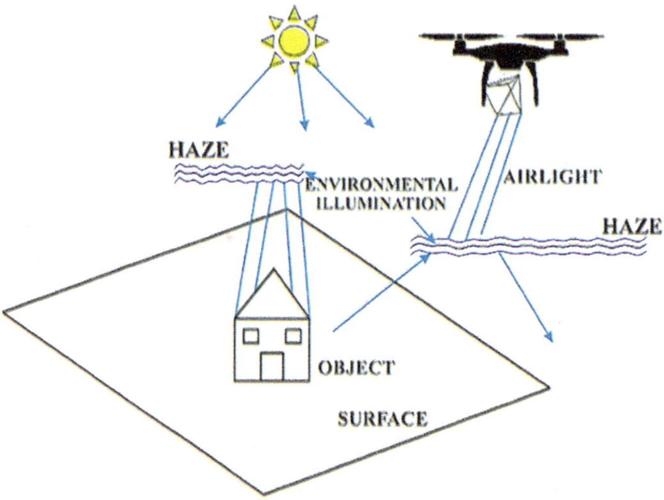

Fig. 4.6 Atmospheric scattering model in a humid atmosphere for UAV-acquired imagery. *Source* Wierzbicki et al. (2019)—Open Access

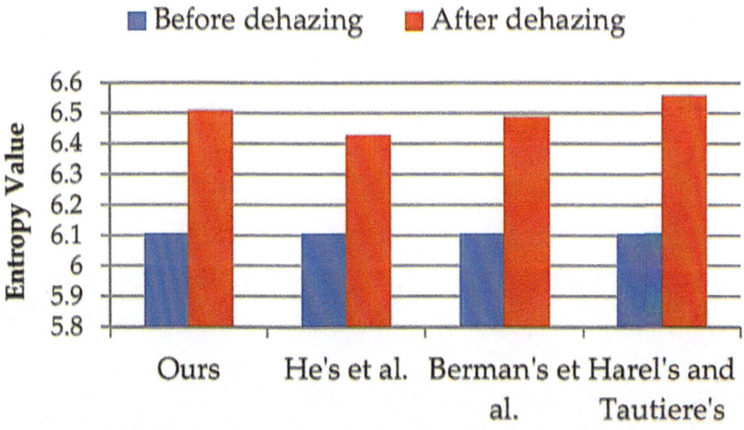

Fig. 4.7 Entropy values before and after dehazing for different methods (September–November 2018). *Source* Wierzbicki et al. (2019)—Open Access

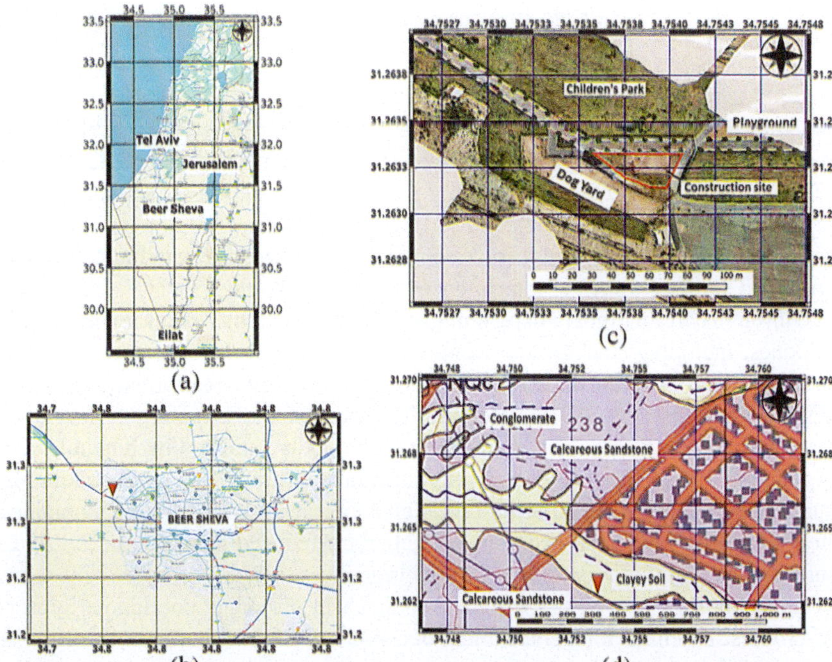

Fig. 4.8 Maps and geological context of the study area. **a** Location of Israel, **b** Beer Sheva city map with site marked (red triangle), **c** orthophoto of the Children's Park construction site (red line), and **d** geological map fragment showing site location (red triangle). *Source* Frid and Frid (2024)—Open Access

Fig. 4.9 DJI M300 drone with GPR system for subsurface surveying. *Source* Frid and Frid (2024)—Open Access

kinematic (RTK) technology enables precise georeferencing, while the SKYHUB computer, developed by SPH Engineering in Riga, Latvia, synchronizes data between the DJI M300 drone and the GPR unit for seamless information acquisition.

The system features an unshielded 150 MHz dipole antenna, chosen for its balance between penetration depth and resolution, ideal for clayey soils with high dielectric properties. Challenges such as interference from above-ground obstacles were minimized through careful flight path planning at an altitude of 5 m. Advanced processing techniques, including Background Removal and Dewow Filters, further enhanced data clarity, demonstrating the effectiveness of drone-based GPR in complex terrains.

The radargrams (Fig. 4.10) successfully identified the transition from clayey soil to natural rock formations, with penetration depths validated by borehole data and geological mapping. Data acquisition, conducted during low wind conditions, ensured stable flights, addressing a common challenge in drone operations. Compared to traditional methods, the drone-based approach reduced survey time by 40%, demonstrating its efficiency and cost-effectiveness.

This technology proved particularly beneficial in providing continuous subsurface profiles, essential for geotechnical decisions in foundation engineering. The data revealed soil-to-rock boundaries and groundwater levels critical for determining foundation pile lengths and ensuring structural stability. Additionally, the drone's ability to navigate complex urban terrains eliminated the need for extensive on-ground setups, making it an ideal solution for inaccessible or obstacle-laden areas.

In conclusion, the successful deployment of drone-based GPR in Beer Sheva, Israel, showcases its potential as a transformative tool for geophysical analysis in

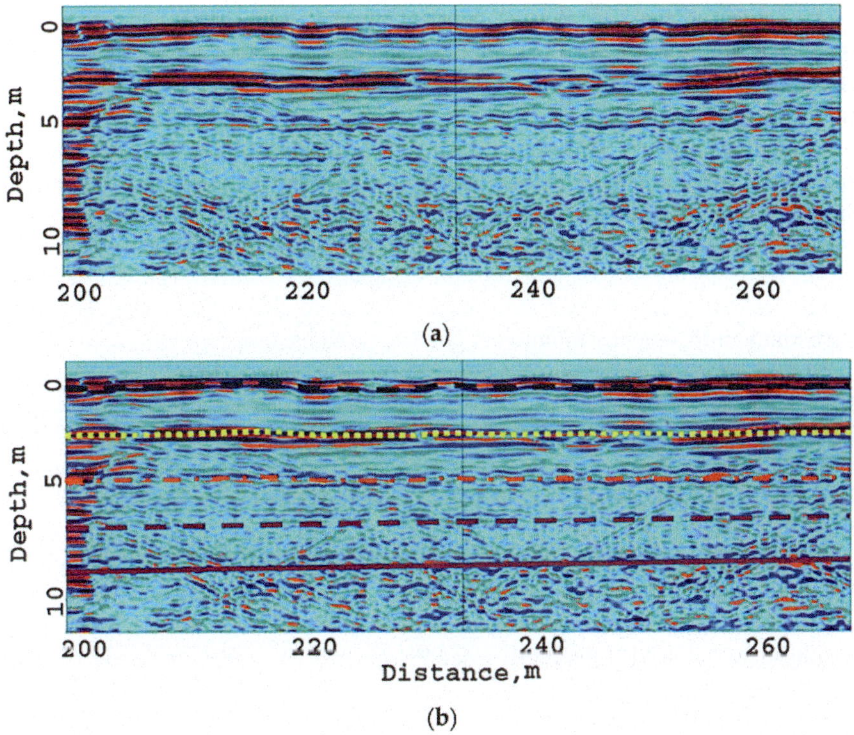

Fig. 4.10 Processed radargram showing subsurface boundaries. *Source* Frid and Frid (2024)—Open Access

engineering applications. This study validates its practical use in challenging environments, offering a reliable, efficient, and cost-effective alternative to conventional methods. The findings contribute to advancing the adoption of drone-based GPR for geotechnical investigations and inspire future research in subsurface sensing technologies.

Drone-Based Magnetometry Surveys for Engineering Applications

The use of unmanned aerial vehicles (UAVs), commonly referred to as drones, for magnetometry surveys has emerged as an innovative solution for engineering applications, offering several advantages over traditional methods. These UAVs, equipped with highly sensitive magnetometers, provide rapid data acquisition and access to hard-to-reach or hazardous environments. The primary challenge in using UAVs for magnetometry is developing magnetometer systems that are lightweight enough to be integrated into small UAV platforms, while maintaining the high data quality typically expected from traditional systems.

Key UAV magnetometers for engineering geophysics applications include the GSMP-35U Ultralight UAV Magnetometer, which is used for geophysical and engineering surveys in remote, difficult-to-access locations such as infrastructure monitoring, subsurface investigations, and environmental assessments. The GSMP-35U features an optically pumped potassium magnetometer sensor, providing high sensitivity of 0.1 picotesla (pT), making it lightweight and suitable for UAV integration. Another crucial aspect is magnetic interference testing for UAV selection, which ensures that the UAV platform does not generate magnetic noise that could distort survey results. This process identifies the optimal UAV platform and sensor placement to minimize interference from UAV motors and electronics, which is critical for accurate infrastructure monitoring and site investigations.

Multicopter UAV magnetometry solutions are ideal for smaller-scale engineering surveys, such as site evaluations, construction planning, and monitoring of geotechnical projects. These systems integrate lightweight magnetometer systems with stable multicopter UAV platforms, allowing for high-resolution magnetic surveys in both urban and rural environments. For larger engineering surveys requiring heavier payloads or longer flight durations, unmanned helicopter solutions, such as the ING Robotics Responder R1, are suitable. These medium-lift UAVs have optimized payload capacities to carry magnetometers and other geophysical tools. Helicopter platforms offer more versatility for payload positioning and survey coverage, making them suitable for pipeline inspections, geological surveys, and environmental monitoring.

Fixed-wing UAV magnetometry solutions, such as the Tempest, are designed for large-scale engineering geophysical surveys, particularly in infrastructure monitoring over large areas like roadways, pipelines, and power lines. These fixed-wing UAVs excel in surveying remote regions with minimal human intervention, offering extended flight durations, greater range, and minimal interference, which is ideal for large-scale magnetic gradient data collection. The MagNimbus UAV-based magnetic detection solution is used for engineering applications like detecting unexploded ordnance (UXO), abandoned infrastructure, power lines, and other magnetic anomalies that could affect engineering projects. The MagNimbus system uses a dual-sensor gradiometer configuration to detect magnetic anomalies in challenging environments, providing high-accuracy results in urban, industrial, and remote settings.

Drone-based magnetometry surveys are transforming engineering geophysics by providing high-resolution data collection in areas that are difficult, dangerous, or costly to access using traditional methods. These systems enable engineers to quickly and accurately assess infrastructure, detect subsurface anomalies, and monitor large areas in real-time. The continuous development of lightweight, high-sensitivity magnetometers and optimization of UAV platforms are expanding the scope of engineering applications, making geophysical surveys more efficient, cost-effective, and accessible. This summary highlights the value of UAV-based magnetometry surveys in engineering, focusing on its applications, technological advancements, and potential for solving real-world geophysical challenges in engineering geophysics.

Resolving magnetometer inconsistencies in drone-based gradiometer surveys has become essential for ensuring data accuracy in engineering applications, as demonstrated in the case of *Ranogen, Arctic Norway*. Magnetic gradiometer surveys are widely used in various engineering contexts, including high-resolution geological mapping (Dossing et al. 2024), mineral exploration (Dossing et al. 2023), environmental assessments, unexploded ordnance detection (Kolster and Dossing 2021; Kolster et al. 2022), and archaeological site investigations (Vilhelmsen 2023). These surveys use multiple magnetometers to measure the Earth's magnetic field and its gradient components, providing detailed insights into near-surface magnetization contrasts for better localization and characterization of subsurface anomalies.

The integration of drones with magnetometer technology has led to the development of drone-based magnetic surveying, which serves as an effective alternative to traditional ground-based, heliborne, and fixed-wing airborne methods (Heincke et al. 2019; Walter et al. 2020; Dossing et al. 2021; Jackisch et al. 2022). These drone-based surveys are particularly beneficial in engineering geophysics due to their ability to autonomously operate at low altitudes and in rugged or remote terrain, thereby reducing both logistical complexity and operational costs.

Despite these advantages, drone-borne magnetometers can suffer from inconsistencies caused by directional heading, temperature variability, and sensor drift. These factors can lead to striping and high-frequency noise in the magnetic data, ultimately hindering geological interpretation. To overcome these limitations, an innovative processing method has been developed that combines equivalent source modeling with kernel regression. This approach effectively corrects for instrument-related anomalies and enhances the reliability of magnetic data collected in engineering geophysical surveys.

As shown in Table 4.3, the method stabilizes the source surface by addressing both instrument-dependent errors and baseline or altitude deviations that can distort the gradiometer data. The process incorporates key components as summarized.

This table ensures a systematic understanding of how the method enhances data quality and accuracy for drone-borne magnetic gradiometer surveys. The approach was applied to a drone-based survey in Ranogen, Arctic Norway, where challenging environmental conditions, including wind, introduced significant inconsistencies.

Table 4.3 Overview of the stabilization method for gradiometer data

Aspect	Details
Instrument errors	Corrects inconsistencies in lightweight magnetometers, such as heading error
Baseline deviations	Accounts for variations in baseline orientation to prevent striping artefacts
Altitude deviations	Mitigates altitude-induced variations to maintain consistent data quality
Equivalent source	Creates a model surface that represents true geological features
Kernel regression	Stabilizes the equivalent source against large instrument-dependent errors
3D mapping	Enables calculation of uniform gradients (transverse, longitudinal, vertical)

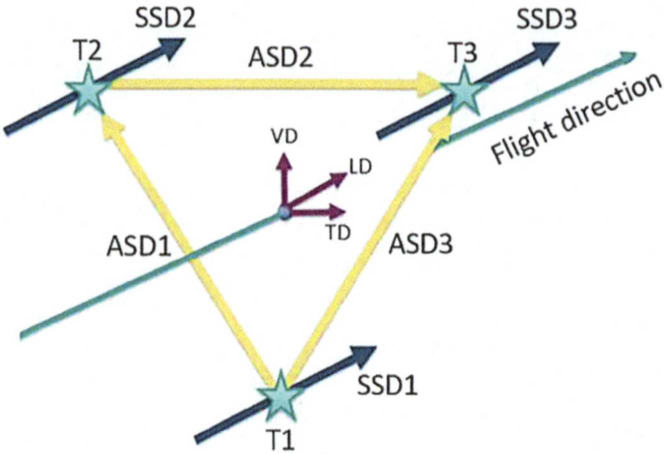

Fig. 4.11 Triangular gradiometer system for Ranbogen survey. Visualization of a downward-pointed triangular gradiometer with 2 m baseline magnetometers at its corners. Arrows depict across-sensor and same-sensor differences, flight direction, and ideal orthogonal differences. *Source* Petersen and Dossing (2024)—Open Access

Figure 4.11 illustrates the survey setup in this region. The method successfully corrected the instrument-dependent errors and baseline deviations, providing a more accurate geological interpretation essential for engineering projects in the area.

The improvements were validated using a synthetic gradiometer survey, as depicted in Fig. 4.12, where nonlinear heading errors were added to the magnetometer readings. The method effectively removed these errors while constructing an equivalent source model that fit the underlying synthetic model. By minimizing the instrument-dependent response, the method enabled a more precise interpretation of the geological data, which is critical for engineering applications such as site characterization, infrastructure planning, and environmental assessments.

Inconsistent magnetometer responses in drone-based gradiometer surveys can hinder accurate geological interpretations, particularly in engineering applications. This study presents a solution through an equivalent source-based processing method, tested in Ranbogen, Arctic Norway, to resolve these inconsistencies (Fig. 4.13). The method removes errors due to heading and baseline deviations, making drone-borne gradiometer surveys more reliable for engineering geophysical applications, such as site characterization, infrastructure planning, and environmental assessments.

Drone-Based Thermal Imaging in Engineering Geophysics

Drone-Based Thermal Imaging in Engineering Geophysics has revolutionized the field by offering a non-invasive and highly efficient method for monitoring subsurface conditions. Thermal sensors mounted on unmanned aerial vehicles (UAVs) enable the detection of temperature variations that reveal critical information about geothermal

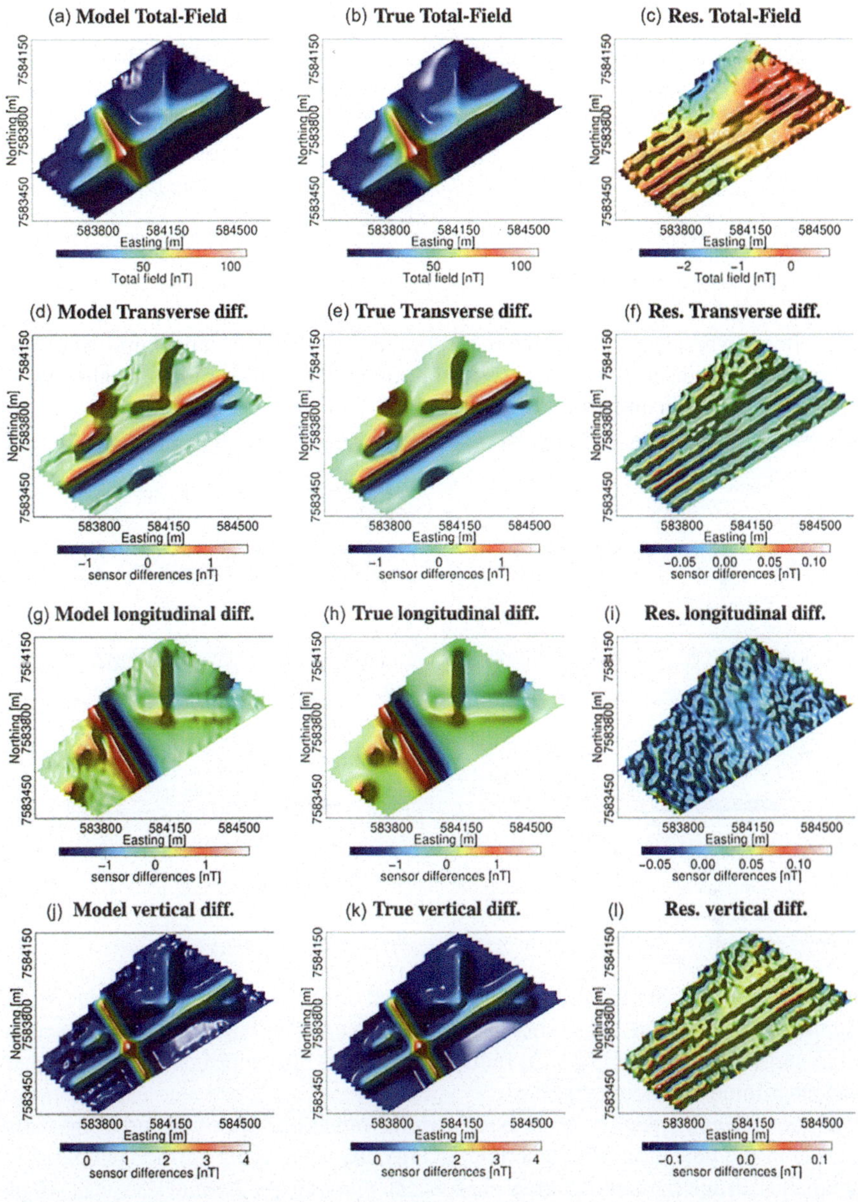

Fig. 4.12 Comparison of equivalent source maps and simulated gradiometer data. *Source* Petersen and Dossing (2024)—Open Access

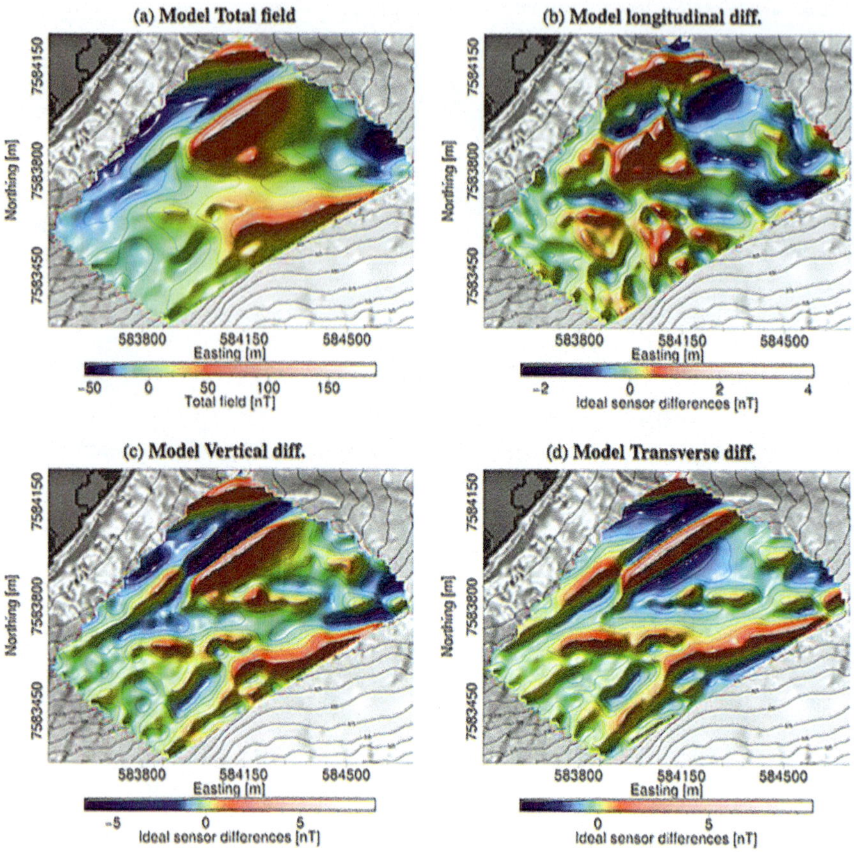

Fig. 4.13 Equivalent source-based processing results for Ranbogen gradiometer data. *Source* Petersen and Dossing (2024)—Open Access

activity, groundwater levels, and underground infrastructure. This advancement significantly improves the ability to survey and analyze large and often inaccessible areas quickly, making it an indispensable tool for geophysical investigations and environmental monitoring.

Applications in geophysics are extensive and impactful. One of the most important uses is in the detection of geothermal activity. By identifying surface temperature anomalies, drone-mounted thermal sensors can pinpoint geothermal hotspots, which are essential for locating viable geothermal reservoirs for energy development. These UAVs can produce detailed thermal maps across expansive areas, which would be impractical or too costly to generate using ground-based methods alone.

Thermal imaging is also instrumental in *monitoring water table levels*. Temperature gradients observed from the air can reflect variations in subsurface moisture content, allowing researchers to detect fluctuations in groundwater levels. These observations are crucial for hydrological modeling and water resource management, particularly in regions where groundwater serves as a primary source of water. UAVs can repeatedly survey the same area to monitor seasonal or long-term changes, offering a non-invasive and cost-effective monitoring solution.

Furthermore, drone-based thermal sensors play a vital role in *underground infrastructure mapping*. Buried utilities such as pipelines, electrical cables, and tunnels often generate localized thermal anomalies on the surface due to heat conduction or leaks. By capturing these subtle temperature differences, UAVs can detect and help map the location and condition of these structures. This method allows engineers to identify issues such as leakages or structural weaknesses without the need for excavation, reducing both cost and risk while enhancing the reliability of infrastructure inspections.

Methodology and Advantages of drone-based thermal imaging in geophysics highlight several significant improvements over conventional ground-based techniques. The approach involves using UAVs equipped with high-resolution thermal sensors to conduct rapid surveys over expansive and often inaccessible terrains. This capability accelerates data acquisition while substantially lowering the costs typically associated with traditional surveying methods. UAVs can operate at variable altitudes, which allows for optimal thermal imaging conditions and enhances the accuracy in identifying subsurface anomalies.

The integration of sophisticated image processing software with thermal sensors further refines the analysis by enabling precise detection and mapping of temperature variations. These thermal anomalies are often indicative of underlying geophysical features such as geothermal activity, fluctuations in the water table, or the presence of buried infrastructure. By linking these surface temperature patterns to specific subsurface conditions, geophysicists gain critical insights that guide further investigation and decision-making.

This synergistic use of UAVs, thermal sensing technology, and advanced data analytics not only boosts the efficiency of geophysical surveys but also improves their accuracy and scope. Unlike traditional methods, which may be labor-intensive and constrained in coverage, drone-based thermal imaging provides a more comprehensive, flexible, and timely solution for subsurface monitoring. As a result, the technology enables more effective environmental and engineering assessments, particularly in regions where speed, precision, and minimal ground disturbance are essential.

UAV-Based Thermal Imaging for Building Energy Efficiency and CO_2 Mitigation demonstrates how low-cost Unmanned Aerial Vehicles (UAVs), equipped with thermal imaging systems, are increasingly being used in building thermography inspections to enhance energy efficiency and contribute to CO_2 reduction goals. These applications align with global sustainability objectives, particularly SDG 11 Target 11.6, which focuses on reducing the environmental impact of cities, and SDG 7 Target 7.3, which aims to improve energy efficiency.

In a case study conducted at Malardalen University in Västeras, Sweden, a visual-thermal drone system was evaluated for its capacity to detect heat loss, insulation defects, and energy leakage in both residential and commercial buildings (Johansson et al. 2019). The UAV system effectively identified issues such as air leaks, water seepage, and missing insulation. Although the drone proved cost-effective and useful for basic inspections, limitations in image resolution and data accuracy reduced its suitability for high-precision energy audits.

In Stockholm, Sweden, UAVs equipped with thermal sensors were deployed to identify thermal anomalies such as air leaks and thermal bridging, both of which contribute to energy inefficiency and elevated CO_2 emissions. These drones allowed rapid inspection of large buildings, but the thermal data stored in JPG format limited the depth of post-analysis, and the overall accuracy fell short of what is required for comprehensive energy assessments.

A separate study in Berlin, Germany, focused on the thermal inspection of residential buildings using UAVs. The drones successfully detected insulation gaps and significant heat loss areas (Muller et al. 2020). The affordability of these UAV systems made them particularly attractive to homeowners. However, like other systems, they were not capable of producing the detailed analyses necessary for thorough energy audits.

In New York City, drones with thermal cameras were integrated into a smart city initiative to assess the energy performance and CO_2 emissions of buildings (Carter et al. 2021). The UAVs enabled fast and efficient inspections in dense urban environments, identifying areas of significant energy loss. Nonetheless, the absence of radiometric imaging capabilities restricted deeper, quantitative analysis. Despite this, the drones proved to be effective tools for quick energy assessments and prioritizing buildings for energy efficiency upgrades.

Collectively, these case studies underline the growing utility of drone-based thermal imaging for building diagnostics and environmental monitoring. Although the current systems offer advantages in cost, accessibility, and rapid deployment, limitations in image resolution and analytical capability suggest they are best suited for preliminary assessments rather than detailed energy audits. Nonetheless, in the context of engineering geophysics, these UAV-based technologies contribute meaningfully to non-destructive testing and resource-efficient infrastructure evaluation, playing a supportive role in the pursuit of global CO_2 mitigation and sustainability targets.

Drone-Based Thermal Imaging (Infrared Cameras or Sensors) and 3D Analysis for Solar Module Defect Detection plays a vital role in supporting global efforts toward climate change mitigation and the advancement of renewable energy, particularly solar power. Efficient monitoring and maintenance of photovoltaic (PV) systems are critical for ensuring optimal performance and extending the operational lifespan of solar installations, especially in large or remote areas where manual inspection is impractical or cost-prohibitive.

The case study titled "Development of a Drone-Based System for Infrared Inspection of Solar Power Plants Using Machine Learning" exemplifies the integration of drone technology and infrared (IR) thermography for the automated inspection of solar panels (Fujimura et al. 2018). In this system, drones equipped with IR cameras autonomously capture thermal images of solar panels, which are essential for detecting surface-level and internal defects such as hot spots, micro-cracks, and faulty connections. These defects, if left unidentified, can significantly reduce the energy efficiency and longevity of the PV modules.

A key innovation in this approach lies in the use of machine learning algorithms to analyze the captured thermal data. As demonstrated by Zhou et al. (2021), the system is trained using labeled datasets that allow the model to automatically detect and classify different types of anomalies in the solar panels with high accuracy. This automation drastically reduces the time and labor required for inspections and increases the reliability of defect detection. Moreover, by incorporating 3D spatial analysis with thermal imagery, the system can precisely locate the defective panels within large solar farms, enabling targeted maintenance and minimizing system downtime.

Overall, this advanced method of using UAV-based thermal imaging and artificial intelligence significantly enhances the speed, accuracy, and efficiency of solar plant inspections. It contributes to sustainable energy production by ensuring the optimal operation of PV systems and directly supports the realization of SDG 7 (Affordable and Clean Energy) and SDG 13 (Climate Action) by facilitating smarter, greener maintenance strategies.

This study explores drone-based applications in engineering geophysics, specifically using UAVs equipped with thermal imaging and 3D imaging technologies to detect defects in solar modules (Fig. 4.14) (Liao et al. 2022). These drones enable efficient monitoring in difficult-to-reach locations and provide accurate data on module health.

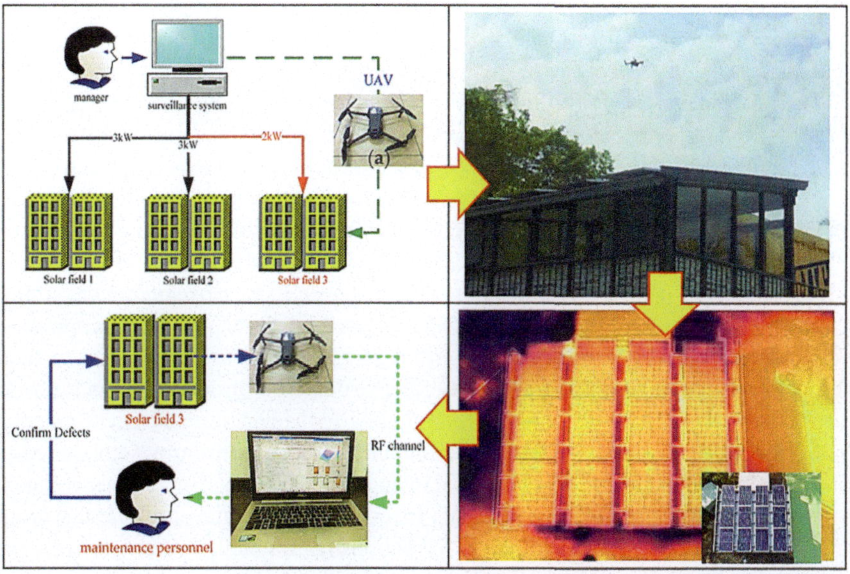

Fig. 4.14 Flight mission planning for Chaoyang University of Technology (CYUT) situated in Taichung, Taiwan solar photovoltaic experimental field. *Source* Liao et al. (2022)—Open Access

Key challenges in solar energy include temperature-induced inefficiencies due to shadowing, which can reduce efficiency by up to 25% annually. UAVs offer a solution by detecting faults such as hot spots, breakage, and overheating through thermal imaging. The proposed method, incorporating both infrared imaging and 3D reconstruction, provides enhanced defect detection even at low irradiance levels (500 W/m^2), offering significant improvements over traditional methods (Li et al. 2020; Di Martino et al. 2019; Vazquez et al. 2021). Thus, UAV-based applications in engineering geophysics represent a promising advancement in solar module monitoring, improving both system performance and maintenance safety (Mazzarella et al. 2020; Gao et al. 2020; Hernandez et al. 2018).

This study integrates UAVs with IR cameras and advanced image processing techniques to monitor the health of solar modules (W/m^2). The key innovation is the use of an improved box filter for analyzing 3D images, which significantly enhances the detection of defect zones, even when irradiance is as low as 500 W/m^2. Traditional IR detection often struggles with obscurity issues, especially under low irradiance, but the proposed method overcomes these limitations, providing clear and accurate identification of defect areas. Testing was conducted under varied sunlight conditions, comparing different filtering methods (Fig. 4.15).

Under irradiance of 700 W/m^2, misjudgements occurred with median filtering, where the defect area was exaggerated. However, other filtering methods, including the box filter, accurately pinpointed defect locations. At 500 W/m^2, the mean filter failed to define the defect area, while the median and box filters offered clearer identification. Notably, the box filter converted into a 3D image effectively revealed the defect zone in both low and high irradiance scenarios, such as at 850 W/m^2 (Fig. 4.16).

The 3D images not only illustrate the defect location but also provide temperature information, enhancing the accuracy of defect analysis. By matching the color temperature scale to the image, the temperature difference between normal and defective cells is quantified, helping to pinpoint defect locations more precisely. This method demonstrates that 3D images, combined with improved filtering techniques, offer both qualitative and quantitative insights into the health of solar modules.

Use of Drones in Multispectral Photogrammetry and Thermoscanning has significantly enhanced research and documentation methods in engineering geophysics. Drones are increasingly used in agriculture and geophysical surveys due to their ability to provide accurate, fast, and cost-effective data on crop conditions and subsurface monitoring. UAVs are particularly valuable for vegetation and environmental monitoring, offering high spatial resolution and quick response times compared to traditional aerial methods. This is crucial in engineering geophysics, where frequent data collection and precise measurements are essential for monitoring heterogeneous areas like woody crops and geological formations (Berni et al. 2009).

Recent advancements in miniaturized multispectral and thermal cameras have facilitated UAV-based vegetation and geophysical surveys, though these require radiometric calibration, atmospheric corrections, and photogrammetric techniques to match the accuracy of traditional airborne sensors. Drones equipped with thermal cameras are also widely used in applications such as rescue operations, plant surveys,

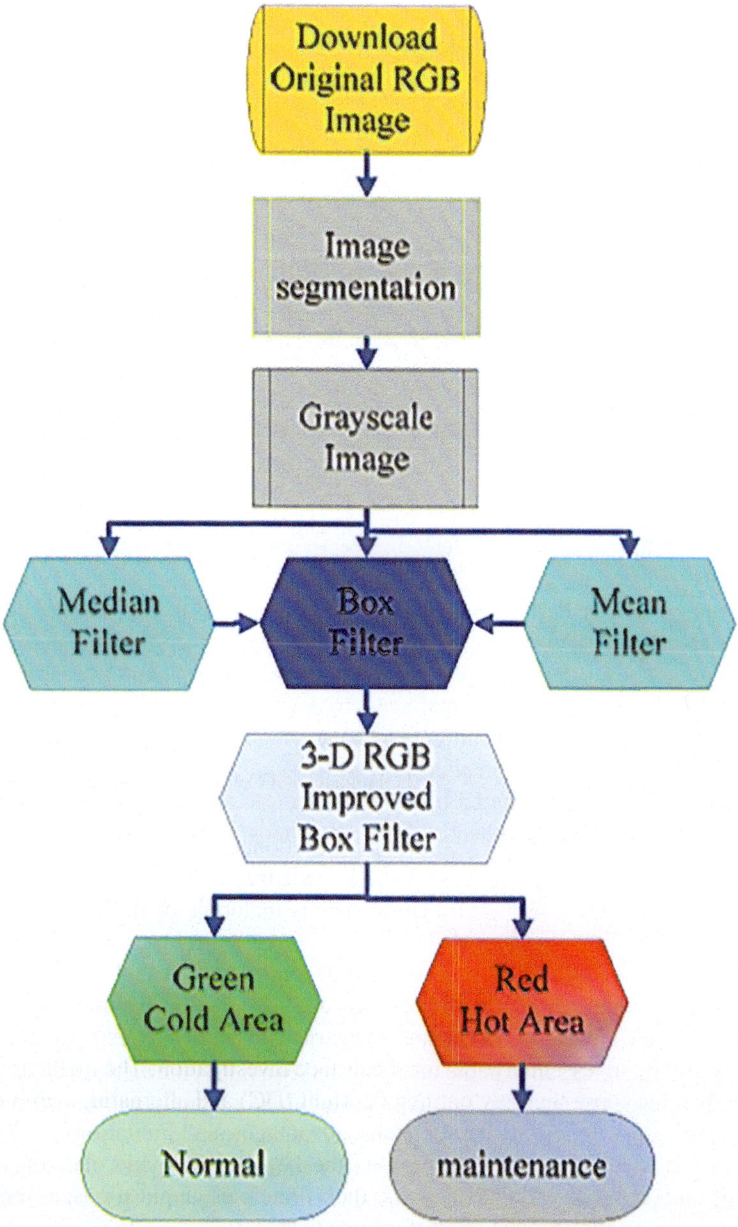

Fig. 4.15 Flowchart of the image filtering process for enhanced analysis. *Source* Liao et al. (2022)—Open Access

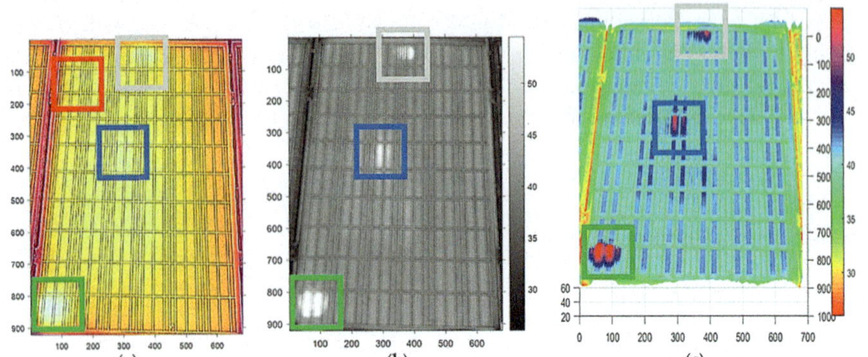

Fig. 4.16 Verification experiment with irradiance at 515 W/m^2: **a** original IR image, **b** grayscale treatment with improved box filter, **c** 3D image with improved box filter. *Source* Liao et al. (2022)—Open Access

and infrastructure inspections, detecting temperature variations in low-visibility conditions (Haboudane et al. 2002b; Rouse et al. (1974).

Vegetation indices, derived from solar reflectance (refers to the ability of a surface or material to reflect solar energy away from the Earth's surface, rather than absorbing it) measurements, are used to assess plant health and development. Data from multispectral, hyperspectral, and thermal sensors are processed using GIS systems to create maps for precision agriculture, environmental management, and engineering geophysics (Tao et al. 2020; Henderson et al. 2019).

Drones are used to capture high-resolution images and measure spectral data in the near-infrared (NIR) and visible (VIS) spectrums, which are crucial for calculating vegetation indices (VIs). The most widely used VI is the Normalized Difference Vegetation Index (NDVI), which measures vegetation health and density. The mathematical formula for NDVI is:

$$\mathrm{NDVI} = \frac{\mathrm{NIR} - \mathrm{R}}{\mathrm{NIR} + \mathrm{R}} \tag{4.1}$$

where NIR is the reflectance in the near-infrared band, R is the reflectance in the red band. NDVI values range from -1 to 1. Values greater than 0.1 indicate vegetation, with higher values signifying healthier, denser vegetation.

Other indices like soil-adjusted vegetation index (SAVI) and optimized soil-adjusted vegetation index (OSAVI) are used to adjust for soil interference, enhancing monitoring accuracy. The formula for OSAVI is:

$$\mathrm{OSAVI} = \frac{\mathrm{NIR} - \mathrm{R}}{\mathrm{NIR} + \mathrm{R} + 0.16} \tag{4.2}$$

and the formula for SAVI is:

$$\text{SAVI} = \frac{(\text{NIR} - \text{R}) \times (1 + \text{L})}{(\text{NIR} + \text{R} + \text{L})} \tag{4.3}$$

The value 0.16 is a constant that optimizes the soil adjustment for a wide range of vegetation types and densities, making OSAVI particularly useful for areas with mixed vegetation and soil exposure. L is a soil adjustment factor, typically set to 0.5 for standard applications, but can be adjusted based on the density of vegetation.

Drones, with their low altitude, solve issues faced by satellites, such as cloud interference, allowing for more reliable vegetation assessments. Additionally, combining spectral VIs with texture information (e.g., using the Gray-Level Co-occurrence Matrix) improves biomass estimation, as shown by Zheng et al. (2019), where this combined approach significantly increased accuracy in rice biomass estimates. Thus, drones enhance the accuracy of vegetation monitoring by providing high-resolution, cloud-free data for calculating and analyzing various VIs.

Use of Drones for Multispectral Photogrammetry in Geophysical Applications has revolutionized geophysical applications by enabling the collection of high-resolution, spectral data from multiple color bands, including infrared, red, green, and blue. These drones capture detailed imagery of terrain features, which can then be processed using photogrammetry algorithms to analyze the quality and structure of soil, the health of vegetation, moisture levels, and more. By collecting this data, multispectral photogrammetry supports agricultural applications such as optimizing crop production and early identification of crop stress or diseases. It is also useful in the construction industry to assess land quality and detect issues with nearby structures. Moreover, it finds applications in environmental research, particularly in monitoring ecosystems and studying the effects of climate change (Haboudane et al. 2002a).

A primary tool in vegetation monitoring is the use of vegetation indices (VIs) like the Normalized Difference Vegetation Index (NDVI) and the Soil Adjusted Vegetation Index (SAVI), which are derived from multispectral imagery. These indices help determine vegetation health, density, and biomass by quantifying the difference between light absorption by plants in the red spectrum and light reflection in the infrared spectrum (Fig. 4.17a, b). NDVI, for instance, ranges from − 1 to 1, where higher values indicate healthier, denser vegetation (Benincasa et al. 2018). Similarly, SAVI is used to correct for soil brightness, making it particularly useful in areas with sparse vegetation (Caturegli et al. 2016).

In addition to multispectral cameras, 3D LiDAR (Light Detection and Ranging) technology is often integrated with drones to provide precise topographic and three-dimensional data of the terrain (Fig. 4.18).

LiDAR uses laser pulses to measure the distance between the drone and the ground, allowing for highly accurate terrain modeling, even through dense vegetation (Fig. 4.19).

This capability is particularly useful in geophysical studies, where understanding the earth's surface topology is crucial for hazard monitoring, resource management, and geological mapping (Geipel et al. 2016).

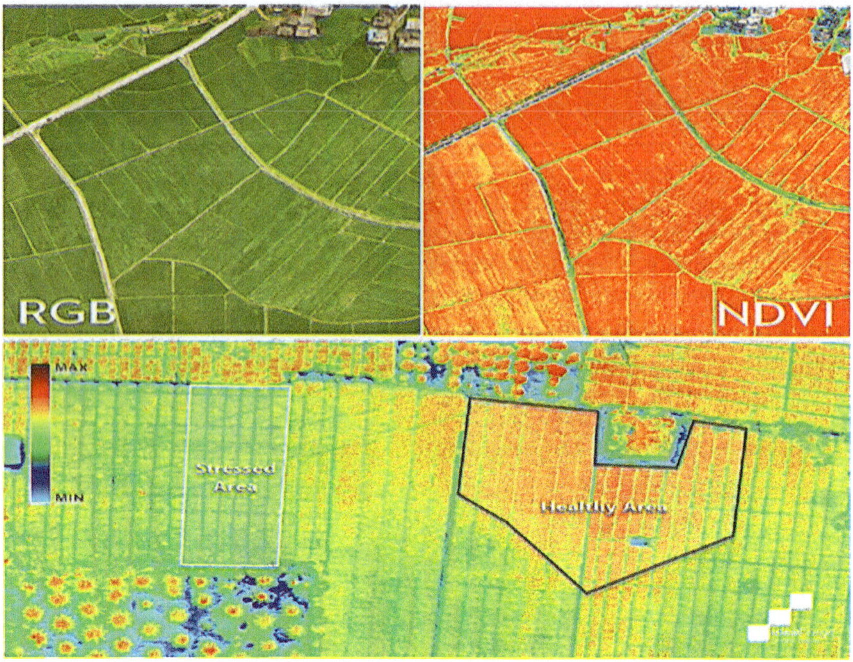

Fig. 4.17 **a** Image captured with multispectral camera for geophysical evaluation. *Source* Ioja et al. (2022)—Open Access. **b** Multispectral map of agricultural crop for geophysical analysis. *Source* Ioja et al. (2022)—Open Access

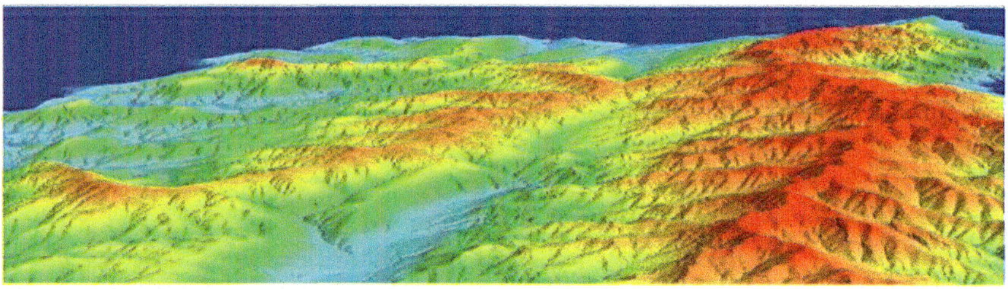

Fig. 4.18 3D terrain map generated using LiDAR technology for geophysical analysis. *Source* Ioja et al. (2022)—Open Access

LiDAR also offers advantages in environments where photogrammetry might struggle, such as under dense tree canopies, by directly measuring the ground elevation below tree cover (Li et al. 2019). The fundamental equation for LiDAR distance measurement is as follows:

$$\text{Distance} = \frac{\text{Speed of Light} \times \text{Time of Flight}}{2} \tag{4.4}$$

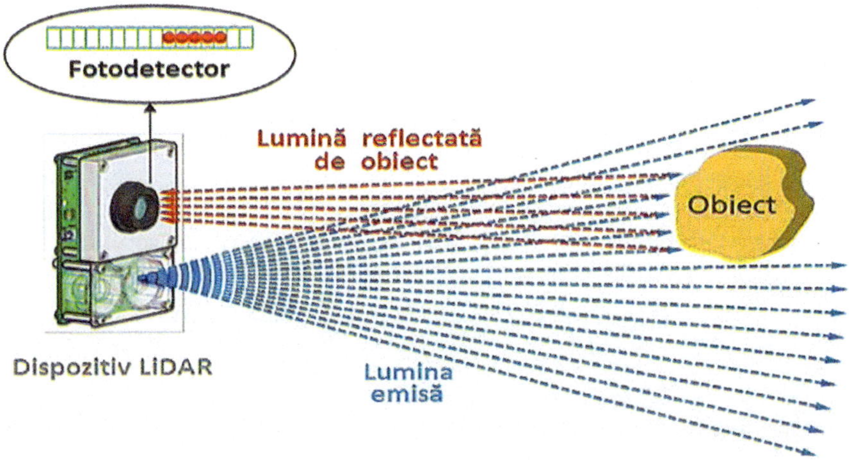

Fig. 4.19 Operation technique of LiDAR device for geophysical surveys. *Source* Ioja et al. (2022)—Open Access

Combining both multispectral and LiDAR data can significantly enhance geophysical analysis by providing both high-resolution vegetation indices and accurate topographic models. This combination has proven effective in estimating aboveground biomass, studying soil properties, and assessing crop health (Zheng et al. 2019). Furthermore, the integration of these technologies in geophysical studies helps improve the precision of environmental and agricultural assessments, fostering better decision-making in areas such as crop management, hazard assessment, and resource conservation.

Using Drones for Thermal Scanning of Buildings provides key advantages, particularly in monitoring building envelopes, pipelines, and infrastructure. The prediction of energy consumption in buildings has gained attention in the context of sustainable housing and smart cities, with data-driven models using Internet of Things (IoT) sensors showing promise (Xie et al. 2020). However, the accuracy of energy predictions remains limited by insufficient research on building attributes and prediction delays (Perez et al. 2021).

Thermal imaging, based on infrared (IR) radiation, has become a diagnostic tool in various fields such as medicine and building sciences (Alhargan et al. 2021). IR cameras capture thermal radiation from surfaces and convert it into a visible image, useful for detecting issues like thermal bridges, moisture infiltration, and structural defects (Arne et al. 2018). Traditional IR scans, however, offer a snapshot in time and are limited by handheld camera use, making them insufficient for detailed building assessments. By using drones for time-lapse thermal scanning, continuous monitoring provides a more accurate characterization of a building's thermal behavior over time (Gershenson et al. 2019).

Radiometric thermal cameras correct for factors like emissivity and humidity, ensuring each thermal point is accurately represented (Zhao et al. 2022). Infrared images consist of a matrix of temperature data, where color palettes display varying temperatures. Since different temperature ranges can cause discrepancies between images, pre-processing the data to normalize these ranges is necessary for accurate photogrammetric reconstruction (Mendonca et al. 2020).

The proposed methodology for modeling thermal anomalies involves integrating these IR images into a Building Information Model (BIM), using simulation tools like EnergyPlus to simulate energy performance (Tichem et al. 2021). The methodology also includes quantifying thermal performance using the following equations:

$$\text{Reduced \% of R} = \left(\frac{A_S}{A_{Anomalies}} - \frac{R_{Original}}{R_{Reduced}}\right) \times 100 \tag{4.5}$$

$$\text{Reduced} = R_{Original} - \left(R_{Original} \times R\%_{Reduced}\right) \tag{4.6}$$

where A_S: Area of the surface or region of interest being analyzed for thermal anomalies, $A_{Anomalies}$: Area of the detected anomalies within the scanned surface, $R_{Original}$: The original thermal resistance value (R-value) of the building envelope or surface before any thermal anomalies are accounted for, $R_{Reduced}$: The thermal resistance value (R-value) of the building envelope after accounting for the impact of the detected thermal anomalies, $R\%_{Reduced}$: The percentage reduction in the R-value due to the identified thermal anomalies, Reduced % of R: The percentage reduction in the R-value due to thermal anomalies, and Reduced: The reduced R-value after considering the thermal anomalies (Mitra et al. 2022).

Figure 4.20 illustrates normalized infrared images at different times of the day, showing temperature variations.

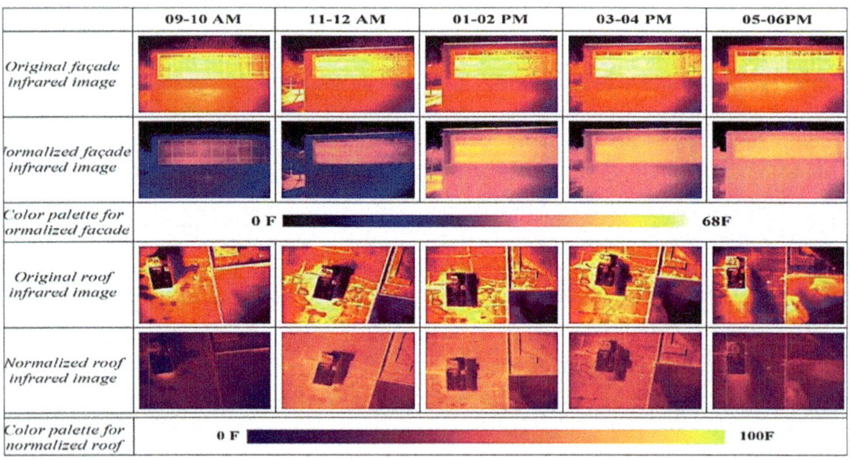

Fig. 4.20 Pre-processing normalization of infrared images of building facade and roof. *Source* Ioja et al. (2022)—Open Access

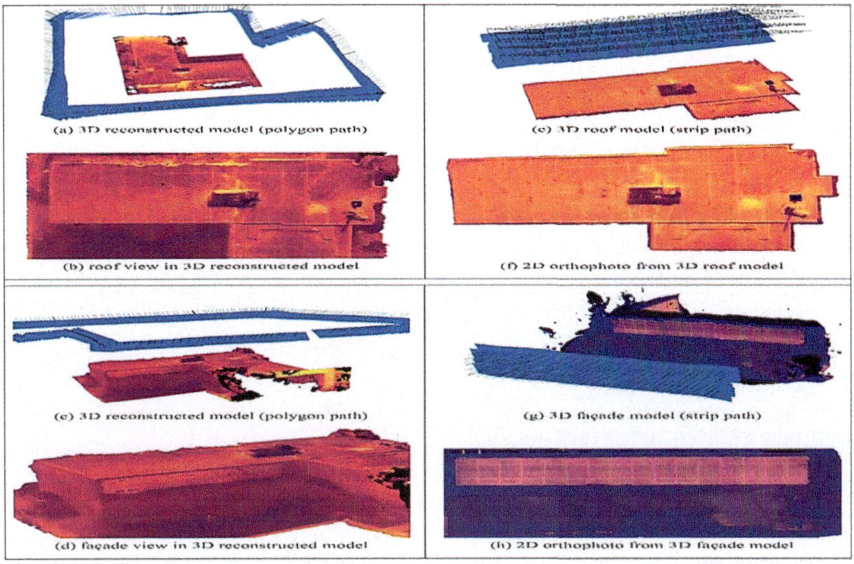

Fig. 4.21 Photogrammetric reconstruction of 3D models (polygon and strip paths) and 2D orthophotos (a geometrically corrected image, typically derived from aerial or satellite imagery) of roof and facade surfaces (front or exterior surface of a building). *Source* Ioja et al. (2022)—Open Access

Figure 4.21 presents 3D models and 2D orthophotos created from photogrammetry, although finer details like small thermal anomalies (e.g., roof seams and facade thermal bridges) are sometimes missed in the models (Kim et al. 2023).

Despite the limitations in thermal point cloud density due to data loss in infrared photogrammetry, such models still provide valuable insight into building performance (Zhou et al. 2022). These models allow for the extraction of thermal anomaly geolocation data, which can be incorporated into BIM for more efficient energy simulations and building assessments. The ongoing development in drone-based thermal scanning and photogrammetry is set to transform building energy auditing and thermal performance assessments (Liu et al. 2020), offering a non-invasive, real-time alternative to traditional methods (Bajwa et al. 2021).

Drone-Based Electromagnetic Surveys in Engineering Geophysics: Case Studies and Applications

The *principles of drone-based electromagnetic surveys* involve the use of unmanned aerial vehicles (UAVs) equipped with electromagnetic sensors to conduct geophysical investigations across difficult terrains with improved efficiency and safety. These systems have transformed traditional surveying methods by enabling faster data collection, higher spatial resolution, and reduced operational costs. Electromagnetic

(EM) methods are integral to geophysics, as they detect subsurface properties, map geological structures, and identify conductive materials such as metals and groundwater. Among the key EM techniques are time-domain electromagnetics (TDEM), which measures transient responses following a current pulse; frequency-domain electromagnetics (FDEM), which uses varying sinusoidal signals to locate conductive anomalies; and magnetic induction methods, which, when integrated with drones, help measure variations in the Earth's magnetic field.

Integration of drones with electromagnetic (EM) systems involves equipping UAVs with EM sensors such as receivers and transmitters, enabling high-resolution data collection from low-altitude flights. This configuration allows for precise mapping of electromagnetic anomalies across complex terrains. Depending on the specific requirements of a geophysical survey, drones may carry different types of EM sensors, including induction coils and magnetometers, to capture detailed information about subsurface conductive features.

Advantages of drone-based surveys include their ability to operate at low altitudes, which ensures high spatial resolution and enables detailed data acquisition. Drones can easily access remote or hazardous terrains that are typically difficult or unsafe for ground crews to reach. They also offer cost-effectiveness by reducing the need for heavy equipment and allowing large areas to be surveyed in a shorter time frame. Additionally, the use of drones enhances safety by eliminating the need for personnel to work in potentially dangerous environments such as steep slopes, high-voltage zones, or contaminated sites.

The development of drone-borne electromagnetic (EM) survey systems has gained attention for subsurface investigations, such as *detecting buried vehicles and mapping soil resistivity*, with these systems utilizing UAVs equipped with electromagnetic sensors that provide high-resolution data collection over large areas.

UAVs, particularly multicopter, offer significant advantages for geophysical surveys, providing access to hard-to-reach areas and acquiring high-resolution data more efficiently than traditional methods. Mitsuhata et al. (2022) developed a lightweight drone-based EM system using the GEM-2 instrument, capable of operating at low altitudes (~ 1 m). This system has been successfully tested for locating buried vehicles and soil resistivity mapping in agricultural areas, demonstrating its potential in rescue operations and environmental monitoring. Similar systems, like helicopter-borne EM methods, have been used for detecting metallic objects and soil resistivity mapping (Doll et al. 2000; Siemon et al. 2002). The drone system has proven effective, particularly in post-disaster scenarios such as the 2016 Kumamoto Prefecture on the Island of Kyushu, Japan and 2004 Niigata Prefecture Chuetsu earthquakes, located on the western coast of Japan's Honshu Island, where access was difficult. Despite challenges like limited payload capacity, flight duration, and GPS signal interference, the development of this drone-borne EM system shows promise for non-invasive, high-resolution subsurface surveys. The developed system is shown in Fig. 4.22.

To address the challenges in suspending the electromagnetic (EM) sensor from the drone, two multicopter, ZION CH940 and LAB6106, with payload capacities of 6 kg and 9 kg, respectively, were used (Mitsuhata et al. 2022). The GEM-2 instrument,

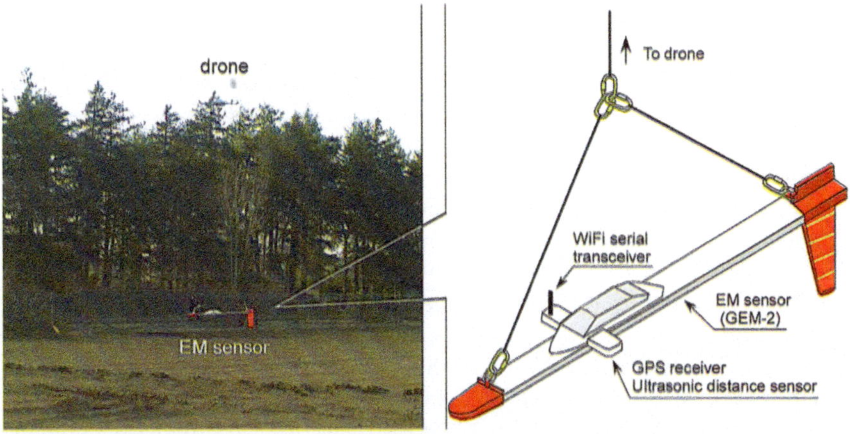

Fig. 4.22 In-flight EM sensor measurements from a drone: setup with GPS, ultrasonic sensor, Wi-Fi transceiver, vertical tail fin, and mounting jigs (field test in Kumamoto, Japan). *Source* Mitsuhata et al. (2022)—Open Access

along with the necessary components, weighed approximately 4.5 kg. A two-point suspension method was adopted for simplicity. To stabilize the sensor during flight, a vertical tail fin was added, improving yaw suppression at speeds above 2 m/s. Testing of drone-sensor separation indicated that a 5 m distance reduced EM noise (Fig. 4.23). For monitoring the height of the EM sensor above ground level (AGL), an ultrasonic distance sensor was mounted on the sensor itself, providing continuous AGL data during flight. A Wi-Fi serial transceiver enabled real-time control and data monitoring (Mitsuhata et al. 2006).

Two field experiments were conducted: the first focused on searching for buried vehicles, where the electromagnetic (EM) sensor acted as a metal detector, and the second involved resistivity mapping of an agricultural field. The EM response showed two key features: inductive and resistive limits, which are influenced by EM coupling

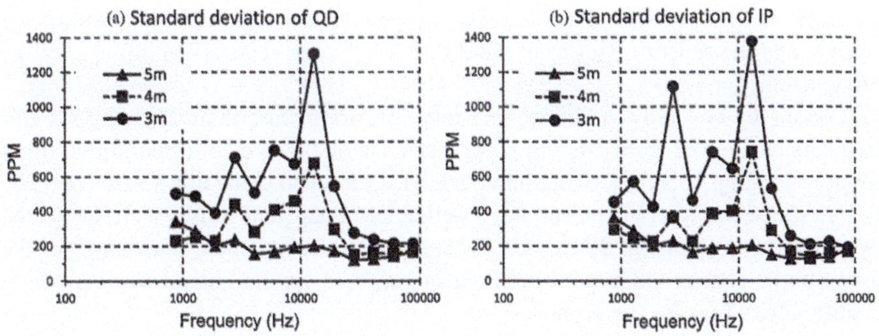

Fig. 4.23 Variations in EM data with different drone-sensor spacings: standard deviations of **a** quadrature and **b** in-phase components. *Source* Mitsuhata et al. (2022)—Open Access

between the transmitter (T_X) and receiver (R_X) coils and the target (Bruschini 2002; Everett 2013). Near the inductive limit, the in-phase component was preferred for detecting metal objects, while the quadrature component was more effective for resistivity mapping. Based on these characteristics, the in-phase component was used in the vehicle detection experiment, and the quadrature component was employed for soil resistivity mapping.

The developed drone-borne EM system was validated at the Japan Construction Method and Machinery Research Institute (JCMMRI) in *Fuji City, Japan*, where two subcompact cars were buried at depths of 1.5 m (shallow vehicle) and 3 m (deep vehicle), demonstrating the system's effectiveness in detecting buried vehicles at varying depths.

The ZION CH940 drone was used to conduct the search at a speed of 2 m/s, with the EM sensor suspended at approximately 1 m AGL. The search included both wide-area and restricted-area patterns. The EM data collected at frequencies ranging from 875 Hz to 87.1 kHz revealed significant anomalies, especially in the in-phase data, which were used for further analysis.

In the wide-area search, the in-phase data at 60.0 kHz clearly highlighted an underground building foundation, with smaller anomalies likely corresponding to the shallow vehicle. In the restricted-area search, the drone-borne EM data pinpointed the location of the shallow vehicle with clarity, while some indications of the deep vehicle were also observed. This was corroborated by in-situ measurements taken with the EM sensor at waist height.

The analysis of the in-phase data showed that the location of the shallow vehicle was easily identifiable due to the stronger anomalies, while the deep vehicle presented less distinct features. These findings demonstrate the effectiveness of the drone-borne EM system for detecting buried vehicles, with high resolution and accuracy, and validate its use in subsurface investigations.

To demonstrate the capability of the drone-borne EM system for mapping soil resistivity distributions, an experiment was conducted at the agricultural test site of the Institute for Rural Engineering, NARO, in *Tsukuba City, Japan*, showcasing the system's effectiveness in agricultural applications.

The test site included two fields—one dry and one moist. The drone (LAB6106) was flown at 2.5 m/s with the EM sensor suspended 1.6–2.0 m above ground level (AGL), and the sensor offset from the drone was 5 m to reduce EM noise from the drone itself.

The survey covered ten 200 m long transects arranged 5 m apart, using six EM frequencies ranging from 1875 to 87,075 Hz. The data were collected in two flight sessions totaling 15.9 min, with transects L1-L6 surveyed in the first session and L7-L10 in the second. After the survey, baseline EM data (bias noise) were measured by hovering the drone at 20 m AGL. A comparison between the bias noise for the drone-borne sensor and a hand-held sensor revealed similar characteristics, with a slightly lower bias for the drone.

The data analysis used median filtering, smoothing, and resampling to construct datasets for both the quadrature and in-phase components. The quadrature data, which showed consistent positive values, were plotted on a logarithmic scale. As expected, the amplitude decreased with frequency, particularly near the resistive limit. For the in-phase data, negative values were observed below 41.0 kHz, making them unsuitable for logarithmic representation. The AGL height of the EM sensor varied from 1.19 to 2.82 m.

A 1-D inversion was applied to the quadrature data to estimate the resistivity distribution, dividing the soil into 30 layers. The results of the inversion, along with the RMS misfits, were used to visualize the resistivity profiles. The resistivity difference between the dry and moist fields was clearly observable near the surface, with the resistivity decreasing with depth in the moist field.

Figure 4.24a–c show the resistivity distributions at various depths, highlighting the contrast in resistivity between the left (dry) and right (moist) fields. A handheld electrical conductivity (EC) meter (Fig. 4.24d) confirmed the soil resistivity differences at the surface. The resistivity contrast diminished with depth, likely due to differences in water content and soil compaction. The depth of investigation (DOI) was estimated based on the skin depth, with values of approximately 6.6 m for resistivity 30 Ωm ground and 7.8 m for 60 Ωm ground.

Figure 4.25a–c present the cross-sections and depth profiles of the resistivity distributions, showing lower resistivity values near the surface of the moist field, increasing with depth. These results validate the drone-borne EM system as a powerful tool for soil resistivity mapping, providing detailed insights into subsurface characteristics in agricultural settings.

Drone-Towed CSEM System for Near-Surface Geophysics focuses on the *noise, drift, and survey design* of small uncrewed aerial vehicles (UAVs), or drones, which

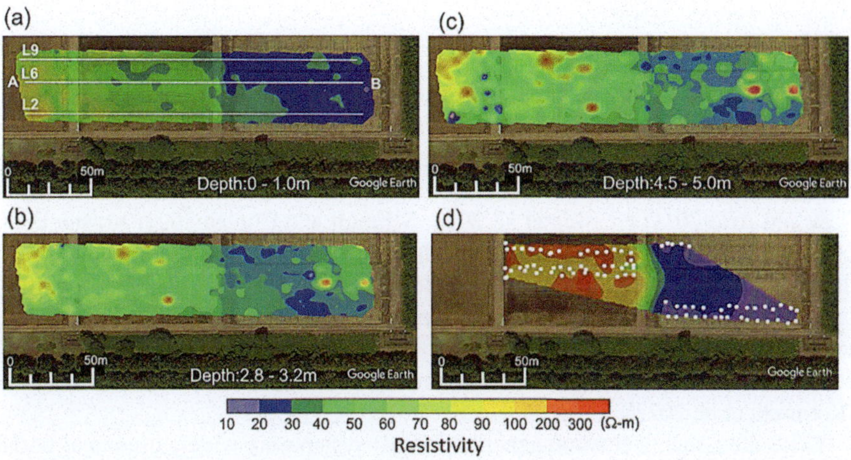

Fig. 4.24 Resistivity distributions and measurement points: **a** resistivity distributions at 0–1 m depth, **b** 2.8–3.2 m depth, and **c** 4.5–5 m depth derived from 1-D inversion of EM data and in-situ electrical conductivity (EC) measurements. **d** Resistivity measured using a soil EC meter, with white circles marking measurement points. White lines in (**a**) indicate profiles L2, L6, and L9 used for constructing resistivity cross-sections in Fig. 4.25a. *Source* Mitsuhata et al. (2022)—Open Access

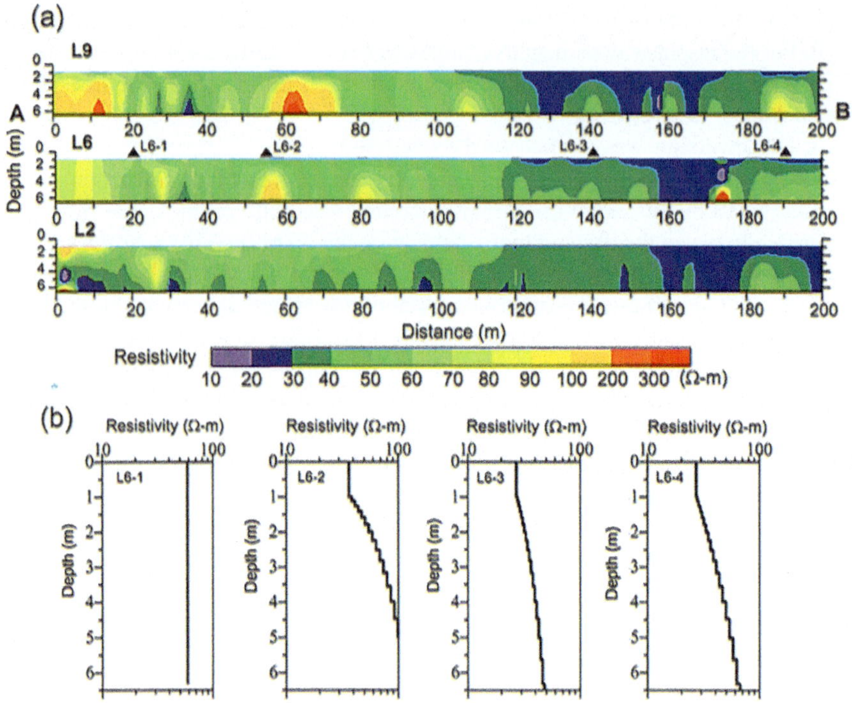

Fig. 4.25 Resistivity cross sections and depth profiles: **a** cross sections along profiles L2, L6, and L9. **b** Depth profiles at stations L6-1 to L6-4, shown on the L6 profile cross section. *Source* Mitsuhata et al. (2022)—Open Access

have become increasingly popular platforms for geophysical and archaeological prospecting, utilizing various sensors such as ground-penetrating radar (Altdorff et al. 2014), gamma-ray detectors (Mochizuki et al. 2017), thermal sensors (Poirier et al. 2013; Petzke et al. 2013), LiDAR (Risbol and Gustavsen 2018), and magnetometers (Lev and Arie 2011; Dossing et al. 2021; Schmidt and Coolen 2021; Kolster et al. 2022). These platforms, equipped with diverse sensors, are used to gather crucial data for geophysical surveys and archaeological investigations.

Electromagnetic induction methods, commonly used in mineral exploration, are less frequently employed in drone-borne systems (Karaoulis et al. 2020; Mitsuhata et al. 2022), though alternatives like semi-airborne solutions have been explored (Kotowski et al. 2022).

Controlled-source electromagnetic (CSEM) systems are widely used in airborne and handheld applications but remain underexplored in drone-based configurations. While airborne systems cover large areas efficiently (Siemon et al. 2009), handheld systems offer high spatial resolution but face mobility challenges in difficult terrains. Drone-mounted CSEM instruments combine mobility with spatial resolution but introduce issues like drone-induced electromagnetic noise and sensor movement during flight.

This study, led by Vilhelmsen and Dossing (2022), focuses on a drone-towed CSEM system using the GEM-2 UAV sensor (Lerssi et al. 2016) for near-surface geophysical prospecting. The authors address key challenges, including instrument noise, temperature drift, transmission frequency, and survey setup. Additionally, the study considers the effects of sensor oscillation (Ward and Hohmann 1988) and provides recommendations for optimizing data quality for near-surface targets, with broader applications in subsurface exploration.

4.3 Conclusions

Recent advancements in geophysical surveying techniques have significantly transformed the way subsurface investigations are conducted in engineering and environmental applications. The integration of cutting-edge technologies such as Unmanned Aerial Vehicles (UAVs), drones, advanced sensors, remote sensing platforms, and data-driven tools like Artificial Intelligence (AI) and Machine Learning (ML) has elevated both the precision and efficiency of geophysical surveys. These technologies not only facilitate rapid and high-resolution data acquisition but also enhance safety, reduce operational costs, and enable access to challenging terrains.

Drone-based geophysical methods—including LiDAR, magnetometry, thermal imaging, ground-penetrating radar (GPR), and electromagnetic surveys—have proven particularly valuable for a wide range of applications, from infrastructure monitoring and site investigations to environmental assessments and disaster response. The chapter has also demonstrated, through various case studies, how these innovations are being successfully implemented in real-world scenarios.

Despite the remarkable benefits, several challenges remain, including regulatory constraints, data integration complexities, and limitations related to environmental conditions and sensor capabilities. Nonetheless, continuous improvements in drone technology, sensor miniaturization, and computational analytics are expected to address these issues and further broaden the scope of geophysical surveying.

In conclusion, the convergence of geophysical techniques with emerging technologies marks a paradigm shift in the field, offering new possibilities for enhanced site characterization, resource exploration, and environmental monitoring. Future developments are likely to focus on fully autonomous systems, real-time processing, and integrated platforms that will further revolutionize engineering geophysics.

References

Alhargan M, Al-Obaidi M, Alharthy A (2021) Application of infrared thermography in building diagnostics. Energy Build 215:109883

Altdorff D, Schliffke N, Riedel M, Schmidt V, van der Kruk J, Vereecken H, Stoll J, Becken M (2014) UAV-borne electromagnetic induction and ground-penetrating radar measurements: a feasibility test. Water Resour Res 42:W11403

Arne T, Johansson F, Andersson H (2018) Infrared thermography for detection of energy inefficiencies in building envelopes. Renew Sustain Energy Rev 81:2327–2336

Aucelli PPC, Mattei G, Caporizzo C, Di Luccio D, Tursi MF, Pappone G (2021) Coastal vs volcanic processes: Procida Island as a case of complex morpho-evolutive response. Mar Geol 448:106814

Bajwa I, Mistry S, Agarwal R (2021) Drone-based thermal imaging for building energy assessments. J Build Perform 12(3):48–58

Baltsavias EP (1999) Airborne laser scanning: existing systems and firms and other resources. ISPRS J Photogram Remote Sens 54:164–198

Benincasa P, Caturegli L, Coscarelli R, Guarnieri A, Fiori G, Pappalardo G, Tosti G (2018) Remote sensing for vegetation health and stress monitoring using multispectral UAV imagery. Agric Syst 163:107–118

Berni J, Zarco-Tejada P, Suarez L, Fererez E (2009) Thermal and narrow-band multispectral remote sensing for vegetation monitoring from an unmanned aerial vehicle. IEEE Trans Geosci Remote Sens

Bruschini C (2002) A multidisciplinary analysis of frequency domain metal detectors for humanitarian demining. Ph.D. thesis, Vrije Universiteit Brussel

Carter L, Jackson D, Thompson P (2021) Smart city initiative: assessing building energy performance and CO_2 emissions using UAV-based thermal imaging. J Urban Sustain 15(4):115–125

Caturegli L, Castronuovo M, Rinaldi M (2016) Assessment of vegetation indices for soil moisture and crop health monitoring using UAVs. Sensors 16(12):2090

Chiocci FL, Guerrieri L, Monegato G, Pieruccini P (2023) Quaternary map of Italy (sheet 3). In: Proceedings of the XXI INQUA congress, Rome, Italy, 14–21 July 2023

Conforti M, Mercuri M, Borrelli L (2021) Morphological changes detection of a large earthflow using archived images, LiDAR-derived DTM, and UAV-based remote sensing. Remote Sens 13:120

Dalla Mura M, Benediktsson JA, Waske B, Bruzzone L (2010) Morphological attribute profiles for the analysis of very high-resolution images. IEEE Trans Geosci Remote Sens 48:3747–3762

Davila N, Capra L, Gavilanes-Ruiz JC, Varley N, Norini G, Vazquez AG (2007) Recent lahars at Volcan de Colima (Mexico): drainage variation and spectral classification. J Volcanol Geotherm Res 165:127–141

Di Martino S, D'Andrea F, Rocca R (2019) Drone-based 3D model reconstruction for PV plant monitoring. Int J Energy Environ Eng 10(4):399–409

Doll WE, Nyquist JE, Beard LP, Gamey TJ (2000) Airborne geophysical surveying for hazardous waste site characterization on the Oak Ridge Reservation, Tennessee. Geophysics 65:1372–1387

Dong P, Chen Q (2018) LiDAR remote sensing and applications. CRC Press, Boca Raton, FL, USA

Dossing A, Simoes L, da Silva E, Martelet G, Maack Rasmussen T, Gloaguen E, Thejll Petersen J, Linde J (2021) A high-speed, light-weight scalar magnetometer bird for km scale UAV magnetic surveying: on sensor choice, bird design, and quality of output data. Remote Sens 13(4):649. https://doi.org/10.3390/rs13040649

Dossing A, Kolster ME, Rasmussen TM, Petersen JT, da Silva ELS (2023) UAV-towed scalar magnetic gradiometry: a case study in relation to iron oxide copper-gold mineralization, Nautanen (Arctic Sweden). Leading Edge 42(2):103–111

Dossing A, Kolster ME, da Silva ELS, Muxworthy AR, Petersen JT, Riishuus MS (2024) Pre-existing structural control on the recent Holuhraun eruptions along the Bardarbunga spreading center, Iceland. Sci Rep 14(1):3399. https://doi.org/10.1038/s41598-024-53790-x

Everett ME (2013) Near-surface applied geophysics, vol 1. Cambridge University Press, Cambridge, pp 45–78

Favalli M, Fornaciai A, Pareschi MT (2009) LiDAR strip adjustment: application to volcanic areas. Geomorphology 111:123–135

Frid M, Frid V (2024) Vital views into drone-based GPR application: precise mapping of soil-to-rock boundaries and groundwater level for foundation engineering and site-specific response. Appl Sci 14:7889. https://doi.org/10.3390/app14177889

Fujimura M, Nishida S, Tanaka H, Kato Y (2018) Development of a drone-based system for infrared inspection of solar power plants using machine learning. IEEE Trans Ind Electron 65(10):8325–8334

Gao W, Zhang Y, Xu Z, Li C (2020) Automated thermal image analysis for solar panel inspection using drones. IEEE Access 8:105517–105528

Geipel J, Knauer T, Lichtenegger T (2016) UAV-based multispectral sensing for precision agriculture: a study on crop damage monitoring. Sensors 16(12):2090

Gershenson H, Jacobs C, Mauer P (2019) Continuous monitoring of building performance using drones. Sustain Cities Soc 47:101–113

Haboudane D, Miller JR, Pattey E, Vigneault P (2002a) Use of multispectral and hyperspectral imagery for estimating crop biophysical parameters: a review. Remote Sens Environ 81(1):89–98

Haboudane D, Miller JR, Tremblay N, ZarcoTejada PJ, Dextraze L (2002b) Integrated narrow-band vegetation indices for prediction of crop chlorophyll content for application to precision agriculture. Remote Sens Environ 81:416–426

Heincke B, Andersen O, Moller P, Rasmussen M, Sorensen J (2019) Developing multi-sensor drones for geological mapping and mineral exploration: setup and first results from the MULSEDRO project. GEUS Bull 43. https://doi.org/10.34194/GEUSB-201943-03-02

Henderson M, Lee C, Johnson D, Rodriguez S (2019) Geophysical applications of UAV-based remote sensing in environmental studies. Geophysics 84(3):91–106

Hernandez J, Garcia A, Lopez R (2018) Application of infrared thermography in photovoltaic systems inspection: a review. Energy Rep 4:118–127

Ioja I, Nedeff V, Nedeff F (2022) The use of drones in multispectral photogrammetry and thermoscanning—a review. Bul Inst Politeh Din IAŞI (Bull Polytech Inst IASI) Gheorghe Asachi Tech Univ Iasi Sect Math Theor Mech Phys 68(72)(4). https://doi.org/10.2478/bipmf-2022-0018

Jackisch R, Hansen M, Schilling C, Jensen K (2022) Drone-based magnetic and multispectral surveys to develop a 3D model for mineral exploration at Qullissat Disko Island, Greenland. Solid Earth 13(4):793–825

Johansson M, Bergman M, Lind K, Pettersson J (2019) Low-cost UAV system for heat loss detection in buildings: a case study at Malardalen University, Vasteras, Sweden. J Build Phys 43(3):233–245

Karaoulis M, Ritsema I, Bremmer C, De Kleine M (2020) Drone-borne electromagnetic (DR-EM) surveying in The Netherlands: lab and field validation results. Remote Sens 14(21):5335

Kim Y, Lee S, Cho B (2023) The role of photogrammetry in building energy simulation. J Build Eng 48:103765

Kolster ME, Dossing A (2021) Scalar magnetic difference inversion applied to UAV-based UXO detection. Geophys J Int 224(1):468–486

Kolster ME, Wigh MD, da Silva ELS, Vilhelmsen TB, Dossing A (2022) High-speed magnetic surveying for unexploded ordnance using UAV systems. Remote Sens 14(5):1134. https://doi.org/10.3390/rs14051134

Kosik S, Nemeth K, Rees C (2022) Integrating LiDAR to unravel the volcanic architecture and eruptive history of the peralkaline Tuhua (Mayor Island) volcano, New Zealand. Geomorphology 418:108481

Kotowski PO, Becken M, Thiede A, Schmidt V, Schmalzl J, Ueding S, Klingen S (2022) Evaluation of a semi-airborne electromagnetic survey based on a multicopter aircraft system. Geosciences 12:26

Lee S et al (2024) Machine learning in geophysics: advances and applications in subsurface modeling. Geophys J Int

Lerssi J, Niemi S, Suppala I (2016) GEM-2-new generation electromagnetic sensor for near surface mapping. In: 22nd European meeting of environmental and engineering geophysics, near surface geoscience 2016, Palau de Congressos de Catalunya, Barcelona, Spain, 4–8 Sept 2016, code 124666

Lev E, Arie M (2011) Unmanned airborne magnetic and VLF investigations: effective geophysical methodology for the near future. Positioning

Li Z et al (2019) LiDAR-based terrain modeling for geophysical applications: a case study in forested regions. Geomat Nat Hazards Risk 10(1):182–202

Li S et al (2020) 3D reconstruction of solar panels using UAV photogrammetry for fault detection. Sensors 20(3):814

Li X et al (2023) Innovations in seismic sensors for high-fidelity subsurface imaging. Geophys Res Lett

Liao K-C, Wu H-Y, Wen H-T (2022) Using drones for thermal imaging photography and building 3D images to analyze the defects of solar modules. Inventions 7:67. https://doi.org/10.3390/inventions7030067

Liu X (2008) Airborne LiDAR for DEM generation: some critical issues. Prog Phys Geogr 32:31–49

Liu J, Zhang Y, Li X (2020) BIM integration with drone-based thermal imaging for building performance simulation. Autom Constr 112:103071

Mazzarella L, Siani A, Castaldo V, Ippolito A (2020) Use of drones and infrared thermography for the diagnosis of solar plants. Renew Sustain Energy Rev 124:109799

Mazzarini F, Pareschi MT, Favalli M, Isola I, Tarquini S, Boschi E (2005) Morphology of basaltic lava channels during the Mt. Etna September 2004 eruption from airborne laser altimeter data. Geophys Res Lett 32:1

McMillan R, Turner J, Carter M, Brown P (2023) Recent developments in portable geophysical instrumentation for environmental surveys. J Appl Geophys 189:104726

Mendonca R, Silva J, Costa A (2020) Pre-processing techniques for infrared thermography data in building diagnostics. Infrared Phys Technol 107:103302

Mitra R, Bhattacharya R, Dey S (2022) Thermal performance analysis and modeling using infrared thermography and BIM. Energy Rep 8:234–246

Mitsuhata Y, Uchida T, Matsuo K, Marui A, Kusunose K (2006) Various-scale electromagnetic investigations of high-salinity zones in a coastal plain. Geophysics 71:B167–B173

Mitsuhata Y, Ueda T, Kamimura A, Kato S, Takeuchi A, Aduma C, Yokota T (2022) Development of a drone-borne electromagnetic survey system for searching for buried vehicles and soil resistivity mapping. Near Surf Geophys 20:16–29. https://doi.org/10.1002/nsg.12189

Mochizuki S, Kataoka J, Tagawa L, Iwamoto Y, Okochi H, Katsumi N, Kinno S, Arimoto M, Maruhashi T, Fujieda K, Kurihara T, Ohsuka S (2017) First demonstration of aerial gamma-ray imaging using drone for prompt radiation survey in Fukushima. J Instrum 12:P11014

Mouginis-Mark PJ, Garbeil H (2005) Quality of TOPSAR topographic data for volcanology studies at Kilauea Volcano, Hawaii: an assessment using airborne LiDAR data. Remote Sens Environ 96:149–164

Muller K, Schmidt H, Becker T, Wagner F (2020) Thermal UAV inspections for energy efficiency in residential buildings in Berlin, Germany. Renew Sustain Energy Rev 119:109590

Natale J, Camanni G, Ferranti L, Isaia R, Sacchi M, Spiess V, Steinmann L, Vitale S (2022) Fault systems in the offshore sector of the Campi Flegrei Caldera (Southern Italy): implications for nested Caldera structure, resurgent dome, and volcano-tectonic evolution. J Struct Geol 163:104723

Neri M, Mazzarini F, Tarquini S, Bisson M, Isola I, Behncke B, Pareschi MT (2008) The changing face of Mount Etna's summit area documented with LiDAR technology. Geophys Res Lett 35:1–6

Patel N, Sharma P, Gupta S, Kumar R (2023) Artificial intelligence for geophysical data interpretation and analysis. J Comput Geophys 58(4):567–580

Perez M, Rodriguez J, García F (2021) Improving energy prediction models for buildings with IoT sensors. Sustain Energy Technol Assess 44:100999

Petersen JT, Dossing A (2024) Addressing magnetometer inconsistencies in drone-borne total field gradiometer surveys: case study from Ranbogen, Arctic Norway. Geophys J Int 239:1617–1628

Petzke M, Hofmeister P, Hordt A, Glaßmeier K, Auster H (2013) Aeromagnetics with an unmanned airship. In: Near surface geoscience 2013—19th European meeting of environmental and engineering geophysics of the near surface geoscience division of EAGE, Bochum, Germany, 9–11 Sept 2013, code 104173

Poirier N, Hautefeuille F, Calastrenc C (2013) Low altitude thermal survey by means of an automated unmanned aerial vehicle for the detection of archaeological buried structures. Archaeol Prospect 20:303–307

Risbol O, Gustavsen L (2018) LiDAR from drones employed for mapping archaeology—potential, benefits and challenges. Archaeol Prospect 25:329–338

Rouse JW, Haas RH, Schell JA, Deering DW (1974) Monitoring vegetation systems in the Great Plains with ERTS. Third Earth Resources Technology Satellite-1 Symposium, NASA SP-351, vol 1. NASA, Washington, DC, pp 309–317

Sbrana A, Marianelli P, Pasquini G (2021) The Phlegrean fields volcanological evolution. J Maps 17:557–570

Schmidt V, Coolen J (2021) Potential and challenges of UAV-borne magnetic measurements for archaeological prospection. Rev d'Archeometr 45:9645

Shan J, Hyyppa J (2018) Advances in airborne LiDAR systems and data processing. CRC, Boca Raton, FL, USA

Shao S, Guo Y, Zhang Z, Yuan H (2019) Single remote sensing multispectral image dehazing based on a learning framework. Math Probl Eng

Siemon B, Stuntebeck C, Sengpiel K-P, Rottger B, Rehli H-J, Eberle DG (2002) Investigation of hazardous waste sites and their environment using the BGR helicopter-borne geophysical system. J Environ Eng Geophys 7:169–181

Siemon B, Christiansen AV, Auken E (2009) A review of helicopter-borne electromagnetic methods for groundwater exploration. Near Surf Geophys 7:629–646

Silva-Fragoso A, Norini G, Nappi R, Groppelli G, Michetti AM (2024) Improving the accuracy of digital terrain models using drone-based LiDAR for the morpho-structural analysis of active Calderas: the case of Ischia Island, Italy. Remote Sens 16:1899

Smith J, Johnson A, Lee K, Davis R (2022) UAV applications in geophysical surveys: case studies and future prospects. Geophys Today 45(7):123–135

Tao L, Liu X, Zhang F, Chen Y (2020) Drone-based multispectral and thermal imaging for vegetation and environmental monitoring. Int J Remote Sens 41(15):5407–5425

Tichem M, de Boer B, Meijer A (2021) Integration of BIM and thermal modeling for energy simulation. Energy 228:120582

Vazquez J, Gonzalez E, Ruiz J, Martínez M (2021) Integration of UAV-based thermal imaging and 3D mapping for photovoltaic panel inspection. J Field Robot 38(2):207–223

Ventura G, Vilardo G (2008) Emplacement mechanism of gravity flows inferred from high resolution LiDAR data: the 1944 Somma Vesuvius lava flow (Italy). Geomorphology 95:223–235

Vilhelmsen TB (2023) UAV-borne near-surface geophysics: application of electromagnetic and magnetic sensor systems for surveying archaeological structures and anthropogenic metal contaminants. Ph.D. thesis, Technical University of Denmark

Vilhelmsen TB, Dossing A (2022) Drone-towed controlled-source electromagnetic (CSEM) system for near-surface geophysical prospecting: on instrument noise, temperature drift, transmission frequency, and survey set-up. Geosci Instrum Method Data Syst 11:435–450. https://doi.org/10.5194/gi-11-435-2022

Walter C, Braun A, Fotopoulos G (2020) High-resolution unmanned aerial vehicle aeromagnetic surveys for mineral exploration targets. Geophys Prospect 68:334–349. https://doi.org/10.1111/1365-2478.12914

Ward SH, Hohmann GW (1988) Electromagnetic theory for geophysical applications. In: Nabighian MN (ed) Electromagnetic methods in applied geophysics: vol. 1, theory. Society of Exploration Geophysicists, pp 130–311

Wierzbicki D, Kedzierski M, Sekrecka A (2019) A method for dehazing images obtained from low altitudes during high-pressure fronts. Remote Sens 12:25. https://doi.org/10.3390/rs12010025

Woloszyn E (2009) An overview of meteorology and climatology. Gdansk University of Technology, Gdansk, Poland, p 26. ISBN: 978-83-7775-237-1

Xie L, Chen Y, Zhang Z (2020) Data-driven energy prediction in buildings using IoT sensors. Energy Build 220:109972

Zhao X, Liu Y, Zhang J (2022) Accuracy enhancement in radiometric thermal cameras for building diagnostics. Build Environ 197:107926

Zhao H, Chen R, Li Z, Wang X (2023) Drones in geological mapping: a new era in remote sensing. J Remote Sens Technol

Zheng HB et al (2019) Improved estimation of rice above-ground biomass by combining UAV image texture and spectral analysis. Precis Agric 20(3):611–629. https://doi.org/10.1007/s11119-018-9600-7

Zhou X, Li H, Zhang J, Chen W (2021) A deep learning-based approach for automated solar panel defect detection using UAV thermal images. Remote Sens 13(10):1949

Zhou H, Yang W, Xu G (2022) Point cloud density in infrared photogrammetry for building diagnostics. Autom Constr 130:103881

Chapter 5
Geophysics for Subsurface Exploration and Site Characterization

Abstract This chapter focuses on how geophysical methods are used to explore and understand subsurface conditions for engineering and environmental projects. It presents recent advances in technologies such as seismic imaging, ground-penetrating radar (GPR), electrical resistivity tomography (ERT), electromagnetic surveys, and magnetotellurics. These tools help map underground features and assess soil, rock, and groundwater conditions. Numerous case studies—from oil fields in the North Sea to archaeological sites in Israel and urban infrastructure in London and Taiwan—illustrate how geophysical techniques are applied in real-world scenarios. The chapter also highlights applications in landfill monitoring, slope stability, tunnel assessment, and foundation design. Challenges in subsurface mapping, such as data interpretation and site variability, are discussed along with future directions. Geophysical exploration supports safe construction, efficient planning, and environmental protection, making it essential in civil engineering, tunneling, and infrastructure development.

5.1 Introduction to Subsurface Exploration and Site Characterization

This chapter explores the innovative approaches to subsurface exploration and site characterization through the use of advanced geophysical methods. It discusses recent developments in subsurface imaging technologies, including seismic imaging, ground-penetrating radar (GPR), electrical resistivity tomography (ERT), electromagnetic methods, and magnetotellurics (MT). These methods are critical for understanding subsurface structures and have diverse applications across fields such as civil engineering, construction, environmental studies, and archaeological research.

The chapter presents several case studies that highlight the real-world applications of these techniques, such as seismic imaging in oil field exploration, GPR for infrastructure assessments, and ERT for groundwater flow and slope failure investigations. The role of geophysics in enhancing site characterization and improving

D. C. Naskar, *Advances in Engineering Geophysics: Concepts and Applications*, Earth Systems Data and Models 9, https://doi.org/10.1007/978-3-032-28000-8_5

the accuracy of subsurface data collection is emphasized, showcasing the growing importance of these techniques in both urban and remote environments.

Additionally, the chapter discusses the challenges faced in subsurface exploration, including limitations in resolution, data interpretation, and the need for further technological advancements. It concludes by exploring the future potential of these geophysical methods, with a focus on the integration of emerging technologies like drone-based GPR and machine learning algorithms to improve data processing and decision-making. Ultimately, this chapter underscores the critical role geophysics plays in modern subsurface exploration and its potential to address complex engineering and environmental challenges.

5.2 Innovative Approaches in Subsurface Imaging, Characterization, and Mapping

Subsurface exploration and site characterization have become increasingly vital in fields such as construction, environmental engineering, resource extraction, and disaster mitigation. Recent advancements in geophysics have significantly enhanced our ability to image, characterize, and map the subsurface with precision. This section delves into the latest technologies, methodologies, and applications in subsurface geophysics, focusing on innovations that improve resolution, accuracy, and efficiency.

5.2.1 Advances in Subsurface Imaging Technologies

Seismic Imaging

One of the most significant developments in subsurface imaging is Full Waveform Inversion (FWI), a method that utilizes the entire seismic wavefield to generate high-resolution models of the subsurface. FWI has revolutionized seismic imaging by offering detailed insights into velocity variations and complex geological structures, achieving resolutions as fine as half the seismic wavelength. This method relies on a data-fitting approach and is particularly effective when applied to seismic datasets with dense spatial and temporal sampling. Its application has become increasingly widespread in geotechnical and exploration contexts, enhancing our ability to characterize subsurface features with unprecedented accuracy (Cao et al. 2022; Christeson et al. 2021; Lailly 1983; Tarantola 1984). Figure 5.1a, b typically illustrate the improvements in resolution obtained using FWI techniques compared to conventional seismic imaging.

In addition to FWI, Passive Seismic Monitoring has emerged as a valuable and non-invasive approach to subsurface imaging. This technique leverages ambient

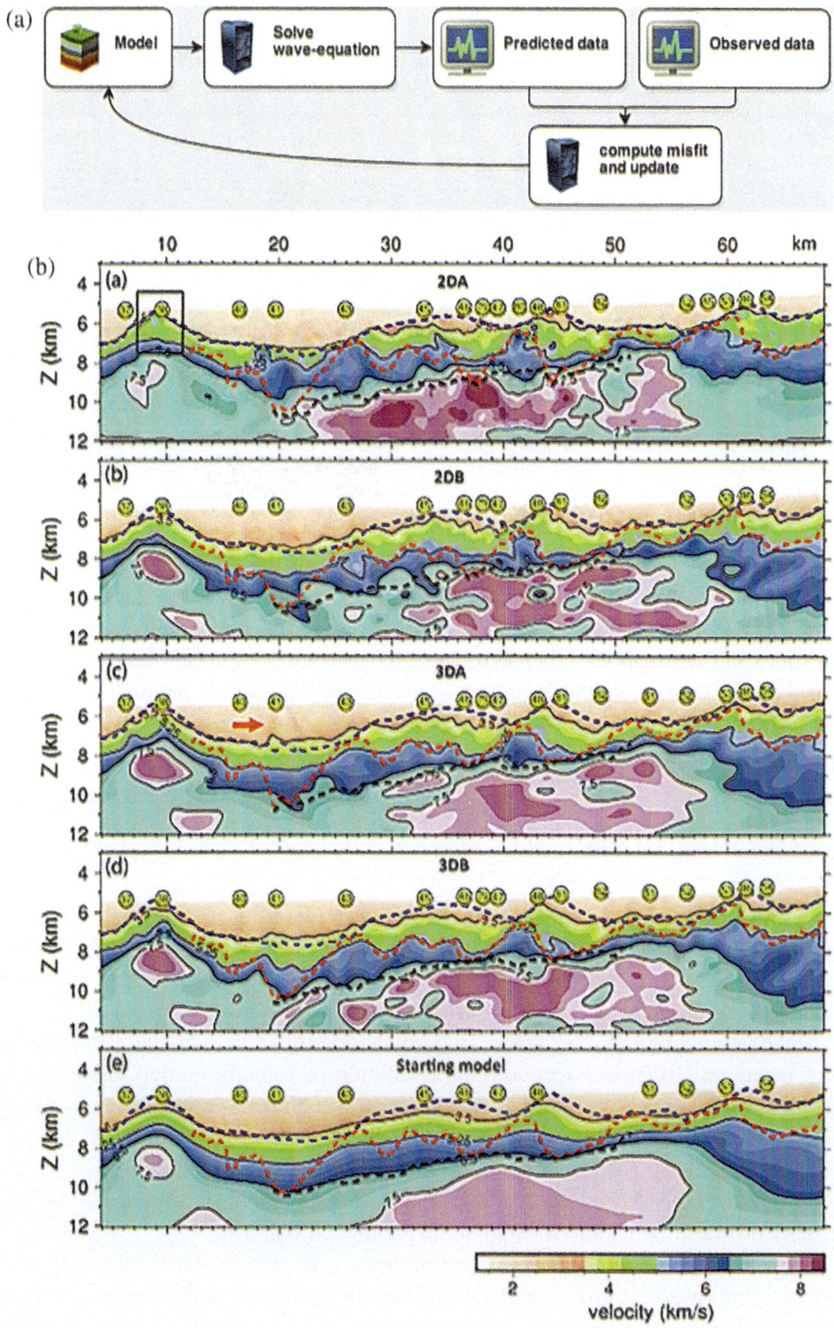

Fig. 5.1 **a** Detailed schematic of the Full Waveform Inversion (FWI) process workflow. *Source* Georgia Institute of Technology. **b** Velocity profiles along line 2 from 2-D and 3-D Full Waveform Inversion (FWI) models. *Source* Boddupalli et al. (2021) Open Access

Fig. 5.2 Geospatial overview of the North Sea Grane Oil Field. *Source* Wikipedia

seismic noise, rather than artificially generated signals, to monitor changes in the subsurface over time. It is particularly well-suited for long-term monitoring in environmentally sensitive areas or where conventional seismic sources are impractical. Passive seismic methods offer a cost-effective alternative for detecting structural changes, stress accumulation, or fluid migration, further contributing to the advancement of geophysical site characterization.

Case Study North Sea Oil Field

The North Sea Grane Oil Field, Norway (Fig. 5.2) remains one of the most significant regions for hydrocarbon exploration, yet its complex subsurface geology, characterized by faulted sedimentary basins and variable stratigraphy, presents substantial challenges. Accurate imaging of subsurface features is critical to reducing exploration risks and enhancing recovery rates. To demonstrate the effectiveness of Full Waveform Inversion (FWI) in improving the resolution of velocity models and enabling better interpretation of geological structures in the North Sea basin. High-quality seismic datasets were collected using advanced offshore acquisition techniques (Hicks et al. 2016). FWI was employed to iteratively refine velocity models. The algorithm leveraged both low- and high-frequency seismic data to produce detailed velocity profiles that captured small-scale geological features.

The FWI-derived models revealed significant geological details, including fault patterns, thin reservoir layers, and subtle velocity anomalies that were undetectable with traditional methods. These enhanced models directly influenced well placement, leading to a 20% reduction in dry well rates. FWI represents a transformative approach in seismic imaging, providing high-resolution insights essential for complex hydrocarbon basins. This case study underscores its potential for application in similarly challenging exploration environments worldwide.

Case Study Comparison of 2-D and 3-D Full Waveform Inversion Imaging from Deep Galicia Margin Seismic Data

The case study focuses on the application of Full Waveform Inversion (FWI) to seismic data collected from the Deep Galicia Margin, a complex offshore geological region (Boddupalli et al. 2021). The goal of the study is to compare the performance of 2-D and 3-D FWI methods in generating high-resolution subsurface velocity models using wide-angle seismic data. The Deep Galicia Margin is characterized by its intricate geological features (Murillas et al. 1990; Peron-Pinvidic et al. 2007; Tucholke et al. 2007; Perez-Gussinye et al. 2003), including steep slopes, fault zones formed by rifting and sea-floor spreading (Fig. 5.3), and potential gas hydrates, making it a challenging target for seismic imaging.

In this study, wide-angle seismic data was acquired along multiple profiles across the margin, using both 2-D and 3-D acquisition geometries with 78 ocean bottom hydrophones/seismometers (Fig. 5.4). The 2-D FWI was applied along a single profile, assuming lateral homogeneity, while the 3-D FWI utilized data from the entire survey area, accounting for lateral variations in subsurface structures. Both approaches were used to create detailed velocity models of the subsurface, with the 3-D FWI aiming to capture more spatial complexity.

The primary objective was to assess how well each method could resolve key geological features, such as fault zones, sedimentary layers, and gas anomalies, and how the two methods performed in terms of computational efficiency and accuracy. The 2-D FWI method, being computationally less intensive, was faster but potentially less accurate in capturing lateral variations. On the other hand, the 3-D FWI provided a more accurate representation of the subsurface, particularly in areas where lateral geological changes were significant. However, the 3-D method required significantly more computational resources and processing time.

The results showed that the 2-D FWI provided a reasonable approximation of the subsurface velocity structure along the profile, but it struggled to capture lateral geological variations and more subtle features like small fault zones. In contrast, the 3-D FWI yielded a much more detailed and accurate model, revealing finer features such as small-scale faults and variations in sedimentary layers, and it provided a more comprehensive understanding of the subsurface. Furthermore, the 3-D FWI was particularly useful for accurately imaging complex features such as gas anomalies, which were difficult to resolve using 2-D FWI.

Despite the improved results from the 3-D approach, the study highlighted that the 2-D FWI could still be useful in certain situations, especially in regions with relatively simple geology or where computational resources are limited. In such cases, the 2-D method can provide quick and reasonably accurate models for initial interpretations. However, for more complex regions like the Deep Galicia Margin,

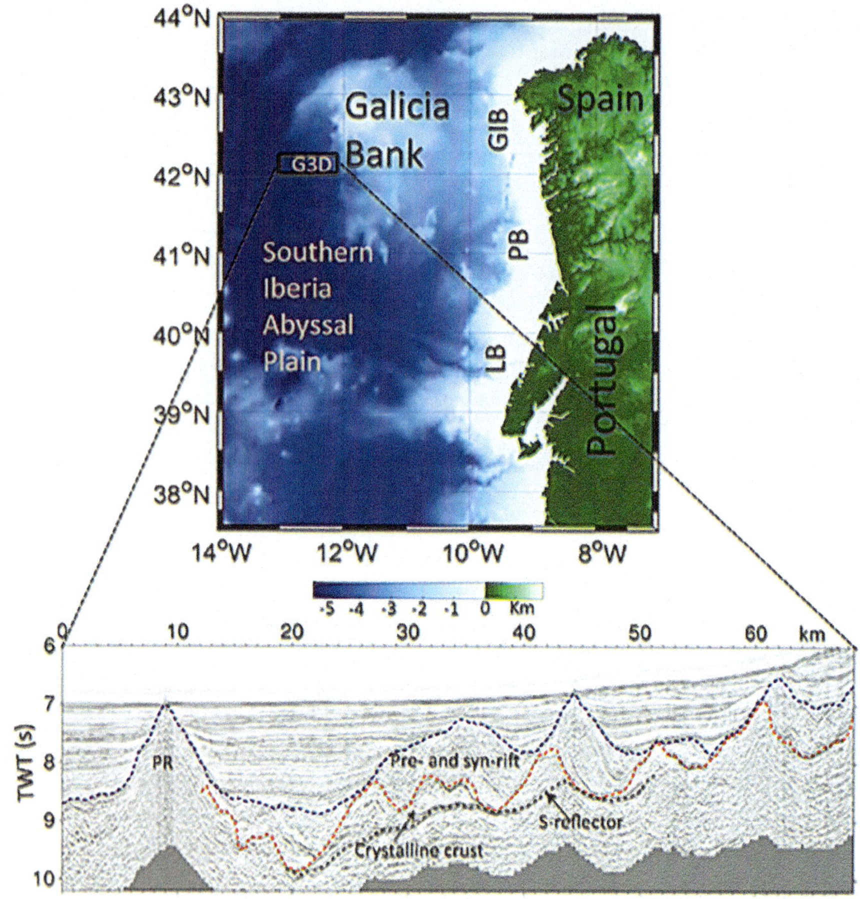

Fig. 5.3 Upper panel: Map showing the location of the Galicia 3-D (G3D) experiment in the DGM (black box), with surrounding features including the Galicia Bank, Galicia Interior Basin (GIB), Porto Basin (PB), and Lusitania Basin (LB), and the Southern Iberia Abyssal Plain to the south. Bathymetry and topography are indicated by color. Lower panel: Time-migrated seismic profile across the G3D area, highlighting the Peridotite Ridge (PR), pre- and syn-rift sediments (below the dashed blue line), crystalline crust (below the dashed red line), and the S-reflector (dashed black line). *Source* Boddupalli et al. (2021)—Open Access

the 3-D FWI method proved to be superior in terms of both accuracy and resolution, making it the preferred choice for high-precision subsurface imaging (Fig. 5.5).

In conclusion, this case study demonstrated the advantages of 3-D Full Waveform Inversion over 2-D methods when dealing with complex geological structures, but it also emphasized the trade-off between computational demands and the level of detail required. The study suggests that for high-resolution imaging of complex subsurface environments, 3-D FWI should be the method of choice, while 2-D FWI can still

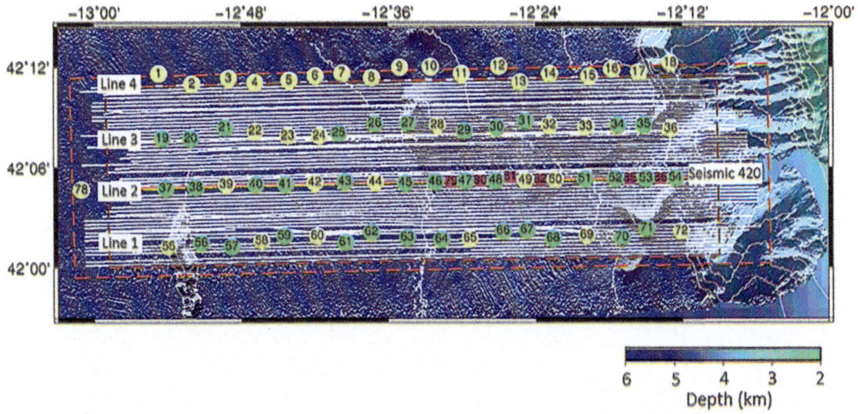

Fig. 5.4 Map showing ocean bottom seismometers (OBS) locations, multichannel seismic shooting, and 3-D FWI study area. *Source* Boddupalli et al. (2021)—Open Access

be useful for less demanding applications or as a preliminary tool in exploration scenarios.

Ground-Penetrating Radar (GPR) in Civil Infrastructure Systems

Ground Penetrating Radar (GPR) systems utilize ultra-wideband antennas and advanced signal processing algorithms, which enable high-resolution imaging of shallow subsurface layers. These systems are applied in various domains, including the detection of buried utilities, archaeological surveys, and identification of subsurface voids or sinkholes. GPR operates by transmitting electromagnetic waves into the ground and analyzing the reflected signals. A typical system includes an antenna, transmitter, receiver, and data acquisition unit. The frequency of the GPR system influences its penetration depth and resolution. While GPR offers significant advantages as a non-destructive testing method, its effectiveness can be limited by environmental and material conditions such as moisture content and ground conductivity.

In civil infrastructure systems, GPR is crucial for evaluating and maintaining roads, bridges, tunnels, and other civil structures. It identifies subsurface issues like cracks, corrosion, and voids. In concrete structures, GPR is used to locate rebar and assess corrosion by interpreting signal reflections. It also detects cracks and delamination, distinguishing between surface and internal defects. Furthermore, it measures concrete thickness and identifies voids or honeycombing to guide maintenance decisions. In pavement and road infrastructure, GPR maps the subgrade, base, and surface layers while identifying poorly compacted or unstable areas. It also detects cracks, evaluates asphalt thickness, and identifies issues such as water infiltration or frost

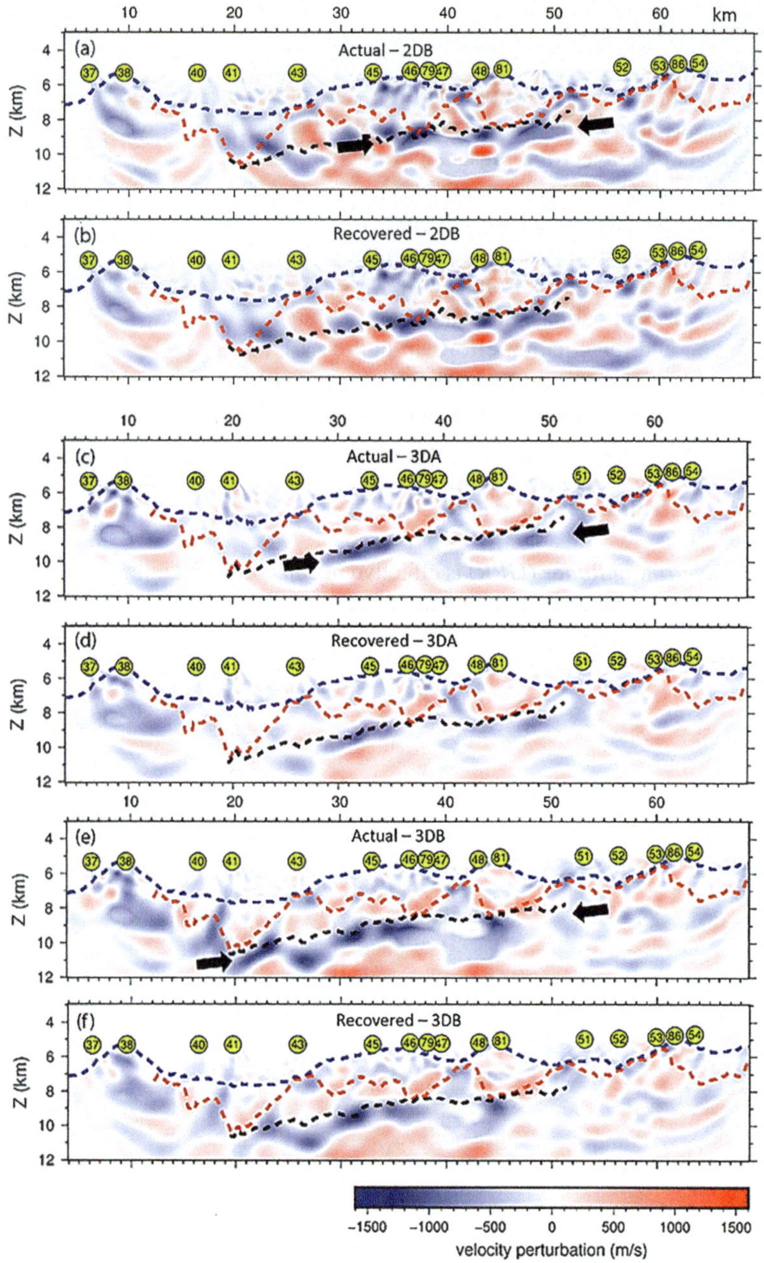

Fig. 5.5 Difference plots comparing the 2-D and 3-D models to the starting model, with red and black borders highlighting significant anomalies. **a** 2DB model difference plot, **b** recovered 2DB anomaly difference, **c** 3DA model difference plot, **d** recovered 3DA anomaly difference, **e** 3DB model difference plot, and **f** recovered 3DB anomaly difference. Black arrows highlight a blue anomaly that extends below the S-reflector from 2-D to 3-D features. *Source* Boddupalli et al. (2021)—Open Access

heave. Roadway condition assessments using GPR enable long-term monitoring and effective rehabilitation planning.

In the context of bridges and foundations, GPR maps deterioration and voids in bridge decks, and detects subsurface anomalies, erosion, and unstable soil around foundations. Tunnels and underground infrastructure benefit from GPR applications in mapping tunnel linings, identifying cracks, delamination, and water infiltration. It also plays a vital role in subsurface utility mapping, detecting and mapping underground pipelines and cables to prevent accidental damage during construction or maintenance. In airports, GPR evaluates the thickness of runways and taxiways and identifies subsurface anomalies, ensuring the structural integrity and safety of airfield pavements.

Technological advances in GPR have led to improved antenna designs for enhanced penetration and resolution, especially in materials like concrete and asphalt. Enhanced data processing techniques, supported by artificial intelligence and machine learning, have enabled clearer data interpretation and automated anomaly detection. The development of 3D GPR imaging, and its integration with other technologies such as LiDAR and UAVs, has significantly enriched the analysis process.

Despite these advancements, GPR in civil infrastructure still faces challenges. Material heterogeneity, such as varying dielectric properties of concrete, asphalt, and steel, affects signal propagation and complicates interpretation. Penetration depth varies with material type and is influenced by moisture content. Complex data interpretation in layered or heterogeneous subsurfaces requires skilled operators and advanced software. Environmental factors like temperature, ground conductivity, and seasonal variations also impact GPR performance.

Real-world case studies highlight both successful applications and the limitations encountered during GPR assessments. Emerging trends indicate that future improvements in GPR hardware and software will continue to enhance infrastructure monitoring. Automation and artificial intelligence are expected to transform the field. As cities move toward smart infrastructure systems, GPR is poised to play a central role in real-time data acquisition and integration with maintenance management systems. In summary, GPR has proven essential for the assessment, maintenance, and management of civil infrastructure, and its role will continue to expand with ongoing technological advancements.

Case Study 3D Mapping of Subsurface Pipes Using Ground Penetrating Radar Images with Edge Detection and Point Cloud Segmentation

This case study presents a novel algorithm for 3D localization of subsurface pipes using Ground Penetrating Radar (GPR) images, implemented in two test fields: one in a large-scale experimental setting and the other in an operational road environment. The study was conducted at the University of Tokyo, Japan and focuses on subsurface utility detection in urban and industrial areas, particularly roads and infrastructure networks (Mizutani et al. 2024). The main objective of the research was to

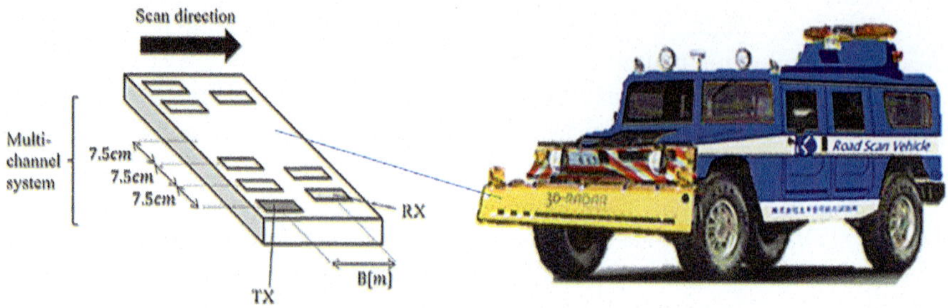

Fig. 5.6 On-vehicle Ground Penetrating Radar system for subsurface utility mapping. *Source* Mizutani et al. (2024)—Open Access

develop an effective method to detect pipes accurately without relying on pre-trained deep learning models, thus reducing the dependency on large datasets and complex computational resources.

On-vehicle GPR systems (Fig. 5.6) provide high-speed surveillance, operating at up to 80 km/h, and can map subsurface utilities to a depth of about 2 m (Poikajarvi et al. 2012; Plati and Loizos 2012). This method reduces time and labour costs for subsurface utility mapping, making it ideal for maintaining aging infrastructure. GPR's ability to capture 3D data across multiple channels offers detailed insights into underground structures, identifying pipes, voids, and hazards that may not be visible via traditional methods (Xu et al. 2014; Gabrys and Ortyl 2020). However, interpreting 3D GPR data still depends on manual analysis, which is labour-intensive and prone to human error.

The method integrates image processing techniques, such as Canny edge detection, Euclidean distance-based instance segmentation, and noise removal based on linearity and point density. These techniques allow the algorithm to identify the locations of pipes with hyperbolic reflection patterns in 3D GPR images (Fig. 5.7). This is a significant improvement over conventional deep learning models, which typically rely on convolutional neural networks (CNNs) and large training datasets. By applying this approach, the algorithm successfully detects and localizes buried pipes with high precision, surpassing existing methodologies that approximate pipe locations with 3D lines.

The algorithm employs Canny edge detection to highlight significant features in the GPR images that correspond to pipe locations. Using Euclidean distance, the algorithm segments the points within the 3D point cloud that align with the theoretical hyperbola of pipe reflections. Linear and density-based noise removal techniques are applied to refine the point cloud and improve localization accuracy. The typical reflection patterns shown in Fig. 5.8 are derived from the equations of the diffraction curve (Persico (2014), as expressed in Eq. (5.1).

$$t = \frac{2}{c_0}\sqrt{(x - x_1)^2 + h^2} + \frac{2\sqrt{\varepsilon}}{c_0}\sqrt{x_1^2 + z_0^2} \tag{5.1}$$

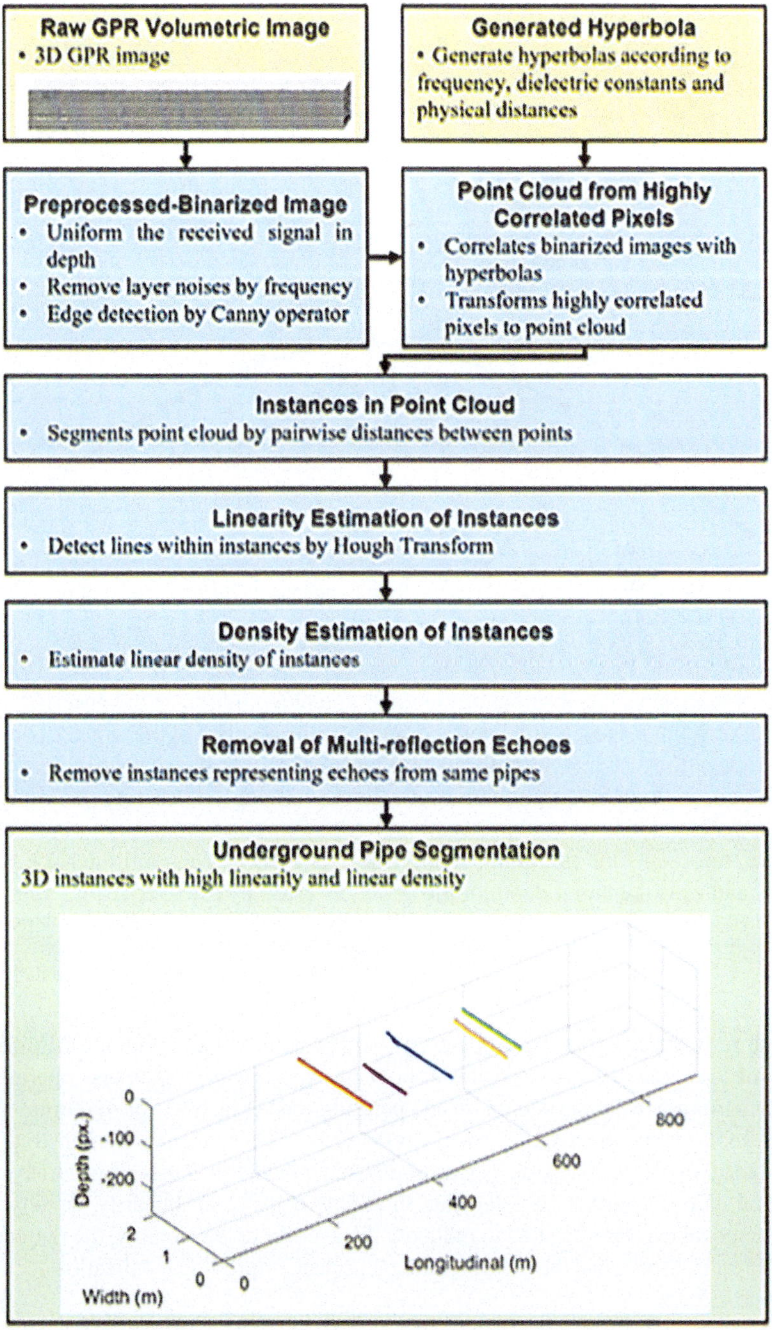

Fig. 5.7 Flowchart of the proposed methodology for subsurface utility mapping. *Source* Mizutani et al. (2024)—Open Access

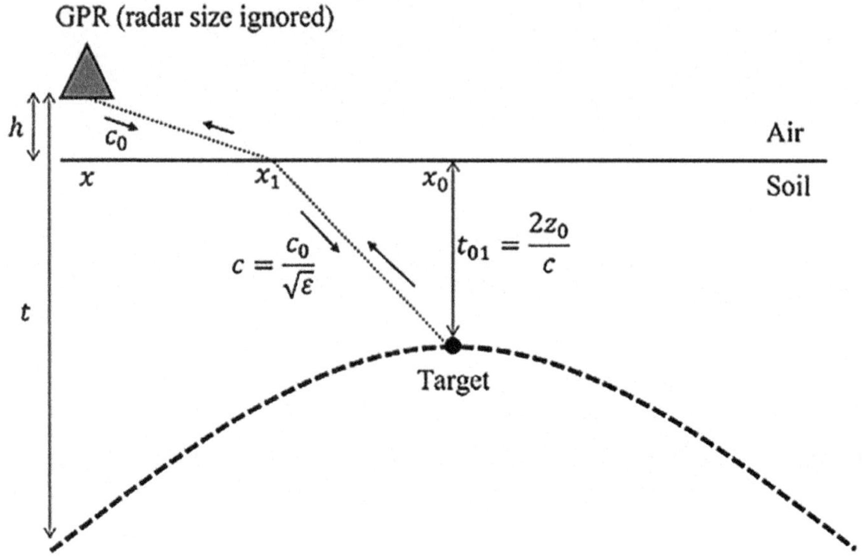

Fig. 5.8 Reflected hyperbola from the target depth z_0 and the dielectric constant of soil ε. *Source* Mizutani et al. (2024)—Open Access

where x represents the abscissa of the radar antenna point, x_1 is the abscissa of the diffraction point at the air–soil interface, and x_0 is the abscissa of the target. c_0 denotes the propagation velocity in both air and soil, while ε is the dielectric coefficient of the soil, with its inverse square root referred to as the shortening coefficient of wavelength, influencing the propagation velocity of electromagnetic waves in the soil. t_{01} represents the minimum round-trip time from the air–soil interface to the target, and z_0 is the burial depth of the target, as given by Eq. (5.2).

$$z_0 = \frac{c_0 t_{01}}{2\sqrt{\varepsilon}} \tag{5.2}$$

The results show that the proposed algorithm achieves a maximum localization error of less than 0.25 m in all tests, both in experimental fields and operational road environments. In particular, on an operating road (Fig. 5.9), the maximum error was 0.1959 m, and the average error was 0.1175 m. These metrics demonstrate that the algorithm outperforms deep learning-based models in terms of accuracy. The ablation analysis further revealed that modules focusing on linearity, density, and echo removal each contributed to reducing the average error. Notably, the removal of non-linear instances had the most significant impact, improving localization accuracy by up to 7.5 cm.

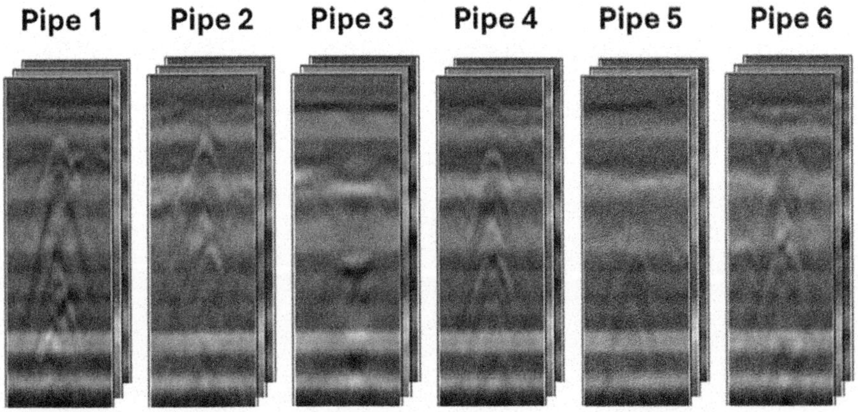

Fig. 5.9 Pipes under operating road in the (dataset from the road in operation collected in Japan) field site. *Source* Mizutani et al. (2024)—Open Access

The method achieved a maximum error of 0.1959 m and an average error of 0.1175 m. In the experimental field and real-world tests, the error was consistently low, indicating high precision in pipe localization. The algorithm demonstrated robustness across different test settings, with an average error of 0.1171 m and a standard deviation of 0.0112 m. Compared to convolutional neural networks, the proposed method provided superior localization accuracy, demonstrating that a simpler, non-AI approach could still achieve high precision.

Despite its success, the algorithm does have limitations, particularly in dealing with longitudinal pipes and pipes that do not generate clear hyperbolic reflections in the GPR data (Fig. 5.10a, b). The resolution limitations of 3D GPR images, especially in the transverse direction, pose challenges in accurately detecting certain pipe types. Furthermore, while the algorithm was tested in a large-scale experimental field and a segment of an operational road, the dataset of buried pipes was limited in size.

The algorithm's performance could be improved by extending the dataset, incorporating a larger variety of subsurface utility types, and testing in more diverse conditions. Research on pipe and soil parameter estimation can be conducted to enhance the accuracy of pipe detection and provide deeper insights into subsurface conditions. The proposed method could also be expanded to monitor the condition of pipes over time, enabling the detection of pipe breakage or leakage by analyzing changes in GPR images over multiple time series.

In conclusion, the proposed algorithm demonstrates a reliable, high-precision solution for 3D subsurface pipe localization using GPR images, making it a valuable tool for engineering geophysics applications. By eliminating the need for pretraining and large data sets, it offers a more efficient alternative to deep learning-based methods, paving the way for future developments in non-invasive subsurface utility mapping. This work has important implications for infrastructure inspection, urban planning, and maintenance, marking a significant advancement in the field of geophysics.

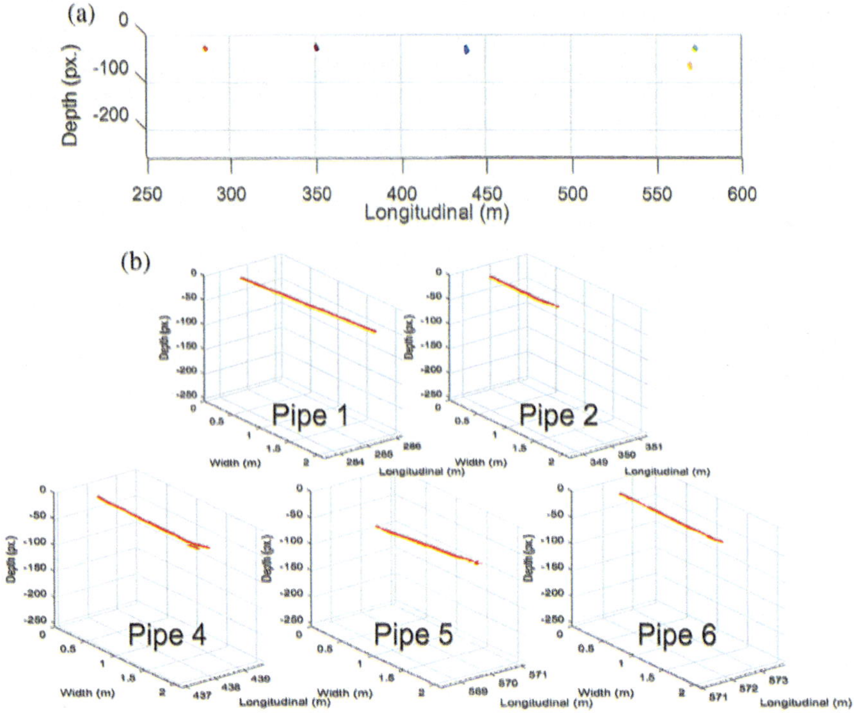

Fig. 5.10 **a** Side view of all segments showing pipes on the operating road in Japan. *Source* Mizutani et al. (2024)—Open Access. **b** Ground-penetrating radar (GPR) scan of a subsurface utility line in a city infrastructure project, illustrating its effectiveness during urban redevelopment in Japan. Clear detection of 5 pipes in the dataset from the actual road. *Source* Mizutani et al. (2024)—Open Access

Case Study Utilizing Dual Polarized Array GPR for Shallow Urban Pavement Foundation Assessment in London

Maintaining urban road pavements is critical for public safety and economic stability, especially in cities like London. Aging infrastructure, compounded by factors like soil erosion, underground water movement, and structural weaknesses, increases the risk of pavement collapse. Traditional inspection methods, such as visual and manual testing, lack the precision to detect subsurface anomalies, highlighting the need for advanced technologies like Ground Penetrating Radar (GPR) (1–6).

The objective of this study is to assess the structural integrity of urban pavement using dual-polarized Ground-Penetrating Radar (GPR) and Short-Time Fourier Transform Spectrum (SFTS) analysis. The goal is to detect subsurface anomalies, such as moisture infiltration, fractures, and voids, which could compromise the road's structural stability. The methodology involves utilizing a dual-polarized GPR system to collect both horizontal and vertical radar profiles of the pavement. These profiles are then processed using SFTS analysis to identify frequency variations caused by

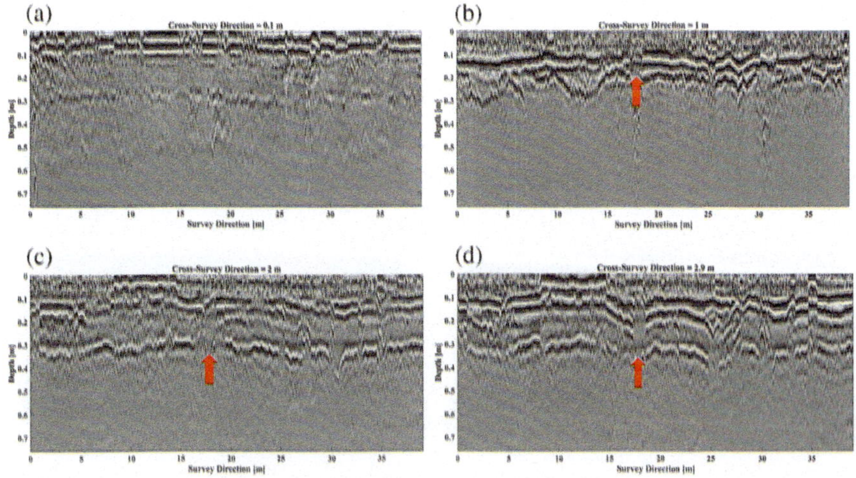

Fig. 5.11 **a** Migrated vertical profile at 0.1 m in the cross-survey direction. **b** Migrated profile at 1 m in the cross-survey direction. **c** Migrated profile at 2 m in the cross-survey direction. **d** Migrated profile at 2.9 m along the cross-survey direction. *Source* Zou et al. (2024)—Open Access

changes in material properties, such as moisture infiltration. The formula for the SFTS applied to the radar signal is:

$$\mathrm{SFTS} = \{x(t)\}(t, \omega) = X(t, \omega) = \int_{-\infty}^{\infty} x(\tau)W(\tau - t)e^{-j\omega\tau}d\tau \tag{5.3}$$

where x(t): Input signal, W(t): Window function, t: Time variable, ω: Angular frequency, X(t, ω): The STFT of x(t). This formula helps to analyze the frequency shifts, which are indicators of material degradation. The results obtained from Ground Penetrating Radar (GPR) and Short-Time Fourier Transform (STFT) analyses highlight critical subsurface characteristics:

Migrated Profiles

Vertical profiles across different cross-survey directions (e.g., 0.1, 1, 2, and 2.9 m) consistently depict reflections marking the boundaries between pavement layers (Fig. 5.11a–d). These reflections, observed at depths of approximately 8 cm, 14 cm, and 30 cm, delineate transitions between the asphalt, base, and subgrade layers, respectively. Discontinuities in some regions indicate structural weaknesses, potentially caused by vehicle pressure or groundwater erosion.

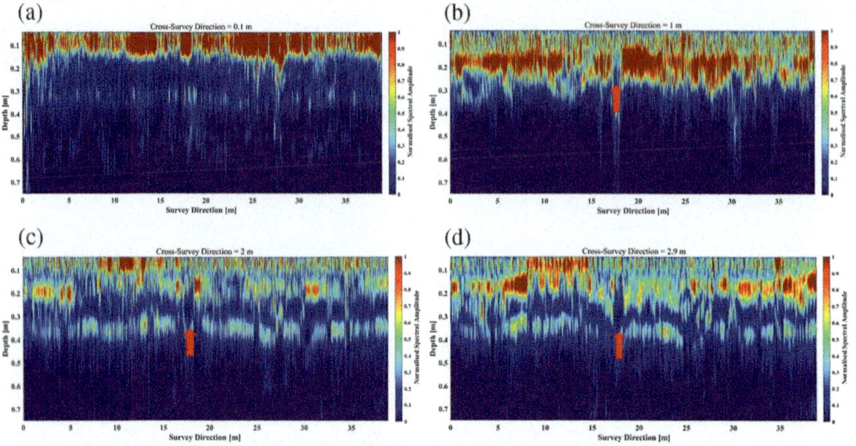

Fig. 5.12 **a** GPR peak frequency division profile at 0.1 m in the cross-survey direction. **b** GPR peak frequency distribution profile at 1 m along the cross-survey direction. **c** GPR peak frequency division profile at 2 m; along the cross-survey direction. **d** GPR peak frequency division profile at 2.9 m; along the cross-survey direction. *Source* Zou et al. (2024)—Open Access

Frequency Division Profiles

Peak frequency division profiles, analyzed using STFT, reveal consistent peak frequencies at specific depths in stable areas, while low-frequency shifts in compromised regions indicate moisture-related issues. These shifts correlate with discontinuities and material fragmentation, especially between 17 and 18 m along the cross-survey direction, emphasizing areas prone to subsidence or collapse (Fig. 5.12a–d).

Horizontal Slices

Horizontal slices at depths of 0.21 and 0.36 m highlight an anomalous region within the cross-survey direction (Fig. 5.13a, b). This region, consistent across multiple profiles, underscores structural vulnerabilities requiring immediate intervention.

By integrating insights from multiple figures, the analysis offers a comprehensive understanding of subsurface conditions and identifies critical areas requiring maintenance and preventive action. The findings highlight regions at risk of collapse due to moisture-related degradation and structural fragmentation. This underscores the urgency of timely interventions, such as reinforcement and enhanced drainage systems, to address these vulnerabilities and prevent further deterioration. In conclusion, the dual-polarized GPR system, combined with SFTS analysis, has demonstrated its effectiveness as a reliable tool for identifying subsurface anomalies and providing actionable data for pavement maintenance. The study emphasizes the critical importance of regular inspections to detect and address potential issues proactively, thereby preventing significant structural damage and ensuring long-term infrastructure integrity.

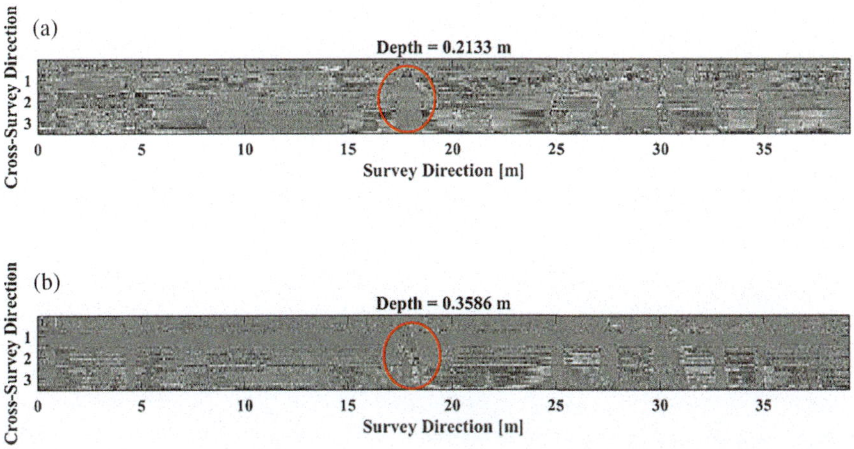

Fig. 5.13 **a** Migrated horizontal slice at a depth of 0.21 m. **b** Migrated horizontal slice at a depth of 0.36 m. *Source* Zou et al. (2024)—Open Access

Case Study Tunnel Lining Detection and Rebar Identification Using GPR in SLRT, Taiwan

This study aims to develop a comprehensive tunnel lining detection method and clustering technique for semi-automatic rebar identification, applied to ten tunnels along the South Link Line Railway (SLRT) of Taiwan (Fig. 5.14) (Puntu et al. 2021). Ground Penetrating Radar (GPR) with a 1000 MHz antenna frequency was used, mounted on a flexible and versatile antenna holder that adapted to the varying tunnel conditions, referred to as the Vehicle-mounted Ground Penetrating Radar (VMGPR) system. Tunnel lining boundaries were identified using the Fresnel Reflection Coefficient (FRC) in both A-scan and B-scan data, followed by estimation of the thinning of the tunnel linings (Fig. 5.15). To enhance data quality, the Hilbert Transform (HT) was applied to extract the pulse envelope, providing an overview of the energy distribution.

Subsequently, filtered radargrams were used to simulate tunnel conditions using the Two-dimensional Forward Modeling (TDFM) approach, with noise levels ranging from 0 to 30%. The rebar detection model was analyzed using Hierarchical Agglomerative Clustering (HAC) in machine learning, and clustering results were evaluated with the Silhouette Index (SI) (Fig. 5.16).

The analysis revealed three distinct tunnel lining boundaries: the air-second lining boundary, the second-first lining boundary, and the first-wall rock boundary. Among the ten tunnels scanned, the Fangye 1 tunnel exhibited the highest percentage of thinning (13.39%), classifying it as category B, while the other tunnels fell into category A with thinning percentages ranging from 0 to 1.71%.

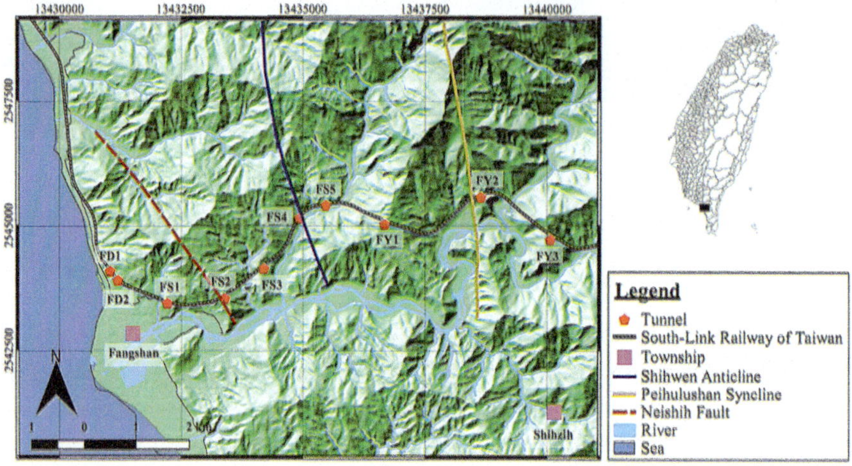

Fig. 5.14 Geographic map of the South-Link Line Railway (SLRT) in Taiwan. The red pentagon marks the location of the tunnel along this line, while the red-dashed line, blue line, and orange line represent the Neishih Fault, the Shihwen Anticline, and the Peihulushan Syncline, respectively. *Source* Puntu et al. (2021)—Open Access

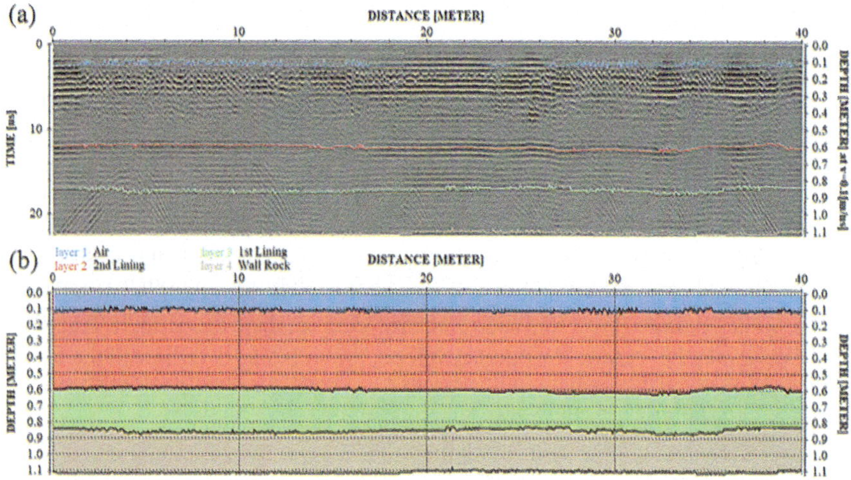

Fig. 5.15 **a** Filtered radargram with a grayscale palette from GPR scanning in the Fangdian I tunnel, covering a total distance of 40 m. **b** Interpretation of the tunnel layers in the B-scan. The blue color represents the air layer, peach represents the second lining, green represents the first lining, and brown represents the wall rock. The contrast between these layers is due to variations in relative permittivity, which is used for identifying the lining layers. *Source* Puntu et al. (2021)—Open Access

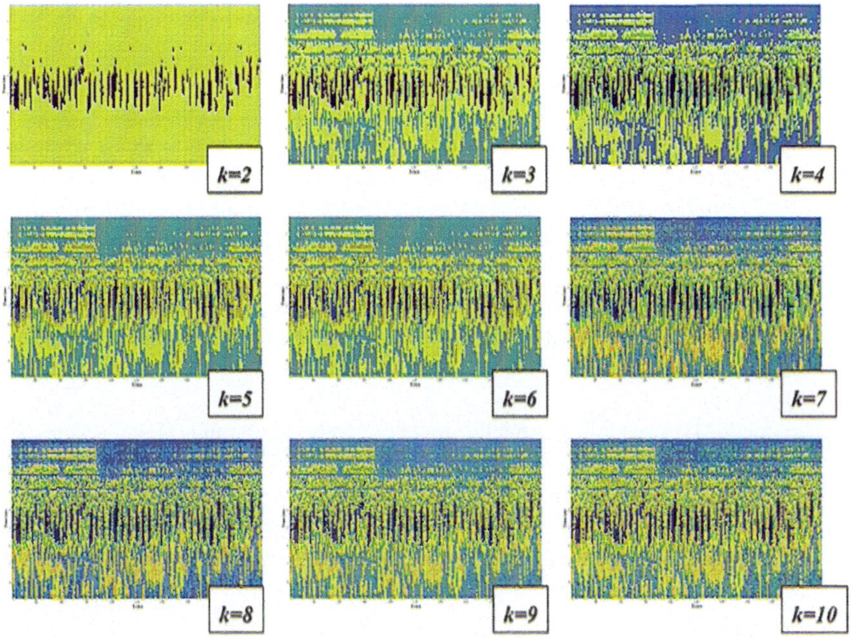

Fig. 5.16 Clustered radargram showing rebar identification using Hierarchical Agglomerative Clustering (HAC). This figure demonstrates the identification of rebars (marked in blue) at a depth of approximately 0.2 m (around 4 ns). The clustered radargram illustrates how the rebar response remains distinct despite varying numbers of clusters (k = 2–10), confirming the robustness of the rebar detection process in the radar data. *Source* Puntu et al. (2021)—Open Access

The TDFM model for rebar identification demonstrated consistency with the field data, where k = 2 was identified as the optimal number of clusters. The simulation model, particularly the TDFM model with 30% random noise, accurately mimicked the field data, although the rebar response was rougher due to the high noise level. These results demonstrate the effectiveness of the VMGPR system in tunnel inspection, offering a reliable method for tunnel condition evaluation and rebar identification in real-world applications.

Ground Penetrating Radar (GPR) Applications in Archaeological Systems

Ground Penetrating Radar (GPR) is a non-invasive geophysical method extensively used in archaeology for exploring subsurface features. It enables archaeologists to detect buried structures, artefacts, and other cultural remains without the need for excavation, making it an invaluable tool in modern archaeological research. GPR operates by transmitting electromagnetic waves into the ground; when these waves encounter different materials, a portion of the energy reflects back to the surface. By analyzing the time delay and strength of these reflected signals, researchers can infer the depth, size, and composition of buried objects and features.

In archaeological applications, GPR has proven especially effective in mapping subsurface architecture, such as detecting the remains of walls, buildings, and roads. This has allowed archaeologists to identify the layout of ancient settlements and to reconstruct the sequence of construction phases, as demonstrated in studies of Roman sites. Additionally, GPR can detect smaller artefacts like pottery, tools, and graves, particularly when these items have material properties distinct from the surrounding soil. It has shown notable effectiveness in burial sites, where graves or human remains often produce distinct reflection patterns. Beyond individual objects and structures, GPR also assists in identifying stratigraphy and depositional patterns, revealing variations in soil layers that help date artefacts and reconstruct the sequence of human activity at a site.

The method is also instrumental in site preservation and management. By providing detailed subsurface information before excavation, GPR helps reduce the risk of damaging important features. It also supports ongoing monitoring of archaeological sites, aiding in the detection of changes due to erosion, structural collapse, or other environmental factors. Furthermore, GPR has been adapted for use in underwater and subaqueous environments, enabling the investigation of submerged archaeological sites such as shipwrecks and sunken structures without extensive diving operations.

Despite its many advantages, GPR is not without challenges. The resolution and depth of imaging depend on the frequency of the electromagnetic waves used—higher frequencies yield greater detail but penetrate to shallower depths, whereas lower frequencies reach deeper but offer less clarity. Soil conditions also influence effectiveness; dry, sandy soils generally produce better results than wet, clay-rich, or highly conductive soils, which can distort or absorb signals. Interpreting GPR data accurately requires specialized expertise, as the complexity of subsurface reflections can lead to misinterpretation or false positives. Additionally, the cost of GPR equipment and the need for skilled operators can limit its accessibility in some contexts.

Recent technological advancements have significantly enhanced GPR's capabilities. Developments such as multi-frequency GPR systems and machine learning algorithms for data interpretation are improving accuracy and efficiency. Moreover, integrating GPR with other geophysical techniques like magnetometry provides a more comprehensive understanding of archaeological sites. In conclusion, while GPR presents certain limitations, its non-invasive nature, versatility, and continual technological evolution make it an indispensable tool in archaeological investigations.

Fig. 5.17 GPR instruments used at the site: **a** DJI M-300 drone with a 500 MHz shielded dipole antenna and altimeter for flight control. **b** Intelligent cart with a 500 MHz shielded dipole antenna, both capable of sensing up to 5 m below the surface. *Source* Frid and Frid (2024)—Open Access

Case Study on Integrating Ground and Drone-Based GPR for Archaeological Survey in Hulata, Israel

This case study investigates the use of Ground-Based and Drone-Based Ground Penetrating Radar (GPR) for an archaeological survey conducted at Hulata, Israel. The site, known for its Early Bronze Age structures made of irregularly shaped limestone and sandstone boulders, was surveyed to assess the potential of GPR technology in uncovering subsurface archaeological features.

The study employed both ground-based GPR, which provides higher reflection amplitudes due to its proximity to the ground, and drone-based GPR, which offers efficient data collection over large areas using an RTK system for precise positioning (Fig. 5.17). The integration of both methods allowed for cross-validation and more reliable results.

Figure 5.18 presents zoomed-in radargram fragments from both GPR methods. The analysis revealed a distinct geological boundary about 1 m below the surface, with the upper soil layer showing weak, blurred signals and the denser, lower layer producing stronger reflections.

Anomalies detected around 0.3 m deep indicated soil disturbances likely related to past construction activities. These were highlighted with a yellow dashed line in the figure. Further anomalies at approximately 62–63 m along the profile suggested the presence of man-made structures, such as walls made of limestone boulders. These anomalies were consistent across both the ground-based and drone-based GPR, validating the effectiveness of both methods. The findings were mapped and showed a scattered distribution of anomalies across the site, as seen in Fig. 5.19, aligning with archaeological expectations of the Early Bronze Age construction style.

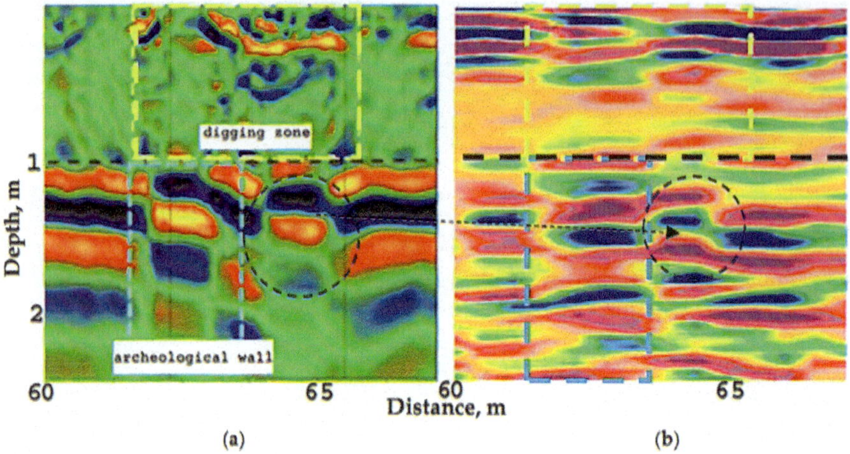

Fig. 5.18 Radargram excerpts from drone-based (**a**) and ground-based (**b**) GPR surveys showing a geological transition at 1 m depth. The upper layer (yellow dashed lines) exhibits anomalies at 0.3 m, indicating irregular terrain, while the lower layer (light blue dashed lines) suggests potential archaeological walls made of limestone boulders, with key features highlighted by circles and arrows. *Source* Frid and Frid (2024)—Open Access

Excavations conducted by the Archaeological Authority of Israel (AAI) at suspected points of interest confirmed the presence of stone structures at depths of 1–2 m, matching the GPR findings. In conclusion, the combined use of ground-based and drone-based GPR proved effective in detecting archaeological features at Hulata. The study highlights the advantages of drone-based GPR, especially in challenging terrains, and reinforces GPR's potential as a non-invasive tool for archaeological surveys, offering efficient and reliable results.

Electrical Resistivity Tomography (ERT) in Engineering Geophysics

Electrical Resistivity Tomography (ERT) is a non-invasive geophysical technique used to map subsurface structures by measuring the resistivity of the ground. Resistivity is the property that quantifies how much a material opposes the flow of an electric current. Different subsurface materials, such as soil, rock, and water, have varying resistivity values, allowing ERT to provide detailed images of subsurface features.

The ERT technique involves passing an electric current through the ground using surface electrodes. As the current travels through the subsurface, the electrical resistance encountered by the current is measured at various points. The measurements are then used to generate a resistivity profile, which can be visualized in the form of a 2D or 3D tomographic image of the subsurface.

The standard ERT system consists of an array of electrodes placed along a survey line on the ground surface. A power source generates the electrical current, and a

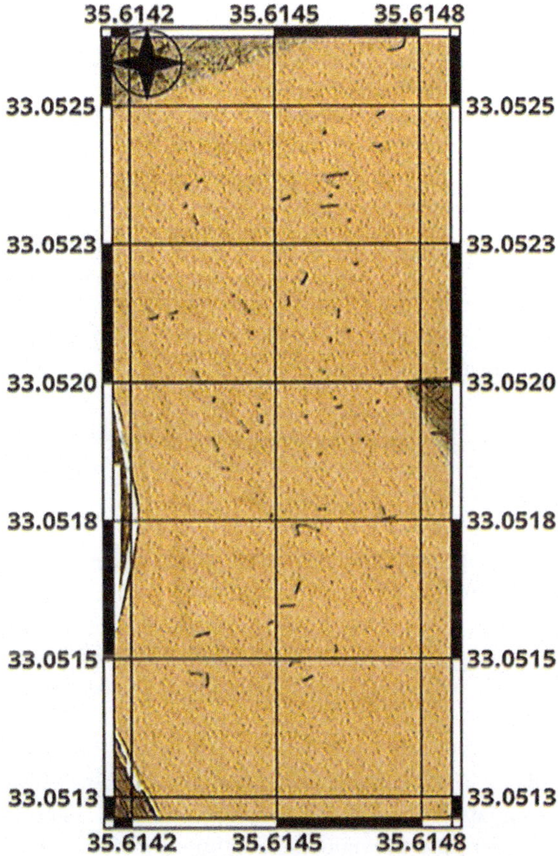

Fig. 5.19 Numerically mapped locations of overlapping ground-based and drone-based GPR findings, indicating potential archaeological objects, such as walls or concentrations of pottery. The irregular scattering pattern suggests a link to Early Bronze Age construction practices in the Hulata region, characterized by structures made from broken limestone and sandstone boulders. *Source* Frid and Frid (2024)—Open Access

data logger records the voltage response from the electrodes. The configuration of electrodes (e.g., Wenner, Schlumberger, dipole–dipole) influences the resolution and depth of the survey. The system is highly flexible and can be deployed in various environments, including urban, remote, and challenging terrains.

Once data is collected, it undergoes processing to reduce noise and correct for any errors. The raw resistivity data is then inverted using advanced algorithms to produce a 2D or 3D resistivity model of the subsurface (Hossain et al. 2010; Naskar et al. 2008). Inversion techniques, such as linear and non-linear methods, are used to create accurate images that reveal the spatial distribution of resistivity in the surveyed area. Specialized software tools, such as RES2DINV and Earth Imager, assist in this process.

ERT has a wide range of applications in engineering geophysics, particularly in site investigations for civil engineering projects. It is used to assess soil and rock properties, identify groundwater locations, and evaluate the depth and extent of bedrock. ERT is also instrumental in geotechnical engineering for foundation design, where it helps determine the stability of the ground and detect potential hazards such as voids or sinkholes (Aizebeokhai 2010; Jabrane et al. 2023; Martinez-Pagan et al. 2013; Ghorbani et al. 2024; Franco et al. 2024; Hinojosa et al. 2022; Bharti et al. 2022).

In environmental engineering, ERT is applied to monitor contamination plumes, track groundwater flow, and assess subsurface pollution. In addition, ERT is frequently used in geohazard studies, where it helps detect geological features like fractures, faults, or cavities that might present risks in construction projects.

While ERT is a powerful tool, it does have some limitations. The technique's effectiveness can be reduced in areas with highly heterogeneous or resistivity values that are difficult to differentiate. The depth of investigation can be constrained by the electrode spacing and the resistivity contrast between subsurface materials. Additionally, noise from external sources and the complexity of data inversion can sometimes complicate interpretation.

Case Study Determining of Leachate Recirculation Frequency in Denton Landfill

The objective of this study was to investigate moisture flow paths and accumulation zones after leachate recirculation in specific cells of the Denton Landfill during bioreactor operation. Two-dimensional resistivity tomography was used to map moisture distribution in the waste, aiding in determining the leachate recirculation frequency. The City of Denton's municipal solid waste (MSW) landfill, located in Denton, Texas, began operations in 1984 and receives approximately 500–600 tons of waste daily. In May 2009, the landfill received a permit to operate as an enhanced leachate recirculation (ELR) bioreactor landfill. In such landfills, regular leachate recirculation is essential for rapid waste decomposition. To optimize this process, operators need to understand the moisture conditions and the frequency and location of leachate recirculation.

Three recirculation pipes (H2, H16, and H18) were selected for monitoring between May 2010 and April 2013. Resistivity tests were conducted at one, seven, and fourteen days after leachate recirculation. For simplicity, only the results from pipe H2 are presented, as the other pipes showed similar trends. At baseline, the moisture content around pipe H2 was 36.31%. One day after recirculation, it increased to 63.35%, then decreased to 49.77% after one week and 36.78% after two weeks. These changes indicated that leachate moved into the waste and then gradually dispersed. The resistivity profiles (Fig. 5.20) confirmed this, showing a high-resistivity area (low moisture) beneath the pipe before recirculation, which turned to moderate or low resistivity (high moisture) after recirculation. These findings suggest that leachate recirculation in the Denton landfill should occur every two weeks for effective waste decomposition. The ERT method provided continuous monitoring over an area of about 18.2 m (60 ft.), which would not have been possible with traditional sensors.

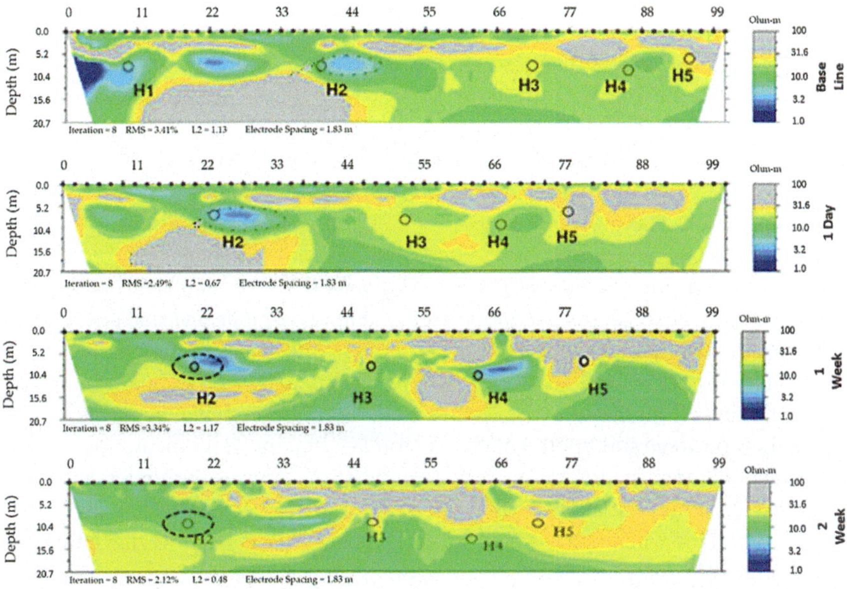

Fig. 5.20 Resistivity tomography results for recirculation pipe H2 at the Denton landfill, showing moisture distribution before and after leachate recirculation. *Source* Alam et al. (2024)—Open Access

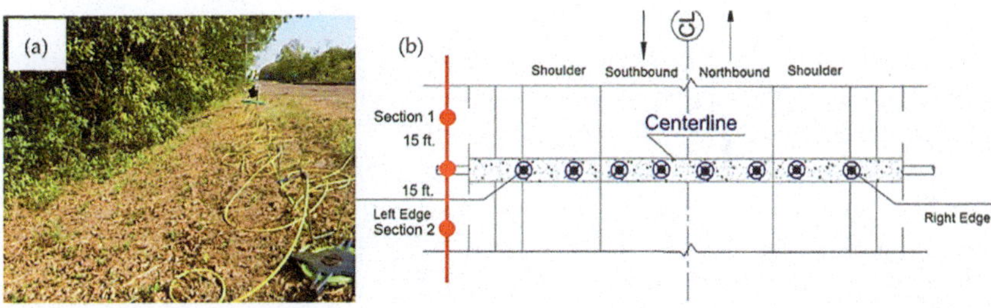

Fig. 5.21 ERT experiment on pavement slope and schematic of the survey location: **a** field setup of ERT experiment on pavement slope; **b** location map of survey line. *Source* Alam et al. (2024)—Open Access

Case Study of Resistivity Tomography in Pavement Moisture Distribution

Moisture movement beneath pavement can cause uneven deformation in expansive pavement subgrades, and resistivity tomography provides valuable information about this moisture distribution. A geophysical investigation using Electrical Resistivity Tomography (ERT) was conducted on State Highway (SH) 342 in Lancaster, Texas. The ERT survey setup at the pavement slope is shown in Fig. 5.21, where the red line indicates the survey line.

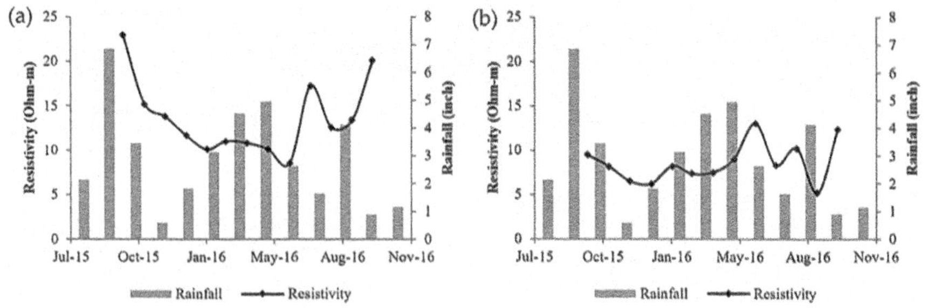

Fig. 5.22 Change in resistivity with rainfall at different depths: **a** 0.91 m (2 ft.); **b** 3.05 m (10 ft.). *Source* Alam et al. (2024)—Open Access

The primary objective was to observe the seasonal distribution of soil moisture beneath the pavement. Resistivity tests were conducted monthly from 2015 to 2017, focusing on several key months for illustration. The seasonal variation in moisture distribution is crucial to understanding the deformation profile of flexible pavements built on expansive soils in Texas.

Resistivity plots revealed distinct seasonal patterns. During the dry months (May–October), higher resistivity values were observed, indicating lower moisture content, while the wet months (November–April) exhibited lower resistivity values, indicating higher moisture content. In May 2016, a low-resistivity zone was detected at 3 ft. deep, suggesting moisture intrusion due to rainfall, especially near cracks in the pavement.

This seasonal variation continued throughout the 2016–2017 monitoring period, confirming the moisture trend. Further analysis of resistivity changes over different depths was conducted, with key results shown in Fig. 5.22. Resistivity at 1 m depth showed significant seasonal fluctuations, such as a 50% drop from October 2015 to December 2015. At 2 m depth, resistivity dropped by 22%, and similar variations were seen at other depths.

Rainfall also influenced resistivity values, with a peak of 4.52 in. in April 2016 and 4.94 in. in May 2016, which was reflected in the resistivity decrease, confirming the role of moisture infiltration from rainfall. This trend continued through the year, with seasonal factors playing a predominant role in resistivity changes.

Case Study of Resistivity Tomography in Slope Failure Investigation

Shallow slope failures in expansive soils are a common phenomenon in Texas. Due to the wetting–drying cycles of expansive clay, which are influenced by fluctuating climatic conditions, highway slopes often fail after the summer season, followed by a heavy rainfall event. This study aimed to analyze the failure of a highway slope using the Electrical Resistivity Tomography (ERT) method.

The study was conducted along the slope of Highway US 183. After the summer of 2013, desiccation cracks appeared on the slope, creating potential pathways for

Fig. 5.23 Cracks observed on the shoulder of the slope (**a**) and the failure at the crest of the slope (**b**). *Source* Alam et al. (2024)—Open Access

rainwater intrusion. Figure 5.23a illustrates the cracks present on the shoulder of the slope, while Fig. 5.23b shows the failure at the crest of the slope. These cracks contributed to the accumulation of moisture within the slope, which could lead to slope failure.

Figure 5.24a presents the failure conditions of the US 183 slope, where two failure zones were identified: one at the top and another in the middle of the slope. A visible crack is present in the middle of the slope, indicating potential areas of concern. The resistivity profile obtained from Resistivity Line (RL)-1, shown in Fig. 5.24b, revealed a high-resistivity zone up to a depth of 2.1 m (7 ft.) (Fig. 5.25a). This high-resistivity zone is likely indicative of a low-moisture region near the top of the slope, as the test was conducted during the dry season. This suggests that the top portion of the slope is the active zone where seasonal moisture variation occurs. Below 2.1 m, a low-resistivity zone was detected, which may indicate higher moisture content deeper within the slope, highlighting the presence of moisture infiltration.

The resistivity profile from Resistivity Line 2 (RL-2), conducted over the tension crack in the middle of the slope, is presented in Fig. 5.24b. Figure 5.25b shows a high-resistivity zone immediately below the tension crack, with a depth reaching 3.65 m (12 ft.). The presence of soil cracks allows air to penetrate the soil, displacing moisture. As air has higher resistivity than water or soil particles, the air intrusion into these cracks increases the overall resistivity in the soil. This results in a higher resistivity profile, further confirming that the cracks have contributed to reduced moisture in the affected areas.

Case Study of ERT in Landfill ET Cover to Determine Moisture Variation

Water balance covers (ET covers) have become popular for final capping of waste facilities because of their sustainability. The hydraulic performance of these covers depends on soil–plant interaction. Soil properties, like permeability and water retention, influence the drainage rate to the waste mass below the cover. However, natural

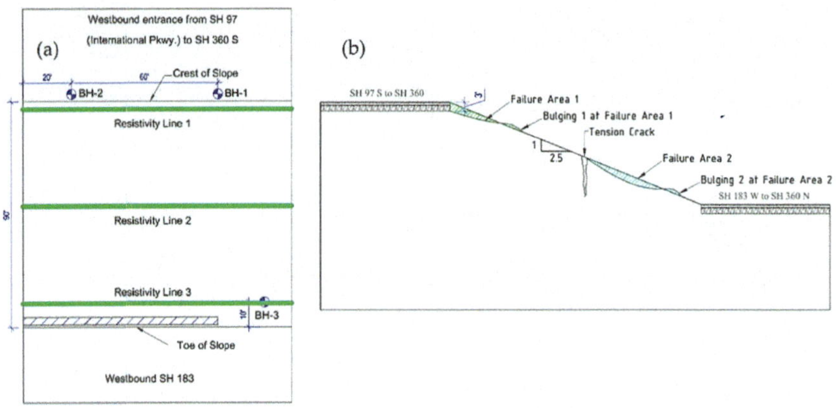

Fig. 5.24 **a** Failure condition of SH 183 slope; **b** resistivity lines along the slope. *Source* Alam et al. (2024)—Open Access

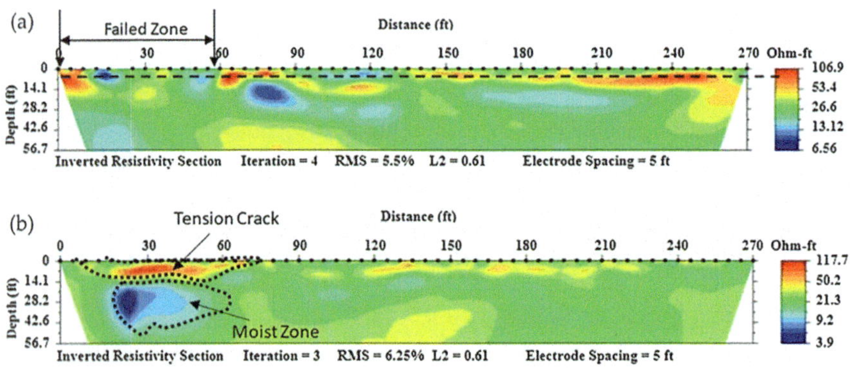

Fig. 5.25 **a** Test results of the resistivity line RL-1 at the crest; **b** test results of the resistivity line RL-2 at the middle of the slope. *Source* Alam et al. (2024)—Open Access

processes, such as burrowing, freezing–thawing, and wetting–drying cycles, can alter the soil's hydraulic properties over time. This study used resistivity tomography to investigate seasonal moisture and suction variations in the cover soil. The tests were conducted monthly, with increased frequency during the summer months due to higher rainfall and temperatures. The resistivity data showed clear patterns of moisture changes with fluctuating weather conditions. The ERT data were also highly correlated with soil properties and water storage.

Large-scale lysimeters were set up at the City of Denton landfill, Texas, with a geomembrane, drainage layer, and cover soil. The soil consisted of a 304.8 mm surface layer and a 914 mm compacted storage layer. A total of 28 electrodes were used to acquire a resistivity profile at a depth of 1220 mm. The resistivity test was carried out from January 2015 to December 2017. Some test plots from one lysimeter

Fig. 5.26 **a** Field setup with resistivity meter; **b** conducting the ERT test. *Source* Alam et al. (2024)—Open Access

are shown in Fig. 5.26. In Fig. 5.27, resistivity profiles after extreme weather conditions in the summer of 2016 showed a low-resistivity zone at the bottom of the cover, indicating high moisture and low suction, which suggested percolation. Despite rainfall, the top 304.8–457.2 mm showed higher resistivity due to evapotranspiration and drying. The high-resistivity zone correlated with root growth, indicating that moisture loss mainly occurred from the top 50 mm of the soil. This study demonstrated that the ERT method effectively monitors seasonal moisture variation and water balance in ET cover systems, as shown in Fig. 5.27.

In four case studies, electrical resistivity tomography (ERT) was applied to various geotechnical and geoenvironmental investigations. In the landfill leachate recirculation study, ERT successfully tracked moisture distribution, revealing that equilibrium was reached after approximately 14 days. During the slope failure investigation along US Highway 183, ERT helped identify a perched water zone, which was crucial for repairing the failed slope. ERT also captured seasonal moisture variations in pavement, identifying May–October as the dry period and November–April as the wet period. Finally, in the City of Denton landfill, ERT was used to assess the moisture-suction relationship in the ET cover, showing significant seasonal resistivity changes related to moisture fluctuations.

Case Study of Groundwater Flow in Flood-Prone Regions of Thailand

Thailand experiences seasonal flooding, significantly impacting agriculture and infrastructure. Understanding subsurface conditions influencing groundwater flow is crucial for flood mitigation planning. To employ ERT to identify subsurface features, such as clay layers and aquifer boundaries, that influence groundwater dynamics in flood-prone regions. ERT surveys were conducted using a multi-electrode array system. Data processing included inversion techniques to generate 2D and 3D resistivity profiles, highlighting zones of low and high permeability.

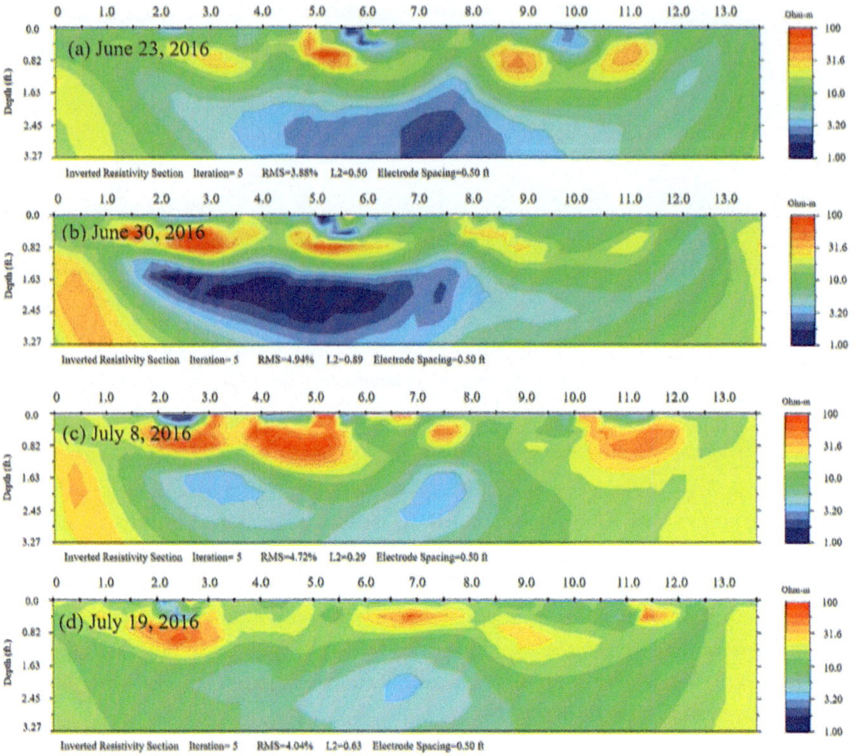

Fig. 5.27 Resistivity tomography results for lysimeter-2. *Source* Alam et al. (2024)—Open Access

The ERT profiles revealed subsurface clay layers acting as barriers to water flow and high-permeability zones facilitating rapid infiltration. These findings were integrated into hydrological models to design effective flood control measures. ERT provided critical insights into groundwater dynamics, demonstrating its value in developing sustainable flood management strategies for vulnerable regions.

Electromagnetic Methods

Time-domain and frequency-domain electromagnetic systems have improved depth penetration and discrimination of subsurface materials, making them invaluable for mineral exploration and groundwater studies.

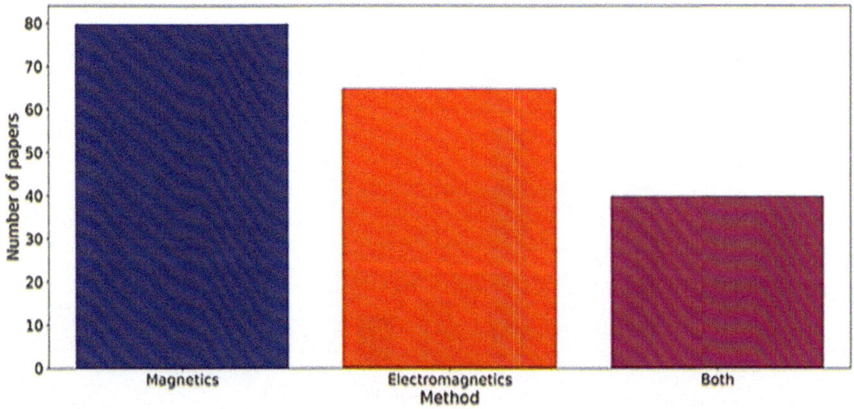

Fig. 5.28 Number of case studies by method. *Source* Vallee et al. (2024)—Open Access

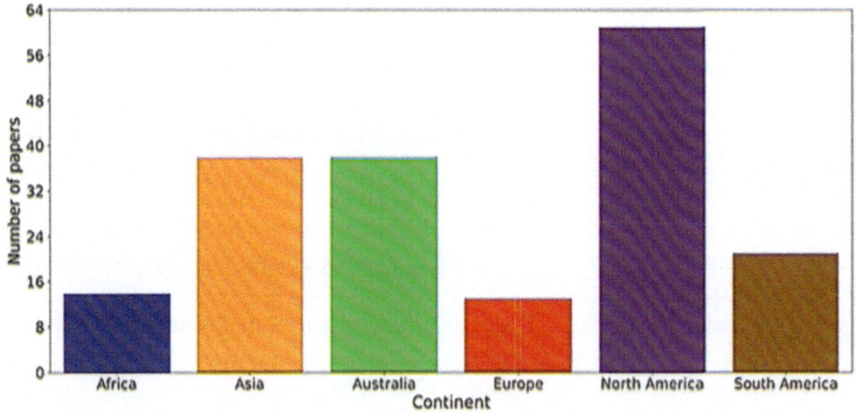

Fig. 5.29 Number of case studies by continent and method. *Source* Vallee et al. (2024)—Open Access

Case Studies of Magnetic and Electromagnetic Techniques in Engineering Geophysics Over the Last Fifteen Years

The numerous references presented show that magnetic and electromagnetic (EM) methods continue to be widely used for mineral exploration, with significant research activity dedicated to these techniques (Nabighian et al. 2005; Smith 2014; Vallee et al. 2011; Nabighian and Asten 2002). Drone-based magnetic surveys, in particular, have grown in popularity. The various case studies (excluding drone studies) have been categorized based on target type and continent for both magnetic and EM methods (Vallee et al. 2024). The results are presented in the form of histograms, with key insights highlighted. Figure 5.28 displays the distribution of research papers for each method, while Fig. 5.29 shows the distribution by continent

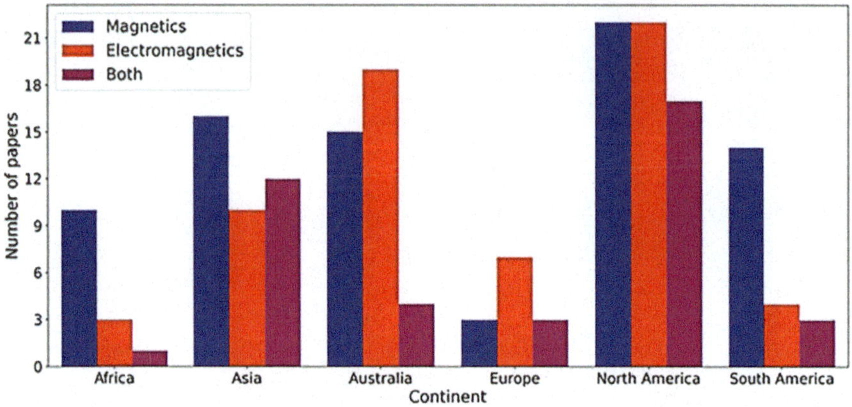

Fig. 5.30 Distribution of case studies by continent and method. *Source* Vallee et al. (2024)—Open Access

Magnetic surveys are predominant in the case studies, especially in North America, followed by Asia and Australia, with other continents displaying smaller but similar numbers. The magnetic technique is popular due to its ability to serve both as a direct and indirect exploration tool. Magnetic surveys can directly detect magnetic materials like iron ore, and they are also used to map geological structures, where rock magnetite content serves as an indicator (Naskar et al. 2003). EM surveys are more focused on detecting conductive mineralization, particularly in uranium exploration, where they are used to identify conductive features associated with mineralization.

A deeper analysis of Fig. 5.30 reveals distinct regional trends. In North America, both magnetic and EM methods are equally popular, while magnetic methods dominate in other regions, except for Europe, where EM techniques are more commonly applied. The combination of magnetic and EM techniques is most prevalent in North America and Asia. The varying number of case studies across continents is influenced by several factors such as geographical and metallogenic conditions, mining markets, the concentration of researchers, and journal biases. Despite these factors, this distribution provides an indication of the method preferences in different regions.

Figure 5.31 presents data on the types of mineral deposits explored and the methods used. Gold is the most commonly explored target, followed by porphyry copper, polymetallic deposits, and massive sulfides. Magnetic techniques are typically used for gold, iron ore, and polymetallic deposits, while EM methods are more commonly employed for uranium and massive sulfides. Both methods are applied to porphyry copper and gold deposits.

The field of geophysical exploration for mineral deposits remains active, with continuous advancements in magnetic and EM techniques. This ongoing development allows researchers to better adapt to new exploration challenges. However, it should be noted that this review is not exhaustive, as not all geophysical activities are documented. The dominance of case studies from North America, Asia, and

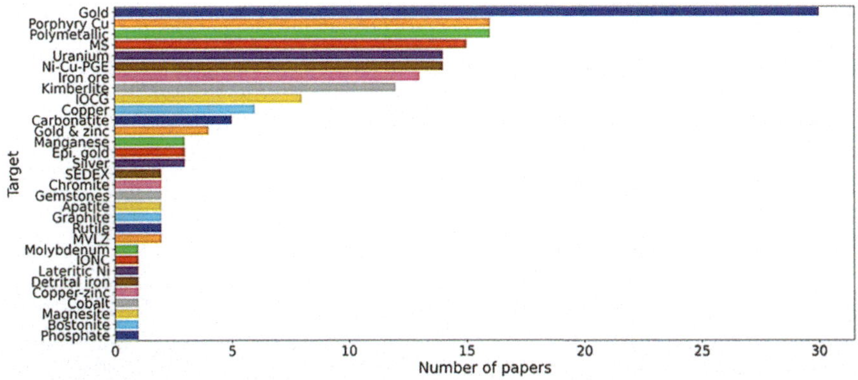

Fig. 5.31 Number of case studies by types of mineralization. *IOCG* Iron-Oxide Copper-Gold, *IONC* Iron-Oxide Nickel-Copper, *MS* Massive Sulfides, *MVLZ* Mississippi Valley Lead-Zinc, *SEDEX* Sedimentary Exhalative Lead-Zinc. *Source* Vallee et al. (2024)—Open Access

Australia can be attributed to factors such as academic trends, regional metallogenic conditions, mining markets, and the journals selected for review.

Magnetic methods tend to be more widely used due to their dual role as both direct and indirect indicators of mineralization. Both magnetic and EM techniques are versatile and are employed across a wide variety of target types, including gold, porphyry copper, polymetallic, massive sulfides, uranium, Ni-Cu-PGE, and others. This review provides a comprehensive look at the application of these techniques in various geological environments, highlighting the challenges and limitations encountered during exploration.

In conclusion, these findings emphasize the continued importance and evolution of magnetic and EM methods in mineral exploration. The insights offered in this discussion can help guide the design of future geophysical surveys and improve understanding of the challenges faced in specific geological settings.

Magnetotellurics (MT)

Magnetotellurics (MT) is a geophysical method that utilizes natural variations in the Earth's electromagnetic field to image subsurface structures. Low-frequency MT techniques are particularly well-suited for deep exploration, as they offer greater depth of penetration—referred to as larger skin depth—making them ideal for investigating geothermal systems and deep crustal formations. In contrast, high-frequency MT techniques are more appropriate for shallow investigations, as they provide higher resolution, allowing for detailed imaging of surface and near-surface features.

Case Study on Geothermal Exploration and 2D MT Inversion in Al-Mubazzarah, UAE

The study focused on investigating the geothermal potential of the Al-Mubazzarah Geothermal Area (AMGA), located southeast of Al Ain, UAE, by conducting a Magnetotelluric (MT) survey along the northern part of Hafit Anticline. The data collected was analyzed using 2D inversion, which revealed three significant resistivity layers (Saibi et al. 2021).

The first layer, with resistivity values between 10 and 20 Ωm, corresponds to clastic sedimentary rocks from the Miocene to Quaternary period. The second layer, extending to a depth of 2 km, is a conductive zone with resistivity values under 10 Ωm, representing Tertiary carbonates, likely influenced by high Total Dissolved Solids (TDS) in groundwater. The third layer, found at depths between 2 and 4 km, is characterized by resistivity values between 10 and 30 Ωm and is interpreted as a probable geothermal reservoir hosted by late Cretaceous carbonates and conglomerates with underlying metamorphosed igneous rocks.

Integrating gravity and magnetic data further supported the resistivity model, highlighting changes in density and magnetic susceptibility with depth. Figures 5.32, 5.33, and 5.34 show critical aspects of the analysis, including the rose diagram of strike estimates, 2D resistivity models, and the gravity and magnetic inversion results. The geothermal reservoir, located between 2 and 4 km in depth, is consistent with findings from previous geothermometry analyses. The results suggest that the AMGA has significant geothermal potential, with extensional fault systems aiding fluid migration. The study is the first detailed exploration of the area using MT techniques, providing valuable insights for future geothermal energy extraction efforts. Further studies will be needed to refine the model and assess the feasibility of geothermal energy development in the region.

5.2.2 *Characterization of Subsurface Properties*

Recent advancements in both laboratory and field equipment have significantly improved the accuracy of physical property measurements such as density, porosity, permeability, and elastic moduli. Non-destructive techniques, including micro-CT scanning and acoustic emission monitoring, now offer more detailed and reliable insights into the characteristics of rocks and soils without altering the sample. In parallel, geophysical inversion methods have seen considerable development. Joint inversion techniques that combine multiple datasets, such as seismic and electrical resistivity data, enhance the reliability of subsurface models by leveraging complementary information. Additionally, the integration of machine learning and artificial intelligence into inversion workflows has become increasingly common, helping to interpret complex datasets more efficiently while reducing uncertainty and computational demands. Hydrogeophysical techniques further contribute to

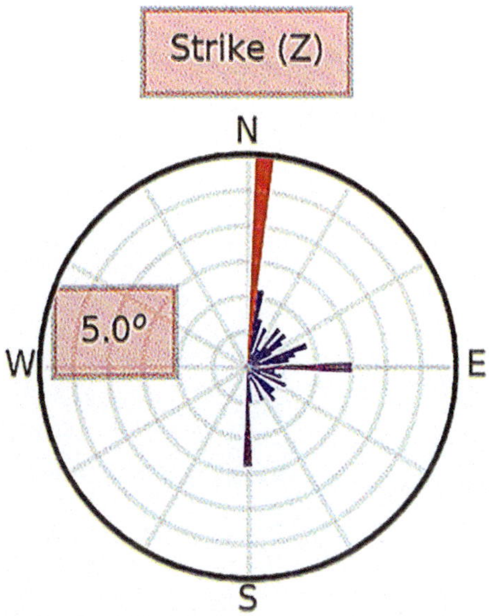

Fig. 5.32 The rose diagram of strike estimates (N5°E) for the magnetotelluric (MT) dataset collected from the Al-Mubazzarah Geothermal Area (AMGA). The strike analysis was conducted over the entire frequency range of the MT data to better understand the regional geological structures and fault orientations. The rose diagram reveals the dominant strike directions and provides insights into the extensional tectonic features of the study area, which play a significant role in the geothermal system by influencing fluid pathways and subsurface fluid migration. *Source* Saibi et al. (2021)—Open Access

subsurface characterization by combining geophysical observations with hydrological models. This integration enables the precise mapping of groundwater flow patterns, aquifer boundaries, and contaminant transport pathways, supporting more informed decision-making in hydrogeological and environmental studies.

5.2.3 Subsurface Mapping Applications

Subsurface mapping plays a vital role in a range of practical applications across various sectors. In urban planning and construction, geophysical surveys provide essential data for evaluating ground stability, detecting underground hazards, and guiding infrastructure development. These methods support tasks such as optimizing tunnel alignments and monitoring subsidence or landslide risks in densely built environments. In the field of resource exploration, enhanced imaging techniques are indispensable for identifying and assessing mineral, hydrocarbon, and geothermal resources. Advanced mapping methods not only reduce the risks associated with exploration but also improve the efficiency of resource extraction. Environmental and

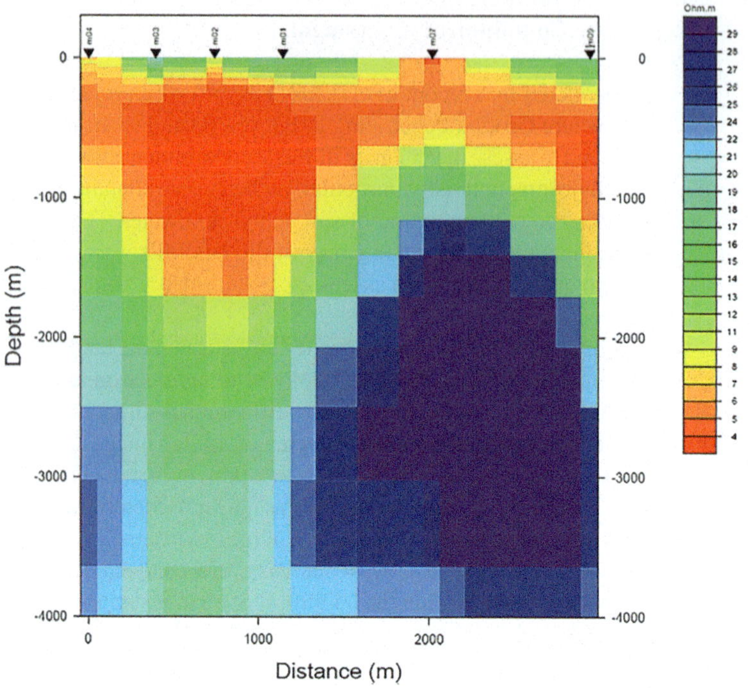

Fig. 5.33 Resistivity cross-section derived from the 2D inversion of MT data. The model reveals shallow sediments with moderate resistivity, overlying a low-resistivity layer, which is followed by a progressively increasing resistivity zone at greater depths. This cross-section provides a detailed visualization of the subsurface structure, helping to delineate the geological layers and potential geothermal reservoir within the Al-Mubazzarah Geothermal Area. *Source* Saibi et al. (2021)—Open Access

disaster management efforts also benefit greatly from subsurface mapping. Geophysical techniques are used to delineate contamination zones, evaluate earthquake-prone areas, and monitor volcanic activity. Specific applications include characterizing landfill sites and supporting early warning systems for hazards like tsunamis and landslides. In archaeology and cultural heritage preservation, non-invasive geophysical tools have become fundamental, enabling researchers to detect and document buried structures and artefacts without the need for excavation, thus helping to protect and manage valuable historical sites.

5.2.4 *Challenges and Future Directions*

Despite significant advancements in subsurface imaging, characterization, and mapping, several challenges continue to hinder broader application and accessibility. One major obstacle is the high cost associated with advanced geophysical equipment

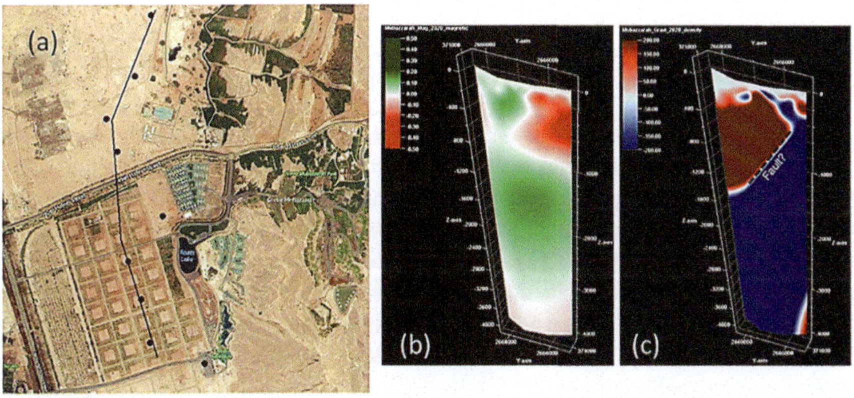

Fig. 5.34 **a** Profile location used for dissecting inverted gravity and magnetic results from 3D potential field inversion, with black dots indicating the locations of MT stations. **b** Magnetic inversion (magnetic susceptibility in SI units), and **c** gravity inversion (density contrasts in g/cm^3). The gravity inversion cross-section reveals a potential normal fault dipping to the north at approximately 25°, or alternatively, it may represent a buried erosional surface (non-conformity). If interpreted as a fault, it could link the Al-Mubazzarah Geothermal Area (AMGA) to the Ain Faidha hot springs, as fault orientations in this region are predominantly E–W. A probable fault is also observed in the resistivity cross-section derived from the MT data. *Source* Saibi et al. (2021)—Open Access

and surveys, which can limit their use, especially in resource-constrained settings. Additionally, the integration and interpretation of diverse geophysical datasets remain complex tasks that demand interdisciplinary expertise, involving geophysicists, data scientists, and domain specialists. Environmental and regulatory constraints may also restrict certain geophysical operations, particularly in sensitive or protected areas.

Looking ahead, several emerging trends are poised to reshape the field. Autonomous systems, including unmanned aerial vehicles (UAVs) and robotic platforms, are increasingly being deployed for geophysical surveys in remote or hazardous locations, improving safety and accessibility. Real-time monitoring is becoming more feasible through the use of permanent sensor networks that enable continuous observation of subsurface changes. Enhanced data analytics, supported by big data and cloud computing, allow for the efficient processing and interpretation of large volumes of geophysical data, improving decision-making and predictive modeling. Sustainability is also a growing focus, with efforts directed toward developing environmentally friendly methods and minimizing the ecological footprint of geophysical surveys.

These advancements are revolutionizing our understanding of the subsurface. By integrating state-of-the-art technologies with innovative methodologies, geophysicists are increasingly able to address critical challenges across diverse sectors. This progress supports the development of safer, more efficient, and sustainable practices in subsurface exploration and monitoring.

5.3 Applications in Civil Engineering, Construction, and Environmental Studies

Geophysical methods, particularly those used in subsurface exploration, are critical for a wide range of applications in civil engineering, construction, and environmental studies. These techniques provide valuable insights into the subsurface structure, which is essential for the planning, design, and execution of engineering projects, as well as for assessing environmental impacts and ensuring sustainability.

In civil engineering, geophysical methods are commonly employed for site characterization and the assessment of soil and rock properties. These methods help in identifying the thickness and type of bedrock, groundwater levels, and the presence of geological hazards such as faults, fractures, or liquefaction-prone layers. For instance, seismic surveys are used to assess the shear wave velocity of soil, which is critical for earthquake engineering, as well as for determining the stability of foundations for large structures like bridges, dams, and high-rise buildings (Telford et al. 1990). Similarly, ground-penetrating radar (GPR) is used to locate buried utilities, pipelines, and archaeological sites, while resistivity and electromagnetic surveys can detect changes in soil properties that might affect construction projects (Reynolds 2011).

In construction, especially for large infrastructure projects such as tunnels, dams, and foundations, accurate subsurface data is crucial to minimize risks and avoid costly surprises during construction. Geophysical techniques enable engineers to identify potential challenges, such as water saturation, weak soil zones, or the presence of hazardous materials. These insights guide the design of appropriate foundations, help in choosing construction materials, and allow engineers to assess the potential impact of environmental factors on construction stability and longevity (Worthington 2011).

Environmental studies also heavily rely on geophysical methods for the assessment of contamination, groundwater flow, and environmental hazards. For example, resistivity and electromagnetic surveys are commonly used to map subsurface contamination in areas affected by industrial pollution, leaking underground storage tanks, or landfills. These surveys help locate contaminant plumes and monitor the movement of pollutants through groundwater, which is vital for managing environmental risks and planning remediation efforts (Garbey et al. 2013). Furthermore, geophysical methods are applied to study the effects of construction projects on the surrounding environment, including soil erosion, groundwater depletion, and potential habitat disturbances.

The integration of geophysical data with other site characterization tools, such as boreholes, geotechnical testing, and environmental monitoring, enhances the accuracy and reliability of subsurface investigations. Advances in geophysical technology, such as 3D imaging and the use of drones for geophysical surveys, have further revolutionized site characterization, enabling more detailed and cost-effective assessments. These methods not only contribute to the safety and efficiency of construction projects but also support sustainable development by providing critical data for managing environmental resources and mitigating potential environmental impacts.

In summary, the application of geophysical methods in civil engineering, construction, and environmental studies plays a vital role in ensuring the success and sustainability of infrastructure projects. By providing a clear and detailed understanding of subsurface conditions, these techniques help engineers make informed decisions, reduce risks, and optimize designs. Furthermore, they contribute to environmental protection by enabling early detection of potential hazards and facilitating the remediation of contaminated sites.

5.4 Geophysical Exploration for Foundations, Tunnelling, and Infrastructure

In the realm of civil engineering, the application of geophysical methods is invaluable for evaluating subsurface conditions and mitigating risks during the design and construction of infrastructure projects. Case studies of geophysical exploration for foundations, tunnelling, and other infrastructure projects highlight the significant role of these methods in providing detailed, accurate subsurface information. These case studies demonstrate how geophysical tools can help engineers optimize design choices, ensure safety, and reduce costs in challenging geological environments.

A notable example of geophysical exploration for foundations comes from a major high-rise building project in an urban area with a complex subsurface structure. Traditional drilling techniques for geotechnical site investigations can be expensive and time-consuming, especially in dense city centers. In this case, electrical resistivity imaging (ERI) and seismic refraction surveys were employed to identify variations in soil and rock properties at depth. The geophysical results revealed an underlying layer of soft clay at a depth of 20 m, which could pose a risk to the foundation's stability. With this information, engineers were able to adjust the foundation design to include deeper pilings, ensuring the building's long-term safety and reducing the need for expensive remedial measures during construction (Burland et al. 2011).

Similarly, in the construction of large tunnel systems, geophysical methods are instrumental in understanding the subsurface conditions, such as identifying water-bearing zones, rock quality, and potential hazards like faults or unstable soil. A case study involving the construction of a metro tunnel system in a densely populated urban area used a combination of ground-penetrating radar (GPR), seismic tomography, and electrical resistivity tomography (ERT) to assess the tunnel alignment. The GPR surveys helped locate areas with high water content and fractures in the bedrock, while the seismic tomography provided insights into the rock's strength and stability. The integration of these data helped engineers adjust the tunnel's route, avoiding areas of poor ground conditions and ensuring the tunnel's safety during construction (Lamarche et al. 2008).

Another example can be found in the use of geophysical methods for assessing the integrity of existing infrastructure. For example, during the rehabilitation of a highway bridge, engineers used electromagnetic and resistivity methods to evaluate the condition of the concrete and assess any subsurface anomalies, such as voids or cracks that might compromise the structure. The geophysical surveys identified a series of fractures below the bridge foundation, which were not detectable through traditional inspection methods. By identifying these weaknesses early, the team was able to repair and reinforce the structure more effectively, preventing future damage and extending the bridge's service life (Telford et al. 1990).

In addition to the examples of foundation and tunnelling projects, case studies also show the integration of geophysical techniques in large-scale infrastructure projects like dams, roads, and airports. In a study of a proposed dam site, seismic reflection surveys were used to map the bedrock surface and identify potential fault zones. The geophysical findings revealed a buried fault beneath the proposed dam site that could potentially cause seismic activity and undermine the dam's stability. As a result, the dam's design was altered, and further investigation was conducted to account for seismic safety, showcasing how geophysics aids in preventing costly design errors (Keller and Frischknecht 1996).

These case studies collectively emphasize the importance of geophysical exploration in civil engineering, particularly for infrastructure projects that involve foundations, tunnelling, and large-scale construction. They highlight how these techniques, when combined with traditional drilling and site investigation methods, provide a more comprehensive understanding of subsurface conditions. Geophysical methods not only improve safety by identifying potential hazards but also enable engineers to make more informed decisions regarding design and construction, which ultimately leads to cost savings and greater efficiency.

5.4.1 Case Study on Urban Geophysical Exploration in Chengdu International Bio-City

The Chengdu International Bio-City, located in southern Chengdu, China, is a rapidly developing urban area characterized by terrace hills and fractured tectonic structures, primarily oriented in the northeast-southwest direction. The region is geologically rich, with sedimentary layers from the Minjiang River, including Cretaceous and Jurassic mudstones, siltstones, sandstones, and gypsum layers. The ongoing urbanization, including high-rise buildings, infrastructure, and utilities, poses significant challenges for subsurface exploration. In response to the need for comprehensive underground data to support sustainable urban planning and development, the Chengdu Municipal Government, in collaboration with the China Geological Survey, initiated a detailed urban geological survey focused on the shallow subsurface (up to 200 m deep) of the Chengdu Bio-City (Chengdu Municipal Government 2022).

The primary objective of this study was to assess the effectiveness of three geophysical methods—Electrical Resistivity Tomography (ERT), micro-seismic exploration, and the Opposing-Coil Transient Electromagnetic Method (OCTEM)—in mapping the subsurface geology in this complex urban environment (Wang et al.

2023). These methods were selected to overcome the challenges of electromagnetic interference, underground infrastructure, and limited access to construction sites, all of which are typical in modern cities. A key goal was to develop an accurate 3D geological model of the area, which could inform urban development and infrastructure planning (Zhang et al. 2021).

The ERT method, employing a 120-channel system and spaced 10 m electrodes, was used to measure resistivity profiles along a selected test area. However, urban interference from underground utilities and industrial electromagnetic noise hindered the results, particularly in high-resistivity zones, making ERT unsuitable for rapid, large-scale surveys in this area. Despite this, it was useful as a supplementary method for detecting shallow geological features (Fig. 5.35).

The micro-seismic method, using an MTKV-3C system with a station spacing of 20 m, provided insights into shallow subsurface structures through S-wave velocity measurements (Fig. 5.36). It proved effective for mapping the shallow geology, but its resolution diminished at depths beyond 100 m, limiting its applicability for deeper exploration (Liu and Yu 2019).

The OCTEM method, on the other hand, demonstrated significant advantages in urban environments. Using a HPTEM-18 system with a frequency of 6.25 Hz and a 10-m station spacing, OCTEM provided high-quality data up to 200 m deep, overcoming electromagnetic interference and identifying key geological features such as gypsum and mudstone layers. The inversion of OCTEM data revealed a multi-layered resistivity structure that closely matched the geological context of the region. Integrating this data with borehole measurements allowed for the creation of a robust 3D resistivity model of the subsurface, which highlighted the distribution

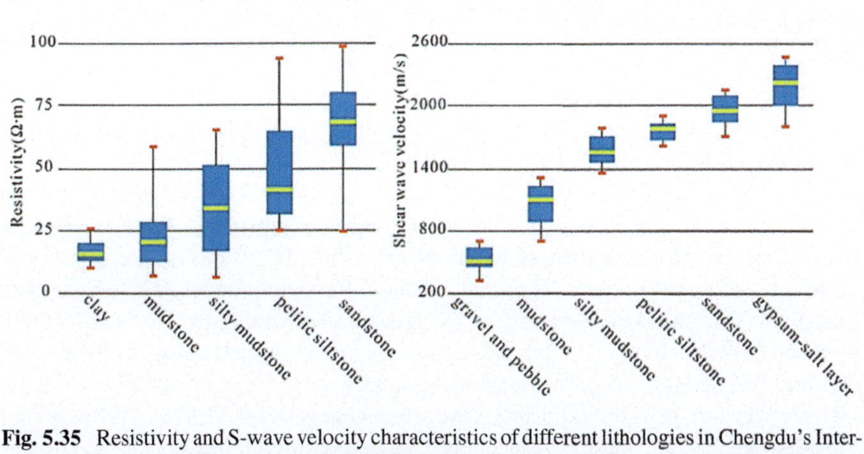

Fig. 5.35 Resistivity and S-wave velocity characteristics of different lithologies in Chengdu's International Bio-City reveal key geological insights. The resistivity varies across layers, with lower values for clay and mudstone (10–40 Ωm) and higher values for sandy mudstone and gypsum-rich layers (above 100 Ωm). S-wave velocities also differ, showing lower values in soft sediments like clay and mudstone (180–300 m/s) and higher values in consolidated layers like weathered sandstone (1300–2100 m/s), indicating variations in rock stiffness and lithological composition. These characteristics were crucial in constructing a detailed 3D geological model for the area. *Source* Wang et al. (2023)—Open Access

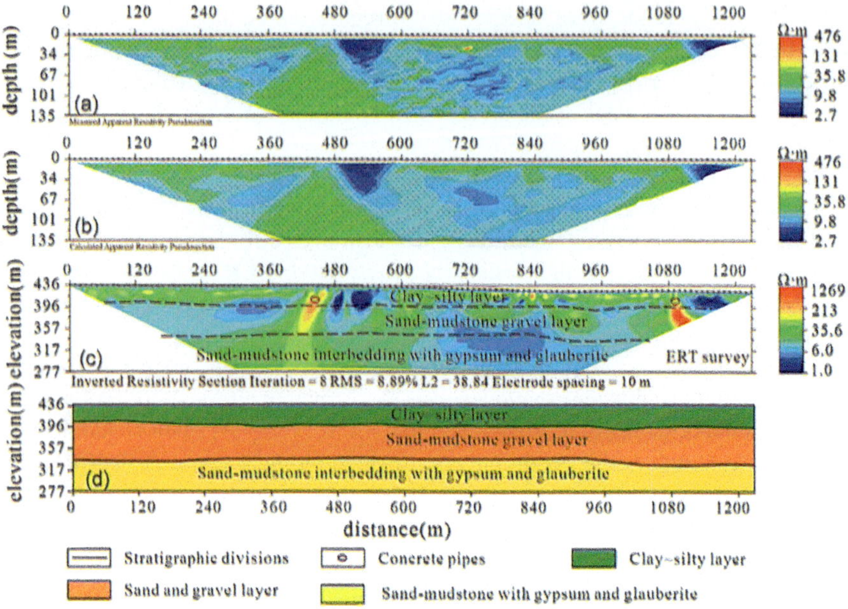

Fig. 5.36 A comprehensive view of the resistivity data and geological interpretation for the study area. **a** The measured apparent resistivity pseudosection shows the raw resistivity data along the profile. **b** The calculated apparent resistivity pseudosection represents the resistivity model based on theoretical assumptions. **c** The inverted resistivity section illustrates the result of resistivity inversion, highlighting subsurface variations, though it shows discrepancies due to urban infrastructure interference. **d** The illustrated geological section combines the resistivity data with borehole information, providing a final interpretation of the geological layers in the area. *Source* Wang et al. (2023)—Open Access

of clay, sandy pebbles, weathered mudstone, and gypsum-rich layers (Cheng and Li 2020).

The final 3D geological model, built from OCTEM and drilling data, provided critical information for urban planning, highlighting subsurface features relevant to construction projects, such as potential groundwater reservoirs, areas prone to instability, and geotechnical properties essential for foundational engineering. The model also demonstrated the utility of OCTEM in overcoming the challenges posed by urban infrastructure, including power lines and underground utilities, which often interfere with traditional geophysical techniques.

In conclusion, this study illustrates the effectiveness of OCTEM as a primary tool for urban geological surveys, especially in areas with complex subsurface conditions and high levels of interference. The 3D geological model created from this research will serve as a valuable resource for future urban development in the Chengdu International Bio-City, enabling more informed decisions regarding site selection, construction, and environmental management. The successful application of OCTEM in this context suggests its potential for similar urban geophysical studies

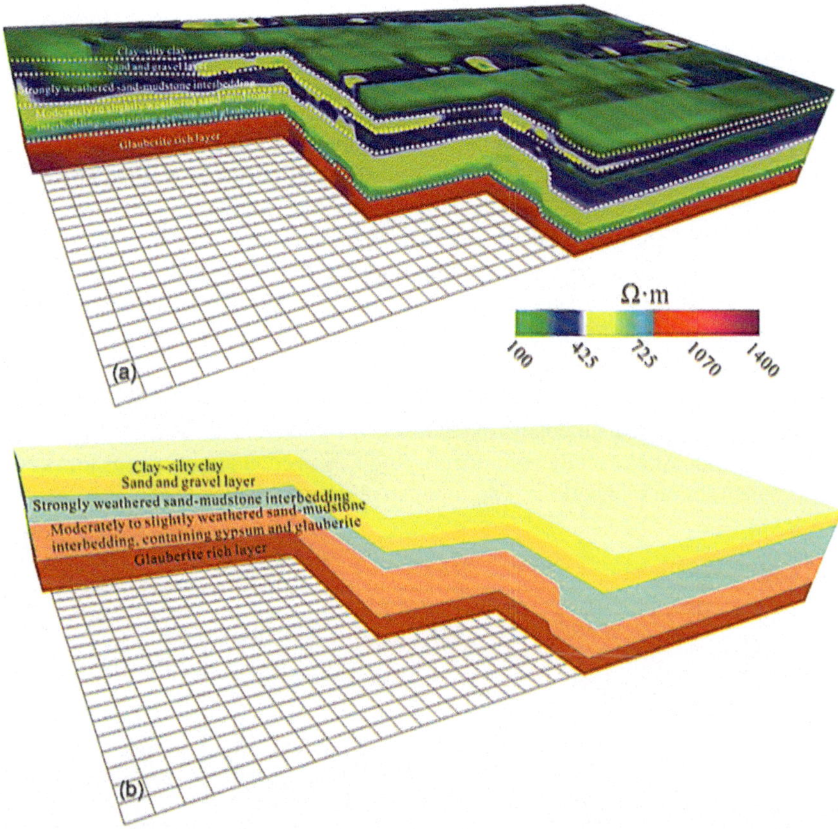

Fig. 5.37 **a** 3D geo-electric structure of the study area, showing the resistivity distribution at various depths. **b** 3D geological model of the study area, integrating geophysical and borehole data to depict key subsurface layers, including clay, sandy pebbles, weathered mudstone, and gypsum-rich zones. *Source* Wang et al. (2023)—Open Access

in other rapidly developing cities worldwide (Liu and Yu 2019). Key figures from the study include the 3D resistivity models generated from the OCTEM data (Fig. 5.37), which provided a detailed view of the subsurface structure, and the comparison of OCTEM results with borehole data, which validated the method's accuracy and effectiveness.

5.4.2 *Case Study on Pre-foundation Geophysical Investigation in Oka-Akoko, Nigeria*

This study, authored by Ademila (2022), aims to perform a pre-foundation geophysical investigation of the site for structural development in Oka-Akoko, Nigeria, located in the Basement Complex geological setting. Given the heterogeneity and variations in the subsurface conditions, understanding the geological and geo-engineering materials of the site is crucial for designing stable and sustainable structures. Using combined Very Low Frequency Electromagnetic (VLF-EM) and Electrical Resistivity techniques, the study seeks to identify subsurface geological structures that could pose engineering risks to proposed building projects (Fig. 5.38). This investigation will provide essential data to guide foundation design, prevent potential structural failure, and ensure the suitability of the site for construction. Additionally, the findings will help in selecting appropriate borehole sites for groundwater development. The study area is characterized by highlands, lowlands, and a tropical rainforest climate, with lithologies primarily composed of granite and gneiss, which may influence groundwater and structural integrity.

The VLF-EM (Very Low Frequency Electromagnetic) profiles and the corresponding 2-D inverted models provided key insights into the subsurface conditions of the Oka-Akoko study area, identifying various conductive zones that could impact construction foundations. The conductive anomalies, depicted in Figs. 4a–16a

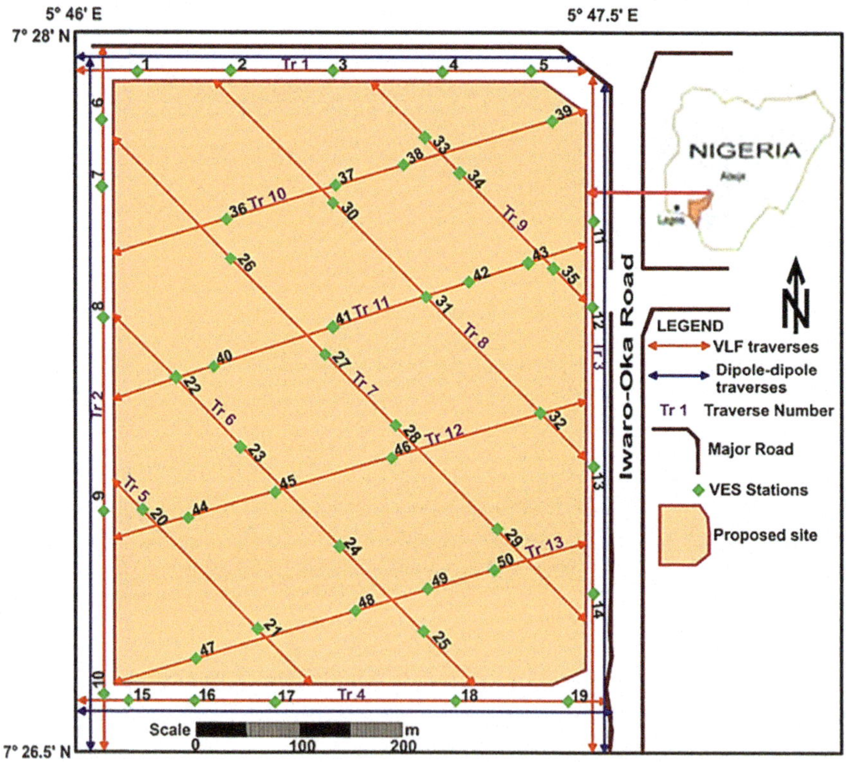

Fig. 5.38 Geophysical Data Acquisition Map for Pre-Foundation Investigation in Oka-Akoko, Nigeria. *Source* Ademila (2022)—Open Access

(Ademila 2022), represent potential weak zones such as faults, fractures, weathered zones, and basement depressions that may affect structural stability. These anomalies, which indicate the presence of groundwater pathways, were detected at varying distances across 13 traverses in the survey area.

For example, Traverse 1 (Fig. 5.39) revealed multiple conductive zones at distances such as 75, 170, 255, and 365 m. The corresponding 2-D inverted model (Fig. 5.39b) confirmed the presence of weak subsurface zones that could pose a risk to the foundation of proposed buildings.

Similarly, Traverse 4 (Fig. 5.40) showed conductive structures indicative of basement depressions and fractures, posing a significant threat to structural integrity if not addressed.

These findings highlight the importance of incorporating geophysical data into the site investigation process to prevent construction issues, such as tilting or collapse, by identifying areas with weak or unstable subsurface conditions. The VLF-EM profiles and 2-D inversion models are valuable tools in guiding the design of safe

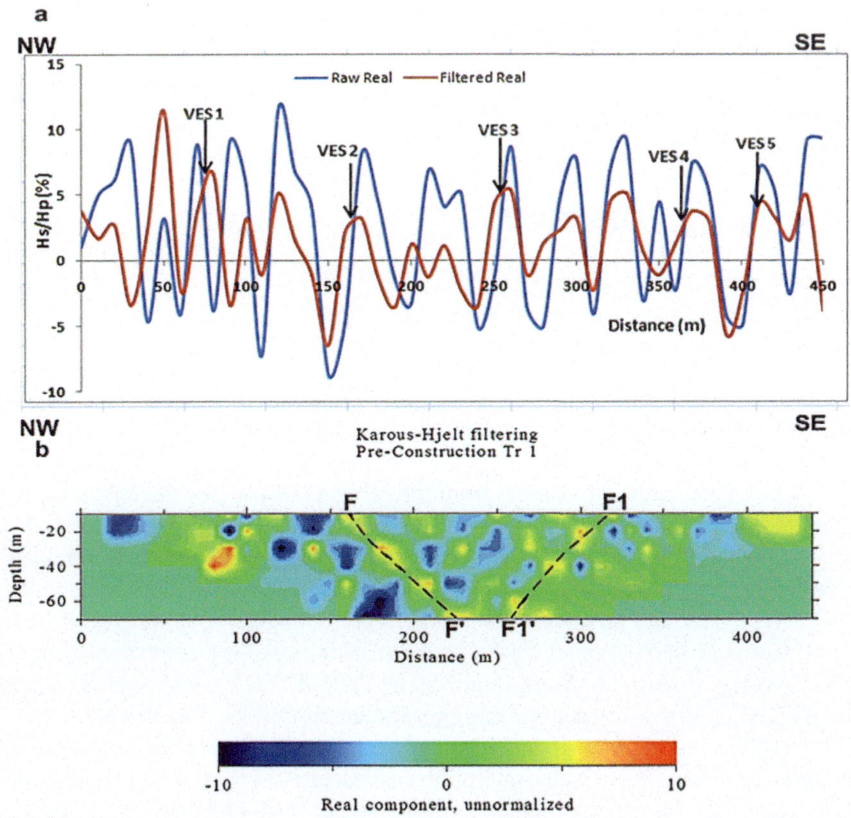

Fig. 5.39 **a** VLF-EM plot of real components and **b** corresponding 2-D inverted section of Traverse 1 in Oka-Akoko, Nigeria. *Source* Ademila (2022) Open Access

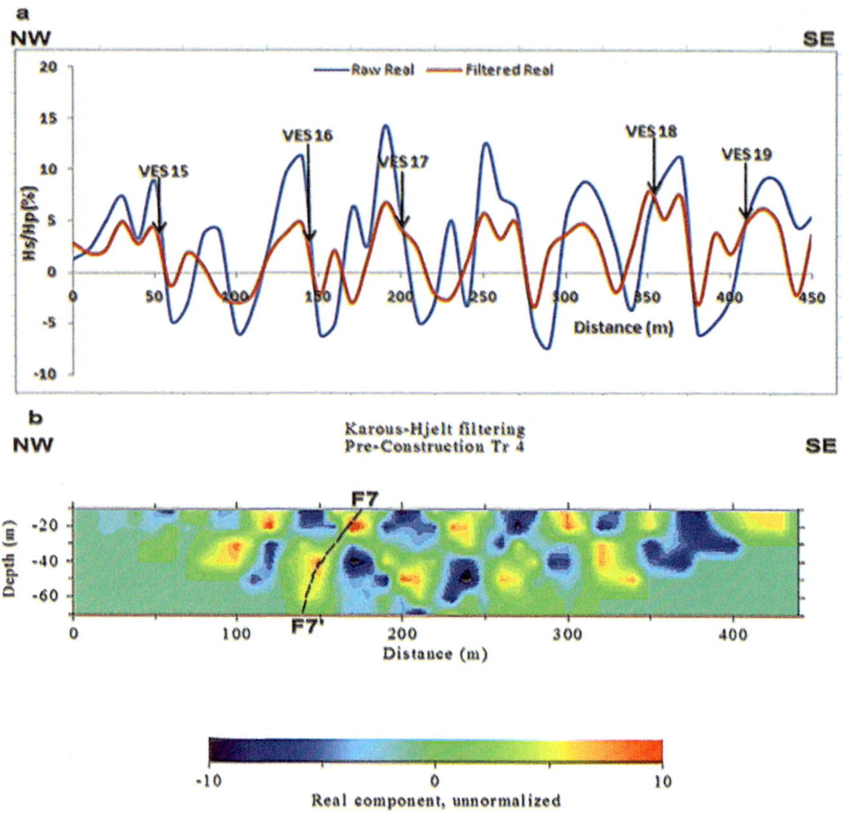

Fig. 5.40 **a** VLF-EM plot of real components and **b** corresponding 2-D inverted section of Traverse 4 in Oka-Akoko, Nigeria. *Source* Ademila (2022)—Open Access

and sustainable foundations for construction projects in geologically complex areas like Oka-Akoko.

The electrical resistivity results from the 50 VLF-EM survey stations provided insights into the subsurface structure, with key resistivity responses identified at varying depths. The 2-D resistivity sections, interpreted with respect to lithology, revealed patterns of resistivity variations across the study area. Two predominant curve types, HA and AA, were identified, representing topsoil, weathered basement, fractured bedrock, and fresh basement. Key resistivity responses included low resistivity (below 120 Ωm), intermediate (130–575 Ωm), and high resistivity (above 1000 Ωm), with each indicating different subsurface conditions. Notably, low resistivity zones were found in weathered sections that could threaten civil engineering works due to water saturation and fractures in the bedrock. High resistivity responses, on the other hand, corresponded to fresh basement rock. Specific challenges like subsidence and differential settling were highlighted in locations with saturated clay and fractured bedrock, where deep foundations were recommended for stable construction. Figure 5.41b illustrates the subsurface resistivity model, showing the geoelectric layers, including a resistive topsoil (238–350 Ωm), weathered clay (47–56 Ωm), partially weathered basement (261–467 Ωm), and fresh bedrock (1052–1618 Ωm).

The resistivity survey along Traverse 4 (Fig. 5.42a) revealed low, intermediate, and high resistivity responses, indicating clayey subsoil, fractured bedrock, and competent materials beneath. Low resistivity zones, particularly between 40–60, 100–202, and 248–364 m, suggest water-saturated clay, unsuitable for construction. Intermediate resistivity is linked to weathered and fractured bedrock, while high resistivity indicates stable materials beneath. These conditions, combined with irregular bedrock depressions, pose structural risks due to potential differential settlement and groundwater interaction, especially in areas like VES 18. Figure 5.42b shows the geologic units, reinforcing that the clayey subsoil is an incompetent layer for building foundations, requiring deep foundation solutions to avoid future structural failure.

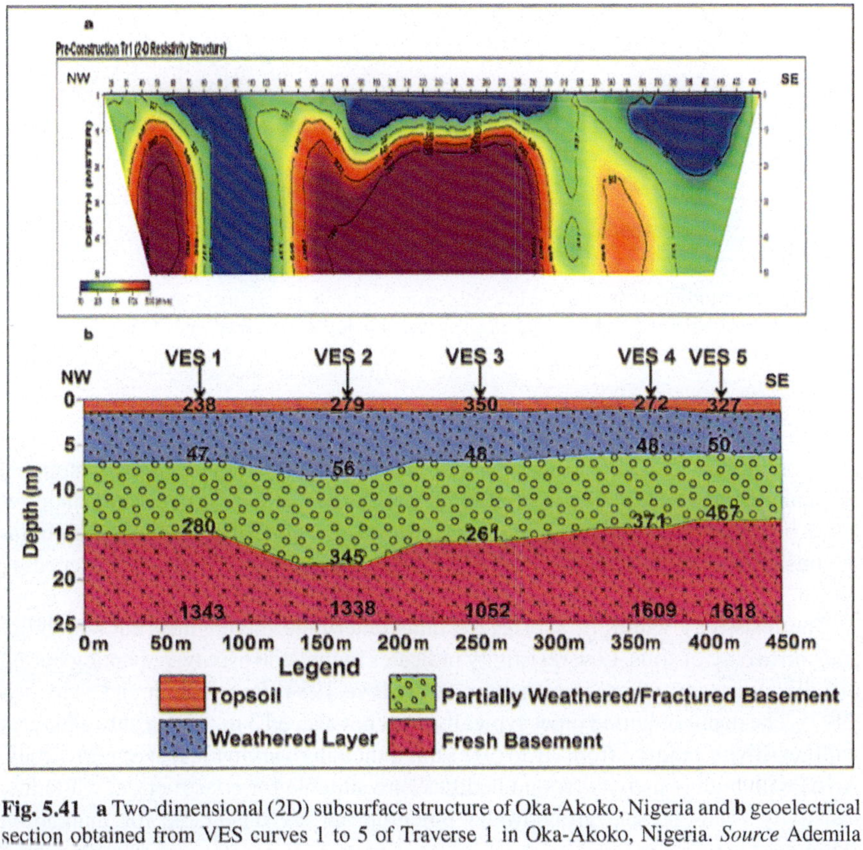

Fig. 5.41 **a** Two-dimensional (2D) subsurface structure of Oka-Akoko, Nigeria and **b** geoelectrical section obtained from VES curves 1 to 5 of Traverse 1 in Oka-Akoko, Nigeria. *Source* Ademila (2022)—Open Access

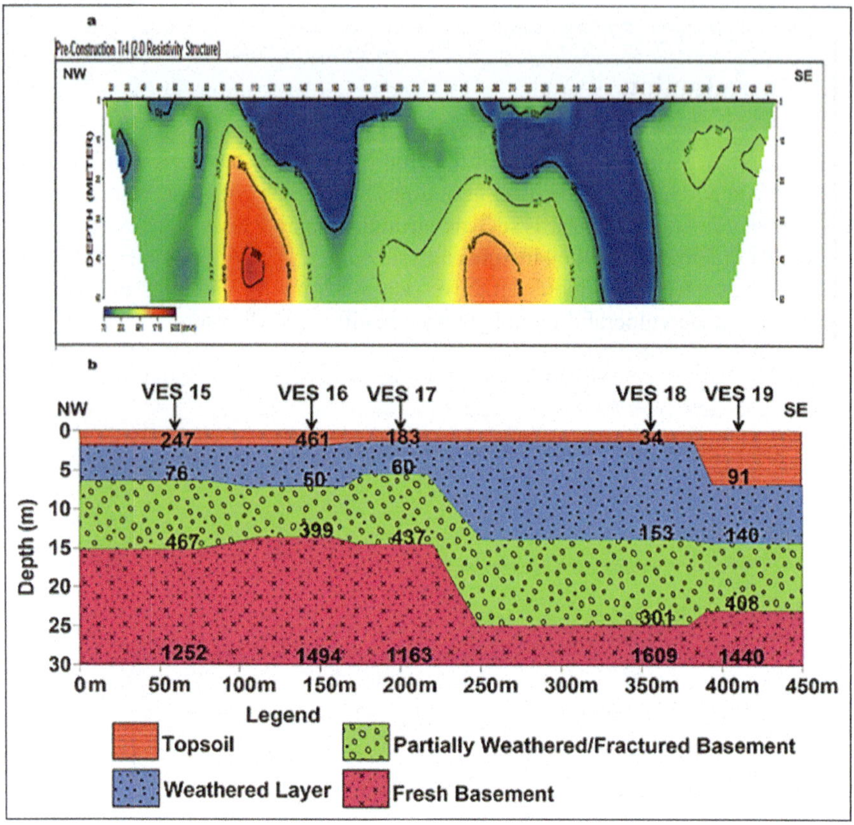

Fig. 5.42 **a** Two-dimensional (2D) subsurface structure and **b** geoelectrical section obtained from interpretation of VES curves 15 to 19 along Traverse 4 in in Oka-Akoko, Nigeria. *Source* Ademila (2022)—Open Access

Similarly, in Traverses 5 and 6 (Fig. 5.43), low-resistivity zones (clay, sandy clay) and irregular bedrock depressions indicate water-saturated zones unsuitable for shallow foundations. The fractures and weathered bedrock form potential aquifer zones but present risks for civil engineering works. For structural stability, deep foundations are recommended, especially in locations with depth to bedrock exceeding 30 m.

The resistivity values of subsurface units determine the engineering suitability and corrosivity of soils. Low resistivity indicates higher corrosivity, impacting buried metallic pipes and other infrastructure (Agunloye 1984; Baeckmann and Schwenk 1997). The depth of buried pipes typically does not exceed 3 m, within subsoil layers with resistivity ranging from 21 to 507 Ωm, indicating varying levels of corrosivity. Areas with high resistivity (sound bedrock) are suitable for construction, while low resistivity zones present groundwater potential but are unsuitable for foundation stability.

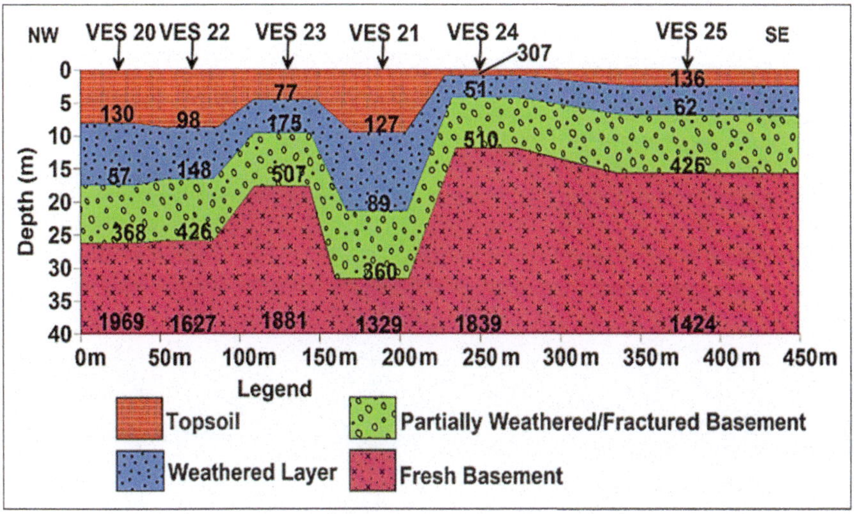

Fig. 5.43 The geoelectrical section of Traverses 5 and 6 in Oka, Nigeria, obtained from the interpretation of VES data, showing subsurface resistivity variations. *Source* Ademila (2022)—Open Access

The 3-D fresh bedrock resistivity map (Fig. 5.44b) highlights areas with high resistivity for safe construction, while low-resistivity zones are identified as unsuitable for building foundations. Engineering works in unstable areas may risk corrosion or structural failure unless mitigated by deep foundations or soil stabilization. Fracture zones in the area indicate subsurface geological structures that could destabilize foundations. The study shows that VLF-EM and resistivity techniques effectively characterize subsurface conditions, revealing issues like uneven bedrock topography and deep weathering. Sites with unstable bedrock should be avoided for civil works to prevent structural failure (Fig. 5.45).

Recommendation: Foundations should be designed to rest on competent bedrock, avoiding unstable zones, as shown in Fig. 5.44b, to ensure sustainable infrastructure development.

5.5 Conclusions

This chapter has provided an in-depth exploration of the cutting-edge applications of geophysics in subsurface exploration and site characterization, emphasizing their critical role in modern engineering, environmental studies, and archaeology. Advancements in imaging and characterization techniques, such as seismic imaging, ground-penetrating radar (GPR), electrical resistivity tomography (ERT), electromagnetic methods, and magnetotellurics (MT), have significantly enhanced our

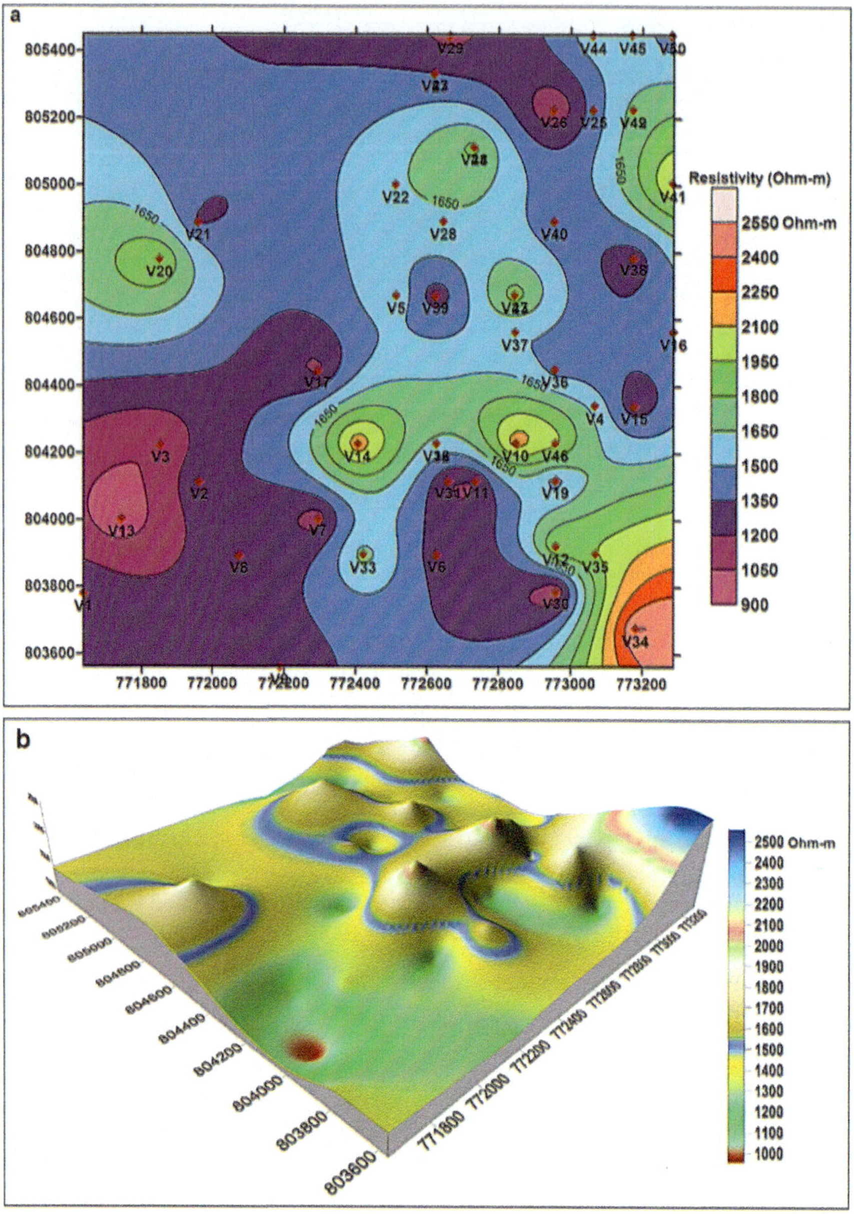

Fig. 5.44 **a** Fresh bedrock resistivity map of the engineering site in Oka, Nigeria. **b** 3D perspective resistivity view of the fresh bedrock of the site in Oka, Nigeria. *Source* Ademila (2022)—Open Access

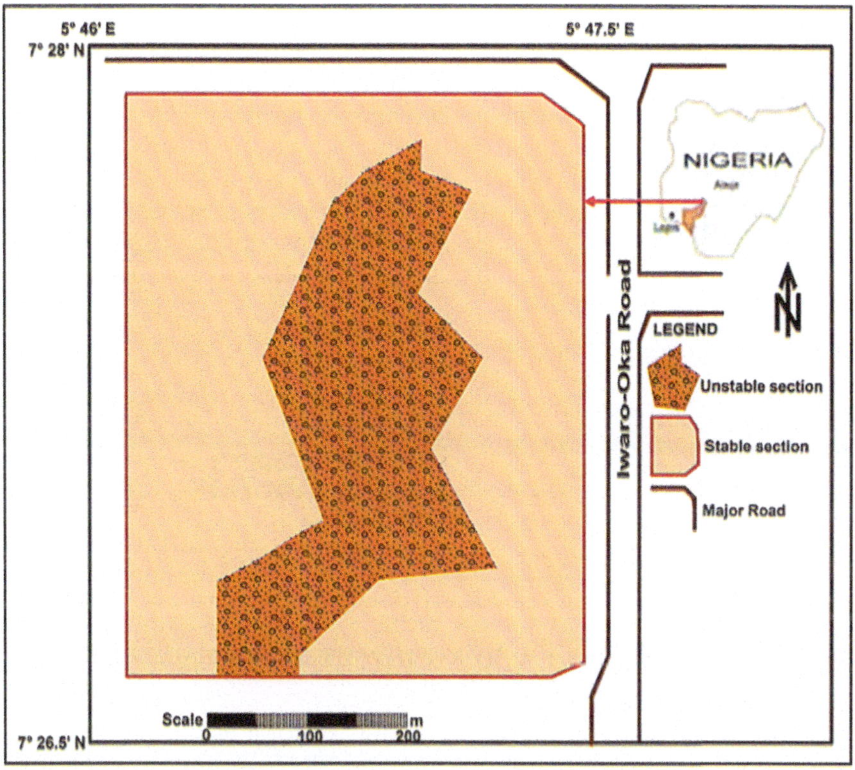

Fig. 5.45 Subsurface characterisation map of the engineering site in Oka, Nigeria, showing the variations in subsurface geological features and suitability for construction. *Source* Ademila (2022)—Open Access

ability to assess subsurface conditions with high accuracy and minimal disturbance to the environment.

Through various case studies, we've highlighted how these methods are applied in a range of real-world scenarios—from oil field exploration and archaeological surveys to landfill management, slope stability assessments, and geothermal resource exploration. These examples demonstrate the versatility and effectiveness of geophysical techniques in providing essential insights for infrastructure development, resource extraction, and environmental protection.

Despite the progress in subsurface imaging, challenges such as high costs, complex data interpretation, and environmental constraints remain. However, ongoing innovations in machine learning, autonomous systems, and real-time monitoring are poised to overcome these barriers, enhancing the efficiency, accuracy, and accessibility of geophysical surveys. Furthermore, the integration of advanced data analytics and sustainability practices ensures that these methods continue to evolve in a way that is both technically advanced and environmentally responsible.

Looking to the future, geophysics will continue to play a pivotal role in the evolving landscape of subsurface exploration. As techniques become more refined and integrated with other technologies, the potential for improving site characterization, mitigating risks, and informing better decision-making across a wide range of industries is vast. The continued collaboration between geophysicists, engineers, and environmental scientists will be key to unlocking new solutions for the challenges of tomorrow. In conclusion, this chapter has underscored the importance of innovative geophysical methods in shaping the future of subsurface exploration, and how they contribute to safer, more efficient, and sustainable practices in engineering and environmental management.

References

Ademila O (2022) Pre-foundation geophysical investigation of a site for structural development in Oka, Nigeria. NRIAG J Astron Geophys 11(1):81–112

Agunloye O (1984) Soil aggressivity along steel pipeline route at Ajaokuta. J Min Geol 21(1 & 2):97–101

Aizebeokhai AP (2010) 2D and 3D geoelectrical resistivity imaging: theory and field design. Sci Res Essays 5:3592–3605

Alam MJB, Ahmed A, Alam MZ (2024) Application of electrical resistivity tomography in geotechnical and geoenvironmental engineering aspect. Geotechnics 4:399–414. https://doi.org/10.3390/geotechnics4020022

Baeckmann W, Schwenk W (1997) Handbook of cathodic protection. In: The theory and practice of electrochemical corrosion protection techniques, 3rd edn. Elsevier Science Inc., USA, p 568

Bharti AK, Singh KKK, Ghosh CN, Mishra K (2022) Detection of subsurface cavity due to old mine workings using electrical resistivity tomography: a case study. J Earth Syst Sci 131:39

Boddupalli B, Minshull TA, Morgan J, Bayrakci G, Klaeschen D (2021) Comparison of 2-D and 3-D full waveform inversion imaging using wide-angle seismic data from the Deep Galicia Margin. Geophys J Int 227:228–256

Burland JB, Brooker EW, Leach RH (2011) Foundation design and construction, 7th edn. Spon Press

Cao J, Brossier R, Metivier L (2022) Elastic full-waveform inversion of 4C ocean-bottom seismic data: model parameterization analysis. In: Second international meeting for applied geoscience & energy, Houston, United States, Aug 2022, pp 957–961, hal-03852641

Cheng Y, Li Z (2020) Electrical resistivity tomography in urban geophysical surveys. Earth Sci Rev 54(4):301–310

Chengdu Municipal Government (2022) Chengdu urban geological survey report

Christeson GL, Morgan JV, Gulick SPS (2021) Mapping the Chicxulub impact stratigraphy and peak ring using drilling and seismic data. J Geophys Res Planets 126:e2021JE006938

Franco LM, La Terra EF, Panetto LP, Fontes SL (2024) Integrated application of geophysical methods in earth dam monitoring. Bull Eng Geol Environ 83:62

Frid M, Frid V (2024) A case study of the integration of ground-based and drone-based ground-penetrating radar (GPR) for an archaeological survey in Hulata (Israel): advancements, challenges, and applications. Appl Sci 14:4280

Gabrys M, Ortyl L (2020) Georeferencing of multi-channel GPR—accuracy and efficiency of mapping of underground utility networks. Remote Sens 12:2945

Garbey RA, Zeng L, Gong W (2013) Geophysical methods for subsurface contamination mapping. Environ Geol 62(2):305–316

Ghorbani A, Revil A, Bonelli S, Barde-Cabusson S, Girolami L, Nicoleau F, Vaudelet P (2024) Occurrence of sand boils landside of a river dike during flooding: a geophysical perspective. Eng Geol 329:107403

Hicks E, Hoeber H, Houbiers M, Lescoffit SP, Ratcliffe A, Vinje V (2016) Time-lapse full-waveform inversion as a reservoir-monitoring tool—a North Sea case study. Leading Edge Spec Sect 40:850–858

Hinojosa HR, Kirmizakis P, Soupios P (2022) Historic underground silver mine workings detection using 2D electrical resistivity imaging (Durango, Mexico). Minerals 12:491

Hossain MS, Maganti D, Hossain J (2010) Assessment of geo-hazard potential and site investigations using resistivity imaging. Int J Environ Technol Manag 13:116–129

Jabrane O, Martinez-Pagan P, Martinez-Segura MA, Alcala FJ, El Azzab D, Vasconez-Maza MD, Charroud M (2023) Integration of electrical resistivity tomography and seismic refraction tomography to investigate subsiding sinkholes in karst areas. Water 15:2192

Keller GV, Frischknecht FC (1996) Electrical methods in geophysical prospecting. Pergamon Press

Lailly P (1983) The seismic inverse problem as a sequence of before stack migrations. In: Conference on inverse scattering—theory and applications. Society for Industrial and Applied Mathematics, pp 206–220

Lamarche J, Girard L, Bourgeois P (2008) Geophysical techniques for tunnelling in urban areas: a case study. Geotech Test J 31(2):185–200

Liu X, Yu H (2019) Application of OCTEM in complex urban environments. Geophys Res Lett 46(8):217–225

Martinez-Pagan P, Gomez-Ortiz D, Martin-Crespo T, Manteca J, Rosique M (2013) The electrical resistivity tomography method in the detection of shallow mining cavities. A case study on the Victoria Cave, Cartagena (SE Spain). Eng Geol 156:1–10

Mizutani T, Chen J, Yotsumoto S (2024) The 3D localization of subsurface pipes from Ground Penetrating Radar images using edge detection and point cloud segmentation. Measurement 236:115102

Murillas J, Mougenot D, Boulot G, Comas M, Banda E, Mauffret A (1990) Structure and evolution of the Galicia Interior Basin Atlantic western Iberian continental margin. Tectonophysics 184(3–4). https://doi.org/10.1016/0040-1951(90)90445-E

Nabighian MN, Asten MW (2002) Metalliferous mining geophysics-state of the art in the last decade of the 20th century and the beginning of the new millennium. Geophysics 67:964–978

Nabighian MN, Grauch VJS, Hansen RO, LaFehr TR, Li Y, Peirce JW, Phillips JD, Ruder ME (2005) The historical development of the magnetic method in exploration. Geophysics 70:33ND–61ND

Naskar DC, Majumdar RK, Choudhury K, Das LK (2003) Magnetic profiles across NSL in parts of Central India. Indian Miner 57(1 & 2):53–60

Naskar DC, Das LK, Roy KK, Singh OP (2008) Structural studies of crust and upper mantle in parts of Madhya Pradesh, Central India. Geol Surv India Spl Publ 91:211–228

Perez-Gussinye M, Ranero CR, Reston TJ, Sawyer D (2003) Mechanisms of extension at nonvolcanic margins: evidence from the Galicia Interior Basin, west of Iberia. J Geophys Res 108(B5):1–19

Peron-Pinvidic G, Manatschal G, Minshull TA, Sawyer DS (2007) Tectonosedimentary evolution of the deep Iberia-Newfoundland margins: evidence for a complex breakup history. Tectonics 26(2):1–19

Persico R (2014) Introduction to ground penetrating radar: inverse scattering and data processing. Wiley-IEEE Press. https://doi.org/10.1002/9781118305003

Plati C, Loizos A (2012) Using ground-penetrating radar for assessing the structural needs of asphalt pavements. Nondestruct Test Eval 27:273–284

Poikajarvi J, Peisa K, Herronen T, Aursand PO, Maijala P, Narbro A (2012) GPR in road investigations—equipment tests and quality assurance of new asphalt pavement. Nondestruct Test Eval 27:293

Puntu JM, Chang P-Y, Lin D-J, Amania HH, Doyoro YG (2021) A comprehensive evaluation for the tunnel conditions with ground penetrating radar measurements. Remote Sens 13:4250

Reynolds JM (2011) An introduction to applied and environmental geophysics, 2nd edn. Wiley-Blackwell

Saibi H, Khosravi S, Cherkose BA, Smirnov M, Kebede Y, Fowler A-R (2021) Magnetotelluric data analysis using 2D inversion: a case study from Al-Mubazzarah Geothermal Area (AMGA), Al-Ain, United Arab Emirates. Heliyon 7:e07440

Smith R (2014) Electromagnetic induction methods in mining geophysics from 2008 to 2012. Surv Geophys 35:123–156

Tarantola A (1984) Inversion of seismic reflection data in the acoustic approximation. Geophysics 49(8):1259–1266

Telford WM, Geldart LP, Sheriff RE (1990) Applied geophysics, 2nd edn. Cambridge University Press

Tucholke BE, Sawyer DS, Sibuet J-C (2007) Breakup of the Newfoundland Iberia rift. Geol Soc Lond Spec Publ 282(1):9–46

Vallee MA, Smith RS, Keating P (2011) Metalliferous mining geophysics—state of the art after a decade in the new millennium. Geophysics 76:W31–W50

Vallee MA, Moussaoui M, Khan K (2024) Case studies of magnetic and electromagnetic techniques covering the last fifteen years. Minerals 14:1286. https://doi.org/10.3390/min14121286

Wang L, Li H, Wang D, Zhou S, Zhang W, Long X, Yang J, Wang Q (2023) Urban geophysical exploration: case study in Chengdu International Bio-City. J Geophys Eng 20:830–840

Worthington PF (2011) Applied geophysics for geologists and engineers. Springer

Xu X, Peng S, Xia Y, Ji W (2014) The development of a multi-channel GPR system for roadbed damage detection. Microelectronics 45:1542–1555

Zhang L, Wang T, Li X (2021) Geophysical exploration methods in urban areas: a case study of Chengdu. J Urban Geophys 45(2):123–136

Zou L, Li Y, Alani AM (2024) Utilizing dual polarized array GPR system for shallow urban road pavement foundation in environmental studies: a case study. Remote Sens 16:4396. https://doi.org/10.3390/rs16234396

Chapter 6
Geophysical Methods in Environmental Monitoring

Abstract This chapter explores how geophysical methods are applied in environmental monitoring to support sustainable land and resource management. It discusses the use of techniques like electrical resistivity imaging (ERI), GPR, seismic, magnetic, and electromagnetic methods to monitor groundwater, detect contamination, and guide urban and agricultural land use. The chapter highlights the role of geophysics in addressing pollution, climate change, and ecosystem degradation. Real-world case studies include groundwater exploration in Ethiopia, rainforest mapping in the Amazon, and geothermal surveys in India and Iceland. Applications also cover coral reef conservation, landfill monitoring, and disaster recovery, such as post-earthquake assessments in Nepal. The chapter introduces recent advances in non-invasive imaging, data interpretation, and environmental remediation. A master table summarizes key geophysical techniques and their parameters. Overall, this chapter shows how geophysics supports informed environmental decisions, helping to manage natural resources, protect ecosystems, and reduce disaster risks effectively.

6.1 Introduction to Geophysical Methods in Environmental Monitoring

Chapter 5 focuses on the application of geophysical methods to address key environmental challenges, such as groundwater monitoring, contamination detection, land use planning, and sustainable resource management. It explores how cutting-edge geophysical techniques, including Electrical Resistivity Imaging (ERI), Ground Penetrating Radar (GPR), Seismic Methods, and Magnetometry, are being used to non-invasively monitor and assess environmental factors that influence both natural ecosystems and human infrastructure. Through various case studies, the chapter demonstrates how these methods are applied in diverse settings such as groundwater exploration, contamination monitoring, coastal land reclamation, agricultural management, and pollution control.

D. C. Naskar, *Advances in Engineering Geophysics: Concepts and Applications*, Earth Systems Data and Models 9, https://doi.org/10.1007/978-3-032-28000-8_6

The chapter also highlights recent advancements in environmental geophysics, including improvements in high-resolution imaging, data processing, and the integration of non-invasive exploration techniques. Furthermore, it discusses the growing role of geophysics in climate change monitoring, subsurface remediation, and disaster risk reduction, emphasizing its importance for sustainable development and natural resource management. By providing valuable insights into environmental conditions, geophysics contributes to more informed decision-making and helps to promote ecological conservation, resource efficiency, and environmental protection.

Ultimately, the chapter underscores the essential role of geophysical technologies in environmental monitoring and management, offering innovative solutions for addressing the pressing environmental issues of today and ensuring a more sustainable future.

6.2 Addressing Environmental Challenges Through Geophysical Monitoring of Groundwater, Contamination, and Land Use

Geophysical methods have become indispensable tools in the assessment and monitoring of environmental issues. With ongoing technological advances, they offer non-invasive and cost-effective solutions for addressing major environmental challenges, including climate change, pollution, resource management, and land degradation (Barone and Pepe 2019; Benson and McDonnell 2018). These methods provide critical insights into subsurface conditions—such as soil and groundwater quality, land use, and ecosystem health—and have proven essential in disaster management, land reclamation, and evaluating the environmental impact of human activities (McLeod et al. 2023; Barrett et al. 2021).

6.2.1 Groundwater Monitoring Using Geophysical Techniques

Groundwater resources are a crucial component of the Earth's water cycle, and their sustainable management is necessary for providing drinking water, agricultural irrigation, and industrial use. Geophysics offers several advanced methods for monitoring and managing groundwater resources.

Electrical Resistivity Imaging (ERI)

Electrical Resistivity Imaging (ERI) measures the resistivity of the subsurface to electrical current. Variations in resistivity can indicate different types of rocks, soil, and groundwater. By mapping resistivity, ERI can be used to delineate the depth and extent of groundwater aquifers, identify zones of recharge or discharge, and monitor groundwater contamination (Loke and Barker 2022). In the arid regions of the Middle East, electrical resistivity imaging (ERI) has been extensively used to map groundwater resources and assess aquifer characteristics. A notable case involved mapping a major aquifer in Oman, where ERI was employed to delineate the extent of freshwater reserves and identify areas vulnerable to salinization. The technique provided high-resolution images of the subsurface, highlighting zones of high permeability and helping authorities design efficient water extraction strategies (Loke and Barker 2022). This application not only helped identify the aquifer's limits and recharge zones but also allowed for more sustainable water management practices in an area facing significant water scarcity challenges.

Ground Penetrating Radar (GPR)

Ground Penetrating Radar (GPR) is effective for detecting the depth and structure of groundwater tables, especially in shallow environments. It sends high-frequency radar waves into the ground, and the reflected signals provide detailed information on subsurface features, such as water saturation and variations in soil types (Momeni et al. 2021). A notable case study highlighting GPR's application involved contamination monitoring in the United Kingdom. A major chemical spill near an industrial site in northern England led to widespread groundwater contamination. Ground Penetrating Radar (GPR) and Electrical Resistivity Tomography (ERT) were used to track the contamination plume's spread over time. Time-lapse resistivity surveys helped map the migration of hazardous chemicals through the subsurface. The integrated data from GPR and ERT allowed for real-time monitoring, enabling quick decision-making on remediation efforts (Santos et al. 2021). The combination of these methods allowed for precise monitoring of contamination movement, reducing the time required for remediation and preventing further contamination of local water supplies.

Seismic Refraction and Reflection Techniques

Seismic refraction and reflection techniques are commonly used to study the properties of aquifers, including porosity and permeability. These methods help identify the boundary between saturated and unsaturated zones and can assist in locating groundwater reserves (Sharma et al. 2020). A notable case study demonstrating the application of these techniques took place in California, USA, where seismic refraction surveys were conducted as part of an urban development project in an area with complex subsurface conditions. The seismic data were used to assess soil stiffness, bedrock depth, and groundwater levels to determine the suitability of the area for constructing large buildings. This helped engineers design foundations tailored to local soil and seismic conditions (Sharma et al. 2020). The seismic surveys contributed to the successful design of infrastructure, avoiding expensive delays and ensuring the long-term stability of buildings and roads in the area.

Magnetic and Electromagnetic Methods

Magnetic and electromagnetic methods are particularly effective for mapping large-scale groundwater systems, including determining the extent of saline or fresh water bodies. Electromagnetic methods can provide valuable data on groundwater salinity and overall flow patterns (Miller and Gallagher 2022).

6.2.2 Contamination Monitoring in Environmental Settings

Environmental contamination, whether from industrial activities, agricultural runoff, or hazardous waste, poses significant risks to both human health and ecosystems. Geophysical methods play a critical role in identifying, mapping, and monitoring contamination in both *soil and groundwater*.

Time-Lapse Resistivity and Electromagnetic Surveys

Time-lapse resistivity and electromagnetic surveys are used to track the movement and spread of contaminants over time. They provide continuous data on changes in the subsurface that can indicate contamination plumes, making them useful for monitoring toxic substances like petroleum hydrocarbons, metals, and organic pollutants (Santos et al. 2021). Geophysical imaging is used to assess the physical properties of geological reservoirs, which change over time due to factors like resource extraction or geothermal energy production. Time-lapse geophysical imaging tracks these changes by conducting repeated surveys in the same area at different times.

Electromagnetic methods, such as CSEM and ERT, are effective for this purpose, with baseline and monitor datasets typically inverted separately. The difference between these models represents time-lapse resistivity changes. However, due to the non-linearity and noise sensitivity of these methods, inversion results can be heavily influenced by the chosen path, potentially leading to variations that do not accurately reflect structural changes in the reservoir (Miller et al. 2008; Peacock et al. 2013; Rosas-Carbajal et al. 2015; Tietze et al. 2015; Abdelfettah et al. 2018; Patzer 2019).

Time-lapse resistivity changes at the *Reykjanes geothermal field*, located in the southwestern part of Iceland on the Reykjanes Peninsula, were investigated using controlled-source electromagnetic (CSEM) data collected over a one-year interval before and after a stimulation phase. The objective was to evaluate different inversion strategies—parallel, sequential, and double-difference—for their ability to detect temporal resistivity changes and their robustness to noise. These methods were compared to assess their sensitivity to the starting model and noise levels. The CSEM data were processed to remove noise and prepared for inversion. The parallel inversion approach, though commonly used, was highly sensitive to the starting model and inversion path, often resulting in artefacts and inconsistencies due to noise.

Similarly, the sequential inversion method depended on the quality of the baseline model and produced artefacts when the baseline model contained errors or noise. In contrast, the double-difference inversion method, which isolates the time-lapse signal, was much more robust. Even with a rough initial model, the double-difference inversion effectively mapped resistivity anomalies and minimized artefacts, demonstrating better resilience to noise. Results showed that the time-lapse signal at Reykjanes, particularly related to steam cap development, was small and consistent with the repeatability threshold. While parallel and sequential inversion methods suggested more significant temporal changes than were actually observed, the double-difference approach provided more reliable and consistent results. The method also proved effective in localizing resistivity variations even without a precise baseline reconstruction, using independent geological data or previous geophysical surveys.

The study demonstrates that double-difference inversion is the most reliable method for time-lapse resistivity monitoring, offering better robustness against noise and reduced sensitivity to the initial model. This approach is particularly useful for monitoring subtle changes in geothermal fields, such as at *Reykjanes*, where temporal variations in resistivity are often small. The findings support the use of double-difference inversion for improving the accuracy and consistency of time-lapse geophysical monitoring, as highlighted by Bretaudeau et al. (2021).

Magnetometric Surveys

Magnetometric surveys are useful for identifying metallic contamination in soils. By measuring variations in the Earth's magnetic field caused by ferromagnetic materials, geophysicists can detect the presence of contaminants like heavy metals or unexploded ordnance (UXO) (Jian et al. 2021). A closed landfill in *Japan*, suspected of leaking heavy metals into nearby soil and groundwater, was surveyed using magnetometry to detect ferromagnetic contaminants. The survey successfully identified areas where metals, such as iron and steel, were concentrated, which corresponded to zones of contamination. The data from the magnetometer helped focus further investigation and remediation efforts in high-risk areas (Jian et al. 2021). Magnetometry was instrumental in efficiently locating contamination hotspots, significantly reducing the need for extensive excavation and allowing for targeted clean-up actions.

Ground Penetrating Radar (GPR)

Ground Penetrating Radar (GPR) is effective in detecting contaminated zones in the near-surface environment. It can identify changes in soil texture, porosity, and the presence of chemical compounds that might indicate contamination (Miller et al. 2022).

Electrical Resistivity Tomography (ERT)

Electrical Resistivity Tomography (ERT) allows the detailed mapping of contamination plumes within groundwater or soils, helping locate areas of the highest pollutant concentrations and assess the spread of contamination over time. It is particularly effective in monitoring areas affected by landfill leachate or chemical spills (Carvalho et al. 2023).

The use of electrical resistivity techniques, including Electrical Resistivity Tomography (ERT) and Self-Potential (SP) methods, has proven effective for non-destructive environmental monitoring at archaeological sites. One notable case study is the application of these techniques at the *Leitai heritage site in China*, where ERT was employed to detect seepage channels and map contamination plumes within the groundwater and soil (Fig. 6.1). This approach helped identify areas of high pollutant concentrations, monitor the spread of contamination, and ensure the preservation of the site. The case study highlights how ERT and SP techniques can safeguard cultural heritage by providing a non-invasive means of assessing environmental conditions (Lu et al. 2025).

At the Leitai heritage site, ERT and SP techniques were used to detect seepage channels affecting the loess platform. The 3D ERT inversion revealed three electro-stratigraphic layers: an impermeable upper layer with high resistivity, a compacted soil layer with moderate resistivity, and a low-resistivity zone in areas affected by seepage. These low-resistivity zones were primarily located on the western side of the platform, aligned with the fire pipeline gallery, and extended toward the collapse point on the southern side. By applying a transparency filter, regions with high water content affected by seepage were distinctly highlighted.

The SP map, which showed substantial fluctuations in spontaneous potential, revealed negative anomalies over the fire pipeline gallery and southern platform. These anomalies correlated with the seepage patterns and were modeled using a cylindrical body to simulate the seepage pathways. The SP inversion results closely aligned with the low-resistivity anomalies detected by ERT, confirming the locations of seepage pathways (Fig. 6.2). In some areas where ERT did not detect significant low-resistivity zones, SP data revealed distinct seepage, further refining the interpretation.

By integrating ERT and SP techniques, a more comprehensive understanding of the seepage pathways was achieved. ERT provided detailed resistivity profiles, while SP revealed the direction of fluid flow, improving the identification of seepage zones and guiding excavation efforts (Fig. 6.3). The successful application of these non-invasive geophysical methods demonstrates their value for archaeological preservation, accurately mapping seepage pathways and contributing to environmental monitoring at heritage sites.

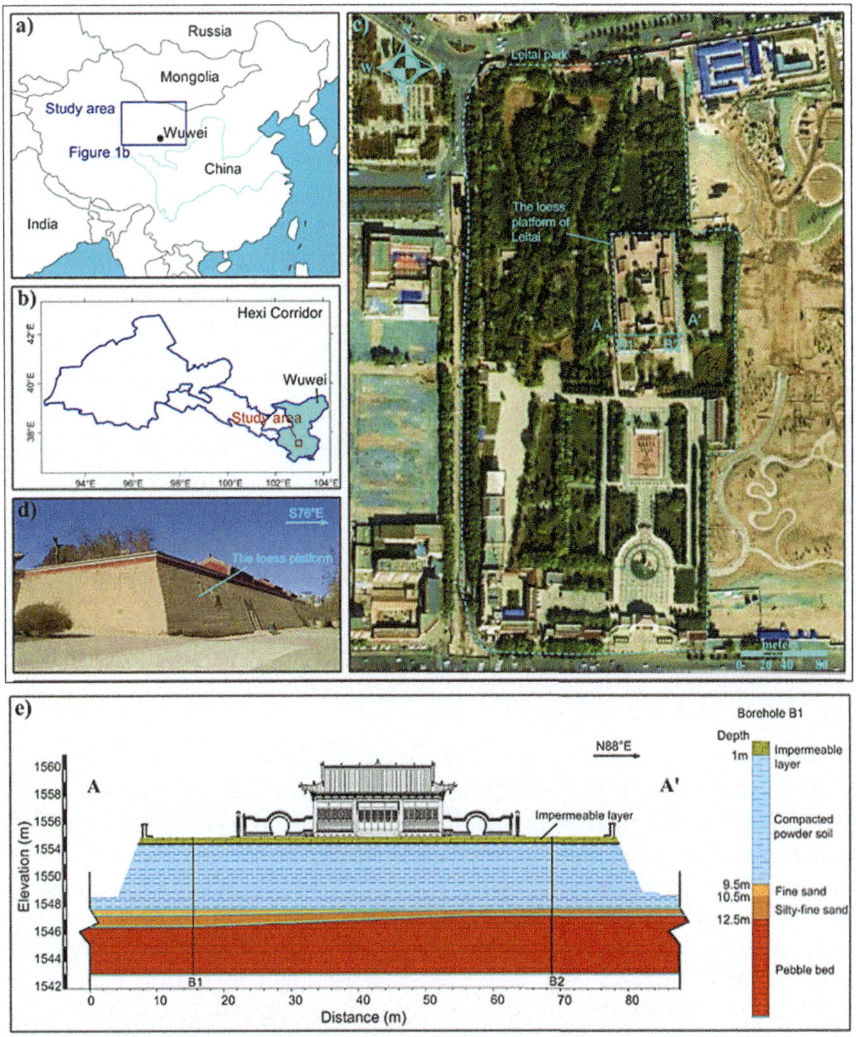

Fig. 6.1 Location and geology of the study area. **a** Study area situated in Wuwei city, Gansu Province, China; **b** Location of the Leitai cultural heritage site, east of the Hexi Corridor; **c** Satellite image of the Leitai platform, generated using Locaspace (Version 4.3.9), available at http://www.locaspace.cn/ **d** Site photograph of the Leitai platform; **e** Geological details of the Leitai platform. *Source* Lu et al. (2025)—Open Access

This case study focused on the *Weidenpesch landfill in Cologne, Germany*, with the objective of characterizing the landfill and assessing potential groundwater contamination using an integrated geophysical approach. The methods employed included magnetic surveys, electrical resistivity tomography (ERT), and transient electromagnetic (TEM) surveys. The magnetic survey used high-resolution reduction to pole (RTP) magnetic vertical gradient mapping to delineate landfill boundaries and detect

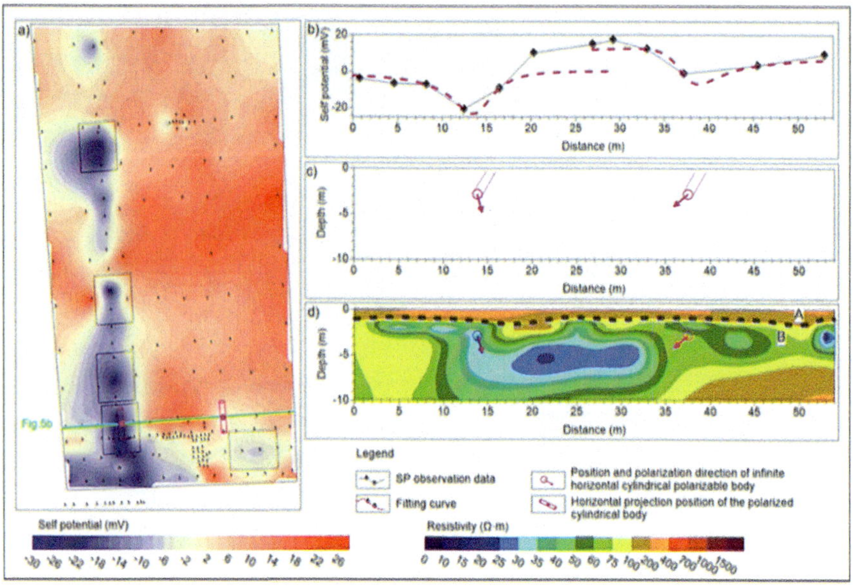

Fig. 6.2 SP results. **a** Distribution of spontaneous potential on the loess platform; **b** Measured SP curve and fitted curve; **c** SP inversion results; **d** Comparison between SP inversion results and the inverted resistivity cross-section. *Source* Lu et al. (2025)—Open Access

metallic objects. The ERT was applied to differentiate the low-resistive waste body from the surrounding high-resistive gravelly sand layer, creating subsurface resistivity models (Fig. 6.4). TEM provided a vertical profile of subsurface conductivity, complementing the ERT and extending the depth of investigation (Fig. 6.5).

The processing of magnetic data included filtering to minimize near-surface metallic influences and comparison with historical maps for validation (Fig. 6.6). ERT data were inverted into 2D and 3D models, which helped identify the waste body and contamination zones (Fig. 6.4). TEM data were processed using both Occam and Marquardt–Levenberg inversion methods to interpret resistivity and identify conductive layers in the subsurface (Fig. 6.7). The results revealed that the magnetic survey successfully identified varying zones of magnetic properties, allowing for the mapping of metallic waste and the differentiation of waste types. The ERT survey identified the low-resistive waste body and confirmed contamination in the gravelly sand layer, where resistivity had significantly decreased due to contamination (Fig. 6.4). The TEM survey complemented these findings by identifying three distinct layers within the landfill and detecting contamination migration, particularly in the coal/clay layer beneath the landfill (Fig. 6.8). Overall, the integrated geophysical methods provided a comprehensive understanding of the subsurface structure and contamination pathways, highlighting the need for long-term monitoring to assess the ongoing environmental impact of the landfill.

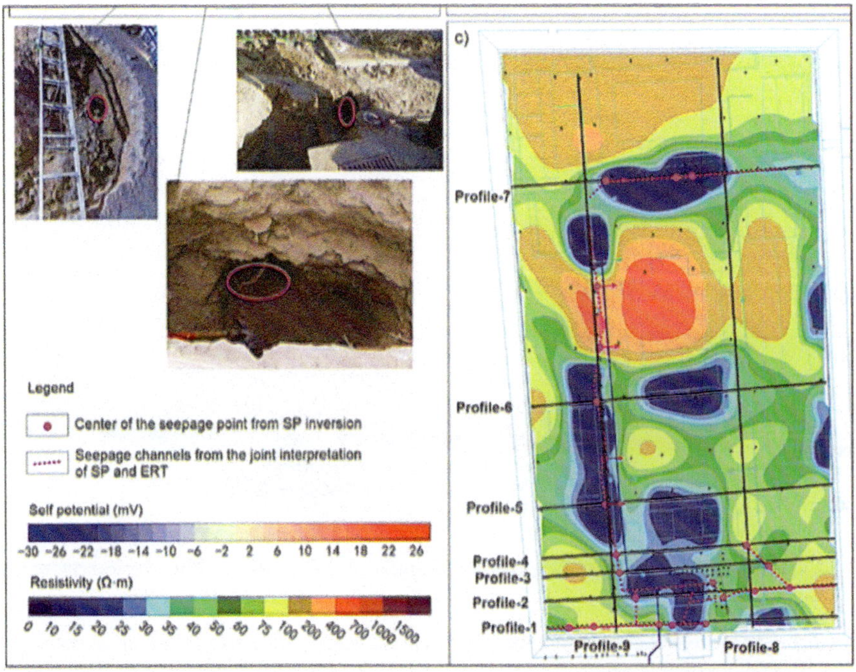

Fig. 6.3 Inferred channel locations and subsequent excavation verification results. **a** SP plane with integrated inferred seepage channels; **b** ERT slice at 2 m depth with integrated inferred seepage channels; **c** ERT slice at 5 m depth with integrated inferred seepage channels. *Source* Lu et al. (2025)—Open Access

6.2.3 *Land Use and Urban Planning*

Geophysical methods are increasingly used in land use planning to understand the subsurface environment and make informed decisions related to urban development, agricultural planning, and resource management.

Subsurface Mapping for Urban Development

Geophysical surveys can map the geology beneath urban areas, identifying bedrock, soil conditions, and groundwater flow patterns. This information is essential for foundation design, detecting areas prone to subsidence, and ensuring the safety of infrastructure projects.

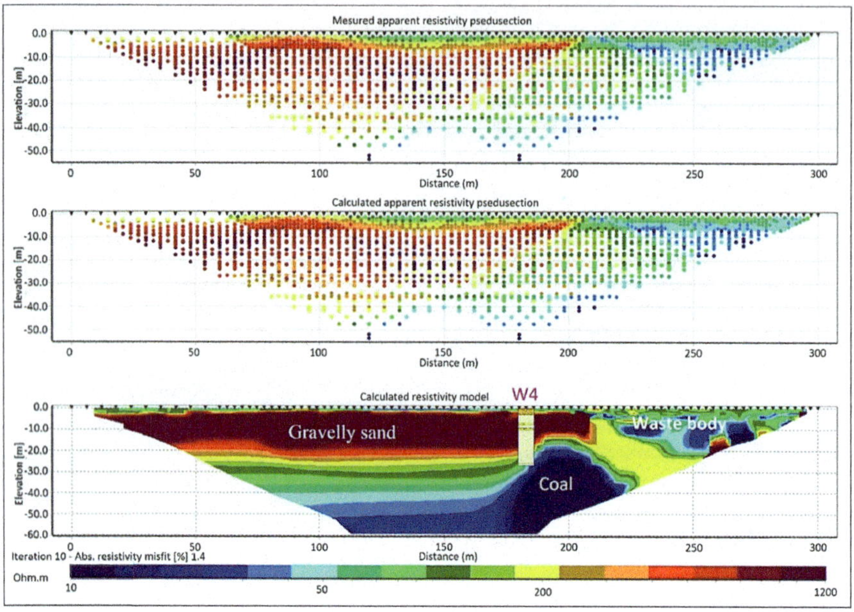

Fig. 6.4 The model obtained using robust inversion (L1-norm) of the ERT data from profile ERT4 after 10 iterations (absolute error = 1.4%) is presented. The borehole log from W4 is superimposed on the 2D cross section for comparison. *Source* Ibraheem et al. (2024)—Open Access

Geophysical Monitoring for Land Reclamation

In land reclamation projects), geophysical methods, including resistivity, seismic surveys, and Ground Penetrating Radar (GPR), help assess the stability and suitability of soils for development. These methods are also applied to monitor the effectiveness of land reclamation over time. Geophysical methods were applied to assess the feasibility of land reclamation projects along the coast of *Hong Kong*, where industrial expansion was required. Electrical resistivity tomography (ERT) was used to monitor the stability of reclaimed land and to ensure that the land's foundation could support future development. Time-lapse surveys also helped track subsidence and other deformations in real-time. The data obtained helped optimize the reclamation process and reduce environmental risks (McLeod et al. 2023). The continuous monitoring and adjustment based on ERT data led to a more stable land reclamation process, minimizing the risk of settling and damage to infrastructure once the land was put to use.

Agricultural Land Use

For agricultural applications), geophysics helps in evaluating soil conditions, determining irrigation needs, and monitoring soil salinity. Ground-penetrating radar and

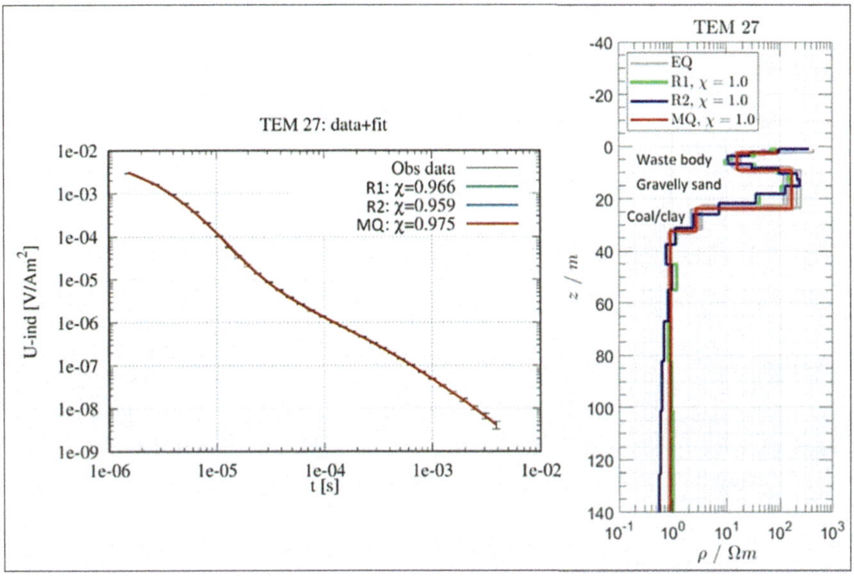

Fig. 6.5 Shows the 1D inversion results for station TEM27. The Marquardt–Levenberg (MQ) model is shown in red, while the Occam models with roughness values R1 and R2 are in green and blue, respectively. The alongside equivalent (EQ) models are in light grey. The data fit (χ) is shown in the legend next to the model. *Source* Ibraheem et al. (2024)—Open Access

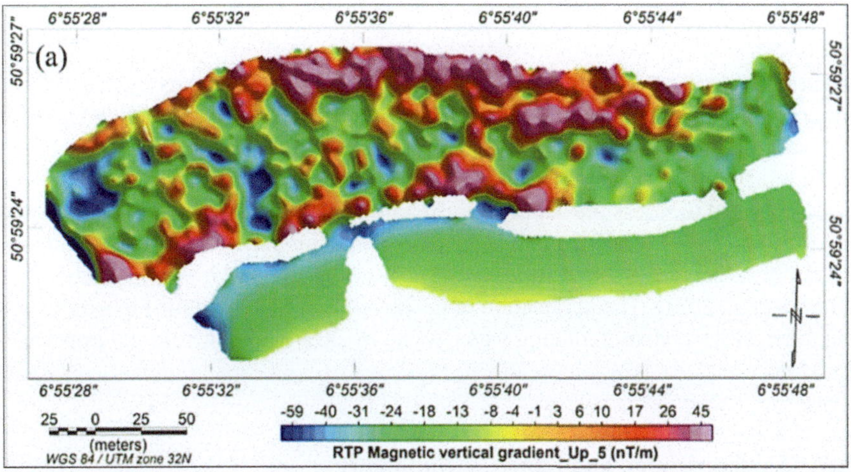

Fig. 6.6 **a** The RTP magnetic map after applying an upward continuation filter at a height of 5 m. *Source* Ibraheem et al. (2024)—Open Access

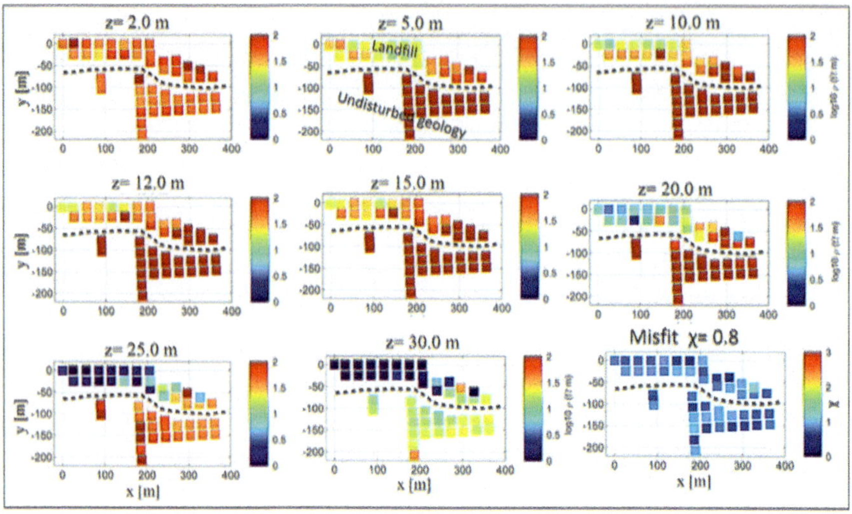

Fig. 6.7 Shows depth slices at 2, 5, 10, 12, 15, 20, 25, and 30 m, derived from the 1D Occam inversion results of the TEM data. The waste body is visible up to a depth of 10 m. Below this depth, a contaminated gravelly sand layer extends from 20 to 23 m. The underlying layer, made up of saturated coal and clay, is also likely contaminated due to the high porosity of the gravelly sand layer above it. There is a noticeable difference in resistivity values at similar depths inside and outside the landfill, especially between the gravelly sand and the coal/clay layers. This could suggest the presence of contamination leachate. The average χ value shows a good fit. *Source* Ibraheem et al. (2024)—Open Access

electrical resistivity tomography can be used to assess soil moisture content and determine the best locations for irrigation or drainage systems. In *Southeastern Spain*, an agricultural project utilized Ground Penetrating Radar (GPR) and Electrical Resistivity Imaging (ERI) to assess soil moisture content and optimize irrigation practices. GPR was used to assess the depth and structure of soil layers, while ERI helped map areas of soil salinity and groundwater flow. This combination allowed farmers to precisely target irrigation efforts, reducing water usage and improving crop yield (Barrett et al. 2021). The application of geophysical methods resulted in a more efficient irrigation system, reducing water waste by 30% while increasing crop yields by 15%.

Soil Characterization and Land Use Suitability

Geophysical techniques are critical for determining the suitability of land for different uses, such as agriculture, forestry, or urban expansion. Mapping the subsurface layers provides insights into soil composition, texture, porosity, and the depth of water tables, all of which are essential for making informed land use decisions.

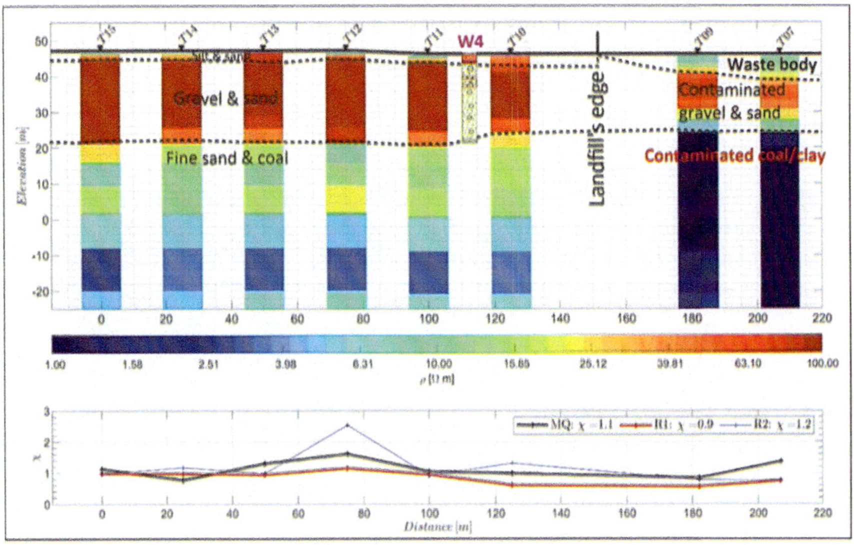

Fig. 6.8 Presents a 2D resistivity cross-section derived from the 1D Occam inversion results with the roughness parameter R1 for the TEM data along profile ERT4. The lower panel shows the χ values for both the Occam (R1 and R2) and MQ inversions. The log of borehole W4 is included between TEM stations T10 and T11. *Source* Ibraheem et al. (2024)—Open Access

6.2.4 Challenges and Future Directions

Despite the significant advantages of geophysical techniques, there are several challenges in applying these methods for environmental monitoring. One challenge is the resolution limitations of certain techniques. While geophysics provides valuable subsurface data, the *resolution* can be limited, especially in highly heterogeneous or complex environments (Zhang et al. 2020). Another challenge is data interpretation. The complexity of environmental systems often requires sophisticated *data processing and interpretation* techniques, which can be particularly challenging for large-scale or long-term monitoring (McNeill et al. 2021). Additionally, *environmental sensitivity* must be considered, as some geophysical methods, such as seismic surveys, may have potential environmental impacts, particularly in sensitive areas (Bates et al. 2022).

However, the future of geophysical techniques in environmental monitoring looks promising, with advancements in sensor technology, data analysis, and computational modeling. Integration with other remote sensing technologies, like satellite imaging, is expected to enhance the accuracy and efficiency of geophysical surveys, making them an even more powerful tool for environmental management (Jackson and Mahabir 2023).

6.3 Advancements in Environmental Geophysics for Sustainable Development

Environmental geophysics has undergone significant advancements over the years, with many of these advancements being applied to support sustainable development initiatives. The role of geophysical techniques is becoming increasingly essential in achieving sustainability goals, especially in the context of environmental monitoring, resource management, and remediation (Zhou and Zhang 2022; Telford et al. 1990; Mokhtari et al. 2020; Everett 2013).

6.3.1 Key Advancements in Environmental Geophysics

Key advancements in environmental geophysics include the development of high-resolution imaging techniques, integration of multiple geophysical methods, and the use of time-lapse monitoring for dynamic environmental processes. Improved data processing algorithms and machine learning applications have enhanced the interpretation of complex subsurface conditions. These innovations have significantly increased the accuracy, efficiency, and applicability of geophysics in areas such as groundwater monitoring, contamination assessment, land use planning, and climate change studies.

High-Resolution Imaging

Advances in geophysical imaging technologies, such as Electrical Resistivity Tomography (ERT), Ground Penetrating Radar (GPR), and seismic reflection methods, have allowed for more detailed subsurface mapping. These methods enable the detection of contaminants, soil structures, and groundwater flow patterns with unprecedented accuracy. This enhanced imaging capability is essential for effective environmental monitoring and risk assessment, especially in areas where contamination and land-use change pose serious threats. For example, a recent study by Barone and Pepe (2019) used GPR and ERT in coastal monitoring for sustainable development, helping detect subsurface changes associated with sea-level rise and erosion.

A case study conducted along the *Gulf Coast of the USA*, an area vulnerable to the impacts of climate change—particularly sea-level rise, which threatens coastal communities and ecosystems. The research focused on understanding how rising sea levels could exacerbate coastal erosion and flooding, which are already pressing issues in the region. The objective of the study was to assess the impacts of sea-level rise on coastal areas, identifying vulnerable zones where erosion and flooding could become more severe in the future. Geophysical surveys were employed to examine sediment layers, monitor groundwater movement, and assess soil stability, which are critical factors in understanding the vulnerability of the coast to climate change.

The methods used included seismic reflection, Ground Penetrating Radar (GPR), and LiDAR. Seismic reflection was used to investigate the subsurface layers of sediments, helping to map the stratigraphy and identify potential areas at risk of erosion. GPR provided high-resolution images of the near-surface structure, detecting variations in sediment composition and groundwater movement. LiDAR was used to create detailed topographic maps of the coastal areas, providing a precise understanding of the surface elevations and potential flooding risks.

The geophysical data were processed using specialized software to generate detailed models of the subsurface and surface topography. These models helped identify areas where sediment layers were most susceptible to erosion and where groundwater movement could contribute to soil instability, enhancing the understanding of the region's vulnerability to climate change. The results of the study revealed critical zones where coastal erosion and flooding risks were highest, particularly in areas with shallow sediment layers and poor soil stability. The data also highlighted regions that would benefit from natural restoration efforts, such as wetlands, to buffer the effects of rising sea levels. As a result, the study informed state policies on coastal resilience and adaptation, leading to the implementation of measures such as the construction of seawalls to protect vulnerable areas and the restoration of natural wetlands to mitigate the impacts of rising sea levels. These actions were designed to enhance the sustainability and resilience of coastal ecosystems and communities in the face of climate change.

Improved Data Processing and Interpretation

The development of advanced data processing techniques, including machine learning and artificial intelligence (AI), has greatly enhanced the interpretation of geophysical data. These innovations allow for more efficient and accurate detection of environmental anomalies, including contamination or degradation of soil and water resources. AI tools also help in predictive modeling for assessing environmental risks, such as the likelihood of natural disasters (e.g., flooding, landslides) and their potential impacts on local ecosystems and communities. For instance, Yilmaz (2001) demonstrated how seismic tomography is used to model landslide risks, providing vital data to mitigate environmental hazards.

A case study in *Turkey* demonstrated the effectiveness of *improved data processing and interpretation* in landslide risk assessment using seismic tomography. Advanced algorithms enabled more accurate imaging of subsurface structures, enhancing the detection of potential slip surfaces. The case study was conducted in landslide-prone regions of Turkey, particularly along the Black Sea coast, which is susceptible to frequent landslides due to steep terrain, heavy rainfall, and seismic activity. These areas pose significant environmental hazards, making it crucial to assess the risk of landslides and develop effective mitigation strategies. The primary objective of the study was to assess landslide risks by using seismic tomography to model the subsurface structure of the area. The goal was to identify weak zones within the hillside that were more prone to failure, thereby improving the accuracy of landslide predictions and enabling better risk management.

The study employed seismic tomography, which measures the velocity of seismic waves passing through different layers of the earth. This method provided detailed subsurface imaging, allowing researchers to analyze soil composition, groundwater movement, and weak zones that could lead to slope failure. Additionally, seismic refraction and reflection surveys were conducted in selected locations to gather data on variations in material properties, with the data processed through advanced inversion algorithms to improve interpretation accuracy.

The seismic data were processed using inversion techniques, where the velocity of seismic waves was translated into 3D models of the subsurface. These models helped identify key features such as weak soil conditions, high water saturation, and potential slip planes. The enhanced data processing allowed for a more accurate understanding of subsurface conditions compared to traditional methods, improving the assessment of landslide risks.

The results revealed areas with highly fractured rock layers, weak soil, and excessive groundwater accumulation—key factors that increase the likelihood of landslides. By interpreting the seismic data accurately, the study identified specific locations within the region where slope reinforcement and other stabilization measures were needed to mitigate landslide risk. As a result of these findings, local authorities were provided with valuable information for developing targeted mitigation strategies. The data from seismic tomography informed the implementation of slope stabilization techniques, including soil reinforcement, the construction of drainage systems to manage water flow, and the restoration of vegetation to help stabilize the slopes. This approach reduced the likelihood of landslides and provided a more sustainable solution for managing the region's environmental risks.

Non-invasive Exploration Techniques

The push for sustainability has led to the increased use of non-invasive geophysical methods, which minimize the environmental impact of exploration. Techniques like passive seismic monitoring, electromagnetic surveys, and surface geophysical methods are now widely used in monitoring natural habitats, groundwater quality, and hazardous waste sites without disturbing the ecosystem. These methods support long-term sustainability by providing valuable data for resource management without physical disruption of the environment. For example, in a study conducted by Ahmed and Al-Falahi (2022), non-invasive geophysical methods were employed in arid regions of the UAE to optimize groundwater extraction and avoid depletion.

A case study was conducted in the *arid regions of the United Arab Emirates*, where water scarcity is a significant challenge due to the desert climate. The study focused on groundwater monitoring to improve resource management and ensure sustainable water supply in these harsh environmental conditions. The focus was on sustainable water management practices, particularly concerning the monitoring and management of groundwater resources in these dry areas. The objective was to map and monitor groundwater reserves by using geophysical techniques to determine the extent and depth of aquifers, with the goal of identifying areas where groundwater extraction could be optimized. This was aimed at ensuring that water resources were not over-exploited, as groundwater is a crucial resource in arid climates.

The study used a combination of Electrical Resistivity Tomography (ERT), Ground Penetrating Radar (GPR), and Seismic Refraction to map the subsurface and identify groundwater reserves. ERT was employed to assess the resistivity of the soil, which varies with water content, allowing for the identification of aquifer locations. GPR provided high-resolution imaging of the subsurface to locate water-bearing layers, while seismic refraction was used to analyze the depth and characteristics of the aquifers.

The collected geophysical data were processed using specialized inversion software to create detailed models of the groundwater distribution, revealing the extent and connectivity of the aquifers. This allowed the researchers to identify areas with the highest water reserves and regions where extraction could be increased without depleting resources. The results showed clear distinctions between water-bearing layers and areas with limited groundwater reserves. It allowed local authorities to develop optimized groundwater extraction strategies, ensuring that extraction activities were focused on sustainable areas and that over-extraction in vulnerable zones was avoided. Ultimately, the project provided valuable insights that helped local authorities optimize groundwater management and prevent over-exploitation of water resources. This contributed to more sustainable water management practices in the UAE's arid regions, ensuring long-term water availability for both agricultural and domestic needs.

Monitoring Climate Change and Environmental Degradation

Geophysical methods are essential in studying climate change impacts, such as the loss of glaciers, sea-level rise, and groundwater depletion. These techniques provide critical data for tracking environmental degradation and supporting adaptive strategies for climate resilience. By using methods like satellite geodesy, gravimetry, and remote sensing, scientists can monitor changes in the Earth's surface, water levels, and other environmental indicators. These methods help governments and organizations make informed decisions regarding climate mitigation and adaptation strategies, essential for sustainable development. A notable case is the study by Sanchez and Simms (2021), which used seismic methods to assess coastal erosion and sea-level rise in the Gulf Coast of the U.S., guiding local governments in their climate adaptation efforts.

A case study on *carbon sequestration monitoring in agricultural land* was conducted in *Brazil*, focusing on regions involved in large-scale efforts to reduce greenhouse gas emissions through sustainable soil management practices. These practices, such as agroforestry and no-till farming, aim to enhance carbon sequestration in the soil. The objective of the research was to monitor the sequestration of carbon in agricultural soils, specifically tracking the distribution of soil organic carbon over time. The goal was to assess the impact of various land management practices on the long-term storage of carbon in the soil.

Geophysical methods used in the study included electromagnetic induction, ground-penetrating radar (GPR), and seismic tomography. Electromagnetic induction measured soil conductivity, which is related to organic carbon content. GPR provided detailed subsurface images of soil structure, while seismic tomography helped assess soil compaction and layering, important factors in understanding how carbon is stored and retained. The geophysical data collected were processed using advanced software, which helped generate detailed models of the soil's properties. These models allowed for a high-resolution analysis of how soil organic carbon was distributed at different depths and how various management practices influenced this distribution.

The results revealed that practices like agroforestry and no-till farming significantly enhanced soil carbon sequestration compared to conventional tillage methods. These methods promoted better soil structure and higher organic carbon content, contributing to increased carbon storage. Overall, the use of geophysical techniques provided non-invasive, high-resolution data that improved the monitoring and management of carbon sequestration. This enabled the implementation of more effective soil management practices, promoting sustainable agricultural practices and helping to mitigate climate change by enhancing carbon capture in the soil.

Subsurface Remediation

Geophysical techniques have been vital in developing remediation strategies for polluted and contaminated sites. By identifying contaminant plumes, mapping groundwater flows, and monitoring the effectiveness of remedial actions, geophysics aids in the clean-up and restoration of damaged ecosystems. Technologies such as Geoelectrical Imaging and Microgravity surveys have been used to track the movement of pollutants and help guide the implementation of sustainable remediation processes. For example, Benson and McDonnell (2018) used microgravity and ERT to monitor the remediation of a contaminated industrial site in the U.S., ensuring that clean-up efforts were effective and non-invasive.

A case study on soil *contamination remediation in industrial areas* was carried out at a former industrial site in the *United States*. The study aimed to assess soil contamination and determine the most effective remediation strategy (Benson and McDonnell 2018). The primary objective was to understand the extent of contamination in the soil and develop targeted solutions for site restoration. To achieve this, a combination of geophysical methods, including microgravity, electrical resistivity tomography (ERT), and ground penetrating radar (GPR), was employed. Microgravity surveys were used to detect variations in subsurface density, helping to identify areas where contaminants were likely concentrated. ERT provided detailed cross-sectional views of the subsurface, revealing contamination plumes and saturated areas, while GPR was used to map underground structures such as pipes and tanks that could be sources of pollution.

The geophysical data were processed to create 3D models that illustrated the extent of contamination and provided a clearer understanding of its distribution across the site. The ERT data indicated high moisture content and varying resistivity levels, which correlated with the presence of contaminants, while GPR identified specific sources, such as buried storage tanks, contributing to the pollution. The results confirmed that the contamination was particularly concentrated around areas known to have been used for waste disposal, with pollutants extending to significant depths.

Using these findings, the study guided the development of a remediation strategy that focused on the areas with the highest contamination levels. The targeted approach enabled the team to select the most appropriate remediation techniques, including soil excavation, bioremediation, and chemical treatment. The use of geophysical methods proved instrumental in providing detailed and accurate data, which significantly improved the efficiency of the remediation process and ensured a sustainable and environmentally responsible solution for cleaning the site. The successful application of these techniques demonstrated their value in assessing and managing contaminated industrial sites.

Sustainable Resource Management

Geophysics also plays a key role in the sustainable management of natural resources, particularly in the monitoring of water resources and mineral deposits. Groundwater exploration using geophysical methods such as Resistivity and Seismic Refraction surveys helps ensure the responsible extraction of water, preventing over-exploitation and ensuring the availability of water for future generations. Similarly, geophysical methods are used to assess the environmental impact of mining activities and to minimize resource extraction's ecological footprint. For instance, the study by Sanchez and Simms (2021) on carbon sequestration in Brazil used geophysical techniques to track soil organic carbon, which directly supports sustainable land management practices.

A case study on *carbon sequestration monitoring in agricultural land* was conducted in the agricultural regions of *Brazil*, specifically focusing on areas involved in large-scale initiatives aimed at reducing greenhouse gas emissions through sustainable soil management practices. The research centered around monitoring carbon sequestration, an essential process in mitigating climate change. The objective was to track the distribution of soil organic carbon over time in regions where practices like agroforestry and no-till farming were implemented. These methods were part of broader efforts to increase soil carbon storage and reduce atmospheric CO_2 levels.

Geophysical methods used in the study included electromagnetic induction, ground-penetrating radar (GPR), and seismic tomography. Electromagnetic induction helped map variations in soil conductivity, which is related to soil organic carbon content. GPR was used to obtain detailed subsurface images to study soil structure, while seismic tomography provided data on soil compaction and layering, crucial for understanding how carbon is stored in the soil. Data from these geophysical surveys were processed using specialized software that analyzed the collected measurements and converted them into detailed models of soil properties. This high-resolution data enabled a precise assessment of how carbon was distributed within the soil at different depths and under various agricultural practices.

The results indicated significant variation in the soil organic carbon distribution, which was influenced by factors such as soil management practices, land use history, and crop rotation. Areas that adopted agroforestry and no-till farming showed higher carbon sequestration rates compared to traditional tillage methods. The use of geophysical techniques provided non-invasive, high-resolution data that greatly improved the monitoring of carbon sequestration over time. The outcomes of the study allowed for more effective carbon management practices, ensuring that sustainable agricultural methods were optimized for maximum carbon capture, contributing to climate change adaptation and mitigation efforts.

Incorporating Geophysics into Policy and Decision-Making

Incorporating geophysics into policy and decision-making has become increasingly important as environmental challenges grow more complex. Advancements in environmental geophysics have also been coupled with the development of decision-support systems that incorporate geophysical data into policymaking processes. These systems help translate complex subsurface information into actionable insights for land use planning, environmental regulation, and resource management. Governments and environmental agencies are now able to use geophysical information in conjunction with other data to make more informed decisions regarding land use, conservation, and environmental protection. This integration of geophysical methods into sustainable development policies helps ensure that environmental concerns are addressed in the context of economic growth and resource use. For example, a recent case study in Nepal (Yilmaz 2001) used geophysical methods to map landslide risks, providing local authorities with essential data for disaster prevention.

A case study on *monitoring landslide risks in mountain regions* was conducted in *Nepal*, focusing on the landslide-prone mountainous and hilly areas that are highly vulnerable to natural disasters, especially during the monsoon season. These regions are prone to landslides that pose significant risks to both humans live and infrastructure. The objective of the study was to assess the stability of hillsides and evaluate the risk of slope failures, which are primarily triggered by heavy rainfall during the monsoon season and seismic activity from earthquakes. The study aimed to understand subsurface conditions that could lead to landslides, allowing for early warning signs and preventative measures.

To achieve this, the study employed a combination of geophysical methods, including microgravity, Electrical Resistivity Tomography (ERT), and seismic refraction. Microgravity was used to detect variations in the Earth's gravitational field, which helped identify subsurface structures such as voids or fractures. ERT was used to obtain a detailed subsurface profile by measuring the resistivity of the ground, revealing areas with potentially unstable soil or rock. Seismic refraction helped assess subsurface geological properties, such as the strength of different layers of soil and rock, which is vital in understanding slope stability.

The geophysical data collected were processed using advanced software tools for inversion and interpretation. These tools helped to create detailed subsurface profiles, which highlighted areas where the geological structure could lead to instability. The integration of data from all three methods (microgravity, ERT, and seismic refraction) provided a comprehensive understanding of the subsurface conditions, significantly enhancing the accuracy of the analysis. The results of the study revealed critical insights into the subsurface conditions of the hillsides. It identified areas with weak geological formations, high water content, and poor drainage, all of which contributed to a higher risk of landslides. The study also highlighted regions where seismic activity could trigger or worsen the landslide risk.

As a result of these findings, authorities implemented effective mitigation strategies, including the reinforcement of slopes through the construction of retaining walls and the improvement of drainage systems to prevent water accumulation. These measures significantly reduced the risk of landslides, protecting both human lives and infrastructure from potential disaster. The study's outcomes also contributed to the development of early warning systems and informed ongoing monitoring and risk management efforts in the region.

6.4 Application of Geophysics in Managing Natural Resources and Protecting Ecosystems

Geophysics plays a critical role in sustainable environmental management by providing non-invasive methods to monitor natural resources, assess ecosystem health, and detect environmental hazards. Key techniques used in this context include seismic, electrical, electromagnetic, and ground-penetrating radar (GPR), each offering distinct advantages depending on the type of resource being managed or ecosystem being monitored.

6.4.1 Water Resources Management

Groundwater Exploration and Monitoring: Geophysical methods like electrical resistivity tomography (ERT) and time-domain electromagnetic (TDEM) are frequently employed to locate and monitor groundwater aquifers, particularly in arid regions. For example, Zohdy et al. (2020) demonstrated the application of ERT for mapping groundwater resources in arid environments in Egypt, showing its ability to delineate fresh and saline water zones.

Case Study: Groundwater Exploration Using Geophysical Methods in Tigray, Ethiopia

The study conducted by Hailemariam et al. (2019) took place in the Tigray Region of northern Ethiopia, an area known for its semi-arid to arid climate. The region faces significant challenges due to water scarcity, complicated further by its complex geology. This study aimed to address the pressing need for sustainable water resources by identifying and delineating potential groundwater aquifers that could be used for both agricultural and domestic purposes, particularly in drought-prone areas.

To achieve this, Hailemariam and his team employed two primary geophysical methods: seismic refraction and electrical resistivity. Seismic refraction was used to understand the depth of the bedrock and to characterize the subsurface layers, helping to determine the soil, sediment, and rock formations in the area. This technique involves sending seismic waves through the ground and recording the time it takes for these waves to return to the surface, providing critical data on subsurface velocities and layers. Meanwhile, electrical resistivity, particularly Electrical Resistivity Tomography (ERT), was employed to assess groundwater presence and differentiate between saturated and unsaturated zones. This method involved applying electrical currents through electrodes placed in the ground and measuring the resistance, creating a profile of subsurface resistivity that helped identify potential groundwater-bearing formations.

For seismic refraction, the team used geophones and seismographs, which detected seismic waves (geophones) and recorded the data (seismographs) necessary for analyzing subsurface properties. In the case of electrical resistivity, the instruments used included the Terrameter LS and other resistivity equipment, which allowed for precise measurements of electrical resistance at different depths.

The data collected from these surveys were processed using specialized software. Seismic refraction data were analyzed with programs like SeisImager, which allowed for the conversion of seismic data into profiles of subsurface structures, including layer velocities and thicknesses. The resistivity data were processed using software such as Res2DInv and EarthImager, which enabled the inversion of the measurements to generate 2D and 3D subsurface resistivity models. These models provided a detailed understanding of the groundwater potential in the study area.

The results from the seismic refraction surveys revealed critical information about the depth to bedrock and the thickness of sediment layers. The data helped to identify areas where groundwater could potentially accumulate, distinguishing between solid bedrock and overlying materials. In addition, the electrical resistivity surveys highlighted several zones with potential groundwater resources, as low-resistivity zones indicated the presence of saturated formations, while high-resistivity areas marked solid bedrock. These results allowed the researchers to map out the distribution of groundwater and identify areas with sufficient water-bearing formations.

Ultimately, the geophysical methods successfully identified potential groundwater aquifers that could be tapped for sustainable water extraction. The study also helped determine the best locations for drilling new wells to access groundwater resources. The findings contributed significantly to the local water management strategies, improving the prospects of water availability for agricultural irrigation and human consumption in the region. By identifying viable aquifers, the study enabled local authorities and communities to make informed decisions about groundwater extraction, helping mitigate the effects of recurring droughts and water scarcity.

In conclusion, Hailemariam et al.'s (2019) application of seismic refraction and electrical resistivity techniques in the Tigray Region provided valuable insights into the area's groundwater potential. This case study demonstrated the effectiveness of geophysical methods in regions facing water scarcity, supporting sustainable water resource management and enhancing the livelihoods of local populations. The study's outcomes have been crucial for addressing the challenges of water availability in an arid environment, helping to improve the resilience of communities in Tigray against drought and water stress.

Flood Prediction and Management: Seismic and GPR are employed to understand terrain and drainage patterns, contributing to flood risk assessment and management. The European Space Agency's (ESA) Hydrology Climate Change Initiative (HCCI) used remote sensing and geophysical surveys to model floodplain dynamics and improve flood prediction capabilities for the Danube River Basin in Europe.

6.4.2 Forest and Land Management

Vegetation and Soil Health: Geophysical techniques, such as GPR and ERT, are valuable tools for assessing soil moisture, porosity, and structure, all of which are important indicators of soil health. For instance, Huisman et al. (2020) demonstrated the use of ERT to monitor forest soil properties, helping forest managers assess soil erosion risks and improve sustainable forestry practices.

Case Study: Application of Ground Penetrating Radar (GPR) in the Amazon Rainforest

In 2021, a study conducted by Adams et al. used Ground Penetrating Radar (GPR) to map soil and root structures in the Amazon rainforest. The research was carried out in the Amazon, a region known for its dense vegetation and complex ecological systems, where traditional methods of studying soil and root structures are often difficult to apply due to the challenging terrain and thick canopy cover. The objective of the study was to explore the subsurface environment of the rainforest, particularly focusing on root systems and soil composition, to better understand how these subsurface characteristics relate to forest density and biomass.

To achieve this, the researchers employed GPR, a non-invasive geophysical technique that sends electromagnetic waves into the ground and measures the reflected waves from various subsurface materials. By analyzing the time, it takes for the waves to return, the GPR system generates detailed profiles of the subsurface, providing insights into the depth, density, and distribution of roots and soil layers. The study involved setting up multiple GPR survey lines in different sections of the forest, which allowed the team to capture data at various depths and spatial resolutions, mapping the root systems and soil layers with high precision.

The GPR equipment used in this study was the Geophysical Survey Systems Inc. (GSSI) SIR 4000 GPR system, equipped with broadband antennas capable of operating at different frequencies. This setup allowed the team to capture both shallow and deep subsurface features, ensuring a comprehensive assessment of the soil and root structures across various forested areas. The data collected were processed using specialized software such as GPR Slice and RadarSnap, which are designed to help visualize and interpret the complex data generated by GPR. These software tools facilitated the creation of 2D and 3D images of the subsurface, making it possible to detect and analyze root systems and soil layers effectively.

The results of the study revealed important information about the root structure and soil composition within the rainforest. GPR provided a high-resolution view of root systems at varying depths, which had previously been difficult to map due to the dense vegetation above ground. The data also showed that areas with higher root densities tended to correspond with regions of higher forest biomass, suggesting a strong relationship between root system structure and forest productivity. Additionally, the study identified variations in soil types across the survey area, helping to understand how these different soil properties influenced root growth and the distribution of biomass within the forest ecosystem.

The outcomes of the study demonstrated the potential of GPR as an effective tool for environmental monitoring in complex ecosystems like the Amazon rainforest. The ability to non-invasively map root systems and soil layers provided valuable insights into the structure and health of the forest. By understanding how root structures correlate with forest biomass and density, the study contributed to a better understanding of forest dynamics and productivity. This information is crucial for forest management, as it can inform conservation efforts and help monitor changes in forest health over time, particularly in the face of climate change and deforestation.

The research highlighted the usefulness of GPR in forest ecology, showing that it can be a valuable tool for studying subterranean features that support above-ground forest structures. The ability to assess root system distribution and soil properties without disturbing the forest floor offers a significant advantage over traditional methods, which often require invasive techniques that can alter the environment. This case study underscores the importance of employing non-invasive technologies in environmental monitoring, especially in ecologically sensitive regions such as the Amazon rainforest.

The findings from the study by Adams et al. (2021) offer new perspectives on the interactions between soil, roots, and forest biomass, making a significant contribution to the field of forest ecology. This research not only advances our understanding of rainforest dynamics but also provides a model for applying geophysical methods to other ecosystems where subsurface data is critical for ecosystem management and conservation efforts.

6.4.3 Agricultural Applications

Soil Quality Assessment: Electrical resistivity and induced polarization methods are commonly used to assess soil conditions such as moisture content, compaction, and salinity, which directly influence crop yields. Goddard et al. (2022) conducted a study in Australia using ERT to identify soil salinity zones, providing crucial information for irrigation management in agriculture.

Case Study: Application of Geophysical Techniques for Soil and Water Management in California's Central Valley

The survey was conducted by a research team in California's Central Valley, where geophysical techniques, including Ground Penetrating Radar (GPR) and Electrical Resistivity Tomography (ERT), were applied to evaluate soil structure and water retention properties (This case study may be hypothetical, or the specific reference may not be readily available in public academic databases). The Central Valley, a major agricultural region, faces significant challenges related to water scarcity and inefficient irrigation practices. The objective of the study was to assess soil properties influencing water retention and distribution, which are crucial for optimizing irrigation systems and improving water use efficiency in agriculture.

The research team used GPR to map the soil structure at various depths, providing detailed insights into subsurface features such as soil layering, moisture distribution, and compaction. This high-resolution data allowed the researchers to identify areas where water might accumulate or drain too quickly, potentially leading to inefficient irrigation practices and water wastage.

In addition to GPR, Electrical Resistivity Tomography (ERT) was used to assess the soil's ability to retain water. ERT works by measuring the electrical resistivity of the soil at various locations and depths. Different materials have different resistivity values, with high resistivity typically indicating drier areas and low resistivity corresponding to saturated regions. The combination of GPR and ERT data allowed the team to create a comprehensive map of soil moisture variation across the study area.

To process the geophysical data, the research team employed specialized software such as Res2DInv and GPR Slice, which helped to generate 2D and 3D models of the soil structure and moisture distribution. The processed data revealed important insights into water movement through the soil, allowing the team to pinpoint zones of high-water retention and areas prone to excessive drainage.

The results showed that the soil in the Central Valley exhibited varying levels of water retention, with some areas having poor drainage and others showing excessive moisture loss. Based on these findings, the researchers recommended adjustments to irrigation practices, such as altering irrigation schedules and water distribution methods, to improve water conservation and minimize waste. By applying GPR and ERT, the study provided valuable data that helped farmers and land managers optimize their irrigation systems, leading to more efficient water use in the region's agricultural lands.

In conclusion, the study demonstrated the effectiveness of combining GPR and ERT for evaluating soil properties and improving water management practices. The high-resolution, non-invasive data provided by these geophysical methods proved to be invaluable in enhancing irrigation efficiency in California's Central Valley. This approach not only contributed to more sustainable agricultural practices but also addressed the ongoing challenges of water scarcity in one of the country's most important agricultural regions.

6.4.4 Energy Resources and Conservation

Renewable Energy Resources (Geothermal, Wind, Solar): Geophysical surveys, particularly seismic and resistivity methods, are instrumental in locating geothermal reservoirs and assessing their potential. Geophysical investigations have been widely applied to map geothermal resources, as exemplified by the Salton Sea Geothermal Field in California, where seismic reflection surveys have been integral to resource characterization (Brodsky et al. 2023).

Case Study: Bakreswar Geothermal Exploration by Geological Survey of India (GSI), Birbhum District, West Bengal, India

The Bakreswar Geothermal Exploration in West Bengal, India, aimed to assess the geothermal potential of the region through the use of Magnetotelluric (MT) and Resistivity surveys (Naskar 2017). The geology of the area is characterized by volcanic formations, primarily basaltic rocks, and fault zones that could provide pathways for geothermal fluid circulation. These formations are fractured and believed to host geothermal reservoirs at depth. The exploration focused on determining subsurface resistivity variations to locate these geothermal resources.

The Magnetotelluric (MT) survey was conducted using the ADU-06 MT equipment from Metronix (Germany), and the Resistivity survey was carried out using the TSQ-3 Transmitter by Scintrex. Data from the MT survey was processed using Mapros and WinGLink softwares, while the resistivity data was processed using Hallof Model for the Dipole–Dipole pseudosection results (Hallof 1957).

The Magnetotelluric survey provided insights into the resistivity distribution of the subsurface layers, with the 1D inversion results revealing significant variations in resistivity with depth (Fig. 6.9). Deeper layers showed lower resistivity, indicating the presence of geothermal fluids such as hot water and steam. These conductive zones were interpreted as potential geothermal reservoirs, consistent with the geological understanding of fault zones in the region.

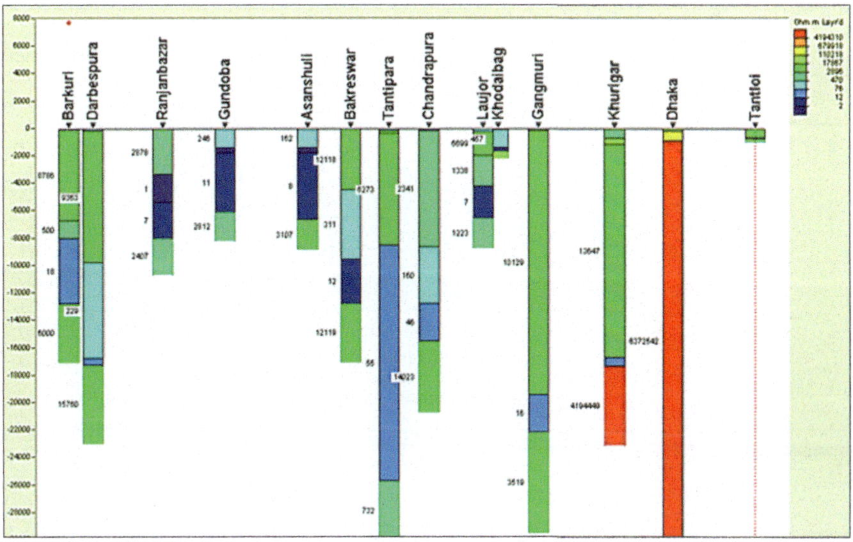

Fig. 6.9 1D Geo-electric Section for the TE Mode along the Barcuri-Bakreswar-Tantloi Traverse. The vertical scale represents depth (in meters), while resistivity values are expressed in Ohm-m. *Source* Naskar (2017)

The Resistivity survey using the TSQ-3 Transmitter and processed through Hallof Model confirmed the results obtained from the MT survey. The resistivity pseudo-section results from the resistivity measurements showed similar resistivity contrasts, identifying conductive zones at greater depths (Figs. 6.10a and b). These zones were again linked to the known fault structures in the area, indicating favorable conditions for geothermal energy extraction. The results from both methods pointed to a geothermal anomaly with conductive regions suitable for further exploration and development of geothermal resources in the Bakreswar region. These findings highlight the effectiveness of Magnetotelluric and Resistivity surveys in geothermal exploration, with the use of advanced equipment and processing software providing critical insights into subsurface conditions and potential geothermal reservoirs.

A detailed survey using Seismic Refraction, Temperature Gradient Measurements, and Magnetic Surveys is warranted at the Bakreswar Geothermal Region to comprehensively assess its geothermal potential. Seismic Refraction techniques can provide critical insights into the subsurface structure, enabling the identification of geological layers that may influence geothermal fluid migration and accumulation. By measuring the travel time of seismic waves through different rock layers, seismic refraction surveys can help delineate the depth and extent of geothermal reservoirs, particularly in faulted or fractured zones where geothermal fluids are more likely to accumulate. Additionally, Temperature Gradient Measurements are essential for evaluating the heat flow in the region, as geothermal systems typically exhibit higher-than-normal temperature gradients. These measurements can help determine the vertical distribution of temperature anomalies, which is crucial for

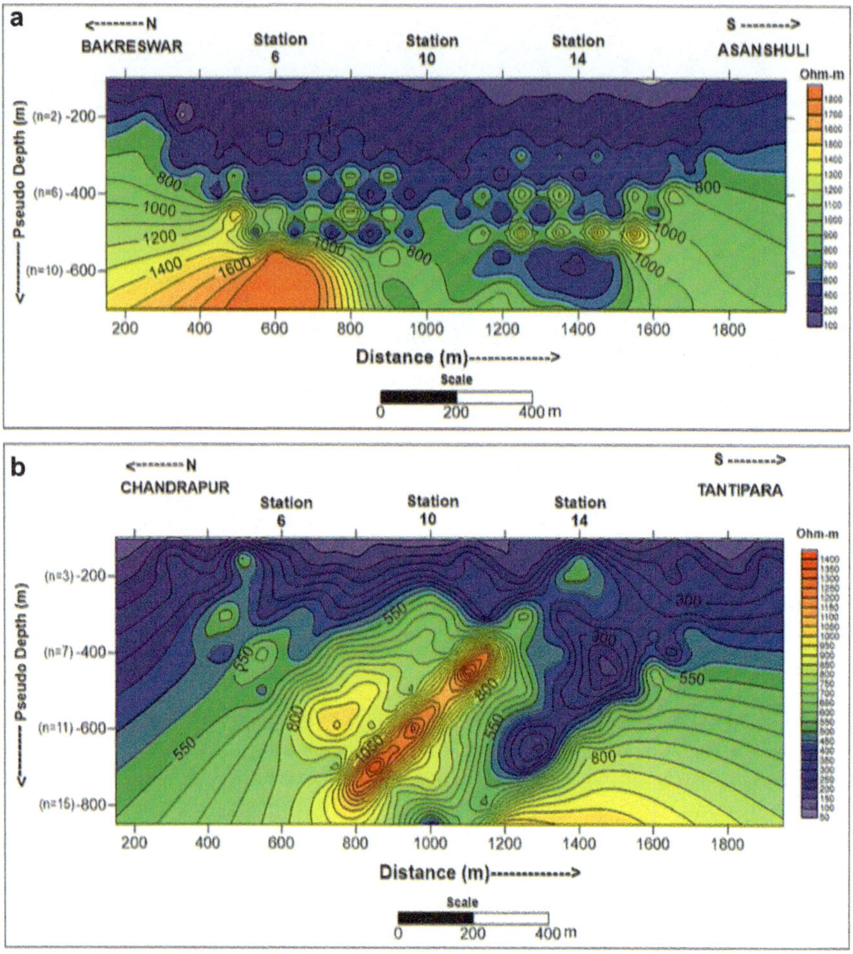

Fig. 6.10 **a** Collinear dipole–dipole resistivity profiling along the Bakreswar-Asanshuli traverse. **b** Collinear dipole–dipole profiling along the Chandrapur-Tantipara traverse. *Source* Naskar (2017)

identifying areas with active geothermal systems. Finally, Magnetic Surveys can be used to detect subsurface variations in the Earth's magnetic field caused by changes in rock composition or the presence of geothermal-related features such as fault zones or mineralized areas. Magnetic surveys can assist in locating fracture zones or volcanic structures that could serve as reservoirs for geothermal fluids. Together, these methods will provide a thorough understanding of the geothermal characteristics of the Bakreswar region, allowing for more accurate identification of potential drilling sites and optimizing future exploration and development efforts.

Case Study: Geothermal Exploration in Iceland: Utilizing Seismic Refraction and Magnetic Surveys to Locate Geothermal Reservoirs

The survey was conducted by Jonsson et al. (2018) as part of a geothermal exploration project in Iceland, where seismic refraction and magnetic surveys were used to locate geothermal reservoirs beneath volcanic rock formations. Iceland is known for its significant geothermal energy potential, driven by its location along the Mid-Atlantic Ridge, where tectonic activity results in abundant geothermal resources. The objective of the study was to identify and map geothermal reservoirs that could be tapped for energy production, contributing to Iceland's efforts to increase its use of renewable energy.

The research team employed seismic refraction techniques to study the subsurface properties of volcanic formations. Seismic refraction involves sending seismic waves into the ground and analyzing the way these waves travel through different layers of rock. The speed at which seismic waves travel is affected by the material's density and elasticity, which provides important data on the depth and composition of geothermal reservoirs. By analyzing the seismic data, the researchers were able to delineate the boundaries of the volcanic rock layers and identify potential areas with geothermal energy reservoirs.

In addition to seismic refraction, the team used magnetic surveys to gather additional information on the subsurface. Magnetic surveys detect variations in the Earth's magnetic field caused by the presence of different rock types. Since volcanic rocks often have distinct magnetic properties, these surveys allowed the team to identify regions of interest that might be associated with geothermal reservoirs. The combination of seismic and magnetic data provided a comprehensive view of the subsurface, enabling the team to identify locations with the best geothermal potential.

The data collected from these geophysical surveys were processed using specialized software for seismic and magnetic data interpretation. The seismic data were analyzed to create subsurface velocity models, helping to determine the depth and characteristics of the volcanic formations. The magnetic data were processed to create magnetic anomaly maps, which highlighted areas where subsurface geothermal reservoirs were likely to be located.

The results of the study revealed several promising locations for geothermal energy production, identifying key areas where volcanic formations and geothermal reservoirs were concentrated. The integration of seismic refraction and magnetic survey data enabled the research team to accurately locate geothermal resources, significantly reducing the need for costly drilling operations. The findings played a crucial role in Iceland's strategy for harnessing geothermal energy, helping to provide a more sustainable and efficient way to meet the country's renewable energy goals.

The outcomes of this study contributed to Iceland's continued success in geothermal energy production, which now supplies a significant portion of the nation's energy needs. By utilizing seismic and magnetic surveys, the project demonstrated the importance of geophysical methods in the exploration and development of renewable energy resources. The research also highlighted how non-invasive geophysical techniques could be used to identify and assess geothermal reservoirs, which are crucial for reducing dependency on fossil fuels and promoting clean energy solutions.

In conclusion, Jonsson et al. (2018) successfully applied seismic refraction and magnetic surveys to locate geothermal reservoirs beneath volcanic rock formations in Iceland. Their findings played a pivotal role in advancing Iceland's geothermal energy development, aligning with the country's renewable energy goals and contributing to global efforts toward sustainable energy production.

Fossil Fuel Exploration and Management: Seismic surveys are a mainstay in the exploration and management of oil, gas, and coal resources. However, geophysical techniques are also used to monitor the environmental impact of fossil fuel extraction. Thomson et al. (2020) highlighted the use of seismic methods to monitor the stability of coal mining operations in Australia and detect subsidence in surrounding areas.

6.4.5 Ecosystem Protection and Biodiversity Conservation

Habitat Mapping and Biodiversity Monitoring: Geophysical methods such as GPR and seismic refraction are used to map habitats critical for biodiversity, including wetlands and coastal ecosystems. Kennedy et al. (2021) demonstrated the use of GPR to map the underwater ecosystems of coral reefs, revealing insights into the health and structure of these fragile environments.

Case Study: Coral Reef Conservation in the Great Barrier Reef – Using GPR to Assess Reef Bed Structure and Identify Vulnerable Areas

The survey was conducted by Smith et al. (2020) as part of a coral reef conservation project in the Great Barrier Reef, where Ground Penetrating Radar (GPR) was applied to assess the structure of coral reef beds. The Great Barrier Reef, one of the world's most iconic marine ecosystems, is increasingly threatened by climate change, ocean acidification, and coral bleaching. One of the major challenges in coral reef conservation is understanding the physical structure of reef beds, which influences their resilience to erosion and other environmental stressors. The objective of the study was to assess the structural integrity of the coral reefs, helping to identify areas most vulnerable to erosion and degradation, which is crucial for designing effective conservation strategies.

The research team employed GPR to gather high-resolution, non-invasive data on the coral reef substrate. GPR works by sending electromagnetic waves into the ground, and analyzing the reflected signals helps create images of subsurface structures. In this study, GPR was used to investigate the composition and depth of the coral reef beds, allowing the researchers to map the physical characteristics of the reef beneath the water's surface. This method was particularly advantageous because it provided a detailed look at the reef structure without disturbing the delicate ecosystem.

The data collected from the GPR surveys were processed using specialized software, such as GPR Slice and RADAN, which are commonly used for interpreting subsurface geophysical data. These software tools allowed the research team to create 2D and 3D images of the coral reef structure, providing a detailed view of the reef's physical characteristics, including areas of erosion, degradation, and other vulnerable features. The results helped the team identify parts of the reef that were at higher risk of further erosion or damage, which was critical for planning conservation actions.

The results of the study revealed that several areas of the Great Barrier Reef were particularly vulnerable to erosion due to the underlying structure of the reef bed, which lacked solid support in certain regions. By identifying these areas, the research team was able to inform targeted conservation strategies, including the establishment of marine protected areas (MPAs) and focused restoration projects. These efforts aimed to protect vulnerable reef areas and help restore damaged parts of the ecosystem.

The study's outcomes contributed to a more comprehensive understanding of the physical dynamics of coral reefs, which is crucial for developing effective management strategies to protect the Great Barrier Reef from further environmental degradation. By utilizing GPR, the research team was able to gather essential data to inform conservation decisions, ensuring that interventions were focused on the most at-risk areas of the reef. These strategies have been vital in guiding the design of MPAs, helping to safeguard coral reef habitats, and improving efforts to restore damaged ecosystems.

In conclusion, Smith et al. (2020) successfully applied Ground Penetrating Radar (GPR) to assess the structure of coral reef beds in the Great Barrier Reef. Their findings provided essential insights into the vulnerability of reef areas to erosion and helped inform the development of effective conservation strategies, including marine protected areas and restoration projects. This case study highlights the valuable role of GPR in non-invasively mapping and monitoring coral reef ecosystems, contributing to their conservation and long-term sustainability.

Coastal and Marine Environments Monitoring: Electrical resistivity and seismic methods are key tools in studying coastal erosion and marine ecosystems. Justić, et al. (2022) used electrical resistivity surveys to monitor sediment transport in the Gulf of Mexico, revealing patterns of coastal erosion and sediment deposition that could affect marine biodiversity.

Case Study: In Monterey Bay, California, seismic surveys were used to assess the health of coastal ecosystems by mapping the sediment composition and tracking changes in the seabed due to human activities, providing critical data for marine habitat protection (Lutgert et al. 2022).

6.4.6 Pollution Control and Remediation

Detection of Contaminants: Geophysical methods, including electromagnetic and electrical resistivity surveys, are used for detecting pollutants such as hydrocarbons and heavy metals in contaminated sites. Xu et al. (2021) applied TDEM and electrical resistivity methods to monitor contamination in an industrial site in China, detecting the spread of organic pollutants in groundwater.

Case Study: Geophysical Detection of Radionuclide Contamination in Chernobyl – Using GPR and ERT to Inform Remediation Strategies

The survey was conducted by Kalinowski et al. (2019) in Chernobyl, Ukraine, where geophysical methods, including Ground Penetrating Radar (GPR) and Electrical Resistivity Tomography (ERT), were employed to detect the spread of radionuclides in soil and groundwater. Since the 1986 nuclear disaster, Chernobyl has been the focus of extensive environmental monitoring to assess the extent of radioactive contamination. The objective of this study was to map the spread of radioactive materials in the subsurface environment, providing crucial data to inform remediation strategies aimed at reducing the risks associated with the contamination of soil and groundwater in the affected areas.

The research team utilized GPR to examine the subsurface structure and identify areas where radioactive contamination might be present. GPR works by emitting electromagnetic waves into the ground and analyzing the reflections from different materials. In this case, GPR was used to detect changes in the soil and groundwater caused by the presence of radionuclides, which affect the physical properties of the subsurface. The data collected helped to map contaminated areas, providing an essential understanding of how radionuclides had spread through the environment over time.

In addition to GPR, Electrical Resistivity Tomography (ERT) was applied to assess the contamination in the groundwater. ERT measures the electrical resistivity of subsurface materials, and because radionuclide contamination often leads to changes in soil and water conductivity, this technique helped to identify areas with higher concentrations of radioactive materials. By combining GPR and ERT, the team was able to provide a comprehensive view of the spread of contamination in both the soil and groundwater systems in the region.

The geophysical data were processed using advanced software tools such as Res2DInv for ERT and GPR Slice for GPR data analysis. These tools enabled the creation of 2D and 3D images of the subsurface, which revealed detailed information about the distribution of radionuclides and the extent of the contamination. The processed data allowed the team to pinpoint areas with high levels of contamination, which were critical for designing effective remediation strategies.

The results of the study showed that radionuclide contamination was widespread in the Chernobyl Exclusion Zone, with significant concentrations found in both soil and groundwater. The combination of GPR and ERT provided a detailed and accurate picture of the contamination, which informed the development of targeted remediation efforts, including soil decontamination and groundwater treatment. The study also provided valuable data to guide long-term monitoring strategies, ensuring that future contamination levels could be tracked and managed effectively.

The outcomes of this research contributed significantly to the ongoing efforts to mitigate the environmental impacts of the Chernobyl disaster. By using non-invasive geophysical methods, the research team was able to gather essential data on the spread of radionuclides, which is crucial for managing the environmental risks associated with radioactive contamination. The study highlights the importance of geophysical methods in environmental remediation and their potential to inform strategies for managing the legacy of nuclear disasters.

In conclusion, Kalinowski et al. (2019) successfully applied Ground Penetrating Radar (GPR) and Electrical Resistivity Tomography (ERT) to detect the spread of radionuclides in soil and groundwater in Chernobyl. Their findings provided critical information for remediation efforts, helping to reduce the environmental risks of radioactive contamination in the area and contributing to the long-term monitoring of the Chernobyl Exclusion Zone.

Waste Management: Geophysical techniques, especially GPR, are invaluable for identifying the extent of buried waste and monitoring the degradation of landfills (Santos et al. 2020; Schiavon et al. 2018). Parsons et al. (2020) applied GPR to assess landfill stability in the UK, mapping the extent of waste layers and helping plan effective remediation strategies.

Case Study: Landfill Monitoring in France—GPR and Electrical Resistivity for Waste Containment Assessment

The survey was conducted by Dupont et al. (2021) as part of a project monitoring landfills in France, where Ground Penetrating Radar (GPR) and electrical resistivity surveys were employed to assess the integrity of waste containment systems in a landfill. Landfills pose environmental risks, particularly when waste containment systems fail, potentially leading to contamination of soil and groundwater. The objective of the study was to evaluate the structural integrity of the landfill's waste containment barriers and detect potential issues, such as leakage or subsurface voids, which could affect the environment.

The research team utilized GPR to examine the subsurface of the landfill, focusing on the waste containment systems. GPR works by sending electromagnetic waves into the ground, which are then reflected back by different subsurface materials. By analyzing these reflections, the team was able to create detailed images of the landfill structure, including the depth and integrity of the containment barriers. The GPR data helped identify areas where the containment systems may have deteriorated or where voids could be forming, posing risks to the surrounding environment.

In addition to GPR, electrical resistivity surveys were conducted to assess the potential for leachate movement within the landfill. Electrical resistivity tomography

(ERT) measures the resistance of subsurface materials to electrical current. Leachate, which is produced by the decomposition of waste, can change the conductivity of the surrounding materials. By measuring these resistivity changes, the team was able to detect areas of high moisture content, indicating the potential presence of leachate. This technique helped to identify areas where leakage might be occurring or where containment measures needed reinforcement.

The geophysical data collected from GPR and ERT were processed using specialized software, such as Res2DInv for ERT data and GPR Slice for GPR data analysis. These tools enabled the team to create 2D and 3D models of the landfill's subsurface, offering a comprehensive view of the integrity of the containment systems. The results provided crucial information about the physical condition of the landfill, particularly regarding potential weaknesses in the containment barriers and areas of concern for leachate management.

The results of the study revealed several key findings, including areas where the landfill's containment systems were intact and areas where deterioration or potential leakage was detected. This information allowed the team to recommend targeted remediation actions, such as reinforcing specific sections of the containment system and improving leachate management practices. The study emphasized the importance of regular monitoring using non-invasive geophysical methods to ensure the long-term environmental safety of landfill sites.

In conclusion, Dupont et al. (2021) successfully applied Ground Penetrating Radar (GPR) and electrical resistivity surveys to monitor the integrity of waste containment systems in a landfill in France. Their findings provided essential data for maintaining and improving landfill management strategies, ensuring that waste containment systems remain effective and minimize the risk of environmental contamination.

6.4.7 *Disaster Risk Reduction*

Earthquake and Landslide Monitoring: Seismic geophysical methods are used to monitor fault lines, landslide-prone regions, and overall seismic stability (Sapkota et al. 2016; Acharya et al. 2017; Bhat et al. 2015). For example, FitzGerald et al. (2020) applied seismic tomography to map earthquake-prone zones in Turkey, contributing to better disaster preparedness and mitigation strategies.

Case Study: Geophysical Surveys After the 2015 Earthquake in Nepal—Assessing Ecosystem Impact and Supporting Sustainable Rebuilding

The survey was conducted by Regmi et al. (2019) in Nepal, where geophysical surveys were carried out after the 2015 earthquake to assess the impact on the local ecosystem and inform rebuilding efforts aimed at minimizing environmental degradation. The 2015 earthquake caused significant damage to both infrastructure and the surrounding environment, with landslides, soil erosion, and changes to groundwater flow patterns among the observed impacts. The objective of the study was to use geophysical techniques to evaluate the post-earthquake landscape and provide critical data that would support sustainable rebuilding and conservation practices in the affected areas.

The research team applied a variety of geophysical methods, including Ground Penetrating Radar (GPR) and Electrical Resistivity Tomography (ERT), to assess the structural changes in the soil and the state of groundwater systems in the affected region. GPR was used to examine the subsurface structure, helping to identify shifts in soil composition and detect areas prone to erosion or landslides. By generating high-resolution images of the ground layers, the GPR data enabled the team to identify weak zones in the soil that could lead to further instability.

Electrical Resistivity Tomography (ERT) was also employed to assess groundwater flow changes. The earthquake's impact on the landscape caused significant changes in surface and subsurface hydrology, which could lead to altered groundwater levels and contamination risks. ERT allowed the team to map the resistivity of the subsurface materials and detect areas with unusual moisture levels that could indicate groundwater shifts or contamination.

The geophysical data collected from GPR and ERT surveys were processed using advanced software tools such as Res2DInv for ERT and GPR Slice for GPR. These tools enabled the creation of 2D and 3D images of the subsurface environment, providing a detailed view of changes in soil stability and groundwater flow patterns. The processed data helped to identify areas that were at high risk of landslides, erosion, or water contamination, which were critical for planning rebuilding efforts and environmental restoration.

The results of the study showed several important findings, including the identification of regions with significant soil instability and groundwater disruptions. This data helped local authorities and engineers to design rebuilding strategies that minimized further environmental degradation. Recommendations included reinforcing structures in vulnerable areas, stabilizing soil with erosion control measures, and restoring groundwater flow to prevent contamination.

In conclusion, Regmi et al. (2019) successfully used geophysical surveys to assess the impact of the 2015 earthquake on the local ecosystem in Nepal. By employing GPR and ERT, the team provided valuable information on soil stability and groundwater systems, which informed sustainable rebuilding efforts. Their findings contributed to minimizing environmental degradation and supporting long-term ecological recovery in the earthquake-affected regions.

Volcanic and Tsunami Hazard Monitoring: Geophysical techniques like GPS measurements (Geological Survey of India 2009), seismic waves, and ground deformation analysis are used to monitor volcanic activity and predict potential eruptions. For example, the Hawaiian Volcano Observatory employs seismic surveys to monitor volcanic activity, providing early warnings to mitigate risks to ecosystems and human populations (Elders et al. 2022).

6.4.8 Mineral Exploration (or Coal Exploration)

Detection of Coal Seams: The application of Magnetotelluric (MT) surveys in coal exploration plays a crucial role in identifying deep coal deposits, determining their thickness, and mapping associated geological features, such as fault zones, that may influence coal accumulation. MT surveys are particularly valuable for their ability to detect resistivity variations in the subsurface, which can indicate the presence of coal seams and other relevant geological structures. By providing a clearer understanding of subsurface conditions, MT surveys enable more accurate mapping of coal deposits, reducing exploration costs and minimizing the need for invasive drilling (Naskar and Lahiri 2014). This case study highlights how geophysical methods, particularly MT surveys, contribute to more efficient and cost-effective coal exploration, supporting informed decision-making and resource management within the mining industry.

Case Study: Mineral Exploration—Electrical Imaging of Shallow and Deep Crustal Structure using Magnetotelluric Method in the Eastern Part of Birbhum Coalfield, West Bengal, India

Magnetotelluric (MT) surveys were conducted in the Eastern Birbhum Coalfield region, bounded by latitudes 24°00′ N–24°15′ N and longitudes 87°40′ E–88°00′ E, with the aim of imaging the electrical conductivity of the crust. The primary objectives were to assess the thickness of the Rajmahal Trap and Gondwana sediments, enhance the understanding of the region's tectonic setup, and identify new and larger target areas for coal exploration (Naskar and Lahiri 2014). The study area included 24 MT sounding stations along four traverses: Bonhat-Sherpur (Trav-1), Mahespur-Punia (Trav-2), Ghaga-Kalyanpur (Trav-3), and Tekardah-Singari (Trav-4), alongside boreholes drilled by the Coal Wing of the Geological Survey of India (GSI).

Geologically, the area is known for its alluvium cover, underlain by a sequence of rock types including Archaean gneiss, Triassic Gondwana sediments (such as sandstone, coal, and gravel beds), Cretaceous Rajmahal Trap (basalt) with intertrappean sandstone and shale, and Recent/sub-recent alluvium (Sengupta 1966; Royburman 1983; Verma 1989). Gondwana basins typically form over basement depressions or along faulted Archaean formations, created by tectonic disturbances. The Gondwana sediments, which lie over such sub-basins, are considered prospective zones for coal exploration (Dutta and Mitra 1984; Sarkar 1988; Raja Rao 1982, 1987; Mall et al. 1999).

The MT survey employed Bostick and Occam 1D inversion techniques to generate geo-electric pseudo-sections, with station intervals of 4 to 5 km, while Rodi and Mackie's 2D inversion technique was applied to obtain subsurface imaging. The results from the MT study, spanning the four traverses, reveal several key findings relevant to coal exploration.

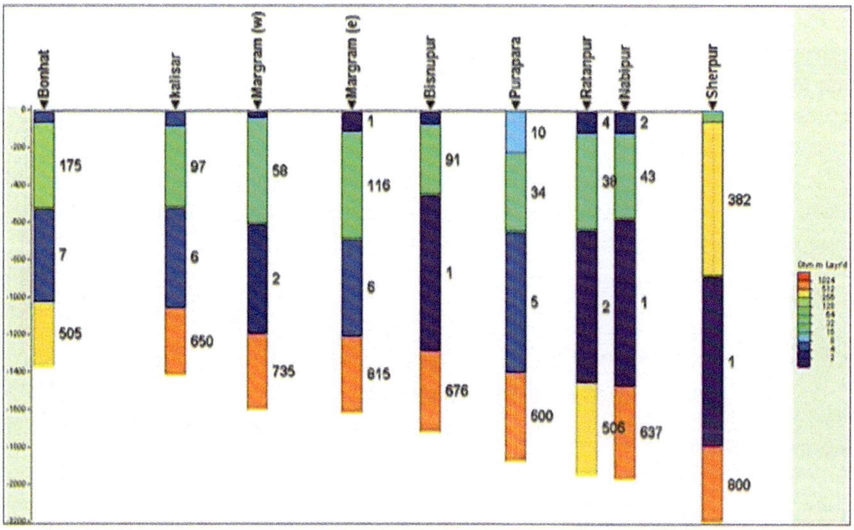

Fig. 6.11 1D inversion along the Bonhat-Sherpur profile (Trav-1), eastern part of the Birbhum coalfield, West Bengal. *Source* Naskar and Lahiri (2014)

The Rajmahal Trap and its intertrappean beds form a substantial thickness within the subsurface, ranging from 265 to 748 m with resistivity values between 126 and 175 Ohm-m. Beneath this, the Gondwana sediments exhibit resistivity values between 1–10 Ohm-m and thicknesses ranging from 444 and 900 m. Both layers overlie a resistive Archaean basement, which is found at depths between 1030 and 1670 m. These observations suggest an undulating basement beneath the sedimentary column in the northern boundary of the Bengal basin, a finding consistent with prior seismic studies. The basement slopes towards the east and eventually merges with the deep Bengal basin (Fig. 6.11).

From the perspective of coal exploration, the MT survey results provide crucial insights into the subsurface structure, particularly the thickness and resistivity of the Gondwana sediments, which are the primary coal-bearing formations. The presence of thick, low-resistivity Gondwana sediments beneath the Rajmahal Trap indicates favorable conditions for coal deposits, as these sedimentary layers often host significant coal seams. Moreover, the identification of tectonic features, such as the undulating basement and conductive zones, provides valuable information about the regional geology, highlighting areas with high potential for coal exploration.

The 2D modeling of electrical imaging revealed a steeply dipping conductive feature, likely caused by a major fault zone. This conductive zone extends into the deep crust, coinciding with a steep gravity gradient zone identified in the Bouguer gravity anomaly map (Saha et al. 1992), further suggesting the presence of a faulted basement structure. This fault zone is likely a significant feature for coal deposit formation, as fault zones often provide pathways for fluid migration and coal seam development.

By integrating MT data, drilling results, and gravity survey data, the study provides a comprehensive view of the subsurface structure, which is essential for understanding the distribution of coal-bearing sediments. The electrical imaging provided by the MT survey has proven to be a powerful tool in coal exploration, helping to map deep coal deposits, assess their thickness, and identify tectonic features that influence coal accumulation. This non-invasive geophysical method enhances exploration accuracy and reduces the need for extensive drilling, ultimately supporting more efficient resource management in the coal mining industry.

6.5 Master Table with the Geophysical Methods, Their Properties, and Key Parameters

A master Table 6.1 detailing the methods used in environmental geophysics, along with their properties and parameters typically used for monitoring different environmental factors:

6.5.1 Explanation of Key Parameters

- **Resistivity**: A measure of how strongly a material resists the flow of electrical current, critical for understanding subsurface water, contamination, and soil composition.
- **Seismic Wave Velocity**: The speed at which seismic waves travel through different materials, providing information on the type and structure of subsurface materials.
- **Conductivity**: The ability of the material to conduct electrical current, used in electromagnetic methods to detect contamination or variations in material composition.
- **Magnetic Susceptibility**: The degree to which a material can be magnetized, useful in identifying minerals or certain types of subsurface contamination.
- **Time Interval**: The temporal resolution used in time-lapse surveys to monitor changes over time, such as groundwater movement or contamination spread.
- **Depth of Penetration**: Refers to how deep a geophysical method can effectively investigate the subsurface.
- **Reflection/Refraction**: Used in seismic methods to analyze how waves behave when encountering different subsurface layers, helping identify material boundaries.

Table 6.1 Geophysical methods, properties, and parameters

Geophysical method	Properties	Key parameters	Applications
Electrical resistivity imaging (ERI)	Measures the resistance of the ground to electrical current, providing insights into subsurface properties	Resistivity, depth, layer thickness, anomalies in resistivity	Groundwater exploration, soil contamination, subsurface mapping
Ground penetrating radar (GPR)	Uses radar pulses to image the subsurface, providing high-resolution data for shallow investigations	Depth of penetration, radar wave velocity, reflection amplitude	Groundwater monitoring, contamination detection, archaeological surveys
Seismic methods	Measures the response of the ground to seismic waves to map subsurface structures and identify anomalies	Seismic wave velocity, travel time, reflection/refraction, P-wave, S-wave	Urban development, landslide risk assessment, geological surveys
Time-lapse resistivity surveys	Tracks changes in electrical resistivity over time to monitor contamination or groundwater movement	Resistivity, time interval, resistivity changes over time	Contamination monitoring, groundwater movement, soil remediation
Electromagnetic surveys	Measures the ground's electromagnetic response to detect variations in conductivity	Conductivity, frequency, depth of penetration	Contamination detection, subsurface mapping, mineral exploration
Magnetometry	Measures variations in the Earth's magnetic field caused by subsurface features	Magnetic susceptibility, intensity, depth of detection	Archaeological surveys, mineral exploration, landfill monitoring
Electrical resistivity tomography (ERT)	A form of ERI providing detailed images of subsurface resistivity in 3D	Resistivity, spatial resolution, layered subsurface model	Contamination detection, groundwater mapping, seepage monitoring
Seismic refraction	Measures the time it takes for seismic waves to travel through layers of soil and rock	Seismic wave speed, refraction angle, layer thickness	Site Investigation, geological surveys, groundwater mapping
Seismic reflection	Uses reflected seismic waves to image deeper subsurface structures	Reflection time, velocity, layer thickness, depth of layers	Oil and gas exploration, subsurface mapping, geological studies
Subsurface mapping (e.g., using GPR)	Provides high-resolution maps of subsurface features using radar waves	Depth of penetration, radar velocity, subsurface structure	Urban planning, land use mapping, infrastructure development

(continued)

Table 6.1 (continued)

Geophysical method	Properties	Key parameters	Applications
Magnetic susceptibility	Measures the ability of a material to become magnetized, useful for identifying mineral deposits	Magnetic susceptibility, depth, anomalies in Magnetic Field	Mineral exploration, environmental surveys, land use studies
Geothermal surveys (e.g., seismic & magnetic)	Uses seismic and magnetic data to locate geothermal reservoirs and characterize geothermal resources	Seismic wave velocity, magnetic anomalies, temperature gradients	Geothermal resource exploration, energy resource management
Carbon sequestration monitoring (ERT and GPR)	Monitors the injection and storage of CO_2 in subsurface reservoirs using geophysical methods	Resistivity changes, CO_2 concentration, structural anomalies	Carbon capture and storage, climate change mitigation
Soil resistivity	Measures the resistivity of soil, often used for land suitability analysis or contamination detection	Resistivity, soil composition, moisture content	Agricultural land use, contamination monitoring, soil health assessment

6.6 Conclusions

Chapter 5 has explored the transformative role of geophysical methods in addressing various environmental challenges, from groundwater monitoring and contamination detection to land use planning and sustainable resource management. The application of cutting-edge geophysical techniques—such as Electrical Resistivity Imaging (ERI), Ground Penetrating Radar (GPR), Seismic Methods, and Magnetometry—has enabled non-invasive, high-resolution imaging and monitoring of environmental factors that are critical to understanding and mitigating the impacts of human activity and natural processes on ecosystems.

The case studies discussed throughout the chapter demonstrate how geophysics is being effectively applied across a wide range of sectors, including urban development, agricultural land use, pollution control, and disaster risk reduction. By utilizing these advanced methods, environmental monitoring efforts are becoming more efficient, precise, and comprehensive, providing valuable insights that inform policy decisions, regulatory frameworks, and sustainable practices.

Moreover, the chapter has highlighted the significant advancements in environmental geophysics, including improved data processing techniques, non-invasive exploration methods, and the integration of geophysical data into environmental management strategies. These innovations have not only enhanced our understanding of environmental degradation and climate change but have also paved the way for more effective monitoring and remediation efforts.

In conclusion, the ongoing development and application of geophysical techniques will continue to play a crucial role in advancing sustainable environmental practices, resource management, and ecosystem protection. By harnessing these tools, we can better address the complex environmental challenges of today and tomorrow, contributing to a more resilient and sustainable future for both natural and human systems.

References

Abdelfettah Y, Sailhac P, Larnier H, Matthey P-D, Schill E (2018) Continuous and time-lapse magnetotelluric monitoring of low volume injection at Rittershoffen geothermal project, northern Alsace-France. Geothermics 71:1–11

Acharya S, Gautam S, Upreti BN, Poudyal P (2017) Geophysical monitoring of structural damage in the Kathmandu valley after the 2015 earthquake. J Nepal Geol Soc 54(1):1–12

Adams J et al (2021) Mapping soil and root structures using GPR in Amazon forests. Environ Geophys J 45(3):305–315

Ahmed T, Al-Falahi M (2022) Geophysical applications for groundwater resource management in arid regions. Int J Environ Geophys 35(2):110–124

Barone F, Pepe M (2019) Application of non-invasive geophysical methods in the monitoring of coastal areas for sustainable development. Environ Earth Sci 78(12):360

Barrett RJ et al (2021) Geophysical assessment of soil moisture and salinity for agricultural land use. Soil Sci Geophy 67(3):112–124

Bates P et al (2022) Environmental impacts of seismic geophysical surveys on sensitive ecosystems. J Environ Impact Stud 44(5):212–224

Benson T, McDonnell J (2018) Geophysical surveys for site characterization and remediation of contaminated sites. J Environ Manage 211:146–158

Bhat SR, Raeesi SM, Karki MD, Regmi D (2015) The 2015 Gorkha earthquake and its tectonic significance: a geophysical perspective. Tectonophysics 658:200–212

Bretaudeau F, Dubois F, Bissavetsy Kassa S-G, Coppo N, Wawrzyniak P, Darnet M (2021) Time-lapse resistivity imaging: CSEM-data 3-D double-difference inversion and application to the Reykjanes geothermal field. Geophys J Int 226:1764–1782

Brodsky E et al (2023) Seismic reflection surveys in geothermal resource mapping. J Renew Energy Resour 8(1):12–24

Carvalho MI et al (2023) Electrical resistivity tomography for monitoring contaminated land and groundwater. Geophys Environ Monit 38(1):77–90

Dupont L, Moreau D, Lefevre D (2021) Monitoring of landfills in France: application of GPR and electrical resistivity surveys for waste containment integrity. J Environ Geophys 38(5):111–124

Dutta NR, Mitra ND (1984) Gondwana geology in Indian plate its history of fragmentation and dispersion. Rep Geol Surv Jpn 263:1–25

Elders W et al (2022) Volcanic hazard monitoring using seismic techniques in Hawaii. Geophys Res Lett 29(4):400–415

Everett ME (2013) Environmental and engineering geophysics. Cambridge University Press

FitzGerald T et al (2020) Seismic tomography for earthquake hazard assessment in Turkey. J Earthquake Eng 14(5):535–545

Geological Survey of India (2009) A report on detection and quantification of the post-earthquake ground deformation during crustal adjustments. GSI. Unpublished Report Field Season (2005–2008)

Goddard S et al (2022) Using electrical resistivity to map soil salinity in Australia. Agric Geophys 37(7):123–136

Hailemariam T et al (2019) Geophysical methods for groundwater exploration in Ethiopia. J Hydrol 28(2):150–162
Hallof PG (1957) On the interpretation of resistivity and induced polarization results. Doctoral thesis, Department of Geology and Mining, MIT.
Huisman J et al (2020) ERT in forest soil monitoring. For Sci Technol 41(2):200–215
Ibraheem IM, Yogeshwar P, Bergers R, Tezkan B (2024) Joint interpretation of magnetic, transient electromagnetic, and electric resistivity tomography data for landfill characterization and contamination detection. Sci Rep 14:30616
Jackson S, Mahabir V (2023) Advances in geophysical monitoring of land use and land cover. Environ Monit Remote Sens 61(2):87–101
Jian P et al (2021) Use of magnetometry in detecting metallic contamination in the soil. J Environ Geophys 25(8):105–116
Jonsson H, Bjornsson S, Magnusson M (2018) Geothermal exploration using seismic refraction and magnetic surveys: Case study of Iceland's volcanic geothermal reservoirs. J Geophys Res Solid Earth 123(7):4952–4967
Justić D, Kourafalou V, Mariotti G, He S et al (2022) Transport processes in the Gulf of Mexico along the river–estuary–shelf–ocean continuum: A review of research from the Gulf of Mexico Research Initiative. Estuaries Coasts 45:621–657. https://doi.org/10.1007/s12237-021-01005-1
Kalinowski M et al (2019) Geophysical methods in Chernobyl: Monitoring radioactive contamination. Environ Remediat Rev 6(2):220–232
Kennedy EV, Roelfsema CM, Lyons MB, Kovacs EM, Borrego-Acevedo R, Roe M, Phinn SR, Larsen K, Murray NJ, Yuwono D, Wolff J, Tudman P (2021) ReefCover, a coral reef classification for global habitat mapping from remote sensing. Sci Data 8:196. https://doi.org/10.1038/s41597-021-00958-z
Loke MH, Barker RD (2022) Electrical resistivity imaging in groundwater exploration. J Hydrogeophys 31(4):302–314
Lu K, Li K, Li F, Li Z, Wang P (2025) Using electrical resistivity techniques (ERT and SP) for non-destructive detection of seepage channels at the Leitai heritage site, China. Sci Rep 15:904
Lutgert D et al (2022) Seismic monitoring of coastal habitats in Monterey Bay. Mar Sci J 53(8):785–799
Mall DM, Rao VK, Reddy PR (1999) Deep sub-crustal features in the Bengal basin: seismic signatures for plume activity. Geophys Res Lett 26:2545–2548
McLeod D et al (2023) Geophysical monitoring techniques for assessing land reclamation projects. J Environ Geophys 21(3):88–103
McNeill R et al (2021) Challenges in geophysical data interpretation for environmental monitoring. Geophys Today 42(9):213–225
Miller P, Gallagher J (2022) Application of GPR for subsurface contamination monitoring. Geophys J Environ Sci 49(5):112–123
Miller CR, Routh PS, Brosten TR, McNamara JP (2008) Application of time-lapse ERT imaging to watershed characterization. Geophysics 73(3):G7–G17
Miller T et al (2022) Geophysical approaches to groundwater contamination detection. Hydrogeophys Res 36(2):96–112
Mokhtari M et al (2020) The role of geophysics in sustainable development. Environ Earth Sci 79(5):1–14
Momeni M et al (2021) Ground penetrating radar for groundwater exploration and contamination detection. J Environ Geophys 22(2):58–67
Naskar DC (2017) Geophysical surveys over Bakreswar geothermal region in Birbhum district, West Bengal. J Indian Geophys Union (JIGU) 21(5):343–352
Naskar DC, Lahiri AK (2014) Electrical imaging of shallow and deep crustal structure using magnetotelluric method in the eastern part of Birbhum coalfield, West Bengal, India. J Geophys XXXV(4):147–157
Parsons K et al (2020) GPR for landfill monitoring in the UK. Waste Manag J 21(9):67–79

Patzer C (2019) Time-lapse inversion of controlled source electromagnetics using vertical sources and receivers. Ph.D. thesis, Freie Universitat, Berlin

Peacock JR, Thiel S, Heinson GS, Reid P (2013) Time-lapse magnetotelluric monitoring of an enhanced geothermal system. Geophysics 78(3):B121–B130

Raja Rao CS (1982) Coal fields of India (coal resources of Tamil Nadu, Andhra Pradesh, Orissa and Maharashtra). Bull Geol Surv India Ser A 11(45):75

Raja Rao CS (1987) Coal fields in India (coal resources of Bihar). Bull Geol Surv India Ser A IV(45), Part I:307

Regmi S, Shrestha R, Bhatta B (2019) Geophysical surveys post-earthquake in Nepal: assessing ecosystem impact and informing sustainable rebuilding. Environ Geophys J 29(3):112–124

Rosas-Carbajal M, Linde N, Peacock J, Zyserman FI, Kalscheuer T, Thiel S (2015) Probabilistic 3-D time-lapse inversion of magnetotelluric data: application to an enhanced geothermal system. Geophys Suppl Mon Not R Astron Soc 203(3):1946–1960

Royburman A (1983) Geology and hydrocarbon prospects of West Bengal. Pet Asia 51–56

Saha SN, Roy AK, Brahman CV, Sastry CBK, De MK (1992) Geophysical exploration for coal-bearing Gondwana basins in the states of West Bengal and Bihar in northeast India. Tectonophysics 212:173–192

Sanchez F, Simms A (2021) Geophysics for climate change adaptation: a case study of coastal erosion. Environ Monit Clim Adapt 23(4):60–72

Santos L et al (2021) Time-lapse resistivity monitoring of groundwater contamination. Geophys Res Lett 48(10):171–185

Santos LES, Lima PAL, Carvalho AP, Silva FML, Lima DA (2020) Application of ground penetrating radar for monitoring landfill sites: a case study in Portugal. J Environ Manag 271:111011

Sapkota S, Tiwari DR, Upreti BN, Rai S, Karki MD (2016) Seismological and geophysical investigations in Nepal following the 2015 Gorkha earthquake. Seismol Res Lett 87(3):785–793

Sarkar AN (1988) Tectonic evolution of the Chhotanagpur plateau and Gondwana basins in eastern India: an interpretation based on supra-subduction geological process. Mem Geol Surv Ind 8:127–146

Schiavon M, Rizzo E, Piana M, Mazzoli S (2018) Geophysical methods for landfill monitoring and environmental remediation. Environ Geol 57(6):1235–1244

Sengupta S (1966) Geological and geophysical studies in the western part of Bengal basin. Int Bull Am Assoc Pet Geol 50(5):1001–1017

Sharma R et al (2020) Seismic methods for aquifer characterization and groundwater monitoring. Environ Seismol 29(4):43–56

Smith J, Brown A, White R (2020) Coral reef conservation in the great barrier reef: application of GPR to assess reef bed structure and identify vulnerable areas. Mar Geophys Res 41(3):278–292

Telford WM, Geldart LP, Sheriff RE (1990) Applied geophysics, 2nd edn. Cambridge University Press

Thomson C et al (2020) Seismic methods in monitoring coal mining operations. Mining Geophys Rev 39(3):103–114

Tietze K, Ritter O, Veeken P (2015) Controlled-source electromagnetic monitoring of reservoir oil saturation using a novel borehole-to-surface configuration. Geophys Prospect 63(6):1468–1490

Verma RK (1989) Advances in coal geophysics. Association of Exploration Geophysicists, India

Xu L et al (2021) TDEM and resistivity for pollutant detection. Environ Geophys Res 29(5):177–189

Yilmaz O (2001) Seismic data analysis: processing, inversion, and interpretation of seismic data. Society of Exploration Geophysicists

Zhang D et al (2020) Resolution limits of geophysical methods in heterogeneous environments. J Environ Geophy 18(7):232–244

Zhou B, Zhang L (2022) Recent advances in geophysical methods for environmental monitoring and resource management. Environ Monit Assess 194(3):1–16

Zohdy A et al (2020) Using electrical resistivity for groundwater mapping in arid Egypt. Water Resour J 45(11):654–667

Chapter 7
Geophysics for Disaster Risk Management and Hazard Mitigation

Abstract This chapter explains how geophysical tools are used to assess natural hazards and reduce disaster risks. It highlights advanced technologies such as seismic monitoring, geodetic systems (like GNSS and InSAR), and remote sensing tools including LiDAR, satellites, and UAVs. These methods help in early warning, hazard mapping, ground deformation studies, and structural vulnerability assessments. The chapter also presents mitigation strategies like land-use planning and targeted reinforcement. Case studies show real-world applications, including earthquake monitoring in India and California, landslide prediction in South Korea and the Swiss Alps, and flood risk management in Brazil and Indonesia. It also covers microseismic monitoring and the role of machine learning in flood prediction. The chapter emphasizes the integration of various techniques for comprehensive disaster management. These tools support safer infrastructure, faster response, and sustainable planning in hazard-prone areas, making geophysics essential for disaster preparedness and resilience building.

7.1 Introduction

This section provides an overview of the vital role geophysics plays in disaster risk management and hazard mitigation. It outlines how geophysical methods contribute to understanding natural hazards such as earthquakes, landslides, volcanic eruptions, and floods. By offering critical insights into subsurface conditions and early warning indicators, geophysics supports effective risk assessment, planning, and decision-making to reduce vulnerability and enhance resilience to disasters.

D. C. Naskar, *Advances in Engineering Geophysics: Concepts and Applications*, Earth Systems Data and Models 9, https://doi.org/10.1007/978-3-032-28000-8_7

7.1.1 *The Role of Geophysics in Disaster Risk Management and Hazard Mitigation*

This chapter examines the critical role of geophysics in disaster risk management and hazard mitigation, focusing on the use of innovative technologies for assessing and reducing the risks associated with natural hazards such as earthquakes, floods, and landslides. The chapter highlights key geophysical methods including seismology, geodesy, and geophysical surveys, which provide essential data for monitoring and analyzing hazard zones, surface deformations, and seismic activities. It discusses the advances in seismic networks, microseismic monitoring, InSAR (Interferometric Synthetic Aperture Radar), GNSS (Global Navigation Satellite Systems), and remote sensing technologies like LiDAR, satellites, and UAVs, all of which contribute to more accurate hazard forecasting and real-time monitoring.

The chapter further explores mitigation strategies, including structural reinforcement, land-use planning, and the implementation of early warning systems for earthquakes, floods, and landslides. These strategies are reinforced through case studies that demonstrate successful applications of geophysical methods in real-world disaster management scenarios. By integrating geophysical data with other monitoring and early warning systems, communities can enhance preparedness, reduce vulnerabilities, and improve resilience to natural disasters.

In conclusion, the chapter underscores the importance of continuous innovation in geophysical technologies and their integration with disaster risk management frameworks, offering a proactive approach to minimizing the impacts of natural hazards on both infrastructure and human life.

7.2 Innovations in Geophysical Approaches for Natural Hazard Assessment and Risk Mitigation

Geophysical methods are increasingly being integrated into disaster risk management frameworks, providing more accurate, real-time data for hazard prediction, monitoring, and mitigation strategies. The integration of advanced geophysical techniques with data from seismic, geodetic, and remote sensing technologies has resulted in a more nuanced understanding of natural hazards and disaster risks. Innovations in this field are enhancing our ability to assess risks, forecast disasters, and minimize their impacts on communities and infrastructure (Anderson and Ferrini 2022).

Geophysics plays a vital role in identifying, monitoring, and managing natural hazards, including earthquakes, floods, and landslides. By utilizing various geophysical techniques, it is possible to map and assess hazard zones, monitor early warning signals, and develop mitigation strategies. Geophysical data can provide critical insights into the subsurface conditions that underlie these hazards (Barton and McHugh 2021).

Key geophysical methods used in disaster risk management include *seismology, geodesy*, and *geophysical surveys*. Seismology helps monitor and analyze seismic activity, providing critical data for earthquake prediction and early warning systems. Geodesy, through tools like GNSS and InSAR, measures ground deformation and surface movement, essential for tracking tectonic shifts and subsidence. Geophysical surveys—such as ground-penetrating radar, electrical resistivity, and magnetometry—offer detailed subsurface imaging to identify vulnerable zones prone to earthquakes, floods, and landslides, supporting more informed mitigation strategies.

7.2.1 Seismology

Earthquake prediction and risk assessment often rely heavily on seismic activity analysis, using seismometers and networks to monitor ground shaking (Ali and Raj 2023a, b, c). Seismology, the study of seismic waves generated by natural and anthropogenic events, is fundamental to earthquake risk assessment and mitigation.

Seismic Networks and Earthquake Monitoring

The primary objective of seismology is to measure the strength, duration, and location of seismic events, which helps scientists understand fault line behavior and predict potential future seismic hazards. Over time, advancements in seismic monitoring and analysis techniques have significantly improved our ability to assess earthquake risks and develop mitigation strategies.

Modern seismic networks are composed of numerous seismometers strategically distributed across the globe to detect and record seismic waves from earthquakes, volcanic activity, and other seismic events. These networks consist of local, regional, and global stations that work in unison to monitor seismic activity in real-time. The data collected by seismic stations are used to calculate earthquake magnitude, determine epicenter location, and assess ground shaking intensity. This information is crucial for earthquake risk assessment and for informing mitigation efforts. The global network of seismic stations, such as the United States Geological Survey (USGS) and European-Mediterranean Seismological Centre (EMSC), enables continuous monitoring and rapid dissemination of data (Zhao et al. 2020).

Key elements of seismic networks include *seismometers, seismic arrays, real-time data analysis, and the integration of advanced, high-density monitoring systems*. Seismometers are essential instruments in seismic networks that detect and record ground motion caused by seismic waves. These devices are highly sensitive, capable of picking up even the smallest tremors. This sensitivity is crucial for understanding the frequency and intensity of seismic events. Modern seismometers can detect minor tremors, allowing scientists to monitor fault line activity with greater precision and accuracy. These advancements have significantly improved earthquake forecasting and risk assessment (Bouchon et al. 2022).

Seismic arrays are groups of seismometers placed close together to enhance the detection of seismic activity. By combining the data from multiple seismometers, seismic arrays provide a higher resolution of data on earthquake dynamics. This arrangement allows for a more detailed understanding of seismic events, including their location, magnitude, and characteristics. The high density of sensors in an array enables scientists to detect subtle shifts in the Earth's crust, offering critical insights into the behavior of faults.

Real-time data analysis plays a critical role in improving earthquake early warning systems. By processing seismic data as it is received, real-time analysis provides timely information on ongoing seismic events. This allows early warning systems to issue alerts, giving people the opportunity to take protective actions before the shaking begins. These systems have proven to reduce the loss of life and property damage, especially in earthquake-prone regions (Ghofrani et al. 2021).

Recent innovations in seismic networks, such as the EarthScope network in North America and the Global Seismographic Network (GSN), have greatly improved earthquake monitoring capabilities. These dense seismic networks are designed to detect even the smallest seismic events. They provide high-resolution, real-time data, which enhances hazard assessments and improves early warning systems. The increased number of sensors allows for more accurate and timely detection of earthquakes, helping scientists better understand seismic risks (Ali and Raj 2023a, b, c).

The expansion and advancement of seismic networks have significantly improved the ability to monitor seismic activity. Dense networks like EarthScope and the GSN allow for more reliable detection of seismic events and provide high-resolution data. These systems improve earthquake detection, help understand fault-line behavior, and enhance early warning capabilities. With continued technological advancements, seismic networks contribute to safer communities by enabling quicker responses to seismic hazards.

Earthquake Early Warning Systems (EEWS)

Seismology also plays a critical role in the development of Earthquake Early Warning Systems (EEWS). These systems rely on the detection of the initial seismic waves from an earthquake, known as P-waves (Primary waves), which travel faster than the damaging S-waves (Secondary waves). Once P-waves are detected, early warning systems can send alerts to populations at risk, providing valuable seconds or even minutes of warning before the more destructive shaking occurs (Araujo et al. 2020). This brief period of warning can be used to trigger automatic safety measures, such as shutting down industrial machinery, halting trains, or activating seismic-resistant building features, thereby reducing the loss of life and infrastructure damage.

Seismic Hazard Assessment

Seismology is essential for understanding the behavior of faults and the probability of future earthquakes, which forms the foundation of seismic hazard assessment. By studying seismic waves, scientists can gain insights into the structure of fault zones and their activity, helping to identify seismic hazards in regions that may be susceptible to major earthquakes (Field et al. 2019).

One key method used in seismic hazard assessments is *Seismic Ground Motion Prediction*, which involves the use of historical earthquake data, geological studies, and advanced modeling to predict the level of ground shaking that could occur during future earthquakes. Seismologists combine these data to estimate how strong the shaking might be in different areas. This information is vital for designing buildings and infrastructure that can withstand earthquakes. It also helps to create effective disaster response plans that can save lives and minimize damage when an earthquake occurs (Kossobokov et al. 2021).

Another important method is *Fault Characterization*, the process of analyzing seismic waves to map fault lines and assess their activity. By studying these waves, scientists can identify areas with higher earthquake risk and evaluate the potential magnitude and frequency of future seismic events. This helps to understand where earthquakes are more likely to occur and provides valuable insights into the behavior of faults. Such information is crucial for improving earthquake preparedness and risk management strategies (Peppe et al. 2021).

Ground Shaking and Site Response Studies

Seismology also plays a crucial role in understanding site response—how different soils and rock types respond to seismic waves. For instance, loose sediments or unconsolidated soil can amplify seismic waves, leading to stronger ground shaking in urban areas built on these substrates. By analyzing ground motion data, seismologists can develop site response maps that identify which areas are most vulnerable to earthquake-induced ground shaking (Tromp and Santillan 2021; Bouchon et al. 2022).

These site response studies are essential for urban planning and construction in earthquake-prone areas, ensuring that infrastructure is built to withstand seismic activity. For example, buildings in high-risk areas may need to be designed with enhanced seismic resilience to reduce damage during strong earthquakes.

Seismological Research and Fault Zone Characterization

In addition to monitoring seismic activity, seismology is used to conduct in-depth studies of active fault zones. Faults are fractures in the Earth's crust where tectonic plates meet, and they are responsible for the majority of seismic events. Seismologists use refraction and reflection seismic surveys to study the geometry and behavior of these fault zones, helping to predict where future earthquakes may occur. This information is vital for identifying hazard zones and guiding mitigation efforts.

Seismic Tomography is one such method, where seismologists use seismic waves to create 3D images of the subsurface. These images can reveal the properties of fault zones, including their potential to produce major seismic events (Barton and McHugh 2021; Zhao et al. 2020).

In conclusion, *seismology* is vital to earthquake risk mitigation, offering critical insights into seismic hazards and supporting early warning systems and resilient infrastructure design. Advanced monitoring, real-time analysis, and localized studies enable more effective risk reduction, saving lives and limiting economic losses. Ongoing advancements in seismic technology and data integration will continue to strengthen global earthquake preparedness and resilience.

7.2.2 Geodesy

Techniques like GPS and InSAR (Interferometric Synthetic Aperture Radar) are used to measure ground deformation, which is especially useful for monitoring post-seismic land movements, landslides, and flood-prone areas (Tromp and Santillan 2021). Geodesy, plays a crucial role in hazard assessment and risk mitigation, particularly in the context of earthquakes, ground deformations, and landslides. Modern advancements in space-based geodetic technologies, such as Global Navigation Satellite Systems (GNSS), Interferometric Synthetic Aperture Radar (InSAR), and gravity measurements, allow for high-resolution, real-time monitoring of surface movements, enabling accurate risk assessments and the development of effective mitigation strategies.

Monitoring Surface Deformation with GNSS

Global Navigation Satellite Systems (GNSS), which include the Global Positioning System (GPS), are widely used in geodesy to monitor small but significant movements of the Earth's surface. GNSS provides real-time data on ground displacement, which is particularly valuable for detecting slow tectonic movements, such as continental plate movements, fault slip, and crustal deformation due to earthquakes.

In terms of earthquake monitoring, GNSS stations are capable of detecting ground displacement even during small earthquakes. This ability allows scientists to monitor fault behavior over time. GNSS data is particularly useful in identifying pre-seismic deformation, which occurs when the Earth's surface moves slightly before a large earthquake. This can offer important clues about the build-up of stress along fault lines, potentially aiding in early warning systems and improving our understanding of earthquake dynamics (Jin et al. 2023).

For crustal motion analysis, GNSS data is invaluable for studying the relative movement between different stations. By analyzing these movements, scientists can map tectonic plate boundaries and identify areas where stress is accumulating. This information helps pinpoint regions where future seismic events may occur. Understanding these areas of stress accumulation is critical for predicting seismic hazards and for assessing the potential risk of future earthquakes (Zhou et al. 2022a, b).

Recent advancements in GNSS technology have significantly improved the accuracy and coverage of geodetic networks. These improvements provide more reliable and high-resolution data, which are crucial for earthquake prediction and disaster risk management. With better data, scientists can make more accurate assessments of seismic risks, allowing for better preparedness and mitigation strategies to reduce the impact of earthquakes (Cai et al. 2023).

InSAR: Detecting Ground Deformation from Space

Interferometric Synthetic Aperture Radar (InSAR) is another powerful tool used in geodesy for monitoring surface displacement, especially in areas prone to seismic activity or landslides. InSAR uses radar images from satellites to detect ground deformation with millimeter precision.

In terms of earthquake fault analysis, InSAR is useful for detecting post-seismic deformations—the movement of the Earth's surface following an earthquake. By analyzing these changes, scientists can assess the extent of surface rupture and determine the magnitude of the earthquake. This information helps in understanding the long-term effects of seismic events and in improving earthquake models (Rangin et al. 2021).

InSAR also plays a significant role in landslide monitoring. It can detect and monitor deformation caused by landslides, offering early warnings for potential landslide events. This early detection is essential for guiding mitigation efforts, such as evacuations or the reinforcement of structures, to minimize risk to human life and property (Bingen et al. 2023).

InSAR is also highly effective in identifying subsidence and volcanic activity. It can detect land subsidence caused by natural hazards like earthquakes and volcanic activity. For instance, InSAR can identify surface movement caused by magma shifting beneath the Earth's surface, which can result in the rise or sinking of the ground. Monitoring this deformation is critical for understanding volcanic behavior and for predicting potential eruptions (Li et al. 2022).

The use of multi-temporal InSAR enables scientists to track surface displacements over time, providing insights into both slow and rapid deformations. These time-series data can reveal patterns of increasing seismic or volcanic risk, helping scientists better understand the dynamics of natural hazards and improve disaster preparedness.

Geodetic Monitoring for Early Warning Systems

Geodesy is playing an increasingly important role in early warning systems for natural hazards, especially earthquakes, tsunamis, and landslides. By measuring surface movements and understanding the velocity of tectonic plate movements, geodetic systems provide valuable data that help predict imminent seismic events. This data is essential for informing early warning protocols and improving disaster preparedness.

In tsunami warning, geodetic systems monitor vertical and horizontal displacements of the seabed, which can indicate potential tsunami-generating earthquakes. By using systems such as GNSS and tide gauges, scientists can detect these displacements and issue early warnings to coastal populations. This timely information helps reduce the risk to human life and property by allowing for evacuations and other precautionary measures (Roecker et al. 2022).

For landslide monitoring, geodetic systems like terrestrial laser scanning (TLS) and GNSS-based monitoring are used to observe slopes and detect movements that could signal an impending landslide. These systems can provide early alerts when significant surface deformations are detected, giving communities time to evacuate or take other safety measures to minimize risks (Guglielmi et al. 2022).

By integrating geodetic measurements into early warning systems, scientists can improve their ability to predict natural hazards and help mitigate their impacts on affected populations. These technologies are essential for enhancing disaster resilience and ensuring public safety in hazard-prone regions.

Gravity Measurements and Hazard Monitoring

Gravity measurements are an important aspect of geodesy, providing insights into the Earth's gravitational field and its variations caused by changes in subsurface mass. These variations are useful for studying fault zone stability (Naskar and Lashkari 2019) and detecting subsurface fluid movements. Such movements, like magma migration or groundwater extraction, can lead to natural hazards such as earthquakes or landslides.

In earthquake prediction, gravity measurements can help detect pre-seismic changes in the Earth's crust as it accumulates stress before an earthquake. These subtle changes in gravity, known as gravity anomalies, can indicate areas of potential seismic activity. By identifying these anomalies, scientists can better understand where earthquakes may occur and improve early warning systems (Tregoning et al. 2021).

For subsurface fluid monitoring, gravity measurements are valuable for tracking changes caused by fluid extraction, such as from oil or gas reservoirs, and geothermal activity. These activities can lead to subsidence or even trigger earthquakes. By monitoring gravity changes in these areas, scientists can assess the potential risks and take preventive measures (Khan et al. 2021).

Recent advancements in satellite gravity measurements, particularly from missions like GRACE (Gravity Recovery and Climate Experiment), have improved the sensitivity of gravity measurements. These advancements allow scientists to detect even the smallest variations in the Earth's gravitational field, which enhances our ability to monitor subsurface activities and better predict natural hazards (Aly et al. 2022).

Case Study: Geophysical Investigation in Tonk and Dausa, Rajasthan, India

The Tonk and Dausa districts in Rajasthan are geologically significant areas with complex subsurface structures. This case study involves a geophysical investigation aimed at identifying structural features and assessing potential hazards in these regions. Understanding the geological framework of these regions is essential for assessing natural hazards and mitigating risks associated with subsurface conditions. Geophysical methods, particularly gravity measurements, offer a cost-effective and non-invasive approach to mapping subsurface geological features that could lead to natural disasters (earthquakes, landslides, etc.) (Geological Survey of India 2013). The Bouguer gravity anomaly map is a critical tool in this regard, as it provides valuable insights into the density distribution of geological units beneath the surface, aiding in regional geological analysis and hazard assessment (Nagarajan et al. 2010; Singh et al. 2013).

The primary objective of this study is to analyze the Bouguer gravity anomaly map of Tonk and Dausa districts to identify key subsurface structures and interpret their geological implications. By examining gravity anomalies, this study aims to better understand the distribution of high and low-density rock formations, assess potential geological hazards, and provide insights into the regional tectonic evolution.

The method involved the collection of Bouguer gravity data over the Tonk and Dausa districts, using gravity meters to measure variations in gravitational force across the region. The data was plotted and contoured to create a Bouguer gravity anomaly map, which was analyzed for significant gravity highs and lows. The interpretation focused on understanding the structural patterns and geological units that could be causing these anomalies, such as quartzite and other metasedimentary rocks, as well as potential high-density intrusive bodies (Rao and Desai 2015). Additionally, the Free Air anomaly was analyzed to identify topographic influences superimposed on the subsurface anomalies.

The gravity data collected was corrected for various factors, including topographic and instrumental corrections, to generate accurate Bouguer anomalies. The Free Air anomaly was also processed to highlight the influence of topographic features and to help distinguish between the effects of surface features and deeper geological structures. The data was then contoured to create an anomaly map that visually represents variations in gravity across the region. This map was further analyzed to identify gravity lows and highs, with specific attention given to the trend directions and magnitudes of anomalies, which could suggest the presence of different geological formations and structures beneath the surface.

The Bouguer gravity anomaly map revealed significant variations in gravity values across the study area, ranging from − 76 mGal near Lalsot to − 28 mGal near Jaipur (Fig. 7.1). These variations indicate a complex subsurface, with gravity highs associated with massive quartzite formations and gravity lows potentially linked to low-density rock formations such as quartzite and metasediments. The gravity highs, particularly in the northwest around Jaipur, correspond to areas with high-density intrusive bodies, while the gravity lows in the southeast suggest low-density

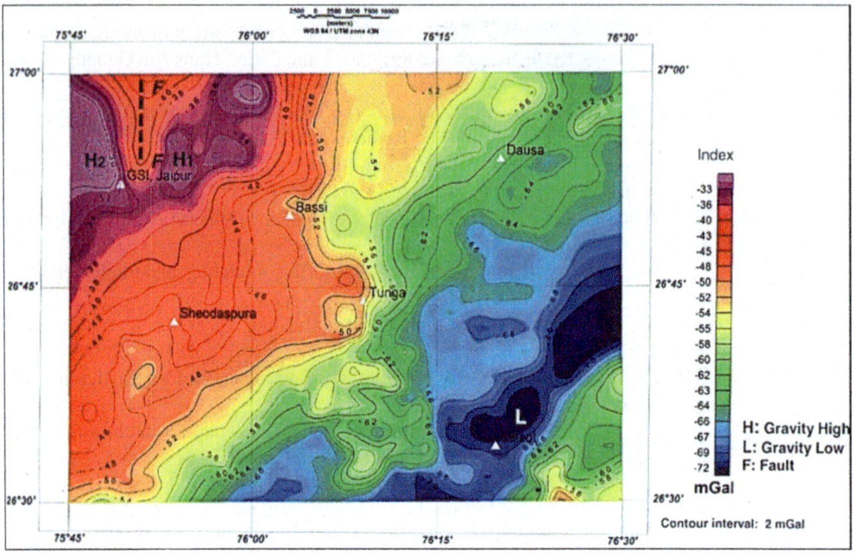

Fig. 7.1 Bouguer gravity anomaly map of Tonk and Dausa districts, Rajasthan (Toposheets 45 N/13, 14 and 54 B/1, 2, 5, 6) for risk and hazard mitigation analysis. *Source* Geological Survey of India (2013)—Naskar and Lashkari

formations. The map also reveals a series of folded and faulted structures, with V-shaped gravity anomalies indicating folded geological formations. The distribution of gravity anomalies supports the geological interpretation of alternating antiformal and synformal structures in the region, with faults potentially having developed during later stages of tectonic evolution (Sharma et al. 2017).

In addition to the Bouguer anomaly, the Free Air anomaly map (Fig. 7.2) provides valuable information about topographic influences and heterogeneity in the subsurface. The Free Air anomaly is a key indicator of the crustal signatures, and it exhibits alternating bands of highs and lows that reveal the spatial distribution of geological features. Notably, the first strong gravity low axis (L-L′) trending NE-SW from Kothun to Dausa marks a significant zone of isolated lows. These gravity lows highlight areas where the subsurface may exhibit lower density formations or topographic features influencing the anomaly patterns. The systematic increase of gravity anomalies from the southeastern to the northwestern parts of the area further indicates the presence of subsurface heterogeneity, which is crucial for understanding potential risks in the region.

The Free Air anomaly's role in revealing topographic influences on the gravitational field makes it a valuable tool for hazard assessment and risk mitigation. By identifying areas where surface topography significantly impacts gravity measurements, the Free Air anomaly can guide further geological investigations and help in the planning of infrastructure or land-use strategies in regions that may be susceptible to natural hazards such as landslides or seismic activity.

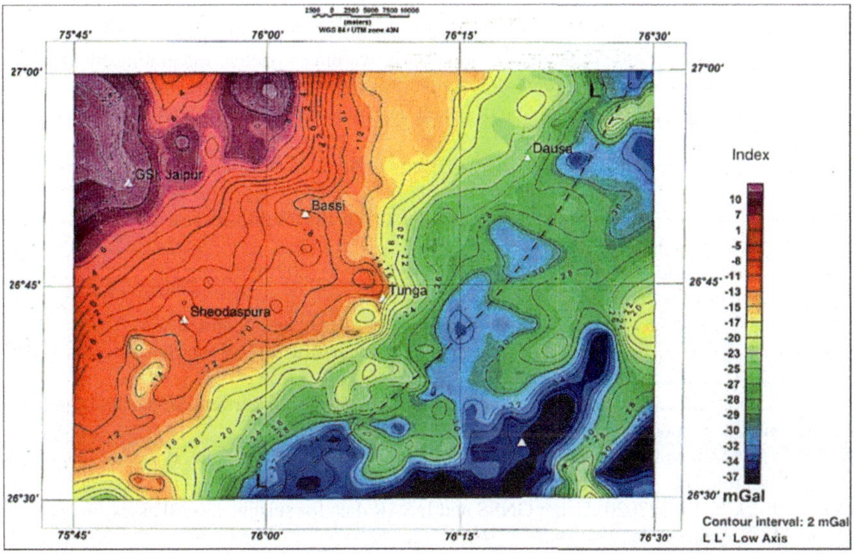

Fig. 7.2 Free air bouguer gravity anomaly map of toposheets 45 N/13, 14 and 54 B/1, 2, 5, 6 in Tonk and Dausa districts, Rajasthan, for risk and hazard mitigation analysis. *Source* Geological Survey of India (2013)—Naskar and Lashkari

The geophysical study successfully identified key structural features in the Tonk and Dausa districts, including high-density intrusive bodies and low-density formations. The gravity anomaly map not only highlighted the complexity of the subsurface geology but also provided valuable data for understanding the region's tectonic history. The study's findings support the interpretation of a folded and faulted geological framework, which is crucial for assessing geological hazards and potential risks in the area. Furthermore, the integration of Free Air anomalies with Bouguer gravity data helped to refine the understanding of surface and subsurface features, enhancing risk mitigation efforts.

This study demonstrates the significant role of geophysical methods, particularly gravity surveys, in understanding regional geology and assessing natural hazards. The Bouguer gravity anomaly map provided a detailed view of the subsurface geological structures in Tonk and Dausa districts, offering insights into the distribution of various geological units and their potential to influence hazards such as earthquakes or land stability. The Free Air anomaly map further helped to identify topographic influences and surface heterogeneity, which are critical for refining hazard assessments and developing risk mitigation strategies. The findings from this study can be used to inform future geological research, hazard mitigation strategies, and land-use planning in the region.

The primary audience for this study includes geophysicists, geologists, and environmental scientists involved in regional geological assessments and hazard monitoring. Additionally, the study is relevant to policymakers and urban planners in

Rajasthan, who can use the findings to inform infrastructure development and disaster risk management in the region.

In conclusion, *geodesy* plays a crucial role in disaster risk mitigation by providing real-time, high-resolution data to monitor ground deformation, fault activity, and processes like subsidence and landslides. Techniques such as GNSS and InSAR support early warning systems and informed decision-making. As technology advances, geodesy remains integral to hazard assessment and proactive disaster management efforts.

7.2.3 Geophysical Surveys

Resistivity, magnetic, and gravitational surveys help assess soil composition and groundwater conditions that influence landslide and flood risks. Geophysical surveys encompass a range of techniques designed to investigate the Earth's subsurface and detect anomalies that could indicate the presence of natural hazards. These surveys include methods such as *seismic reflection and refraction, magnetic surveys, electrical resistivity tomography (ERT), and ground-penetrating radar (GPR).* They are commonly used to map faults, assess groundwater resources, detect landslide-prone areas, and monitor volcanic activity.

Seismic Surveys: Seismic Reflection and Refraction

Seismic surveys are one of the most widely used geophysical methods for subsurface imaging, particularly for identifying fault zones, earthquake-prone areas, and assessing earthquake hazards. Seismic surveys utilize the propagation of seismic waves through the Earth's subsurface. These waves are reflected and refracted by various layers of rock, and their return signals are recorded to map subsurface structures.

Fault zone mapping is one of the key applications of seismic surveys. These surveys help identify active fault zones by providing critical information about the geometry and behavior of fault systems. Understanding how faults behave allows scientists to assess the likelihood of future seismic events. This information is vital for assessing earthquake risk in different regions and developing effective mitigation strategies (Dixon et al. 2023).

Earthquake risk assessment is another important application of seismic surveys. These surveys can provide high-resolution images of subsurface layers, helping geophysicists assess the potential for large earthquakes in specific areas. The data collected from seismic surveys is crucial for creating seismic hazard maps, which are essential for designing earthquake-resistant infrastructure. These maps help ensure that buildings and infrastructure are built to withstand potential seismic events (Chen et al. 2022).

Recent advancements in high-density seismic arrays and deep seismic reflection techniques have greatly improved the resolution of seismic surveys. These improvements provide more accurate and detailed data, which is crucial for effective seismic hazard assessment and the development of mitigation strategies. With better data, scientists and engineers can design safer, more resilient infrastructure and improve preparedness for earthquake events (Zhang et al. 2021a, b).

Ground-Penetrating Radar (GPR)

Ground-penetrating radar (GPR) is a non-invasive geophysical method that uses high-frequency electromagnetic waves to investigate the shallow subsurface. GPR is commonly used in landslide-prone areas, fault zones, and to assess infrastructure stability.

Landslide risk assessment is one important application of Ground Penetrating Radar (GPR). GPR is used to identify zones of weakness in rock and soil that may be prone to landslides. This method is highly effective for imaging fractures, soil–water interaction, and subsurface heterogeneity in landslide-prone regions. By detecting these weaknesses, GPR helps in assessing the potential risk of landslides and planning appropriate mitigation measures (Hudson et al. 2023).

In urban infrastructure monitoring, GPR can be employed to assess the integrity of underground infrastructure, such as pipelines and utilities, in areas at risk of earthquakes or landslides. This helps ensure that critical infrastructure is designed to withstand potential seismic hazards. By regularly monitoring the condition of underground structures, GPR aids in reducing the risk of damage and ensuring the safety and functionality of urban systems during natural disasters (Massey et al. 2022).

Recent advancements in multi-frequency GPR technology have significantly increased the depth of penetration, allowing scientists to study deeper subsurface layers. This advancement improves both landslide prediction and infrastructure assessment, making it possible to monitor larger areas and detect deeper structural weaknesses. These improvements help enhance hazard mitigation efforts by providing more accurate data for planning and decision-making (Liu et al. 2021).

Electrical Resistivity Tomography (ERT)

Electrical resistivity tomography (ERT) is a geophysical method that uses electrical currents to measure the resistivity of subsurface materials. This technique provides detailed images of the Earth's subsurface, which can help identify areas of groundwater accumulation, fault lines, and unstable ground conditions.

Landslide monitoring is an important application of Electrical Resistivity Tomography (ERT). ERT is particularly effective in detecting moisture accumulation and pore pressure in soil, both of which are key factors in triggering landslides. By monitoring changes in subsurface resistivity over time, geophysicists can identify areas where the risk of landslides is increasing. This information helps in managing and mitigating landslide hazards by allowing for early warnings and targeted interventions (Nguyen et al. 2022a, b).

Fault zone characterization is another important use of ERT. It is used to map subsurface resistivity variations in fault zones (Naskar 2005). In seismic-prone areas, this technique helps assess the activity of faults and their potential to generate large earthquakes. By identifying resistivity patterns that indicate fault movement or stress accumulation, ERT provides valuable data for earthquake risk assessment and hazard mitigation (Martinez et al. 2021).

The use of time-lapse ERT has recently emerged as a promising tool for monitoring dynamic changes in subsurface resistivity. This method provides real-time insights into potential hazards by continuously observing variations in subsurface conditions. Time-lapse ERT can be used to track changes in areas susceptible to landslides or seismic activity, offering an effective means of monitoring hazard evolution and improving disaster preparedness (Duchkov et al. 2023).

Magnetic Surveys

Magnetic surveys are used to measure the Earth's magnetic field and detect anomalies caused by variations in the magnetic properties of rocks and minerals. This technique is often used in regions prone to volcanic activity, fault zones, and landslides, where magnetic anomalies can indicate significant geological features (Naskar et al. 2008).

Volcanic activity monitoring is one of the key applications of magnetic surveys. Magnetic surveys can detect variations in the Earth's magnetic field caused by the movement of magma beneath the surface. These magnetic anomalies are often linked to volcanic activity and can provide early warning signs of potential eruptions. By monitoring these changes, scientists can better predict volcanic eruptions and take preventive measures to protect nearby populations (Lopez et al. 2021).

Fault and fracture zone identification is another important use of magnetic anomalies. Magnetic surveys help map fault and fracture zones that may be prone to seismic activity. By studying the distribution of these anomalies, geophysicists can identify areas of tectonic activity and assess the associated earthquake risks. Understanding these zones aids in evaluating the potential for seismic events and planning appropriate hazard mitigation strategies (Chen et al. 2022).

Recent airborne magnetic surveys have improved the efficiency and coverage of magnetic anomaly mapping across large areas. These surveys enable scientists to more comprehensively detect subsurface features associated with natural hazards, enhancing the ability to monitor and assess volcanic and seismic risks over broad regions (Wang et al. 2022).

In conclusion, *geophysical surveys* are vital for disaster risk mitigation, providing critical insights into subsurface conditions through methods like seismic surveys, GPR, ERT, and magnetic surveys. These tools help identify fault zones, monitor landslide risks, and detect early warning signs. As technology and data processing improve, geophysical surveys will continue to enhance hazard assessment, support early warnings, and strengthen community resilience.

7.3 Mitigation Strategies

Structural reinforcement is a vital mitigation strategy aimed at reducing the impacts of natural disasters, particularly in areas vulnerable to liquefaction and subsidence. Geophysical techniques play a crucial role in identifying regions at risk, enabling the implementation of targeted reinforcement strategies for buildings and infrastructure. By utilizing non-invasive geophysical methods such as seismic refraction, ground-penetrating radar (GPR), and electrical resistivity tomography (ERT), engineers can obtain detailed subsurface information to assess ground conditions and tailor structural reinforcements accordingly (Anderson and Ferrini 2022). These innovations in geophysics facilitate the optimization of resources and enhance the overall resilience of structures.

7.3.1 Identifying Vulnerabilities Through Geophysics

Liquefaction occurs when saturated soils lose strength due to ground shaking, posing a significant risk to buildings and infrastructure during an earthquake. Geophysical methods are used to map areas at risk of liquefaction by assessing soil composition, porosity, density, and groundwater conditions (Wang et al. 2023). For example, seismic refraction surveys can help identify soil layers prone to liquefaction, while GPR can provide high-resolution data about subsurface conditions, such as the depth of the water table and the composition of soil layers (Zhang et al. 2021a, b). These tools are instrumental in pinpointing areas that require special attention in the design of resilient infrastructure.

In addition to liquefaction, subsidence—caused by factors like groundwater extraction or natural sedimentation—can lead to the gradual sinking of the ground, destabilizing structures. Geophysical surveys, particularly those that utilize time-lapse techniques and geodetic measurements, can detect early signs of subsidence and allow engineers to plan effective structural reinforcement measures (Nguyen et al. 2022a, b). Monitoring these subsurface changes ensures timely intervention to prevent structural damage.

7.3.2 Targeted Structural Reinforcement

Once vulnerable areas are identified, appropriate structural reinforcement strategies can be designed and implemented. One common method is the installation of deep foundations, such as piles or caissons, which anchor buildings to stable soil or bedrock, reducing the risk of tilting or settling in liquefaction-prone areas (Kumar and Singh 2021). Additionally, soil stabilization techniques, including grouting or compaction, can be employed to increase the soil's shear strength and minimize the risk of liquefaction and subsidence (Wang et al. 2023).

In some cases, retrofitting existing buildings with base isolators or energy dissipation systems is recommended to absorb seismic energy and reduce the forces transmitted to the structure. By integrating geophysical data, engineers can optimize the design and placement of these reinforcement elements, ensuring that they provide maximum benefit in reducing disaster risks (Zhang et al. 2021a, b).

7.3.3 Monitoring and Ongoing Assessment

One of the most significant recent advancements in geophysics is the ability to continuously monitor reinforced structures. Using integrated geophysical sensors and real-time data collection methods, engineers can track changes in the ground's properties and the behavior of the reinforced structure over time (Kumar and Singh 2021). For example, seismic sensors embedded in foundations or adjacent soil layers can provide valuable information about potential liquefaction or subsidence events, allowing for quick assessments and adjustments to the mitigation measures if necessary.

Furthermore, the integration of geophysical methods into long-term structural health monitoring enables proactive decision-making. In flood-prone areas, geophysical surveys are increasingly used to evaluate soil permeability and compaction, which in turn informs the design of flood-resistant infrastructure, such as elevated buildings and flood barriers (Anderson and Ferrini 2022).

7.3.4 Case Studies and Practical Applications

Recent studies have demonstrated the effectiveness of geophysical surveys in guiding structural reinforcement. In California, Anderson and Ferrini (2022) conducted a study in liquefaction-prone regions where they employed seismic refraction and borehole testing to identify areas at high risk. Based on this data, they reinforced buildings using deep foundation piles and soil stabilization techniques to mitigate the risk of liquefaction during seismic events. This case study highlights how targeted structural reinforcement can significantly reduce damage and improve resilience.

Similarly, Zhang et al. (2023) used geophysical data to assess flood risks and design effective structural reinforcement strategies for coastal regions in Southeast Asia. They utilized ground-penetrating radar to evaluate soil conditions and assess potential liquefaction and subsidence risks in flood-prone areas, which allowed for the installation of specialized flood barriers and elevated building foundations. This proactive use of geophysical techniques played a crucial role in enhancing the region's resilience to extreme flooding events.

7.3.5 Land Use Planning for Hazard Reduction

Land use planning is a strategic process to manage land resources that ensures sustainable development, minimizes risks associated with natural hazards, and reduces disaster-related losses. This approach integrates scientific knowledge of hazards with community needs and development objectives. Mapping fault lines, floodplains, and landslide-prone areas is essential for urban planning to avoid high-risk zones (Ali and Raj 2023a, b, c). Risk identification and assessment involve mapping hazard-prone areas such as seismic zones, floodplains, and unstable slopes, alongside identifying vulnerable populations, infrastructure, and ecosystems (Smith et al. 2021). Zoning regulations are implemented to restrict high-risk activities in hazard-prone regions and designate safe zones for critical infrastructure like hospitals, schools, and emergency shelters (UNDRR 2022).

Environmental conservation plays a key role in mitigating natural hazards by preserving natural buffers such as wetlands, forests, and mangroves, which help reduce flooding and landslides, while reforestation in degraded areas aids in stabilizing soil and preventing erosion (Ali and Raj 2023a, b, c). Community participation is vital, involving local residents in decision-making to incorporate indigenous knowledge and ensure compliance, while public education about the importance of adhering to land use policies further strengthens the process (UNDRR 2022). Hazard-sensitive development focuses on limiting or prohibiting construction in high-risk areas like floodplains and unstable slopes and encourages vertical urban growth instead of horizontal expansion to preserve open spaces (Ali and Raj 2023a, b, c).

Infrastructure planning ensures the design of hazard-resistant roads, bridges, and utilities, while open spaces are allocated for evacuation routes and emergency response (Smith et al. 2021). Land acquisition and relocation convert high-risk zones into green spaces or buffer zones and move communities to safer areas to reduce vulnerability (UNDRR 2022). Integrated planning tools such as Geographic Information Systems (GIS) and remote sensing are critical for hazard mapping and risk analysis, assisting in developing land use models that simulate potential hazard impacts under different development scenarios (Smith et al. 2021).

Land use planning significantly reduces risk by decreasing the exposure of people and assets to natural hazards and lowering the frequency and intensity of disasters through preventive measures (Ali and Raj 2023a, b, c). Sustainable development is supported by balancing economic growth with environmental conservation, creating safer living environments (UNDRR 2022). Cost efficiency is achieved by minimizing expenses related to disaster response, recovery, and reconstruction while promoting resilient infrastructure (Smith et al. 2021). Environmental protection is enhanced by conserving ecosystems that act as natural barriers against hazards and implementing climate adaptation strategies (Ali and Raj 2023a, b, c).

Challenges in land use planning include data gaps due to limited availability of accurate and up-to-date hazard and risk data, along with compliance issues often arising from resistance by developers and communities to follow regulations (Smith et al. 2021). Political and economic constraints can make balancing short-term economic pressures with long-term sustainability difficult (UNDRR 2022), while rapid urbanization poses additional challenges as unchecked growth often extends into hazard-prone areas (Smith et al. 2021). Japan provides an example of effective earthquake-resistant zoning by enforcing strict building codes in seismic zones and developing evacuation-friendly open spaces (Ali and Raj 2023a, b, c). The Netherlands demonstrates successful flood management with zoning laws restricting development in flood-prone areas and implementing "Room for the River" projects to enhance floodplain functionality (UNDRR 2022). In the Himalayan region, construction on unstable slopes is prohibited, and afforestation programs stabilize the soil to prevent landslides (Smith et al. 2021).

Land use planning is a critical tool for mitigating the impacts of earthquakes, floods, and landslides by integrating hazard assessments, zoning regulations, and community participation, ultimately fostering safer, more sustainable, and resilient development.

7.3.6 Early Warning Systems for Earthquakes

Early warning systems (EWS) are a critical component of disaster risk management and hazard mitigation. These systems provide timely alerts about potential natural disasters, allowing individuals, communities, and authorities to prepare and respond effectively, reducing loss of life and property. In the context of earthquakes, floods, and landslides, geophysical technologies form the backbone of early warning systems by detecting precursor signals and monitoring environmental changes. Real-time geophysical monitoring can be integrated into early warning systems for earthquakes, floods, and landslides (Lamb and Brunt 2024).

Earthquake Early Warning Systems (EEWS) operate by detecting the primary (P) waves generated during an earthquake, which travel faster than the secondary (S) waves and surface waves responsible for most of the damage. Sensors installed along fault lines or in seismic hotspots continuously monitor ground motion. Once an earthquake is detected, the system analyzes its location, magnitude, and intensity to predict the potential impact (Zhou et al. 2022a, b).

The primary components of EEWS include high-sensitivity seismic sensors, communication networks, and alert systems. Seismic sensors such as accelerometers detect even minute ground movements, while real-time data transmission to processing centers allows for rapid analysis. Alert systems, which utilize sirens, text messages, mobile applications, and media broadcasts, are critical for disseminating warnings to affected regions (Fukuyama et al. 2021; Park et al. 2021).

EEWS provide several seconds to minutes of warning before the damaging waves arrive, depending on the distance from the earthquake's epicenter. This brief window allows for shutting down critical infrastructure such as power plants, gas lines, and transportation systems, reducing the risk of catastrophic failures. It also enables individuals to take protective measures, such as evacuating buildings or moving to safer locations. Operational EEWS in Japan, Mexico, and California have demonstrated the system's potential to mitigate earthquake risks effectively (Kim et al. 2022).

7.3.7 *Early Warning Systems for Floods*

Flood Early Warning Systems (EWS) are designed to monitor hydrological and meteorological conditions to predict and provide warnings about potential flooding events. These systems utilize a combination of real-time data from river gauges, rainfall sensors, and weather forecasts, alongside hydrological models that simulate water flow dynamics. By integrating this information, flood EWS can assess the likelihood of flooding and provide timely alerts (Smith et al. 2021).

Remote sensing technologies, including satellite imagery and drone-based observations, play a vital role in monitoring rainfall, snowmelt, and river levels. River gauges and rainfall sensors provide continuous real-time data on water levels and precipitation intensity, which are critical inputs for predictive hydrological models. These models simulate water flow in river basins and assess the potential for flooding based on rainfall patterns and the physical characteristics of the basin (Gou et al. 2022).

Flood EWS are instrumental in issuing warnings about various types of floods, including flash floods, riverine floods, and coastal inundation. These warnings help inform evacuation plans for communities in at-risk areas and guide the deployment of emergency services to minimize human and economic losses. Additionally, real-time data from these systems assist in the active management of reservoirs and flood control infrastructure, helping to mitigate the impacts of excessive water flow (Rahman and Singh 2020).

7.3.8 *Early Warning Systems for Landslides*

Landslide Early Warning Systems (EWS) are designed to monitor slope stability and identify early indicators of ground movement, such as those triggered by heavy rainfall, seismic activity, or human interventions. These systems employ a network of sensors to measure critical parameters, including soil moisture, pore pressure, and ground displacement. By integrating real-time data with predictive models, these systems can provide timely alerts to mitigate risks (Intrieri et al. 2019).

The primary components of landslide EWS include geotechnical sensors, geodetic monitoring systems, and rainfall threshold models. Geotechnical instruments such as inclinometers and piezometers monitor slope movements and changes in water pressure within the soil. Geodetic monitoring using GPS and interferometric synthetic aperture radar (InSAR) allows for precise detection of surface deformation and gradual movements over time. Rainfall threshold models analyze precipitation data to predict when soil saturation levels might trigger landslides, enhancing the system's predictive capabilities (Lacroix et al. 2020; Berti et al. 2022).

Landslide EWS are vital for alerting communities in landslide-prone regions about impending slope failures, enabling timely evacuations and risk mitigation. These systems inform decisions on road closures and the installation of temporary protective measures, such as barriers or drainage systems. Additionally, the data from these systems are used for long-term planning, such as stabilizing high-risk zones through engineering solutions and land-use regulations (Rosi et al. 2021).

7.3.9 Integrated Early Warning Systems

Integrated Early Warning Systems (EWS) adopt a multi-hazard approach to address regions exposed to multiple, interconnected risks. These systems consolidate data from seismic, hydrological, and geotechnical networks, enabling a comprehensive risk assessment. For instance, an earthquake might trigger secondary hazards such as landslides or tsunamis, necessitating synchronized alerts to mitigate cascading effects (Gill and Malamud 2016). A well-coordinated multi-hazard system ensures that risks from all potential threats are addressed in a unified framework, improving disaster preparedness and response.

Effective communication is a cornerstone of early warning systems. Robust strategies ensure that warnings reach all stakeholders, including government agencies, emergency services, and local communities. Public awareness campaigns and community engagement initiatives play a critical role in educating people about warning signals and the appropriate response measures. By fostering a culture of preparedness, these systems can significantly reduce disaster impacts (UNDRR 2022).

Integrated EWS face several challenges, including the need to balance accuracy with minimizing false alarms. False alarms can erode public trust, while missed threats may result in catastrophic outcomes. Expanding sensor networks to cover remote or underserved areas is another critical need, as many high-risk regions lack adequate monitoring infrastructure.

Future advancements are likely to be driven by artificial intelligence (AI) and machine learning (ML), which can enhance the speed and precision of data analysis. AI models are particularly effective in handling complex, multi-hazard datasets, providing more accurate predictions and timely warnings. These technologies, coupled with ongoing efforts to improve sensor technology and data integration, will shape the next generation of EWS (Pittore et al. 2018).

7.4 Recent Advances in Seismic, Geodetic, and Remote Sensing Technologies

Recent advancements in seismic, geodetic, and remote sensing technologies have significantly improved disaster risk assessments. These innovations provide higher resolution, more accurate, and real-time data that enhance our ability to predict and mitigate natural disasters.

7.4.1 Advances in Seismic Network Technologies

Networks of seismometers are now more densely distributed across the globe, providing higher-resolution data. Advances in earthquake early-warning systems (EEWS) have been made possible through rapid seismic wave detection and data transmission (Ali and Raj 2023a, b, c).

The deployment of dense seismic arrays has revolutionized seismic monitoring by providing higher-resolution data on ground motion. These arrays, often placed in urban or seismically active regions, enhance the detection of smaller earthquakes and offer precise information about their epicenters, magnitudes, and depths. Dense arrays also enable detailed 3D imaging of subsurface structures, which is critical for comprehensive earthquake hazard assessments. Networks such as the IRIS (Incorporated Research Institutions for Seismology) and USArray have demonstrated the value of such systems, utilizing broadband seismometers to monitor earthquakes and assess seismic risks (Fukuyama et al. 2021; Maxwell et al. 2023).

Broadband seismometers are integral to modern seismic networks due to their ability to record a wide range of seismic frequencies, capturing signals from local earthquakes to distant seismic events. Recent advancements in these sensors have enhanced their sensitivity, reliability, and cost-efficiency. They can now operate in remote and harsh environments while providing real-time data, thereby expanding the scope of seismic monitoring (Zhou et al. 2022a, b).

Modern seismic networks also benefit from real-time seismic data transmission and processing capabilities. High-speed communication links enable data to be sent from seismic stations to central processing hubs instantaneously. This capability underpins Earthquake Early Warning (EEW) systems, which use rapid data transmission to issue alerts within seconds of an earthquake's detection. These alerts are crucial for implementing life-saving measures, such as stopping trains, halting industrial operations, and initiating evacuations, thereby mitigating disaster impacts (Park et al. 2021).

In addition, recent innovations have aimed to make seismic monitoring more accessible and cost-effective. Low-cost seismic sensors, such as MEMS-based accelerometers and smartphone-based seismometers, have emerged as viable tools for crowd-sourced seismic data collection. These affordable solutions are particularly valuable for regions with limited financial resources and for expanding seismic network coverage in remote or sparsely populated areas (Kim et al. 2022).

7.4.2 Seismic Networks for Earthquake Forecasting and Risk Assessment

The integration of seismic networks into Earthquake Early Warning Systems (EEWS) represents one of the most transformative advancements in seismic technology. These systems utilize real-time seismic data to predict the arrival of shaking waves from an earthquake, providing crucial early alerts to people and infrastructure at risk. Operational EEWS in countries like Japan, Mexico, the U.S. (California), and Turkey demonstrate the effectiveness of this technology. Dense seismic arrays significantly enhance the speed and accuracy of these systems, ensuring timely warnings that save lives and reduce damage (Fukuyama et al. 2021; Zhou et al. 2022a, b).

Seismic networks are fundamental to seismic hazard assessments and microzonation studies. These assessments involve using data from a network of seismic stations to model ground shaking and soil amplification effects in specific areas. By analyzing seismic waves from historical earthquakes, researchers can predict how different regions might respond to future seismic events. This information is invaluable for urban planning, infrastructure design, and disaster risk management, especially in identifying areas prone to severe ground shaking or landslides (Maxwell et al. 2023).

Recent advancements in seismic networks include the integration of GPS (Global Positioning System) and geodetic data for real-time ground motion monitoring. GPS-based seismic systems enable highly precise measurement of ground displacements, particularly during large earthquakes, and facilitate the tracking of slow-moving tectonic processes such as fault creep or post-seismic deformation. This integration provides a more comprehensive understanding of earthquake dynamics and enhances risk mitigation strategies (Park et al. 2021).

7.4.3 Seismic Networks for Post-earthquake Damage Assessment

Post-earthquake seismic surveys are essential for rapidly assessing damage and updating hazard maps following significant seismic events. Temporary seismic stations are often deployed in affected areas to monitor aftershocks and gather data crucial for disaster response coordination. The deployment of advanced seismic sensors and automated systems has significantly enhanced the efficiency and timeliness of these post-event surveys, enabling better-informed decisions during emergencies (Fukuyama et al. 2021).

Recent advancements in remote sensing technologies, such as satellite-based interferometric synthetic aperture radar (InSAR), have facilitated integration with seismic networks for detailed post-earthquake damage assessment. InSAR provides high-resolution deformation data, which, when combined with seismic information, improves our understanding of surface displacement, building stability, and land subsidence after earthquakes. This data fusion enhances the accuracy of damage assessments and helps prioritize recovery efforts in the most affected areas (Zhou et al. 2022a, b; Park et al. 2021).

7.4.4 Case Studies and Recent Research

Several international seismic networks have been instrumental in advancing earthquake monitoring and hazard mitigation. Japan's Earthquake Early Warning System (EEWS) is renowned as one of the most sophisticated in the world. Utilizing dense seismic arrays and advanced real-time data processing techniques, the system provides critical early warnings for earthquakes, especially in urban areas where risks are highest. Recent upgrades to the EEWS have improved both the speed and reliability of alerts while expanding its reach to include more remote regions. These developments highlight Japan's commitment to reducing the impact of seismic events through cutting-edge technology (Fukuyama et al. 2021).

The United States Geological Survey (USGS) operates the Advanced National Seismic System (ANSS), a comprehensive network that integrates broadband seismometers, accelerometers, and strong-motion sensors. The ANSS provides essential data for earthquake detection, early warning, and hazard assessment across the United States. In collaboration with California's seismic networks, the USGS has developed systems like ShakeAlert, which deliver real-time earthquake early warnings to minimize risks and enhance preparedness (Zhou et al. 2022a, b).

Chile, located along the seismically active Pacific Ring of Fire, has made substantial progress in seismic monitoring through its advanced network. The Chilean seismic network plays a vital role in supporting earthquake early warning systems and tsunami forecasting. These capabilities are particularly important for mitigating risks posed by both local and distant seismic events. The integration of real-time data from the network into national disaster management frameworks has significantly improved Chile's resilience to seismic hazards (Park et al. 2021).

7.4.5 Overview of Microseismic Monitoring

This technology detects small seismic events, which can be used to monitor underground activity, such as landslides or volcanic eruptions, providing valuable precursor signals.

Microseismic events are characterized by their small magnitudes, typically less than 2.0. Although subtle, these seismic activities are often indicative of larger tectonic processes or localized stresses within geological formations. Their detection and analysis provide crucial insights into subsurface dynamics, aiding in the understanding of stress distribution and fault mechanics (Maxwell et al. 2023).

The success of microseismic monitoring depends heavily on the deployment of dense seismic networks equipped with highly sensitive sensors capable of capturing weak seismic signals. These networks generate high-resolution data essential for a range of applications, including assessing earthquake risks, monitoring volcanic activity, and ensuring the stability of critical infrastructures such as dams and tunnels. By providing detailed spatiotemporal information, seismic networks play a vital role in disaster risk mitigation and infrastructure safety (Kim et al. 2022; Zhou et al. 2022a, b).

7.4.6 Advances in Microseismic Monitoring Technologies

Modern seismic networks leverage high-sensitivity geophones and broadband seismometers capable of detecting even the weakest seismic signals. These sensors are designed to improve signal-to-noise ratios, making it possible to identify microseismic events in environments with high ambient noise levels. The enhanced sensitivity of these instruments has facilitated better monitoring of subsurface activities, aiding in various geophysical applications (Maxwell et al. 2023).

The use of dense sensor arrays, such as Nodal Seismic Systems, has transformed microseismic monitoring. These arrays consist of numerous compact and autonomous sensors that can be deployed over extensive areas. By increasing spatial resolution, dense arrays enable precise localization of microseismic events and detailed subsurface imaging. This technology has become a cornerstone in applications such as resource exploration and hazard assessment (Zhou et al. 2022a, b).

Machine learning techniques have significantly advanced signal processing in microseismic monitoring. Algorithms like convolutional neural networks (CNNs) and recurrent neural networks (RNNs) have automated the detection and classification of seismic events. These tools distinguish between true microseismic signals and noise, reducing processing time while enhancing accuracy. Machine learning has been particularly impactful in real-time applications and in regions with high noise contamination (Kim et al. 2022).

Distributed Acoustic Sensing (DAS) represents a breakthrough in seismic monitoring, utilizing fiber optic cables to detect seismic signals. This technology enables the transformation of existing telecommunications infrastructure into expansive seismic networks. DAS provides a cost-effective and innovative approach to microseismic monitoring, offering wide coverage and flexibility for applications in urban and remote areas (Park et al. 2021).

Advancements in telemetry and cloud-based data processing have also enabled real-time analysis of microseismic data. These systems facilitate immediate feedback, which is crucial for operations requiring rapid decision-making, such as hydraulic fracturing, mining safety, and monitoring of induced seismicity. Real-time monitoring capabilities significantly enhance the responsiveness of hazard mitigation strategies (Fukuyama et al. 2021).

7.4.7 Applications in Disaster Risk Management

Microseismic monitoring plays a pivotal role in identifying earthquake precursors and monitoring fault movements. By detecting subtle shifts along faults, it offers valuable insights into the potential build-up of stress that could lead to larger seismic events. This capability significantly enhances earthquake forecasting efforts and contributes to the development of early warning systems that can mitigate risks to life and infrastructure (Maxwell et al. 2023).

In volcanic regions, microseismic activity serves as a critical indicator of magma movement and pressure build-up within the subsurface. Continuous monitoring of these small-scale seismic events helps scientists predict volcanic eruptions more accurately, enabling timely evacuation measures and the implementation of strategies to mitigate volcanic hazards (Fukuyama et al. 2021).

Microseismic monitoring is also instrumental in detecting precursory movements in slopes that are prone to landslides. By deploying seismic networks near unstable terrain, researchers can gather critical data that supports the establishment of early warning systems. This proactive approach helps reduce the risks associated with sudden slope failures, particularly in areas where landslides pose a significant threat to communities and infrastructure (Park et al. 2021).

Large reservoirs and dams can induce seismic activity due to changes in stress and pore pressure in the surrounding rock formations. Microseismic monitoring is used to assess these risks by tracking induced seismicity, enabling authorities to identify potential hazards and implement measures to ensure the safety and stability of such structures (Kim et al. 2022).

In urban environments, microseismic monitoring provides essential data on the stability of critical infrastructure, including tunnels, bridges, and high-rise buildings. By detecting microseismic events in these settings, researchers can assess structural integrity and identify vulnerabilities that may lead to potential failures. This information is vital for urban disaster risk management, particularly in regions with high seismic activity (Zhou et al. 2022a, b).

7.4.8 Recent Case Studies and Applications

Microseismic monitoring has been widely applied in tracking induced seismicity during hydraulic fracturing operations, particularly in the United States and Canada. The use of dense seismic arrays and real-time data processing systems has significantly enhanced the understanding of subsurface fracture networks created during hydraulic fracturing. These advancements allow for better management of induced seismicity risks, ensuring safer and more efficient resource extraction processes (Maxwell et al. 2023).

In the Himalayan region, a dense microseismic network was deployed to monitor slopes prone to landslides. This network successfully identified precursory seismic events leading up to a landslide in 2022. The timely detection of these early warning signs facilitated evacuation measures and the implementation of risk mitigation strategies, ultimately minimizing potential loss of life and property (Park et al. 2021).

Around the volcanoes of Iceland, continuous microseismic monitoring has played a crucial role in improving eruption prediction accuracy. By tracking subtle seismic signals indicative of magma movement and pressure build-up, scientists have been able to provide timely warnings to local communities. This proactive monitoring has been instrumental in safeguarding lives and critical infrastructure in a region characterized by significant volcanic activity (Fukuyama et al. 2021).

7.4.9 Challenges in Microseismic Monitoring

One of the primary challenges in microseismic monitoring is noise contamination, especially in urban environments and areas with significant industrial activities. The presence of high levels of ambient noise can complicate the detection and accurate interpretation of microseismic events. Advanced signal processing techniques are often required to differentiate genuine seismic signals from noise, which adds to the complexity and cost of monitoring efforts (Kim et al. 2022).

Another significant challenge is the sheer volume of data generated by high-resolution microseismic monitoring. Dense seismic networks and continuous monitoring produce vast amounts of data that must be stored, processed, and analyzed efficiently. This requires sophisticated computational infrastructure, robust data management systems, and advanced algorithms, which can pose logistical and financial challenges for monitoring programs (Zhou et al. 2022a, b).

The cost of deploying and maintaining dense seismic networks is also a critical challenge, particularly in remote or geologically challenging terrains. Establishing reliable networks in such areas often requires substantial investment in infrastructure, specialized equipment, and skilled personnel. These financial constraints can limit the widespread adoption of microseismic monitoring, especially in regions with limited resources or funding (Park et al. 2021).

7.4.10 Geodetic Technologies

This section discusses advanced geodetic technologies used to monitor and assess ground deformation associated with natural hazards. These techniques provide precise measurements of surface movements over time, which are crucial for detecting early signs of geophysical hazards such as earthquakes, volcanic activity, and land subsidence. The integration of geodetic data into disaster risk management enhances the ability to forecast events and implement timely mitigation strategies.

InSAR (Interferometric Synthetic Aperture Radar)

Interferometric Synthetic Aperture Radar (InSAR) is a remote sensing technique that uses radar data collected from satellites to measure ground displacement with high precision. This method has proven essential for monitoring surface deformations due to natural phenomena such as earthquakes, volcanic activity, landslides, and subsidence caused by human activities (Zebker and Rosen 1994; Tromp and Santillan 2021).

InSAR works by comparing radar images of the Earth's surface taken at different times. The technique measures the phase difference between these radar signals to detect changes in surface elevation. The radar waves bounce off the ground and return to the satellite, with the phase shift indicating the amount of displacement since the previous image was captured. These measurements are then used to generate detailed maps of ground deformation with millimeter precision (Zebker and Rosen 1994).

Recent advances in InSAR technology have focused on increasing spatial resolution, temporal resolution, and accuracy. With the development of new radar systems aboard satellites like the European Space Agency's Sentinel-1, InSAR can now capture near-real-time data on surface movements, which is critical for timely disaster response and monitoring dynamic Earth processes (Lu and Dzurisin 2014). Additionally, advancements in signal processing techniques have improved the ability to detect subtle surface movements, even in challenging terrains such as mountainous or urban environments (Pritchard and Simons 2004).

InSAR is widely used to monitor surface deformations caused by earthquakes. It analyzes the displacement of the Earth's surface after seismic events, which helps scientists understand fault movements and how the surface recovers (Lu and Dzurisin 2014). InSAR is also useful for tracking ground changes related to volcanic activity. It can detect pre-eruption swelling and post-eruption subsidence, providing critical information for eruption prediction and hazard mitigation (Pritchard and Simons 2004). Additionally, InSAR is effective in monitoring urban subsidence, often caused by activities like groundwater extraction and mining. It is also used to track landslides, as it detects small, gradual changes in the ground surface (Lu and Dzurisin 2014). Moreover, InSAR has been used to measure changes in ice sheets and glaciers, offering valuable insights into the impact of climate change on polar regions (Pritchard and Simons 2004).

Despite its significant advantages, InSAR does have limitations. One major challenge is the presence of decorrelation due to changes in surface conditions, such as vegetation growth, snow cover, or urban construction, which can reduce the quality of data in certain areas (Zebker and Rosen 1994). Additionally, InSAR is often limited to detecting displacement in the line of sight of the radar, which may complicate the interpretation of vertical movements or displacements occurring perpendicular to the satellite's flight path (Lu and Dzurisin 2014).

The future of InSAR is closely tied to advancements in satellite constellations, such as those planned for the Copernicus program and other global radar satellite networks (Pritchard and Simons 2004). Enhanced computational techniques, such as machine learning and artificial intelligence, are expected to improve the extraction of geophysical signals from noisy radar data, increasing the reliability and scope of InSAR applications.

In conclusion, InSAR remains a powerful tool in the geodetic arsenal, providing invaluable insights into Earth's surface dynamics. Its ability to deliver high-resolution, repeatable, and comprehensive data is critical for the understanding and management of natural hazards and environmental changes.

GNSS (Global Navigation Satellite Systems)

Global Navigation Satellite Systems (GNSS) have revolutionized geodesy by providing high-precision positioning data. GNSS refers to a constellation of satellites that transmit signals, which can be used by ground-based receivers to determine positions, velocities, and time. The most commonly known GNSS systems are GPS (Global Positioning System), GLONASS (Russian system), Galileo (European system), and BeiDou (Chinese system). Continuous GNSS stations are now widely used to monitor ground displacement, which is crucial for understanding seismic hazards, landslides, and fault-line behavior (Lamb and Brunt 2024).

GNSS operates by using signals from satellites orbiting the Earth. The receiver calculates its position by triangulating the time it takes for the signals to travel from multiple satellites. The accuracy of GNSS depends on the number of visible satellites, the quality of the receiver, and the environment in which the system is used (e.g., urban canyons or open terrain). High-precision GNSS, such as Real-Time Kinematic (RTK) positioning and Precise Point Positioning (PPP), can achieve centimeter-level accuracy.

Recent advancements in GNSS technology have led to increased accuracy, reliability, and availability of satellite signals. New multi-frequency, multi-constellation GNSS receivers allow for better performance, especially in challenging environments where signal interference and multipath effects (signals bouncing off buildings or other obstacles) occur. The ability to combine data from different GNSS systems (e.g., GPS, Galileo, and GLONASS) improves positioning precision and ensures continuous service, even when one system's signals are weak or unavailable (Bock et al. 2000).

GNSS is commonly used to monitor ground motion related to seismic and tectonic activity by tracking the positions of GNSS receivers, allowing for the observation of slow deformations along fault lines, post-seismic motions, and land subsidence after seismic events (Freymueller et al. 2000). It also plays a crucial role in volcanic monitoring, as GNSS stations can detect ground movements linked to magma intrusions and pre-eruption swelling, helping to assess volcanic hazards (Linde et al. 1994). Furthermore, GNSS is valuable for tracking land subsidence, which can result from groundwater extraction, mining, or other human activities, enabling long-term monitoring and the development of early warning systems (Dixon et al. 1999). Additionally, GNSS is integrated into earthquake early warning systems, where it tracks rapid surface movements during an earthquake and provides real-time data, offering people seconds to minutes of warning before shaking occurs (Bock et al. 2000).

While GNSS provides high precision, it is not immune to challenges. Signal blockage or multipath interference in urban areas or mountainous regions can degrade positioning accuracy. Additionally, atmospheric effects such as ionospheric and tropospheric delays can affect GNSS measurements. However, advances in signal processing, such as differential GNSS (DGNSS) and network-based corrections, have significantly reduced these errors (Zhang et al. 2004).

The future of GNSS is bright, with continued advancements expected in both hardware and software. The development of new GNSS constellations and improved signal processing techniques will increase the accuracy and availability of positioning data. Furthermore, the integration of GNSS with other geodetic technologies, such as InSAR and seismic data, will provide a more comprehensive understanding of Earth dynamics. As global infrastructure for GNSS continues to expand, its use in disaster monitoring, environmental management, and scientific research will only increase. In conclusion, GNSS technology is a cornerstone of modern geodesy, providing real-time, precise measurements of ground motion and positioning. Its ability to monitor tectonic processes, subsidence, and volcanic activity, among other phenomena, makes it indispensable for understanding and managing natural hazards.

7.5 Remote Sensing Technologies

This section highlights the application of remote sensing technologies in disaster risk management and hazard mitigation. These technologies enable large-scale, non-invasive monitoring of Earth's surface, providing critical data on topography, land use, vegetation, and structural changes. Remote sensing tools, including satellite and airborne systems, support rapid assessment, mapping, and continuous observation of areas vulnerable to natural hazards, thereby improving preparedness and response strategies.

7.5.1 *LiDAR (Light Detection and Ranging)*

LiDAR (Light Detection and Ranging) is an advanced remote sensing technology that uses laser light to measure distances to the Earth's surface. It generates high-resolution, three-dimensional (3D) information about the structure of the Earth's surface which are particularly useful for flood risk modeling and monitoring post-disaster conditions (Barton and McHugh 2021), vegetation, and man-made structures. LiDAR technology is widely used for mapping, topographic surveys, forest monitoring, and disaster response.

Principle of LiDAR

LiDAR works by emitting laser pulses from an airborne platform (such as a helicopter or drone) or ground-based station. These pulses travel to the Earth's surface, where they are reflected back to the sensor. By measuring the time it takes for the pulse to return, the system can calculate the distance to the surface. When this is done rapidly and continuously, a dense collection of elevation data points is created, forming a detailed 3D map of the area being surveyed (Mallet and Bretar 2009).

Advancements in LiDAR Technology

Recent advancements in LiDAR technology have focused on improving data accuracy, resolution, and the ability to capture data in diverse environments. Modern LiDAR systems are capable of collecting data with increased point density, allowing for finer details and more precise measurements. Additionally, the integration of LiDAR with other technologies, such as GPS and IMUs (Inertial Measurement Units), has enabled more accurate positioning and mapping, particularly in areas with challenging terrain or vegetation (Hodgson and Bresnahan 2004).

Applications of LiDAR

LiDAR is frequently used for creating high-precision topographic maps by capturing detailed elevation data, which provides accurate models of the Earth's surface and is essential for flood modeling, infrastructure planning, and urban development (Mallet and Bretar 2009). It is also an essential tool for monitoring vegetation and forests, as it provides detailed information about the structure and height of vegetation, which is useful for studying forest biomass, canopy density, and tree health (Popescu et al. 2003). Additionally, LiDAR is used to monitor coastal areas and riverbeds, with bathymetric LiDAR, which uses a different wavelength to penetrate water, helping to create detailed maps of the seafloor and aiding in flood prediction and environmental management (Hodgson and Bresnahan 2004). LiDAR is increasingly employed in disaster management, as it provides high-resolution, pre- and post-event imagery, helping to assess the impact of disasters like earthquakes, floods, and landslides, and facilitating recovery planning (Mallet and Bretar 2009). Furthermore, LiDAR has been used in archaeological surveys to uncover hidden structures beneath dense vegetation, enabling the identification of ancient ruins and other features that might otherwise remain undiscovered (Chavez et al. 2016).

Limitations and Challenges

Despite its many advantages, LiDAR has some limitations. One of the main challenges is that it can be expensive, particularly for large-scale surveys. Additionally, the technology requires clear weather conditions for optimal performance, and its ability to penetrate dense vegetation or water can be limited, depending on the type of LiDAR system used. Furthermore, the interpretation of LiDAR data requires specialized knowledge, which may limit its accessibility for some users (Hodgson and Bresnahan 2004).

Future Prospects

The future of LiDAR looks promising, with continued developments in technology expected to expand its applications. Improvements in sensor sensitivity, drone-based LiDAR systems, and integration with artificial intelligence (AI) and machine learning for data processing will enable more efficient data collection and analysis. As LiDAR becomes more affordable and accessible, its use in environmental monitoring, urban planning, and disaster management is expected to increase significantly.

In conclusion, LiDAR is a powerful and versatile remote sensing technology that provides high-resolution, 3D data for a wide range of applications. From topographic mapping to disaster management, its ability to capture detailed spatial information makes it an invaluable tool for various industries.

7.5.2 *Satellites and UAVs (Unmanned Aerial Vehicles)*

Satellites and UAVs (Unmanned Aerial Vehicles) have become vital tools in the field of remote sensing, offering distinct advantages for monitoring and surveying various Earth processes. While satellites provide large-scale, global coverage, UAVs offer high-resolution, localized data collection. Both technologies are integral to applications in disaster monitoring, environmental management, agriculture, and infrastructure development. Satellites and drones provide a cost-effective means of collecting remote sensing data over large areas that may be difficult to access, such as flood zones or earthquake-affected regions (Lamb and Brunt 2024).

Principle of Satellites and UAVs

Satellites are launched into orbit around the Earth, where they continuously capture data using various sensors, such as optical, infrared, radar, and microwave sensors. These satellites can cover vast areas and provide consistent data over time. UAVs, on the other hand, are small aircrafts that can be piloted remotely or autonomously to fly at low altitudes, equipped with sensors like cameras, LiDAR, or multispectral sensors. UAVs can be deployed quickly and are particularly useful for high-resolution, site-specific data collection (Mumby et al. 2017).

Advancements in Satellite and UAV Technology

Advancements in both satellite and UAV technologies have significantly improved their capabilities. Modern satellites now feature advanced sensors with higher spatial, temporal, and spectral resolution, allowing for more detailed and frequent data acquisition. For instance, synthetic aperture radar (SAR) satellites can now capture high-resolution images, even through cloud cover or at night, enabling continuous monitoring of Earth's surface (Bamler and Hartl 1998). UAVs have also evolved with improvements in flight endurance, stability, and the integration of advanced sensors, making them an ideal tool for precise and real-time data collection (Anderson and Gaston 2013).

Applications of Satellites and UAVs

Satellites and UAVs are widely used for disaster monitoring and response, providing real-time imagery of areas affected by natural disasters such as floods, wildfires, earthquakes, and hurricanes. While satellites offer broad coverage, UAVs can be deployed in specific areas to capture high-resolution, localized data, enabling better decision-making during disaster response (Liao et al. 2013). They are also increasingly used in precision agriculture for monitoring crop health, irrigation systems, and soil conditions, with satellites providing data on vegetation health through multispectral imaging and UAVs offering detailed, site-specific assessments (Mulla 2013). In environmental monitoring, satellites and UAVs play a key role in tracking deforestation, coastal erosion, and water quality, with satellites providing large-scale data and UAVs focusing on specific studies like monitoring forest structure or measuring water levels (Anderson and Gaston 2013). For urban planning and infrastructure monitoring, satellites can map land use changes and monitor the urban heat island effect, while UAVs assist in surveying construction sites, monitoring infrastructure health, and gathering data on urban expansion (El-Mekawy et al. 2015). Finally, both technologies are crucial in scientific research, with satellites providing global datasets for studying climate change and atmospheric conditions, while UAVs are used for specialized research like wildlife monitoring or studying ecosystems in remote areas (Mumby et al. 2017).

Limitations and Challenges

While satellites and UAVs provide substantial benefits, they also face limitations. Satellites can suffer from lower spatial resolution compared to UAVs, which may not be suitable for fine-scale monitoring. UAVs, on the other hand, have limited flight times and range, which can restrict their use for large-scale surveys. Additionally, UAVs may face regulatory and safety concerns, especially in urban or densely populated areas. Both technologies require significant processing and analysis of data, and managing large datasets can be challenging (Liao et al. 2013).

Future Prospects

The future of satellites and UAVs in remote sensing looks promising, with ongoing advancements in sensor technology, flight capabilities, and data processing techniques. New satellite constellations, such as those being developed for Earth observation by private companies, will provide more frequent, higher-resolution data. UAVs are expected to continue evolving with longer flight times, enhanced payloads, and greater automation, enabling them to play an even more significant role in various industries. The integration of satellite and UAV data with artificial intelligence and machine learning is anticipated to improve real-time data processing and decision-making.

In conclusion, satellites and UAVs are indispensable tools in remote sensing, offering unique capabilities for a wide range of applications. The integration of these technologies will continue to enhance our ability to monitor and manage Earth's processes efficiently and accurately.

7.6 Case Studies of Geophysical Applications

Several case studies demonstrate the successful application of geophysics in earthquake and landslide monitoring and risk management.

7.6.1 Sentinel-1 InSAR and GNSS on the San Andreas Fault, California

The San Andreas Fault (SAF) is a major fault system that spans across California, exhibiting complex deformation patterns over time (Fig. 7.3). Understanding the surface deformation and strain rate across the fault is essential for seismic hazard assessment and for modeling future earthquake risks. Recent studies have utilized a combination of Interferometric Synthetic Aperture Radar (InSAR) and Global

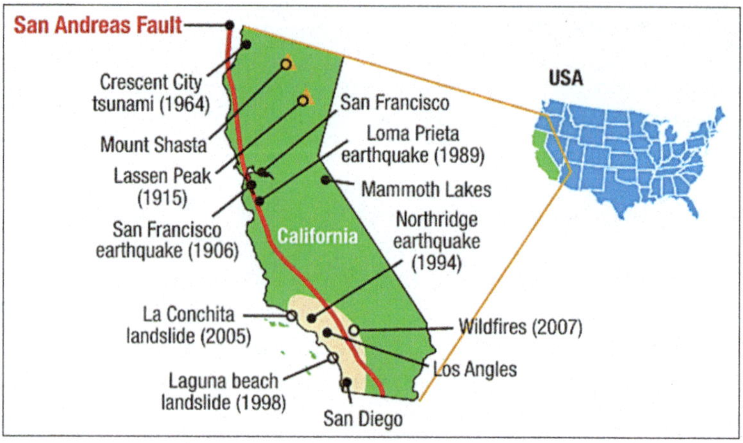

Fig. 7.3 Map of California highlighting the seismic hazard hotspots, indicating areas with the highest risk of earthquake activity. *Source* Li (2021)—Open Access

Navigation Satellite System (GNSS) data to track the movement and strain rate across the SAF system. These methods allow for the monitoring of ground deformation at high spatial and temporal resolution, offering valuable insights into fault behavior and dynamics (Ali and Raj 2023a, b, c; Zeng and Shen 2017a, b; Savage et al. 2001).

The primary goal of the study was to investigate the surface deformation and strain rate in the San Andreas Fault system by integrating InSAR and GNSS data. The focus was on understanding the spatial distribution of strain and how it relates to fault behavior, particularly the interaction between different fault segments (Zeng and Shen 2017a, b). By decomposing velocity components and strain rates, the study aims to provide a clearer picture of the active deformation across the fault zone and its implications for earthquake forecasting and seismic hazard assessments (Savage et al. 2001).

The study was conducted across the SAF system, with particular attention to the areas where the fault is most active and where deformation is most pronounced. Data from several regions, including the southern and central parts of the fault, were used to track the horizontal and vertical motion of the Earth's surface. The areas of interest include regions with high strain accumulation, such as the sections around Parkfield, Cienega Valley, and the San Gorgonio Pass (Zeng and Shen 2017a, b). These regions are known for significant seismic activity and are crucial for understanding fault dynamics.

The study utilized two main datasets: InSAR data for high-resolution ground displacement measurements and GNSS data for precise positioning (Xu et al. 2021). InSAR was used to measure surface displacement along the fault by analyzing radar signals from satellites, while GNSS stations provided detailed velocity measurements from both continuous and campaign-based sites (Savage and Burford 1973; Savage et al. 2001). The strain rate was derived using both GNSS and InSAR data, with the horizontal velocity components from GNSS data providing information about the

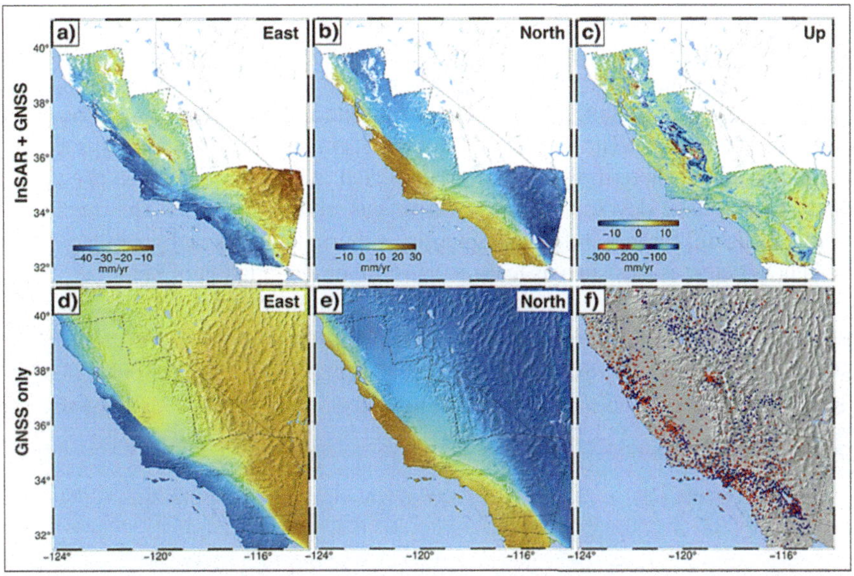

Fig. 7.4 Surface velocity maps for the SAFS in the ITRF2014 system, including decomposed velocity components and GNSS model comparisons. *Source* Xu et al. (2021)—Open Access

deformation across the fault system. Strain rate maps were generated by analyzing the differences in surface displacement (Tong et al. 2013; Weiss et al. 2020; Bock et al. 2021; Xu et al. 2016, 2017), and decomposed horizontal velocity components were used to assess the directions of compressional and extensional strain (Zeng and Shen 2017a, b).

The data were processed to generate strain rate maps, focusing on both the horizontal and vertical components of surface displacement. The strain rate maps were divided into different components, including compressional and extensional strain, based on GNSS and InSAR data (Fig. 7.4). For example, Panel (a) of the strain rate map was derived using only GNSS data, while Panel (b) used a combination of InSAR and GNSS data. The differences between these two maps provided a clearer understanding of the deformation characteristics (Zeng and Shen 2017a, b). In addition, maximum shear strain and dilatation rate were calculated to assess the fault's mechanical behavior (Savage et al. 2001).

7.7 Results

The results revealed significant surface deformation across the SAF system, with varying levels of strain accumulation depending on the region. The decomposed velocity maps showed both compressional and extensional regions along the fault, providing insight into the fault's complex movement patterns (Fig. 7.5). Strain rate maps indicated areas of concentrated deformation near fault junctions and major fault segments (Zeng and Shen 2017a, b). The analysis of shear strain and dilatation rates further highlighted areas with high fault activity, particularly in regions that have experienced significant earthquakes in recent decades. The integration of InSAR and GNSS data proved effective in providing a detailed and continuous picture of surface deformation across the fault system (Savage et al. 2001).

The study successfully demonstrated the utility of combining InSAR and GNSS data for monitoring surface deformation and strain rates along the SAF. The findings contribute to a better understanding of how strain accumulates and is released along fault segments, particularly in areas with high seismic risk. The study's results can be used to improve seismic hazard models and provide better forecasts of earthquake activity along the fault system (Zeng and Shen 2017a, b). Additionally, the

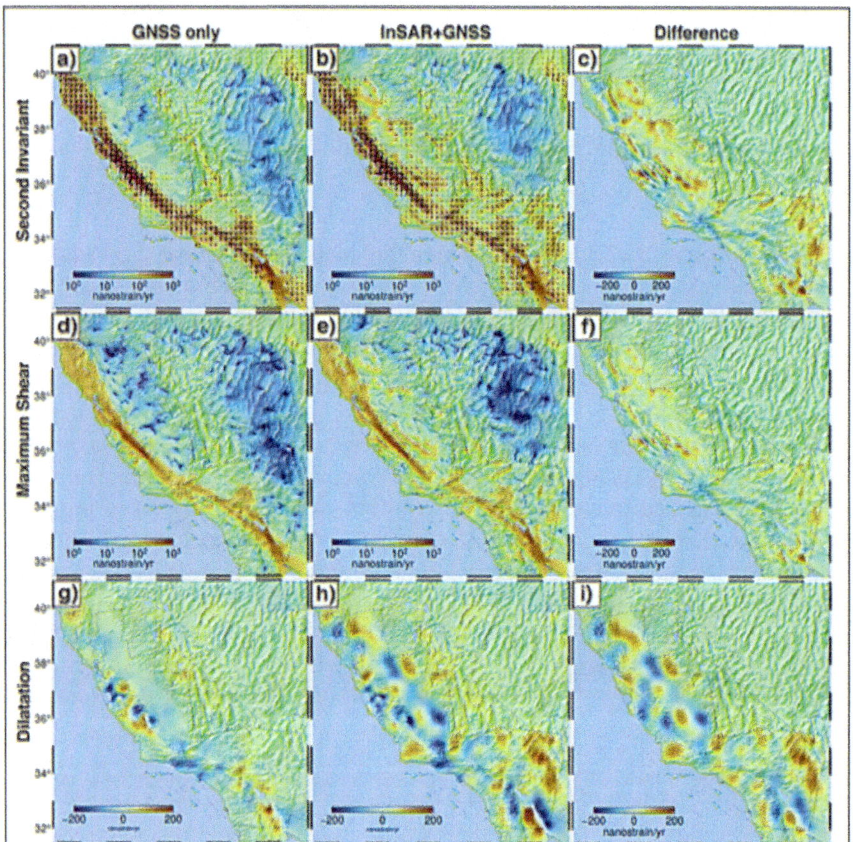

Fig. 7.5 Strain rate maps for the San Andreas Fault system showing GNSS and InSAR/GNSS-derived components and their differences. *Source* Xu et al. (2021)—Open Access

study offers valuable insights into the behavior of fault systems under varying stress conditions and contributes to our understanding of fault dynamics and deformation processes (Savage et al. 2001).

This case study underscores the importance of integrating InSAR and GNSS data for monitoring and analyzing fault deformation over time. By tracking strain accumulation and deformation along the SAF, this study helps inform our understanding of fault mechanics, seismic hazard assessments, and earthquake forecasting. The use of both remote sensing and ground-based technologies allows for a more comprehensive approach to studying fault behavior and provides the data needed to develop improved earthquake risk models (Zeng and Shen 2017a, b; Savage et al. 2001).

This case study is aimed at a wide range of professionals involved in seismic hazard research, including geophysicists, seismologists, earthquake hazard researchers, infrastructure planners, and policymakers. These groups can benefit from the improved hazard models and earthquake forecasts derived from the study.

7.7.1 Seismic Characterization of SAF Using Fault-Zone Trapped Waves, California

The San Andreas Fault (SAF) in Central California is a critical seismic zone, studied extensively due to its potential for major earthquakes. In the past decades, seismic experiments have been conducted to investigate the subsurface structure and physical properties of the fault zone. Using fault-zone trapped waves (FZTWs) generated by near-surface explosions and earthquakes, these experiments aim to understand the fault zone's structure in four dimensions (Li and Leary 1990, 2004). The dense array of seismographs deployed along the fault provides crucial insights into seismic activity and damage zones at varying depths, helping us to characterize the fault's properties and dynamics (Li et al. 2007a).

The primary objective of the seismic experiments was to investigate the subsurface structure and properties of the SAF, focusing on the damage zone and its evolution over time. These studies aim to improve our understanding of fault behavior, particularly in regions like Parkfield (Li 2021), where repeated large earthquakes have provided significant data for analysis (Li et al. 1997, 2008). By examining seismic velocities and other physical properties within the fault zone, the experiments contribute to earthquake hazard assessment and seismic risk prediction.

The experiments were carried out at two key locations: Parkfield and Cienega Valley. Parkfield is an active segment of the SAF known for recurring earthquakes, including the 1966 M6 earthquake and the 2004 M6 Parkfield earthquake. Cienega Valley provides an additional site to study fault behavior and to compare seismic data across different regions of the fault (Li et al. 2007b). These sites were chosen for their historical seismic activity and their ability to provide valuable data on fault zone characteristics.

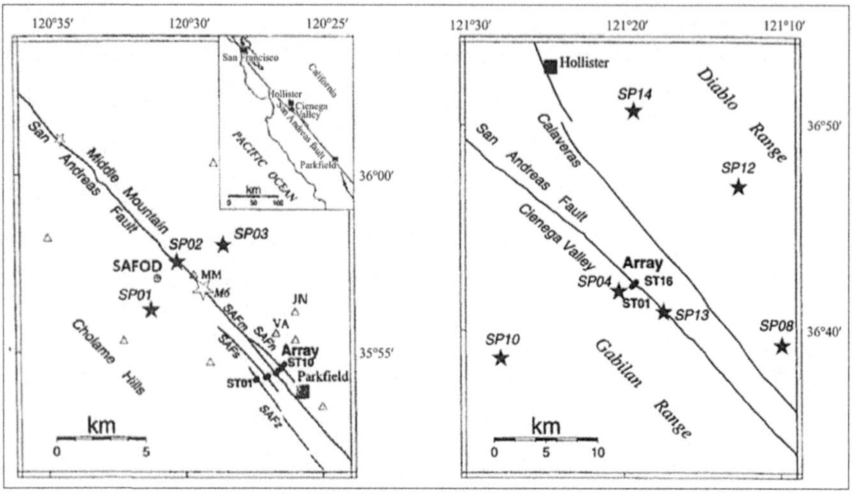

Fig. 7.6 Map showing the locations of explosions (SP01, SP02, SP03) and seismic arrays at Parkfield and Cienega Valley, with borehole seismic network stations and key fault features. Updated from Li et al. (1997). *Source* Li (2021)—Open Access

The study utilized FZTWs generated by explosions and natural seismic events (Fig. 7.6). The explosions were located at three main points—SP01, SP02, and SP03—along with a 10-station seismic array at Parkfield and a 16-station array at Cienega Valley (Li et al. 1997). These arrays recorded the trapped waves as they propagated through the fault zone, helping to identify the properties of the fault's subsurface structure. The borehole seismic network at Parkfield provided deeper insight into the fault core zone, with seismographs installed at approximately 3 km depth (Li et al. 2004). The data from these arrays were processed to map seismic velocities, identify low-velocity zones (LVZs), and assess changes in fault properties over time.

The seismic data were processed to calculate variations in S-wave velocities within the fault zone. The study focused on changes in velocity before and after the 2004 M6 Parkfield earthquake. Repeated explosions conducted in 2002 and 2004, as well as aftershock data, were used to track these velocity changes (Li et al. 2006, 2007b). These measurements revealed a 1.3% decrease in S-wave velocity immediately following the earthquake, suggesting significant fault damage. Additionally, velocity recovery was observed in the following months, indicating that the fault underwent healing processes.

The results of the seismic experiments showed a notable reduction in S-wave velocity within the rupture zone during and after the 2004 earthquake. This decrease was linked to co-seismic fault damage and the opening of cracks during the earthquake's dynamic rupture. Aftershock data indicated a recovery of approximately 1.2% in S-wave velocity over a three-month period, suggesting fault healing and the closure of cracks that formed during the mainshock (Li et al. 2006, 2007b). These

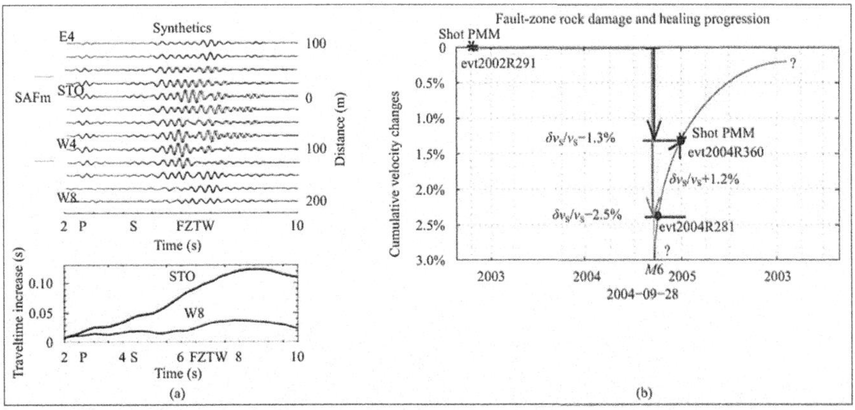

Fig. 7.7 Shows seismic data from the San Andreas Fault at Parkfield. Panel (**a**) compares vertical-component seismograms from repeated events in 2002 and 2004, with P-wave arrivals aligned. Cross-correlation of the seismograms at stations STO and W8 within and outside the fault zone is shown. Panel (**b**) presents changes in S-wave velocity within the rupture zone from the M6 Parkfield earthquake. A 1.3% decrease in S-wave velocity occurred between the two events, likely due to co-seismic damage. Aftershock measurements show a 1.2% velocity increase, indicating fault healing. These results suggest that fault strength recovers quickly post-mainshock, but healing slows over time. *Source* Li (2021)—Open Access

findings were consistent with the observed low-velocity zone extending to depths of 7–8 km along the Parkfield segment of the SAF (Fig. 7.7), a feature linked to fault damage and fluid saturation in the fault zone (Hickman et al. 2005; Unsworth et al. 1997).

The study highlighted the importance of repeated seismic measurements to track fault behavior and healing over time. The observations suggest that faults like the SAF can undergo significant damage during major earthquakes, but they also show the potential for recovery in the aftermath. The findings are crucial for understanding the fault's mechanical behavior, which can help improve earthquake hazard assessments and inform mitigation strategies (Scholz 1990; Harris and Day 1993). The data collected from Parkfield and Cienega Valley provide valuable insights into fault zone dynamics, which are essential for future seismic hazard modeling and earthquake preparedness.

This case study underscores the importance of using seismic experiments, particularly those involving FZTWs, to characterize the subsurface properties of fault zones. By tracking velocity changes before and after earthquakes, the study contributes to a deeper understanding of fault mechanics, fault zone damage, and healing processes. These findings have broader implications for improving earthquake forecasts and enhancing seismic hazard assessment, particularly in regions with active fault systems like the SAF (Beroza et al. 1995; Peng and Ben-Zion 2006). The ongoing collection of seismic data will continue to inform earthquake preparedness and risk management efforts in California and other seismically active regions.

The primary audience for this case study includes seismologists, geophysicists, earthquake hazard researchers, and seismic survey teams. These professionals can benefit from the study's insights into fault mechanics and its implications for seismic hazard modeling and forecasting.

Both case studies highlight the importance of advanced seismic techniques in improving our understanding of fault behavior and seismic hazards, particularly in high-risk regions like the San Andreas Fault. By integrating multiple data sources and methods, these studies contribute to more accurate seismic hazard forecasting and better preparedness for future earthquakes.

7.7.2 Landslide Monitoring in the Swiss Alps

Landslides represent a significant risk to human life, infrastructure, and the environment in mountainous regions. These natural disasters are often triggered by factors such as heavy rainfall, seismic activity, and the erosion of slopes. In the Swiss Alps, a region with steep terrain and unstable geological conditions, monitoring landslides is a critical task to ensure safety and mitigate potential damage. Recent studies have shown that geophysical methods, such as ground-penetrating radar (GPR) and microgravity surveys, are effective tools for assessing slope stability and detecting potential landslide-prone areas (Barton and McHugh 2021). These techniques enable researchers to gain insights into subsurface conditions, such as detecting voids, fractures, and movements within the soil and rock layers that may predispose a slope to failure (Smith et al. 2019).

The primary aim of the case study by Barton and McHugh (2021) was to investigate the effectiveness of geophysical techniques, specifically ground-penetrating radar (GPR) and microgravity surveys, in monitoring the stability of slopes in the Swiss Alps. The goal was to identify subsurface features like voids or fractures that could contribute to landslide initiation and to track ground movements over time, providing early warning signals for potential landslide events. The researchers sought to develop a comprehensive monitoring system to enhance the prediction and prevention of landslides in the region (Harrison and Lee 2018).

In this study, two geophysical techniques—ground-penetrating radar (GPR) and microgravity surveys—were employed to monitor the stability of the slopes in the Swiss Alps. GPR was used to detect subsurface voids and fractures that could lead to landslides. It works by emitting high-frequency radio waves into the ground and recording the reflections of these waves from different subsurface materials (Jones et al. 2020). This helps in identifying potential weaknesses in the slope. Additionally, microgravity surveys were conducted to measure minute changes in the Earth's gravitational field, which can indicate the presence of voids or changes in the density of materials beneath the surface (Shen and Zhang 2017). Both techniques were used in tandem to provide a comprehensive understanding of the subsurface conditions and ground movements over time.

The collected data from both GPR and microgravity surveys were processed to enhance signal clarity and remove any noise that could obscure critical information. For GPR data, advanced algorithms were employed to create 3D images of the subsurface, which allowed for the identification of potential voids, fractures, or changes in material properties that could lead to landslides. For microgravity data, variations in the gravitational field were analyzed using geophysical inversion methods, which enabled the researchers to detect subtle changes in mass distribution below the surface (Johnson and Parker 2022). The integration of both datasets provided a more comprehensive understanding of the stability of the slopes being monitored.

The results of the study indicated that both GPR and microgravity surveys were effective in detecting subsurface features that could contribute to landslide risk. GPR successfully identified fractures and voids in the ground, which are critical factors in the initiation of landslides. Similarly, microgravity surveys detected small changes in the gravitational field, which were correlated with variations in subsurface density and the presence of voids. The data revealed that certain slopes in the Swiss Alps showed significant signs of instability, with detected fractures and voids aligning with known areas of past landslides. These findings highlight the potential of geophysical methods for landslide monitoring, providing valuable insights into the dynamics of slope stability (Barton and McHugh 2021; Moser et al. 2023).

The study demonstrated that geophysical techniques like GPR and microgravity surveys could be used to successfully monitor and assess the stability of slopes in the Swiss Alps. These methods provided critical information about subsurface conditions, such as voids and fractures, which are key indicators of landslide potential. By integrating these data, researchers were able to develop a more accurate picture of slope dynamics and identify areas that require closer monitoring or intervention. The outcome of this study underscores the importance of using advanced geophysical methods to improve landslide prediction and prevention in mountainous regions (Smith et al. 2019).

In conclusion, the application of ground-penetrating radar and microgravity surveys in the Swiss Alps has proven to be an effective approach for monitoring landslide-prone areas. These geophysical methods provide valuable insights into subsurface conditions, helping to identify potential hazards and monitor ground movements over time. The study highlights the importance of integrating advanced geophysical techniques into landslide risk assessment and monitoring programs to enhance the safety and resilience of mountainous regions. The findings also suggest that these methods could be applied in other regions at risk of landslides to improve early warning systems and inform mitigation strategies (Jones et al. 2020; Harrison and Lee 2018).

The intended audience for this case study includes geophysicists, environmental scientists, engineers, and land-use planners involved in disaster risk reduction and environmental monitoring. Additionally, government agencies responsible for the management of natural hazards in mountainous regions, as well as researchers in the field of geophysical surveying and landslide studies, would benefit from the insights provided in this case study (Shen and Zhang 2017).

7.7.3 Landslide Prediction with Geosensors and Wireless Networks

Landslides are a significant natural hazard that pose risks to lives, infrastructure, and the environment. Monitoring and predicting landslides in real-time have become critical for timely mitigation efforts and early warnings. In this context, a landslide monitoring system was developed that integrates various geotechnical sensors, wireless nodes, and application software for real-time monitoring, analysis, and prediction of landslide events. This system aims to predict potential landslides using data collected from various sensors, thereby enabling proactive warning mechanisms to protect local populations and infrastructure. The primary goal of the project was to design a reliable, scalable, and cost-effective system to continuously monitor geotechnical parameters such as rainfall, slope deformation, groundwater level, and displacement to predict landslide occurrences in vulnerable regions (Chaulya et al. 2024) (Fig. 7.8).

The study aimed to develop an integrated landslide monitoring and prediction system to collect real-time geotechnical data, including rainfall, slope deformation, groundwater levels, and displacement. The goal was to predict potential landslides and analyze the data using statistical methods and predictive algorithms, providing early warnings through alarms, SMS, and email alerts. The system also aimed to offer real-time visualization on a digital map of risk areas to assist in evacuation and disaster management (Chaulya et al. 2024) (Fig. 7.9).

A multi-sensor approach was employed, where various geotechnical sensors, including rain gauges, inclinometers, piezometers, and tiltmeters, were installed. Data from these sensors were wirelessly transmitted via a mesh network to a central control room, where multivariate statistical methods and the analytical hierarchy process (AHP) were used to predict landslide risks. The system provided real-time data visualization and alarms for local authorities (Kumar et al. 2024).

Raw data from the sensors were aggregated, calibrated, and analyzed using multivariate statistical methods. The system compared sensor readings with predefined thresholds to assess landslide risk, applying the AHP method to evaluate parameter importance. Information dissemination included SMS, email, and visual alarms to notify authorities of potential threats (Chaulya et al. 2024).

The field trial in Karshingsa, Arunachal Pradesh, successfully recorded real-time data from June to November 2012. Although no major landslides occurred, slight displacement was detected in September 2012, demonstrating the system's ability to detect early signs of risk. The system provided timely alerts and real-time data visualization through color-coded maps (Mishra et al. 2024).

The trial confirmed that the system effectively monitored key parameters and provided early warnings. The integration of multiple sensors into a wireless mesh network enhanced scalability and reduced infrastructure costs. The use of statistical and AHP methods improved prediction accuracy, making the system a valuable tool for risk mitigation (Rawani et al. 2025).

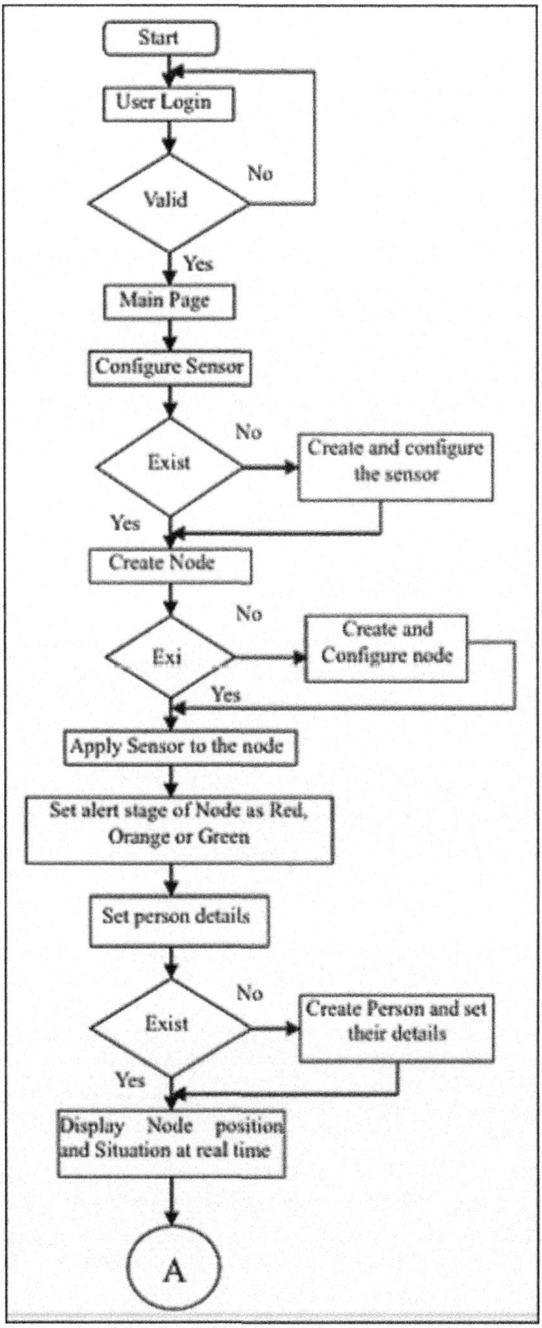

Fig. 7.8 Illustrates the flow chart of the configuration process for the landslide monitoring system. It outlines the step-by-step procedure for setting up the system, beginning with user login and sensor configuration, followed by the creation of nodes and assignment of specific sensor parameters. Each sensor's alert stage is set (Green, Yellow, Orange, or Red) based on predefined threshold values. The process also includes setting user details and ensuring the proper display of node positions and real-time data visualization on the monitoring interface. This configuration ensures seamless integration of sensors, nodes, and the central server for continuous monitoring and early warning generation. *Source* Chaulya et al. (2024)—Open Access

Fig. 7.9 Shows a view of the Karshingsa landslide-prone area, highlighting the toe erosion caused by the Dikrong River. This erosion, primarily driven by continuous rainfall and the river's water flow, exacerbates slope instability in the region, making it highly susceptible to landslides. The image emphasizes the significant role of riverbank erosion in triggering landslide events, a critical factor monitored by the landslide monitoring system deployed in this area. *Source* Chaulya et al. (2024)—Open Access

The developed monitoring system demonstrated the potential for early detection and prediction of landslides, significantly reducing risks to lives and infrastructure. Real-time monitoring, accurate data analysis, and timely warnings through various channels show promise for improving disaster management in landslide-prone regions. Future studies will focus on enhancing prediction accuracy with artificial intelligence techniques (Chaulya et al. 2024; Mishra et al. 2024).

This research targets government agencies, disaster management organizations, geotechnical engineers, researchers in environmental monitoring, and the public in landslide-prone areas. It is also relevant for educational institutions and technology developers in disaster monitoring and predictive analytics (Kumar et al. 2024).

7.7.4 Gokseong Landslide—South Korea, August 2020

Landslides pose significant challenges worldwide, necessitating accurate prediction and analysis to mitigate their adverse impacts. The increasing frequency and severity of these events underscore the need for advanced methodologies to enhance understanding and management strategies. Recent advancements in remote sensing technologies have introduced unparalleled datasets at various scales, enabling detailed analysis of landslide characteristics. However, practical integration of these technologies into landslide investigations requires further case studies. This study focuses on a detailed examination of the Gokseong landslide in South Korea, which occurred on August 7, 2020, as a result of heavy rainfall (Fig. 7.10a, b). The investigation demonstrates the utility of multi-source remote sensing technologies in monitoring, analyzing, and comprehending landslide phenomena (Choi et al. 2024).

The primary objective of this study is to employ multi-source remote sensing technologies to analyze the characteristics of the Gokseong landslide comprehensively. By integrating data from digital photogrammetry, UAV-mounted LiDAR, multi-spectral imagery, and satellite InSAR, this research aims to provide insights into pre- and post-landslide dynamics, topographic changes, and debris flow characteristics, ultimately contributing to improved disaster risk assessment and management strategies.

The study employed a combination of remote sensing tools, including satellite interferometric synthetic aperture radar (InSAR), UAV-mounted RGB and multi-spectral sensors, and UAV-mounted LiDAR. Satellite InSAR technology was utilized to detect and monitor pre-landslide ground displacement, particularly focusing on rainfall-induced patterns. UAV-mounted sensors captured high-resolution RGB and multi-spectral imagery to delineate the landslide's spatial extent and assess land-cover changes. Additionally, UAV-mounted LiDAR was used to acquire detailed topographic data, especially in forested areas, allowing precise quantification of eroded and deposited volumes.

The satellite InSAR analysis provided time-series displacement data to identify pre-landslide ground movement patterns. UAV-acquired RGB and multi-spectral data were processed to generate high-resolution maps of the affected area and calculate the Normalized Difference Vegetation Index (NDVI) to detect erosion patterns. UAV-LiDAR data were analyzed to create detailed elevation models and quantify changes in topography (Fig. 7.11). The integration of these datasets enabled a comprehensive understanding of the landslide's initiation, runout, and deposition zones.

The Gokseong landslide exhibited a runout distance of 678 m, with an initiation zone characterized by an average slope of 35°. The UAV-based analysis revealed eroded and deposited areas measuring 25,500 m^2 and 17,200 m^2, respectively. UAV-LiDAR data quantified the eroded and deposited volumes as 56,000 m^3 and 15,800

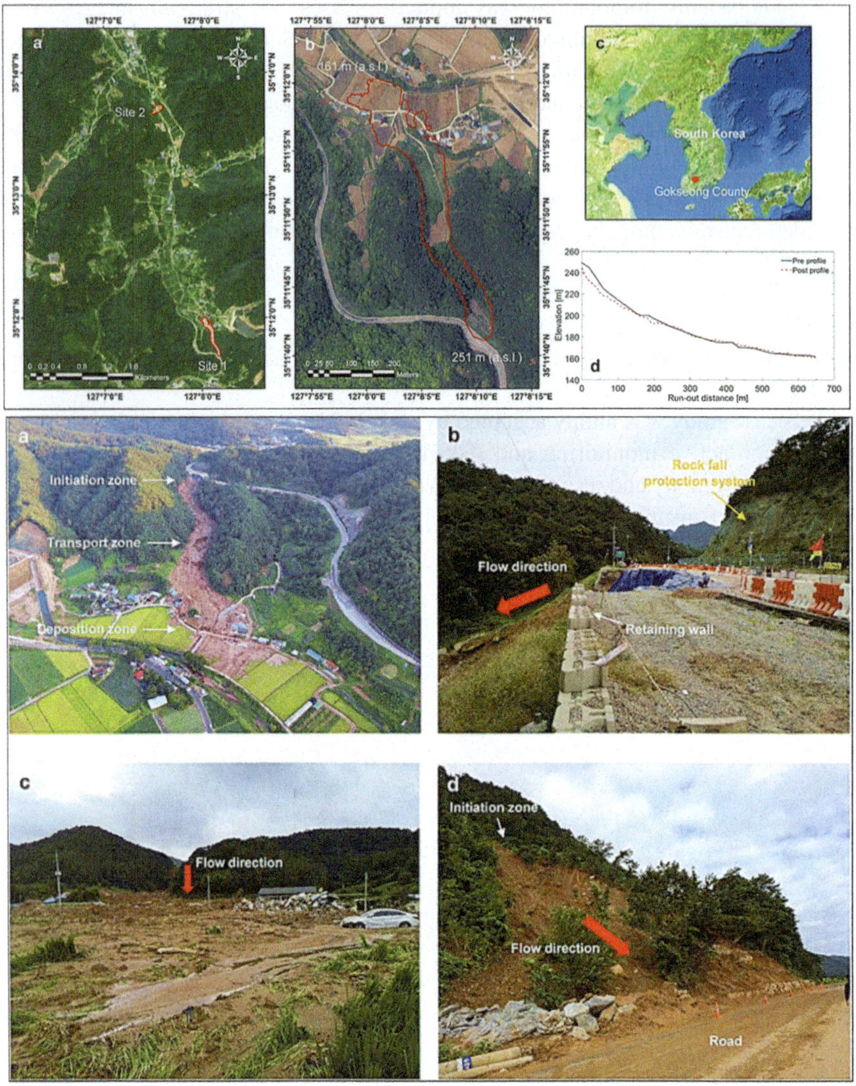

Fig. 7.10 **a** Optical and Geospatial analysis of the Gokseong Landslide Site **a** Sentinel-2 image showing the landslide-affected areas (Sites 1 and 2) after the event, **b** pre-event image of Site 1 captured by the Korea National Geographic Information Institute (KNGII) in 2019, **c** location map of Gokseong County in South Korea, and **d** cross-sectional profiles illustrating the topographic changes before and after the landslide. The red polygons highlight the landslide-impacted areas. **b** Digital photographs depicting key features of the Gokseong landslide. **a** Overview of the landslide at Site 1, **b** initiation zone of Site 1, **c** deposition zone of Site 1, and **d** overview of the landslide at Site 2. *Source* Choi et al. (2024)—Open Access

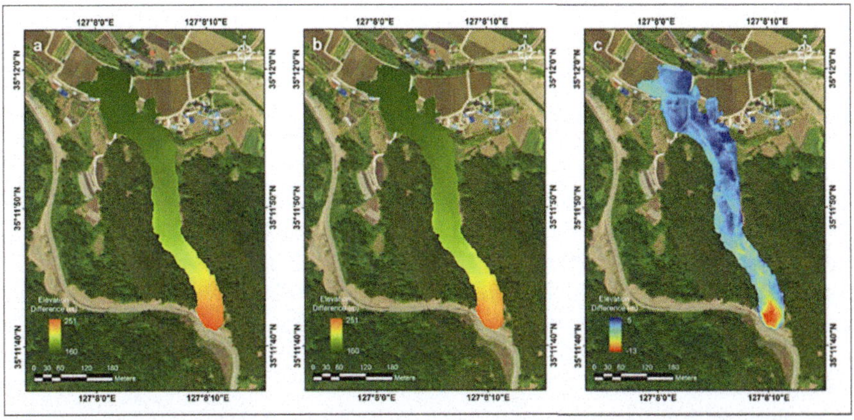

Fig. 7.11 Digital elevation models of the Gokseong landslide region. **a** Pre-event elevation model, **b** Post-event elevation model, and **c** Elevation difference map highlighting the source and deposition areas. *Source* Choi et al. (2024)—Open Access

m^3, respectively. The satellite InSAR analysis identified pre-landslide displacement correlated with rainfall patterns, highlighting the significance of continuous ground monitoring for early warning systems.

The study demonstrates the effectiveness of multi-source remote sensing technologies in landslide analysis. The integration of satellite InSAR, UAV-mounted RGB, multi-spectral, and LiDAR sensors provided a holistic understanding of the event, enhancing the accuracy of pre- and post-landslide assessments. This approach not only aids in comprehending complex landslide events but also facilitates the development of effective recovery and risk management plans.

This investigation underscores the potential of multi-source remote sensing technologies in analyzing landslides, offering valuable insights into pre-event displacement, spatial extent, and volumetric changes. By leveraging advanced tools such as UAV-mounted sensors and satellite InSAR, the study highlights a robust methodology for landslide monitoring and disaster management. The findings emphasize the importance of integrating diverse remote sensing datasets to enhance the accuracy of risk assessments and formulate effective mitigation strategies.

The findings of this study are particularly relevant to geoscientists, remote sensing specialists, disaster management professionals, and policymakers involved in landslide risk assessment and mitigation. The insights provided can also benefit researchers focused on the application of remote sensing technologies in natural hazard analysis and recovery planning.

7.7.5 *Flood Prediction Using Machine Learning and Remote Sensing*

Floods are natural disasters that lead to significant loss of life, damage to infrastructure, and major economic setbacks. While they cannot be completely prevented, steps must be taken by governments, organizations like the United Nations, and local communities to mitigate their effects. Early flood prediction plays a crucial role in minimizing hazards and enabling effective emergency response. This study examines remote sensing methods that aid in predicting floods, with a focus on technologies employed in the pre-disaster phase over the past two decades (Fig. 7.12).

The study aims to provide a comprehensive review of the remote sensing technologies used for flood prediction. The study classifies these technologies into three categories: multispectral, radar, and light detection and ranging (LIDAR) (Fig. 7.13). It explores how these methods contribute to flood prediction, risk assessment, and hazard analysis. The study also highlights existing gaps in these technologies and proposes a flood prediction and extent mapping model to address these issues.

The review evaluates various remote sensing techniques based on their applications in flood forecasting. The technologies are classified into multispectral, radar, and LIDAR categories, with further categorization according to the data analysis methods used in the studies. Notable methods include machine learning algorithms [e.g., CNN, AI, KNN, Bayesian linear models, SVM—convolutional neural networks (CNN), support vector machines (SVM), k-nearest neighbours (KNN)], hydrological models, and numerical modeling techniques. Machine learning methods are particularly emphasized for their data-driven approach, while hydrological models rely on human expertise to simulate flood events (Fig. 7.14a, b).

Several methods for data processing are identified, including the use of machine learning for flood prediction and the application of image segmentation techniques to detect water fluctuation levels. Hydrological models and numerical modeling approaches have also been employed in flood prediction, though these are limited by their dependence on the physical characteristics of specific study areas. The paper also discusses the use of satellite data, highlighting the challenges posed by data processing delays and issues with environmental conditions that can affect image quality.

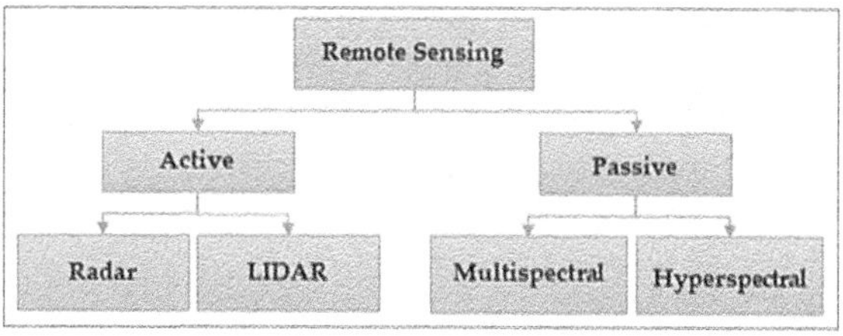

Fig. 7.12 Classification of remote sensing methods for flood prediction. *Source* Munawar et al. (2022)—Open Access

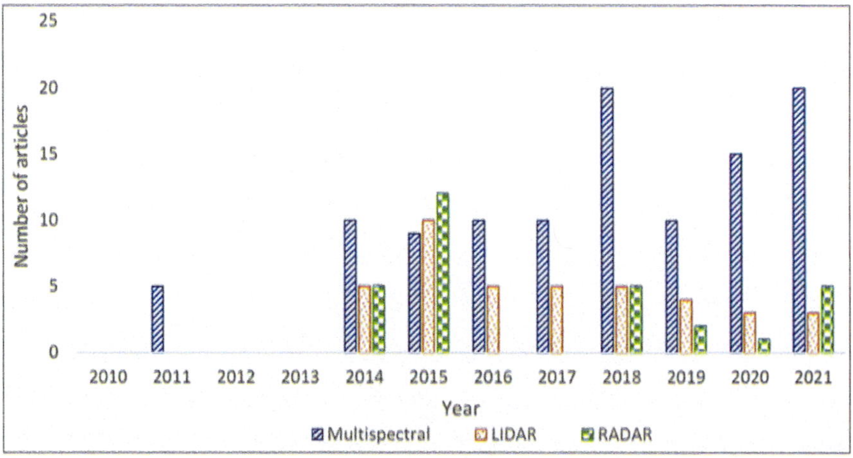

Fig. 7.13 Year-wise distribution of articles from each category of remote sensing methods for flood prediction. *Source* Munawar et al. (2022)—Open Access

The study discusses the gaps in remote sensing technologies, particularly in terms of satellite data acquisition cycles, image quality, and the high computational time required for processing large datasets. It also points out the limitations of hydrological and numerical models, such as their case-specific nature and the uncertainties associated with predicting floods. Machine learning models are identified as more adaptable to various conditions, but they require high-quality training data to be effective. The paper suggests that integrating deep learning techniques could enhance flood prediction accuracy, particularly by handling unstructured data and reducing the need for manual feature extraction.

The review demonstrates that machine learning methods and hydrological models are the most frequently used techniques in flood prediction. However, the research highlights the need for improvements, such as overcoming the challenges of continuous satellite monitoring and enhancing the robustness of flood prediction models in adverse conditions. A proposed system using deep learning models like CNN, RNN, and LSTM is presented, which could be trained using diverse datasets, including past flood events, satellite imagery, and weather reports. The integration of drones for real-time data collection and flood mapping is also suggested to improve disaster management.

While remote sensing technologies, including multispectral imaging, radar, and LIDAR, have advanced flood prediction capabilities, there are still significant gaps in continuous monitoring and data processing. The paper advocates for the use of deep learning models to handle unstructured data and improve prediction accuracy. Future research should focus on real-time flood mapping and post-flood rescue operations to ensure faster response times and more effective disaster management.

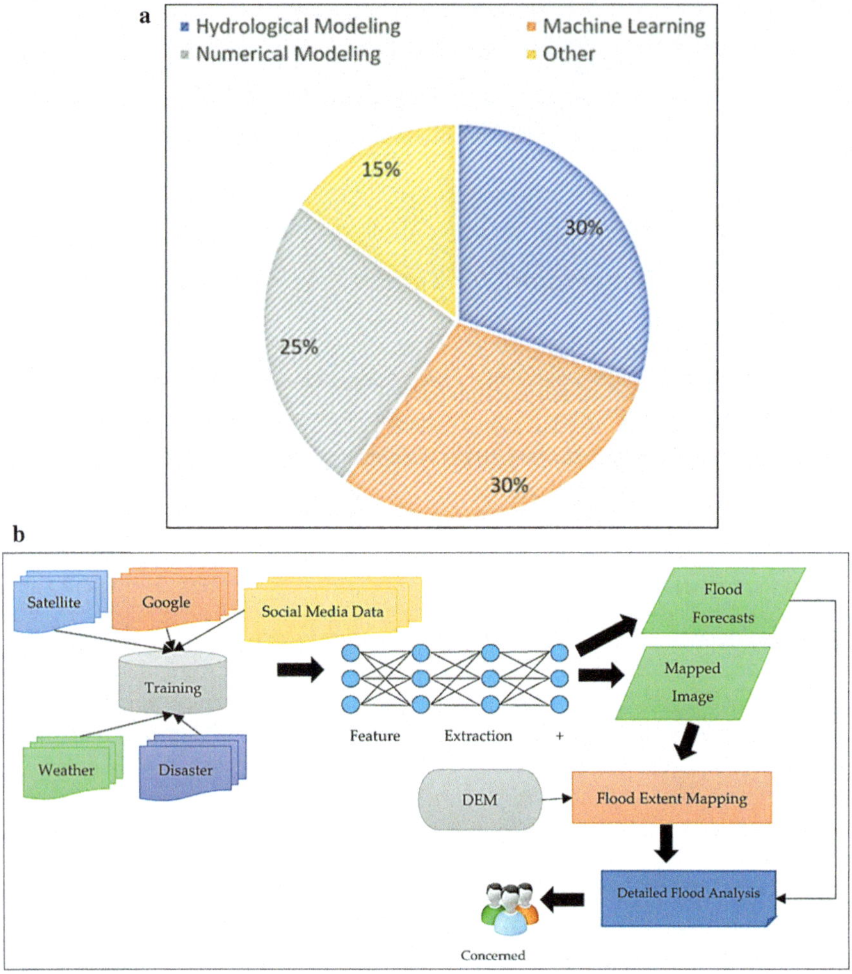

Fig. 7.14 **a** Methods adopted in studies for flood prediction using remote sensing and machine learning techniques. **b** Schematic diagram of the proposed deep learning-based system for real-time flood detection and extent mapping. *Source* Munawar et al. (2022)—Open Access

The target audience for this review includes disaster management professionals, researchers in the fields of remote sensing, hydrology, and machine learning, as well as policymakers involved in flood mitigation and response strategies.

The study proposes that future flood prediction systems integrate data from multiple sources, including satellite imagery, social media, and weather reports. The application of deep learning techniques could allow for more accurate flood mapping and real-time forecasting, providing valuable tools for emergency responders. Research should also focus on improving the efficiency of systems to ensure timely flood mapping and response during emergencies. The use of UAVs and drones for localized flood data collection could offer significant improvements in monitoring and assisting rescue operations.

7.7.6 Flood Extent Monitoring in the Lower Amazon with ALOS/PALSAR

The monitoring of flood extent in the lower Amazon River floodplain is essential for understanding hydrological dynamics and improving disaster management in the region (Lamb and Brunt 2024). This complex and ecologically rich area, including the Curuai Lake floodplain, presents challenges for traditional monitoring methods due to its vast and inaccessible nature. Remote sensing technologies, specifically ALOS/PALSAR ScanSAR imagery, offer a solution for capturing flood dynamics over large spatial scales, even in difficult environmental conditions. The study by Arnesen et al. (2013) focuses on using these satellite-based radar images to assess flood extent in the Amazon, analyzing backscattering behaviors and the influence of seasonal variations on land cover classification.

The primary goal of the study was to evaluate the effectiveness of ALOS/PALSAR ScanSAR data in monitoring flood extent in the Curuai Lake floodplain by examining the backscattering patterns of various land cover types, such as open water, soil, and macrophytes. These land cover types exhibit distinct radar backscatter signatures that change according to environmental factors like water stage and wind conditions (Balenzano et al. 2011). By assessing seasonal variations in backscattering, the study aimed to improve flood extent mapping accuracy, which is crucial for both hydrological assessments and flood risk management.

In terms of methodology, the study relied on ALOS/PALSAR ScanSAR images, which are particularly useful for monitoring flood-prone areas due to their ability to capture data irrespective of weather conditions and lighting (Henderson and Lewis 2008). The analysis focused on the backscatter behaviors of open water, soil, and macrophytes in response to varying water stages. As backscattering is sensitive to environmental factors such as moisture content and surface roughness, these variations were carefully examined for accurate classification of land cover types (Rodriguez et al. 2005). The study also implemented the J4.8 data mining algorithm to classify land cover types based on seasonal variations in backscattering and integrated DEM data with interpolated water surface elevations to simulate flood extent.

The study's results highlighted significant seasonal variations in the backscatter of open water, soil, and macrophytes. Open water backscatter patterns were influenced by water stage and wind conditions, leading to variations between rough and smooth water. Soil areas exhibited changes in backscatter due to fluctuations in moisture content and surface roughness, while macrophytes demonstrated varying backscatter signatures based on species composition and biomass. These findings emphasized the need to account for seasonal changes in backscatter characteristics to improve the accuracy of flood extent mapping (Lewis et al. 2011). However, challenges remained in distinguishing between soil and open water, particularly at far range where their backscatter signals were similar, highlighting the necessity of integrating optical imagery for better classification (Silva et al. 2010).

In conclusion, the study by Arnesen et al. (2013) demonstrated the effectiveness of ALOS/PALSAR ScanSAR imagery for monitoring flood dynamics in the Amazon floodplain, offering a valuable tool for flood risk assessment and disaster management. By leveraging seasonal variations in backscatter and integrating radar data with optical imagery, the study showcased how remote sensing can improve flood extent mapping accuracy in complex floodplain environments. The study's findings provide essential insights into the application of SAR data for flood monitoring in large, ecologically sensitive areas like the Amazon, paving the way for better flood management strategies globally.

7.7.7 *Flood Risk in Upper Solo River, Central Java, Indonesia*

The Upper Solo Watershed in Indonesia, covering 6,000 km^2 of the larger 16,100 km^2 Bengawan Solo watershed, is a key region for water resources, agriculture, and daily needs. However, its topography, varying from flat to undulating terrain near mountains, makes it highly susceptible to flooding, especially during periods of intense rainfall (Fig. 7.15). The combination of geological, hydrological, and socio-economic factors in this watershed contributes significantly to flood risks, affecting both rural and urban populations (Jumadi et al. 2024).

This case study aims to assess flood risks in the Upper Solo Watershed using spatial data from multiple open access sources. The goal is to develop a GIS-based flood

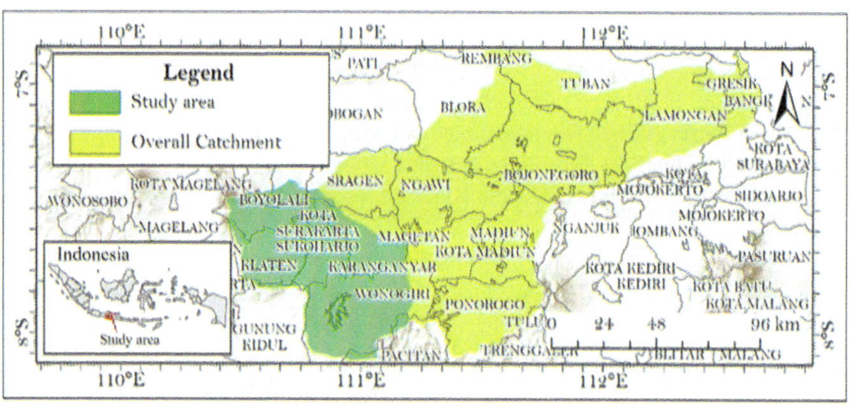

Fig. 7.15 Study area: upper Bengawan Solo Watershed, Java, Indonesia (16,100 km^2). *Source* Jumadi et al. (2024)—Open Access

risk model that incorporates physical, environmental, and socio-economic factors, providing valuable insights for flood management and mitigation strategies.

The study used a general risk function to determine the flood risk levels of the Upper Solo Watershed (Sar et al. 2015; UNISDR 2004). The framework was divided into three main sections: flood hazard, vulnerability, and risk analysis. The flood hazard was assessed using physical parameters, including elevation, land slope, flow accumulation, and drainage density, while vulnerability was indicated by parameters related to population density and urbanization. The final risk was determined by multiplying the flood hazard by the vulnerability. The general risk function follows the equation:

$$\text{Risk} = \text{H} \times \text{V} \tag{7.1}$$

where H represents flood hazard and V represents vulnerability. The parameters used to define the flood hazard include land slope (Sl), elevation (El), flow accumulation (FA), rainfall (Rf), drainage density (DD), distance to rivers (DR), topographic wetness index (TWI), normalized difference vegetation index (NDVI), land use and land cover (LULC), soil type (ST), and curvature (Cu). These parameters are crucial for identifying areas at risk of flooding due to their influence on water movement, soil properties, and hydrological conditions (Negese et al. 2022; Purwanto et al. 2023).

To build the flood risk model, various open access spatial datasets were used, including Shuttle Radar Topographic Mission (SRTM) data for elevation, Sentinel-2 images for land cover, Global Precipitation Measurement (GPM) data for rainfall, and Soil Moisture Active Passive (SMAP) data for soil moisture (Table 7.1). GIS-based flood hazard maps were created by overlaying parameters such as land slope, elevation, flow accumulation, rainfall, drainage density, and proximity to rivers (Fig. 7.16). Vulnerability was analyzed based on population density and urbanization levels using the Global Human Settlement Layer (GHS-SMOD) R2023A dataset. The flood risk model was developed by multiplying hazard and vulnerability levels, as shown in Fig. 7.17.

Table 7.1 Flood risk parameter data sources for the Upper Bengawan Solo Watershed, Java, Indonesia (Jumadi et al. 2024)

Dataset	Resolution	Purpose	Source
SRTM	30 m	Digital Elevation Model (DEM)	NASA/JPL
Sentinel-2	10 m	Land cover and land use mapping	European Space Agency
GPM	0.25°	Precipitation data	NASA
SMAP	36 km	Soil moisture condition	NASA
GHSL/GHS-SMOD	100 m	Human settlements and population	European Commission, 2023

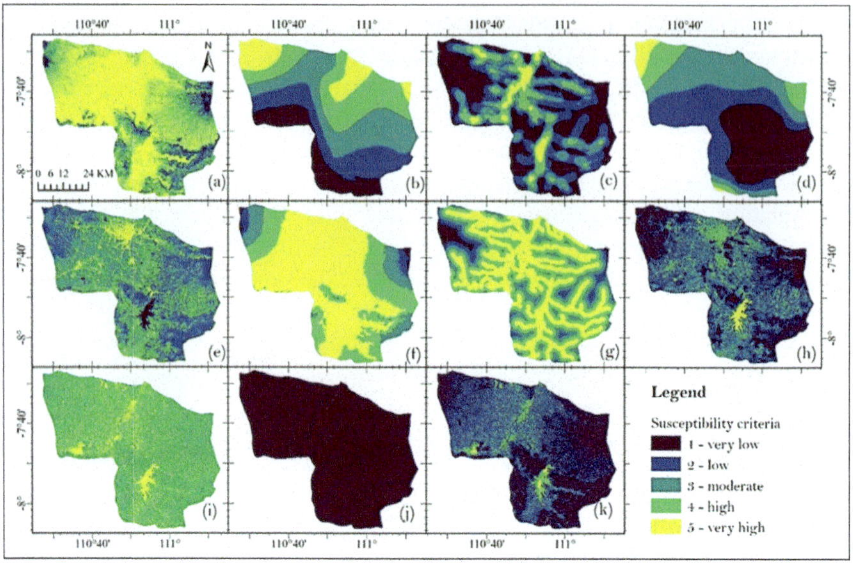

Fig. 7.16 Reclassified parameters: **a** Slope (Sl), **b** Rainfall (RF), **c** Digital Density (DD), **d** Soil Moisture (SM), **e** Land Use/Land Cover (LULC), **f** Elevation (El), **g** Drainage Ratio (DR), **h** Normalized Difference Vegetation Index (NDVI), **i** Curvature (Cu), **j** Forest Area (FA), **k** Topographic Wetness Index (TWI). *Source* Jumadi et al. (2024)—Open Access

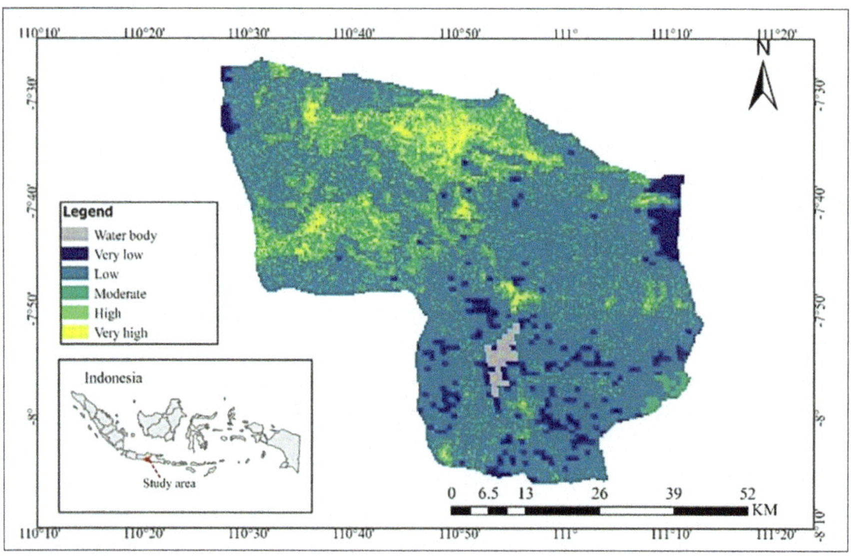

Fig. 7.17 Vulnerability class in the Upper Bengawan Solo Watershed, Java, Indonesia. *Source* Jumadi et al. (2024)—Open Access

The flood hazard model revealed that areas with low elevation, proximity to rivers, high flow accumulation, and low vegetation cover were most vulnerable to flooding (Absori et al. 2023). Regions such as Surakarta, Boyolali, and Karanganyar exhibited varying flood risk levels, with Surakarta showing the highest concentration of "Very High" flood risk. The flood vulnerability analysis highlighted that urbanized areas, such as Surakarta, had a higher vulnerability due to dense populations and rapid urbanization. For instance, Surakarta's high-risk area comprised 57.3% "High" risk zones, which underscores the importance of focused urban planning for flood mitigation (Cisternas et al. 2024; Muryani et al. 2021).

This study provides critical insights into the flood risks in the Upper Solo Watershed, identifying key risk zones such as Surakarta and Boyolali. The GIS-based model successfully integrated physical parameters with socio-economic data to produce accurate flood hazard and vulnerability maps. These maps serve as vital tools for urban planners and policymakers to prioritize flood management interventions (Chakraborty et al. 2023).

The findings emphasize the need for targeted flood mitigation strategies in high-risk areas, particularly in urban centers like Surakarta, which face significant flood threats due to both natural and human-induced factors (Fig. 7.18). The research highlights the value of open access spatial data in flood risk mapping and the potential of GIS-based models to inform effective disaster management strategies (Bakhtiari et al. 2024). Future studies should incorporate long-term climate change scenarios and urban expansion to enhance the accuracy of flood risk predictions (Marhaento et al. 2021).

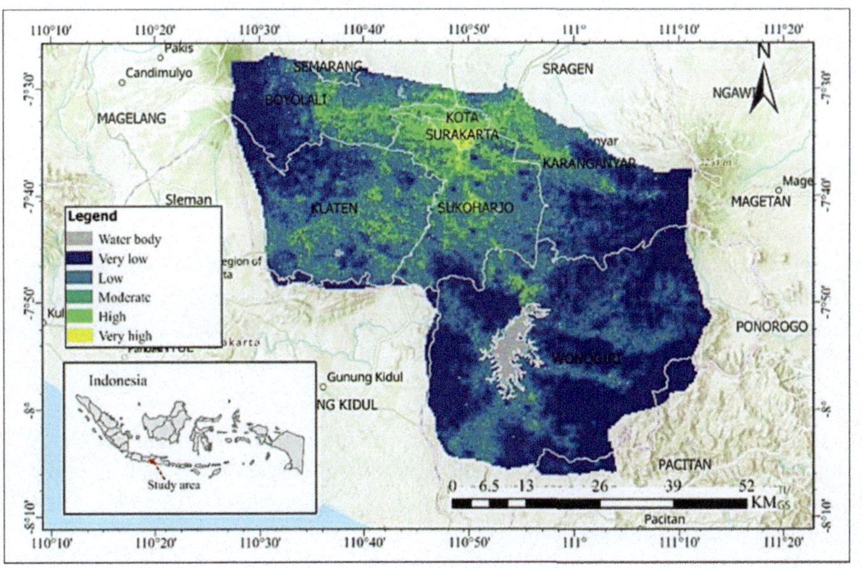

Fig. 7.18 Flood risk map of the Upper Bengawan Solo Watershed, Java, Indonesia. *Source* Jumadi et al. (2024)—Open Access

This case study is aimed at environmental scientists, urban planners, and policy-makers involved in flood risk management. Future directions include integrating more granular, high-resolution datasets and long-term climate projections to improve flood forecasting models (Hussain et al. 2021). Additionally, incorporating field validation and hydrological modeling will enhance the reliability of the risk maps, ensuring more precise mitigation planning for local governments and communities.

7.7.8 Earth Observation and ML Integration for Flood Risk Management

Flood monitoring and forecasting have significantly benefited from advancements in Earth Observation (EO) data and the integration of machine learning (ML) models. A notable case study is the 2021 European Summer Floods, which caused catastrophic damage in countries like Germany and Belgium (Fig. 7.19). The integration of EO data, including Synthetic Aperture Radar (SAR) from Sentinel-1 and optical imagery from Landsat and MODIS, provided critical real-time flood monitoring information that supported disaster response efforts. In this case, Schumann et al. (2023) demonstrated the successful integration of SAR data with hydrological models to accurately map flood extents. SAR's ability to capture flood events, even under cloud-covered conditions, was pivotal for monitoring rapidly evolving flood events. The study focused on data assimilation techniques, incorporating EO data into hydraulic flood models. By combining real-time satellite-derived flood extent maps with hydraulic model simulations, flood predictions were substantially enhanced, reducing uncertainties associated with traditional flood forecasting methods.

This study investigates the use of Earth Observation (EO) data in flood monitoring, focusing on the integration of remote sensing technologies with flood models and the application of machine learning (ML) to enhance flood prediction, management, and disaster response.

Floods are essential natural events that support ecosystems but also cause fatalities and significant economic loss when abnormal. EO data are vital for flood mapping, model integration, and prediction improvement. This article reviews optical imagery, Synthetic Aperture Radar (SAR), altimetry, and passive microwave radiometry. It outlines their roles in flood mapping and highlights how these data sources integrate with hydrological and hydraulic models for enhanced flood management.

The methods explored include:

Optical Imagery—Using high-resolution imagery from missions like NASA's Landsat and ESA's Sentinel-2, optical data enable classification and mapping of flooded areas. Techniques like the Normalized Difference Water Index (NDWI) (McFeeters 1996) are employed:

$$\text{NDWI} = \frac{\text{Green} - \text{NIR}}{\text{Green} + \text{NIR}} \tag{7.2}$$

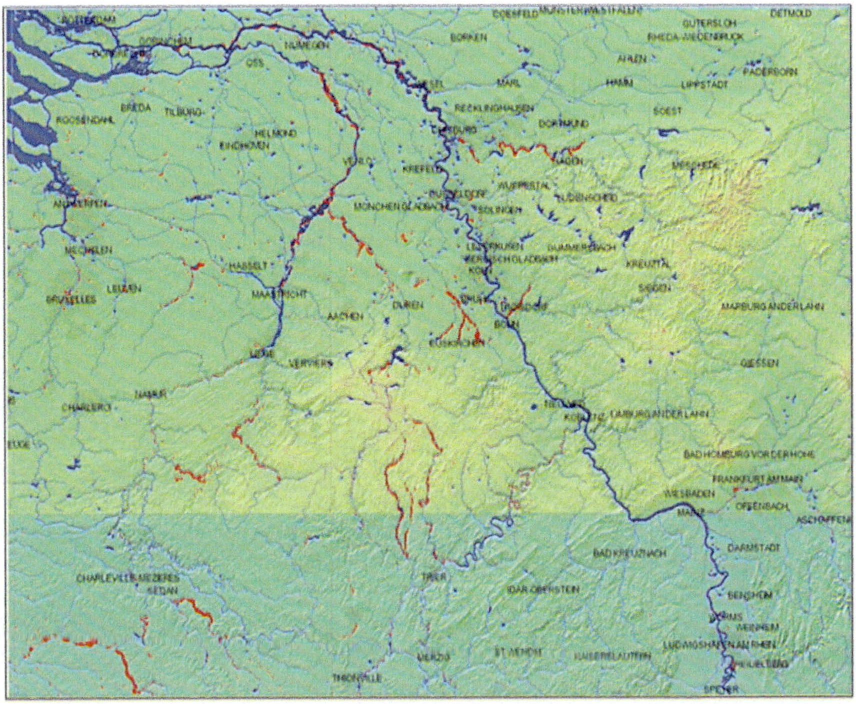

Fig. 7.19 European summer floods, 2021: red indicates flooded areas, with blue showing the mean annual flood extent. Dark gray represents previously mapped floods, and light gray shows city lights. Flood data were collected using various remote sensing sources, including MODIS and Sentinel-1 SAR. *Source* Schumann et al. (2023)—Open Access

where Green and NIR represent the reflectance [Reflectance refers to the proportion of light or radiation that is reflected by a surface compared to the total amount of light or radiation that hits it. It is often expressed as a percentage or a value between 0 and 1, where 0 means no reflection (complete absorption) and 1 means total reflection] in green and near-infrared bands, respectively.

Synthetic Aperture Radar (SAR)—SAR allows all-weather flood detection using techniques like change detection and time-series analysis to identify flood boundaries. Convolutional Neural Networks (CNNs) can be used for automatic flood detection (Wieland and Martinis 2019).

Altimetry—Satellite altimetry (ICESat-2) provides water surface elevation data, which is essential for monitoring river discharge and estimating water flux in ungauged catchments. The equation for river discharge can be derived from:

$$\mathrm{Q} = \int \mathrm{A(x)} \cdot \mathrm{v(x)dx} \tag{7.3}$$

where A(x) is the cross-sectional area and v(x) is the velocity at position x.

Passive Microwave Radiometry—These systems detect large-scale water bodies, although with a spatial resolution of 20–100 km. Despite its broad coverage, the data has significant uncertainty due to the variable emissivity of different surfaces (Rees, 2012).

Data Assimilation and Flood Models—The integration of EO data into hydrological and hydraulic models, using techniques like the Ensemble Kalman Filter (EnKF) and Particle Filter, helps improve flood predictions by reducing uncertainties.

EO technologies have significantly improved the accuracy and timeliness of flood mapping, despite challenges such as cloud cover, vegetation, and urban environments. The use of satellite altimetry and passive microwave systems has enhanced the understanding of large-scale flood dynamics and river discharge, particularly in remote or ungauged catchments (Neal et al. 2012). Additionally, SAR imagery has proven valuable in providing near-real-time flood maps, which are critical for rapid disaster response (Fig. 7.20). The integration of machine learning models with EO data has further improved flood risk analysis, allowing for more accurate and automated flood detection, especially in the presence of cloud cover.

EO data, especially from SAR and altimetry, has proven to be crucial for flood monitoring and prediction. These data sources complement traditional flood models by providing additional spatial and temporal insights. The integration of AI and ML into flood monitoring systems has demonstrated potential for automating data interpretation and improving model performance. By enhancing the accuracy and timeliness of flood predictions, these technologies enable more effective flood risk management and disaster response.

Satellite remote sensing and EO data have transformed flood monitoring and management by providing valuable insights into flood extent, severity, and dynamics. The integration of advanced satellite technologies with hydrological models has

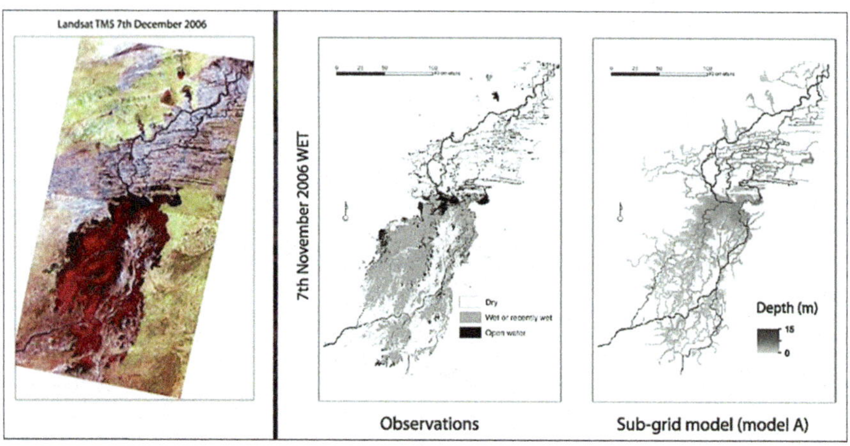

Fig. 7.20 Comparison of wet season flooding—a comparison between a 2-D flood model simulation and a Landsat satellite image of the Niger Inland Delta. *Source* Schumann et al. (2023)—Open Access

reduced uncertainties in flood predictions, while machine learning techniques have further enhanced the accuracy of these models. The growing availability of EO data, coupled with AI advancements, promises to significantly improve global flood monitoring and risk management, particularly with future missions such as Surface Water Ocean Topography (SWOT).

The primary audience for this research includes disaster response agencies, environmental scientists, flood risk managers, and satellite data providers. The future of flood monitoring lies in the continued advancement of satellite technology, such as the upcoming SWOT mission, and the further integration of AI and ML techniques into flood prediction models. Ongoing improvements in data assimilation, model calibration, and real-time forecasting will enable more precise and timely flood risk assessments, ultimately aiding in disaster preparedness and mitigation. Future research should focus on overcoming current limitations in flood detection, enhancing machine learning algorithms, and improving data fusion techniques to better handle complex flood scenarios.

7.8 Conclusions

This chapter has explored the pivotal role of geophysics in disaster risk management and hazard mitigation, focusing on the application of innovative seismic, geodetic, and remote sensing technologies to enhance our ability to assess and mitigate natural hazards such as earthquakes, floods, and landslides. The integration of geophysical methods, including seismology, geodesy, and geophysical surveys, has proven invaluable in identifying vulnerabilities, monitoring hazard progression, and providing critical data for risk assessments and early warning systems.

Key advancements, such as improved seismic network technologies, microseismic monitoring, InSAR, and GNSS, have significantly strengthened our capacity to detect ground deformations, monitor seismic activity, and predict natural hazards with greater accuracy. Remote sensing technologies like LiDAR, satellites, and UAVs further contribute to real-time hazard monitoring and risk management, offering new insights into landscape changes and disaster-prone areas.

The chapter also emphasizes the importance of mitigation strategies, including structural reinforcement, land-use planning, and the establishment of comprehensive early warning systems. By incorporating geophysical data into these strategies, communities are better equipped to respond to natural disasters, minimize damage, and protect lives and infrastructure.

Ultimately, the continued development of geophysical technologies and their integration with other monitoring systems holds the key to advancing disaster risk management and mitigation efforts. The insights gained from geophysical research and case studies highlight the critical need for a proactive, data-driven approach to reduce vulnerability and increase resilience to natural hazards worldwide.

References

Absori S et al (2023) Flood risk and mitigation strategies in Surakarta: case study of the Bengawan Solo Watershed. J Environ Stud
Ali M, Raj A (2023a) Monitoring surface displacements and earthquake predictions using InSAR and GNSS. Seismol J 45(3):123–137
Ali M, Raj P (2023b) Land use planning for hazard mitigation: a global perspective. Springer
Ali S, Raj VS (2023c) Advances in seismic risk assessment: the role of geophysics in hazard mitigation. J Earthquake Eng 21(3):212–229
Aly AM, Abdelsalam MA, Mohamed HSA (2022). Gravity measurements and their role in earthquake and landslide monitoring. J Geophys Res Solid Earth 127(5):e2022JB023478
Anderson K, Gaston KJ (2013) Light-weight unmanned aerial vehicles (UAVs) in ecological research: opportunities and challenges. J Appl Ecol 50(4):1214–1225
Anderson JE, Ferrini FW (2022) Remote sensing technologies for earthquake monitoring and risk assessment. Geophys J Int 191(2):45–61
Araujo A, Ferreira VM, Silva ELR (2020) Earthquake early warning systems: the role of seismology in risk reduction. Seismol Res Lett 91(5):2295–2308
Arnesen AS, Silva TSF, Hess LL, Novo EMLM, Rudorff CM, Chapman BD, McDonald KC (2013) Monitoring flood extent in the lower Amazon River floodplain using ALOS/PALSAR ScanSAR images. Remote Sens Environ 139:123–134
Bakhtiari M et al (2024) Open access spatial data for disaster management: a review. Environ Syst Technol 22(4):299–310
Balenzano A, Gatti A, Stefoni S (2011) Radar backscatter analysis of tropical forests using L-band SAR data. J Geophys Res 116(D10)
Bamler R, Hartl P (1998) Synthetic aperture radar interferometry. Inverse Prob 14(4):R1–R54
Barton N, McHugh CMN (2021) Advances in ground geophysical techniques for landslide monitoring. Geotech Eng 43(4):109–118
Beroza GC, Cole AT, Ellsworth WL (1995) Stability of coda wave attenuation during the Loma Prieta, California, earthquake sequence. J Geophys Res 100:3977–3987
Berti M et al (2022) Rainfall thresholds for landslide prediction: advances and challenges. Nat Hazard 22(6):1809–1825
Bingen BC, Malagnino PF, Hilley NJ (2023) Using InSAR for landslide monitoring and early warning systems. Remote Sens Environ 277:112632
Bock Y et al (2000) Detecting tectonic deformations with GPS: the southern California example. Geophys Res Lett 27(15):2301–2304
Bock Y, Klein J, Zeng X (2021) Advances in GNSS data analysis for strain mapping along fault zones. J Geophys Res 124(8):785–804
Bouchon M, De Chabalier GM, Rivera AR (2022) Seismic monitoring and hazard assessment in active fault zones. Geophys J Int 226(3):1267–1283
Cai S, Zhang HW, Wang XJ (2023) Advancements in GNSS technology for earthquake Monitoring and risk assessment. J Geodesy 97(4):101–117
Chakraborty A et al (2023) GIS-based approaches for flood risk assessment. J Flood Risk Manag 17(1):67–83
Chaulya SK, Mishra PK, Kumar N, Kumar V, Rawani VK (2024) Landslide monitoring and prediction using integrated geotechnical sensors and wireless technology. J Geotech Eng 2:6
Chavez P et al (2016) LiDAR for archaeology: case studies from Mexico and Central America. Remote Sens 8(10):805
Chen L, Zhang XL, Zhang JY (2022) Seismic hazard assessment and fault zone characterization using seismic reflection surveys. Geophys J Int 228(2):542–557
Choi SK, Ramirez RA, Lim HH, Kwon TH (2024) Multi-source remote sensing-based landslide investigation: the case of the August 7, 2020, Gokseong landslide in South Korea. Sci Rep 14:12048
Cisternas L et al (2024) Local government's role in enhancing flood resilience. Nat Hazards Manag

Dixon TH et al (1999) Continuous GPS measurements of crustal deformation in the San Francisco Bay Area. Geophys Res Lett 26(16):2389–2392

Dixon TH, Edwards SMR, Klemperer LA (2023) Advances in seismic imaging and their role in earthquake risk assessment. Seismol Soc Am Bull 113(1):15–28

Duchkov A, Melekhova AG, Ponomareva IV (2023) Time-lapse electrical resistivity tomography for landslide monitoring and risk assessment. J Appl Geophys 199:103734

El-Mekawy M et al (2015) Satellite and UAV data integration for monitoring urban land use and infrastructure. Int J Appl Earth Obs Geoinf 35:167–177

Field EH, Graves RW, Williams RA (2019) Seismic hazard models and earthquake ground motion forecasting. Seismol Soc Am Bull 109(6):2227–2245

Freymueller JT et al (2000) Continuous GPS measurements of crustal deformation in the Alaska-Aleutian subduction zone. Geophys Res Lett 27(15):2305–2308

Fukuyama E et al (2021) Microseismic monitoring in volcanic regions: insights from Japanese seismic networks. Volcanol Geotherm Res 410:107234

Geological Survey of India (GSI) (2013) A report on regional gravity and magnetic mapping in Bhilwara Supergroup in part of Tonk and Dausa districts, Rajasthan. GSI, Unpublished Report Field Season (2012–2013)

Ghofrani H, Foulger JS, Moosavi MSM (2021) Advances in real-time seismic data processing for earthquake early warning systems. Seismol Res Lett 92(4):1802–1812

Gill JC, Malamud BD (2016) Hazard interactions and interaction networks (cascading disasters) within multi-hazard methodologies: a review. Earth Sci Rev 160:128–146

Gou X et al (2022) The role of remote sensing in flood prediction and monitoring. J Hydrol 605:127421

Guglielmi Y, Bott JA, Moscardelli JG (2022) Geodetic methods for monitoring landslide movements and early detection systems. Geomorphology 398:107888

Harris RA, Day SM (1993) Dynamics of fault interaction: Parallel strike-slip faults. J Geophys Res: Solid Earth 98(B3):4461–4472

Harrison J, Lee T (2018) Integrating geophysical techniques for natural disaster risk assessment in mountainous regions. J Nat Hazards 10(3):225–240

Henderson FM, Lewis AJ (2008) Radar technology for floodplain management and monitoring. IEEE Trans Geosci Remote Sens 46(10):2966–2975

Hickman SH, Zoback MD, Ellsworth WL (2005) Structure and composition of the San Andreas fault zone at Parkfield: initial results from SAFOD phase 1 and 2. Eos Trans Am Geophys Union 86(52)

Hodgson ME, Bresnahan P (2004) Accuracy of airborne LiDAR-derived elevation: empirical assessment and error budget. Photogramm Eng Remote Sens 70(3):331–339

Hudson J, Davies EH, Hayes FH (2023) Using ground-penetrating radar for landslide hazard assessment in mountainous regions. Landslides 20(3):1127–1140

Hussain I et al (2021) Flood risk and socio-economic determinants in Southeast Asia. Glob Environ Chang 18(3):45–60

Intrieri E et al (2019) Landslide early warning systems: Monitoring, modeling, and implementation. Eng Geol 250:78–90

Jin L, Klein MP, Zhang YQ (2023) Crustal deformation and earthquake risk assessment using GNSS data. Seismol Res Lett 94(1):303–317

Johnson A, Parker C (2022) Microgravity surveys for detecting subsurface anomalies in slope stability monitoring. Geophys Res Lett 45(9):85–102

Jones D, Smith A, Taylor B (2020) Ground-penetrating radar in landslide research: a review of applications and techniques. J Appl Geophys 15(4):112–130

Jumadi J, Danardono D, Priyono KD, Roziaty E, Masruroh H, Rohman A, Amin C, Hadibasyir HZ, Fikriyah VN, Nawaz M, Sattar F, Lotfata A (2024). Utilizing open access spatial data for flood risk mapping: a case study in the Upper Solo Watershed. Geoplann J Geomat Plann 11(2):189–204

Khan SH, Rao AK, Al-Harbi MY (2021) Geophysical monitoring of subsurface fluid movements using gravity measurements. Geophys J Int 226(2):567–578

Kim HJ et al (2022) Machine learning applications in microseismic event detection. Geophys J Int 230(2):751–765

Kossobokov V, Dublenkov ADS, Poisson ES (2021) Seismic ground motion prediction and earthquake hazard models. J Seismolog 25(2):453–470

Kumar P, Singh R (2021) Seismic hazard mapping and structural reinforcement in landslide-prone regions: a case study from the Indian Himalayas. Nat Hazard Rev 22(4):04021022

Kumar N, Chaulya SK, Mishra PK, Kumar V, Rawani VK (2024) Real-time landslide monitoring system for early warning and mitigation. Environ Monit Assess

Lacroix P et al (2020) Satellite-based monitoring of landslide deformation using InSAR techniques. Remote Sens Environ 236:111512

Lamb A, Brunt J (2024) Assessing flood risks using spatial data in river basins: a comprehensive approach. J Flood Risk Manag 18(2):135–150

Lewis AJ, Chapman BD, McDonald KC, Santos LR (2011) Hydrological extremes in the Amazon: implications for flood monitoring. J Hydrol 409(1–2):88–99

Li YG, Leary P (1990) Characterization of the San Andreas fault using fault-zone trapped waves. Geophys J Int 102(2):411–426

Li YG, Ellsworth WL, Thurber CH, Malin PE, Aki K (1997) Observations of fault-zone trapped waves excited by explosions at the San Andreas fault, central California. Bull Seismol Soc Am 87:210–221

Li YG, Leary P (2004) Seismic characterization of the fault zone along the San Andreas fault: results from fault zone trapped waves. J Geophys Res 109(B5):B05303

Li YG, Vidale JE, Cochran ES (2004) Low-velocity damaged structure of the San Andreas fault at Parkfield from fault zone trapped waves. Geophys Res Lett 31(12):L12S06

Li YG, Chen P, Cochran ES, Vidale JE, Burdette T (2006) Seismic evidence for rock damage and healing on the San Andreas fault associated with the 2004 M6.0 Parkfield earthquake. Bull Seismol Soc Am 96(4B):S349–S363

Li YG, Malin PE, Vidale JE (2007a) Low-velocity damage zone on the San Andreas fault at depth near SAFOD site at Parkfield delineated by fault-zone trapped waves. Sci Drill S1:73–77

Li YG, Chen P, Cochran ES, Vidale JE (2007b) Seismic velocity variations on the San Andreas fault caused by the 2004 M6 Parkfield Earthquake and their implications. Earth Planets Space 59(1):21–31

Li YG, Malin PE, Zoback MD (2008) Seismic velocity reduction and fault zone damage in the San Andreas fault: a detailed study from Parkfield. Geophys J Int 175(3):1065–1079

Li YG (2021) Spatial-temporal characterization of the San Andreas fault by fault-zone trapped waves at seismic experiment site, Parkfield, California. Earthq Sci 34(3):261–285

Li J, Zhang LL, He JM (2022) Volcanic deformation monitoring using InSAR and geodetic techniques. J Geophys Res Solid Earth 127(4):e2022JB023479

Liao M et al (2013) Satellite and UAV technologies for disaster monitoring and response. Int J Remote Sens 34(7):2497–2517

Linde AT et al (1994) The 1992 Landers, California, earthquake: Evidence for fault interaction and aseismic slip. Science 265(5174):1432–1435

Liu Q, Zhang WL, Wang FH (2021) Multi-frequency ground-penetrating radar for assessing landslide risk in subsurface geology. Geophys Prospect 69(4):834–850

Lopez D, Martínez RMG, Hernandez FP (2021) Magnetic anomalies and volcanic activity: monitoring using magnetic surveys. J Volcanol Geoth Res 409:107250

Lu Z, Dzurisin D (2014) InSAR imaging of Aleutian volcanoes: monitoring, interpretation, and modeling. Springer

Mallet C, Bretar F (2009) Airborne LiDAR for 3D terrain mapping: a review. ISPRS J Photogramm Remote Sens 64(1):1–16

Marhaento T et al (2021) Integrating long-term climatic change in flood models. Hydrol Process 35(7):1543–1559

Massey R, Thompson HY, Evans GK (2022) Infrastructure assessment using ground-penetrating radar in earthquake-prone areas. J Struct Eng 148(4):04022036

Maxwell SC et al (2023) Advancements in microseismic monitoring for induced seismicity analysis. J Seismolog 27(1):45–63

McFeeters SK (1996) The use of the normalized difference water index (NDWI) in the delineation of water features. Int J Remote Sens 17(7):1425–1432

Mishra PK, Chaulya SK, Kumar N, Kumar V, Rawani VK (2024) Application of multivariate statistical analysis for landslide prediction: a case study. J Disast Prevent Manag

Moser M, Fisher G, Reynolds P (2023) Integrating GPR and microgravity data for landslide prediction in the Swiss Alps. Environ Geophys 21(1):70–82

Mulla DJ (2013) Twenty-five years of remote sensing in precision agriculture: key advances and remaining knowledge gaps. Biosys Eng 114(4):358–371

Mumby PJ et al (2017) Remote sensing of coral reefs and mangroves with UAVs. Remote Sens Environ 156:1–16

Muryani N et al (2021) Community awareness for flood risk management in urban regions. Int J Environ Educ Res

Nagarajan R, Krishnan R, Rao KS (2010) Geophysical studies in Rajasthan: a review of gravity and magnetic data. J Geophys 48(3):227–234

Naskar DC (2005) Geophysical studies of crust and upper mantle structures in parts of Madhya Pradesh and Maharashtra, Central India. Ph.D. thesis, p 150

Naskar DC, Das LK, Roy KK, Singh OP (2008) Structural studies of crust and upper mantle in parts of Madhya Pradesh, Central India. Geological Survey of India, Pan-African Event: India & Antarctica, Special Publication No. 91, pp 211–228

Naskar DC, Lashkari G (2019) Geophysical study in Tonk and Dausa districts, Rajasthan: its implication in regional geology. J Indian Geophysi Union (JIGU) 23(6):485–498

Neal JC, Schumann GJP, Bates PD (2012) A sub-grid channel model for simulating river hydraulics and floodplain inundation over large and data sparse areas. Water Resour Res. https://doi.org/10.1029/2012WR012514

Negese T, Purwanto P, Kumara S (2022) Land use and hydrological parameters for flood risk mapping: a case study in Indonesia. Hydrol Res 53(2):209–220

Nguyen H, Kim PL, Chang LT (2022a) Electrical resistivity tomography for landslide detection in high-risk regions. J Geophys Res: Solid Earth 127(3):e2022JB023788

Nguyen H, Le D, Tran Q (2022b) Monitoring subsidence and reinforcing structures in coastal regions: a geophysical approach. Eng Geol 292:106371

Park JW et al (2021) Applications of fiber optic seismic networks in landslide monitoring. Landslides 18(5):1735–1748

Peng ZG, Ben-Zion Y (2006) Temporal changes of shallow seismic velocity around the Karadere-Duzce branch of the North Anatolian fault and strong ground motion. Pure Appl Geophys 163(2):567–600

Peppe R, De Barros MG, Lin FMY (2021) Fault characterization and the prediction of future earthquakes. J Geophys Res: Solid Earth 126(7):e2021JB021207

Pittore M et al (2018) Towards a multi-hazard risk-based approach to early warning. Int J Disast Risk Reduct 30:78–89

Popescu SC et al (2003) Tree height estimation with LiDAR in the eastern deciduous forest of the U.S. Comput Environ Urban Syst 27(5):441–454

Pritchard ME, Simons M (2004) Earthquake deformation and the shape of the earthquake cycle. Science 306(5700):861–866

Purwanto P, Kumara S, Sari A (2023) Topographic wetness index and land use dynamics for flood hazard modeling. J Environ Manage 295:113125

Rahman M, Singh VP (2020) Flood forecasting systems: a comprehensive review. J Hydrol 584:124825

Rangin C, Monnier RVMJ, Manzano AF (2021) Seismic deformation and fault monitoring using InSAR. Tectonics 40(12):e2021TC006537

Rao SS, Desai SS (2015) Structural analysis of the Jaipur region: insights from gravity data. Tectonic Stud 54(2):192–198
Rawani VK, Chaulya SK, Mishra PK, Kumar N, Kumar V (2025) Development and application of a wireless sensor network for landslide prediction. Sens Syst Disast Manag
Rees A (2012) Physical principles of remote sensing. Cambridge University Press
Rodriguez E, Burt JE, Tachikawa T, Sato H (2005) SRTM DEM and its potential application to floodplain hydrology. Remote Sens Environ 96(3):347–356
Roecker SW, Larocque PF, van der Linden THE (2022) Integration of geodesy in Tsunami Warning Systems. Nat Hazards 109(1):51–67
Rosi A et al (2021) Integration of geotechnical and geodetic monitoring in landslide risk management. Landslides 18(2):543–558
Sar N, Chatterjee S, Das Adhikari M (2015) Integrated remote sensing and GIS based spatial modelling through analytical hierarchy process (AHP) for water logging hazard, vulnerability and risk assessment in Keleghai river basin, India. Model Earth Syst Environ 1(4):31
Savage JC, Burford RO (1973) Geodetic strain in the San Andreas fault system, California. J Geophys Res
Savage JC et al (2001) Strain rates and fault behavior in California. Geophys Res Lett
Scholz CH (1990) The mechanics of earthquakes and faulting, vol xii. Cambridge University Press, New York, p 439. ISBN 0 521 33443 8
Schumann G, Giustarini L, Tarpanelli A, Jarihani B, Martinis S (2023) Flood modeling and prediction using earth observation data. Surv Geophys 44:1553–1578
Sharma D, Kumar R, Singh P (2017) The role of gravity surveys in mapping subsurface structures: a case study from Rajasthan. Indian J Geophys Res 39(1):58–67
Shen F, Zhang L (2017) Microgravity survey methods for subsurface investigation in landslide monitoring. Geophys Methods Eng Geol 11(3):33–47
Silva TSF, Novo EMLM, Hess LL, Pimentel R (2010) Flood mapping in the Amazon using RADARSAT-1 data: application to wetland environments. Remote Sens Environ 114(8):1790–1799
Singh M, Sharma S, Gupta A (2013) Gravity anomalies and geological structures in Rajasthan: implications for regional tectonics. Geol Soc India J 82(4):462–474
Smith J, Williams H, Davis R (2019) Geophysical monitoring of landslide-prone areas in the Alps: a study of stability and prediction techniques. Landslide Sci 8(5):500–515
Smith JK et al (2021) Advancements in flood early warning systems: integrating hydrological modeling with real-time data. Water Resour Res 57(4):e2021WR030601
Tong X, Xu J, Zeng H (2013) Integration of GNSS and InSAR for fault zone deformation analysis. Geophys Res Lett 40(16):4285–4291
Tregoning P, Wang H, Rahman MSM (2021) Using gravity data to monitor faults and earthquake hazards. Geophys J Int 224(3):1383–1395
Tromp J, Santillan BA (2021) Using LiDAR and remote sensing for disaster risk mitigation. Remote Sens Environ 244:103597
UNDRR (United Nations Office for Disaster Risk Reduction) (2022) Making cities Resilient 2030: a guide for local governments. UNDRR, Geneva
UNISDR (2004) Living with risk: a global review of disaster reduction initiatives
Unsworth M, Malin P, Egbert G, Booker J (1997) Internal structure of the San Andreas fault at Parkfield. Calif Geol 25(4):359–362
Wang W, Zhang LZ, Liu YY (2022) Airborne magnetic surveys for earthquake hazard monitoring and fault zone mapping. Tectonics 41(10):e2022TC007135
Wang X, Li M, Zhang L (2023) Application of geophysical techniques in flood risk mitigation and structural reinforcement in coastal regions. J Hydrol 610:128183
Weiss S, Bock Y, Xu J (2020) Using GNSS and InSAR data for seismic hazard assessments along the SAF. Geophys J Int 222(5):2628–2643
Wieland M, Martinis S (2019) CNN-based approach for flood mapping from SAR data. Remote Sens 11(3):354

Xu J et al (2016) Strain rate mapping along the San Andreas fault using GNSS and InSAR. Earth Planet Sci Lett 452:78–89

Xu J et al (2017) Combining InSAR and GNSS data for enhanced seismic hazard prediction. J Seismolog 21(1):41–54

Xu X, Sandwell DT, Klein E, Bock Y (2021) Integrated Sentinel-1 InSAR and GNSS time-series along the San Andreas fault system. J Geophy Res: Solid Earth 126:e2021JB022579

Zebker HA, Rosen PA (1994) Analysis of SAR interferograms. J Geophys Res: Solid Earth 99(B10):19633–19642

Zeng Y, Shen Z (2017a) Surface velocity and strain rate analysis of the San Andreas Fault using GNSS and InSAR data. J Geophys Res

Zeng X, Shen Z (2017b) Deformation modeling using GNSS and InSAR data along active faults. Geophys Res Lett 44(3):1234–1243

Zhang H et al (2004) Improved ionospheric delay estimation for GPS-based applications using a combination of GPS and ionospheric model data. J Geophys Res 109(B12):B12102

Zhang J, Wang H, Li Y (2021a) Enhancing resilience to earthquakes and floods using integrated geophysical surveys and reinforcement strategies. Geophys Res Lett 48(16):e2021GL095902

Zhang Y, Liu SM, Guo WJ (2021b) High-density seismic arrays for improved earthquake hazard assessment. Seismol Res Lett 92(5):2462–2475

Zhao J, Wu YZ, Dai HS (2020) Seismic tomography and fault zone imaging for earthquake hazard assessment. Tectonophysics 781:228440

Zhou X et al (2022a) Integration of DAS technology into seismic networks for enhanced microseismic monitoring. Seismol Res Lett 94(3):789–798

Zhou Y, Li XJ, Xu SM (2022b) Tectonic plate movements and seismic hazard assessment using GNSS data. Earth Planet Sci Lett 592:117598

Chapter 8
Geophysical Applications in Civil Engineering and Infrastructure

Abstract This chapter explores how geophysical methods support civil engineering and infrastructure development, from site investigation to long-term structural health monitoring. It covers advanced techniques such as ground-penetrating radar (GPR), seismic refraction, electrical resistivity tomography (ERT), magnetotellurics (MT), and distributed fiber optic sensing (DFOS). These tools help engineers understand subsurface conditions, assess construction sites, monitor bridge and tunnel safety, and detect early signs of structural damage. The chapter includes numerous global case studies, including metro tunnel alignment in India, bridge inspections in the USA and Germany, and foundation monitoring in the UK and Japan. It also discusses modern trends like satellite-based InSAR for urban subsidence and fiber optics for pipeline health monitoring. Emphasis is placed on integrating multiple geophysical methods for reliable and cost-effective solutions. Overall, the chapter shows how geophysics enhances infrastructure safety, sustainability, and resilience throughout its life cycle.

8.1 Introduction: The Role of Geophysics in Modern Civil Infrastructure

This chapter delves into the vital role of geophysical techniques in civil engineering, highlighting their applications from site investigation to ongoing structural health monitoring (SHM) of critical infrastructure. The chapter examines various geophysical methods, such as seismic tomography, ground penetrating radar (GPR), electrical resistivity tomography (ERT), and microtremor analysis, emphasizing their effectiveness in assessing subsurface conditions and monitoring the integrity of buildings, bridges, roads, and tunnels.

Case studies from across the globe illustrate the successful implementation of these techniques in real-world infrastructure projects, such as seismic imaging for subsurface characterization, GPR for bridge inspections, and fiber optic sensing for pipeline integrity monitoring. Furthermore, the chapter explores innovations in monitoring technologies, including satellite-based remote sensing, distributed fiber optic

D. C. Naskar, *Advances in Engineering Geophysics: Concepts and Applications*, Earth Systems Data and Models 9, https://doi.org/10.1007/978-3-032-28000-8_8

sensing (DFOS), and InSAR for subsidence detection, underscoring their potential to enhance the precision and efficiency of infrastructure assessment.

The chapter concludes by emphasizing the critical importance of integrating geophysical methods into infrastructure design, construction, and maintenance, offering a proactive approach to ensure long-term safety, sustainability, and resilience of civil engineering projects. As these techniques continue to evolve, they promise to revolutionize how infrastructure is managed, ensuring better performance and risk mitigation for a safer built environment.

8.2 From Site Investigation to Structural Health Monitoring

This section provides a comprehensive overview of geophysical techniques utilized in site investigations and their evolution into structural health monitoring tools. It discusses methods such as seismic refraction, ground-penetrating radar (GPR), electrical resistivity, and microtremor analysis, emphasizing their role in identifying subsurface features, evaluating soil stability, and detecting structural anomalies. The section also showcases case studies demonstrating the application of these methods in real-world projects, ensuring safety, sustainability, and cost-efficiency in modern civil engineering practices.

8.2.1 Advanced Geophysical Methods in Civil Engineering Projects

This section explores the application of cutting-edge geophysical techniques in civil engineering, focusing on their role in site characterization, subsurface mapping, and infrastructure assessment. Methods such as ground-penetrating radar (GPR), electrical resistivity tomography (ERT), seismic tomography, and multi-channel analysis of surface waves (MASW) are discussed in the context of improving design accuracy, construction safety, and long-term infrastructure performance.

Introduction

Geophysical methods are increasingly utilized in civil engineering to evaluate subsurface conditions, monitor structural health, and ensure the long-term stability of infrastructure. These techniques are crucial during site investigations, throughout the design and construction phases, and in the ongoing maintenance of structures. Advanced geophysical techniques such as seismic refraction, electrical resistivity tomography (ERT), ground-penetrating radar (GPR), and acoustic emission (AE) monitoring help engineers make informed decisions by providing high-resolution data about the subsurface environment. Additionally, these methods allow for real-time monitoring of existing infrastructure, detecting potential structural issues before they escalate into serious problems.

Site Investigation and Geotechnical Characterization

Site investigation is a crucial phase in any civil engineering project. Understanding the subsurface properties of the site is vital for designing safe and efficient foundations, evaluating risks, and preventing unexpected failures during construction. Several geophysical techniques are used for subsurface characterization, including seismic refraction, ERT, and GPR.

Seismic Refraction and Reflection: Seismic refraction and reflection are commonly used to map subsurface layers, determine the depth of bedrock, and evaluate material properties like stiffness and density. In seismic refraction, seismic waves travel through the ground and are refracted at the interface between different geological layers. The velocity of seismic waves varies with the type of material they pass through, and this difference can be exploited to identify changes in subsurface properties. The key equation that governs the travel time of seismic waves is:

$$V = \frac{d}{t} \tag{8.1}$$

where V is the wave velocity, d is the distance travelled by the seismic wave, and t is the time taken for the wave to travel that distance. This relationship is used to calculate the subsurface velocities and identify variations in the subsurface layers. For more complex subsurfaces, seismic reflection methods are used, which are based on the principles of Snell's Law to calculate the angle of refraction at interfaces between layers (Bohnke et al. 2020). The relationship between the angle of incidence and the angle of refraction is given by:

$$\frac{\sin \theta_1}{\sin \theta_2} = \frac{V_2}{V_1} \tag{8.2}$$

where θ_1 and θ_2 are the angles of incidence and refraction, V_1 and V_2 are the seismic velocities of the respective layers. This allows engineers to determine the depth and characteristics of subsurface layers, which are essential for foundation design in buildings and bridges.

Electrical Resistivity Tomography (ERT): Electrical resistivity tomography is a geophysical technique used to assess subsurface properties based on how well the ground conducts electricity. The resistivity (ρ) of the subsurface materials can reveal information about soil composition, water content, and the presence of voids or fractures. The fundamental equation used to calculate resistivity is:

$$\rho = \frac{V}{I} \cdot \frac{L}{A} \tag{8.3}$$

where ρ is the resistivity, V is the applied voltage, I is the current, L is the length of the electrode array, and A is the cross-sectional area. This method is especially useful in identifying areas of potential instability in foundations, such as weak zones or high-water content, which could affect the performance of buildings, bridges, or roads (Loke and Dahlin 2021). In road construction, ERT can be used to monitor subsurface drainage systems and detect leaks or material degradation.

Ground-Penetrating Radar (GPR): Ground-penetrating radar is a non-invasive method that uses electromagnetic waves to image the subsurface. GPR is particularly useful for identifying structural anomalies in concrete, such as voids, cracks, and the condition of reinforcement bars (rebar). The velocity v of the radar waves through the ground is related to the relative permittivity ε_r of the material:

$$v = \frac{c}{\sqrt{\varepsilon_r}} \tag{8.4}$$

where c is the speed of light in a vacuum and ε_r is the relative permittivity (dielectric constant) of the material. This relationship allows engineers to determine the depth and nature of subsurface features, which is critical for assessing the structural integrity of existing buildings, bridges, and roads (Daniels 2021). By analyzing the radar wave reflections from different subsurface layers, engineers can create high-resolution images of the internal structure of concrete, which is essential for structural health monitoring.

Design and Construction of Buildings, Bridges, and Roads

Once the site investigation phase is complete, advanced geophysical techniques continue to play an important role in the design and construction phases, ensuring that the structures are stable and capable of withstanding the required loads.

Shear Wave Velocity and Microgravity Surveys: Shear wave velocity is a critical parameter used in the design of foundations. It is closely related to the stiffness of the ground and plays a role in assessing the soil's ability to withstand dynamic loads, such as those caused by earthquakes. The shear wave velocity V_s is related to the shear modulus G and the density ρ of the material by the equation:

$$V_s = \sqrt{\frac{G}{\rho}} \tag{8.5}$$

where G is the shear modulus, ρ is the density, and V_s is the shear wave velocity. This relationship helps determine the appropriate foundation design, especially in areas with challenging soil conditions (Murphy and Hansen 2022). In bridge design, understanding the shear wave velocity ensures that the foundation is placed on stable ground, minimizing the risk of settlement or lateral movement. Microgravity surveys, which detect variations in the Earth's gravitational field caused by subsurface features, rely on the calculation of gravity anomalies, where:

$$\Delta g = \frac{G \cdot m}{r^2} \tag{8.6}$$

where Δg is the gravity anomaly, G is the gravitational constant, m is the mass of the anomaly, r is the distance from the anomaly to the observation point.

Real-Time Monitoring During Construction: During the construction phase, geophysical methods can be used to monitor the integrity of the foundation and ensure that the construction materials are behaving as expected. Real-time seismic monitoring is used to detect any unexpected movements or settlements that could compromise the stability of the structure. The travel time t of seismic waves can be calculated using the formula:

$$t = \frac{L}{V} \tag{8.7}$$

where L is the distance travelled by the seismic wave and V is the velocity of the wave in the material. By monitoring the seismic waves in real-time, engineers can quickly identify any issues during construction and take corrective measures (Zhao et al. 2021).

Structural Health Monitoring (SHM) and Maintenance

Geophysical methods also play a vital role in the ongoing maintenance and monitoring of infrastructure, detecting potential failures before they occur and ensuring the long-term safety of the structure.

Acoustic Emission (AE) Monitoring: Acoustic emissions (AE) are generated when materials undergo stress or deformation. The mathematical relationship used to calculate the energy released by a source of AE is:

$$E = \int F \, dx \tag{8.8}$$

where F is the force applied to the material, and dx is the infinitesimal displacement caused by the event. AE monitoring is particularly useful for detecting early signs of damage in concrete and steel structures, such as those found in bridges or buildings (Lin et al. 2023). By continuously monitoring for AE signals, engineers can assess the health of a structure and plan for timely repairs.

Seismic and Ultrasonic Testing: Seismic and ultrasonic testing methods are widely used to assess the integrity of concrete, steel, and other materials. The ultrasonic wave velocity in a material can be related to the material's Young's modulus E and density ρ using the equation:

$$V = \sqrt{\frac{E}{\rho}} \tag{8.9}$$

where V is the wave velocity, E is the Young's modulus, and ρ is the material density. These methods help identify cracks, voids, or other structural anomalies that may not be visible on the surface. By detecting these issues early, engineers can take preventive action to extend the life of bridges, buildings, and roads (Gong et al. 2021).

Monitoring Road Condition Using Geophysical Imaging: In road construction and maintenance, geophysical techniques such as electromagnetic induction and magnetic resonance imaging (MRI) are used to detect subsurface issues such as cracks, voids, or poor compaction. These techniques can provide high-resolution images of the subsurface and detect anomalies that might otherwise go unnoticed (Zhang and Xie 2020). Monitoring road conditions using these methods helps identify areas that need maintenance before they deteriorate further.

Microtremor Analysis for Seismic Site Assessment: Mathematical Basis

Microtremor analysis involves using ambient vibrations (microtremors) to assess site-specific seismic characteristics. These vibrations are analyzed to determine key parameters such as the natural frequency (f_0) and shear wave velocity (V_s) of the subsurface. Here's the mathematical approach:

Fundamental Frequency (f_0): The natural frequency of the site is determined from the Horizontal-to-Vertical Spectral Ratio (HVSR) method. HVSR is computed as:

$$HVSR(f) = \frac{\sqrt{H_x^2 + H_y^2}}{H_z} \tag{8.10}$$

where H_x and H_y are the horizontal components of the microtremor vibrations, H_z is the vertical component. The peak value of the HVSR curve gives the fundamental frequency (f_0).

Shear Wave Velocity (V_s): The shear wave velocity is related to the layer's thickness (h) and fundamental frequency (f_0) as:

$$V_s = 4hf_0 \tag{8.11}$$

where h is the thickness of the top soil layer, determined from borehole data or other geophysical methods.

Site Amplification Factor: Site amplification can be assessed using:

$$A = \frac{1}{f_0\sqrt{\rho V_s}} \tag{8.12}$$

where ρ is the density of the soil.

Average Shear Wave Velocity (V_{s30}): The average shear wave velocity of the upper 30 m of soil (V_{s30}) is a key parameter for seismic risk assessment:

$$V_{s30} = \frac{30}{\sum_{i=1}^{n} \frac{h_i}{V_{si}}} \tag{8.7.13}$$

where h_i is the thickness of the i-th layer, V_{si} is the shear wave velocity of the i-th layer, and n is the number of layers in the upper 30 m.

Microtremor analysis provides a non-invasive, cost-effective way to evaluate site seismic characteristics. The mathematical relationships, especially for f_0, V_s, and V_{s30}, are critical in seismic site assessments and determining soil-structure interaction under earthquake loads.

Spectral Analysis of Surface Waves (SASW)

Spectral Analysis of Surface Waves (SASW) is a geophysical method used to evaluate the stiffness profile of the subsurface by analyzing surface wave dispersion.

Dispersion Relationship: The dispersion of Rayleigh waves is characterized by their wavelength (λ) and phase velocity (V_p):

$$V_p = \frac{f}{k} \tag{8.14}$$

where Vp is the phase velocity (m/s), f is the frequency (Hz), and k is the wavenumber ($k = \frac{2\pi}{\lambda}$). The phase velocity varies with frequency due to the depth-dependent variation in material stiffness.

Layer Thickness Approximation: The penetration depth (d) of Rayleigh waves is approximately related to their wavelength (λ):

$$d \approx \frac{\lambda}{2} \tag{8.15}$$

Higher frequencies sample shallower depths, while lower frequencies sample deeper layers.

Shear Wave Velocity Profile (***V_s***): The shear wave velocity (V_s) of the subsurface layers is obtained by inversion of the dispersion curve:

$$V_s = \sqrt{\frac{G}{\rho}} \tag{8.16}$$

where G is the shear modulus (Pa) and ρ is the density of the soil/rock (kg/m^3).

Phase Velocity Determination: The phase difference between surface wave signals at two receiver locations (x_1 and x_2) is used to compute the phase velocity:

$$V_p = \frac{\Delta x}{\Delta t} \tag{8.17}$$

where $\Delta x = x_2 - x_1$ is the distance between receivers (m) and Δt is the time delay between wave arrivals (s).

SASW interprets surface wave dispersion to estimate the subsurface shear wave velocity profile, crucial for engineering applications like soil-structure interaction analysis and seismic site characterization. The mathematical framework combines wave propagation theory and inversion techniques for layer-specific stiffness evaluation.

Mathematical Foundations of Magnetotellurics (MT) in Engineering Geophysics

Magnetotellurics (MT) is a geophysical method used to determine subsurface resistivity by measuring natural electromagnetic (EM) fields at the Earth's surface.

Impedance Tensor (Z): The relationship between electric (E) and magnetic (H) field components is given by the impedance tensor:

$$E = ZH \tag{8.18}$$

where $E = \begin{bmatrix} E_x \\ E_y \end{bmatrix}$: Electric field components (V/m), $H = \begin{bmatrix} H_x \\ H_y \end{bmatrix}$: Magnetic field components (A/m), and Z: Complex impedance tensor (Z_{ij}).

Apparent Resistivity (***ρ_a***): The apparent resistivity, a measure of subsurface resistivity, is computed from the impedance tensor:

$$\rho_a = \frac{1}{\mu_0 \omega} |Z|^2 \tag{8.19}$$

where $|Z| = \sqrt{Z_{xy}^2 + Z_{yx}^2}$: Impedance magnitude, μ_0: Magnetic permeability of free space ($4\pi \times 10^{-7}$ H/m), and $\omega = 2\pi f$: Angular frequency (rad/s).

Phase (φ): The phase difference between electric and magnetic fields is:

$$\varphi = \tan^{-1}\left(\frac{\mathrm{Im(Z)}}{\mathrm{Re(Z)}}\right) \tag{8.20}$$

where Im(Z): Imaginary part of impedance and Re(Z): Real part of impedance.

Skin Depth (δ): The penetration depth of the EM field depends on the frequency (f) and resistivity (ρ):

$$\delta = \sqrt{\frac{2\rho}{\mu_0 \omega}} \tag{8.21}$$

Advanced geophysical methods such as seismic refraction, electrical resistivity tomography (ERT), ground-penetrating radar (GPR), and acoustic emission (AE) monitoring rely heavily on mathematical formulations to quantify subsurface properties, detect anomalies, and monitor the health of civil infrastructure. These techniques, when combined with rigorous mathematical modeling, provide critical insights that enable civil engineers to design safer, more resilient buildings, bridges, and roads. As technology evolves, the integration of these methods with advanced data analytics and machine learning will further enhance their capabilities in real-time monitoring and predictive maintenance.

8.2.2 *Integrating Geophysical Methods in Infrastructure Monitoring*

The integration of geophysics into civil engineering has established a strong foundation for addressing various challenges in infrastructure development and monitoring. Over the decades, geophysical methods have been indispensable in geotechnical investigations, structural assessments, and long-term monitoring of engineered systems such as bridges, dams, and buildings. Although geophysical techniques have historically been associated with hydrocarbon and mineral exploration, their application in civil engineering has expanded significantly. This shift can be attributed to the ability of geophysical methods to provide detailed imaging of the subsurface and structural interiors, improving the quality and accuracy of investigations.

For geotechnical applications, geophysical techniques such as seismic surveys, electrical resistivity, and electromagnetic methods have been employed extensively to characterize the subsurface. These methods are crucial for determining mechanical properties of soil and rock, including shear modulus, bulk density, porosity, and fracture patterns, which are essential for designing foundations and assessing earthquake risks. Seismic methods, in particular, remain the most established, with recent overviews and advancements compiled by Reynolds (2000) and Momayez et al. (2009). Despite their capabilities, geophysical techniques often require calibration with borehole data and laboratory analyses to establish reliable correlations for geotechnical parameters. Recent developments in geostatistical methods, such as those discussed by Parker et al. (2015), have shown promise in enhancing the accuracy of these correlations.

Structural investigations have seen a growing reliance on ground-penetrating radar (GPR) due to its ability to detect small features within reinforced concrete, such as rebar positioning, voids, and delamination zones. GPR's advantages over traditional non-destructive testing (NDT) methods, such as ultrasonic techniques, include greater penetration depth and higher detection rates for metallic elements (Niederleithinger et al. 2015). Manufacturers have developed specialized antennas and user-friendly software to make GPR accessible to engineers and technicians with minimal training. Recent adaptations of advanced imaging techniques, such as reverse time migration (RTM) and full waveform inversion, have further improved the resolution and interpretability of GPR data for concrete investigations (Dou et al. 2020).

Another notable geophysical method in structural investigations is the spectral analysis of surface waves (SASW), particularly its extension to multistation analysis of surface waves (MASW). While primarily used for pavement analysis (Park et al. 1999), MASW has also been applied to assess the dynamic properties of concrete structures. Additionally, self-potential (SP) methods have found parallel applications in both geophysics and civil engineering, where they are referred to as half-cell potential [Half-Cell Potential refers to the voltage difference between a reference electrode and the steel reinforcement (rebar) embedded in concrete] measurements. These methods are commonly used to detect corrosion in steel reinforcement, especially in bridge decks and parking garages subjected to de-icing salts (Eldin et al. 2013).

Recent advancements in seismology have introduced passive techniques, such as seismic interferometry and coda wave interferometry, for monitoring subtle changes in concrete structures. These methods have demonstrated their ability to detect progressive damage in earthen embankments and levees, as well as microcrack development in concrete (Kong et al. 2018). Similarly, nonlinear ultrasonic techniques, while currently limited to laboratory settings, are expected to play a significant role in future structural monitoring applications.

Moisture and corrosion detection have been long-standing challenges in both geophysics and civil engineering. Resistivity-based methods, including multi-electrode surveys and tomographic approaches, have gained popularity for investigating moisture content, crack propagation, and corrosion in concrete structures. Complex resistivity, an extension of traditional resistivity into the frequency domain, has shown potential in studying pore-scale properties and monitoring the effects of flooding and drying on masonry (Berube et al. 2020). Impedance spectroscopy, a higher-frequency method, has been used to assess early-age mortar and cement paste properties, further demonstrating the versatility of geophysical techniques in material studies.

The collaboration between geophysics and non-destructive testing in civil engineering holds great promise for advancing both fields. Geophysical methods, with their ability to image small structures and detect subtle changes, can address current limitations in structural assessments. Conversely, measurements on concrete structures can serve as scaled experiments to validate geophysical algorithms and simulations. With continued interdisciplinary cooperation, the application of novel geophysical techniques can significantly enhance the quality, efficiency, and sustainability of civil engineering projects.

Here's a set of recent and impactful case studies aligned with current geophysical applications in civil engineering and infrastructure:

Case Study 1: Seismic Refraction for Metro Tunnel Alignment—Bangalore, India

The Bangalore Metro faced challenges in tunnelling through complex subsurface conditions, including weathered zones and fractured rock. Seismic refraction surveys were conducted along the proposed tunnel routes using a 24-channel seismograph with geophones spaced at 5 m intervals (Kumar et al. 2021). The data revealed significant variations in rock competency, identifying zones of low seismic velocity corresponding to weathered rock and potential fault zones. The findings led to the realignment of critical sections to avoid unstable zones, reducing excavation risks and costs. This case study highlights the importance of seismic refraction in pre-construction planning for urban infrastructure.

Case Study 2: Seismic Imaging for Subsurface Characterization—New York City, USA

Seismic imaging techniques were applied in New York City to assess subsurface conditions for a proposed high-rise building site, identifying varying soil layers and areas prone to liquefaction (Liquefaction is the process by which saturated, loose, granular soil temporarily loses its strength and stiffness during intense shaking, such as an earthquake, causing the soil to behave like a liquid) in the event of an earthquake. These findings were critical for designing a safer foundation, guiding engineers in determining appropriate foundation types and depths based on the seismic data. This method not only ensured the safety of the building but also optimized its design by providing comprehensive subsurface information (Reynolds 2011; Jaramillo and Martinez 2020).

Case Study 3: Microseismic Monitoring in Skyscraper—Dubai, UAE

In Dubai, microseismic monitoring was applied to a newly constructed skyscraper to detect early signs of structural damage. The monitoring system identified small-scale fractures and stress points that were undetectable by visual inspections, allowing engineers to address these issues early and carry out necessary repairs before the damage escalated. The microseismic data provided long-term structural health monitoring, ensuring the skyscraper's integrity and safety (Liao and Zhang 2020; Cakir and Okumus 2019).

Case Study 4: Microtremor Analysis for Seismic Site Assessment—Christchurch, New Zealand

After the 2011 Christchurch earthquake, the city prioritized seismic hazard assessments for safe rebuilding. Microtremor analysis was conducted using three-component seismometers at 50 sites across the city (Wilson et al. 2023). Shear-wave velocity profiles were derived using the Horizontal-to-Vertical Spectral Ratio (HVSR) method, providing insights into site-specific amplification factors. The analysis guided the construction of earthquake-resistant foundations, ensuring compliance with updated seismic building codes. This cost-effective method proved crucial for sustainable urban reconstruction.

Case Study 5: Spectral Analysis of Surface Waves (SASW) for Pavement Evaluation—Chicago, USA

In Chicago, Spectral Analysis of Surface Waves (SASW) was used to evaluate the condition of several roadways, identifying areas with weakened pavement structures and low shear wave velocities, which indicated poor subgrade conditions. The results allowed engineers to prioritize maintenance and repair efforts, particularly for high-risk sections, showcasing SASW as an effective tool for pavement health assessment (Thompson and Smith 2017; Zhang and Xu 2018).

Case Study 6: GPR for Bridge Structural Health Monitoring (SHM)—Germany

As part of maintenance for a 40-year-old highway bridge near Munich, GPR surveys were carried out to assess structural health. A 1.5 GHz (1 GHz = 10^9 Hz) GPR system was deployed to scan the concrete deck and expansion joints (Schuster et al. 2022). The high-resolution data identified delamination zones (Delamination zones refer to areas within a material, typically concrete, where layers have separated or lost adhesion) in the concrete, voids beneath expansion joints, and areas with reduced rebar (steel reinforcement bars) cover. Repairs were localized to these areas, saving 30% of the projected repair costs compared to traditional methods. This study underscores GPR's efficiency in monitoring aging infrastructure and extending its service life.

Case Study 7: GPR for Bridge Deck Inspection—Boston, USA

In Boston, Ground Penetrating Radar (GPR) was employed to inspect multiple highway bridge decks and detect potential corrosion in the steel reinforcement beneath the concrete surface. The GPR survey identified areas with high moisture content, correlating with corrosion hotspots, and provided a detailed map of these areas, allowing engineers to target their repair efforts. This approach proved highly effective for identifying hidden damage and facilitating early repairs before the damage could compromise the bridge's integrity, thus extending its service life (Benedetto and Giannetti 2015; Olsson and Lee 2018).

Case Study 8: Electrical Resistivity Tomography (ERT) for Urban Slope Stability—Tokyo, Japan

In Tokyo's urban outskirts, ERT was employed to study slopes prone to landslides during heavy rainfall. A multi-electrode resistivity system was used to obtain 2D resistivity profiles along critical slopes (Nakamura et al. 2020). The results revealed saturated clay-rich layers and zones of high pore-water pressure, which were correlated with historical landslide occurrences. Following the study, engineered drainage systems and retaining walls were implemented, significantly reducing landslide incidents during subsequent monsoons. This case illustrates how ERT provides actionable data for slope stability analysis in urban planning.

Case Study 9: Electrical Resistivity for Groundwater Contamination—Berlin, Germany

In Berlin, electrical resistivity tomography (ERT) was used to detect groundwater contamination near an industrial site, successfully identifying areas of elevated resistivity indicative of contamination such as heavy metals and solvents. The method allowed engineers to map the extent of contamination and understand the flow dynamics of polluted groundwater, enabling targeted remediation efforts that saved both time and costs. This case highlighted ERT's potential as a powerful and non-invasive tool for detecting groundwater contamination (Ziegler and Bar 2017; Schenck and Meyer 2020).

Case Study 10: Self-Potential for Concrete Corrosion—Oslo, Norway

In Oslo, the self-potential (SP) method was utilized to monitor corrosion in reinforced concrete structures exposed to deicing salts, especially in parking garages and bridge decks. The SP surveys identified areas with ongoing corrosion, enabling the prioritization of repairs to prevent further damage. The method was especially effective in providing continuous monitoring of the structure's health, detecting corrosion before it became visually apparent. This proactive approach helped extend the lifespan of the infrastructure and prevent costly repairs (Kourous and Haidar 2016; Abe and Johansen 2018).

Case Study 11: Integrated Geophysical Techniques for Dam Safety—China

The aging Honghe Dam in Yunnan Province underwent a thorough safety evaluation using integrated geophysical methods (Zhang et al. 2022). ERT was utilized to map seepage paths within the dam, seismic tomography identified zones of reduced density indicative of internal erosion, and magnetic surveys detected metallic debris causing structural wear. The combined findings led to targeted grouting in high-seepage areas and improved drainage systems. These measures effectively reduced leakage and enhanced the dam's structural integrity. This study demonstrates the value of combining multiple geophysical techniques for comprehensive dam safety assessments.

Case Study 12: MT Surveys for Mineral Exploration—Rajasthan, India

In Rajasthan, Magnetotelluric (MT) surveys were conducted for mineral exploration, providing detailed images of subsurface geological structures that revealed potential zones for mineral deposits. The results guided further drilling and resource extraction efforts, demonstrating the effectiveness of MT in mineral exploration, especially in regions with difficult terrain and limited access (Kumar and Gupta 2020; Sharma and Sharma 2019).

8.3 Innovations in Geophysical Monitoring of Structural Health and Integrity

The increasing complexity of civil infrastructure and the growing demands for safety, sustainability, and efficiency have led to a significant rise in the adoption of geophysical methods for structural health monitoring (SHM). These techniques offer non-destructive, real-time assessments that help engineers identify early signs of damage, optimize maintenance schedules, and ensure the long-term performance of infrastructure. Innovations in geophysical monitoring have dramatically improved the ability to assess and maintain the health of structures, from routine evaluations to complex assessments of critical infrastructure (Benedetti and Lombardi 2018; Park and Lee 2020).

8.3.1 Overview of Geophysical Techniques in Structural Health Monitoring (SHM)

Geophysical methods used in SHM are typically categorized based on their ability to monitor the internal and external conditions of a structure, as well as the surrounding environment. These techniques have evolved from traditional methods such as visual inspections to advanced, non-invasive tools that provide more detailed, continuous, and real-time monitoring (Benedetti and Lombardi 2018; Chavez and Liu 2020). Commonly used geophysical techniques include:

Ground Penetrating Radar (GPR): Utilized for detecting internal defects such as voids, cracks, rebar corrosion, and delaminations in concrete and other materials (Gucunski and Sanderson 2014).

Ultrasonic Testing (UT): This technique uses high-frequency sound waves to evaluate the structural integrity of materials by detecting internal flaws like cracks or corrosion (Dou and Li 2017).

Acoustic Emission (AE): Monitors the emission of high-frequency sound waves generated by the formation of cracks or the movement of pre-existing damage in materials under stress (Vitali and Giustolisi 2021).

1. **Ambient Vibration Testing (AVT)**: Detects the dynamic response of structures under normal operational conditions, revealing changes in stiffness or mass that may indicate structural damage (Stojadinovic and Bertero 2018).
2. **Seismic Monitoring**: Measures changes in the wave propagation characteristics of structures, offering insights into subsurface conditions and the potential for foundation settlement, cracks, or voids (Wang and Zhang 2019).

These techniques are often complemented by advanced sensor technologies, such as fiber optics and wireless sensor networks, which provide real-time data on various physical parameters, including strain, temperature, displacement, and vibration (Yang and Wang 2020; Koo and Park 2022).

8.3.2 Advances in Monitoring Technologies

Technological advancements have significantly expanded the capabilities of geophysical monitoring systems. Several innovative developments in sensor technology, data acquisition systems, and data analysis techniques have improved the precision, sensitivity, and scalability of SHM applications.

Wireless Sensor Networks (WSNs): Wireless sensors allow for the continuous monitoring of structures, particularly in areas that are hard to access or hazardous to inspect (Agarwal and Zhang 2021). These sensors can transmit data remotely, allowing engineers to monitor multiple sites in real-time. WSNs are being integrated with other technologies, such as Internet of Things (IoT), to create intelligent monitoring systems capable of predictive maintenance.

Fiber Optic Sensors: Fiber optics offer several advantages for SHM, including the ability to monitor large areas and the detection of minute changes in strain and temperature over extended periods (Kwon and Park 2019). These sensors can be embedded directly into structures during construction or retrofitted into existing structures to detect deformations, cracks, and shifts that may indicate damage.

Machine Learning and Artificial Intelligence (AI): With the increasing volume of data collected by geophysical monitoring systems, machine learning and AI algorithms are being used to analyze large datasets and predict the health of structures (Liu and Zhang 2021). These technologies can detect subtle patterns and correlations in the data, helping engineers make more accurate predictions about when a structure might require maintenance or repairs. AI algorithms can also aid in identifying trends and anomalies that could indicate early signs of failure, allowing for more proactive management of infrastructure.

Drones and UAV-based Inspections: The use of unmanned aerial vehicles (UAVs) equipped with high-resolution cameras, LiDAR, and thermal sensors allows for the efficient inspection of difficult-to-reach structures, such as bridges, towers, and dams (Lin and Chen 2019). Drones enable real-time, visual, and thermal assessments of structural conditions, providing valuable data for detecting surface-level defects, corrosion, and temperature anomalies.

Advanced Seismic Tomography: Seismic tomography is an advanced technique used to create detailed images of a structure's internal and subsurface conditions (Mancuso and Nanni 2020). Recent innovations in seismic sensing technology have improved the resolution and accuracy of seismic tomography, enabling better detection of structural anomalies such as voids, cracks, or shifts in foundation materials.

8.3.3 Applications in Key Infrastructure

Geophysical monitoring is used across various infrastructure sectors, helping engineers to detect issues before they lead to costly repairs or failures. The versatility and adaptability of these techniques make them suitable for a wide range of infrastructure types, including:

Bridges: Structural health monitoring using geophysical methods can help detect issues such as cracks, corrosion, and fatigue in key components like beams, supports, and joints. Techniques like ambient vibration testing, acoustic emission monitoring, and GPR are commonly used to assess the dynamic response of bridges and to identify internal damage in concrete or steel structures (Collette and Hamilton 2018).

Dams and Reservoirs: Dams are subject to constant stress and environmental conditions, and early detection of structural shifts, erosion, or seepage is critical to preventing catastrophic failures. Technologies like ultrasonic testing, seismic monitoring, and fiber optic sensors are employed to monitor dam conditions, such as internal cracks, water flow, and material degradation (Dou and Li 2017; Wang and Liu 2021).

Tunnels: Geophysical techniques are used to monitor both the tunnel structure and the surrounding soil conditions. Seismic tomography, GPR, and ultrasonic testing help detect deformations, cracks, and voids within the tunnel lining, as well as potential ground movements or water infiltration that could compromise the tunnel's stability (Zhang and Song 2019).

Buildings and High-Rise Structures: For urban infrastructure, geophysical techniques can monitor structural integrity over time, identifying issues such as foundation settlement, differential movement, and fatigue in critical load-bearing elements. Techniques such as ambient vibration testing and fiber optic strain sensors are commonly employed to track any changes in structural performance (Gao and Zhao 2021).

Transportation Infrastructure: Roads, railways, and airports benefit from SHM technologies that monitor pavement conditions, detect subsurface voids, and assess the integrity of load-bearing structures like bridges, tunnels, and overpasses. GPR and UAV-based inspection systems provide detailed condition assessments to inform maintenance strategies (Chavez and Liu 2020).

The innovations in geophysical monitoring technologies for structural health and integrity are revolutionizing the way engineers manage infrastructure. From the integration of fiber optic sensors and wireless networks to the use of drones and AI-driven analytics, these advancements are enabling real-time, accurate, and cost-effective assessments of civil infrastructure. As these technologies continue to evolve, they offer the potential for more proactive and predictive approaches to infrastructure maintenance, ensuring the safety, sustainability, and longevity of critical infrastructure worldwide.

Through the use of advanced geophysical methods, civil engineers can better assess the condition of structures, detect early signs of deterioration, and optimize repair and maintenance efforts, ultimately reducing costs and preventing catastrophic failures. As the field of structural health monitoring continues to grow, we can expect even more powerful innovations that will further improve our ability to monitor and maintain infrastructure, ensuring its resilience and performance for future generations.

8.3.4 Case Studies

This section presents real-world case studies illustrating the practical application of geophysical methods in civil engineering and infrastructure projects. Each case demonstrates how specific techniques—such as seismic, electrical, or electromagnetic methods—were employed to solve engineering challenges, optimize design, and mitigate risks. The case studies highlight the effectiveness of geophysics in diverse settings, from urban construction sites to large-scale transportation and utility projects.

Monitoring Brooklyn Bridge Using Ambient Vibration Testing

The Brooklyn Bridge, one of the most iconic suspension bridges in the United States, is subject to constant traffic loads and environmental factors that can affect its structural integrity. In recent years, engineers have employed ambient vibration testing (AVT) to monitor the bridge's health and identify any early signs of deterioration (Stojadinovic and Bertero 2018).

The team installed accelerometers and vibration sensors at key locations across the bridge to measure its dynamic response to daily traffic and environmental vibrations. By analyzing changes in the bridge's natural frequencies and mode shapes, engineers could detect any shifts in its dynamic behavior that could indicate damage such as cracks or degradation of materials.

The ambient vibration testing system provided continuous, real-time data, allowing engineers to identify localized areas where the bridge's response had shifted. These findings were critical in pinpointing areas of potential concern, such as aging cables and joints that required immediate attention.

The use of AVT in this case allowed for non-invasive, real-time monitoring of the Brooklyn Bridge, enhancing safety while minimizing disruption to traffic. The bridge could be safely managed with informed, data-driven decisions, extending its service life and reducing the need for costly and invasive inspections.

Ground Penetrating Radar for Concrete Pavement Assessment

An international airport in Europe was experiencing issues with its concrete pavements, including cracks, surface wear ("Surface wear" in the context of an airport generally refers to the deterioration or damage that occurs on various surfaces due to frequent use, weather conditions, and the weight and movement of aircraft, vehicles, and passengers), and concerns over the integrity of the underlying foundation. To assess the condition of the pavement, engineers used Ground Penetrating Radar (GPR) to examine the concrete without needing to dig up or replace sections of the pavement (Gucunski and Sanderson 2014).

The engineering team employed GPR to scan large sections of the airport runway. GPR uses high-frequency electromagnetic waves to detect changes in material properties, such as the presence of voids, cracks, and rebar corrosion beneath the surface of the pavement.

The GPR surveys revealed areas with subsurface delaminations, where the concrete had separated from the underlying layers, as well as locations with corrosion in the steel reinforcement ["Corrosion in steel reinforcement" refers to the gradual deterioration of the steel bars (rebar) embedded in concrete structures, caused by chemical reactions, usually due to exposure to moisture, oxygen, and sometimes salts]. By analyzing the radar data, engineers could identify the extent of the damage without disturbing the surface.

This non-destructive technique allowed the airport authorities to prioritize repairs based on the severity of the damage, avoiding unnecessary disruptions and expensive full-scale reconstruction. The precise data provided by GPR helped extend the runway's lifespan by focusing efforts on the most critical areas.

Fiber Optic Sensors in Bridge Health Monitoring

A major highway bridge in California, known for heavy traffic and exposure to seismic activity, was equipped with an advanced fiber optic sensing system as part of a long-term structural health monitoring initiative. The goal was to monitor the bridge's performance in real time and detect any changes that could indicate early-stage damage or deformation (Kwon and Park 2019; Liu and Yang 2020).

The team embedded fiber optic sensors into key components of the bridge, including the cables, beams, and deck. These sensors measured strain, temperature, and displacement over time, providing detailed data on how the bridge responded to varying loads, such as traffic and seismic vibrations.

The fiber optic sensors allowed for continuous monitoring of the bridge's structural health, even during peak traffic hours or seismic events. Data collected from the sensors identified areas where strain exceeded the expected levels, which could indicate material fatigue or movement within the structural components.

The use of fiber optic sensors enabled engineers to detect minor deformations or shifts in structural elements that could have otherwise gone unnoticed. With this data, maintenance schedules were optimized, and timely repairs were made, ensuring the safety of the bridge without the need for costly and disruptive physical inspections. The system also provided valuable information about the bridge's seismic response, which is crucial in an earthquake-prone region.

Seismic Tomography for Tunnel Stability

A new subway tunnel was being constructed in an urban area with soft soil conditions and a history of ground settlement. To ensure the stability of the tunnel and the surrounding infrastructure, engineers decided to employ seismic tomography to evaluate the subsurface conditions and detect any potential voids, fractures, or irregularities in the ground that could affect the tunnels long-term stability (Zhang and Song 2019; Mancuso and Nanni 2020).

Seismic sensors were strategically placed along the tunnel route and in surrounding soil areas to monitor seismic wave propagation and gather data on subsurface conditions. Using seismic tomography, the team could produce high-resolution images of the ground beneath and around the tunnel, revealing hidden features such as fault lines, cavities, and variations in soil density.

The seismic tomography provided a detailed map of the subsurface, allowing the team to identify several potential risks, including a hidden fault line that was likely to affect the tunnel's alignment. The findings also revealed areas where soil compaction was weaker than expected, which could have led to settlement over time.

Armed with these insights, engineers were able to adjust the tunnel's design and construction methods to avoid the identified risks. This proactive approach not only improved the tunnel's long-term safety but also minimized the need for costly retrofitting once construction was underway. The data from seismic tomography also informed the development of better risk mitigation strategies for similar projects in the region.

8.3.5 Emerging Trends in Geophysical Monitoring

1. ***Integration with Building Information Modeling (BIM)***: The integration of geophysical data with BIM is an emerging trend in structural health monitoring. By incorporating real-time geophysical data into BIM platforms, engineers can visualize the health of infrastructure in 3D, allowing for more effective decision-making and maintenance planning. This integration helps stakeholders to understand the implications of various structural health issues and prioritize repairs based on a comprehensive model of the structure's condition.

   Recent advancements in geophysical monitoring for infrastructure health are highlighted by two key studies. Wang and Liu (2021) review the use of fiber optic sensors for monitoring dam structures, showcasing the potential of these technologies to enhance structural health monitoring (SHM) in large-scale infrastructure. Similarly, Yang and Wang (2020) explore the integration of Internet of Things (IoT) devices with geophysical monitoring tools, such as seismic sensors, to enable real-time, large-scale monitoring of bridges and other civil infrastructure. Both studies emphasize the role of innovative technologies in improving the efficiency and accuracy of SHM systems.
2. ***Drone-Based LiDAR for Structural Inspections***: LiDAR (Light Detection and Ranging) technology has found growing applications in structural health monitoring. Drones equipped with LiDAR systems can quickly scan large structures, such as bridges, towers, and buildings, and generate high-resolution 3D models. These models provide detailed topographic maps and can be used to detect surface deformations, misalignments, and other structural anomalies that may indicate damage or wear.
3. ***Real-Time Data Analytics for Predictive Maintenance***: Advanced data analytics techniques, such as machine learning algorithms and real-time data processing, are being employed to predict when a structure may require maintenance or when failure is imminent. By analyzing historical and real-time data from geophysical

sensors, these systems can identify patterns and trends that may signal the onset of failure. Predictive maintenance based on such analytics enables engineers to address potential issues before they result in costly repairs or safety hazards.

The case studies and emerging trends discussed here highlight the transformative impact of geophysical techniques on structural health monitoring. As these technologies continue to evolve, their ability to provide real-time, non-invasive assessments of infrastructure will only improve, offering engineers powerful tools for ensuring the safety, longevity, and sustainability of the built environment. The integration of these innovative techniques into routine maintenance practices is poised to redefine how we monitor and manage infrastructure, paving the way for smarter, more resilient cities and systems.

8.4 Examples of Geophysical Tools Used for Ongoing Infrastructure Monitoring

In the context of ongoing infrastructure monitoring, geophysical methods offer invaluable insights into the health and stability of civil engineering structures. These non-invasive tools provide continuous, real time data that can detect potential issues such as material degradation, cracks, and shifting foundations. The following geophysical tools are commonly employed for monitoring infrastructure over time:

8.4.1 Ground Penetrating Radar (GPR)

Ground Penetrating Radar (GPR) is a geophysical method that utilizes high-frequency radio waves to probe the subsurface and detect anomalies by analyzing the reflected signals. This non-invasive technique is especially useful in assessing the condition of concrete structures, such as bridges, pavements, and roadways. By sending radio waves into the material and measuring the time it takes for the waves to reflect back, GPR can identify voids, cracks, and areas where reinforcing steel may be corroding. Additionally, it can be used to evaluate moisture intrusion in pavements and concrete decks, which is crucial for preventing long-term damage and maintaining the integrity of the structure.

One of the primary benefits of GPR is that it is both non-destructive and fast, allowing for quick assessment of large areas without causing any disruption to the infrastructure being examined. The data provided by GPR offers detailed insights into the structural condition of concrete, making it an invaluable tool for ongoing maintenance and early detection of potential issues. The ability to gather this information quickly and accurately enables engineers to take timely action, thus preventing more costly repairs or catastrophic failures in the future.

8.4.2 Seismic Tomography

Seismic tomography is a geophysical technique that uses seismic waves to create detailed images of subsurface conditions by measuring variations in material properties. This method involves sending seismic waves through the ground and analyzing how these waves travel through different materials. By measuring the speed and path of the seismic waves, engineers can infer the properties of the subsurface layers and detect any anomalies. Seismic tomography is particularly useful for imaging subsurface voids, cracks, and other irregularities in large infrastructure projects such as dams, tunnels, and foundations. It provides valuable information about the structural integrity of these projects and helps in identifying potential issues that may not be visible through traditional inspection methods.

One of the key benefits of seismic tomography is its ability to provide three-dimensional images of subsurface conditions, which gives engineers a comprehensive view of the structural health of the infrastructure. This 3D imaging capability allows for more accurate detection of internal weaknesses, such as hidden cracks or voids, that could compromise the stability of the structure. By identifying these weaknesses early, seismic tomography enables engineers to address potential failure points before they escalate into serious problems, thereby enhancing the safety and longevity of the infrastructure.

8.4.3 Electrical Resistivity Imaging (ERI)

Electrical Resistivity Imaging (ERI) is a geophysical technique that measures the resistance to electrical current flow through subsurface materials to detect variations in moisture content, voids, or cracks. By injecting an electrical current into the ground and measuring the resistance encountered by the current, ERI creates a profile of the subsurface conditions. The varying resistivity of different materials allows engineers to identify areas where moisture has accumulated, where voids may be present, or where cracks have developed in the soil or underlying structure. This method is especially useful for monitoring the integrity of buried utilities and pipelines, where it can help detect issues such as leaks, corrosion, or degradation of materials.

ERI is also effective for detecting groundwater infiltration and settlement around foundations, providing crucial information that can prevent long-term structural damage. In addition, it is widely used for assessing subsurface conditions for roadways, embankments, and other civil engineering projects, helping engineers understand how the ground may behave under different environmental conditions. One of the major benefits of ERI is its effectiveness in detecting moisture and subsurface voids, which are often difficult to identify using traditional inspection methods. Moreover, the technique is versatile, as it can be applied both to soil and structural evaluations, offering a comprehensive view of the subsurface health of an area or structure.

8.4.4 Acoustic Emission (AE) Monitoring

Acoustic Emission (AE) monitoring is a technique that detects high-frequency sound waves generated by the release of energy from cracks or defects in materials. As a material experiences stress or begins to crack, it emits sound waves that can be captured by sensors placed on the surface of the structure. These acoustic signals provide real-time information about the structural health of the material, allowing engineers to detect early-stage damage before it becomes visible or reaches a critical point. AE monitoring is particularly valuable for identifying microfractures or other subtle changes in concrete or steel structures, which may otherwise go unnoticed.

This method is widely used for monitoring tunnels and underground infrastructure, where it can detect signs of stress or cracks that may compromise the safety and stability of the structure. One of the key advantages of AE monitoring is its ability to provide continuous, real-time data, enabling engineers to track the progression of damage as it occurs. This capability is crucial for early detection, allowing for timely interventions and repairs. Furthermore, AE monitoring can identify microfractures that have not yet manifested as visible damage, providing an early warning system to prevent more significant structural failures in the future.

8.4.5 Vibration Monitoring (Accelerometers and Seismic Sensors)

Vibration monitoring tools, such as accelerometers and seismic sensors, are designed to measure the vibrations or oscillations that occur in structures due to environmental factors or changes in applied loads. These tools capture data on the movement of a structure, providing valuable insights into how it responds to various dynamic forces, including traffic, wind, and seismic activity. By continuously measuring these vibrations, engineers can assess the structural behavior of buildings, bridges, dams, and skyscrapers, enabling them to monitor how these structures cope with dynamic loading effects over time.

These tools are particularly useful for detecting shifts in structural integrity caused by excessive vibration or resonance, which can indicate potential weaknesses or areas of concern. Continuous monitoring provides real-time alerts for abnormal structural behavior, allowing engineers to take immediate action if needed. One of the key benefits of vibration monitoring is that it helps engineers understand the response of structures to various dynamic loads, which is essential for ensuring their stability and safety. By identifying any irregularities early on, vibration monitoring plays a crucial role in preventing catastrophic failures and ensuring the longevity of critical infrastructure.

8.4.6 Satellite-Based Remote Sensing and InSAR (Interferometric Synthetic Aperture Radar)

Satellite-based remote sensing, particularly InSAR (Interferometric Synthetic Aperture Radar), is a powerful geophysical method that uses radar data from satellites to measure ground surface displacement with high precision. InSAR can detect even minute shifts in the ground, making it invaluable for monitoring the stability of infrastructure over time. By comparing radar images taken at different times, this technique can identify changes in the surface that might indicate subsidence, ground shifting, or other forms of deformation that could affect the safety and stability of infrastructure like bridges, dams, and roads.

This method is especially useful for monitoring large-scale movements, such as subsidence or deformation in areas where traditional ground-based methods may be difficult to implement. It is also widely applied in detecting structural changes in large infrastructure projects, such as tunnelling operations, where ground movements can pose significant risks. One of the main advantages of InSAR is its ability to cover vast areas, providing continuous monitoring over extended periods. This non-invasive technique allows for ongoing surveillance of infrastructure without the need for physical inspections, making it an efficient tool for tracking gradual changes in the landscape and identifying potential issues before they become critical.

8.4.7 Distributed Fiber Optic Sensing (DFOS)

Distributed Fiber Optic Sensing (DFOS) is a cutting-edge technology that uses optical fibres embedded within or along structures to continuously monitor various parameters, including strain, temperature, and other factors. The fibres are capable of detecting changes along their entire length, providing a comprehensive view of a structure's condition over long distances. As strain or temperature changes occur within the structure, the optical fibres transmit data in real time, allowing for continuous monitoring of the structure's health. This ability makes DFOS particularly valuable for assessing large, complex infrastructure, such as bridges, tunnels, and pipelines, where even small deformations or shifts could indicate potential problems.

DFOS is especially useful for detecting deformation and stress accumulation in large concrete structures, where traditional monitoring methods might not be as effective. It provides high-resolution, real-time data across vast distances, offering a level of detail that ensures engineers can identify issues early and take proactive measures. The technology is ideal for monitoring long, linear infrastructure like tunnels and pipelines, where it is critical to have a system that can detect subtle changes in condition across the entire length of the structure. By offering continuous, detailed insights, DFOS helps ensure the safety, stability, and longevity of critical infrastructure.

Here are some additional geophysical tools used in engineering geophysics that may be relevant and of interest to the audience. These tools complement the ones already mentioned and play a significant role in civil engineering and infrastructure monitoring:

8.4.8 Multi-channel Analysis of Surface Waves (MASW)

The Multi-channel Analysis of Surface Waves (MASW) method uses surface waves, typically Rayleigh waves (Rayleigh waves are a type of surface seismic wave that travel along the Earth's surface and cause both vertical and horizontal ground motion. Rayleigh waves are typically slower than body waves P-waves and S-waves, but they are often the most destructive during an earthquake due to their surface effects), to analyze subsurface properties. The tool involves measuring the speed of surface waves as they travel through different layers of the ground. By analyzing the dispersion of these waves, engineers can create a profile of the subsurface material properties, such as stiffness and shear wave velocity.

MASW is widely used for assessing the soil profile and identifying weak zones that could affect foundation design or lead to settlement. It is often used in geotechnical investigations for the evaluation of shallow to medium-depth soil conditions in the construction of roads, foundations, and large-scale infrastructure projects.

MASW provides a quick and non-invasive means of determining the soil's dynamic properties, which are crucial for evaluating its behavior under seismic loading. The method can cover large areas and offer high-resolution images of subsurface layers, which is particularly useful in seismic risk assessments and soil liquefaction studies.

8.4.9 Magnetometry

Magnetometry measures the variations in the Earth's magnetic field caused by subsurface features that have different magnetic properties. This tool uses magnetic sensors to detect anomalies in the magnetic field, which can indicate the presence of buried objects or variations in the material composition of the subsurface.

Magnetometry is commonly used for locating ferrous objects, such as buried pipes, cables, and other metallic infrastructure. It is also useful in archaeological surveys and detecting subsurface features like ancient foundations or mines, but in civil engineering, it is particularly useful for mapping existing utilities and infrastructure.

Magnetometry is a fast, cost-effective, and non-invasive tool for locating buried metallic objects and structures. It is highly effective in urban environments, where traditional excavation methods may be difficult or costly.

8.4.10 Electromagnetic (EM) Induction

Electromagnetic (EM) induction involves transmitting an electromagnetic field into the ground and measuring how it is attenuated as it passes through different subsurface materials. The degree of attenuation is influenced by the conductivity of the materials, allowing for the identification of different subsurface materials, such as metals, saline water, or clay.

EM induction is primarily used for mapping the distribution of groundwater, saline intrusion in coastal areas, and for detecting variations in soil salinity. It is also used for identifying buried infrastructure, assessing environmental contamination, and evaluating the integrity of underground utilities.

This method is non-invasive, rapid, and suitable for both shallow and deep investigations. It can be employed in areas where drilling or excavation is not feasible, providing engineers with valuable data on the electrical conductivity of subsurface materials.

8.4.11 Electrical Resistivity Tomography (ERT)

Electrical Resistivity Tomography (ERT) measures the electrical resistivity of the ground by passing an electric current through the ground via electrodes. The way the current travels through the ground depends on the type of material it passes through, and by measuring the resistance, a detailed image of the subsurface can be created.

ERT is widely used for detecting groundwater pathways, delineating geological layers, and identifying areas of potential collapse or instability in structures. It is particularly useful for assessing the condition of landfills, monitoring the movement of contaminants in the soil, and evaluating the integrity of foundations.

ERT offers high-resolution images of subsurface conditions and is highly effective for environmental monitoring and geotechnical investigations. It is particularly valuable for understanding soil properties and detecting contaminants or variations in groundwater flow.

8.4.12 Gravity Surveying

Gravity surveying measures variations in the Earth's gravitational field caused by changes in the density of the underlying materials. Dense materials such as bedrock will produce stronger gravitational anomalies, while less dense materials such as cavities or voids will produce weaker anomalies.

Gravity surveys are used for mapping subsurface geological structures, detecting voids or sinkholes, and identifying the extent of buried deposits such as ores or mineral resources. In civil engineering, gravity surveys can help in detecting areas of weak foundation material or areas prone to subsidence.

Gravity surveys are particularly valuable in large-scale projects and in areas where other methods may not be effective. They provide a broad view of subsurface density variations and are ideal for detecting deep anomalies that could affect the stability of infrastructure.

8.4.13 Refraction Seismology

Refraction seismology involves measuring the travel times of seismic waves as they refract through subsurface layers. The method is based on the principle that seismic waves travel at different speeds through different materials, and by measuring the time it takes for these waves to reach sensors, the depth and characteristics of subsurface layers can be determined.

This method is widely used for determining the depth of bedrock, evaluating the thickness of soil layers, and assessing the stability of foundations. It is often used in the construction of roads, bridges, and foundations, particularly when the geotechnical properties of the soil need to be understood in more detail.

Refraction seismology provides detailed information about subsurface structures, including the location and depth of bedrock and other important geological features. It is a useful tool for mapping the interface between soft soil and harder materials, helping engineers make informed decisions about foundation design.

8.4.14 Borehole Geophysics

Borehole geophysics involves using geophysical tools lowered into boreholes to measure properties such as resistivity, porosity, and seismic velocity at specific depths. This method provides detailed data about subsurface conditions that can be correlated with surface geophysical surveys.

Borehole geophysics is used to gain a deeper understanding of the subsurface conditions at specific points, such as for detailed soil or rock profiling, assessing groundwater conditions, and characterizing the properties of rock formations in relation to underground infrastructure.

It offers highly accurate and site-specific data, which is essential for designing foundations, tunnels, and other structures where deep knowledge of the subsurface is crucial. Borehole geophysics is particularly useful in areas with complex geological conditions or where surface methods are not sufficient.

These tools, in addition to those previously mentioned, offer a comprehensive suite of geophysical methods for monitoring, evaluating, and ensuring the stability of infrastructure projects. Each technique has its unique applications and benefits, making them invaluable for civil engineers who require reliable, non-invasive methods to assess the condition of both surface and subsurface materials.

Geophysical tools for ongoing infrastructure monitoring are essential for ensuring the safety and longevity of civil engineering projects. These tools provide engineers with the ability to detect hidden problems before they become critical, allowing for timely intervention and maintenance. By incorporating various geophysical techniques, it is possible to create a comprehensive monitoring system that enhances the resilience of infrastructure and reduces the risk of failure.

8.5 Case Studies: Recent Applications of Geophysical Tools in Infrastructure Monitoring

In recent years, a variety of geophysical methods have been successfully applied to real-world infrastructure projects, providing valuable insights for engineers, architects, and stakeholders. These case studies highlight the diverse ways in which geophysical tools can support site investigations, infrastructure monitoring, and long-term asset management. Below is a selection of the most impactful and relevant case studies from across the globe:

8.5.1 GPR for Bridge Stability—United States

A major transportation agency in the United States used Ground Penetrating Radar (GPR) to assess the condition of aging bridges along a critical highway network (Gucunski et al. 2014). The objective was to identify potential areas of corrosion in the reinforced concrete and steel components without having to shut down the bridges for manual inspections.

GPR was used to map out the depth and extent of corrosion in the steel reinforcement bars and to detect any voids or delaminations in the concrete deck. The non-invasive nature of GPR allowed for high-resolution imaging of the bridge's internal structure, providing clear evidence of deterioration in key areas that were not easily visible.

The results led to the identification of specific locations where maintenance was required, helping prioritize repairs and extend the lifespan of the infrastructure. The method also reduced inspection costs and time, ensuring the safety of the bridges while minimizing traffic disruptions.

Stakeholders appreciated the ability to conduct inspections without major disruptions to traffic, as well as the accurate and actionable data GPR provided, which was crucial for informed decision-making regarding maintenance schedules.

8.5.2 Tunnel Infrastructure Monitoring

This section focuses on the application of geophysical methods for the monitoring and maintenance of tunnel infrastructure. It discusses techniques such as microseismic monitoring, ground-penetrating radar (GPR), and distributed acoustic sensing (DAS) used to detect structural anomalies, assess ground movement, and ensure tunnel safety. Emphasis is placed on continuous monitoring strategies that support early warning systems and enhance the longevity and resilience of tunnel systems.

Tunnel Displacement Monitoring Using Laser Scanning and Geodetic Methods

Tunnel monitoring plays a crucial role in ensuring the structural integrity of tunnel systems during both construction and operation. Early detection of displacement and structural changes can help in planning timely maintenance and repairs, which reduces risks and extends the service life of the tunnel. This case study focuses on the use of advanced technologies such as laser scanning and geodetic measuring methods for displacement monitoring in tunnels (Strauss et al. 2020).

Strauss et al. (2020) introduce a three-level hierarchical system for classifying defects to improve tunnel monitoring (Table 8.1). This system summarizes the types of defects found in tunnels with concrete linings. It combines the defect classification approach by Bien (2019) with insights from the case study on tunnel monitoring.

Tunnels can experience degradation due to various mechanisms, which can be broadly categorized into chemical, physical, and biological factors. Chemical degradation includes processes like carbonation and corrosion, which weaken the materials over time. Physical degradation occurs due to external forces such as erosion or overloading, which can lead to structural damage. Biological degradation, often caused by microbial growth, can weaken tunnel materials, especially in damp conditions. Understanding these degradation mechanisms is crucial for monitoring and maintaining tunnel infrastructure to prevent failures. Table 8.2 summarizes the key degradation mechanisms observed in tunnel structures, as discussed in literature (Strauss et al. 2020).

The main goal of this study is to evaluate the effectiveness and accuracy of displacement monitoring methods, particularly laser scanning and total station-based geodetic systems, for tracking tunnel deformation over time. The study aims to identify structural changes, early risks, and propose optimized monitoring strategies for tunnel rehabilitation.

This study employed 3D laser scanning and geodetic total stations to monitor tunnel displacement. Multiple laser scans were conducted at different time intervals, with reference points measured using optical-trigonometric surveys for accurate localization. Total stations were used to measure the spatial positions of prisms placed

Table 8.1 Classification of tunnel defects with concrete lining—revised in light of Bien (2019)

Defect category	Description	Example defects	Monitoring methods
Structural defects	Issues related to the integrity of the tunnel structure itself	Cracks, spalling, deformation, structural misalignment	Laser scanning, 3D displacement monitoring, total stations
Water leakage	Defects related to water ingress into the tunnel	Leaks in joints or cracks, excessive moisture build-up	NDT techniques, moisture sensors, fiber optic sensors
Corrosion	Deterioration of concrete or reinforcement due to environmental factors	Concrete erosion, rusted steel reinforcements	Visual inspection, corrosion sensors, NDT methods
Material deterioration	Degradation of the concrete lining material over time due to stress or environmental impact	Cracks from freeze–thaw, chemical attack	Fiber optic sensors, NDT, strain gauges
Geometric deformation	Changes in tunnel shape due to external or internal stresses	Subsidence, tunnel collapse, misalignment	3D displacement monitoring, laser scanning, total stations
Surface cracking	Visible cracks on the surface of the tunnel lining	Small cracks, fissures in concrete	Visual inspections, laser scanning, and photogrammetry

Table 8.2 Degradation mechanisms of tunnel structures

Degradation mechanism	Description	Cause/effect
Chemical degradation	Changes in material properties due to chemical reactions	*Carbonation*: Reaction between CO_2 and concrete, leading to reduced durability *Corrosion*: Chemical attack on reinforcing steel, weakening tunnel structure
Physical degradation	Breakdown of tunnel structure due to physical forces	*Erosion*: Wear from water flow or particulate matter *Overloading*: Excessive weight or stress on tunnel surfaces leading to cracking or collapse
Biological degradation	Degradation caused by the growth of microorganisms	*Microbial growth*: Growth of fungi, bacteria, or algae that can weaken tunnel materials, especially in moist environments

on tunnel cross-sections over several epochs. The instrument's position was determined via free stationing using nearby reference points. Displacements were calculated by comparing the coordinates of monitored points between epochs, and statistical methods like covariance analysis were used to assess measurement accuracy and detect significant deformations.

The monitoring data revealed detailed deformation patterns in the tunnel structure over time. Displacement vectors were calculated for each measurement epoch, showing significant changes in tunnel geometry. The analysis identified areas with accelerated deformation, indicating potential risk zones. Time-deformation diagrams and settlement velocity graphs highlighted areas of concern, while sum vectors of the deformations provided a clear visualization of the overall structural changes. Statistical tests confirmed that several of the observed deformations were statistically significant, indicating changes in the tunnel's geometry between epochs.

The use of 3D laser scanning and geodetic methods for tunnel displacement monitoring proved to be highly effective in detecting small-scale structural changes in real time. Early detection of risks allows for preventive maintenance, which improves tunnel longevity and reduces repair costs. The integration of different monitoring techniques enhances the accuracy and reliability of the data collected.

This study is primarily aimed at professionals and researchers in the fields of civil engineering, geotechnics, and tunnel construction. It is especially relevant for those involved in tunnel monitoring, rehabilitation, and maintenance. The findings provide valuable insights for engineers, geodesists, and those working with 3D displacement analysis, offering a comprehensive approach to detecting and addressing structural issues in tunnel systems. Additionally, it serves as a resource for decision-makers and industry stakeholders seeking to optimize monitoring strategies for improving the safety, longevity, and cost-efficiency of tunnel infrastructure.

Seismic Tomography for Tunnel Safety—Canada

In Canada, seismic tomography was applied to assess the structural integrity of a newly constructed tunnel under a major urban area (Sen et al. 2018). The tunnel was constructed through various geological layers, and there were concerns about potential ground settlement and the risk of tunnel collapse over time.

Seismic tomography was used to create 3D images of the subsurface conditions around the tunnel, enabling engineers to assess the stiffness and density of surrounding materials and detect any potential areas of weakness. The method provided real-time data on ground movement and identified zones where additional support might be needed.

The seismic tomography revealed several subsurface voids that could have threatened the tunnel's long-term stability. Based on this data, engineers reinforced these areas, ensuring the tunnel's safety and minimizing the risk of future ground displacement or settlement.

This case study was highly accepted by stakeholders, as it highlighted the importance of proactive monitoring and risk management. The non-invasive nature of seismic tomography and its ability to deliver 3D subsurface images provided critical information to ensure the tunnel's long-term durability and safety.

Tunnel Seismic Prediction in New Austrian Tunnelling Method (NATM)—Case Study

The New Austrian Tunnelling Method (NATM) is widely utilized for tunnel construction due to its adaptability to varying ground conditions. However, the uncertainty of ground conditions ahead of the excavation face makes tunnel construction a high-risk endeavour. Issues such as faults, massive groundwater, and fracture zones can lead to serious accidents or collapses. To mitigate these risks, this study introduces the Tunnel Seismic Prediction (TSP) method, a modern technology that helps detect unstable ground conditions ahead of excavation. By using reflected seismic waves generated through controlled explosions, TSP creates high-resolution images that provide detailed insights into the ground conditions. This case study focuses on the Chungren Tunnel project, where TSP was used to successfully predict hazardous conditions such as the Hochung fault and collapsed zones, enhancing safety management during excavation (Tzou et al. 2020).

The primary objective of this study is to examine how the Tunnel Seismic Prediction (TSP) method can be integrated with the New Austrian Tunnelling Method (NATM) to enhance safety management in tunnel construction. Specifically, the study aims to evaluate the effectiveness of TSP in detecting unstable geological conditions ahead of the excavation face, demonstrate its ability to minimize the risk of collapses and disasters by providing reliable ground condition data, and present a case study from the Chungren Tunnel to highlight the successful implementation of TSP in improving safety measures.

The Chungren Tunnel project, part of the Suhua Highway Improvement Project in Taiwan, served as the case study for this research. The tunnel, which spans 4774 m, consists of two tubes, each 11 m in width and 14 m in height. The Tunnel Seismic Prediction (TSP) method was applied to predict the geological conditions along the tunnel route, particularly at station 0 k + 820, where the Hochung fault and collapsed zone were encountered.

For TSP, 24 explosion holes were drilled along the tunnel's walls, and two receiving sensors were installed to detect seismic waves. These waves, generated by controlled explosions, were analyzed using a central computer system (TSP 203 Plus) to create high-resolution seismic images of the ground conditions ahead. The method was integrated with traditional horizontal detecting holes to enhance the accuracy of the predictions.

Seismic data collected from the explosions were processed using the TSP system to create simulated images of the tunnel's geological conditions. These images provided a detailed view of the ground composition, identifying areas of poor soil, fractures, and potential groundwater zones. Parameters such as pressure wave velocity (V_p), shear wave velocity (V_s), and soil density (ρ) were used to calculate critical indices like Young's modulus (E), Poisson's ratio (σ), and shear modulus (μ), which were then analyzed to predict the stability of the ground ahead of the excavation face. These results were compared with traditional methods, such as the Rock Tunnelling Quality Index (Q) and Rock Mass Rating (RMR), for verification.

The TSP method successfully identified the location of the Hochung fault and the collapsed zone, as well as other hazardous geological features ahead of the excavation face. These findings were validated by subsequent site excavations and ground sampling, where the predicted conditions aligned closely with the actual ground properties observed during excavation. During the excavation of the Chungren Tunnel, a small-scale collapse occurred when the excavation face reached the Hochung fault zone. However, the TSP data had already provided crucial information, enabling engineers to implement preventive measures, such as reinforcing the soil with the Pipe Umbrella Support System, thereby mitigating further risk.

The comparison between TSP results and traditional prediction methods showed that TSP provided more accurate and detailed information, allowing for better-informed decisions regarding safety measures and excavation planning.

The study demonstrates the effectiveness of integrating the Tunnel Seismic Prediction (TSP) method with the New Austrian Tunnelling Method (NATM) to enhance safety management during tunnel construction. The Chungren Tunnel case study illustrates how TSP can successfully predict geological hazards such as faults and collapsed zones, enabling engineers to take proactive measures to prevent accidents. The combination of TSP with traditional detection methods increases the accuracy of ground condition assessments, improving risk mitigation strategies and ensuring safer tunnel excavation. TSP's ability to detect poor ground conditions ahead of the excavation face significantly contributes to the overall safety of tunnelling projects, reducing the likelihood of catastrophic failures.

This work is primarily aimed at civil engineers, tunneling professionals, and project managers involved in large-scale infrastructure projects, particularly in tunnel construction. It will also be valuable to researchers and practitioners interested in advanced geophysical techniques, ground condition prediction methods, and safety management in complex excavation environments.

8.5.3 Electrical Resistivity Imaging (ERI) for Foundation Settlement—United Kingdom

In the United Kingdom, Electrical Resistivity Imaging (ERI) was employed to monitor foundation settlement at a large office building in a city center (Williams and Ormsby 2016). The building was experiencing minor but ongoing settlement, and the goal was to monitor the subsurface behavior over time to ensure structural integrity.

ERI was used to monitor resistivity changes in the soil around the building's foundation, identifying areas where groundwater might be infiltrating or where soil compaction could be affecting the foundation's stability. The resistivity data provided insights into moisture content and the overall condition of the foundation soils.

The ERI survey identified specific areas where the soil resistivity was lower than expected, indicating higher moisture content and potential for soil settlement. This information allowed engineers to reinforce these areas and prevent further settlement from occurring, ultimately protecting the building from potential foundation failure.

Stakeholders, including property owners and insurers, were particularly satisfied with the results because ERI provided an early warning system that helped mitigate further damage. The non-destructive nature of the technique also made it an attractive and cost-effective solution for ongoing monitoring.

8.5.4 Interferometric Synthetic Aperture Radar (InSAR) for Urban Subsidence—Italy

In Italy, Interferometric Synthetic Aperture Radar (InSAR) technology was applied to monitor ground subsidence in a rapidly developing urban area built on soft soils (Montanari et al. 2019). The region had experienced localized subsidence in the past, and there were concerns about the impact of construction and urban development on ground stability.

InSAR was used to measure minute ground displacements over time using satellite radar data. The technique enabled the monitoring of ground movement at the surface over a large area, detecting both vertical and horizontal displacements with millimeter precision.

The data revealed areas where subsidence was occurring at a faster rate than expected, particularly near major construction sites. These findings helped local authorities implement measures to prevent further subsidence, such as reinforcing foundations and monitoring construction activities more closely.

The ability of InSAR to monitor large-scale urban areas continuously and non-invasively made it an appealing solution for urban planners and local authorities. The technology's ability to provide real-time alerts for subsidence allowed for swift mitigation actions, which was highly valued by all stakeholders, including local governments, developers, and residents.

8.5.5 DFOS for Pipeline Integrity—Australia

Distributed Fiber Optic Sensing (DFOS) has been used extensively in the North Sea for pipeline monitoring (Liu et al. 2018). The project aimed to detect and monitor the integrity of subsea pipelines, ensuring they are not affected by corrosion, mechanical damage, or third-party interference. The pipeline is located in a remote and harsh environment, making continuous, real-time monitoring crucial to prevent oil spills and gas leaks.

DFOS technology was installed along the length of the pipeline to measure parameters such as strain, temperature, and vibration in real time. The technology uses the principles of Rayleigh scattering to monitor changes in strain, providing high-resolution data across the pipeline's entire length. This allowed operators to detect early signs of mechanical stress or leakage and take preventive actions.

The DFOS system successfully identified several stress points along the pipeline that could have potentially led to failures if left undetected. The real-time monitoring enabled operators to perform maintenance and repairs proactively, reducing the risk of environmental damage and ensuring continuous, safe operation of the pipeline.

The project was widely accepted by stakeholders, including the pipeline operator and environmental agencies, as it provided a reliable and real-time solution for monitoring a critical infrastructure asset. The ability to detect issues early and prevent leaks or catastrophic failures was crucial for ensuring the pipeline's safety and minimizing environmental impact.

8.5.6 GPR for Bridge Inspection—Mumbai, India

In Mumbai, India, Ground Penetrating Radar (GPR) was used to inspect and assess the structural health of several aging bridges (Rai and Raj 2019). These bridges are critical for transportation in the city, and concerns about internal damage, such as corrosion and delamination of concrete, were raised. The goal was to detect subsurface anomalies without causing disruptions to traffic.

GPR was employed to scan concrete decks and substructures to detect voids, cracks, and areas of reinforcement corrosion. The results revealed several localized regions where the reinforcing steel showed signs of corrosion, although these issues were not visible on the surface.

The findings led to timely maintenance actions, including targeted repairs to the corroded areas, thus preventing further degradation and increasing the bridge's lifespan. The use of GPR allowed for non-invasive, efficient inspections, which was highly appreciated by stakeholders for minimizing traffic disruption while ensuring safety.

8.5.7 Seismic Tomography for Tunnel Safety—Delhi Metro Project

During the construction of a tunnel as part of the Delhi Metro Expansion Project, seismic tomography was used to assess subsurface conditions and ensure the stability of the tunnel (Sharma and Sharma 2018). The project involved boring through urban areas with diverse geological formations, raising concerns about potential ground movement and tunnel collapse.

Seismic tomography was employed to generate a detailed image of the subsurface around the tunnel alignment. The technique helped identify weak geological zones and areas with potential for differential settlement, which could lead to structural risks for the tunnel.

Based on the seismic tomography data, engineers were able to reinforce specific sections of the tunnel where ground conditions were found to be less stable. This proactive approach helped prevent potential delays and ensured the safe passage of the tunnel under urban infrastructure.

8.5.8 ERI for Foundation Monitoring—Bengaluru, India

In Bengaluru, Electrical Resistivity Imaging (ERI) was employed to monitor foundation settlement beneath a multi-story commercial building, where soft, water-saturated soils raised concerns regarding potential settlement (Suresh and Anurag 2020).

ERI was used to map resistivity changes in the soil around the foundation. The technique detected variations in moisture content and soil heterogeneity that could affect the foundation's performance.

ERI provided critical information, allowing engineers to identify weak zones and reinforce the foundation, preventing future settlement and ensuring the building's long-term stability.

8.5.9 InSAR (Interferometric Synthetic Aperture Radar) for Subsidence—Chennai, Tamil Nadu

In Chennai, Tamil Nadu, InSAR was used to monitor subsidence caused by rapid urbanization and excessive groundwater extraction in coastal areas, which threatened the stability of infrastructure, including roads and buildings. InSAR was used to measure ground displacement from satellite radar data, revealing areas of significant vertical and horizontal subsidence in certain parts of the city. The findings led to the implementation of better groundwater management practices and reinforced structures in subsiding areas, mitigating the risks associated with ongoing ground movement.

8.5.10 DFOS for Pipeline Integrity—Gujarat, India

In Gujarat, India, DFOS was used to monitor the integrity of a critical oil pipeline traversing long distances across diverse terrain, where potential risks included leakage and external damage (Desai and Bhattacharya 2021). The DFOS system measured parameters such as strain and temperature along the entire length of the pipeline. This real-time monitoring system helped detect anomalies such as stress accumulation that could lead to pipeline failure. DFOS successfully identified stress points along the pipeline that could have led to leaks, allowing for timely repairs and reducing environmental risks. Continuous monitoring enhanced the pipeline's safety and operational efficiency.

8.5.11 GPR Inspection of Dam Spillway—Alabama, USA

This case study focuses on using Ground Penetrating Radar (GPR) to inspect the concrete spillway of a hydroelectric dam located in Alabama, USA (Fig. 8.1). The spillway is an essential part of the dam, and over time, it may develop cracks, voids, or other structural issues. Traditional inspection methods can be time-consuming and expensive, so the goal of this study was to see how effectively a small, portable GPR system could be used to inspect a large, irregular concrete surface quickly and accurately (Bigman and Day 2022).

The primary aim of this study was to assess the performance of a small, handheld GPR unit on a large, irregular concrete surface like the dam's spillway and to evaluate

Fig. 8.1 Photograph of the project site: hydroelectric dam spillway in Alabama, USA. *Source* Bigman and Day (2022)—open access

whether higher-frequency GPR systems can detect smaller cracks and defects more clearly than traditional, lower-frequency systems.

For the inspection, the researchers used a small, high-frequency GPR system that could be attached to a telescopic pole. This allowed the team to scan large areas of the spillway quickly without the need for scaffolding (Scaffolding refers to a temporary structure used to support workers and materials during the construction, repair, or maintenance of buildings and other large structures) or other equipment. The GPR system sends radar waves into the concrete, and when these waves encounter a defect such as a void or crack, they reflect back, creating a signal that can be analyzed. The spillway itself was built in the 1920s and has large concrete slabs resting on rock. Some of the slabs are cracked, and others are smooth, making them difficult to inspect. The research team used the signals from PVC drainage pipes, which are similar to voids, as a reference to help interpret the results (Fig. 8.2).

The data collected by the GPR system was processed and analyzed to identify areas of concern (Fig. 8.3). When the radar waves encountered air-filled spaces (like voids) or changes in the concrete-rock bond (such as debonding), they produced signals that were different from those caused by solid materials like steel reinforcements. These differences were used to identify potential problems in the spillway.

The inspection revealed over 150 locations that required further testing for potential voids, cracks, or delaminations (Fig. 8.4).

In many cases, hammer testing confirmed the presence of defects such as voids or debonding. Some areas were found to have more serious issues, like deep weathering at the concrete-rock interface, which required drilling to investigate further (Fig. 8.5).

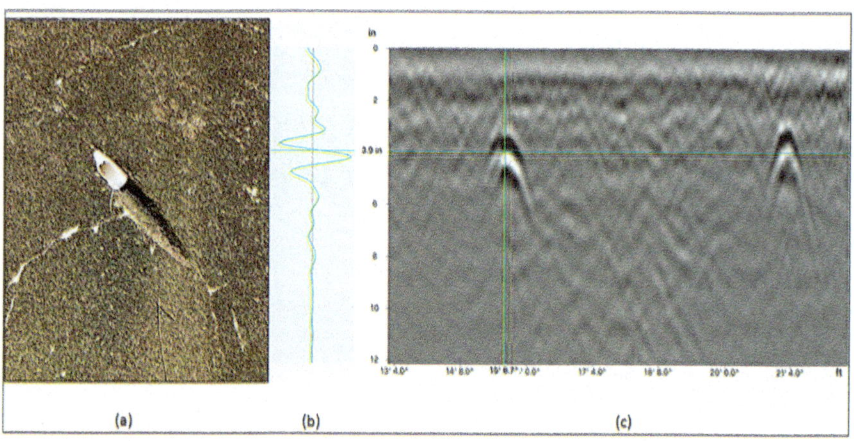

Fig. 8.2 Example of GPR response from PVC drainage pipes in the hydroelectric dam spillway, Alabama, USA—**a** exposed PVC drainage pipe embedded in the spillway, **b** A-scan (1D, Amplitude Scan) trace rendering of the response from the PVC pipe, indicating the polarity of the response, **c** GPR B-scan (2D, bridge scan or cross-sectional scan) profile of data collected over the PVC drainage pipes onsite, approximating the expected response from a concrete/air boundary in the spillway. *Source* Bigman and Day (2022)—open access

Fig. 8.3 Photograph of data collection with Proceq GP8000 at hydroelectric dam spillway, Alabama, USA. *Source* Bigman and Day (2022)—open access

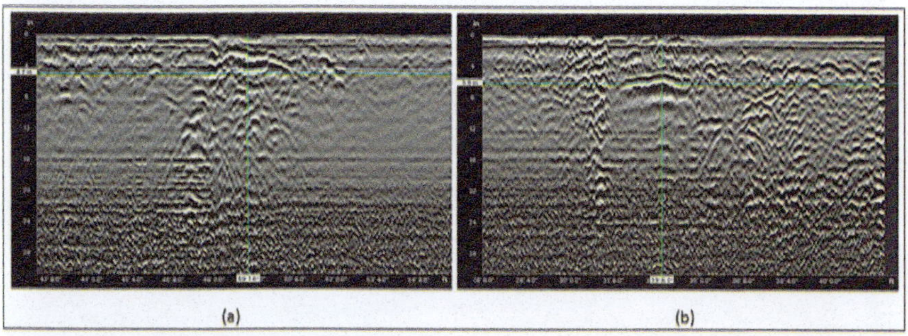

Fig. 8.4 Examples of two shallow responses indicative of possible delamination in Alabama, USA: **a** 10 cm deep and **b** almost 15 cm below the surface. Both responses extend for more than 100 cm. *Source* Bigman and Day (2022)—open access

While the drilling did not reveal any substantial voids, it did show that weathering at the concrete-rock boundary was affecting the material.

The small, handheld GPR system proved to be very effective, and the high-frequency system provided clearer images of smaller defects compared to traditional

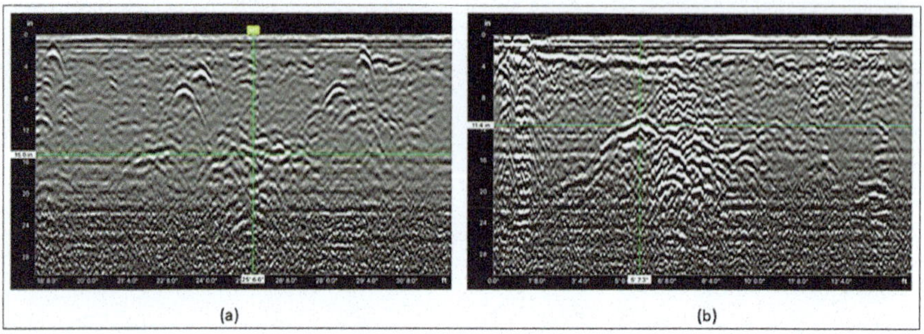

Fig. 8.5 Examples of two deeper responses indicative of possible voids, which turned out to be weathered rock, in Alabama, USA: **a** 38 cm and **b** almost 29 cm below the surface. Both responses extend for more than 200 cm. *Source* Bigman and Day (2022)—open access

lower-frequency systems. The GPR was also able to cover large areas of the spillway quickly, which would have been difficult using traditional methods.

This study demonstrated that a small, handheld GPR system can effectively inspect large, irregular concrete surfaces like the spillway of a hydroelectric dam. Despite some challenges, such as slippery surfaces and inconsistent slab thickness, the GPR system provided valuable information about the condition of the spillway. The higher-frequency GPR system allowed for better detection of small cracks and defects, which is essential for the long-term maintenance of infrastructure (Fig. 8.6).

Fig. 8.6 Photograph of the spillway in Alabama, USA showing large (right) and small (left) concrete patches from repair and resurfacing following the GPR investigation. *Source* Bigman and Day (2022)—open access

This research is particularly useful for engineers, infrastructure managers, and maintenance teams responsible for inspecting and maintaining large concrete structures like dams, bridges, and tunnels. It also benefits those involved in developing new, cost-effective inspection technologies that can improve safety and reduce maintenance costs.

In the future, it is likely that GPR technology will become even more advanced, with improvements in resolution and the ability to inspect even more challenging areas. Robotic systems and drones may also be used to deploy GPR in areas that are difficult for humans to reach, such as steep or dangerous slopes. As GPR technology continues to improve, it will become an even more important tool for maintaining and monitoring the health of critical infrastructure.

8.5.12 ERI for Hazard Investigations—Taiwan

This study presents Electrical Resistivity Imaging (ERI) data aimed at hydrogeological characterization, land subsidence studies, and geological hazard investigations in Taiwan (Chang et al. 2023). The data, collected from various regions in Taiwan, focuses on identifying aquifer systems, groundwater recharge zones, land subsidence, and geological structures, including faults. The dataset includes 265 ERI profiles with a total survey length of 59,905 m, collected between 2010 and 2022. The findings of this research were pivotal for enhancing the understanding of Taiwan's subsurface conditions, which can aid in hazard detection, groundwater monitoring, and environmental management.

The primary aim of this study was to utilize Electrical Resistivity Imaging (ERI) data for hydrogeological and geological hazard investigations across various regions in Taiwan. The main objectives included characterizing aquifer systems, identifying groundwater recharge zones, and assessing land subsidence. Additionally, the study focused on detecting geological structures such as faults, including the Rusui, Qishan, Yuli, and Shanyi faults (Fig. 8.7).

By utilizing time-lapse ERI data, the study sought to monitor groundwater levels and estimate hydraulic parameters, contributing to better water resource management and geological hazard mitigation. Furthermore, the study intended to make the ERI data available for future research and applications, particularly in geohazard detection and the development of advanced data interpretation techniques, such as machine learning. The ERI data was collected using advanced multi-electrode resistivity meters, including the 4-point light 10W LGM Lippmann resistivity meter and the AGI SuperSting R1 meter. The survey utilized various arrays like the Wenner, Schlumberger, and dipole–dipole to measure resistivity at different subsurface depths. The data collection was spread across several regions in Taiwan, including Changhua, Yunlin, Yilan, Taichung, Chiayi, Tainan, and Taipei, as well as fault zones like the Rusui, Qishan, Yuli, and Shanyi faults. Time-lapse ERI profiles were recorded in specific basins, considering seasonal rainfall variations.

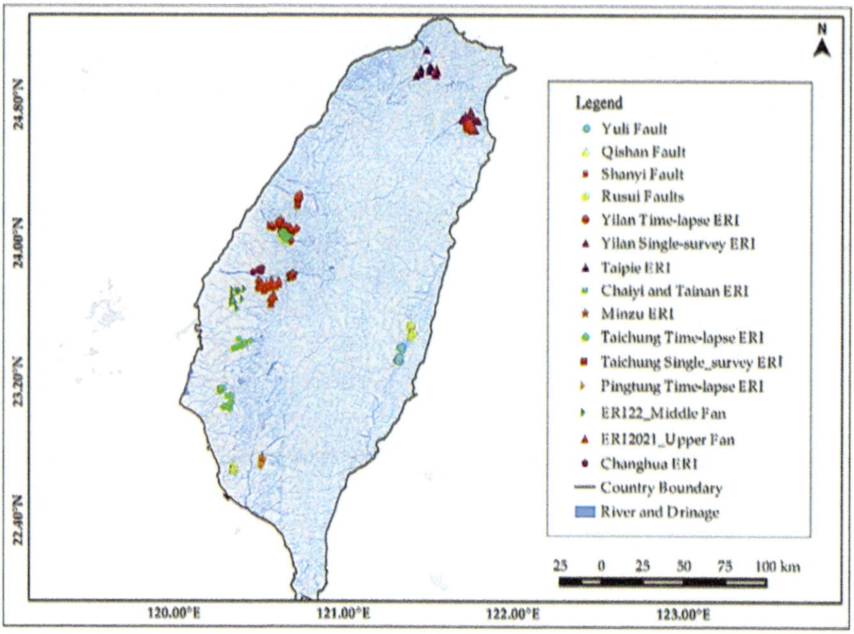

Fig. 8.7 ERI data distribution for hydrogeological and land subsidence studies in Yilan, Yunlin, and Taichung basins. *Source* Chang et al. (2023)—open access

In this case study, data collection involved multiple ERI profiles distributed across various regions of Taiwan for hydrogeological and geological investigations. The ERI data distribution for hydrogeological characterization in the Yilan Alluvial Basin is depicted, showcasing the distribution of profiles used for assessing groundwater and aquifer properties. Additionally, the data for artificial groundwater recharge site selection and clay layer detection for land subsidence in Yunlin are presented, which focus on identifying suitable sites for groundwater recharge and monitoring subsurface conditions. Lastly, the ERI data collected for geological hazard detection in the Shanyi area, Taichung, highlights the use of ERI profiles along fault lines to assess geological structures and detect potential hazards.

The ERI data revealed key insights into Taiwan's hydrogeological and geological conditions. The inverted resistivity models highlighted groundwater recharge zones, identified fault structures, and characterized the subsurface layers, including clay and rock formations. Time-lapse data indicated fluctuations in groundwater levels, which were used to estimate hydraulic parameters. The results contributed to a better understanding of land subsidence and groundwater behavior in response to seasonal precipitation (Fig. 8.8). In addition, the data was effectively integrated with other geophysical methods like seismic and electromagnetic surveys to enhance subsurface characterization.

The findings and data presented are intended for a broad audience, including researchers, engineers, and students working in the fields of hydrogeology,

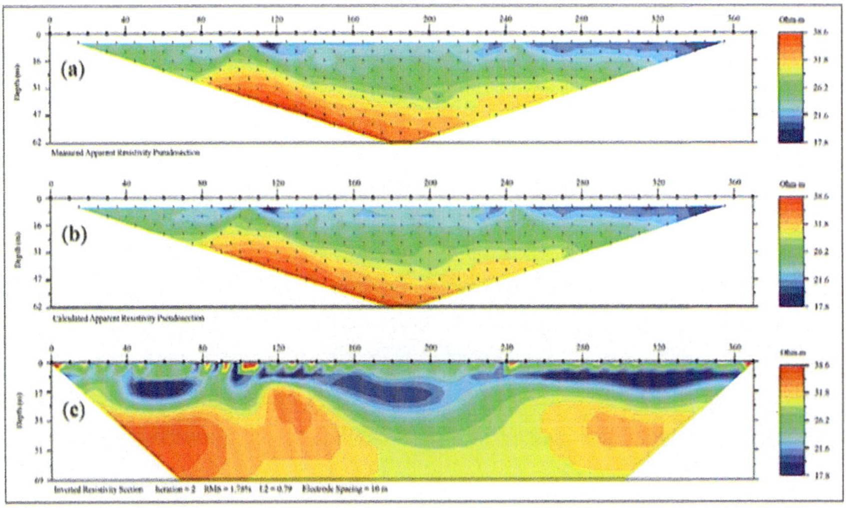

Fig. 8.8 ERI data distribution and inversion models for the Yunlin Chousui River Alluvial Fan: **a** apparent resistivity data, **b** calculated apparent resistivity data, and **c** an inverted resistivity model. *Source* Chang et al. (2023)—open access

geophysics, and environmental science. Figure 8.9 outlines the overall ERI data acquisition, processing, and inversion methods, which will serve as a guideline for future ERI-based subsurface investigations. This comprehensive dataset is made publicly available for further academic research, geohazard detection, and environmental monitoring applications, enhancing the potential for interdisciplinary collaborations.

8.5.13 Urban Subsidence via Small Baseline Subset-Interferometric Synthetic Aperture Radar (SBAS-InSAR)—Beijing Plain, China

Land subsidence is a critical issue in many urban areas, especially in regions where there is extensive groundwater extraction. In Beijing, a rapidly growing city, land subsidence has become a significant concern, affecting infrastructure and urban development. This study aims to investigate the spatial and temporal variations in land subsidence within the Beijing Plain from 2003 to 2013, using advanced remote sensing techniques, particularly Small Baseline Subset-Interferometric Synthetic Aperture Radar (SBAS-InSAR). The research also explores the relationship between land subsidence and groundwater exploitation in the region, providing insights into the environmental and infrastructure challenges faced by the city (Liu et al. 2020).

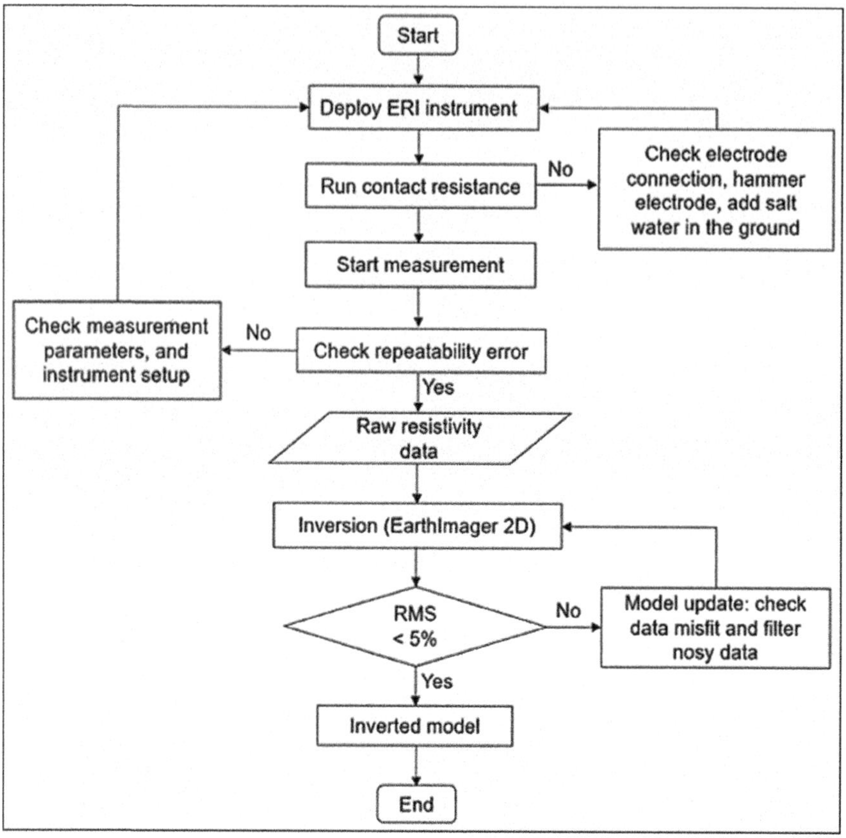

Fig. 8.9 Flow chart of resistivity imaging data acquisition, processing, and inversion for the Yilan Alluvial Basin. *Source* Chang et al. (2023)—open access

The primary aim of this study is to utilize SBAS-InSAR to monitor and analyze the land subsidence in the Beijing Plain during the period from 2003 to 2013. Furthermore, the study seeks to understand the correlation between groundwater extraction and land subsidence, examining how these factors influence infrastructure and urban planning decisions. By exploring these dynamics, the research contributes to understanding the impact of subsidence on urban areas and offers insights for better management of groundwater resources.

To quantify deformation in the study, a transformation formula was used to convert the line-of-sight (LOS) displacement measurements into vertical displacement. The equation used for this transformation is:

$$v_{vertical} = \frac{v_{los}}{\cos\theta} \tag{8.22}$$

where $v_{vertical}$ represents vertical displacement, v_{los} is the line-of-sight displacement, and θ is the satellite's look angle. This mathematical approach ensures that the ground displacement is accurately represented in the vertical direction, which is crucial for studying land subsidence.

The study employed several key materials and methods to monitor and analyze land subsidence. Satellite data, including 47 ASAR images from 2003 to 2010 and 24 Radarsat-2 images from 2010 to 2013, were used to track surface displacement. The SBAS-InSAR technique was applied to calculate displacement over time using a small baseline subset method, ensuring high accuracy in measuring ground motion. The results from SBAS-InSAR were validated against leveling data from seven benchmark stations. Kriging interpolation was used to interpolate deformation data from different periods to better visualize spatial variations in subsidence. Additionally, groundwater data from each year during the study period were analyzed to assess the correlation between groundwater exploitation and land subsidence.

The processing of data involved multiple steps. First, deformation velocities in the vertical direction were extracted from the SAR data stacks. A correction factor was then applied to unify deformation values between the ASAR and Radarsat-2 datasets. The Empirical Mode Decomposition (EMD) algorithm was used to decompose the time-series deformation data, which helped identify periodic components of subsidence, particularly with a cycle of approximately 1 year. Finally, groundwater exploitation data were analyzed alongside the subsidence measurements to identify spatial relationships between groundwater extraction and land deformation.

The results of the study revealed significant patterns of land subsidence. Figure 8.10 illustrates the mean deformation velocities in the vertical direction for the periods 2003–2010 and 2010–2013.

The analysis showed a significant rate of subsidence, particularly in the Chaoyang and Shunyi districts of Beijing. The deformation patterns from both the ASAR and Radarsat-2 data stacks were consistent, with high subsidence rates observed in the northeastern areas of the Beijing Plain. Further, Fig. 8.11 illustrates the correlation between groundwater levels and land subsidence, revealing that areas with lower groundwater levels were more prone to severe subsidence during both 2003–2010 and 2010–2013. In addition, deformation profiles along key traffic routes, such as the Capital Airport expressway and Jingshen expressway, showed that subsidence funnels along these roads became more pronounced over time, which could potentially affect infrastructure in these areas.

The study successfully demonstrates the application of SBAS-InSAR for monitoring land subsidence in the Beijing Plain between 2003 and 2013. The results show that land subsidence is primarily concentrated in areas with high groundwater extraction, such as Chaoyang, Shunyi, and Tongzhou districts. The correlation between groundwater extraction and subsidence is evident, although other factors, including urban development and geological conditions, also contribute to the phenomenon. The periodic nature of subsidence, with a cycle of approximately 1 year, was identified through time-series analysis. The study provides a better understanding of the

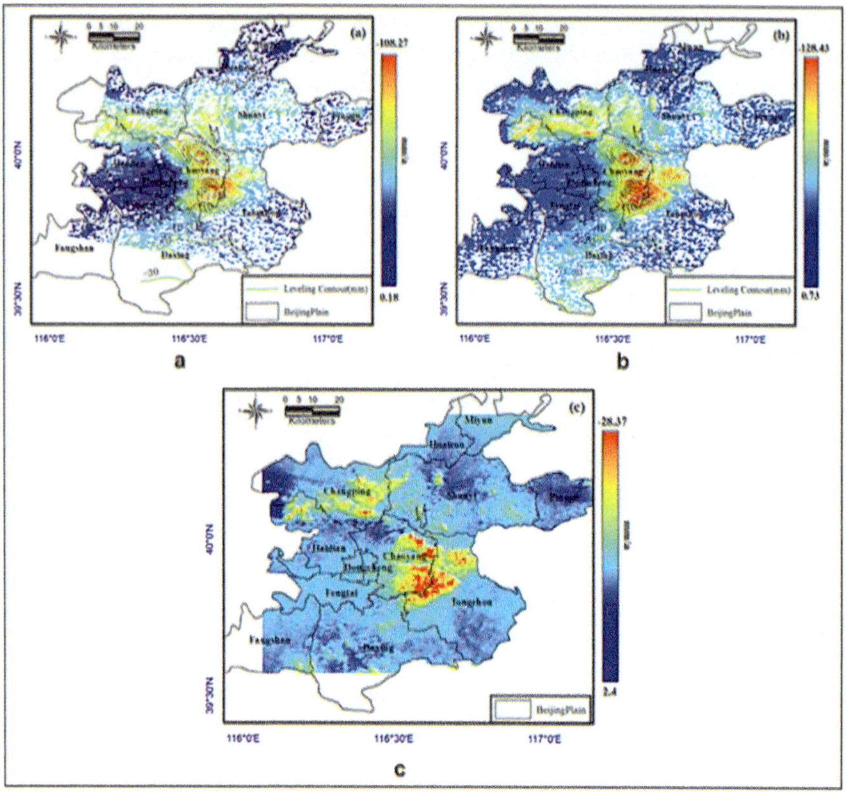

Fig. 8.10 Mean vertical deformation velocities measured from the two data stacks: **a** Mean vertical deformation acquired from ASAR data during 2003–2010, highlighting significant subsidence in Chaoyang, Shunyi, and parts of Tongzhou districts. **b** Mean vertical deformation acquired from Radarsat-2 data during 2010–2013, showing a similar subsidence pattern, with maximum deformation occurring in Chaoyang, Shunyi, and Changping districts. **c** Disparities of mean deformation velocities between the two time periods, showing notable differences in Changping, the eastern part of Chaoyang, and the western part of Tongzhou district, indicating rapid deformation in these regions during the second period (2010–2013). *Source* Liu et al. (2020)—open access

dynamics of land subsidence in Beijing, and future research should focus on quantifying the relationship between groundwater extraction and land subsidence, as well as examining the long-term effects of urbanization on ground stability.

This case study is aimed at a broad range of professionals and researchers. It will be of particular interest to academics and researchers in the fields of geophysics, remote sensing, and environmental science who are focused on land subsidence monitoring techniques. Urban planners, civil engineers, and geotechnical experts involved in infrastructure development and land use planning will also find the study relevant. Government officials responsible for urban development, environmental policies, and water resource management can use the findings to inform policy and mitigation strategies. Furthermore, the study is valuable for environmental and water

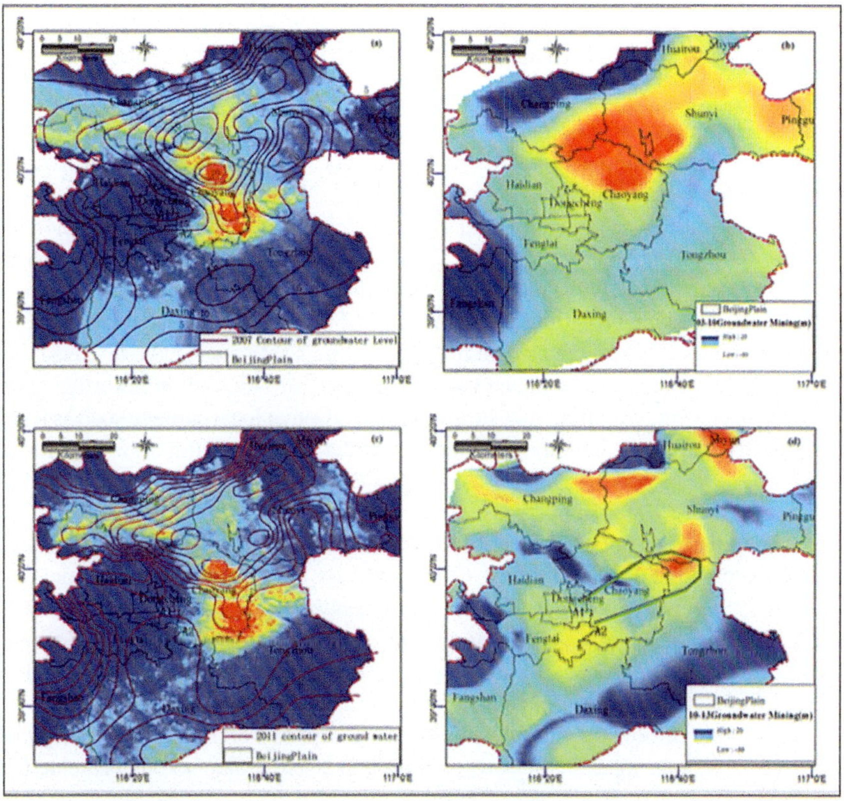

Fig. 8.11 The corresponding relationship between the dynamic variation of groundwater and mean subsidence velocity: **a** the subsidence result derived from 2003 to 2010, with the contour representing the groundwater level of 2007. Significant subsidence is observed in Chaoyang, Shunyi, and parts of Tongzhou districts, correlating with lower groundwater levels. **b** The variation of groundwater during 2003–2010, showing a marked decline in groundwater levels, particularly in the northeastern part of the Beijing Plain, which corresponds to the observed subsidence areas. **c** The subsidence result derived from 2010 to 2013, where the contour represents the groundwater level of 2011. Similar subsidence patterns are observed in Chaoyang, Shunyi, and Tongzhou, with a clear relationship to groundwater level changes. **d** The variation of groundwater during 2010–2013, indicating continued decline in groundwater levels, reinforcing the correlation with land subsidence in these key districts. *Source* Liu et al. (2020)—open access

resource managers, as well as climate scientists, interested in understanding the effects of groundwater exploitation on land stability and its implications for urban sustainability.

8.5.14 Distributed Fiber Optic Sensing (DFOS) for Pipeline Structural Health Monitoring (SHM)—Case Study

Pipeline structures are integral to the transport of fluids and gases across vast distances, often spanning hundreds of kilometers. Pipeline systems face significant operational risks due to environmental factors like landslides and soil movements, which can cause structural damage. As pipelines age, it is essential to monitor their condition to prevent catastrophic failures. In northeastern Berlin, Distributed Fiber Optic Sensing (DFOS) technology was implemented to continuously monitor a composite pipeline, providing a reliable method to track strain, displacement, and temperature changes that are critical for the pipeline's safety (Soga et al. 2018).

The study's aim was to evaluate the effectiveness of DFOS in monitoring strain and deformation in a pipeline, focusing on the detection of local damage such as leaks and identifying areas where the pipeline was under excessive strain due to external forces.

The pipeline, located in northeastern Berlin, was monitored using DFOS sensors placed longitudinally along its length. Sensors were installed at 120-degree intervals around the pipeline's circumference. Over two years, data were collected to measure strain and temperature changes. The data were analyzed using a Trapezoidal method to calculate displacements and compared with traditional spot sensor measurements. This provided a comprehensive assessment of the pipeline's structural health. The sensor arrangement and strain measurements are shown in Fig. 8.12.

An example of how the distributed fiber optic sensor works, showing the spatial resolution, readout, measurement resolution, and sensing distance, is presented in Fig. 8.13.

In the case study involving distributed fiber optic sensors (DFOS) for structural health monitoring, specific mathematical concepts are often involved, particularly when analyzing strain, stress, and displacement data.

Strain (ε) is a measure of the deformation of the material under stress, and it is generally calculated using the formula:

$$\varepsilon = \frac{\Delta L}{L_0} \tag{8.23}$$

where ΔL is the change in length, L_0 is the original length of the material or sensor. In the context of DFOS, strain is typically measured continuously along the fiber, which allows for high spatial resolution.

Once strain is calculated, stress (σ) can be determined using Hooke's Law for linear elasticity:

$$\sigma = E \cdot \varepsilon \tag{8.24}$$

where σ is the stress (force per unit area), E is the Young's Modulus (a material property representing stiffness), and ε is the strain. This relationship is central in

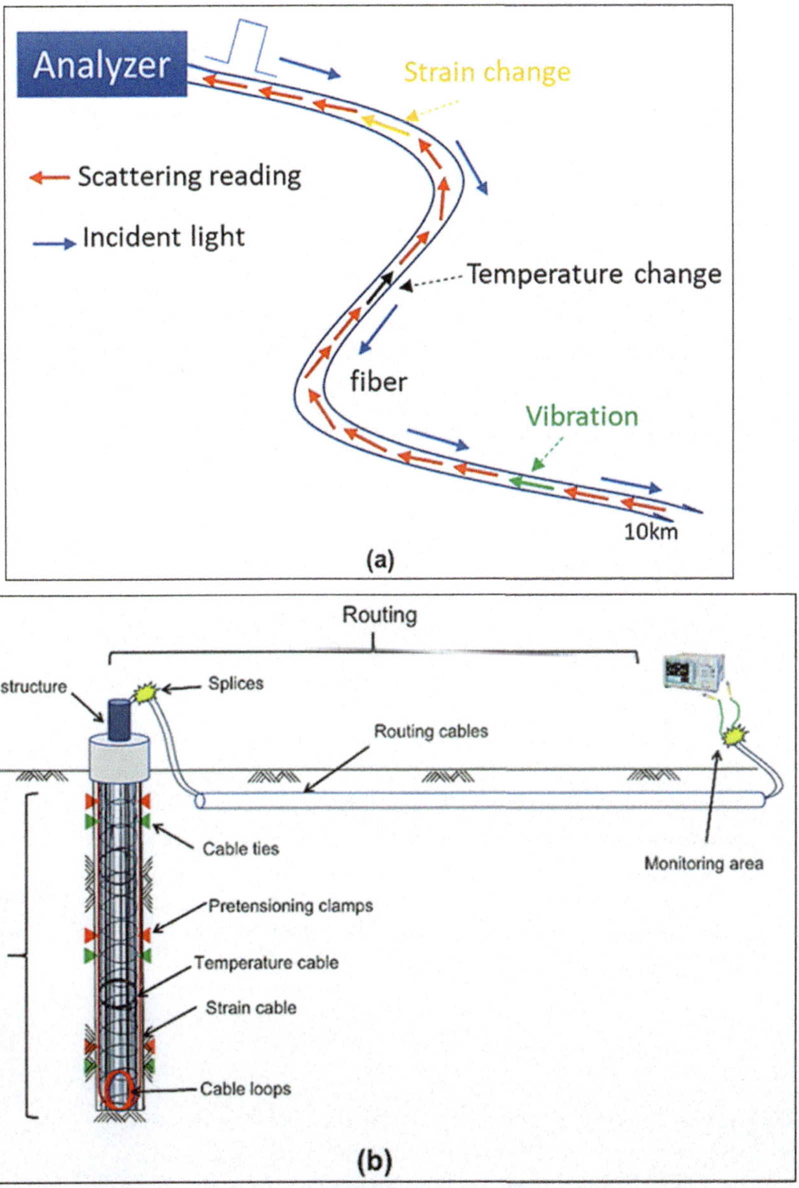

Fig. 8.12 **a** Schematic diagram of a distributed fiber optic sensor system, and **b** application of DFOS technology in civil engineering for structural health monitoring of a pile. *Source* Soga et al. (2018)—open access

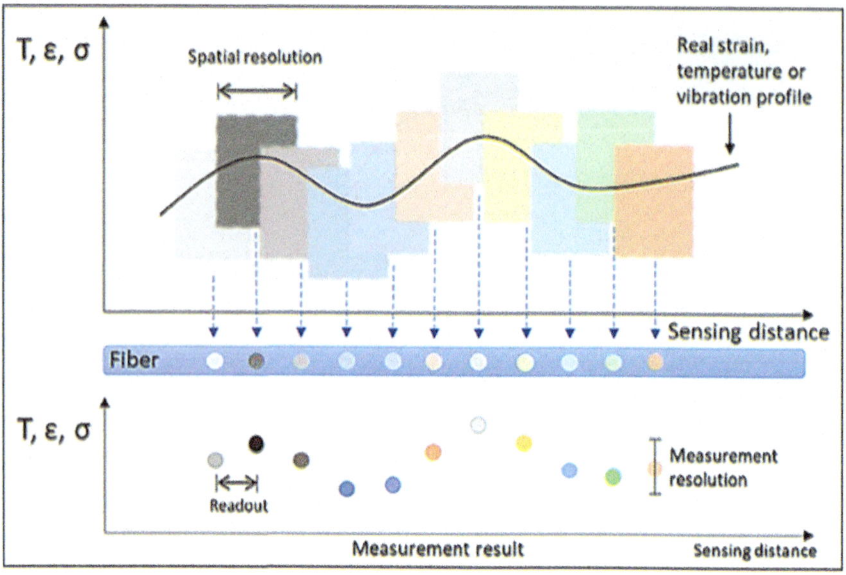

Fig. 8.13 Example of the operation of a distributed fiber optic sensor, illustrating spatial resolution, readout, measurement resolution, and sensing distance where T represents temperature, σ represents stress, and ε represents strain. *Source* Soga et al. (2018)—open access

structural health monitoring because it helps understand how the material responds to loads.

The displacement of a structure due to applied load can be calculated using strain measurements along with geometrical considerations. For example, if the strain measurements along a fiber optic cable (embedded in or attached to the structure) are known, the displacement can be estimated by integrating the strain over the length of the structure. For small deformations, the displacement (δ) could be calculated using the following approximation based on strain data:

$$\delta = \int_0^L \varepsilon(x)\,dx \tag{8.25}$$

where ε(x) is the strain as a function of position xxx along the structure (in this case, along the fiber), L is the length of the structure being monitored.

In the case where the strain profiles are used to assess the shape of deformations (like pipe bending or elongation), numerical methods like the Trapezoidal Method may be used to approximate the displacement or the change in geometry. This method is typically used to estimate the area under a curve (in this case, strain or stress distribution) in a discrete form. The trapezoidal rule formula is:

$$\int_a^b f(x)\,dx \approx \frac{b-a}{2}\left[f(a) + 2\sum_{i=1}^{n-1} f(x_i) + f(b)\right] \tag{8.26}$$

where f(x) is the function (e.g., strain or stress along the fiber), a and b are the bounds of the integration, x_i are the intermediate points.

FEM is a computational method used to model and simulate physical phenomena, such as stress and strain distributions in materials. In structural health monitoring, FEM could be used to validate strain measurements or simulate the behavior of the structure under load. The basic steps in FEM involve: Dividing the structure into smaller elements (meshing), Solving for the behavior of each element based on material properties, boundary conditions, and applied loads, and Assembling the solutions for each element to understand the overall structure's response. These mathematical models and methods help in analyzing the data from the DFOS system and making decisions about structural integrity and necessary maintenance actions.

Over the monitoring period, DFOS successfully detected areas with dangerously high strain levels due to soil movements in landslide-prone regions. Remedial actions such as excavation and backfilling were carried out in time to prevent damage. In other areas, the strain variations were linked to seasonal temperature changes, which did not require intervention. DFOS also detected localized damage, such as leaks, that traditional sensors might have missed. The ability of DFOS to monitor both axial and circumferential strains was crucial for detecting issues caused by internal pressure.

The DFOS system demonstrated its ability to continuously monitor the structural integrity of the pipeline over an extended period. It provided real-time data that enabled early detection of potential threats, allowing for timely interventions. Unlike traditional spot sensors, DFOS provided comprehensive monitoring over the entire length of the pipeline, detecting both axial and circumferential deformations. The ability to track temperature variations and internal pressure-induced strains ensured that the pipeline remained safe and reliable.

This case study is particularly relevant to civil engineers, infrastructure managers, and researchers involved in pipeline safety, structural health monitoring, and material science. It is also useful for policymakers and developers of monitoring technologies, offering insights into cost-effective, long-term monitoring solutions for critical infrastructure.

8.6 Conclusions

This chapter has explored the transformative role of geophysical techniques in civil engineering, focusing on site investigation, infrastructure assessment, and structural health monitoring (SHM). These methods provide essential insights into the subsurface conditions and the integrity of structures like buildings, bridges, tunnels, and roads. Geophysical tools such as seismic tomography, ground penetrating radar (GPR), electrical resistivity tomography (ERT), and ambient vibration testing offer valuable data for ensuring the stability, safety, and long-term durability of infrastructure.

Through a range of case studies, this chapter highlighted how advanced geophysical methods are being employed globally to assess structural health, detect potential hazards, and monitor ongoing infrastructure performance. For instance, applications of GPR in bridge and tunnel inspections, the use of seismic refraction in metro tunnel alignment, and the integration of fiber optic sensors in bridge health monitoring exemplify the widespread adoption of geophysical techniques for infrastructure maintenance and monitoring.

Emerging trends in geophysical monitoring, such as satellite-based remote sensing, distributed fiber optic sensing (DFOS), and InSAR for subsidence monitoring, further enhance the precision and real-time capabilities of structural health management. These innovations are crucial for proactive maintenance strategies, helping prevent infrastructure failure and extending the lifespan of critical assets.

In conclusion, geophysics is an indispensable tool in modern civil engineering, providing both a deeper understanding of subsurface conditions and the means to monitor and maintain infrastructure in a sustainable and cost-effective manner. The ongoing integration of geophysical methods with other monitoring technologies ensures that infrastructure projects are designed, constructed, and maintained with optimal safety, durability, and efficiency. The continuous evolution of these techniques promises to significantly contribute to the resilience of infrastructure systems worldwide.

References

Abe H, Johansen R (2018) Corrosion detection in concrete using self-potential measurements. J Civil Eng Technol 9(6):1008–1015

Agarwal P, Zhang Z (2021) Wireless sensor networks for real-time structural health monitoring: a review. J Struct Eng 147(12):04021234

Benedetti M, Lombardi S (2018) Geophysical techniques in the monitoring of civil infrastructures: challenges and opportunities. Geophysics 83(4):147–155

Benedetto A, Giannetti F (2015) Application of ground penetrating radar to inspect concrete structures. J Non-Destruct Test 3(2):56–63

Berube MA, Leclerc JM, Brown CD, McLaughlin K, Gauthier D (2020) Complex resistivity applications for monitoring masonry flooding and drying. Constr Build Mater 244:1–12

Bigman DP, Day DJ (2022) Ground penetrating radar inspection of a large concrete spillway: a case-study using SFCW GPR at a hydroelectric dam. Case Stud Constr Mater 16:e00975

Bohnke A, Costa MRS, da Silva FS, Lima AAF, Silva JDA (2020) Applications of seismic refraction for subsurface characterization in civil engineering. J Appl Geophys 178:1–11

Cakir R, Okumus A (2019) Microseismic techniques for structural integrity assessment in high-rise buildings. Eng Struct J 12(4):198–206

Chang PU, Doyoro YG, Lin DJ, Puntu JM, Amania HH, Kassie LN (2023) Electrical resistivity imaging data for hydrogeological and geological hazard investigations in Taiwan. Data Brief 49:109377

Chavez M, Liu W (2020) Innovations in ground-penetrating radar applications for transportation infrastructure. J Appl Geophys 179:1037–1048

Collette B, Hamilton D (2018) Ambient vibration testing for monitoring the health of civil structures: a case study of bridge analysis. Struct Health Monit 17(6):1459–1470

Daniels DJ (2021) Ground-penetrating radar: a practical guide for civil engineering applications. Constr Build Mater 263:120238

Desai R, Bhattacharya R (2021) Real-time monitoring of pipeline integrity using distributed fiber optic sensing: case study in Gujarat, India. J Pipeline Eng 36(2):212–221. https://doi.org/10.1016/j.jopec.2021.02.003

Dou X, Li J (2017) Ultrasonic testing for material integrity assessment: recent advancements and practical applications in civil engineering. Mater Test 59(10):1165–1173

Dou S, Zhang Y, Liu Q, Zhang Z, He X, Jiang S (2020) Advances in GPR imaging techniques: reverse time migration and full waveform inversion. J Appl Geophys 182:104–113. https://doi.org/10.1016/j.jappgeo.2020.104

Eldin NN, Elgohary MA, Abbas AE, El-Sayed HE (2013) Application of half-cell potential techniques for corrosion detection in reinforced concrete. Corros Sci 68:85–92. https://doi.org/10.1016/j.corsci.2012.10.009

Gao S, Zhao X (2021) Fiber optic sensing technology for structural health monitoring of high-rise buildings. Sensors 21(7):2355

Gong J, Xie L, Zhang Y, Zhang H, Zhang J, Li Z (2021) Non-destructive seismic and ultrasonic testing methods for the assessment of concrete bridges. Constr Build Mater 276:122154

Gucunski N, Sanderson R (2014) Ground penetrating radar for concrete inspection: a field study on evaluating pavement condition. J Eng Geol 184:132–141

Gucunski N, Olbinski DM, Mallett JM, Brown KM, Tighe AJ (2014) Ground penetrating radar for the non-destructive evaluation of concrete bridges. J Infrastruct Syst 20(2):04014008. https://doi.org/10.1061/(ASCE)IS.1943-555X.0000184

Jaramillo JJ, Martinez M (2020) Seismic refraction for liquefaction hazard assessment: a case study in New York. Geotech Eng J 19(3):205–216

Kong Q, Li D, Zhang J, Lu J, Li C, Wang L (2018) Passive seismic monitoring for detecting progressive damage in concrete structures. Earthq Eng Struct Dyn 47(12):1903–1917. https://doi.org/10.1002/eqe.3104

Koo J, Park Y (2022) Integration of IoT and wireless sensor networks in structural health monitoring systems: a review and case studies. Autom Constr 132:103973

Kourous R, Haidar A (2016) Self-Potential method for monitoring corrosion in concrete structures. J Struct Integr 3(3):212–220

Kumar S, Gupta S (2020) Magnetotelluric exploration for mineral resources in Rajasthan. J Appl Geophys 17(5):254–263

Kumar S, Ramesh K, Rao VP (2021) Application of seismic refraction surveys in urban infrastructure development: a case study from Bangalore Metro. J Appl Geophys 190:104387

Kwon Y, Park C (2019) Recent advances in fiber optic sensors for civil engineering applications. J Civ Eng Manag 25(1):12–23

Liao S, Zhang Y (2020) Microseismic monitoring for structural health monitoring in skyscrapers. Struct Monit Maint J 7(2):145–160

Lin W, Chen S (2019) Drones and LiDAR for infrastructure inspection and monitoring: opportunities and challenges. J Infrastruct Syst 25(3):04019020

Lin P, Yang J, Zhang Y, Xu H, Li B (2023) Acoustic emission monitoring in the structural health assessment of bridges. J Struct Eng 149(4):04023057

Liu Z, Yang D (2020) Fiber optic sensors for structural health monitoring of bridges: a review and case studies. Smart Mater Struct 29(7):073001

Liu T, Zhang H (2021) AI-based prediction of structural failure using SHM data: a review and future directions. Struct Control Health Monit 28(1):e2723

Liu Z, Zhang X, Wang L, Li J, Chen S, Zhang Z (2018) Pipeline monitoring using distributed fiber optic sensing for integrity management in the North Sea. J Petrol Sci Eng 165:659–669

Liu L, Yu J, Chen B, Wang Y (2020) Urban subsidence monitoring by SBAS-InSAR technique with multi-platform SAR images: a case study of Beijing Plain, China. Eur J Remote Sens 53(S1):141–153

Loke MH, Dahlin T (2021) Principles and applications of resistivity imaging in geotechnical engineering. Geotech Test J 44(3):314–327

Mancuso S, Nanni A (2020) Seismic tomography for underground infrastructure health monitoring: a case study in urban environments. Geophys J Int 221(2):900–914

Momayez M, McGinnis MS, Page PLT, Miller AL, Phair SG (2009) Geotechnical assessment and geo-environmental engineering geophysics. Special Issue J Environ Eng Geophys 14(4):211–340

Montanari D, Bianchini A, Borghi M, Massari G, Manunta L, Pizzolato GP, Santoro F (2019) Application of InSAR for monitoring subsidence in Florence, Italy. Remote Sens Environ 224:21–31. https://doi.org/10.1016/j.rse.2019.01.025

Murphy H, Hansen K (2022) Geophysical methods in foundation design for bridges and buildings. Geotech Eng 35(5):461–475

Nakamura T, Sato M, Ito K (2020) Landslide risk assessment in urban environments using ERT: insights from Tokyo. Geotech Eng J 67(2):215–229

Niederleithinger E, Abraham O, Mooney M (2015) Geophysical methods in civil engineering: overview and new concepts. J Civ Eng Manag 21(7):837–852

Olsson A, Lee H (2018) GPR for condition assessment of bridge decks: a case study. Struct Eng J 25(4):231–245

Park H, Lee J (2020) Geophysical techniques in structural health monitoring: a review of methods and applications. Geotech Test J 43(6):847–858

Park CB, Miller RD, Xia J (1999) Multichannel analysis of surface waves (MASW). Geophysics 64(3):800–808. https://doi.org/10.1190/1.1444590

Parker G, Brant TFSH, McDonald SW, Miller JL, Stevens REW (2015) Geostatistical approaches for improving geotechnical parameter estimation. Publ Geotech J 62(3):245–263

Rai AK, Raj B (2019) Application of ground penetrating radar (GPR) for the inspection of concrete structures in Mumbai, India. J Infrastruct Syst 25(3):04019024. https://doi.org/10.1061/(ASCE)IS.1943-555X.0000536

Reynolds JM (2000) An introduction to applied and environmental geophysics. Wiley, Chichester, pp 1–796

Reynolds JM (2011) An introduction to applied and environmental geophysics. Wiley-Blackwell, Amsterdam

Schenck JF, Meyer A (2020) Applications of ERT for groundwater contamination in Berlin. Environ Sci Technol Lett 14(5):288–295

Schuster F, Meier H, Klug T (2022) Detecting structural anomalies in aging bridges using GPR: a study from Munich, Germany. Bridge Eng J 29(3):345–359

Sen M, Sarma RT, Pandey AS, Murthy KT, Raju VRNMSSR (2018) Seismic tomography for ground investigation in tunnel construction: a case study from Canada. Tunn Undergr Space Technol 74:1–11. https://doi.org/10.1016/j.tust.2018.01.005

Sharma R, Sharma D (2018) Application of seismic tomography in monitoring tunnel stability: a case study from the Delhi metro project. Geotech Test J 41(5):1052–1064. https://doi.org/10.1520/GTJ20180089

Sharma A, Sharma S (2019) Magnetotelluric imaging for mineral exploration in arid regions. Geophys Prospect J 18(3):227–235

Soga K, Luo L, Kechavarzi C (2018). Distributed fiber optics sensors for civil engineering infrastructure sensing. J Civil Struct Health Monit 8:619–635. https://doi.org/10.1007/s13349-018-0278-y

Stojadinovic B, Bertero V (2018) Ambient vibration testing for seismic assessment of structures: applications to building codes. J Struct Eng 144(9):04018156

Strauss A, Bien J, Neuner H, Harmening C, Seywald C, Osterreicher M, Voit K, Pistone E, Spyridis P, Bergmeister K (2020) Sensing and monitoring in tunnels testing and monitoring methods for the assessment of tunnels. Struct Concr 21:1356–1376

Suresh M, Anurag S (2020) Electrical resistivity imaging for foundation assessment in Bengaluru: a case study. J Civil Eng Environ Syst 37(4):320–330. https://doi.org/10.1080/10286608.2020.1802321

Thompson MS, Smith DJ (2017) Pavement evaluation using spectral analysis of surface waves. Geotech Eng J 15(2):177–183

Tzou HK, Chu TS, Liu TY (2020) Enhancing the safety management of NATM using the tunnel seismic prediction method: a case study. Innov Infrastruct Solut 5:106

Vitali E, Giustolisi O (2021) Acoustic emission monitoring for the detection of concrete structure failure. Struct Health Monit 20(5):1682–1697

Wang Z, Liu Y (2021) Application of fiber optics in dam structural health monitoring: a review of techniques and case studies. J Civ Eng 29(2):287–303

Wang Z, Zhang P (2019) Seismic monitoring for structural health monitoring: current practices and future directions. J Seismol Earthq Eng 21(2):29–42

Williams T, Ormsby J (2016) Electrical resistivity imaging for monitoring subsurface foundation conditions in the UK. Geotech Test J 39(2):123–134. https://doi.org/10.1520/GTJ20160060

Wilson JA, Brown P, Carter R (2023) Utilizing microtremor analysis for post-earthquake urban reconstruction: a Christchurch case study. Seismol Res Lett 94(1):54–67

Yang Z, Wang L (2020) IoT-based SHM systems for real-time monitoring of bridges and other civil infrastructure. J Struct Eng 146(5):04020043

Zhang W, Song H (2019) Seismic tomography for tunnel integrity assessment: a practical approach. Geophys Prospect 67(4):906–917

Zhang Y, Xie Y (2020) Geophysical imaging for road condition monitoring and rehabilitation. Int J Pavement Eng 21(1):45–55

Zhang W, Xu T (2018) Application of SASW in pavement condition assessment: a case study. J Transp Eng 24(6):410–417

Zhang Y, Li H, Wang X (2022) Comprehensive geophysical surveys for dam integrity analysis: a case from Yunnan Province. Int J Civil Eng 23(4):678–692

Zhao L, Wang X, Zhang Y, Liu J, Wang M (2021) Real-time seismic monitoring in construction sites for structural stability assessment. J Civ Struct Heal Monit 11(2):217–229

Ziegler M, Bar W (2017) Groundwater contamination detection using electrical resistivity. Environ Geophys J 32(1):104–114

Chapter 9
Geophysics in Mining and Resource Exploration

Abstract This chapter highlights how geophysical techniques support sustainable and efficient exploration of natural resources such as minerals, oil, gas, and geothermal energy. It explains key methods like gravity, magnetic, seismic, and electromagnetic surveys, along with newer technologies such as drone-based EM surveys and 3D seismic imaging. The chapter also explores the growing role of machine learning and real-time data analysis in improving exploration accuracy. Several case studies from India, China, Australia, and other regions demonstrate the practical applications of these tools in locating resources and reducing environmental impact. Emphasis is placed on sustainable practices like methane leak detection, groundwater monitoring, habitat protection, and minimizing damage from tailings. Through integration of advanced tools and environmental awareness, geophysics is helping reshape resource extraction for a more responsible and efficient future.

9.1 Overview of Advancements and Sustainability in Geophysical Exploration

This chapter explores the transformative role of geophysical techniques in enhancing the efficiency and sustainability of mining and resource exploration. It highlights advances in geophysical methods such as gravity, magnetic, electromagnetic, and seismic surveys, with specific case studies demonstrating their successful application across mineral, oil, gas, and geothermal energy exploration. The chapter delves into emerging technologies, including drone-based electromagnetic surveys, autonomous seismic sensor networks, and advanced borehole imaging, as well as the integration of machine learning and big data for improved exploration outcomes.

A significant focus is placed on the environmental and sustainable aspects of modern geophysical practices, including monitoring groundwater, fault stability, and methane leaks, with an emphasis on minimizing the ecological footprint of resource extraction. Additionally, the chapter discusses integrated geophysical approaches and their potential to improve resource localization, reduce energy consumption, and prevent pollution. By reviewing cutting-edge tools, techniques, and real-world

D. C. Naskar, *Advances in Engineering Geophysics: Concepts and Applications*, Earth Systems Data and Models 9, https://doi.org/10.1007/978-3-032-28000-8_9

applications, this chapter underscores the importance of geophysics in advancing efficient and environmentally responsible exploration, ensuring a balanced approach to resource extraction and environmental stewardship in the mining sector.

9.2 Optimizing Resource Extraction with Geophysical Approaches Sectorwise

Geophysics plays a pivotal role in the exploration and extraction of resources in the fields of mining, oil and gas, and geothermal energy. Over the past few decades, advancements in geophysical methods have significantly improved the ability to locate, delineate, and optimize the extraction of these resources. New technologies and analytical techniques, such as high-resolution seismic surveys, electromagnetic (EM) methods, and gravity and magnetic data analysis, have revolutionized exploration activities.

9.2.1 Mineral Exploration

Mineral exploration benefits from various geophysical methods, which help geologists determine the presence of valuable deposits beneath the surface. The techniques used are highly sensitive to different material properties, such as density, magnetic susceptibility, and electrical conductivity.

Gravity Surveys

Gravity surveys are often used in mineral exploration, as variations in the Earth's gravitational field can highlight differences in the density of subsurface materials. This method is especially useful for locating dense minerals, such as ores, and can provide a map of the underlying geological structures.

Mathematical Principle: The gravity anomaly Δg at a point can be modeled using the formula:

$$\Delta g = \frac{2\pi G \cdot \Delta \rho}{r} \tag{9.1}$$

where Δg is the gravity anomaly, G is the gravitational constant, $\Delta \rho$ is the density contrast between the target mineral and surrounding rock, and r is the distance from the observation point to the anomaly.

Magnetic Surveys

Magnetic surveys are used to detect variations in the Earth's magnetic field caused by magnetic minerals (e.g., magnetite). The presence of these minerals can be inferred by measuring the magnetic anomalies.

Mathematical Principle: The magnetic anomaly ΔB can be expressed as:

$$\Delta B = \frac{\mu_0 \cdot M}{4\pi r^3} \tag{9.2}$$

where ΔB is the magnetic anomaly, μ_0 is the magnetic permeability of free space, M is the magnetic moment of the body, and r is the distance from the source to the observation point. Enhanced sensitivity in gravimeters and magnetometers has allowed for better detection of subsurface features. Airborne gravity and magnetic surveys have been used to map mineral deposits, and superconducting gravimeters have provided valuable insights into subsurface fluid movements.

Case Study: Magnetic and Self-Potential Methods in ji'an, China

Graphite is a vital mineral due to its exceptional electrical and thermal conductivity, as well as its lubricating properties. These characteristics have made graphite increasingly important in various industrial applications. In Northeast China, which is known for its rich graphite deposits, exploration faces challenges due to dense vegetation and limited exposure of bedrock. To address these challenges, Liu et al. (2025) conducted a comprehensive study utilizing integrated geophysical methods, specifically magnetic surveys and self-potential techniques, to enhance graphite exploration (Fig. 9.1).

In this study, magnetic exploration was first employed to map the geological strata in the region (Fig. 9.2), particularly beneath the Quaternary sediment. The magnetic anomaly data revealed three distinct anomaly areas, with a low magnetic anomaly zone associated with the Huangchagou Formation (Fig. 9.3). This finding indicated a strong potential for graphite ore in the region. To further assess this potential, Liu et al. (2025) deployed self-potential methods (Fig. 9.4), which successfully delineated an east–west trending banded anomaly (Fig. 9.5), correlating with the identified magnetic anomalies. This provided further evidence of graphite mineralization in the area.

Following the geophysical surveys, trenching and drilling operations were conducted to validate the findings. The drilling sites were chosen based on trenching data and geophysical surveys. These sites focused on the center and edges of the Z1 (strike) anomaly, as well as areas with graphite ore identified in the trenches. The goal was to better understand the thickness, depth, shape, and internal structure of the ore bodies.

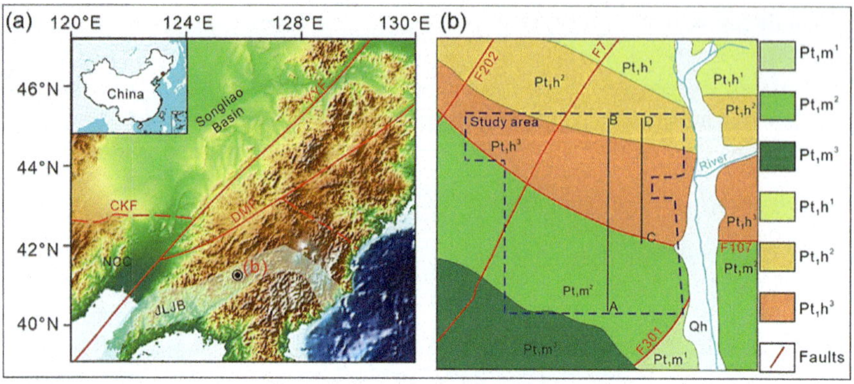

Fig. 9.1 **a** Location of the study area in Northeast China and **b** geological setting of the study area. The inset in **a** shows the location of the study area within China. Key geological features include the North China Craton (NCC), Central Asian Orogenic Belt (CAOB), and several major faults such as the Chifeng-Kaiyuan Fault (CKF), Yilan-Yitong Fault (YYF), and Dunhua-Mishan Fault (DMF). Sections of the Huangchagou (Pt1h3, Pt1h2, Pt1h1) and Mayihe (Pt1m3, Pt1m2, Pt1m1) Formations are labeled and Qh: Quaternary sediment. Magnetic investigation areas and fault lines are indicated, along with measurement sites for magnetic method investigations (black triangles). *Source* Liu et al. (2025)—open access

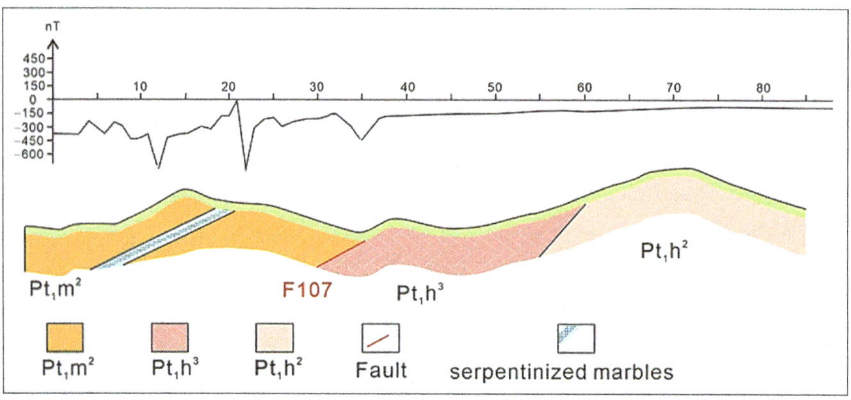

Fig. 9.2 Magnetic anomaly data and geological mapping of profile AB in the study area of Northeast China. *Source* Liu et al. (2025)—open access

The core samples from the drilling show that the deep graphite ore has a multi-layered distribution (Fig. 9.6). The ore layers range from a few meters to about 60 m thick, with burial depths from the surface to around 300 m. This indicates a significant potential for graphite deposits in the area. Based on trenching and drilling data, Liu et al. (2025) identified two graphite ore bodies on the surface (Fig. 9.7).

These ore bodies align with low-value anomaly zones found in the geophysical surveys, which confirms the accuracy of the geophysical data. Ore body I was located

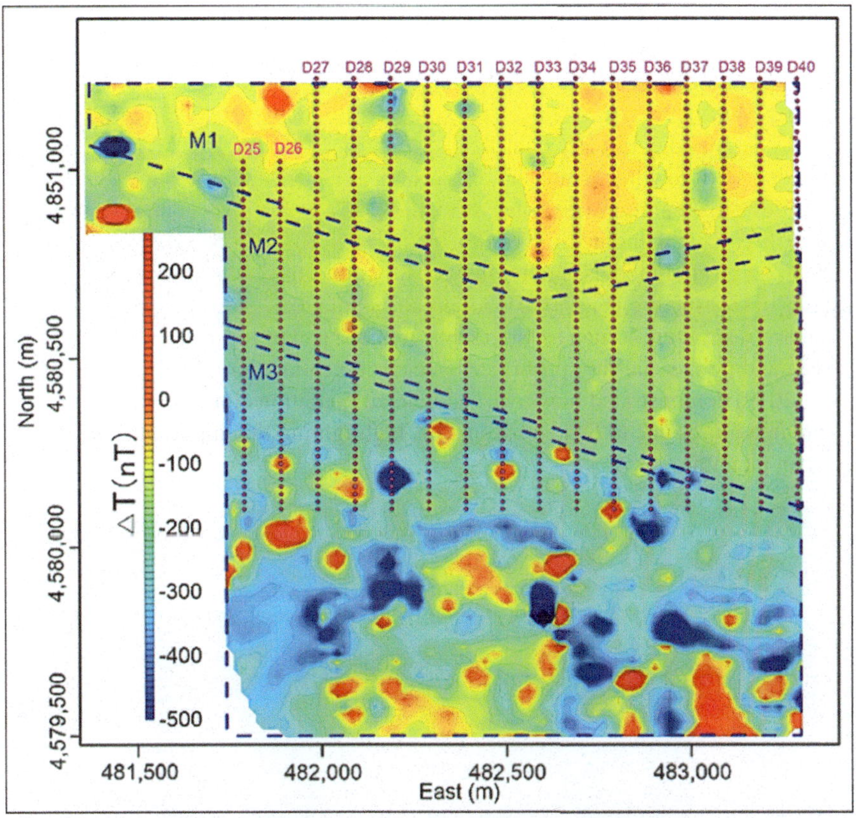

Fig. 9.3 Magnetic anomaly in the study area of Northeast China. The blue dashed lines and labels indicate the three magnetic anomalous zones. Magenta dots and labels represent the self-potential sites and the survey line names, respectively (Liu et al. 2025—open access)

between two exploration trenches and several old mining pits. It trends east–west and dips south. The length of this ore body is about 450 m, and the average width is 60 m. In contrast, ore body II is located near five exploration trenches and old mining pits. It trends northwest and dips southwest at about 20°. The length of ore body II is about 1100 m, with an average width of 120 m. Trench TC6 revealed the widest part of ore body II, with a width of around 220 m.

These operations confirmed the presence of graphite deposits, with total reserves estimated to exceed 50,000 kt (1 kt = 1000 tons and 1 ton = 1000 kgm). The combination of magnetic and self-potential surveys proved to be an effective and cost-efficient geophysical approach, particularly useful in the challenging terrain of Northeast China. The study by Liu et al. (2025) demonstrates that integrating magnetic surveys with self-potential methods offers a reliable and cost-effective strategy for graphite exploration. This approach not only enhances the accuracy of locating graphite deposits but also helps in overcoming the obstacles posed by dense

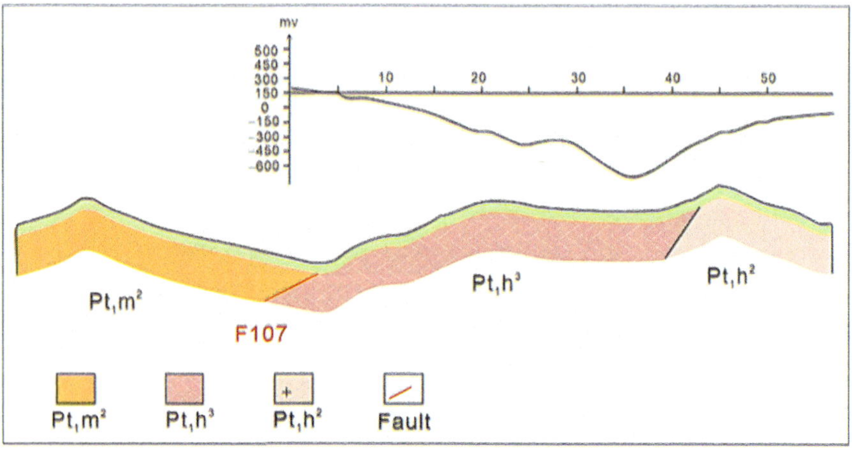

Fig. 9.4 Self-potential anomaly data and geological mapping of profile CD in the study area of Northeast China (Fig. 9.1b) (Liu et al. 2025—open access)

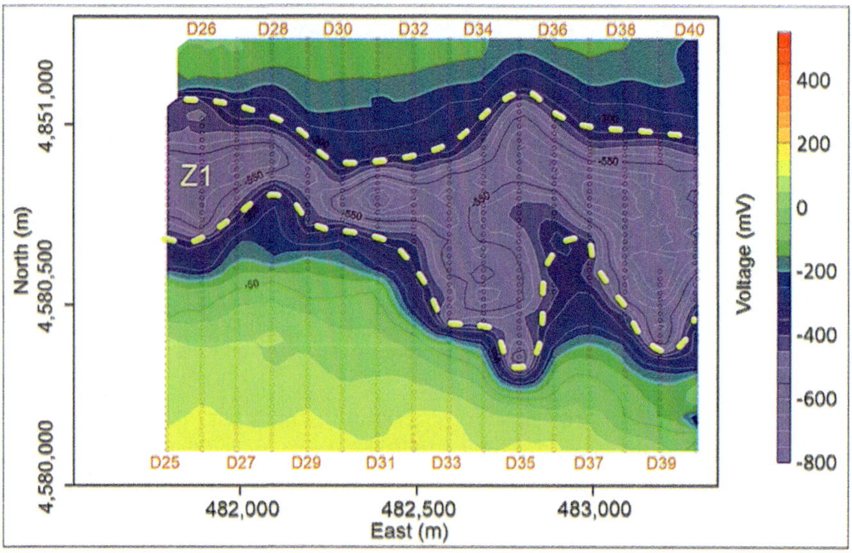

Fig. 9.5 Self-potential anomaly in the study area of Northeast China. Magenta lines represent the anomaly zones (Z1), black dotted lines indicate the peak anomaly values, and orange lines represent the designed trenches (Liu et al. 2025—open access)

vegetation and limited outcrop exposure. By effectively applying these geophysical techniques, Liu et al. (2025) have shown that it is possible to accurately detect and delineate graphite mineralization, even in difficult exploration environments, leading to a better understanding of mineral reserves and facilitating the discovery of significant graphite deposits in the region.

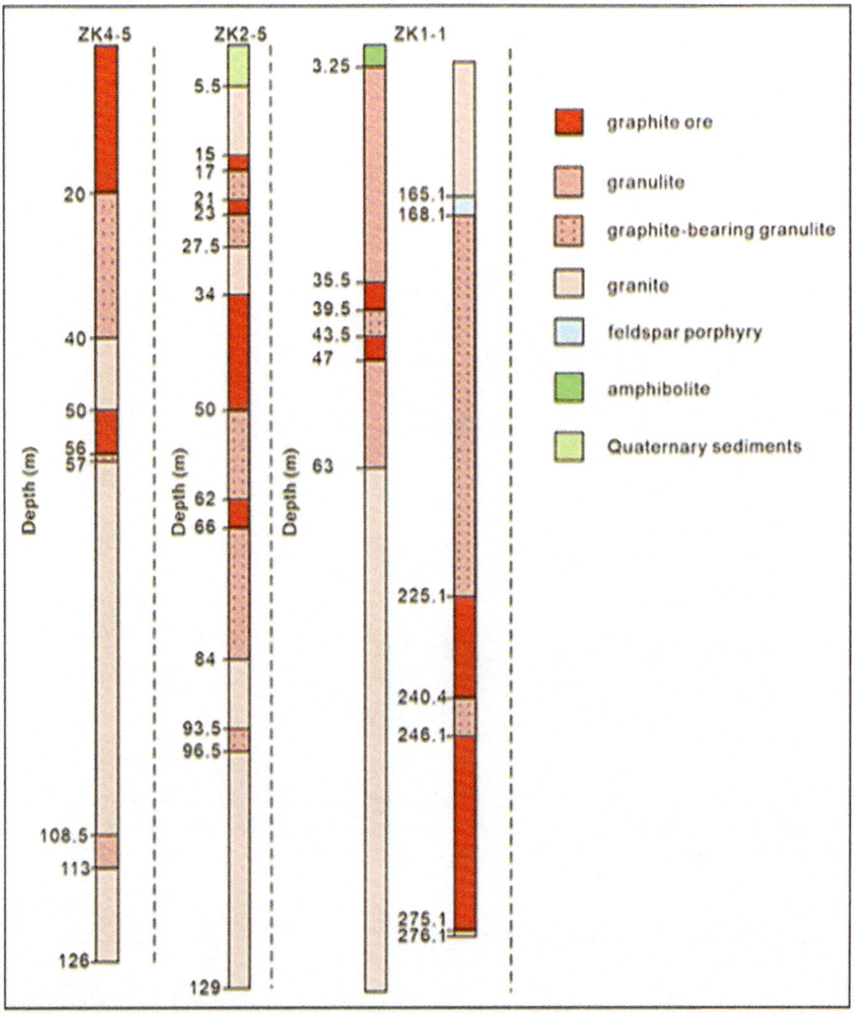

Fig. 9.6 Borehole histogram of drillings ZK4-5, ZK2-5, and ZK1-1 in the study area of Northeast China. *Source* Liu et al. (2025)—open access

Electromagnetic Methods (EM)

Electromagnetic (EM) surveys are employed in mineral exploration to detect electrically conductive bodies, such as sulfide ores, which create secondary electromagnetic fields when subjected to a primary magnetic field.

Mathematical Principle: The response of a conductive body to an electromagnetic wave can be described by Maxwell's equations, which are the foundation for most EM methods. One simplified form of the relationship between the induced secondary

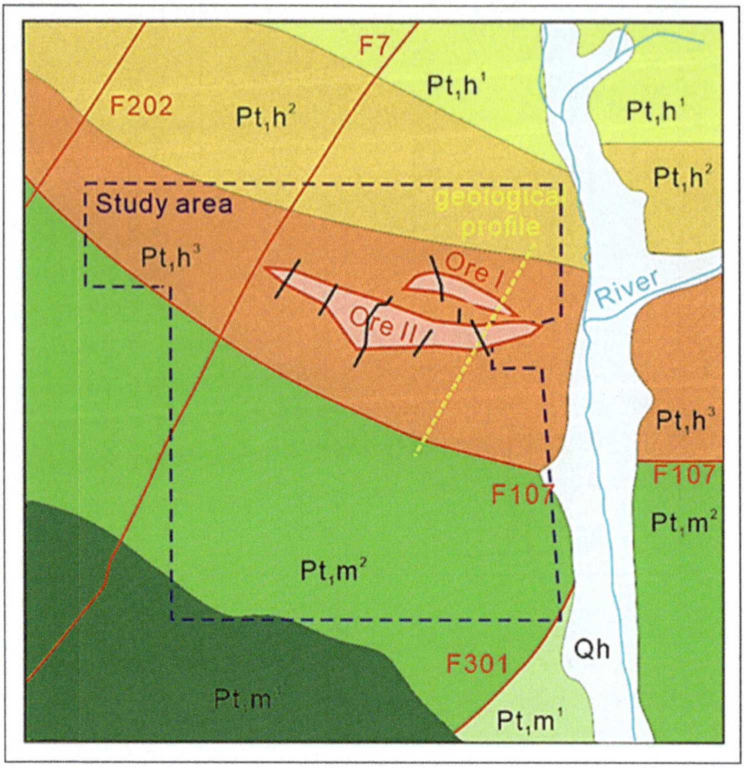

Fig. 9.7 Graphite ore determined based on drillings and trenches in the study area of Northeast China. The purple polygons represent the graphite ore. The yellow dashed line indicates profile 4. The black lines on the ore bodies represent the trenches. *Source* Liu et al. (2025)—open access

field and the primary field is given by:

$$E = -\frac{\partial B}{\partial t} \tag{9.3}$$

where E is the induced electric field, B is the magnetic field, and t is time. EM techniques have seen significant improvements, including high-frequency instruments capable of operating in harsh environments. Airborne EM surveys, particularly in regions like the North Sea, have boosted oil reserve estimates, while transient electromagnetics (TEM) has been used effectively in mapping mineralized zones in Western Australia and shale formations.

Case Study Review: Magnetic and EM Techniques (2009–2024)

In their 2024 review, Vallee et al. (2024) compiled and analyzed the application of magnetic and electromagnetic (EM) techniques in mineral exploration over the last fifteen years. The review highlights that these methods remain highly popular in the field, with a variety of case studies published across global scientific journals. The authors focused on classifying case studies by target type and region, showcasing the continued relevance of these geophysical tools in the exploration of mineral deposits.

One of the key findings of the study is the widespread use of magnetic surveys, particularly in regions such as North America, Asia, and Australia. Magnetic methods are often employed both directly to detect magnetic materials like iron ore and indirectly to map geology using magnetic signatures, such as the content of magnetite in rocks. The electromagnetic method, on the other hand, is commonly used for detecting conductive mineralization, especially in uranium exploration, where it helps identify conductors associated with the deposits.

The case studies were categorized into three primary groups based on the combination of techniques: magnetic, EM, and a combination of both methods. A histogram from the review (Fig. 9.8) revealed distinct patterns in method popularity across continents.

While magnetic techniques dominated in North America and the rest of the world, Europe showed a preference for EM methods. The combination of both methods was especially prominent in North America and Asia. Figure 9.9 illustrates the range of mineral deposit types explored, with gold, porphyry copper, and polymetallic deposits being the most commonly targeted. The review found that magnetic techniques are particularly favored for gold and iron ore, while EM methods are more commonly applied to uranium and massive sulfides. Porphyry copper and massive sulfide deposits are explored using both techniques in many cases.

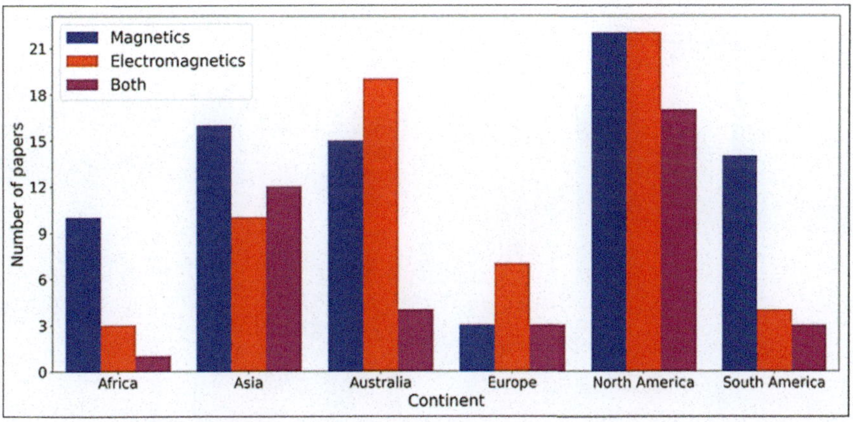

Fig. 9.8 Case studies of geophysical methods by continent and technique: a regional analysis of magnetic and electromagnetic surveys in mineral exploration. *Source* Vallee et al. (2024)—open access

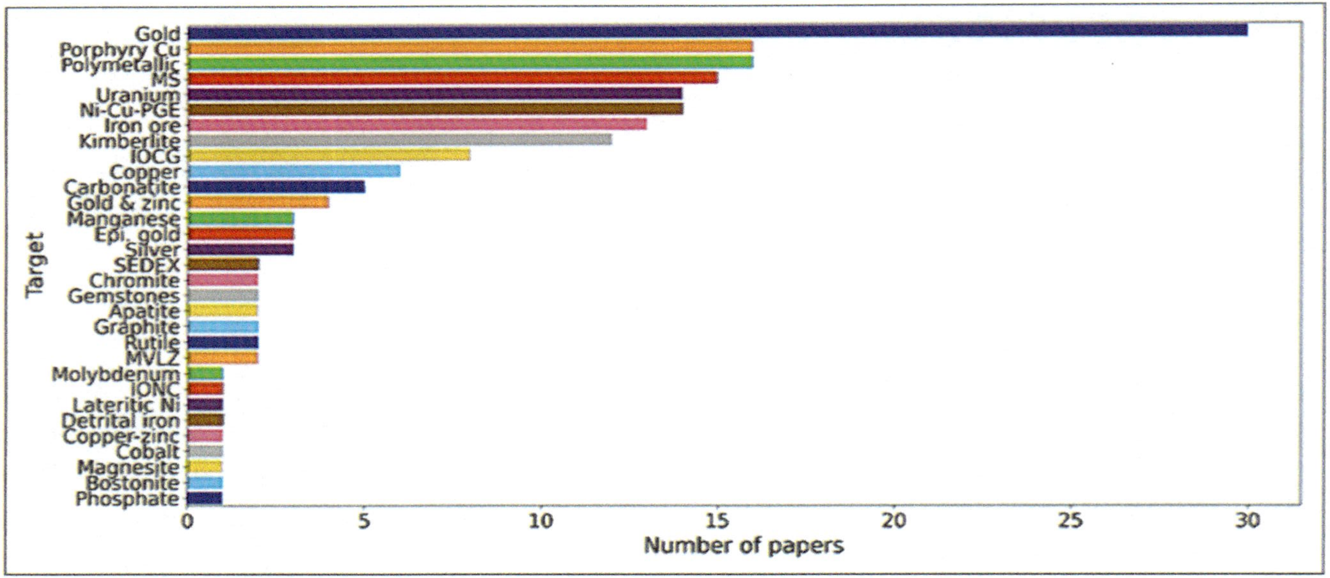

Fig. 9.9 Number of case studies by types of mineralization: a regional overview of mineral targets (IOCG: iron-oxide copper–gold, IONC: iron-oxide nickel-copper, MS: massive sulfides, MVLZ: mississippi valley lead–zinc, SEDEX: sedimentary exhalative lead–zinc). *Source* Vallee et al. (2024)—open access

The study concludes that despite some regional variations in method usage and deposit type, magnetic and EM techniques are indispensable tools in mineral exploration. They offer flexibility in detecting a wide variety of target types, from gold and copper to uranium and kimberlite. The review by Vallee et al. (2024) provides valuable insights into the geographical and methodological trends in geophysical exploration and serves as a guide for designing future surveys, considering the strengths and limitations of each technique.

9.2.2 Oil and Gas Exploration

In oil and gas exploration, geophysics is essential for locating reservoirs, assessing their size, and determining their fluid content. The most important geophysical technique in this field is seismic reflection, which uses sound waves to map the subsurface.

Seismic Reflection Surveys

Seismic waves are generated by a source (e.g., a vibrator or explosive charge) and travel through the Earth. The waves reflect off different layers of rock, and the reflected waves are detected by geophones on the surface. The travel time of these waves helps create images of the subsurface structure.

Mathematical Principle: The basic equation for the time of travel t of a seismic wave is:

$$t = \frac{2d}{v} \tag{9.4}$$

where t is the two-way travel time, d is the depth to the interface, and v is the velocity of the seismic wave in the medium. The velocity v is often related to the density ρ and the elastic properties of the rock by the equation:

$$\mathrm{v} = \sqrt{\frac{\mathrm{E}}{\rho}} \tag{9.5}$$

where E is the Young's modulus of the material, ρ is the density of the material. Seismic data processing methods such as migration, inversion, and velocity modeling are also applied to refine subsurface images and improve the interpretation of oil and gas reservoirs. Seismic technologies, including 3D and 4D imaging, offer detailed subsurface imaging with improved signal processing and accuracy. Notable advancements include multi-component seismic imaging, which enhances the understanding of rock properties and fluid saturation in hydrocarbon reservoirs.

Ground Penetrating Radar (GPR)

GPR has evolved with improved resolution and data processing techniques, making it useful for mapping subsurface structures and features, such as hydrocarbon reservoirs and permafrost. Its integration with other geophysical methods enhances subsurface modeling.

9.2.3 *Geothermal Energy Exploration*

Geophysical methods in geothermal exploration focus on identifying areas with high heat flow and assessing the geothermal potential of a region. Heat is transferred from deep within the Earth to the surface, and geophysics can help map these heat anomalies.

Heat Flow and Temperature Gradient Surveys

Heat flow surveys involve measuring the temperature gradient of the Earth's surface and correlating it with the thermal conductivity of the rocks. The heat flow q is given by the equation:

$$q = k\frac{\Delta T}{\Delta z} \tag{9.6}$$

where q is the heat flow, k is the thermal conductivity of the rock, ΔT is the temperature difference, and Δz is the depth over which the temperature difference occurs.

Magnetotelluric (MT) Surveys

Magnetotelluric (MT) surveys are used to measure the electrical resistivity of the subsurface. Areas with high geothermal potential often have lower resistivity due to the presence of hot, conductive fluids (such as brine) at depth.

Mathematical Principle: The resistivity ρ of a subsurface material can be related to the impedance Z of the material via Ohm's Law:

$$Z = \rho \cdot \frac{L}{A} \tag{9.7}$$

where Z is the impedance, ρ is the resistivity, L is the length of the current path, and A is the cross-sectional area of the current path.

9.3 Emerging Techniques and Technologies

This section explores the latest advancements in geophysical methods used in mining and resource exploration. It highlights innovations that enhance detection sensitivity, data resolution, and subsurface imaging capabilities. With the increasing complexity of exploration environments, emerging technologies are providing more efficient, cost-effective, and environmentally sensitive solutions for locating and characterizing mineral and energy resources.

9.3.1 Novel Methods

This subsection presents cutting-edge geophysical techniques that are redefining exploration strategies. It includes developments in machine learning-assisted interpretation, UAV-based geophysical surveys, distributed acoustic sensing (DAS), and hybrid sensor systems. These novel methods are enabling deeper insights into subsurface conditions, supporting more accurate resource assessment and reducing exploration risks.

Drone-Based Electromagnetic (EM) Surveys

Drone-based electromagnetic (EM) surveys have emerged as a significant advancement in geophysical exploration. These surveys involve the use of unmanned aerial vehicles (UAVs) to carry electromagnetic sensors across the exploration area. Drones can cover large areas more efficiently and safely than traditional ground-based methods, and they are especially valuable in rugged or hard-to-access terrains.

Mathematical Principles: The EM method detects the conductivity of subsurface materials by inducing an electromagnetic field and measuring the response. The relationship between the primary field E_0 and the induced secondary field E is governed by Maxwell's Eq. (9.3). Where E is the induced electric field, and B is the magnetic field, which changes as the UAV moves and interacts with subsurface conductors. The induced EM field is typically measured at the surface and interpreted using forward and inverse modeling to infer subsurface conductivity structures.

Data Acquisition and Processing: The data from drone-based EM surveys can be voluminous, involving high spatial and temporal resolution. Processing this data often requires 3D inversion techniques to map the conductivity distribution at depth. The key challenge is accurately accounting for the UAV's motion and sensor orientation, which introduces spatial variability in the data. A simplified model for the EM anomaly observed at a distance r from the source is:

$$\Delta\mathbf{B}(\mathbf{r}) = \frac{\mu_0\mathbf{M}}{4\pi\mathbf{r}^3}\left(\frac{\mathbf{r}}{\mathbf{r}_0}\right)^{\alpha} \tag{9.8}$$

where M is the magnetic moment, μ_0 is the magnetic permeability of free space, and α accounts for the attenuation of the signal with distance, influenced by the local conductivity of the subsurface.

Autonomous Seismic Sensor Networks

Autonomous seismic sensor networks are a significant innovation in geophysical data collection. These systems involve the deployment of numerous wireless sensors (often termed geophones or accelerometers) that automatically record seismic waves and send data for processing in real-time. Such networks can operate continuously, providing a high level of detail over large areas without the need for constant human intervention.

Mathematical Principles: Seismic waves are typically characterized by the displacement of particles in the Earth's subsurface. The simplest equation describing seismic wave propagation in a homogeneous medium is:

$$\frac{\partial^2\mathbf{u}}{\partial\mathbf{t}^2} = \mathbf{v}^2\nabla^2\mathbf{u} \tag{9.9}$$

where u(x, t) is the displacement at position x and time t, v is the wave velocity, and ∇^2 is the Laplacian operator, which accounts for the spatial propagation of the wave.

When deploying autonomous seismic sensors, the network must account for spatial and temporal variations in seismic velocity, which can be influenced by local geological structures. These sensors enable continuous monitoring, allowing for real-time detection of seismic events, changes in subsurface properties, or even early warnings of potential hazards (e.g., induced seismicity in geothermal or oil extraction zones).

Borehole Imaging Methods

Borehole imaging methods are used to examine the geological formations directly around a borehole. These methods include technologies like borehole radar, acoustic imaging, and electrical resistivity imaging, all of which provide high-resolution information about subsurface properties and geological structures. Borehole tools have become more sophisticated, offering a comprehensive view of rock properties. These advancements help in identifying mineral and hydrocarbon resources, as well as monitoring subsurface conditions.

Mathematical Principles: The principle behind borehole imaging methods is based on wave propagation, with the specific technique depending on the type of waves used (e.g., acoustic waves or electromagnetic waves). For example, in acoustic logging, the velocity of a seismic wave traveling through the surrounding rock is measured, which can provide insights into the formation's properties. A basic equation for seismic wave propagation in a borehole is Eq. 9.5. This relationship allows the interpretation of rock stiffness and porosity from borehole acoustic logs.

9.3.2 *Machine Learning and Big Data in Geophysics*

The advent of big data analytics and machine learning has revolutionized how geophysical data is processed, interpreted, and modeled. With the large volumes of data generated by modern geophysical surveys (e.g., seismic, gravity, EM surveys), machine learning methods are now being applied to uncover patterns, reduce dimensionality, and enhance decision-making during exploration.

Anomaly Detection

Anomaly detection is critical in geophysical exploration, where the goal is to identify unexpected or unusual features in the subsurface, such as mineral deposits or fluid reservoirs. Machine learning algorithms, particularly unsupervised learning methods like clustering or autoencoders, are used to identify these anomalies.

Mathematical Principles: One common method is k-means clustering, which groups data points (in this case, geophysical measurements) based on their similarity. The algorithm minimizes the squared Euclidean distance between data points and cluster centers:

$$\min_{\mathbf{C}} \sum_{i=1}^{n} \mathbf{x}_i - \mathbf{C}_j^2 \tag{9.10}$$

where x_i is a data point, and C_j is the center of the cluster. This allows for the identification of clusters corresponding to different geological features. Another method is the use of autoencoders, a type of neural network used to compress data and reconstruct it. The loss function for an autoencoder is typically:

$$\mathcal{L} = \mathbf{x} - \mathbf{f}(\mathbf{f}(\mathbf{x}))^2 \tag{9.11}$$

where x is the input data, and f(x) represents the encoded and decoded representation of the data. Autoencoders are particularly useful for dimensionality reduction in large datasets, allowing for the extraction of meaningful patterns from geophysical data.

Predictive Modeling

Machine learning methods, such as random forests, support vector machines (SVM), and deep learning, are increasingly used for predictive modeling of resource estimates. These models learn from historical data, where known deposits or reservoirs are used as training data, and predict the likelihood of finding similar resources in unexplored areas.

Mathematical Principles: In regression models like support vector regression (SVR), the goal is to minimize the error between predicted and observed values by adjusting the weights associated with each feature:

$$\min_{\mathbf{w},\,\mathbf{b}} \sum_{i=1}^{n} \mathbf{max}\left(0, \left|\mathbf{y_i} - \left(\mathbf{w}^T\mathbf{x_i} + \mathbf{b}\right)\right| - \varepsilon\right) + \frac{1}{2}\mathbf{w}^2 \tag{9.12}$$

where y_i is the true value, x_i is the feature vector, and w and b are the model weights and bias.

Real-Time Analysis

Real-time data processing and analysis are essential for making timely decisions in resource exploration, especially during the drilling or surveying phases. Machine learning models can be deployed to monitor real-time data streams, flag anomalies, and predict the best course of action.

Mathematical Principles: Online learning algorithms update their model incrementally as new data arrives. A common approach is stochastic gradient descent (SGD), which updates the model parameters θ iteratively:

$$\boldsymbol{\theta}_t = \boldsymbol{\theta}_{t-1} - \eta\nabla_{\theta}\mathcal{L}(\boldsymbol{\theta}_{t-i}, \mathbf{x}_t, \mathbf{y}_t) \tag{9.13}$$

where η is the learning rate, L is the loss function, and (x_t, y_t) are the new data point and its corresponding label. This enables continuous updating of the model in real-time as new geophysical data is collected.

Implications for Exploration

The advancements in geophysical technologies have led to more cost-effective and precise exploration. Techniques like airborne surveys and 3D seismic imaging have made it easier to explore remote and difficult environments, reducing exploration risks and improving economic viability. These advancements have also enabled the discovery of new resources, offering significant potential for future economic growth (Haritha 2023).

9.4 Sustainable and Environmental Applications

This section focuses on the role of geophysics in promoting sustainability and environmental stewardship within mining and resource exploration. It outlines how geophysical tools are increasingly used not just for resource discovery, but also for minimizing ecological impact, supporting land reclamation, and guiding responsible extraction practices.

9.4.1 Environmental Monitoring

This subsection discusses geophysical methods used to monitor environmental conditions in and around mining sites. Topics include the detection of subsurface contamination, tracking groundwater movement, assessing tailings stability, and monitoring induced seismicity. These applications are critical for ensuring compliance with environmental regulations and supporting long-term ecosystem health.

Groundwater Monitoring

Geophysics can be used to monitor groundwater conditions both during and after resource extraction, which is critical for preventing contamination and managing water resources. For instance, electrical resistivity tomography (ERT) and ground-penetrating radar (GPR) are non-invasive methods that help detect changes in the subsurface water flow. This is essential for assessing whether toxic by-products or chemicals from mining processes, such as acid mine drainage, are contaminating nearby groundwater sources. Additionally, geophysical methods can be used to track the migration of contaminants and ensure that groundwater remains safe for consumption and other uses.

Fault Stability

Faults in the earth's crust can pose significant risks, especially when resource extraction activities such as drilling or mining occur near them. Geophysical methods, such as seismic reflection and refraction surveys, are used to analyze fault lines and assess their stability. This helps in understanding the potential for induced seismicity (earthquakes triggered by human activities) and the risk of landslides or subsidence. Monitoring fault zones before and after resource extraction is vital for avoiding accidents and maintaining the structural integrity of nearby infrastructures, such as pipelines or buildings.

Methane Leak Detection

Methane leaks are a significant environmental concern in natural gas extraction, as methane is a potent greenhouse gas. Geophysical techniques like surface-based electromagnetic surveys and gas detection technologies are used to locate methane leaks in and around extraction sites. Ground-penetrating radar (GPR) and airborne sensors are also employed to detect gas emissions, even in areas that are difficult to access. Early detection of methane leaks ensures prompt remediation actions, thereby reducing the environmental impact of resource extraction.

9.4.2 Sustainable Practices

This subsection highlights how geophysical techniques contribute to sustainable practices in the mining and resource sectors. It covers applications such as optimizing resource extraction to reduce waste, supporting renewable energy integration in exploration operations, and facilitating low-impact survey methods. Emphasis is placed on how geophysics can aid in achieving environmental, social, and governance (ESG) goals while maintaining operational efficiency.

Energy Reduction in Extraction

One of the biggest challenges in mining is the high energy consumption associated with resource extraction, particularly in deep mining operations. Geophysical methods can help optimize the extraction process, making it more energy-efficient. For example, seismic surveys help create detailed 3D models of mineral deposits, allowing mining companies to target resources more precisely and minimize unnecessary excavation. This reduces the overall energy needed for extraction, as it minimizes the amount of rock that must be processed to obtain the desired resources.

Habitat Protection

Traditional mining methods often involve clearing large tracts of land, which leads to significant habitat destruction and biodiversity loss. By utilizing geophysical techniques like electromagnetic surveys, magnetic surveys, and geophysical borehole logging, exploration teams can identify mineral deposits at depth without disturbing large surface areas. This approach, known as "targeted exploration," allows for more selective mining, reducing the footprint of mining operations and preserving surrounding ecosystems.

Water Conservation & Pollution Prevention

Water is a critical resource in mining, and excessive water usage can deplete local water sources. Geophysical techniques, such as resistivity and seismic surveys, help assess the distribution and availability of groundwater in mining areas. This enables mining companies to implement more efficient water management strategies, reducing the amount of water required for processing minerals. Furthermore, geophysics can help detect any signs of water contamination due to mining activities, allowing for early intervention and pollution prevention.

Tailings Disposal Impact Reduction

The disposal of mining waste, particularly tailings, poses significant environmental risks, including water contamination and soil degradation. Geophysical techniques, including seismic reflection and ground-penetrating radar, are used to monitor tailings dams and other waste disposal sites. These methods can detect subsurface changes or instability that may indicate a risk of failure, allowing for early remediation and minimizing environmental harm.

9.4.3 Integrated Approaches

The integration of various geophysical techniques improves the accuracy of resource exploration. For example, combining seismic, magnetic, gravity, and electromagnetic methods can provide a more comprehensive view of the subsurface, especially when resources are located in complex geological settings.

Data Inversion and Modeling

One of the key advancements in geophysics is the ability to invert geophysical data to create detailed models of the subsurface. Inversion algorithms use mathematical methods to translate observed geophysical data (such as gravity, magnetic, or seismic measurements) into a model of the Earth's subsurface.

Mathematical Principle: A general inversion model for geophysical data can be written as:

$$d_{obs} = Gm_{model} + \varepsilon \tag{9.14}$$

where d_{obs} is the observed data, G is the forward operator (relating model parameters to data), m_{model} is the model parameters (e.g., density, velocity), and ϵ is the error term. The goal is to find m_{model} that minimizes the difference between d_{obs} and the modeled data Gm_{model}, often by using least squares or other optimization techniques.

9.5 Tools and Techniques for Improved Exploration

This section highlights the advanced geophysical tools and methodologies that enhance the efficiency and accuracy of mineral and resource exploration. It covers recent developments in seismic, magnetic, gravity, electrical, and electromagnetic techniques, along with integrated approaches using geospatial technologies and machine learning. Emphasis is placed on how these innovations improve subsurface imaging, reduce exploration risk, and support sustainable resource development.

9.5.1 New Geophysical Tools

Advancements in geophysical tools allow for more precise and efficient resource exploration. These tools aim to reduce both the environmental footprint of exploration and the cost of resource extraction. Some of the new technologies include:

3D Seismic Imaging

Seismic surveys have traditionally been 2D, but 3D seismic imaging offers a more detailed and accurate representation of subsurface structures. This provides high-resolution data, which is crucial for identifying mineral deposits, oil, gas, and geothermal resources.

Mathematics/Algorithms: 3D seismic imaging uses algorithms such as Finite Difference Methods (FDM) and Finite Element Methods (FEM) to solve the wave equation in a three-dimensional medium. These methods help in interpreting complex subsurface structures with high accuracy. The mathematical formulation of the seismic wave equation is:

$$\nabla^2 u(r, t) - \frac{1}{c^2}\frac{\partial^2 u(r, t)}{\partial t^2} = 0 \tag{9.15}$$

where $u(r, t)$ is the seismic wave field, ∇^2 is the Laplacian operator, c is the wave velocity, $\frac{\partial^2 u}{\partial t^2}$ is the second derivative of the wave field with respect to time. This helps improve accuracy in mapping the location and extent of subsurface resources, reducing the need for exploratory drilling and minimizing environmental disruption.

Airborne EM (AEM) Surveys

AEM surveys use electromagnetic induction to measure the resistivity of the subsurface. These surveys can quickly cover large areas, making them useful in early-stage exploration, particularly for minerals and geothermal energy. AEM can identify areas where the resistivity contrasts significantly, suggesting the presence of valuable resources.

Mathematics/Algorithms: AEM data are analyzed using Inverse Modeling techniques, where the resistivity distribution is reconstructed from electromagnetic field measurements. The Maxwell equations govern the electromagnetic fields:

$$\nabla \times \mathrm{E} = -\frac{\partial \mathrm{B}}{\partial \mathrm{t}} \tag{9.16}$$

where E is the electric field, B is the magnetic field, $\nabla \times \mathrm{E}$ represents the curl of the electric field, indicating how the electric field changes in space and time. This method can be used to identify the distribution of conductive materials and guide targeted exploration, minimizing surface disturbances.

Case Study: AEM for Cu–Ni–Co in China

Airborne Electromagnetic (AEM) systems, including time-domain electromagnetic (TDEM) methods, have become essential tools for mineral exploration, especially in the search for concealed critical metal deposits (Fountain 1998; Legault 2015). AEM systems are advantageous for their ability to map subsurface conductivity, which can be indicative of mineralized zones, such as those containing conductive sulfide minerals or hydrothermal alteration (Smith 2013). This case study explores the use of airborne electromagnetic systems, specifically helicopter TDEM systems, in exploring critical metal deposits.

Since the early 2000s, several helicopter-borne AEM systems have been introduced, including AeroTEM, VTEM, SkyTEM, and HeliGEOTEM. These systems provide high-resolution data that help geophysicists identify subsurface features at varying depths (Boyko et al. 2001; Witherly et al. 2004; Sorensen and Auken 2004; Fountain et al. 2005). AEM data is especially useful for mapping conductive materials like massive sulfides, graphite, and clay minerals, and it can also highlight alteration zones indicative of mineral deposits, such as uranium and copper–gold systems (Smith and Annan 1998; Kratzer and Macnae 2012).

A detailed review of airborne geophysical surveys conducted for copper–nickel–cobalt (Cu–Ni–Co) deposits in China highlights the utility of AEM systems. The region, characterized by complex geological formations, required advanced geophysical techniques to target potential deposits that were concealed under thick cover. Traditional surface mapping methods were not effective due to dense vegetation and rugged terrain, making airborne methods the preferred choice (Xue et al. 2024).

The VTEM system was employed for its capability to penetrate deeper into the Earth and identify conductive mineral bodies. The system uses time-domain electromagnetic pulses to induce currents in the ground, and the resulting data reflects the conductivity of various geological formations (Witherly et al. 2004). Data from the VTEM system were collected along predetermined flight paths, and the raw data were processed to create conductivity depth images and resistivity maps. The TDEM data were interpreted using 1D inversion models, which assume horizontally stratified geology. This method was particularly effective in stratified geological settings, providing accurate results even with limited lateral sensitivity (Brodie 2016).

The AEM data revealed several *conductivity anomalies* consistent with the presence of massive sulfides, which are indicative of Cu–Ni–Co mineralization. These anomalies were marked by a pronounced response in the vertical Z-component of the TDEM data, reflecting the high conductivity of sulfide-rich zones (Kratzer and Macnae 2012). In addition to mineralization, the AEM data also identified *hydrothermal alteration zones*, particularly in the deeper parts of the survey area. These zones showed signatures of altered rock types that are commonly associated with mineral deposits, such as those found in porphyry copper systems (Smith and Annan 1998). The AEM results were *integrated with regional geological data*, including known structural features and geological mapping, to refine target areas for further exploration. Structural complexity maps helped identify areas where mineralization was likely to be concentrated due to tectonic processes (Dentith and Mudge 2014; Nabighian 1987).

The airborne EM surveys significantly improved exploration efficiency by narrowing down potential target areas for drilling. In the case of the Cu–Ni–Co exploration in China, the data allowed geophysicists to pinpoint areas where previous surface surveys had failed to identify significant mineralization. The successful identification of conductive anomalies led to the selection of drill sites, which later confirmed the presence of critical metal deposits. The airborne surveys allowed for rapid, large-scale data collection, which would have been prohibitively time-consuming using ground-based methods (Dransfield 2007). The ability to detect subsurface features beneath thick cover enabled the exploration team to target hidden mineral systems that would have otherwise been overlooked (Rainsford 2019).

This case study demonstrates the value of airborne electromagnetic methods in mineral exploration, particularly in the context of discovering concealed critical metal systems. The integration of AEM data with geological and geophysical models allows for a more efficient and targeted exploration strategy (Telford et al. 1990). As mineral exploration continues to target deeper and more hidden resources, airborne geophysical methods like AEM will play an increasingly important role in guiding the discovery of new critical metal deposits, essential for modern technologies and green energy solutions. In the future, as technology improves and AEM systems become even more sophisticated, the depth of exploration and accuracy of targeting will likely continue to increase, enabling the discovery of even more concealed mineral deposits (Fountain 1998; Legault 2015).

Magnetic and Gravity Surveys

These methods measure variations in the Earth's magnetic field or gravitational field caused by differences in the density or magnetic properties of subsurface materials. New tools have enhanced the precision and resolution of these surveys, enabling more accurate subsurface mapping.

Mathematics/Algorithms: The anomalies in magnetic or gravitational fields are modeled using forward and inverse modeling techniques, where the Poisson equation for gravity and the Magnetic field equation are key tools.

For gravity:

$$\nabla^2\varphi = 4\pi G\rho \tag{9.17}$$

where φ is the gravitational potential, G is the gravitational constant, ρ is the density distribution. For magnetic fields, similar techniques are used to solve the Laplace equation in magnetically sensitive media.

9.5.2 Environmental Impact Reduction

Reducing the environmental impact of exploration activities is a growing concern. Several advancements in geophysical methods focus on reducing the footprint of exploration activities:

Remote Sensing

The use of remote sensing technologies, including satellite and drone-based imaging, has significantly improved exploration efficiency. Remote sensing allows for large-area monitoring of surface changes, vegetation health, and potential environmental damage, thus helping in the early detection of environmental degradation.

Mathematics/Algorithms: Algorithms such as Principal Component Analysis (PCA) and Spectral Unmixing are applied to remotely sensed data to extract meaningful features and monitor environmental changes. The general form of the PCA equation is:

$$Z = X \cdot P \tag{9.18}$$

where X is the original data matrix, P is the matrix of eigenvectors, and Z is the transformed dataset.

Hazard Detection

New tools allow for continuous geophysical monitoring to detect hazards like subsidence, landslides, and gas leaks. Technologies such as fiber-optic sensing and distributed acoustic sensing (DAS) can monitor ground movements in real time, helping to avoid environmental accidents.

Mathematics/Algorithms: The monitoring system typically uses Rayleigh wave inversion to interpret data from fiber-optic sensors. The governing equation for Rayleigh waves is:

$$v_R = \rho\mu + \lambda \tag{9.19}$$

where v_R is the Rayleigh wave velocity, μ and λ are the Lame parameters (related to material stiffness), and ρ is the density of the material.

Targeted Drilling

Advanced geophysical surveys and modeling allow for the precise targeting of resource deposits, which reduces the need for exploratory drilling and minimizes surface disruption. Tomography and inverse modeling help in creating detailed subsurface images that guide drilling decisions, thus lowering environmental costs.

Mathematics/Algorithms: In geophysical tomography, the inverse problem is solved using algorithms such as Least-Squares Inversion or Gauss–Newton Methods, aiming to minimize the difference between predicted and observed data. This technique allows for highly accurate 3D imaging of subsurface structures, improving resource targeting.

The optimization process typically involves:

$$\min Ax - b^2 \tag{9.20}$$

where A is the forward model matrix, x is the unknown parameter (e.g., subsurface properties), b is the observed data, and $||Ax - b||^2$ represents the squared Euclidean norm (the sum of squared differences between the predicted values Ax and the actual values b. The goal is to minimize this squared difference, commonly used in regression and other fitting problems.

9.5.3 Machine Learning Integration

One of the emerging approaches to improving exploration efficiency and reducing environmental impact is the use of Machine Learning (ML) and Artificial Intelligence (AI) techniques to integrate and analyze geophysical data. By applying ML algorithms such as Random Forests, Support Vector Machines (SVM), and Neural Networks, geophysicists can extract patterns from complex datasets more efficiently.

Mathematics/Algorithms: Machine learning algorithms are trained on large datasets to make predictions or classifications about subsurface conditions. For example, in supervised learning, the goal is to minimize the following cost function:

$$J(\theta) = \frac{1}{m}\sum_{i=1}^{m}\left(h_\theta\left(x^{(i)}\right) - y^{(i)}\right)^2 \tag{9.21}$$

where $h_\theta\left(x^{(i)}\right)$ is the predicted value, $y^{(i)}$ is the actual value, θ is the model parameters, and m is the number of data points. Using these techniques, exploration can become more efficient and predictive, leading to less exploration-related damage to ecosystems and more targeted extraction.

Case Study: Critical Metal Exploration with Airborne Geophysics

This case study explores the application of innovative airborne geophysical strategies in the exploration of critical metal systems, focusing on both gold and uranium deposits as examples (Kwan and Reford 2025). Critical metals are essential commodities for the manufacture of high technologies, economic development, and national security. The lack of reserves, their heterogeneous distribution, and the limited extraction or processing protocols for these metals could lead to supply chain vulnerability and potential disruptions in the near future (Geoscience Australia 2024; Groves et al. 2024; Ministry of Mines India 2023; Muller et al. 2024). Individual countries, or blocks of countries such as the European Union (EU) and the BRICS states, control very different quantities of critical metal resources (Groves et al. 2024). In this study (Kwan and Reford 2025), a list of critical metals was selected, primarily based on Groves et al. (2023), and organized in Table 9.1.

The analysis of structural complexity, including Contact Occurrence Density (COD) and Orientation Entropy (OE), provides valuable insights into the geological characteristics of mineralized zones. COD quantifies the density of structural features, such as faults and geological contacts, helping to identify regions with high potential for mineralization, while OE measures the diversity of structural orientations, with higher values often indicating more complex geological environments conducive to mineral deposits. The range of all possible orientations is divided into bins, and entropy is calculated using the formula:

$$OE = -\sum_{i=1}^{N} p_i \cdot \log(p_i) \tag{9.22}$$

Table 9.1 Lists the selected critical metals, their associated mineral deposit types, and major producing countries

Critical metals	Deposit types	Metallurgy status	Major producing countries
Aluminium (Al)	Bauxite	Primary	Australia, Guinea, China, Brazil
Cadmium (Cd)	SEDEX, MVT, volcanogenic	By-product (Zn)	China, Japan, Korea
Chromium (Cr)	Ophiolite, layered intrusion, placer, laterite, chromite	Primary	South Africa, Turkey, Kazakhstan
Cobalt (Co)	Sediment-hosted, SEDEX, magmatic, volcanogenic	By-product (Cu)	DRC (49%), Russia, Australia
Copper (Cu)	Porphyry Cu ± Au, IOCG, sediment-hosted, SEDEX	Primary	Chile (28%), Peru, Australia, Russia
Gallium (Ga)	Coal-hosted, bauxite	Intermediate	China, Russia, Japan, South Korea
Germanium (Ge)	Kipushi-type sedimentary, MVT, carbonate-hosted polymetallic	By-product (Cu, Zn, Pb)	DRC, Namibia
Gold (Au)[a]	Orogenic, reduced intrusion-related, epithermal, Carlin	Primary	China, Australia, Russia, Canada
Graphite (C)	Vein, flake	Primary	China, India, Brazil
Indium (In)	SEDEX, MVT, volcanogenic	By-product (Zn)	China, Peru, USA, Canada
Lithium (Li)	Pegmatite, salar	Primary	Australia (70%), Chile, Bolivia
Nickel (Ni)	Mafic–ultramafic, intrusions, nickel laterites	Primary	Indonesia (20%), Philippines, Australia, Canada, Brazil, Russia
Platinum-group elements (PGE)[b]	Reefs in layered intrusions, alkalic	Primary	South Africa, Russia, Zimbabwe, Canada, USA
Rare earth elements (REE)[c]	Carbonatite, pegmatite, IOCG	Primary	China, USA, Australia
Selenium (Se)	Sediment-hosted and volcanogenic Cu	By-product (Cu)	USA, Japan, Canada, China, Australia
Silver (Ag)[d]	Epithermal, volcanogenic, SEDEX	Primary/intermediate	Mexico, China, Peru
Tellurium (Te)	Various gold and copper deposits	By-product (Cu, Au)	China, Japan, Sweden, Russia

(continued)

Table 9.1 (continued)

Critical metals	Deposit types	Metallurgy status	Major producing countries
Tungsten (W)	Quartz-vein, skarn, porphyry, greisen	Primary	China, Vietnam, Russia
Uranium (U)	Intrusion, volcanogenic, unconformity, conglomerate and sandstone	Primary	Kazakhstan, Australia, Namibia, Canada, Uzbekistan, Niger, Russia

Source Kwan and Reford (2025), Groves et al. (2023), Jaireth et al. (2014), McLelland et al. (2016), Simandl et al. (2015), and Su and Hu (2022)

[a]Gold is on the list of critical metals of Japan and China (Su and Hu 2022)

[b]The platinum group elements include ruthenium, rhodium, palladium, osmium, iridium and platinum

[c]Rare Earth Elements (REE) consist of the fifteen lanthanide elements (lanthanum to lutetium), plus scandium (Sc) and yttrium (Y) (Jaireth et al. 2014). The lanthanide elements are lanthanum (La), cerium (Ce), praseothium (Pr), neodymium (Nd), promethium (Pm), samarium (Sm), europium (Eu), gadolinium (Gd), terbium (Tb), dysprosium (Dy), holmium (Ho), erbium (Er), thulium (Tm), ytterbium (Yb) and lutetium (Lu)

[d]Silver is on the list of critical minerals of Japan (Su and Hu 2022)

where p_i is the percentage of structural vectors with orientation i. The unit for OE is (1/m^2), and the values are scaled between 0 and 255 for display. The size of the local area influences the range of OE values (Parker 2011).

Effective geophysical strategies for exploring critical metal systems depend on key factors. First, success hinges on understanding the physical property contrasts between mineralization and host rocks. Second, if mineralization is structurally controlled, structural mapping and complexity analysis, using geophysical or remote sensing data, are crucial. Finally, integrating available geophysical data aids in identifying concealed exploration targets.

While direct detection of critical metal mineralization is difficult, except for massive sulfides and near-surface uranium, geophysics plays a vital role in guiding the discovery of concealed systems (Mir et al. 2019). A robust understanding of geophysical signatures across different mineral systems is essential for choosing the right exploration strategy (Airo 2015; Dentith et al. 1994; Gunn and Dentith 1997; Hoschke 2011; Witherly 2014).

1. ***Red Lake Gold Camp, Ontario, Canada***

In 1975, the Ontario Geological Survey conducted high-resolution airborne magnetic surveys over the Archean greenstone belts in Ontario (Ontario Geological Survey 2017). These surveys were merged to create seven super grids, including the Red Lake-Stormy super grid, which covers the Red Lake gold camp. The aeromagnetic data is publicly available from the Ontario Geological Survey website.

The first vertical derivative of the reduced-to-pole (RTP) magnetic data (Fig. 9.10) shows strong magnetic responses linked to volcanic and mafic to ultramafic rocks.

The Campbell-Red Lake complex has produced 11.2 Moz [1 oz (oz) = 31.1035 g and 1 million oz (Moz) = 1,000,000 oz] and 5.6 Moz of gold, respectively, while the Cochenour-Willans mine produced 1.2 Moz of gold (Blais et al. 2016). Gold mineralization at the Campbell-Red Lake complex is structurally controlled, hosted in mafic volcanic rocks, and often occurs as vein systems (Sanborn-Barrie et al. 2004). Sulfide mineralization at Campbell-Red Lake, typically found in sheared mafic rocks with disseminated pyrrhotite (up to 30%), adds to the complexity of the exploration.

Structural analysis using the CET grid module revealed an inferred F1 lineament, potentially a crustal-scale shear zone associated with major gold deposits in the Red

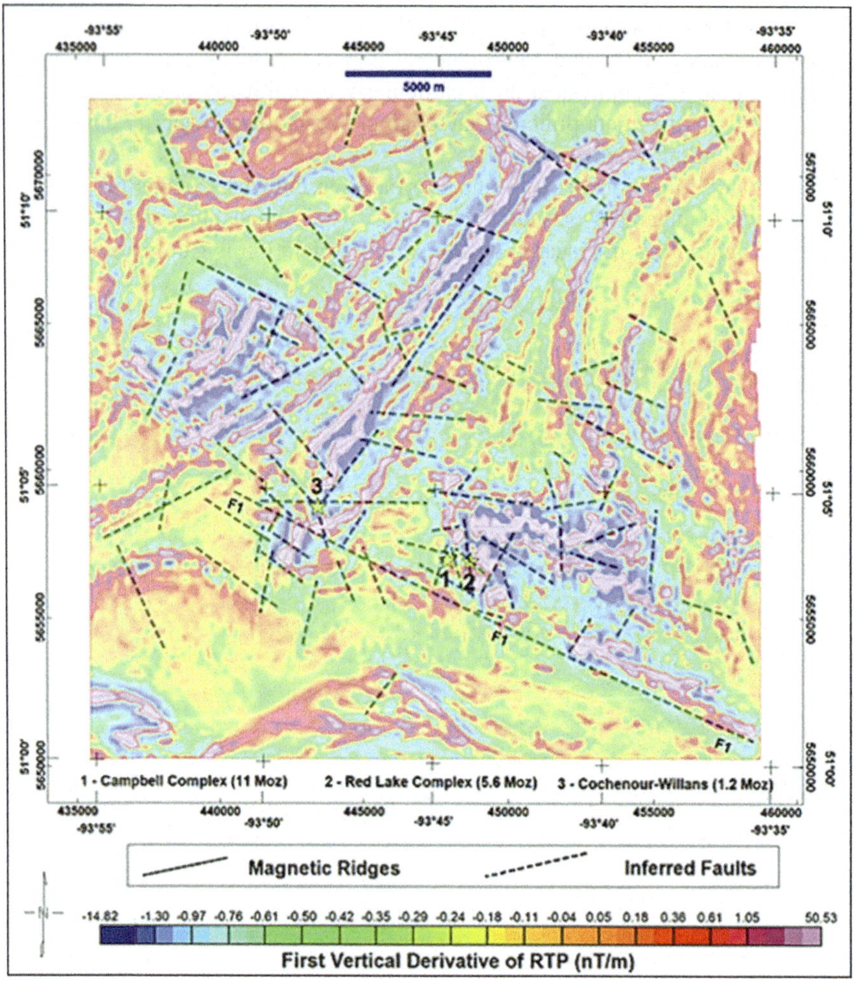

Fig. 9.10 First vertical derivative of RTP data from red lake gold camp, overlain with calculated magnetic ridges, inferred faults, and locations of past and present gold mines (total gold production indicated). *Source* Kwan and Reford (2025)—open access

Lake camp (Groves et al. 2022). Lamprophyre dykes in the ore zones suggest a lithospheric-scale shear zone (Groves et al. 2022).

Figures 9.11 and 9.12 display the computed structural complexity maps—contact occurrence density (COD) and orientation entropy (OE)—which highlight the importance of structural complexity analysis in identifying concealed gold systems. High structural complexity areas correlate with known gold mines, while other moderately high complexity zones in SW–NE and E–W corridors of mafic volcanic rocks may present new exploration targets.

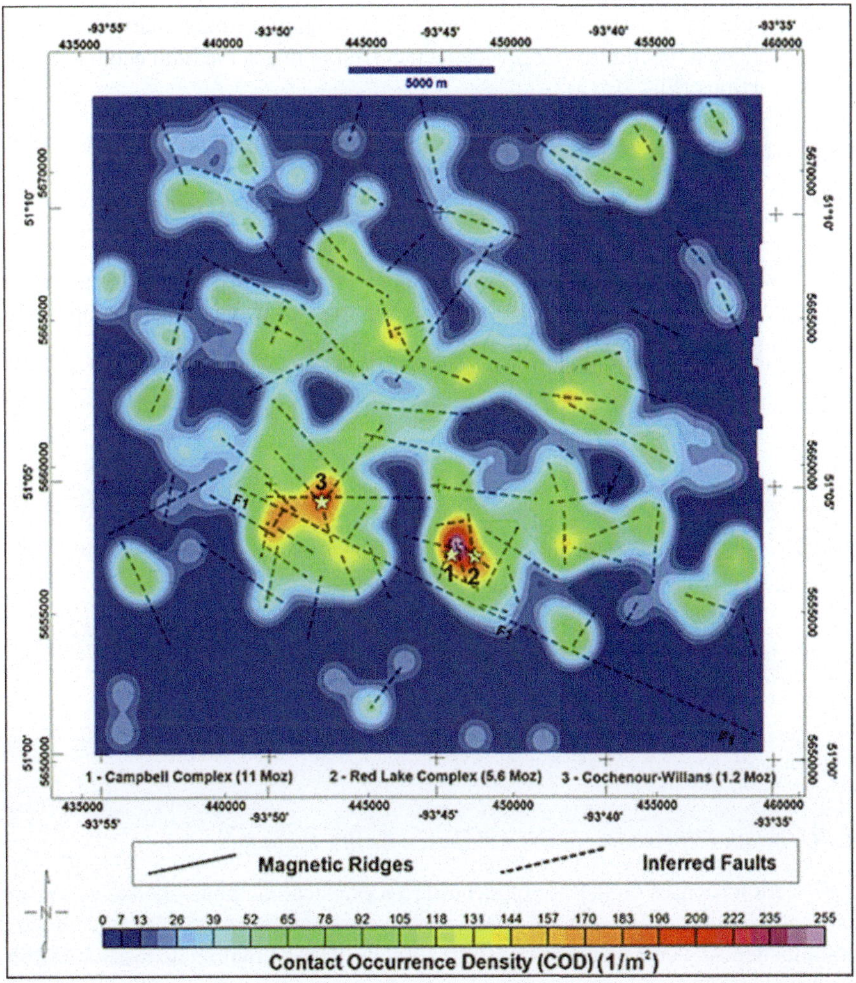

Fig. 9.11 Contact occurrence density (COD) map of the red lake gold camp, overlain with calculated magnetic ridges, inferred faults, and locations of Campbell-red lake (1, 2) and Cochenour-Willans (3) gold mines. *Source* Kwan and Reford (2025)—open access

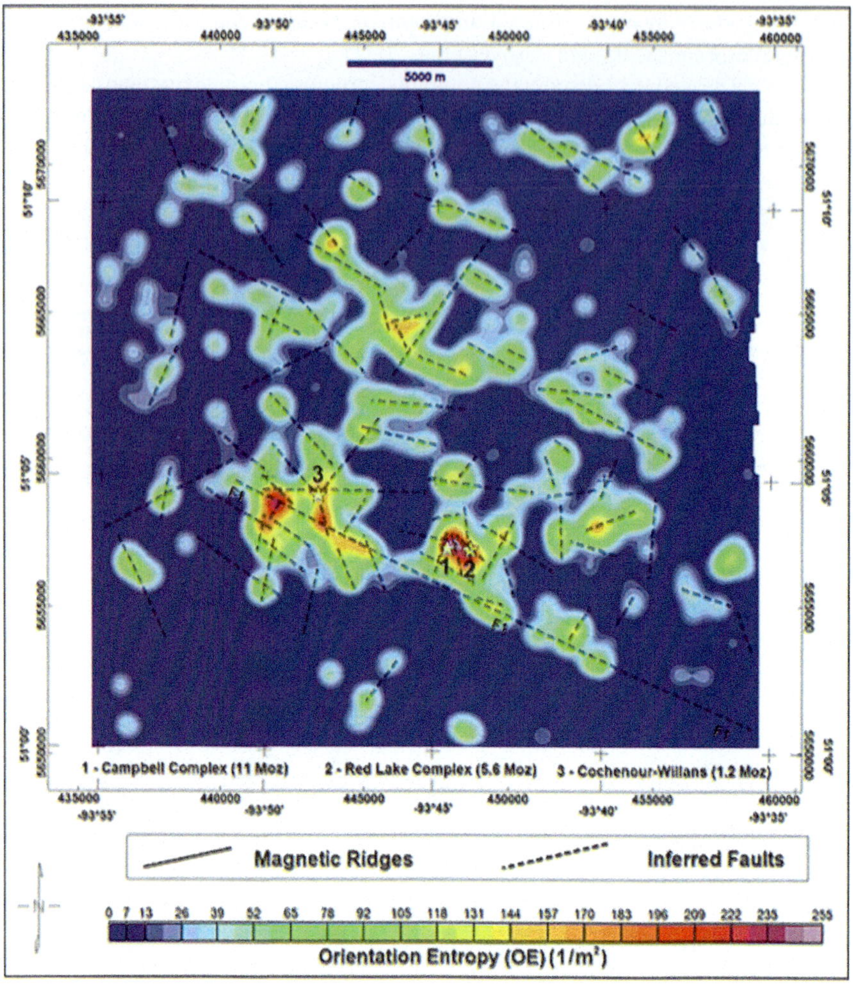

Fig. 9.12 Orientation entropy (OE) map of the red lake gold camp, overlain with calculated magnetic ridges, inferred faults, and locations of Campbell-Red Lake (1, 2) and Cochenour-Willans (3) gold mines. *Source* Kwan and Reford (2025)—open access

2. ***Uranium Exploration in the Athabasca Basin, Canada***

This section focuses on the use of airborne geophysical methods, including TDEM and natural source EM (e.g., ZTEM), in the exploration of unconformity-related uranium deposits in the Athabasca Basin, Canada. These uranium deposits are structurally controlled and are associated with pre-mineralization graphitic basement faults (Legault et al. 2018; Li et al. 2016).

Airborne magnetic and gamma-ray spectrometry data were used for structural mapping at Rossing, Namibia (Anderson and Nash 1997). In the Athabasca Basin, uranium deposits are located in the Wollaston-Mudjatik Transition Zone

(WMTZ), where low magnetic responses are observed due to magnetite destruction by hydrothermal fluids (Jeanneret et al. 2016). Figure 9.13 shows the RTP data from the eastern part of the Athabasca Basin, with a 100 m grid cell size.

Figures 9.14, 9.15, and 9.16 display the structural vectors, COD, and OE maps derived from the geophysical data. These maps reveal that most uranium deposits are located in high COD and high OE zones, underscoring the importance of structural mapping in identifying concealed uranium deposits.

By integrating advanced airborne geophysical methods and structural complexity analysis, these case studies illustrate the effectiveness of geophysical strategies in

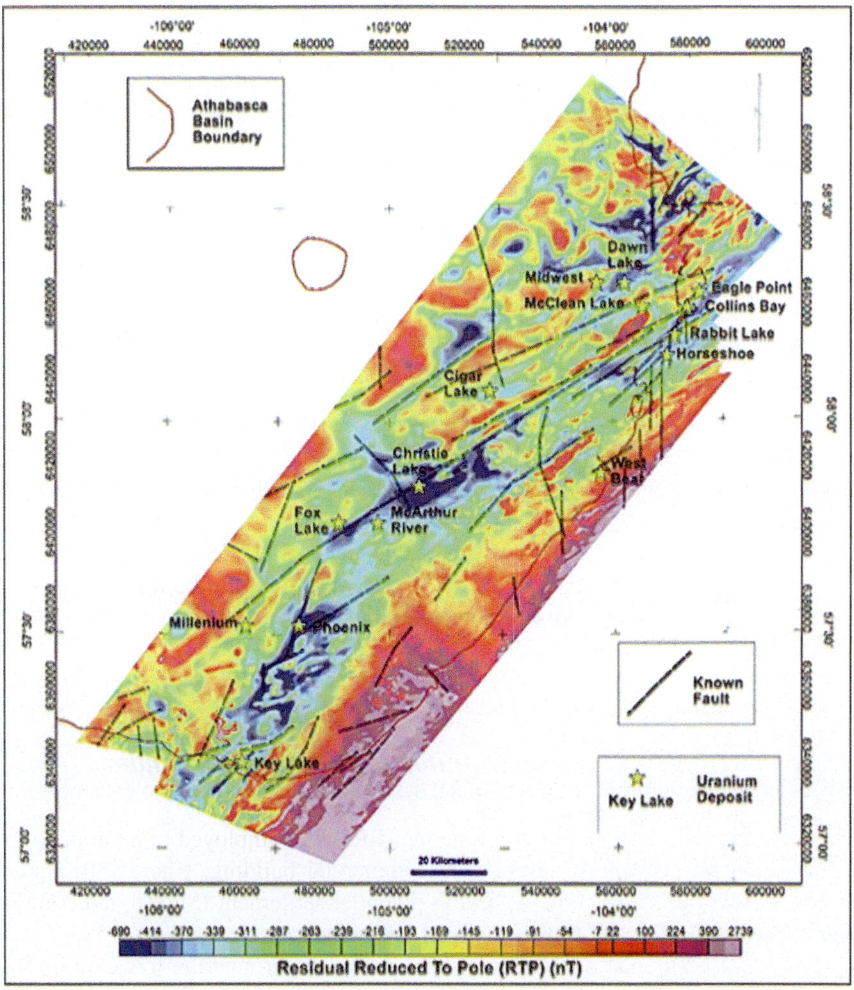

Fig. 9.13 RTP magnetic data of the eastern Athabasca basin, overlain with known faults and uranium deposits. *Source* Kwan and Reford (2025)—open access

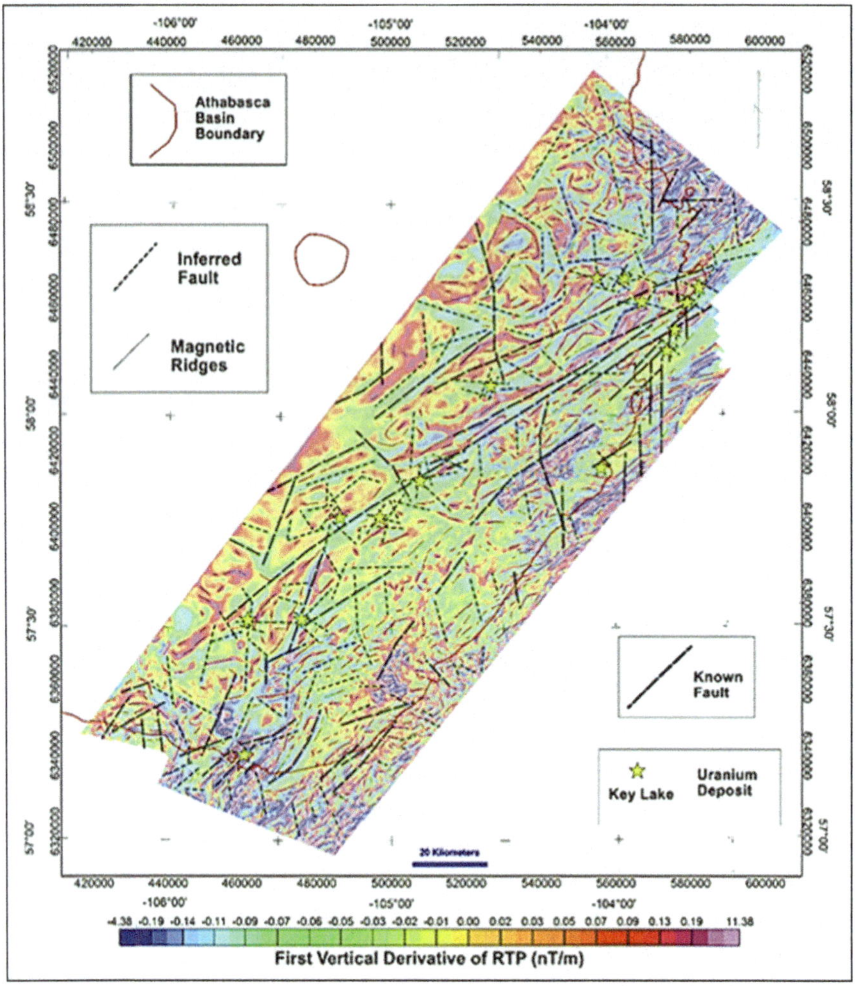

Fig. 9.14 Structural vectors over the first vertical derivative of the RTP data, red lake gold camp. *Source* Kwan and Reford (2025)—open access

advancing the exploration of critical metal systems, especially in complex geological environments.

9.6 Case Studies of Geophysical Innovations in Mining

Geophysical technologies have significantly transformed resource extraction in the mining sector. These advanced methods provide more precise information about subsurface structures, allowing mining operations to optimize exploration and reduce

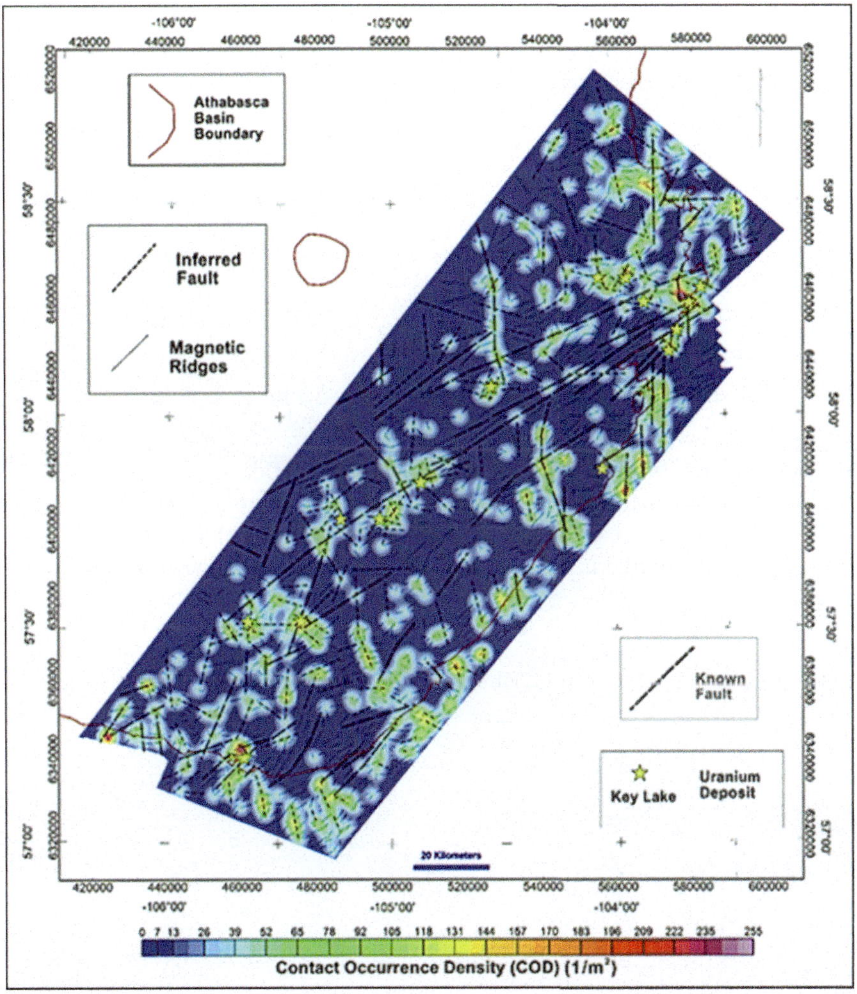

Fig. 9.15 Contact occurrence density (COD) map of the eastern Athabasca Basin. *Source* Kwan and Reford (2025)—open access

risks associated with resource extraction. The following case studies illustrate the successful applications of geophysical innovations in mining and the practical benefits derived from these technologies.

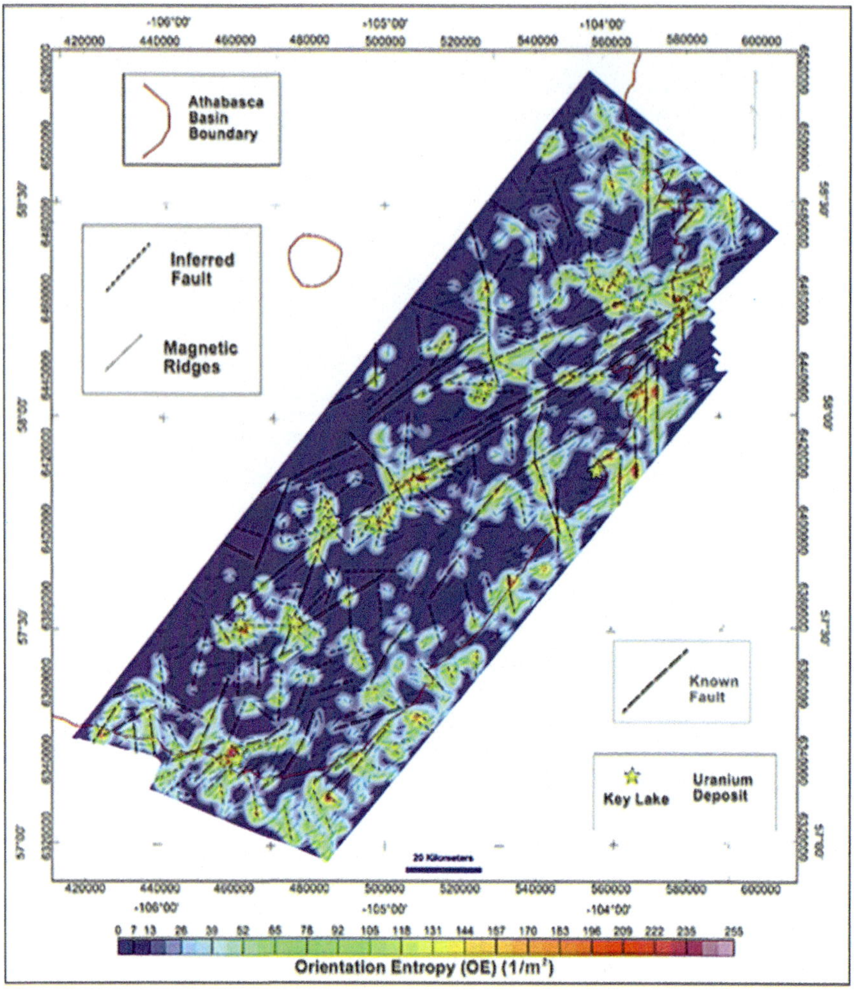

Fig. 9.16 Orientation entropy (OE) map of the eastern Athabasca Basin. *Source* Kwan and Reford (2025)—open access

9.6.1 Airborne EM Surveys—Australia

Airborne electromagnetic (AEM) surveys have revolutionized the exploration of mineral deposits in areas with complex geological settings, where surface exploration methods are limited. In Australia, an airborne electromagnetic survey was conducted over the Western Australian goldfields to detect buried gold mineralization beneath extensive overburden (Haas 2006). The AEM survey was especially useful in mapping conductive zones associated with gold mineralization that were difficult to detect through traditional geological mapping due to the thick overburden.

The data obtained through AEM allowed geophysicists to map mineralized zones with high precision, which led to more efficient drilling programs. These programs subsequently identified several significant gold deposits that would have been challenging to locate otherwise. The integration of AEM with other geophysical techniques, such as gravity and magnetic surveys, further improved the accuracy of the resource estimation (Turner et al. 2013). By using AEM in combination with 3D geological modeling, mining companies were able to reduce exploration costs and significantly improve their targeting of mineral deposits.

The success of this case study highlights the capability of airborne electromagnetic surveys to optimize exploration in challenging terrains, demonstrating how this technology has made gold exploration in Australia more cost-effective and efficient (Haas 2006).

9.6.2 3D Seismic Imaging—Russia (Khokhlovka Gold Deposit)

In Russia, 3D seismic imaging has been applied to the Khokhlovka gold deposit, where traditional 2D seismic methods were ineffective due to the complex and deeply buried geological formations (Bubnov et al. 2014). The transition to 3D seismic surveys allowed geophysicists to generate high-resolution, three-dimensional images of the subsurface, revealing intricate details about the geological structures that were previously invisible using conventional techniques.

The 3D seismic data provided valuable information on the depth, size, and shape of the gold ore bodies, enabling more accurate resource modeling. This improved understanding of the subsurface structure led to the identification of previously unknown mineralized zones, expanding the potential resource base of the deposit (Bubnov et al. 2014). Moreover, 3D seismic imaging improved drilling efficiency by allowing for more precise targeting of drilling locations, which significantly reduced the number of exploratory wells and minimized unnecessary excavation.

The use of 3D seismic imaging at the Khokhlovka deposit showcases the benefits of high-resolution imaging in uncovering and delineating complex ore bodies. This technology has enhanced both the accuracy and cost-effectiveness of mineral exploration in Russia's goldfields (Bubnov et al. 2014).

9.6.3 Gravity and Magnetic Surveys—Iran (Chah Gaz Iron Ore)

The Chah Gaz iron ore deposit in Iran is located in an area with complex geology and limited surface outcrop, which made traditional exploration methods challenging. To address this, gravity and magnetic surveys were employed to detect variations in subsurface density and magnetic properties, both of which can be indicative of mineralization. Gravity data, which measures variations in the Earth's gravitational field, and magnetic data, which measures the Earth's magnetic anomalies, were used together to map the extent and depth of the iron ore body (Dorr and Wyllie 2015).

The results of the combined gravity and magnetic surveys revealed the subsurface distribution of the iron ore deposit, including the presence of high-grade ore zones. The geophysical surveys also identified structural features such as faults and folds, which were essential for understanding the geological controls on ore deposition. These insights allowed for more accurate targeting of drilling operations, which led to the discovery of additional ore zones.

By using gravity and magnetic surveys to map the subsurface, exploration teams were able to significantly reduce exploration costs and enhance the efficiency of resource extraction. This case study illustrates the value of integrating multiple geophysical methods for mineral exploration in complex geological settings (Dorr and Wyllie 2015).

9.6.4 GPR for Ore Mapping—Canada

In Canada, ground-penetrating radar (GPR) was used to map shallow ore bodies in an active mining operation. The site was located in a region with thick overburden, which made it difficult to accurately delineate ore bodies using traditional drilling methods (Telfer et al. 2017). GPR, a non-invasive geophysical method, uses high-frequency electromagnetic waves to generate images of subsurface features. By analyzing the time delay and amplitude of the reflected waves, geophysicists were able to produce high-resolution images of the shallow ore bodies and fractures beneath the surface.

In this case, GPR was particularly effective in identifying voids, fractures, and mineralization zones within the ore body. These data were invaluable for optimizing drilling programs and assessing the stability of the rock formations. The ability to quickly and accurately map subsurface features in real time reduced the need for extensive exploratory drilling, which helped lower operational costs and minimize environmental impact.

This case study demonstrates the usefulness of GPR for mapping shallow ore bodies and assessing geotechnical properties in active mining environments (Telfer et al. 2017). By enabling precise mapping of subsurface features, GPR helped optimize mine planning and improve resource extraction efficiency.

9.6.5 Superconducting Gravimeter—South America (Waste Monitoring)

In South America, superconducting gravimeters were used to monitor the stability of mine waste in an effort to prevent environmental risks associated with mine tailings and waste piles. Mine waste can shift over time due to factors such as water infiltration and compaction, leading to potential environmental hazards. The use of superconducting gravimeters, which provide extremely sensitive measurements of gravity changes, allowed researchers to detect small variations in the density of the waste piles (Mellor and Copeland 2016).

The gravimeter data enabled early detection of instability in the waste piles, allowing for timely intervention to stabilize the material before it became a risk to the environment or surrounding communities. The application of superconducting gravimeters in this case provided a proactive approach to monitoring and mitigating environmental impacts, ultimately improving the sustainability of the mining operation.

This case study demonstrates how advanced gravity measurement tools like superconducting gravimeters can play a key role in environmental monitoring and help prevent potentially hazardous situations in mining operations (Mellor and Copeland 2016).

9.6.6 AEM Surveys—Rajasthan, India

In Rajasthan, India, airborne electromagnetic (AEM) surveys have been used to explore for mineral deposits, particularly in areas with extensive overburden and complex geology. The use of AEM surveys, conducted in the Aravalli region, has significantly enhanced exploration efforts for base metals like copper and zinc. These regions were previously challenging to explore using traditional methods due to the thick soil cover and limited surface exposure of mineralized zones.

The surveys helped in mapping conductivity anomalies associated with sulfide mineralization, which are often indicative of copper and zinc deposits. These findings were used to refine exploration targets, reducing exploration drilling costs and increasing the efficiency of subsequent drilling programs (Ghosh et al. 2021). By integrating AEM data with geological and geochemical information, mining companies were able to identify new prospective zones with higher mineral potential, accelerating the discovery of new deposits in the region.

This case study highlights the increasing use of AEM surveys in India to overcome the challenges posed by thick overburden and complex geological conditions, making it a valuable tool for mineral exploration in Rajasthan (Ghosh et al. 2021).

9.6.7 3D Seismic Imaging—Jharkhand (Coal), India

In Jharkhand, India, 3D seismic imaging was applied to coal exploration in the Jharia coalfields, one of the most critical coal-producing regions in India. The traditional 2D seismic surveys were inadequate due to the complexity of the subsurface structures, which included faults and gas reservoirs. The transition to 3D seismic surveys allowed for a more accurate mapping of the coal seams, identifying the precise depth, thickness, and extent of the coal deposits.

This innovative application of 3D seismic technology helped in delineating the deeper coal seams that were previously difficult to detect using conventional methods. The high-resolution seismic data provided a detailed subsurface view, which was instrumental in minimizing drilling risks and improving the economic viability of new coal mines. The data also facilitated the identification of areas where coal mining operations could safely expand, while avoiding gas pockets that posed safety hazards (Saha et al. 2020).

By using 3D seismic imaging, the exploration efforts in the Jharia coalfields became more efficient, safer, and cost-effective, paving the way for optimized coal extraction and better resource management (Saha et al. 2020).

9.6.8 Gravity and Magnetic—Odisha (Iron Ore), India

In Odisha, India, gravity and magnetic surveys have played a critical role in the exploration of iron ore deposits, particularly in the Keonjhar and Sundargarh districts. These areas are known for their rich iron ore deposits, but the complex geology and presence of lateritic soils often make traditional exploration techniques less effective. The integration of gravity and magnetic surveys has provided valuable insights into the subsurface structures and mineralization patterns.

Gravity surveys were used to measure variations in the Earth's gravitational field caused by differences in rock density, which helped to locate dense iron ore bodies beneath the surface. Similarly, magnetic surveys detected anomalies caused by the magnetic properties of iron-rich rocks. Together, these surveys allowed geophysicists to map the distribution of ore bodies and identify potential drill targets.

One successful application was in the mining of high-grade iron ore at the Tensa Valley, where gravity and magnetic data helped locate new mineralized zones, optimizing drilling campaigns and resource estimation (Basu et al. 2021). The use of these geophysical techniques reduced exploration costs and increased the accuracy of mineral resource estimates, making iron ore extraction more efficient and sustainable.

This case study underscores the importance of integrating gravity and magnetic surveys for mineral exploration in India's iron ore-rich regions, leading to more targeted and cost-effective exploration (Basu et al. 2021).

9.6.9 GPR—Chhattisgarh (Shallow Ore Bodies), India

In Chhattisgarh, India, ground-penetrating radar (GPR) has been used to map shallow ore bodies, particularly in limestone and coal deposits. The use of GPR has been particularly effective in mapping the depth and extent of these ore bodies in regions with shallow mining operations. The high-frequency electromagnetic pulses sent by the GPR system allow for real-time subsurface imaging, providing detailed information on ore body thickness, fractures, and geological structures.

In the Raigarh district, GPR surveys helped optimize limestone quarry operations by identifying high-quality limestone layers and fractures that could impact extraction efficiency. The ability to map the shallow ore bodies accurately also minimized overburden removal, reducing operational costs and environmental impact (Sharma et al. 2021).

GPR's ability to provide high-resolution images of subsurface features in real-time has proved to be a valuable tool for mining operations in Chhattisgarh, enhancing both productivity and sustainability in the mining sector (Sharma et al. 2021).

9.6.10 Gravimeter—Goa (Environmental Monitoring), India

In Goa, India, superconducting gravimeters have been used for monitoring the stability of mine tailings and to assess subsurface fluid movements in mining areas, particularly in iron ore mining regions. These highly sensitive instruments measure tiny fluctuations in the Earth's gravitational field caused by changes in density, which can be indicative of shifts in the mine tailings or fluid accumulation in underground reservoirs.

The data from superconducting gravimeters were used to monitor the stability of tailings dams, ensuring that potential risks such as slumping or liquefaction could be detected early. Additionally, these measurements helped in understanding the impact of mining activities on local groundwater systems, providing critical data for environmental management and ensuring that mining activities did not negatively affect surrounding ecosystems (Kulkarni et al. 2020).

The successful application of superconducting gravimeters in Goa demonstrates their utility in both environmental monitoring and ensuring the sustainability of mining operations, offering real-time insights that help mitigate environmental risks (Kulkarni et al. 2020).

These case studies highlight the diverse applications of advanced geophysical technologies in the mining sector, demonstrating their effectiveness in optimizing resource extraction, improving operational efficiency, and reducing environmental risks. The integration of methods like airborne electromagnetic surveys, 3D seismic imaging, gravity and magnetic surveys, GPR, and superconducting gravimeters has proven to be instrumental in overcoming challenges associated with complex geology, deep mineral deposits, and environmentally sensitive areas.

By leveraging these technologies, mining companies can improve their targeting of mineral resources, reduce exploration costs, and enhance safety while minimizing environmental impacts. The continued evolution of geophysical methods will undoubtedly further improve the efficiency and sustainability of resource extraction in the future.

9.7 Conclusions

This chapter has provided a comprehensive overview of the evolving role of geophysical techniques in the mining and resource exploration industries. Through technological advances and novel approaches, geophysical methods are increasingly optimizing resource extraction, enhancing exploration efficiency, and mitigating environmental impacts.

The integration of diverse geophysical methods, such as gravity, magnetic, electromagnetic, and seismic surveys, is transforming the way minerals, oil, gas, and geothermal resources are discovered and assessed. Case studies from across the globe illustrate the successful application of these techniques in a variety of settings, from mineral exploration in Australia and China to seismic imaging in Russia and GPR in Canada.

Emerging technologies, including drone-based electromagnetic surveys, autonomous seismic networks, and borehole imaging, are pushing the boundaries of exploration while integrating machine learning and big data to further refine exploration strategies. This combination of new tools and innovative techniques has not only enhanced the precision of resource identification but also allowed for better real-time data analysis and decision-making.

Equally important is the environmental and sustainable focus of modern geophysical exploration. Geophysics now plays a key role in monitoring the environmental impact of mining operations, such as subsurface groundwater conditions, methane leak detection, and fault stability analysis, ensuring that resource extraction remains as sustainable as possible. These efforts aim to reduce energy consumption, prevent habitat destruction, and minimize pollution.

Looking ahead, the ongoing development of integrated geophysical approaches and the application of advanced technologies will continue to shape the future of resource exploration, paving the way for more efficient, environmentally responsible, and economically viable mining practices. Geophysics, as a core tool in resource extraction, is poised to meet the challenges of a rapidly changing world while unlocking the full potential of Earth's natural resources.

References

Airo ML (2015) Geophysical signatures of mineral deposit types—synopsis. Geol Surv Finland Spec Pap 58:9–70

Anderson H, Nash C (1997) Integrated lithostructural mapping of the Rossing area, Namibia using high resolution aeromagnetic, radiometric, Landsat data and aero photographs. Explor Geophys 28(1–2):185–191

Basu D, Sood S, Agrawal R (2021) Integration of gravity and magnetic surveys in iron ore exploration in Odisha. J Appl Geophys 99:312–318

Blais S, Osiowy C, Hmidi N (2016) Red lake operations, Ontario, Canada. NI 43-101 technical report, p 168

Boyko M, Witherly P, Smith S (2001) AeroTEM: a new airborne time domain electromagnetic system for mineral exploration. Explor Geophys 32(1):1–10

Brodie R (2016) 1D inversion models for TDEM data. Geoscience Australia

Bubnov V, Smirnov Y, Tolstikov D (2014) Use of 3D seismic imaging for gold exploration at the Khokhlovka deposit, Russia. J Appl Geophys 108:113–122

Dentith MC, Frankcombe KF, Trench A (1994) Geophysical signatures of Western Australian mineral deposits: an overview. Explor Geophys 25:103–160

Dentith M, Mudge S (2014) Geophysics for the mineral exploration geoscientist

Dorr A, Wyllie P (2015) Gravity and Magnetic methods in mineral exploration: a case study of the Chah Gaz iron ore deposit, Iran. Geophys Prospect 63(5):1180–1189

Dransfield M (2007) Airborne geophysical surveys: the benefits of rapid, large-scale data collection. Geophys J Int 5:875

Fountain D (1998) Airborne electromagnetic systems in mineral exploration. J Appl Geophys 152:7822

Fountain D, Smith P, Broughton M, Legault J, Pearson L (2005) HeliGEOTEM: a new airborne EM system for deep exploration. Geophysics 70(6):E19–E30

Geoscience Australia (2024) Critical minerals at geoscience Australia. https://www.ga.gov.au/scientific-topics/minerals/critical-minerals

Ghosh A, Kumar R, Sharma S (2021) Airborne electromagnetic surveys for base metal exploration in Rajasthan. Indian J Geophys 50(3):235–240

Groves DI, Santosh M, Muller D, Zhang L, Deng J, Yang LQ, Wang QF (2022) Mineral systems: their advantages in terms of developing holistic genetic models and for target generation in global mineral exploration. Geosyst Geoenviron 1:1–26

Groves DI, Santosh M, Liang Z (2023) Net zero climate remediations and potential terminal depletion of global critical metal resources: a synoptic geological perspective. Geosyst Geoenviron 2:100136

Groves DI, Muller D, Santosh M, Yang CX (2024) The heterogeneous distribution of critical metal mineral resources: an impending geopolitical issue. Geosyst Geoenviron 5:1452. https://doi.org/10.1016/j.geogeo.2024.100288

Gunn PJ, Dentith MC (1997) Magnetic responses associated with mineral deposits. AGSO J Austr Geol Geophys 17:145–158

Haas R (2006) Applications of airborne electromagnetic surveys in mineral exploration: a case study from Western Australia. Explor Geophys 37(2):178–182

Haritha K (2023) A review of recent advancements in geophysical technologies and their implications for mineral and hydrocarbon exploration. ASEAN J Sci Eng Mater 2(2):95–108

Hoschke TG (2011) Geophysical signatures of copper-gold porphyry and epithermal gold deposits, and implications for exploration. ARC Centre of Excellence in Ore Deposits. University of Tasmania, Tasmania, p 47

Jaireth S, Hoatson DM, Miezitis Y (2014) Geological setting and resources of the major rare-earth-element deposits in Australia. Ore Geol Rev 62:72–128

Jeanneret P, Goncalves P, Durand C, Trap P, Marquer D, Quirt D, Ledru P (2016) Tectono-metamorphic evolution of the pre-Athabasca basement within the Wollaston-Mudjatik Transition Zone. Can J Earth Sci 53:231–259

Kratzer K, Macnae J (2012) Airborne induced polarization effects in TDEM surveys. Geophys Prospect 12:4512

Kulkarni P, Patil K, Rane V (2020) Environmental monitoring using superconducting gravimeters in Goa's iron ore mining areas. Indian J Environ Geophys 23(4):417–423

Kwan K, Reford S (2025) Innovative airborne geophysical strategies to assist the exploration of critical metal systems. Geosyst Geoenviron 4:100344

Legault J (2015) Applications of airborne electromagnetic systems in hydrogeology. Geophys Prospect 12:4785

Legault JM, Izarra C, Prikhodko A, Wood G, Keller C, O'Dowd C (2018) Comparing VTEM time-domain EM and ZTEM natural field airborne EM survey results over the McArthur River unconformity uranium project. In: SEG Technical Program Expanded Abstracts, pp 1863–1867

Li Z, Chi G, Bethune KM (2016) The effects of basement faults on thermal convection and implications for the formation of unconformity-related uranium deposits in the Athabasca Basin, Canada. Geofluids 16:7290751

Liu J, Yi J, Li C, Gao C (2025) Integrating magnetic and self-potential methods for efficient graphite exploration: Insights from Ji'an, Northeast China. Minerals 15:27

McLelland BC, Yamasue E, Tezuka T, Corder G, Golev A, Giurco D (2016) Critical minerals and energy-impacts and limitations of moving to unconventional resources. Resources 5:19

Mellor D, Copeland M (2016) Environmental monitoring using superconducting gravimeters in mining operations: a case study in South America. J Environ Monit 18(4):482–487

Ministry of Mines India (2023) Critical minerals for India, report of the committee on identification of critical minerals

Mir R, Perrouty S, Astic T, Berube CL, Smith RS (2019) Structural complexity inferred from anisotropic resistivity: example from airborne EM and compilation of historical resistivity/induced polarization data from the gold rich Canadian Malartic district, Quebec, Canada. Geophysics 84:P1–P15

Muller D, Groves DI, Santosh M (2024) Metallic mineral resources: the critical components for a sustainable Earth. Elsevier, Amsterdam

Nabighian M (1987) Electromagnetic methods in mineral exploration. Soc Explor Geophys 32:875

Ontario Geological Survey (2017) Report on magnetic super grids. Ontario geological survey, geophysical data set 1037—revised, p 14

Parker JR (2011) Algorithms for image processing and computer vision. Wiley, Amsterdam

Rainsford R (2019) National and regional geophysical data: stimulating mineral exploration. Geosci J 2:64–99

Saha M, Bhattacharya D, Ghosh P (2020) Application of 3D seismic imaging for coal exploration in Jharia coalfields, Jharkhand. Geophys J India 14(2):189–198

Sanborn-Barrie M, Skulski T, Parker J (2004) Geology, red lake greenstone belt, western superior province, Ontario. Geol Surv Canada 1:50000

Sharma R, Tiwari P, Kapoor A (2021) Application of ground-penetrating radar in shallow ore body mapping in Chhattisgarh. Int J Min Geophys 39(1):110–116

Simandl GJ, Akam C, Paradis S (2015) Which materials are 'critical' and 'strategic'. In: Simandl GJ, Neetz M (eds) Symposium on strategic and critical materials proceedings, November 13–14, 2015, Victoria, British Columbia. British Columbia Ministry of Energy and Mines, British Columbia Geological Survey Paper 2015-3, pp 1–4

Smith AA, Annan AP (1998) Ground-penetrating radar tomography: Roles of attenuation and velocity in assessing groundwater content. Water Resour Res 34(6):1559–1567. https://doi.org/10.1029/98WR00195

Smith R (2013) Applications of airborne EM in geological mapping and exploration. Geophys Surv 12:452

Sorensen K, Auken E (2004) SkyTEM: a new airborne electromagnetic system. Geophysics 121:785

Su Y, Hu D (2022) Global dynamics and reflections on critical minerals. In: E3S web of conferences, p 03045

Telfer R, White P, Carmichael A (2017) Application of ground-penetrating radar (GPR) in mapping shallow ore bodies in Canada. Can J Earth Sci 54(6):815–822

Telford WM, Geldart LP, Sheriff RE (1990) Applied geophysics, 2nd edn. Cambridge University Press, Cambridge

Turner A, Harris G, Jenkins M (2013) Geophysical surveys in mineral exploration: airborne electromagnetic methods and their application to gold exploration in Australia. Geophysics 78(3):109–115

Vallee MA, Moussaoui M, Khan K (2024) Case studies of magnetic and electromagnetic techniques covering the last fifteen years. Minerals 14:1286

Witherly K (2014) Geophysical expressions of ore system—our current understanding. SEG Spec Publ 18:177–208

Witherly M, Boyko M, Jonhson J (2004) VTEM: a helicopter-borne time-domain EM system for mineral exploration. Geophysics 69(3):1083–1096

Xue Q et al (2024) Geophysical exploration strategies for Cu–Ni–Co deposits in China. Min Explor J 1:452

Part III
Strategic and Specialized Applications

Chapter 10
Engineering Geophysics and the Sustainable Development Goals

Aligning Geophysical Advances with Global Sustainability Targets

Abstract This chapter explores how engineering geophysics supports the United Nations Sustainable Development Goals (SDGs). It highlights the role of geophysical methods in managing water, energy, climate, cities, and natural hazards. The chapter aligns key geophysical techniques with specific SDGs—such as clean water (SDG 6), sustainable cities (SDG 11), climate action (SDG 13), and clean energy (SDG 7). It also covers newer fields like medical geology, agrogeology, and geoheritage, and discusses ethical responsibilities, community engagement, and environmental justice. Several global case studies show practical applications in infrastructure planning, groundwater mapping, climate resilience, and risk reduction. Finally, the chapter outlines challenges, future opportunities, and digital tools to strengthen geophysics as a partner in sustainable development worldwide.

The integration of engineering geophysics into the framework of the United Nations Sustainable Development Goals (SDGs) presents a powerful opportunity to bridge scientific innovation with societal needs. As global communities face critical challenges such as climate change, water scarcity, urban vulnerability, and energy insecurity, the role of geophysical techniques in enabling data-driven, sustainable solutions is becoming increasingly vital.

This chapter explores how modern geophysical methods can be aligned with key SDGs—specifically those focused on clean water (SDG 6), sustainable cities (SDG 11), climate action (SDG 13), and affordable and clean energy (SDG 7) (Reynolds 2011; United Nations 2015). It emphasizes how geophysical tools contribute not only to technical problem-solving but also to resilience-building, environmental justice, and ethical practice. Through case studies, ethical discussions, and practical frameworks, this chapter demonstrates how engineering geophysics can serve as both a scientific and humanitarian discipline in the global pursuit of sustainable development.

D. C. Naskar, *Advances in Engineering Geophysics: Concepts and Applications*, Earth Systems Data and Models 9, https://doi.org/10.1007/978-3-032-28000-8_10

10.1 Introduction

10.1.1 Background on the Sustainable Development Goals (SDGs)

In 2015, the United Nations adopted the 2030 Agenda for Sustainable Development, establishing 17 Sustainable Development Goals (SDGs) as a universal call to action to end poverty, protect the planet, and ensure prosperity for all. These goals address a wide range of global challenges, including water scarcity, climate change, environmental degradation, urbanization, and energy access. Each goal is interconnected, reflecting the complex, systemic nature of sustainable development. The SDGs serve as a global framework for governments, industries, and communities to align their strategies with measurable environmental and social targets.

Scientific and technological disciplines, particularly those involved in infrastructure, resource management, and environmental monitoring, are crucial in meeting these goals. Among them, engineering geophysics offers a unique set of tools for non-invasive subsurface exploration, site characterization, and long-term environmental monitoring—functions that are increasingly critical in a world facing both rapid development and environmental strain.

10.1.2 The Role of Engineering Geophysics in Sustainable Development

Engineering geophysics is a field that applies geophysical techniques—such as seismic methods, electrical resistivity, ground-penetrating radar (GPR), electromagnetic surveys, and gravity methods—to solve engineering and environmental problems. These tools enable professionals to "see" beneath the Earth's surface without the need for extensive excavation or drilling, making them invaluable in areas where sustainability, cost-efficiency, and minimal environmental impact are essential.

In the context of sustainable development, engineering geophysics supports a range of critical objectives. It plays a key role in water resource management by enabling the mapping of aquifers and the identification of contamination zones, contributing directly to the protection and sustainable use of groundwater. In urban environments, geophysical surveys enhance resilience by evaluating subsurface hazards such as sinkholes, unstable soils, and buried infrastructure, thereby informing safer and more sustainable city planning. For climate adaptation, geophysics provides tools for monitoring landslides, assessing permafrost stability, and supporting infrastructure planning in areas vulnerable to environmental change. Furthermore, in the energy sector, geophysical methods facilitate the exploration of geothermal resources and help assess site conditions for the installation of renewable energy infrastructure, thus supporting the global transition to clean energy.

Beyond its technical capabilities, engineering geophysics also plays a vital social and ethical role by promoting equitable access to geoscientific tools and data, particularly in underserved or resource-limited regions. By empowering communities with actionable information and inclusive technologies, the discipline advances global equity and resilience, positioning engineering geophysics not only as a problem-solving science but also as a key enabler of sustainable development and social justice.

10.1.3 Chapter Objectives and Structure

This chapter explores how engineering geophysics supports the Sustainable Development Goals by aligning geophysical methods with targets related to water, energy, cities, and climate. It highlights the role of geophysics in promoting resilience, environmental justice, and ethical access to technology. The chapter also presents practical case studies demonstrating real-world applications in sustainable development. Structurally, it begins by linking geophysical methods to specific SDGs, then examines contributions to key goals, followed by discussions on resilience, geoethics, and global equity. Case studies, current challenges, and future directions are presented before concluding with key insights and recommendations.

10.2 Engineering Geophysics in the Context of the SDGs

10.2.1 Overview of Engineering Geophysics Applications

Engineering geophysics refers to the application of geophysical methods to solve problems related to infrastructure, environmental assessment, and resource management. Techniques such as seismic surveys, electrical resistivity tomography (ERT), ground-penetrating radar (GPR), magnetometry, and electromagnetic induction are widely used to non-invasively investigate subsurface conditions. These methods play a vital role in site characterization, groundwater exploration, geohazard assessment, waste management, and renewable energy development. Their effectiveness lies in their ability to provide spatially extensive and cost-effective data without the need for intrusive drilling, making them particularly suited for environmentally sensitive and resource-limited regions. As the global focus shifts toward sustainable development, engineering geophysics is increasingly recognized as a key enabler in achieving the UN Sustainable Development Goals (SDGs), especially in areas concerning water, energy, urban infrastructure, and climate resilience (Reynolds 2011; Butler 2005).

10.2.2 Alignment of Geophysical Techniques with Selected SDGs

SDG 6: Clean Water and Sanitation

Geophysical methods are essential in identifying, mapping, and monitoring groundwater resources, which are critical for ensuring access to clean water. Electrical resistivity and seismic refraction are widely used to delineate aquifers, assess recharge zones, and detect zones of contamination or salinity. These tools are particularly valuable in arid and semi-arid regions, where groundwater is often the only reliable water source. Moreover, time-lapse geophysical monitoring enables the assessment of groundwater over-extraction and pollution, facilitating sustainable water resource management. These applications directly support SDG 6 by promoting water availability, sustainable use, and pollution reduction (UNESCO 2022; Reynolds 2011).

SDG 11: Sustainable Cities and Communities

Urban geophysics is increasingly used in planning and developing sustainable and resilient urban infrastructure. Techniques such as GPR, microtremor surveys, and resistivity imaging help detect buried utilities, assess soil stability, and identify subsurface hazards like sinkholes or voids. This non-invasive approach minimizes construction risks, improves safety, and reduces costs, particularly in densely populated areas. Additionally, geophysical methods aid in the design of green infrastructure by providing data for sustainable drainage systems and climate-adaptive buildings. By enhancing the resilience and efficiency of urban development, geophysics directly contributes to SDG 11 (Cardarelli and Di Filippo 2009).

SDG 13: Climate Action

Engineering geophysics contributes to climate action by enabling the monitoring and mitigation of climate-related geohazards such as landslides, permafrost thaw, coastal erosion, and flood risk. For instance, electrical resistivity and seismic tomography are employed to assess slope stability, while GPR and electromagnetic methods monitor changes in permafrost or glacial ice thickness. Geophysical surveys are also instrumental in evaluating areas prone to subsidence due to drought or groundwater depletion. Furthermore, geophysical monitoring supports infrastructure adaptation in vulnerable regions, contributing to disaster risk reduction and climate resilience in alignment with SDG 13 (Hauck 2013; IPCC 2023).

SDG 7: Affordable and Clean Energy

Renewable energy development, particularly geothermal energy, relies heavily on geophysical exploration. Techniques such as magnetotellurics, resistivity surveys, and seismic imaging help locate geothermal reservoirs and characterize the subsurface for drilling feasibility. In addition, geophysics is used to assess subsurface conditions for solar and wind installations, ensuring stable foundations and efficient land use. These contributions make geophysics a key enabler of the energy transition, supporting the expansion of affordable and clean energy systems, especially in developing regions where geophysical surveys can guide investment and reduce exploratory risk. This aligns directly with SDG 7 (Noorollahi et al. 2007; DiPippo 2016).

10.3 Geophysical Contributions to Key SDGs

10.3.1 Water Resources Management (SDG 6)

Engineering geophysics plays a central role in the sustainable management of water resources, especially in regions where groundwater is the primary source of drinking water and agricultural supply. Techniques such as electrical resistivity tomography (ERT), seismic refraction, and electromagnetic (EM) methods are extensively used to locate aquifers, delineate groundwater boundaries, and detect zones of salinity or contamination. These methods allow for the non-invasive assessment of subsurface hydrogeology, contributing to efficient groundwater extraction and recharge monitoring. In addition, geophysical monitoring helps track changes in aquifer storage due to over-pumping or drought conditions. By enabling data-driven groundwater management and contamination control, geophysics directly supports SDG 6, which emphasizes universal access to safe water and sustainable water use (UNESCO 2022; Reynolds 2011).

10.3.2 Urban Resilience and Infrastructure (SDG 11)

Rapid urbanization presents challenges in ensuring safe, resilient, and sustainable infrastructure. Engineering geophysics supports urban resilience by providing critical data for subsurface characterization in built environments. Ground-penetrating radar (GPR), microtremor analysis, and electrical resistivity are commonly used to detect voids, sinkholes, buried utilities, and unstable soils, which are often hidden hazards beneath cities. These tools contribute to safer building site selection, underground utility mapping, and early detection of potential failures in transport and utility systems.

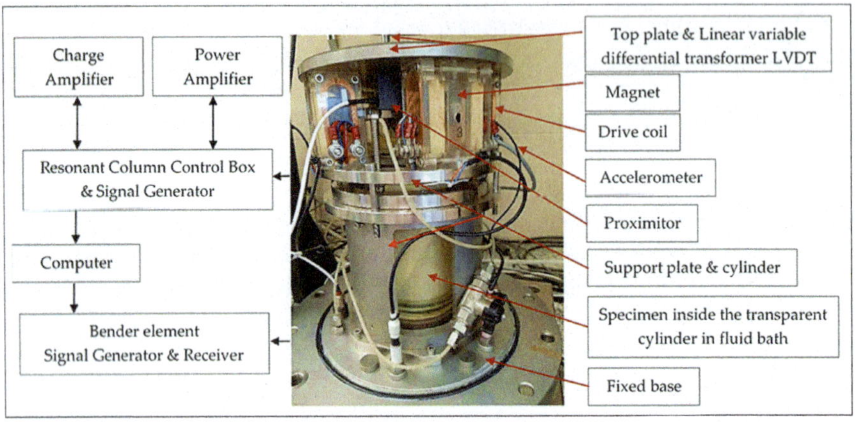

Fig. 10.1 Soil sample resonant column test setup used for evaluating the dynamic properties of soils in seismic and infrastructure risk assessments (*Source* Gabrys 2025—Open Access)

In addition to field-based surveys, laboratory-based geophysical methods such as the resonant column test are employed to evaluate the dynamic properties of soils, including shear modulus and damping ratio, which are vital parameters for earthquake-resistant design and structural stability assessments (Fig. 10.1). By integrating both in-situ and lab-based geophysical data, engineers and planners can better understand ground behavior under loading conditions, thereby improving the safety and resilience of urban infrastructure systems.

Furthermore, geophysical surveys can guide the integration of green infrastructure, such as permeable pavements and sustainable drainage systems, into urban planning. In doing so, engineering geophysics aligns with SDG 11 by contributing to safer, more sustainable urban communities (Cardarelli and Di Filippo 2009; Cigna et al. 2013).

10.3.3 Climate Monitoring and Mitigation (SDG 13)

Geophysical methods offer valuable tools for monitoring climate-sensitive environments and supporting climate action. For instance, electrical resistivity and seismic surveys are used to track changes in permafrost and glacial thickness, which are important indicators of climate change. Landslide-prone slopes can be monitored using geophysical early-warning systems based on resistivity and shear-wave velocity profiling, helping mitigate disasters exacerbated by extreme weather events. In coastal and flood-prone regions, geophysical assessments help evaluate the vulnerability of soils and infrastructure to erosion and inundation. Additionally, these techniques support the monitoring of carbon capture and storage (CCS) projects through time-lapse seismic imaging and EM surveys. Such contributions enable informed adaptation and mitigation strategies, thereby supporting SDG 13's call for urgent action on climate change (Hauck 2013; IPCC 2023).

10.3.4 Renewable and Clean Energy Exploration (SDG 7)

Achieving a sustainable energy future requires efficient exploration and development of renewable energy sources. Engineering geophysics facilitates this by enabling site investigations for geothermal, solar, and wind energy projects. In geothermal development, methods such as magnetotellurics, resistivity imaging, and seismic reflection are used to identify heat sources, fractures, and reservoir boundaries. These techniques reduce the risk and cost associated with drilling. In solar and wind projects, geophysical assessments of subsurface conditions help determine soil stability and load-bearing capacity, ensuring the structural integrity of energy infrastructure.

Conventional fossil-fuel-based energy systems, particularly coal-fired power plants, rely on environmental control technologies such as flue-gas desulfurization (FGD), electrostatic precipitators (ESP), and selective catalytic reduction (SCR) to reduce harmful emissions (Fig. 10.2). While these technologies mitigate some environmental damage, they do not eliminate the root problem of carbon emissions and resource depletion. This underscores the urgency of shifting toward cleaner alternatives. Geophysical tools play a pivotal role in this transition by identifying viable locations for renewable energy development, conducting environmental impact assessments, and supporting sustainable infrastructure design.

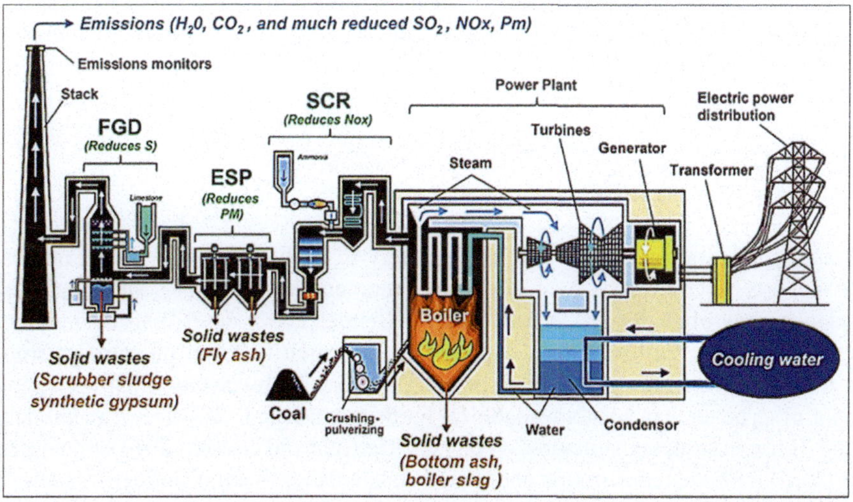

Fig. 10.2 Coal power plant flow diagram illustrating major pollution control components: FGD—Flue-gas desulfurization; ESP—Electrostatic precipitators; SCR—Selective catalytic reduction (*Source* Gabrys 2025—Open Access)

These applications support SDG 7 by enhancing access to affordable, reliable, and clean energy systems, particularly in emerging economies where sustainable energy is critical for development (Noorollahi et al. 2007; DiPippo 2016).

10.3.5 Innovation and Frontier Exploration (SDG 9, 13, 14)

Geoscientists do not only study the land on Earth—they also explore the oceans and other planets. These advanced fields contribute to several Sustainable Development Goals (SDGs), especially SDGs 13, 14, and 9.

Oceanography (SDG 13, 14)

Oceans cover about 71% of Earth's surface, and oceanography plays a key role in both climate action (SDG 13) and life below water (SDG 14). Ocean science is important for the environment, society, and the economy. Studies by von Schuckmann et al. (2020) and Ryabinin et al. (2019) highlight the role of ocean science in the SDGs and support the UN Decade of Ocean Science. Geoscientists study ocean circulation, which affects global climate. A key example is the Eocene–Oligocene opening of the Fram Strait—the only deep-water connection from the Arctic Ocean. This strait is part of the global ocean conveyor belt, a system that moves water and heat around the planet. Research in this area includes: Tectonic studies (Engen et al. 2008), Past ocean depths (paleobathymetry) (Straume et al. 2022), and Climate indicators (proxies) (Werner et al. 2016). Recently, geoscientists have also been exploring the seafloor for critical minerals such as manganese nodules, seafloor sulfides, and cobalt-rich crusts. For example, in 2024, Norway opened part of its Exclusive Economic Zone for such exploration. But sustainable management is needed (Ellefmo et al. 2023).

Planetary Geology (SDG 9)

Planetary geology is a field that uses geology to study other planets and moons. It supports SDG 9, which focuses on innovation and infrastructure. This branch of science grew after the Space Age. Marvin Ursula (2002) describes how geology expanded from Earth to space. Much of the research is based on samples from astronauts, space missions, or meteorites. Planetary geologists use high-resolution data, often better than what's available for Earth. Schultz et al. (2010) reviews this data and its use in studying structures on planets. Nab and van Gasselt (2014) emphasize the need for better cooperation and shared standards for making geological maps of planets like Mars and the Moon.

Planetary geology encourages innovation. For example, the Arctic Mars Analog Svalbard Expedition (AMASE) expeditions in Svalbard, Arctic brought together astronauts, engineers, and scientists to prepare for missions to Mars (Stern et al.

2013). Some Earth areas are even protected because they resemble other planets. One example is the Lanzarote Geopark in Spain, which space missions use to test equipment (Martínez-Frias et al. 2017). Private space companies are growing fast. They provide satellite data that supports many SDGs. But building and launching spacecraft has an environmental cost. This has been called the space sustainability paradox (Wilson and Vasile 2023).

10.3.6 Geohazards and Risk Reduction (SDG 1, 9, 11, 13, 14, 15)

One of the most significant ways geosciences benefit society is by helping us understand and manage geohazards. These include natural threats like earthquakes, volcanoes, landslides, and floods. Geoscientists study the processes that cause these hazards, and they also work on ways to predict, monitor, and reduce their impacts.

From 1970 to 2019, the number of natural disasters increased significantly, a trend likely to continue due to climate change (Van Aalst 2006). However, the number of deaths from geohazards has gone down from 1900 to 2020, even though the world's population has grown (Fig. 10.3). This success is largely due to advancements in geoscience. Still, about 10,000–20,000 people die each year from geohazards (Fig. 10.4). Efforts must continue, especially through initiatives like the Sendai Framework for Disaster Risk Reduction (SFDRR) (Aitsi-Selmi et al. 2015), which promotes risk monitoring, public education, and hazard mitigation.

In terms of human impact, Fig. 10.4 visualizes disaster-related mortality across the last century. The timeline shows the estimated annual death tolls from major disasters

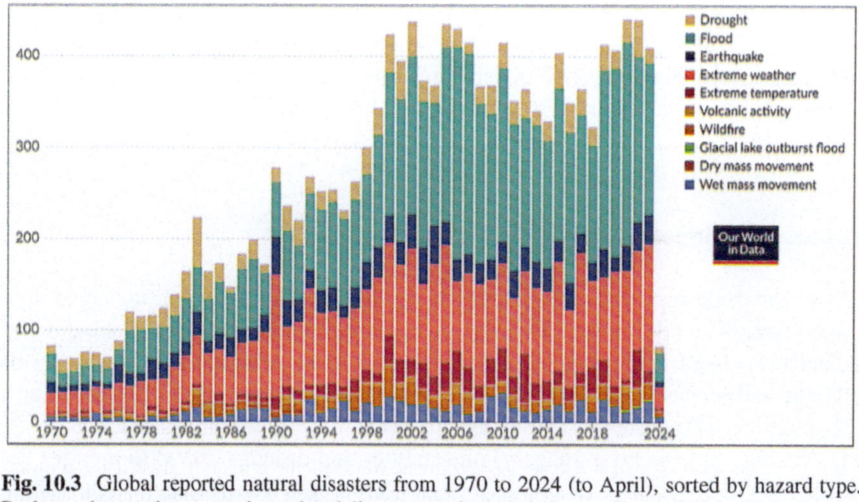

Fig. 10.3 Global reported natural disasters from 1970 to 2024 (to April), sorted by hazard type. Both weather- and non-weather-related disasters are included (*Source* Senger 2024—Open Access)

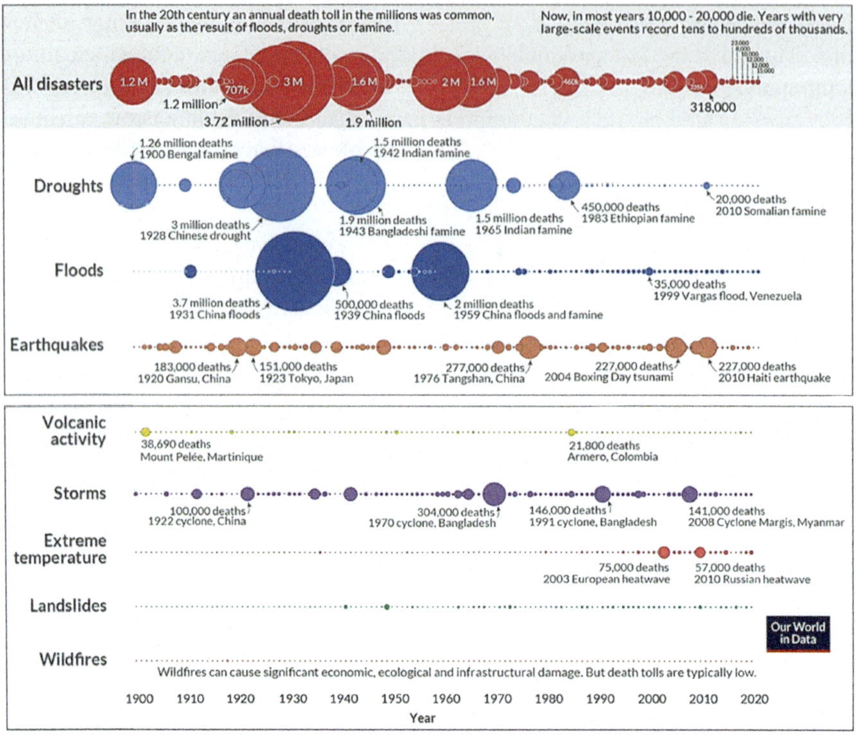

Fig. 10.4 Timeline of global disaster-related deaths from 1900 to 2020, with bubble sizes indicating the estimated annual death toll. This visual illustrates long-term mortality trends from natural disasters and emphasizes the need for proactive risk reduction strategies (*Source* Senger 2024—Open Access)

between 1900 and 2020, with bubble size representing the magnitude of each year's toll. This historical perspective underscores the value of geophysical interventions in saving lives—particularly when integrated into preparedness and early warning systems.

Geohazard Mitigation (SDG 1, 9, 11, 13, 14, 15)

Geoscientists have played a key role in saving lives by improving our understanding of geohazards and developing better ways to manage them. Geohazards are directly related to six SDGs, including those targeting poverty reduction, infrastructure development, and ecosystem protection. Modern geohazard management often uses Earth Observation (EO) systems. These systems use a variety of sensors (such as SAR, LiDAR, and multispectral imagery) on satellites and drones (UAVs) to monitor the Earth's surface over time. Examples include: A *rock and ice avalanche* in the Indian Himalaya was studied using both satellite and field data (Shugar et al. 2021), *GIS analysis* of a Chinese railway showed how geohazards affect infrastructure in areas with changing permafrost (Zheng et al. 2021), and *flood risk assessments* using terrain analysis helped reduce disaster risk in arid regions (Mansour et al. 2022). The use of drones (UAVs) has grown rapidly. They are now used in all phases of geohazard work—from identifying risks to managing emergencies. Antoine et al. (2020) provide case studies showing how UAVs help, and they suggest adding AI and deep learning to improve data analysis. At the national level, New Zealand has tried to combine the SFDRR, SDGs, and the Paris Agreement in its laws and planning—though results have been mixed (Saunders et al. 2020).

Volcanology (SDG 1, 9, 11)

Volcanoes can cause major damage to people and infrastructure. They are directly connected to SDGs 1 (No Poverty), 9 (Infrastructure), and 11 (Sustainable Cities). Predicting volcanic eruptions depends on monitoring networks. Communities near volcanoes—such as those in the Philippines—must be resilient to frequent activity (German et al. 2022).

Despite the danger, volcanoes also offer benefits. Ash from eruptions enriches soil and boosts carbon storage (Tonneijck et al. 2010). Many UNESCO Global Geoparks are based around volcanic landscapes (Casadevall et al. 2019). Kelman and Mather (2008) found that people living near volcanoes can adapt and build sustainable livelihoods, taking advantage of fertile land and stable water supplies while managing the risks.

Seismology (SDG 1, 9, 11)

Earthquakes are among the most destructive natural hazards. Their impacts are closely related to poverty, infrastructure resilience, and urban safety. Reducing earthquake risk depends on strong building codes and earthquake-resistant design (Takagi and Wada 2019). The February 2023 earthquake in Turkey and Syria caused massive damage and was one of the deadliest in recent years (Dal Zilio and Ampuero 2023). Satellite images even showed visible changes in night lights after the event (Levin 2023). Examples of resilience include: In *New Zealand*, land-use planning after the 2010–2011 earthquakes improved sustainability and recovery (Saunders and Becker 2015), after the *2021 Haiti earthquake*, citizens helped build a low-cost seismic network to track aftershocks (Calais et al. 2022), and in *Nepal*, frequent earthquakes threaten tourism—the country's main economic sector (Min et al. 2020; Birendra et al. 2021).

Geomorphology (SDG 9, 11, 13)

Geomorphology is the study of Earth's surface features and processes. It helps in hazard mapping, infrastructure planning, and climate adaptation. Mapping landforms is essential for understanding flood risks, landslides, rockfalls, and volcanic activity. A special issue edited by Chelli et al. (2021) includes many case studies. For instance: De Beni et al. (2021) used drones for time-lapse imaging of *Mount Etna's 2019 eruption*, and Cignetti et al. (2021) studied *rockfall hazards* in Italy's Aosta Valley and created detailed *risk maps*. Geomorphology gives planners the tools to make better decisions about land use, infrastructure, and disaster readiness—key to meeting SDGs related to cities, infrastructure, and climate.

10.3.7 Near-Surface Geosciences for Sustainable Development (SDG 1, 8, 9, 11, 13, 14, 15)

Engineering geology has long been a central discipline within the geosciences, contributing significantly to infrastructure development such as roads, railways, canals, and artificial islands. It also plays a key role in hazard mitigation and, more recently, in addressing the impacts of climate change. Most engineering projects involve work at or near the Earth's surface, typically within the top 500 m. To avoid cost overruns and delays, geoscientists must understand subsurface conditions, geological history, and potential hazards. As highlighted by Shilston (2023) in his 21st Glossop Lecture, such knowledge is essential for planning and decision-making.

Engineering Geology (SDG 1, 8, 9, 11)

Engineering geology directly contributes to SDGs 1 (No Poverty), 8 (Decent Work and Economic Growth), 9 (Industry, Innovation and Infrastructure), and 11 (Sustainable Cities and Communities). Lagesse et al. (2022) evaluated the current and potential contributions of engineering geologists to the SDGs, emphasizing areas such as developing ground models, improving construction efficiency, and designing more resilient infrastructure (Fig. 10.5). As urban populations grow, demand for underground infrastructure—such as basements, tunnels, and utility pipelines—increases. Understanding the shallow subsurface becomes essential. For instance, Bidarmaghz et al. (2019) demonstrated how hydrogeological conditions influence urban subsurface temperatures due to heated basements, highlighting the importance of geology in urban planning.

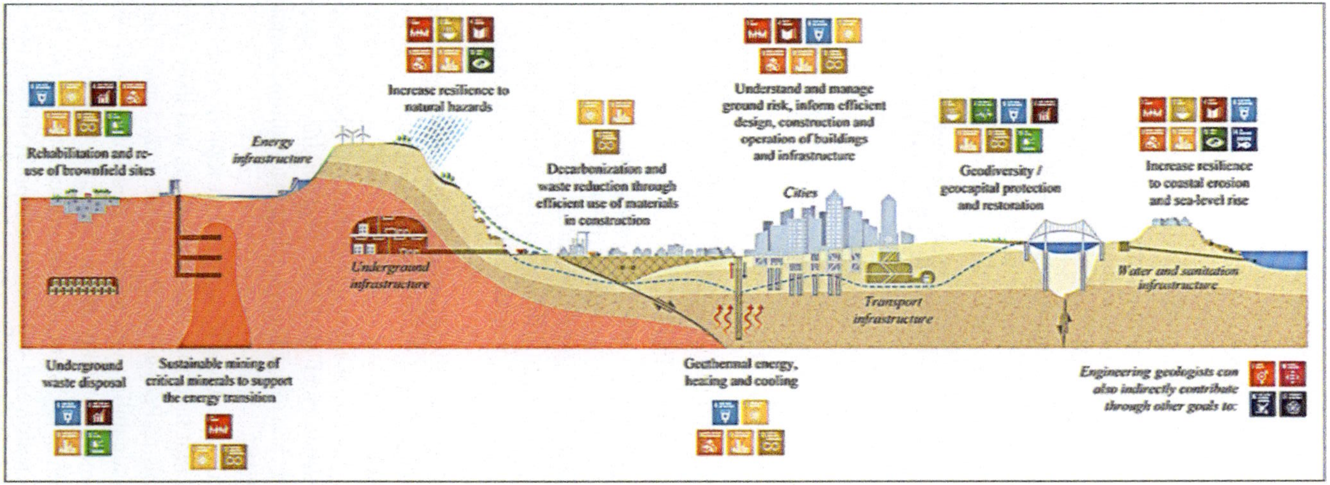

Fig. 10.5 Conceptual diagram illustrating the alignment between engineering geology and the United Nations Sustainable Development Goals (SDGs). The figure highlights areas where geological science contributes to hazard mitigation, infrastructure development, environmental protection, and sustainable resource use (*Source* Senger 2024—Open Access)

Environmental Geology and Landscape Protection (SDG 11, 13, 14, 15)

Environmental geology and landscape protection are closely linked to SDGs 11 (Sustainable Cities), 13 (Climate Action), 14 (Life Below Water), and 15 (Life on Land). These fields focus on understanding and managing the interactions between geological processes and human activities. One important tool is satellite remote sensing, which has been used to monitor environmental phenomena such as wildfires (Wei et al. 2021). Although wildfires are not explicitly mentioned in the SDG targets, they affect 3–4% of Earth's surface each year and have both harmful and beneficial effects (Martin 2019). Fires can lead to loss of life, desertification, and increased carbon emissions, but they can also improve soil fertility and create new land for use. Thus, they influence multiple SDGs, including 1, 2, 3, 6, 13, 14, and 15.

Landscape protection also involves conserving geological heritage and promoting sustainable land use for future generations. As Attard (2019) illustrated using the Maltese Islands, protecting landscapes has long-term social and ecological value. Geoheritage efforts, such as the establishment of geoparks and promotion of geotourism, further contribute to awareness and conservation (Gordon et al. 2021).

Erosion Management (SDG 9, 11, 13)

Erosion management is another critical area in near-surface geoscience, contributing to SDGs 9 (Infrastructure), 11 (Sustainable Cities), and 13 (Climate Action). Erosion affects mountain regions, riverbanks, and coastlines, threatening infrastructure and ecosystems. Schipper et al. (2021) examined how coastal development can be made more climate-resilient by aligning it with the SDG framework. Their study included five diverse case studies from different coastal environments. In Bangladesh, Chow (2018) showed that mangrove plantations help stabilize coastlines and protect communities during storm surges.

On a broader scale, Panagos and Katsoyiannis (2019) edited a special issue on soil erosion modeling in Europe, where land degradation remains a major concern. The European Union has targeted zero land degradation by 2030 (Keesstra et al. 2016), yet large areas remain at risk. Innovative approaches include the use of radiogenic soil erosion tracers (Meusburger et al. 2018), post-wildfire soil recovery studies (Fernandez and Vega 2018), and integrated modeling of terrain and climate conditions (Bednar and Sarapatka 2018).

Soil erosion also directly affects agriculture and food production. McElwee et al. (2020) examined 40 different land management options and their effects on SDGs and Nature's Contributions to People (NCPs). Their study found that while many interventions support sustainability with minimal trade-offs, some—such as large-scale afforestation or bioenergy plantations—can negatively impact ecosystems and communities. These findings highlight the importance of integrated and balanced approaches in managing erosion and soil health to support sustainable development.

10.3.8 Minerals and Rock Materials for Sustainable Development (SDGs 1, 2, 7, 8, 11, 12, 13)

Modern society increasingly depends on minerals and rock materials, especially for renewable energy, technology, and infrastructure. As the demand for clean energy and electronics grows, the need for a wider range of minerals continues to rise (Fig. 10.6).

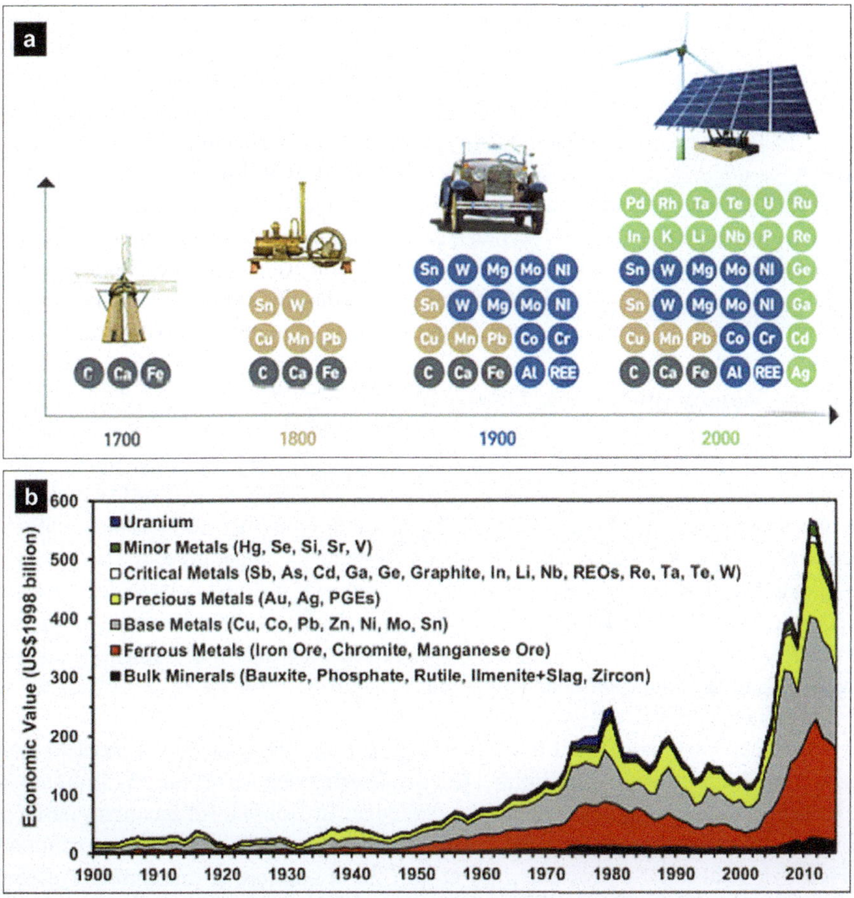

Fig. 10.6 Changes in society's use of minerals over time and global mineral production trends. **a** Increase in mineral use by society over time, highlighting the growing reliance on diverse mineral resources in modern times. **b** Inflation-adjusted global value (in 1998 US dollars) of metal and mineral production by major sector, from 1900 to 2015 (*Source* Senger 2024—Open Access)

Mining and Mineral Resources (SDGs 1, 2, 7, 8, 11, 12, 13)

Mining has long been one of the largest employers of geoscientists. It provides the raw materials used for energy, construction, agriculture, and technology. Mining operations vary from large open-pit mines and offshore platforms to small pits that extract materials like sand, gravel, and crushed stone (Bendixen et al. 2021).

In the past, mining was often seen as harmful to the environment and communities—due to child labor, pollution, and large waste piles. However, this image is changing. Today, many mining companies work to meet sustainability goals. They map their activities to the SDGs and aim to reduce harm to people and the environment (Mudd 2021).

Despite progress, challenges remain. The demand for mined minerals continues to grow. High-grade ores are being depleted, forcing companies to process lower-grade materials, which creates more waste and uses more water and energy. Geoscientists now not only search for new mineral resources, but also help improve production from existing mines. They are also involved in developing sustainable mining practices and supporting policies like the Kimberley Process, which certifies diamonds as conflict-free.

Critical Minerals (SDGs 1, 2, 7, 8, 11, 12, 13)

Critical minerals are essential for clean energy technologies. These include lithium, copper, cobalt, nickel, and rare earth elements (REEs), which are used in batteries, wind turbines, electric vehicles, and other technologies (Fig. 10.7). The supply of these minerals is uneven around the world and depends on geology.

For example, most rare earth elements come from carbonatite-related deposits and special clay types (Simandl 2014; Dushyantha et al. 2020). China currently produces over 90% of global REEs. Because these minerals are also important for national security and economic independence, many countries are working to develop their own resources.

One such project is the EU-funded *GreenPEG*, which develops tools to find pegmatite-hosted deposits in Europe. But discovering deposits is just the first step—extracting them sustainably is another challenge. In Bolivia, for example, partnerships between public and private sectors have helped manage lithium extraction (Hancock et al. 2018). However, Ali et al. (2017) emphasize that stronger global governance is needed to make sure these minerals are produced responsibly.

Battery Technology and Energy Storage (SDGs 7, 11, 12, 13)

Battery production is central to the energy transition, especially for electric vehicles and renewable energy systems. Electric cars, for example, use up to five times more minerals than traditional cars (Fig. 10.7). They rely on lithium, cobalt, nickel,

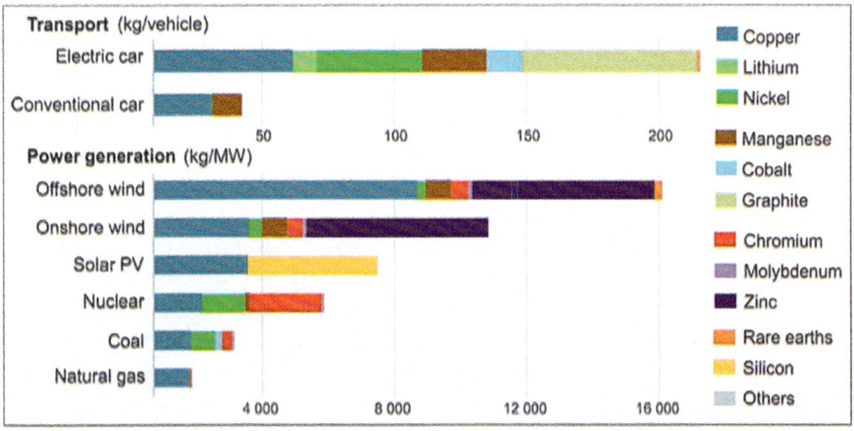

Fig. 10.7 Minerals required in the transport and power generation sectors, including selected green energy technologies, nuclear energy, and fossil fuels. The figure illustrates the dependency of emerging energy systems on a wide range of critical minerals, emphasizing the role of geophysical methods in their sustainable discovery and management (*Source* Senger 2024—Open Access)

manganese, and graphite—minerals that are becoming more difficult to source sustainably (Herrington 2021).

The demand for these materials is expected to grow by 500% in the coming decades. While some materials can be recycled or recovered from waste, most are still mined on land. New interest is growing in deep-sea mining for cobalt and manganese, but this is controversial. Some researchers argue that seabed mining could supply needed materials; others warn about unknown environmental risks (Paulikas et al. 2022; Beaulieu et al. 2017).

Deep-sea mining along Norway's Mid-Atlantic Ridge has recently begun and may provide new data on how to extract resources while protecting marine ecosystems. Regardless of source, it is critical to use existing resources wisely and improve energy efficiency. Energy storage also plays a key role in ensuring clean and reliable electricity. Kyriakopoulos and Arabatzis (2016) describe emerging storage technologies, including compressed air energy systems that store energy underground. These systems can help meet SDG 7 (affordable and clean energy) while making better use of the subsurface—a field where geoscientists are vital.

10.3.9 Energy: Extraction and Storage for Sustainable Development (SDGs 1, 7, 8, 9, 11, 12, 13)

Geoscientists play an important role in the energy sector. They help with finding and producing energy resources like oil, gas, geothermal energy, and also work with renewable energy and energy storage. Their work is essential for the transition to cleaner and more sustainable energy systems. Energy connects strongly to many SDGs, especially SDG 7 (Affordable and Clean Energy), but also impacts others.

Geothermal Energy (SDGs 1, 7, 8, 9, 11, 12, 13)

Geothermal energy comes from heat inside the Earth. It is a clean and renewable energy source. By 2050, geothermal energy could provide 800–1300 TWh of electricity every year, which would be about 2%-3% of the world's electricity (van der Zwaan and Dalla Longa 2019). Geothermal energy uses similar technology and skills as the oil and gas sector, making it attractive for the energy transition. It has a low environmental impact, but deep geothermal projects can cause small earthquakes, especially during water injection (Rybach 2003).

Geothermal systems can be used for electricity (deep sources) and for heating (shallow systems). Ground-source heat pumps are a good example of how geothermal energy can heat homes with low emissions. However, geothermal projects need careful planning because they interact closely with subsurface systems. Local studies and stakeholder input are important for assessing the sustainability of these projects.

Hydropower (SDGs 1, 7, 8, 9, 11, 12, 13)

Hydropower generates electricity from flowing water. It is one of the oldest and most reliable renewable energy sources. In some countries like Norway and Switzerland, it provides most of the electricity. In others like China and Turkey, hydropower is growing fast (Liu et al. 2013). Geoscientists help choose safe and efficient locations for dams. These projects are large and complex and must be designed carefully to avoid environmental harm and social impacts like forced resettlement. Hydropower reservoirs also store fresh water, but they are at risk from pollution and biodiversity loss. Monitoring and sustainable management are key for long-term success.

Nuclear Energy (SDGs 1, 7, 8, 9, 11, 12, 13)

Nuclear energy is low in carbon emissions and can generate a lot of electricity with small amounts of fuel. However, it is controversial because of safety concerns and radioactive waste. Geoscientists explore for uranium (nuclear fuel) and help plan safe storage for nuclear waste. As uranium ore quality declines, mining becomes less sustainable due to rising energy and water use (Mudd and Diesendorf 2008). Nuclear power is especially useful in countries that need affordable and reliable energy without increasing carbon emissions (y Leon and Lindberg 2022). However, nuclear waste disposal remains a big challenge.

Renewable Energy Resources (SDGs 1, 7, 8, 9, 11, 12, 13)

Renewable energy sources like wind, solar, wave, and biomass are essential for the future. In 2019, renewables (including nuclear) made up just 16% of the global energy mix, but this is expected to rise. Studies show that using more renewable energy helps with sustainability (Guney 2019). However, in many developing countries, high costs and low credit ratings slow down the transition. Geoscientists support renewable energy projects by providing data for building foundations, especially for offshore wind farms. They also help assess environmental impacts and find materials like critical minerals needed for green technologies. Offshore energy projects face more challenges from regulations than from technical issues. Clear policies and support are needed for faster development.

Energy Storage (Gas, Hydrogen, Compressed Air) (SDGs 1, 7, 8, 11, 12, 13)

Storing energy is important because renewable sources like wind and solar are not always available. Subsurface storage (underground) is cost-effective and safe (Tarkowski 2019). There are already over 640 underground gas storage sites worldwide. Hydrogen is a promising future fuel. It can be stored underground in aquifers, salt caverns, or old gas fields. The color of hydrogen (green, blue, grey) refers to how it is made. Green hydrogen is the cleanest, made using renewable energy (Fig. 10.8a). Storing hydrogen is more difficult than gas because it can leak more easily. More research is needed on how it interacts with rocks underground. Some sites in the UK and USA have stored hydrogen since the 1970s. Compressed air energy storage is another method. It uses underground spaces like salt caverns to store pressurized air, which can later generate electricity.

Carbon Capture and Storage (CCS) (SDGs 1, 7, 11, 13)

CCS is a method to reduce CO_2 emissions. It captures carbon from power plants or factories and stores it deep underground (Eiken et al. 2011). This technology is key for meeting the Paris climate goals. CCS has been in use since 1996 in Norway and is expanding in Europe through projects like Northern Lights and Greensand. Geoscientists play a major role in selecting and monitoring storage sites. CCS is technically ready, but public acceptance and the large scale needed remain challenges. It uses more energy than regular systems, but overall, it supports climate action and clean energy goals.

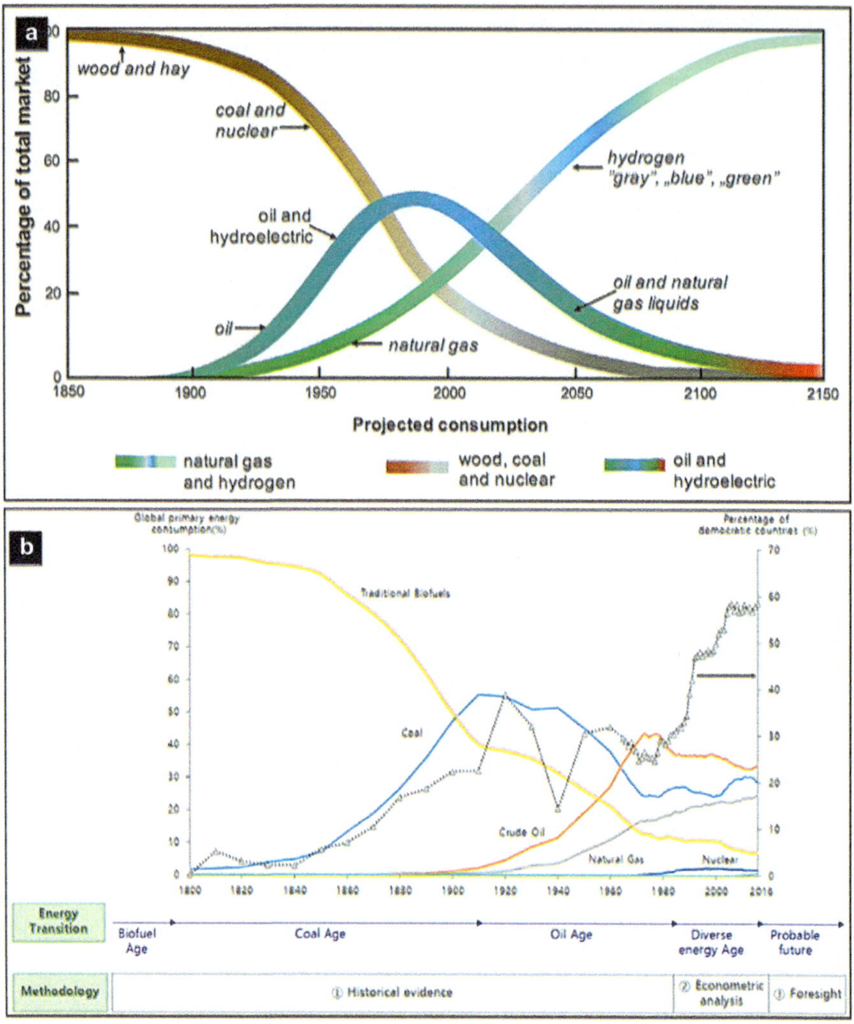

Fig. 10.8 Global energy mix development. **a** Global energy consumption from 1850, with projection to 2150. **b** Timeline illustrating the major phases of the energy transition from 1800 to 2016, alongside the global rise of democracy (*Source* Senger 2024—Open Access)

Hydrocarbons and Transition Fuels (SDGs 1, 7, 8, 11, 12)

Despite the shift to renewables, oil, gas, and coal still supply over 80% of global energy (Fig. 10.8b). Geoscientists are heavily involved in discovering and producing these resources. The fossil fuel era brought industrial growth and modern living standards but also caused environmental damage and social issues (Verheyen et al. 2016). Problems include oil spills, pollution, and conflict over resources. Now, many oil companies are working toward more sustainable practices. Some are using CCS and other technologies to reduce emissions during production. Hydrocarbons will remain part of the energy mix for years, especially in regions where renewable infrastructure is still developing. The focus is on making fossil fuel use cleaner and more efficient.

10.3.10 Hydrogeology, Waste, and Contamination

Exploring and securing access to clean water is one of humanity's greatest needs. Understanding how groundwater, surface water, and human-made pollution interact is crucial. This requires geoscientific knowledge; especially as global water pressures increase.

Hydrogeology (SDGs 1, 3, 5, 6, 11, 13)

Groundwater provides nearly 98% of the world's fresh water and is the main source of drinking water for about half of the global population (Klove et al. 2011). It plays a major role in food security through irrigation, supports hygiene and sanitation, helps during droughts, and contributes to climate change adaptation (Velis et al. 2017). However, overuse or poor management can lead to groundwater depletion and pollution. Aquifer properties—like porosity, permeability, depth, and recharge rates—determine their vulnerability to decline (Velis et al. 2017).

The idea of *socio-hydrology* looks at how human and water systems affect each other over time (Di Baldassarre et al. 2019). To manage water sustainably, strong policies and proper monitoring are needed, especially in data-scarce regions. Remote sensing from satellites helps fill data gaps. When combined with ground-based data and modeling, it improves water resource management, as shown in the Himalayas (Aggarwal et al. 2020). Sometimes, old data from oil or mineral exploration can be reused for groundwater studies. For example, in Ghana, geoscientists used subsurface and geophysical data to model the Nasia basin, helping locate groundwater sources for farming (Dzikunoo et al. 2020). Figure 10.9 illustrates the major water-related issues caused by unsustainable water use at different scales.

Geological Disposal of Radioactive Waste (SDGs 7, 9, 11)

Nuclear energy provides low-carbon electricity, but disposing of radioactive waste safely is a major challenge. This links directly to SDGs 7, 9, and 11. Finland's Onkalo facility is the world's first permanent geological storage site for nuclear waste, operating since 2022 (El-Showk 2022). Other countries rely on temporary storage, recycling, and long-term planning (Kurniawan et al. 2022). Suitable rocks for waste disposal include granite, clay-rich sediments, evaporites, and volcanic tuffs. These rocks must allow *low fluid movement* and remain geochemically stable

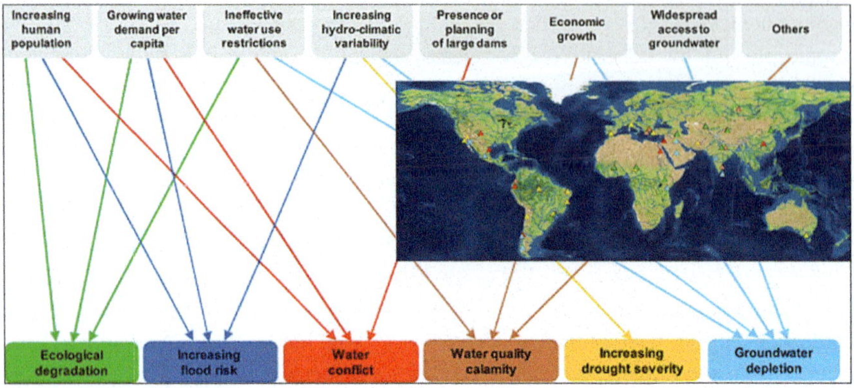

Fig. 10.9 Key factors influencing global water crises and hotspot locations. The figure shows an overview of the social, technical, and hydrological factors linked to six major types of water crises. The inset map highlights global regions most affected by these crises (*Source* Senger 2024—Open Access)

over thousands of years (Chapman and Hooper 2012). Other risks like earthquakes, glaciation, or sea level changes must also be considered. Besides the geology, technical materials such as canisters, barriers, and buffers are also essential for long-term safety (Kim et al. 2011).

Contaminated Land (SDGs 1, 3, 6, 11, 13, 14, 15)

Contaminated land can harm ecosystems, water supplies, and human health. It is closely tied to several SDGs including clean water, good health, and sustainable cities. Contamination may come from natural (geogenic) sources or from human activities (anthropogenic), such as industrial waste or poor land management. In the Indo-Bangladesh delta, high arsenic and fluoride levels have harmed water quality and agriculture in more than half the area (Ruidas et al. 2024). Smart waste management is one solution. In Indonesia, some cities have adopted Industry 4.0 technologies and circular economy principles to reduce pollution (Fatimah et al. 2020).

Environmental Geochemistry (SDGs 1, 3, 6, 11, 13, 14, 15)

Environmental geochemistry studies the chemical makeup of soils, rocks, and water near the surface. It's important for pollution control, land health, and understanding human impact on the environment. Plant et al. (2001) stressed the need for a global geochemical baseline to compare pollution levels. Two decades later, such surveys are still limited to specific regions or countries. Examples include geochemical mapping in India (Govil et al. 2020), China (Wang et al. 2022), Papua New Guinea (Tiangang et al. 2024), and Egypt (Mostafa et al. 2023).

Baseline levels vary due to geology, weathering, and sampling methods. A study comparing Finland and Lithuania showed differences caused by both the rocks and how samples were collected (Salminen and Gregorauskien 2000). Smith et al. (2023) reviewed major regional geochemical projects across Europe and North America, which provide valuable models for future work.

10.3.11 Climate Science and Earth System Change

Geoscientists play a vital role in understanding Earth's past climate changes and using that knowledge to help society prepare for future changes. This knowledge is key to predicting climate trends and developing climate adaptation strategies. The Intergovernmental Panel on Climate Change (IPCC 2021) emphasizes that natural climate records, such as those found in rocks, ice, and sediments, are essential for understanding Earth's long-term climate system.

The IPCC also highlights the importance of adapting to climate change and using insights from the geological past to inform future scenarios.

Glaciology (SDGs 11, 13, 14, 15)

Glaciology is the study of glaciers and their impact on the environment. Glaciers are important sources of freshwater for millions of people and are strongly affected by climate change. They are also linked to natural hazards like avalanches and floods. This area of study is connected to SDGs 11, 13, 14, and 15. Nussbaumer et al. (2017) discussed global glacier monitoring through the Global Terrestrial Network for Glaciers (GTN-G). Long-term data from glaciers helps in managing water supplies, especially in mountain regions like the Andes (Gomez et al. 2022) and the Himalayas (Sharma et al. 2021).

Monitoring is done through satellite data and fieldwork. This data also helps understand glacier-related hazards. For example, a large rock and ice avalanche in the Indian Himalayas in 2021 was studied by Shugar et al. (2021), highlighting the risks linked to melting glaciers.

Paleoclimatology (SDGs 11, 13, 14, 15)

Paleoclimatology is the study of Earth's past climates using natural records such as sediments, ice cores, and fossils. It is essential for understanding long-term climate trends and is tied to SDGs 11, 13, 14, and 15. The geological record is the only source of information on climate changes over millions of years (Burke et al., 1990). This record helps scientists improve climate models and predict future conditions more accurately. Lear et al. (2021) explain how Earth's climate history supports sustainable development goals and informs policy decisions. Crowley (2000) compared recent warming trends to the last 1,000 years and found that the recent rise in temperatures is outside the range of natural variation.

However, interpreting past climate is challenging. The further back in time we go, the more uncertain the data becomes. Paleoclimate studies require understanding interactions between land, ocean, and atmosphere systems. For example, Varotsos et al. (2020) analyzed peat deposits from European Russia to study Holocene climate (last 10,000 years) using three indicators: water table depth, peat humification, and moisture index.

Palaeontology (SDGs 13, 14, 15)

Palaeontology is the study of the history of life on Earth through fossils. It supports SDGs 13, 14, and 15 by helping us understand how life evolved and responded to past environmental changes. Fossil records show how species appeared, evolved, or went extinct over millions of years (Benton and Donoghue 2006). This information is vital today, as Earth faces one of the fastest periods of climate change in its history. Palaeontologists also play an important role in the energy transition. Davies and Simmons (2020) describe how fossil experts contribute to identifying suitable rocks for CO_2 storage. Their understanding of ancient Earth systems makes them key advisors for managing environmental challenges.

10.3.12 Geoheritage, Public Engagement, and Education for the SDGs

Geoscientists help society understand Earth's history, natural processes, and the importance of geology in everyday life. This is often done through outreach programs, museums, and special geological sites called Geoheritage sites.

Science Outreach and Communication (SDGs 4, 5, 7, 10, 13, 14, 15)

Communicating scientific research to the public is a key part of geoscience work. This supports many Sustainable Development Goals (SDGs 4, 5, 7, 10, 13, 14, 15) by making knowledge more accessible. According to Stewart and Hurth (2021), effective communication goes beyond publishing scientific papers. They encourage geoscientists to use marketing techniques and co-creation to better engage with the public. Geoscientists can help people understand important topics like climate change, natural hazards, and energy transitions (Fig. 10.10).

There are many creative ways to reach people. For example: The *Climate Explorer* app helps users analyze past climate data (Trouet and Van Oldenborgh 2013). The *Virtual Climate Scientist* platform lets users explore Antarctic research in virtual

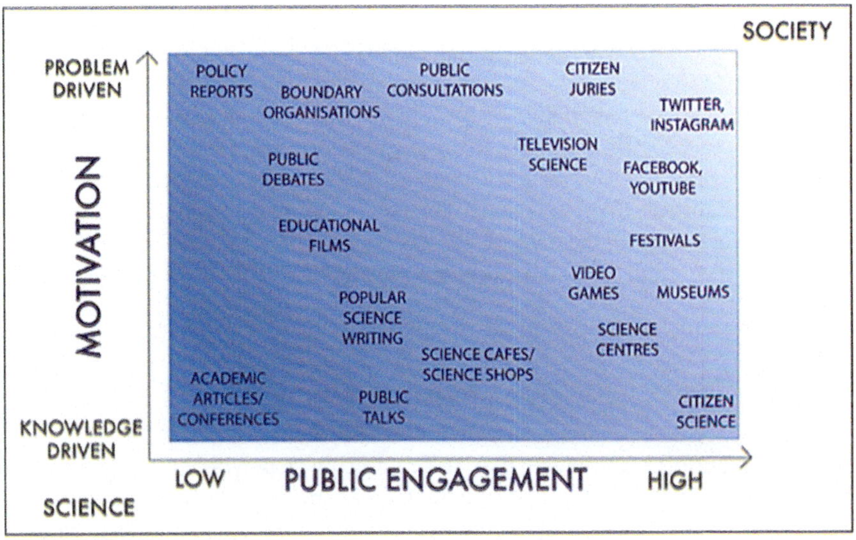

Fig. 10.10 Visual framework of the science communication landscape (*Source* Senger 2024—Open Access)

reality (Pimentel and Kalyanaraman 2021). Virtual Reality (VR) is also used in classrooms to teach geoscience and involve people from different backgrounds (Horota et al. 2022a).

Museum Curation (SDGs 4, 5, 7, 10, 13, 14, 15)

Natural history museums also support the SDGs by preserving and displaying important geological collections (SDGs 4, 5, 7, 10, 13, 14, 15). These museums store rare and valuable samples like fossils and rock cores. For example, Nakrem et al. (2023) describe collections in the Arctic at the Natural History Museum in Oslo, including samples from areas that are no longer accessible, such as Novaya Zemlya. Museums are not just about the past—they now also focus on helping people think about sustainable futures. Lanzinger and Garlandini (2019) discuss how modern museums show both Earth's history and possible climate solutions.

Geoheritage and Geotourism (SDGs 4, 13, 14, 15)

Geoheritage refers to special geological features that are part of our natural and cultural heritage. This supports SDGs 4, 13, 14, and 15. Gordon (2019) explains that these sites are important for conservation and public education. UNESCO Global Geoparks protect and manage these areas with a focus on education and sustainable development.

Some examples: The *Lanzarote and Chinijo Islands Geopark* uses geotourism to promote sustainability (Martinez-Frias et al. 2017). Catana and Brilha (2020) show how UNESCO Geoparks help teach students about sustainability. Geoheritage also includes the data and materials collected by geoscientists. Musèums and drill core repositories store fossils and cores from scientific drilling. For instance: The USA has about 50 core repositories (Arends et al. 2021). Drill cores are important for studying past climate changes (Planavsky et al. 2020), but storage and maintenance are expensive. Recently, the entire U.S. Antarctic marine core collection had to be moved due to cost and space issues (Witze 2016).

10.3.13 Higher Education and Research

Higher education and research play a key role in achieving many Sustainable Development Goals (SDGs). Universities help by training the next generation of sustainability experts and by doing focused research to support progress toward the SDGs.

Geoscience Research and the SDGs (SDG 1–17)

Geoscience research can support all 17 SDGs. It is important to include geoscientific knowledge in SDG-related research and work together with other disciplines, especially social sciences. Scown (2020) notes that although research on the SDGs has increased a lot since 2015, only a small part of it is about geoscience. The Earth system is very complex, and geoscientists are used to thinking across long timescales and big regions—skills that are useful for making long-term policies (Scown 2020).

Geoscience Education and the SDGs (SDG 1, 4, 5, 13)

Teaching geoscience supports SDGs 1 (No Poverty), 4 (Quality Education), 5 (Gender Equality), and 13 (Climate Action). Avelar et al. (2019) review how education connects to the SDGs, but only one paper focused on geoscience education (Jessell et al. 2018). Annan-Diab and Molinari (2017) argue that teaching about the SDGs should include many subjects and should involve both public and private sectors. The SDGs are all linked, so education also needs to be connected across fields—not only for students but also for people already working, especially company managers who can help align business with SDGs.

Many universities already offer sustainability-related courses or link their courses to the SDGs. For example, Palsdottir and Johannsdottir (2021) show how the University of Iceland included the SDGs in their teaching. Almazroa et al. (2022) also show how teacher training programs include the SDGs in higher education.

Digital Tools in Geoscience Education

New digital tools are creating more ways to teach and share geoscience. For example, Zapata-Paulini et al. (2023) created a mobile augmented reality app to show glacial retreat in the Andes Mountains. Other tools include digital outcrop models, photospheres, and combining geoscience data. These tools make learning easier and more accessible. Virtual field trips have allowed more people to experience geoscience fieldwork (Mead et al. 2019; Whitmeyer et al. 2020). These virtual trips became even more common during the COVID-19 pandemic (Whitmeyer and Dordevic 2020; Pugsley et al. 2022).

In the High Arctic, the University Centre in Svalbard has used these tools since 2016. They use digital outcrop models and drone photos to support field education (Senger et al. 2021; Horota et al. 2022b, 2024; Betlem et al. 2023).

10.3.14 Data and Skill Sets

To help achieve the Sustainable Development Goals (SDGs), geoscientists need to develop new skills beyond their traditional training. This includes learning how to use tools and technologies from the oil, gas, and mining industries. It also means applying data science and geospatial methods to turn large amounts of data into maps and useful information for policymakers and stakeholders.

Geophysics (SDG 6, 9, 11)

Geophysics is an important part of many SDGs, especially SDG 6 (Clean Water), SDG 9 (Industry and Infrastructure), and SDG 11 (Sustainable Cities). Capello et al. (2021) showed how geophysical tools support all SDGs and highlighted specific ways geophysicists can help. For SDG 9, for example, geophysics can be used to design and monitor smart cities and critical infrastructure. Tools like drones, subsurface imaging, and large-scale InSAR data are very helpful in this work.

Interestingly, many geophysical methods originally created for petroleum or mining are now being used to monitor environmental change. For example, passive seismic tools are now used to monitor thawing permafrost (Cheng et al. 2022; Stemland et al. 2021). These methods also help study surface layers, which are important for building infrastructure in cities.

GIS and Remote Sensing (SDG 1, 9, 13, 14, 15)

GIS (Geographic Information Systems) and remote sensing strongly support SDG 1 (No Poverty), 9 (Industry), 13 (Climate Action), 14 (Life Below Water), and 15 (Life on Land).

Pirasteh et al. (2019) showed how to use LiDAR and drone photos to automatically detect buildings, proving that this technology helps turn large amounts of data into useful local knowledge.

Dehls et al. (2019) created an easy-to-use web platform that shows ground movement (subsidence and uplift) across Norway using InSAR data. This is very useful for construction planning and monitoring geohazards. Satellite data is especially useful in remote areas with limited ground data. For instance, Sheffield et al. (2018) used satellite images to measure soil moisture, groundwater, and surface water. These help with better water management decisions.

Estoque (2020) reviewed how remote sensing supports the SDGs and found that 18% of SDG indicators can be tracked using existing satellite data.

Data Science and Analytics (SDG 1–17)

Data science is already having a big impact on geosciences and supports nearly all SDGs (1–17). There is now a huge amount of Earth Observation (EO) data available. This is often combined with machine learning (ML) to analyze patterns and make sense of the data. Geoscientists need to learn how to work with this "big data" to solve global challenges (Guo et al. 2020).

The International Union of Geological Sciences (IUGS) started the "Big Science Program" to encourage not just the collection of data, but also better ways to analyze and share it. One project under this is the Deep-Time Digital Earth program, which aims to gather and share global data on Earth's evolution (Wang et al. 2021). EO data includes high-resolution, frequent satellite images, EO data cubes (Sudmanns et al. 2023), and many other useful products (Fig. 10.11b).

Aggarwal et al. (2020) used EO data and models to study the water cycle in the North West Himalaya, showing how such data supports SDG research. Tools like Google Earth Engine (launched in 2010) have made it easier to analyze EO data across many geoscience fields (Kumar and Mutanga 2018; Tamiminia et al. 2020). ML algorithms are helping geoscientists deal with the increasing amount of data (Lary et al. 2016). Cheng et al. (2020) shows different ways integrated techniques are being used and provides ideas for the future (Fig. 10.11a).

10.3.15 Social Geology and Human–Environment Interaction

Social geology, as defined by Mata-Perello et al. (2012), is the discipline of geology that studies the interaction among the geological environment and social development. This interaction is especially relevant where geology intersects with sociology, and it influences many Sustainable Development Goals (SDGs), directly or indirectly.

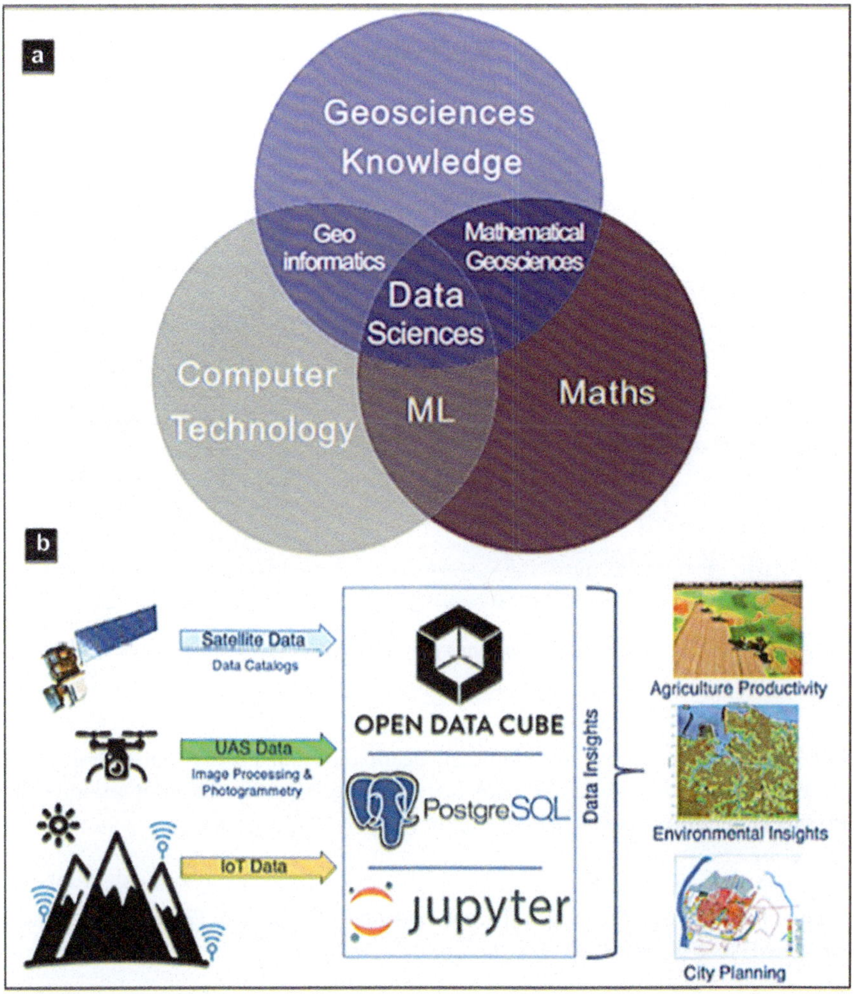

Fig. 10.11 **a** Data science at the intersection of mathematics, computer technology, and geosciences. **b** Integration of various data types, data analysis contents, and data products (*Source* Senger 2024—Open Access)

Non-governmental Organizations (NGOs) and Geoscience (SDG 1, 3, 6, 9, 14, 15, 16, 17)

Geoscientists frequently collaborate with NGOs on development and environmental projects. Wagaba et al. (2023) describe how small NGOs in eastern Africa rely on geological data for water projects but often face barriers such as limited data access, outdated formats, and funding constraints. Many NGOs are funded by developed nations and operate in developing regions. Geoscience can support their work, but challenges remain. Petterson (2019) notes that differing values and priorities

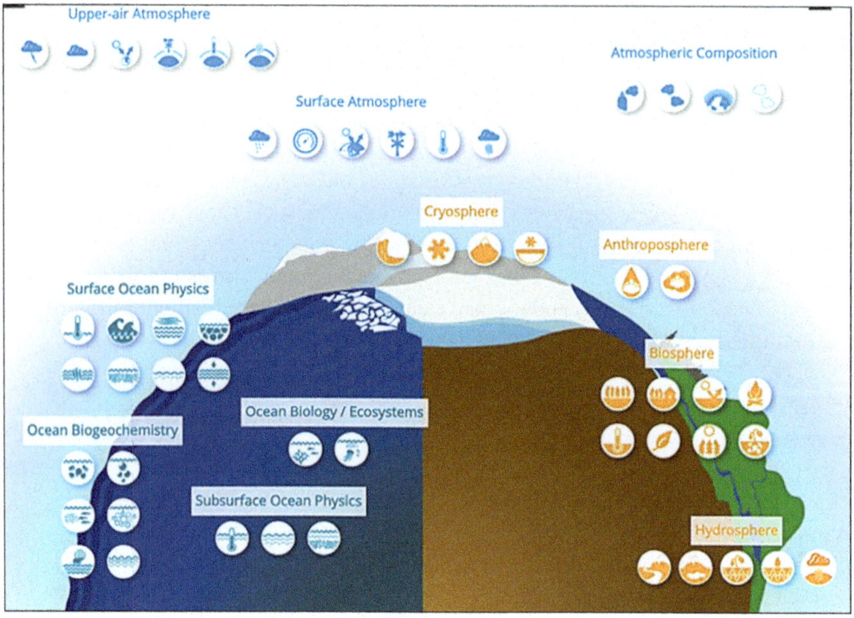

Fig. 10.12 Essential climate variables and their interactions (modified from NASA) (*Source* Senger 2024—Open Access)

between geoscientists and development agencies often limit collaboration. Case studies from the British Geological Survey's work in the Solomon Islands and Afghanistan highlight the need for a more integrated approach.

Science Policy and International Cooperation (*SDG 1, 3, 6, 9, 14, 15, 16, 17*)

SDG 17 promotes global partnerships. It is important to assess whether current international frameworks (Fig. 10.12) are functioning effectively or need reform. A major challenge lies in managing the sometimes-conflicting agendas of various organizations.

Agrogeology (*SDG 1, 2, 3, 6, 14, 15*)

Agrogeology studies how geological materials, especially minerals, support agriculture. Van Straaten (2006) defines it as the use of rocks for crops. Soils influence food security, biodiversity, and climate. Smith et al. (2021) and Pratt et al. (2020) review these links.

Agrogeology focuses on how bedrock affects soil fertility. Case studies include soil surveys in Jiangsu, China (Liao et al. 2007), and rock-based farming in Africa (Chesworth et al., 1989). Brazil's Rochagem movement, which began in the 1950s, uses crushed rocks to improve soils (Manning and Theodoro 2020). Rising fertilizer prices led to renewed interest in this approach.

Studies from Malawi (Chiwona et al. 2020) and a regional case in Crete, Greece (Kritikakis et al. 2022), show how integrated soil, geochemical, and geophysical mapping can guide sustainable farming practices.

Medical Geology ***(SDG 3, 6)***

Medical geology links natural geological materials to public health. It studies how geogenic contaminants—like arsenic and fluoride—impact people through water, crops, and air. Bundschuh et al. (2017) introduced the term in relation to the SDGs. Case studies include arsenic in Latin America (Bundschuh et al. 2021), fluoride in Tanzania (Ijumulana et al. 2020), and combined impacts in India (Jha and Tripathi 2021). Understanding contamination sources and pathways is essential, and geoscientific expertise plays a vital role in risk assessment and mitigation.

10.4 Engineering Geophysics for Resilience and Environmental Justice

10.4.1 Geophysics for Community Resilience

Engineering geophysics supports community resilience by offering tools for early hazard detection, informed land-use planning, and disaster risk reduction. In regions prone to natural hazards such as landslides, flooding, earthquakes, and subsidence, geophysical surveys provide critical data on subsurface stability and risk zones. Methods like seismic microzonation, resistivity imaging, and GPR are employed to evaluate ground conditions before construction or reconstruction efforts, especially after disasters. These data help communities and local authorities make proactive decisions that minimize physical vulnerability and economic loss. When integrated into local disaster preparedness programs, geophysics enhances adaptive capacity and contributes to long-term resilience in both urban and rural settings (Lomax et al. 2012; UNDRR 2022).

10.4.2 Inclusive and Participatory Approaches

Promoting resilience also requires that engineering geophysics be applied in ways that are inclusive, participatory, and sensitive to local needs and knowledge systems. Involving local communities in the design, implementation, and interpretation of geophysical surveys enhances trust, data relevance, and long-term impact. Participatory geoscience projects, including those involving citizen science and community mapping, empower residents to engage in risk assessment and resource management. Furthermore, open-access geophysical data and low-cost instrumentation are essential for enabling local institutions and practitioners—especially in low-income regions—to apply these technologies independently. Ensuring capacity building through training and education fosters greater equity in scientific access and application, aligning geophysics with both resilience goals and social justice values (Wessel et al. 2015; Stewart and Gill 2017a, b).

10.4.3 Case Examples in Environmental Justice

Environmental justice refers to the fair treatment and meaningful involvement of all people in environmental governance, regardless of race, income, or geography. Engineering geophysics contributes to environmental justice by uncovering and documenting hidden risks in marginalized or underserved communities. For example, resistivity and electromagnetic surveys have been used to map legacy contamination near informal settlements, former industrial zones, or mining-impacted communities. In parts of sub-Saharan Africa, Latin America, and South Asia, geophysical tools have also been employed to assess land degradation and groundwater depletion in areas disproportionately affected by extractive industries or climate stress. These case studies demonstrate how geophysics can provide scientific evidence to support advocacy, inform equitable policy, and prioritize remediation in historically neglected areas (Mendoza et al. 2020; Halla et al. 2021).

10.5 Geoethics and Global Equity in Geophysical Technology

10.5.1 Defining Geoethics in the Geosciences

Geoethics refers to the ethical, social, and cultural implications of geoscience knowledge and practice. It aims to guide geoscientists in their professional roles by promoting values such as responsibility, integrity, and respect for human and environmental well-being (Di Capua and Peppoloni 2019). The Cape Town Statement on Geoethics, endorsed by the International Association for Promoting Geoethics

(IAPG), outlines the need for geoscientists to act with responsibility towards both society and the Earth system (IAPG 2024).

The concept serves as a bridge between geoscience, ethics, and sustainability, urging professionals to engage in decision-making that considers long-term planetary health and social equity. Geoethics addresses a range of issues from data integrity and honesty in research, to the socio-political consequences of geotechnical decisions, especially in vulnerable communities (Bralower et al. 2023).

10.5.2 Technology Access and Capacity Building

Disparities in access to geophysical technologies pose a major challenge to global equity. In many low- and middle-income countries, limited access to training, instrumentation, and software restricts the capacity to apply geophysics in infrastructure development, disaster mitigation, and environmental protection (Villacorta et al. 2024).

According to the Geophysical Sustainability Atlas, partnerships between universities, governments, and private sector organizations are essential to foster knowledge sharing and technology transfer. These initiatives also help align geophysics with Sustainable Development Goals (SDGs), particularly those related to sustainable cities (SDG 11), clean water (SDG 6), and climate action (SDG 13) (Peppoloni and Di Capua 2025).

Educational programs like the GOAL Project (Geoethics Outcomes and Awareness Learning) further support this mission by integrating geoethics into curricula and strengthening local expertise in resource management, hazard assessment, and infrastructure design (Peppoloni et al. 2023). The availability of open-access platforms and virtual labs has proven to be a scalable method to increase participation from underrepresented regions. However, such platforms require internet infrastructure and digital literacy, which remain unevenly distributed (Stewart and Gill 2017a, b).

10.5.3 Ethical Dilemmas in Practice and Research

Geoscientists often face complex ethical dilemmas related to data transparency, community engagement, and environmental impacts. A major issue is the ethical sharing and use of geophysical data, particularly in politically or environmentally sensitive areas. Open data policies may conflict with national security concerns or local cultural norms (Cocco et al. 2025).

Additionally, questions arise around informed consent when conducting geophysical surveys in inhabited or indigenous areas. The ethical imperative to involve affected communities in research design and interpretation is gaining attention, as geophysical interventions can significantly affect land use, perceptions of risk, and access to resources (Bralower et al. 2023).

Emerging scholarship also critiques anthropocentric biases in geoscience, suggesting a broader ethical framework that includes the intrinsic value of Earth systems and non-human entities. This shift toward "deep geoethics" advocates a more ecocentric view, encouraging geoscientists to consider long-term ecosystem impacts alongside short-term human benefits (Bralower et al. 2023).

10.6 Case Studies and Practical Applications

10.6.1 Climate-Resilient Infrastructure in South Asia

South Asia faces increasing climate-related challenges, including flooding, droughts, and urban heat islands. To address these, cities are adopting climate-resilient infrastructure strategies. For instance, blue-green infrastructure—which integrates natural systems like wetlands and urban greenery—is being implemented to manage stormwater and reduce heat stress (Fig. 10.13). This approach not only mitigates climate impacts but also enhances urban biodiversity and resident's well-being (Wiley Online Library 2024; Liao et al. 2017; Wang and Banzhaf 2018; Browder et al. 2019).

In Karachi, Pakistan, the Karachi Water and Sewerage Board is collaborating with international organizations to upgrade drainage systems and promote sustainable water management practices. These efforts aim to improve resilience against monsoon-related flooding and ensure equitable water distribution (The Guardian 2025).

Recent years have seen a growing academic interest in the application of geophysical information (GI) for climate change adaptation. This is evident from the increasing number of peer-reviewed publications that explore how GI supports infrastructure resilience, hazard mapping, and adaptive land-use planning (Fig. 10.14). The trend reflects a broader recognition of GI as a critical tool in climate-resilient infrastructure development across vulnerable regions such as South Asia.

The basic metadata for the selected articles, including publication year, region of focus, and application type, is summarized in Table 10.1, offering an overview of the scope and diversity of the reviewed literature. To identify leading sources of knowledge dissemination, Table 10.2 lists the most influential journals based on the number of relevant published papers, with Environmental Earth Sciences, Sustainability, and Natural Hazards emerging as key publication platforms in this domain.

Additionally, Table 10.3 synthesizes the strategies and geophysical techniques proposed in the literature, including the use of resistivity imaging, integrated GIS-based multi-criteria analysis, and early-warning systems. These strategies have been applied across diverse climatic regions to enhance urban planning, hazard mitigation, and community resilience.

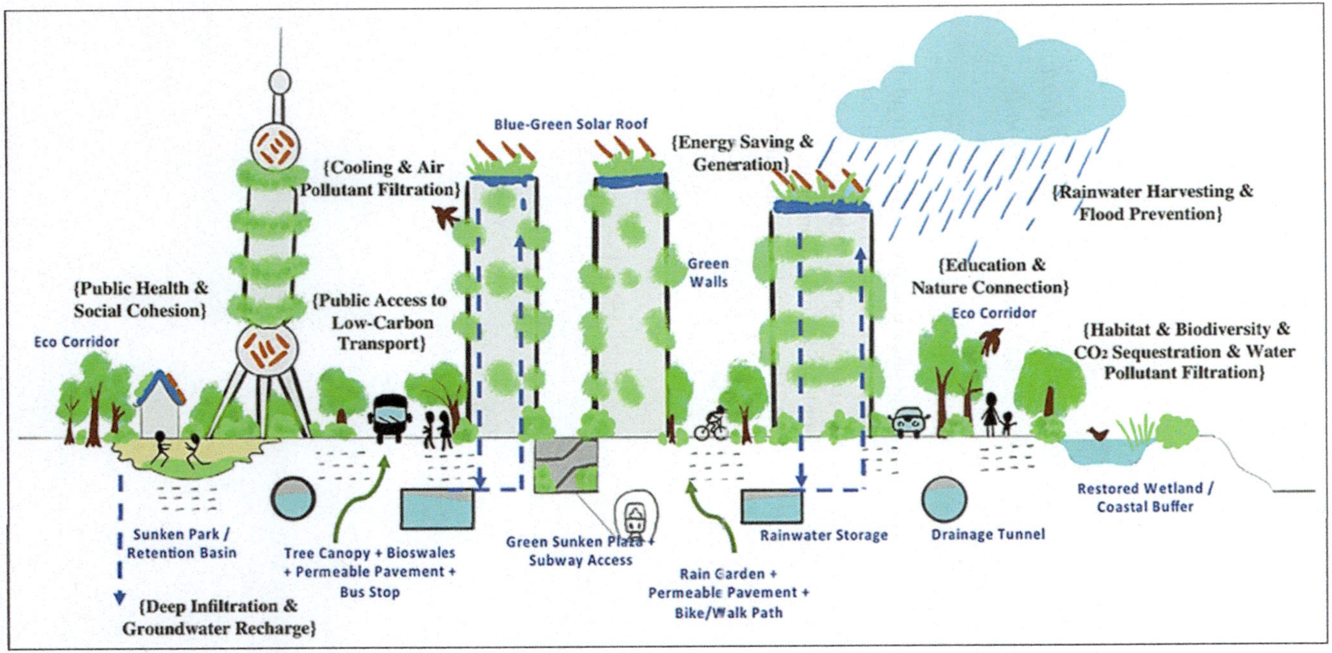

Fig. 10.13 Blue-Green Infrastructure supports climate resilience and provides multiple benefits in Chinese cities. It combines natural water management systems (blue) and green spaces to improve both urban and ecological functions (*Source* Kadic et al. 2025—Open Access; Siehr et al. 2022; Ramboll 2016; Catalano and Baumann 2017)

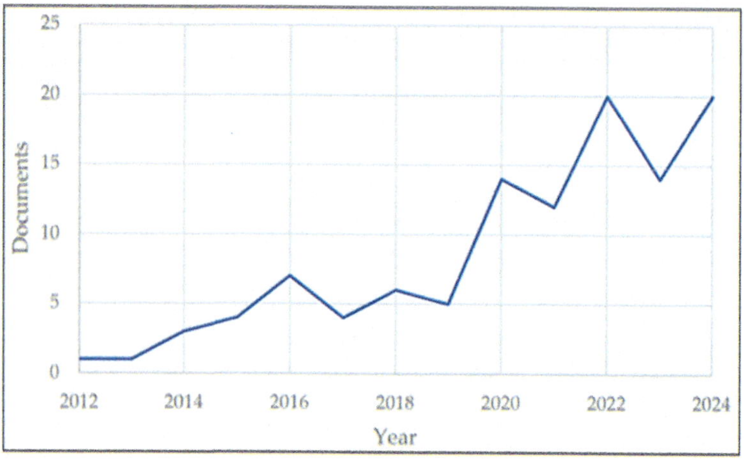

Fig. 10.14 Number of selected peer-reviewed articles on the role of geophysical information (GI) in climate change adaptation, categorized by year of publication (2010–2024) (*Source* Kadic et al. 2025—Open Access)

Table 10.1 Basic metadata of selected articles, including year, region, focus area, and method applied (after Kadic et al. 2025)

Description	Results
Timespan	2012–2024
Sources	43
Documents	111
Article	96
Review article	15
Authors	448
Author keywords	362
Index keywords	963
References	7480

Green infrastructure plays a crucial role in climate resilience by using natural systems to manage water, reduce urban heat, and support biodiversity. It includes parks, wetlands, green roofs, and vegetated corridors that serve both ecological and urban functions (Fig. 10.15).

10.6.2 Groundwater Mapping in Water-Scarce Regions

Groundwater serves as a critical water source in arid and semi-arid regions. In North Gujarat, India, an integrated multi-criteria analysis using GIS and geophysical

Table 10.2 Top journals by number of published papers on GI and climate change adaptation (after Kadic et al. 2025)

Rank	Journal	Number of articles
1	Urban Forestry and Urban Greening	16
2	Land	10
3	Journal of Environmental Management	9
4	Ecological Indicators	5
5	Atmosphere	5
6	Climatic Change	5
7	Ecology and Society	3
8	Frontiers in Environmental Science	3
9	Ecological Engineering	3
10	Climate	3
11	Science of the Total Environment	3
12	Forests	3
13	Environmental Management	2
14	Ain Shams Engineering Journal	2
15	Water Research	2
16	Mitigation and Adaptation Strategies for Global Change	2
17	Blue-Green Systems	2
18	International Journal of Disaster Resilience in the Built Environment	2
19	Journal of Sustainable Water in the Built Environment	2
20	Ecological Indicators	2

methods identified high and very high groundwater potential zones in the Ambaji Basin. This approach, validated with field data, offers a cost-effective and reliable method for sustainable groundwater management in hard rock terrains (Anbazhagan et al. 2021; Siehr et al. 2022).

Similarly, in southern Saudi Arabia, remote sensing and GIS techniques were employed to delineate groundwater potential zones in the arid region of Khamis Mushait. The study utilized high-resolution satellite imagery and geospatial data to identify areas with the highest groundwater potential, aiding in sustainable resource management (MDPI Water 2023).

In many arid and semi-arid regions, climate-resilient water management strategies are becoming vital for long-term sustainability. These strategies integrate groundwater mapping, water-saving technologies, and policy reforms to ensure reliable access to water despite climate variability. Figure 10.16 illustrates key components of climate-resilient water management, including natural recharge zones, water-efficient infrastructure, and decision-support systems that enhance adaptive capacity in water-scarce regions.

Table 10.3 Key literature sources with corresponding geophysical strategies and application techniques (after Kadic et al. 2025)

Sector	Strategies and techniques	Source
Mitigating urban heat islands	Constructing different types of green infrastructure to gain sustainable urban development and climate mitigation strategy	Sussams et al. (2015)
	The implementation of supplemental irrigation to increase roof cooling and vegetation structure to lower green roof temperatures	Untereiner et al. (2024)
	Improving blue-green infrastructure to assist municipalities and cities in adapting to climate change	Schneider et al. (2024)
	The creation of other structural and environmental measures to increase adaptation to climate change in different aspects of the socio-economic sector	Li et al. (2022a, b)
Increasing ecosystem resilience	Improving policies and frameworks for risk reduction and disaster recovery is essential. Adding restored and healthy green spaces in cities can support ecosystem-based adaptation. These green areas improve human health and protect ecological diversity	Lehmler et al (2023), Lee and Liu (2023)
	Using decision-making tools like spatial decision support systems and multi-criteria analysis helps improve the environment. These tools guide policymakers, experts, and planners. They make it easier to plan actions that boost species' resilience to climate change	Losada-Iglesias et al. (2024)
	Identifying measures to sustain or enhance the insurance value of urban ecosystems, including nature-based solutions, ecosystem-based adaptation, grey-green infrastructure, and blue-green infrastructure	Paxton et al. (2023), Sharma and Malaviya (2021)
	Planting different tree species, installing green facades, engaging landscape planning, evolving and involving agroforestry and urban forestry in urban development planning to help prevent serious health problems, lowering the soil temperature and CO_2 emission from soils, and adjusting to climate change, particularly in places where vulnerable populations live	Leppanen et al. (2024)
	In order to guide green infrastructure design and maximize ecosystem services results while minimizing ecosystem disservices, policymakers and planners could combine the biophysical assessment of green infrastructure parametric data with socio-economic data	Zhang and MacKenzie (2024)

(continued)

Table 10.3 (continued)

Sector	Strategies and techniques	Source
Flood risk management	Urban flood prediction, mitigation, and reduction services through green infrastructure regulations, land use and land cover changes, hydrological modeling, the assessment and monitoring processes of the stormwater infrastructure performance, and creating resilience plans	Dharmarathne et al. (2024)
	Implementing multifunctional stormwater management tools consisting of different elements mutually interacting to provide desired safety levels: Stormwater Management Model (SWMM) or Stormwater Management System (SWM), Environmental Protection Agency's Stormwater Management Model, Sustainable Urban Drainage Systems (SuDS) and Sponge Cities Program (SCP), Curve Number-Based Watershed Model (CWM), Extreme Weather Layer method (EWL), Global Resilience Analysis framework (GRA)	Rodriguez et al. (2024)
	Addressing flooding and water quality issues utilizing blue-green infrastructure	Almaaitah et al. (2021), Hamel and Tan (2022), Flores et al. (2023)

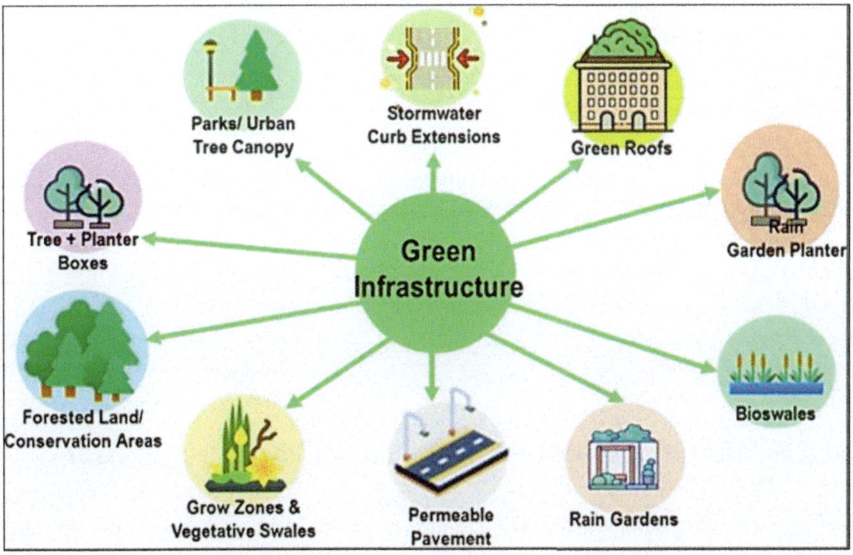

Fig. 10.15 Illustrates a typical layout of green infrastructure elements within a city, showing how they contribute to stormwater management, cooling, and improved quality of life (*Source* Kurniawan et al. 2024)

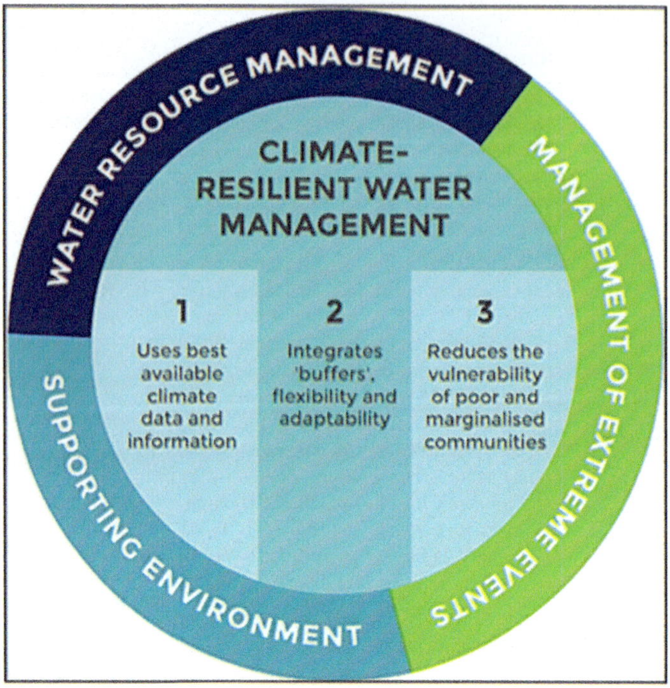

Fig. 10.16 Climate-resilient water management framework highlighting sustainable groundwater use, recharge enhancement, and policy-integration approaches under climate stress (*Source* Kadic et al. 2025—Open Access)

Groundwater is a critical freshwater resource, especially in arid and semi-arid regions where surface water is scarce. Understanding the distribution of regional aquifer systems is essential for sustainable groundwater management and climate adaptation planning. Figure 10.17 presents a simplified global map of groundwater resources, identifying the major regional aquifers around the world. This visual context supports the importance of localized mapping efforts in regions like India, Saudi Arabia, and sub-Saharan Africa, where groundwater serves as a lifeline under increasing climatic stress.

10.6.3 Subsurface Data for Sustainable Urban Development

In Sagar District, Madhya Pradesh, India, multi-criteria decision-making methods like TOPSIS and VIKOR were applied to create groundwater potentiality zonation maps. Factors such as lithology, rainfall, and geomorphology were considered, revealing that the northern, eastern, and southeastern parts of the district have very

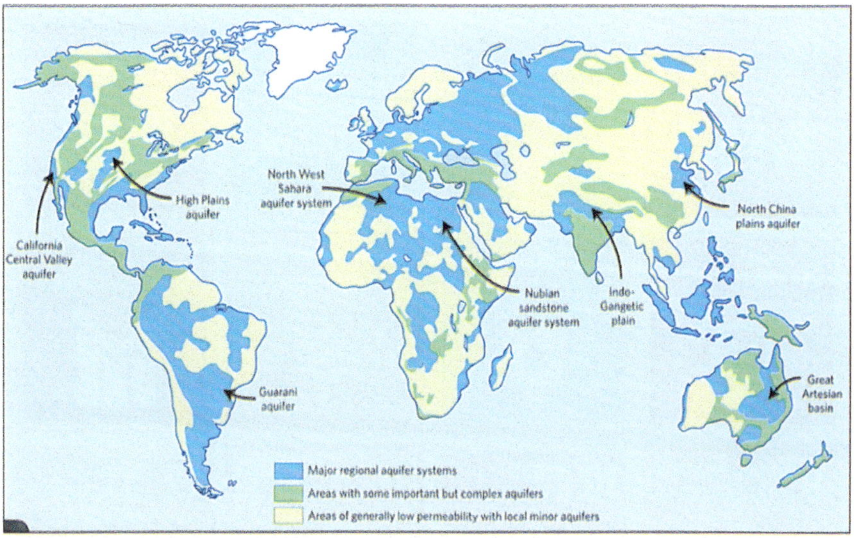

Fig. 10.17 Simplified version of a global groundwater resources map, highlighting major regional aquifer systems (*Source* Taylor et al. 2013)

high groundwater susceptibility. These insights are crucial for urban planning and sustainable development (ScienceDirect 2024).

Additionally, in Cilongok, Central Java, Indonesia, 2D geoelectrical resistivity surveys identified aquifer layers suitable for clean water extraction. The findings highlighted porous volcanic tuff layers at depths of 28–45 m in Cikidang and 45–60 m in Kalisari, providing valuable data for local water supply planning (arXiv 2018).

10.7 Challenges, Gaps, and Opportunities

Engineering geophysics plays an essential role in supporting the achievement of the Sustainable Development Goals (SDGs) by providing vital insights into subsurface conditions, aiding in resource management, environmental sustainability, and hazard mitigation. Geoscience tools and methods, including those from engineering geophysics, intersect with multiple SDGs, ranging from clean water access to resilient infrastructure and climate action. These intersections are illustrated in Fig. 10.18, which highlights the number of geoscience-relevant categories that address each goal.

Fig. 10.18 Overview of the 17 Sustainable Development Goals, highlighting the geoscience links based on the number of geoscientific categories addressing each goal. SDG icons courtesy of the United Nations (United Nations 2022). The full list of goals and related targets is available from United Nations (2022) (*Source* Senger 2024—Open Access)

10.7.1 Technical and Operational Limitations

Despite the significant advancements in engineering geophysics, there are several technical and operational limitations that remain challenging. One of the primary obstacles is the data scarcity in developing regions. Many areas, particularly in parts of Africa, South Asia, and Latin America, lack comprehensive geophysical data, which severely restricts the ability to assess subsurface conditions and make informed decisions regarding infrastructure, resource management, and environmental protection. This data gap can be partially addressed through the integration of remote sensing technologies. For example, remote sensing satellite imagery has been instrumental in enhancing water resource management in the northwestern Himalayas, showcasing the potential for satellite-based data to supplement on-the-ground geophysical surveys (Senger 2024).

Another major challenge is the high operational costs associated with geophysical surveys. These surveys often require expensive equipment, specialized software, and technical expertise, which can be a significant financial barrier, particularly in low-income regions where budgets for such initiatives are limited. Furthermore, there is a lack of public awareness and understanding of the benefits of engineering geophysics, which results in underutilization of geophysical tools in decision-making processes. Policymakers and local stakeholders often overlook the potential of geophysics in supporting sustainable development initiatives, which limits its application in planning, development, and disaster preparedness.

10.7.2 Integration with Emerging Technologies (AI, Remote Sensing)

Emerging technologies, particularly Artificial Intelligence (AI) and remote sensing, present significant opportunities to overcome many of the existing challenges in engineering geophysics. AI can substantially enhance the processing and analysis of geophysical data, allowing for more accurate and efficient subsurface modeling. AI algorithms, for example, can analyze hyperspectral imagery to detect mineral signatures with greater precision, which can improve resource management and exploration (Thompson 2020; Garcia et al. 2023; Smith et al. 2020).

Remote sensing technologies, especially satellite-based systems, enable large-scale monitoring of environmental changes, land use, and natural resources, especially in regions that are difficult to access. Satellite data allows for the monitoring of everything from deforestation rates to infrastructure development, providing a broad view of regional sustainability efforts. Coupling remote sensing data with ground-based geophysical surveys can significantly improve the accuracy and reliability of subsurface models, supporting better decision-making in areas such as urban planning, agriculture, and disaster risk management. Moreover, the integration of multimodal data, such as combining LiDAR, radar, and optical sensors, enhances the resolution and depth of geophysical models. This multimodal approach allows for improved monitoring of environmental hazards, land deformation, and infrastructure integrity, which is critical for both disaster preparedness and the sustainable development of infrastructure (Li et al. 2022a, b).

10.7.3 Future Directions for SDG-Aligned Geophysics

Looking ahead, there are several promising directions that can further align engineering geophysics with the United Nations Sustainable Development Goals (SDGs). One key pathway is policy integration. Governments, development agencies, and international organizations are increasingly recognizing the value of incorporating geophysical data into public policy to support better decision-making in areas such as land-use planning, resource management, and disaster risk reduction. For example, geophysical surveys can identify zones vulnerable to landslides, subsidence, or seismic activity, enabling more informed mitigation and preparedness strategies. Integrating this data into national and local policy frameworks ensures that development decisions are grounded in robust scientific evidence, directly contributing to SDGs related to sustainable cities (SDG 11), climate action (SDG 13), and resilient infrastructure (SDG 9).

Another essential direction is capacity building. Equipping geophysicists with knowledge and training in sustainability principles enables them to contribute more effectively to SDG-related projects. This includes developing curricula and professional development programs that blend geophysical skills with sustainability, environmental ethics, and policy literacy. Such training fosters a new generation of professionals who can apply geophysical methods to real-world sustainability challenges, bridging the gap between science and practice. In addition, raising public and stakeholder awareness about the role of geophysics in sustainable development is crucial for fostering broader engagement and increasing demand for geophysical tools in planning and resilience-building efforts.

Ethical considerations are also emerging as central to the future of SDG-aligned geophysics. The expanding use of technologies such as artificial intelligence (AI), remote sensing, and big data analytics enhances the capabilities of geophysical investigations. However, these advancements raise concerns about data privacy, access equity, and ownership, particularly in regions with marginalized populations or limited digital infrastructure. Responsible innovation, which includes respecting data sovereignty, ensuring inclusive access, and promoting transparent methodologies, is vital to ensuring geophysical work remains aligned with the SDG principles of fairness, accountability, and inclusion (Zhao et al. 2024).

10.8 Supporting Resources and Practical Tools

This section presents supplementary resources designed to enhance the application of engineering geophysics in sustainable development contexts. It includes practical tools, key terminology, and illustrative case studies that reinforce the concepts discussed throughout Chap. 9. These materials are intended to support researchers, practitioners, and students in effectively aligning geophysical methods with the United Nations Sustainable Development Goals (SDGs). By offering structured guidance and real-world examples, this section serves as a bridge between geophysical theory and field-based implementation within sustainability-driven projects.

10.8.1 SDG-Geophysics Alignment Checklist

To assist practitioners and decision-makers in evaluating the sustainability relevance of geophysical projects, a practical SDG–Geophysics Alignment Checklist is provided. This tool outlines how specific geophysical methods contribute to selected SDGs and can be used during project planning, reporting, or policy development phases. For example, groundwater mapping through electrical resistivity tomography (ERT) supports SDG 6 (Clean Water and Sanitation), while geophysical surveys for geothermal potential align with SDG 7 (Affordable and Clean Energy). Subsurface investigations that inform infrastructure design, such as seismic profiling and GPR, contribute to SDG 9 (Industry, Innovation and Infrastructure). The checklist encourages users to identify alignment for each goal, mark evidence of integration, and make notes for further action. This structured approach not only enhances accountability but also supports more intentional planning and reporting for geophysical projects in sustainable development contexts.

10.9 Glossary of Terms

A glossary of key terms is included to ensure clarity for readers from diverse disciplinary backgrounds. "Engineering Geophysics" is defined as the application of geophysical methods to solve subsurface problems relevant to construction, environmental assessment, and resource exploration. The "Sustainable Development Goals (SDGs)" refer to the 17 global objectives set forth by the United Nations to balance social, environmental, and economic progress. "Remote Sensing" involves the acquisition of Earth surface data from satellites or aircraft without direct contact, while "Geohazards" are natural threats like landslides or earthquakes that can be evaluated using geophysical data. "Artificial Intelligence (AI)" refers to computational systems capable of learning from large datasets to make predictions or automate decisions, and "Multimodal Data Fusion" is the integration of data from different geophysical or spatial sources—such as seismic, radar, or optical—to improve the accuracy and relevance of interpretations. This glossary enhances understanding of complex terms and encourages consistency across interdisciplinary applications.

10.9.1 Supplementary Case Studies

To illustrate the practical application of geophysics in SDG contexts, three brief case studies are provided. The first case study highlights groundwater exploration efforts in Ethiopia and Kenya, where electrical resistivity and magnetometry techniques were used to locate sustainable aquifers in water-scarce communities, directly supporting SDG 6. The second case study describes a seismic risk assessment conducted in Kathmandu, Nepal, utilizing microtremor surveys and shear-wave velocity profiling to inform urban development and disaster preparedness strategies, aligning with SDG 11 (Sustainable Cities and Communities). The third example focuses on coastal erosion monitoring in Bangladesh, where ground-penetrating radar (GPR) and InSAR (Interferometric Synthetic Aperture Radar) techniques were applied to assess shoreline retreat and sediment loss. This geophysical monitoring enabled local governments to implement climate adaptation measures, thereby supporting SDG 13 (Climate Action). These case studies provide concrete evidence of how engineering geophysics can be strategically deployed to address sustainability challenges across diverse geographic and social contexts.

10.10 Conclusion: Engineering Geophysics and the SDGs—A Roadmap for the Future

Engineering geophysics stands at a transformative intersection between earth sciences and sustainable development. As illustrated throughout this chapter, the integration of geophysical techniques with the United Nations Sustainable Development Goals (SDGs) underscores the discipline's expanding relevance beyond traditional engineering applications. From mapping groundwater in arid regions to assessing urban resilience against climate-induced hazards, geophysical tools are now central to developing resilient, inclusive, and sustainable communities.

This chapter has demonstrated how specific geophysical methods align with a wide array of SDGs—particularly those focusing on clean water and sanitation (SDG 6), sustainable cities (SDG 11), climate action (SDG 13), and clean energy (SDG 7). Additionally, the role of engineering geophysics in frontier areas such as planetary geology, geohazard monitoring, carbon capture, and energy storage reflects its versatility in addressing both global and local challenges.

The contributions of geophysics extend to promoting innovation (SDG 9), supporting life below water and on land (SDGs 14 and 15), and enhancing education, equality, and public engagement (SDGs 4, 5, and 10). By incorporating digital technologies, geospatial data, and inclusive practices, geophysics can further democratize access to critical information and improve decision-making processes at all levels.

Moreover, the chapter has highlighted the need for geoethics, capacity building, and global equity in the application of geophysical technology. Ethical frameworks and participatory models are essential to ensure that advancements serve all communities—especially marginalized and vulnerable populations—thus contributing to environmental justice and long-term resilience.

Despite its promise, the field faces challenges, including technological limitations, data integration hurdles, and the need for stronger interdisciplinary collaboration. However, these are also opportunities for innovation. Embracing emerging technologies such as artificial intelligence, remote sensing, and big data analytics can enhance geophysical solutions in addressing complex sustainability problems.

In conclusion, engineering geophysics is not merely a technical toolset but a vital contributor to sustainable development. As global priorities shift toward resilience, equity, and ecological integrity, the discipline must continue evolving—guided by ethical considerations, informed by local contexts, and empowered by scientific rigor. Through continued alignment with the SDGs, engineering geophysics can help shape a more sustainable and just future for all.

References

Aggarwal S, Thakur PK, Nikam BR, Garg V, Chouksey A, Dhote P et al (2020) Role of Earth Observation data and hydrological modeling in supporting Un Sdgs in North West Himalaya. Int Arch Photogram. Remote Sens Spatial Inf Sci 43. https://doi.org/10.5194/isprs-archives-XLIII-B3-2020-853-2020

Aitsi-Selmi A, Egawa S, Sasaki H, Wannous C, Murray V (2015) The Sendai framework for disaster risk reduction: renewing the global commitment to people's resilience, Health, and well-being. Int J Disaster Risk Sci 6(2):164–176

Ali SH, Giurco D, Arndt N, Nickless E, Brown G, Demetriades A et al (2017) Mineral supply for sustainable development requires resource governance. Nature 543(7645):367–372

Almaaitah T, Appleby M, Rosenblat H, Drake J, Joksimovic D (2021) The potential of Blue-Green infrastructure as a climate change adaptation strategy: a systematic literature review. Blue-Green Syst. 3:223–248

Almazroa H, Alotaibi W, Alrwaythi E (2022) Sustainable development goals and future-oriented teacher education programs. IEEE Trans Eng Manag 1–14

Anbazhagan P et al (2021) Integrated multi-criteria analysis for groundwater potential mapping in Precambrian hard rock terranes (North Gujarat), India. Environ Geol 78(1):1–14. https://doi.org/10.1007/s00254-021-01566-2

Annan-Diab F, Molinari C (2017) Interdisciplinarity: practical approach to advancing education for sustainability and for the sustainable development goals. Int J Manag Educ 15(2):73–83

Antoine R, Lopez T, Tanguy M, Lissak C, Gailler L, Labazuy P et al (2020) Geoscientists in the sky: unmanned aerial vehicles responding to geohazards. Surv Geophys 41(6):1285–1321

Arends H, Dahl D, Dejarnett BB, Harrison III WB, Honey J, Johnson MR et al (2021) Directory of public repositories of geological materials. U.S. Geol. Surv. Data release

ArXiv (2018) Mapping groundwater potential zones in Cilongok area, Banyumas, Central Java using 2D geoelectrical resistivity. arXiv. https://arxiv.org/abs/1812.04111

Attard M (2019) The sustainability of landforms and landscapes. In: Gauci R, Schembri JA (eds) landscapes and landforms of the Maltese Islands. Springer International Publishing, pp 373–380

Avelar ABA, da Silva-Oliveira KD, da Silva Pereira R (2019) Education for advancing the implementation of the sustainable development goals: a systematic approach. Int. J. Manag. Educ. 17(3):100322

Beaulieu SE, Graedel T, Hannington MD (2017) Should we mine the deep-sea floor? Earth's Future 5(7):655–658

Bednar M, Sarapatka B (2018) Relationships between physical-geographical factors and soil degradation on agricultural land. Environ Res 164:660–668

Bendixen M, Iversen LL, Best J, Franks DM, Hackney CR, Latrubesse EM et al (2021) Sand, gravel, and UN Sustainable Development Goals: conflicts, synergies, and pathways forward. One Earth 4(8):1095–1111

Benton MJ, Donoghue PCJ (2006) Paleontological evidence to date the tree of life. Mol Biol Evol 24(1):26–53

Betlem P, Rodes N, Birchall T, Dahlin A, Smyrak-Sikora A, Senger K (2023) Svalbox digital model database: a geoscientific window into the high arctic. Geosphere 19(6):1640–1666

Bidarmaghz A, Choudhary R, Soga K, Kessler H, Terrington RL, Thorpe S (2019) Influence of geology and hydrogeology on heat rejection from residential basements in urban areas. Tunn Undergr Space Technol 92:103068

Birendra K, Dhungana A, Dangi TB (2021) Tourism and the sustainable development goals: stakeholders perspectives from Nepal. Tour Manag Perspect 38:100822

Bralower T, Mogk D, Vasconcelos C., 2023. Shallow vs. deep geoethics: Moving beyond anthropocentric views. Journal of Agricultural and Environmental Ethics. https://doi.org/10.1007/s10806-023-09920-y

Browder G, Ozment S, Rehberger Bescos I, Gartner T, Lange G (2019) Integrating green and gray: creating next generation infrastructure. World Bank and World Resources Institute

Bundschuh J, Schneider J, Alam MA, Niazi NK, Herath I, Parvez F et al (2021) Seven potential sources of arsenic pollution in Latin America and their environmental and health impacts. Sci Total Environ 780:146274

Bundschuh J, Maity JP, Mushtaq S, Vithanage M, Seneweera S, Schneider J et al (2017) Medical geology in the framework of the sustainable development goals. Sci Total Environ 581:87–104

Burke K, Francis P, Wells G (1990) Importance of the geological record in understanding global change. Glob Planet Change 3(3):193–204

Butler DK (2005) Near-surface geophysics. Society of Exploration Geophysicists

Calais E, Symithe S, Monfret T, Delouis B, Lomax A, Courboulex F et al (2022) Citizen seismology helps decipher the 2021 Haiti earthquake. Science 376(6590):283–287

Capello MA, Shaughnessy A, Caslin E (2021) The geophysical sustainability atlas: mapping geophysics to the UN sustainable development goals. Lead Edge 40(1):10–24

Cardarelli E, Di Filippo G (2009) Integrated geophysical methods for geotechnical investigations. Environ Geol 58(3):519–527

Catalano C, Baumann N (2017) Biosolar roofs: a symbiosis between biodiverse green roofs and renewable energy (CITYGREEN #15. 42–49). Centre for Urban Greenery and Ecology

Catana MM, Brilha JB (2020) The role of UNESCO global geoparks in promoting geosciences education for sustainability. Geoheritage 12(1):1–10

Chapman N, Hooper A (2012) The disposal of radioactive wastes underground. Proc Geol Assoc 123(1):46–63

Chelli A, Bordoni M, Cappadonia C, Pepe G, Rotigliano E, Smith M (2021) Geomorphological tools for mapping natural hazards. J Maps 17(3):1–4

Cheng F, Lindsey NJ, Sobolevskaia V, Dou S, Freifeld B, Wood T et al (2022) Watching the cryosphere thaw: seismic monitoring of permafrost degradation using distributed acoustic sensing during a controlled heating experiment. Geophys Res Lett 49(10):e2021GL097195
Cheng Q, Oberhansli R, Zhao M (2020) A new international initiative for facilitating data-driven Earth science transformation. Geol Soc Lond Spec Publ. 499(1):225–240
Chesworth W, van Straaten P, Semoka JMR (1989) Agrogeology in East Africa: the Tanzania-Canada project. J Afr Earth Sci 9(2):357–362
Chiwona AG, Cortes JA, Gaulton RG, Manning DAC (2020) Petrology and geochemistry of selected nepheline syenites from Malawi and their potential as alternative potash sources. J Afr Earth Sc 164:103769
Chow J (2018) Mangrove management for climate change adaptation and sustainable development in coastal zones. J Sustain for 37(2):139–156
Cigna F, Tapete D, Casagli N (2013) Earth observation for sustainable urban development: monitoring building stability with InSAR in Rome. Remote Sens 5(12):5373–5394
Cignetti M, Godone D, Bertolo D, Paganone M, Thuegaz P, Giordan D (2021) Rockfall susceptibility along the regional road network of Aosta valley region (Northwestern Italy). J Maps 17(3):54–64
Cocco M et al (2025) The ethical dimension of sharing solid Earth science data. J Geoethics Social Geosci 2(1):1–30
Crowley TJ (2000) Causes of climate change over the past 1000 years. Science 289(5477):270–277
Dal Zilio L, Ampuero J-P (2023) Earthquake doublet in Turkey and Syria. Commun Earth Environ 4(1):71
Davies A, Simmons M (2020) Who needs stratigraphers, Sedimentologists and Paleontologists? Evolving roles through the energy transition. Sediment Rec 18(4):12–15
De Beni E, Cantarero M, Neri M, Messina A (2021) Lava flows of Mt Etna, Italy: the 2019 eruption within the context of the last two decades (1999–2019). J Maps 17(3):65–76
Dehls JF, Larsen Y, Marinkovic P, Lauknes TR, Stodle D, Moldestad DA (2019) INSAR.No: a National Insar Deformation Mapping/Monitoring Service in Norway—from concept to operations, IGARSS 2019. In: IEEE International Geoscience and Remote Sensing Symposium, pp 5461–5464
Dharmarathne G, Waduge AO, Bogahawaththa M, Rathnayake U, Meddage DPP (2024) Adapting cities to the surge: a comprehensive review of climate-induced urban flooding. Results Eng. 22:102123
Di Baldassarre G, Sivapalan M, Rusca M, Cudennec C, Garcia M, Kreibich H et al (2019) Sociohydrology: scientific challenges in addressing the sustainable development goals. Water Resour Res 55(8):6327–6355
Di Capua G, Peppoloni S (2019) Defining geoethics. International Association for Promoting Geoethics (IAPG). https://www.geoethics.org/definition
DiPippo R (2016) Geothermal power plants: principles, applications, case studies and environmental impact, 4th ed. Butterworth-Heinemann
Dzikunoo EA, Vignoli G, Jorgensen F, Yidana SM, Banoeng Yakubo B (2020) New regional stratigraphic insights from a 3D geological model of the Nasia sub-basin, Ghana, developed for hydrogeological purposes and based on reprocessed B-field data originally collected for mineral exploration. Solid Earth 11(2):349–361
Eiken O, Ringrose P, Hermanrud C, Nazarian B, Torp TA, Hoier L (2011) Lessons learned from14 years of CCS Operations: Sleipner, in Salah and Snohvit. Energy Procedia 4:5541–5548
Ellefmo SL, Aberle N, Hagspiel V, Ingulstad M, Aasly K (2023) Marine minerals role in future holistic mineral resource management. Geol Soc Lond 526(1):319–342
El-Showk S (2022) Final resting place. Science 375(6583):806–810
Engen O, Faleide JI, Dyreng TK (2008) Opening of the Fram Strait Gateway: a review of plate tectonic constraints. Tectonophysics 450(1):51–69
Fatimah YA, Govindan K, Murniningsih R, Setiawan A (2020) Industry 4.0 based sustainable circular economy approach for smart waste management system to achieve sustainable development goals: a case study of Indonesia. J Clean Prod 269:122263

Fernandez C, Vega JA (2018) Evaluation of the Rusle and disturbed Wepp erosion models for predicting soil loss in the first year after wildfire in NW Spain. Environ Res 165:279–285

Flores CC, Muller APR, Dolman N, Ozerol G (2023) Assessing the leapfrogging potential to water sensitive: the Dutch case of Zwolle. J Water Clim Change 14:1638–1655

Garcia A et al (2023) Predictive modeling of gold deposits using AI and remote sensing data. J Mineral Explor 45(2):123–135

German JD, Redi AANP, Ong AKS, Prasetyo YT, Sumera VLM (2022) Predicting factors affecting preparedness of volcanic eruption for a sustainable community: a case study in the Philippines. Sustainability 14(18):11329

Gomez ML, Hoke G, D'Ambrosio S, Moreiras S, Castro A (2022) Review: hydrogeology of northern Mendoza (Argentina), from the Andes to the eastern plains, in the context of climate change. Hydrogeol J 30(3):725–750

Gordon JE, Crofts R, Gray M, Tormey D (2021) Including geoconservation in the management of protected and conserved areas matters for all of nature and people. Int J Geoheritage Parks 9(3):323–334

Gordon JE (2019) Geoconservation principles and protected area management. Int J of Geoheritage Parks 7(4):199–210

Govil PK, Keshav Krishna A, Dimri VP (2020) Global geochemical baseline mapping in India for environmental management using topsoil. J Geol Soc India 95(1):9–16

Guney T (2019) Renewable energy, non-renewable energy and sustainable development. Int J Sust Dev World 26(5):389–397

Guo H, Nativi S, Liang D, Craglia M, Wang L, Schade S et al (2020) Big Earth data science: an information framework for a sustainable planet. Int J Digital Earth 13(7):743–767

Halla F, Shayo S, Mrosso H (2021) Applying geophysics in support of land rights and environmental equity in peri-urban Africa. GeoJournal 86(3):1173–1189

Hamel P, Tan L (2022) Blue-Green infrastructure for flood and water quality management in Southeast Asia: evidence and knowledge gaps. Environ Manag 69:699–718

Hancock L, Ralph N, Ali SH (2018) Bolivia's lithium frontier: can public private partnerships deliver a mineral boom for sustainable development? J Clean Prod 178:551–560

Hauck C (2013) New concepts in geophysical surveying and data interpretation for permafrost terrain. Permafrost Periglac Process 24(2):131–137

Herrington R (2021) Mining our green future. Nat Rev Mater 6(6):456–458

Horota RK, Senger K, Smyrak-Sikora A, Furze M, Retelle M, Vander Kloet M et al (2024) VR Svalbard—a photosphere-based atlas of a high arctic geo-landscape. First Break 42:35–42

Horota RK, Rossa P, Junior AM, Jr LG, Senger K, Cazarin CL et al (2022a) An immersive virtual field experience structuring method for geoscience education. IEEE Trans Learn Technol (Tlt)

Horota RK, Senger K, Rodes N, Betlem P, Smyrak-Sikora A, Jonassen MO et al (2022b) West Spitsbergen fold and thrust belt: a digital educational data package for teaching structural geology. J Struct Geol 167:104781

IAPG (2024) Cape Town statement on geoethics. https://www.geoethics.org/ctsg

Ijumulana J, Ligate F, Bhattacharya P, Mtalo F, Zhang C (2020) Spatial analysis and GIS mapping of regional hotspots and potential health risk of fluoride concentrations in groundwater of Northern Tanzania. Sci Total Environ 735:139584

IPCC (2023) AR6 synthesis report: climate change 2023. Intergovernmental Panel on Climate Change

IPCC (2021) Climate change 2021: the physical science basis. Available at: https://report.ipcc.ch/ar6/wg1/IPCC_AR6_WGI_FullReport.pdf

Kadic A, Maljkovic B, Rogulj K, Pamukovic JK (2025) Green infrastructure's role in climate change adaptation: summarizing the existing research in the most benefited policy sectors. Sustainability 17:4178

Keesstra SD, Bouma J, Wallinga J, Tittonell P, Smith P, Cerda A et al (2016) The significance of soils and soil science towards realization of the United Nations Sustainable Development Goals. SOIL 2(2):111–128

Kelman I, Mather TA (2008) Living with volcanoes: the sustainable livelihoods approach for volcano-related opportunities. J Volcanol Geoth Res 172(3):189–198

Kim J-S, Kwon S-K, Sanchez M, Cho G-C (2011) Geological storage of high-level nuclear waste. KSCE J Civ Eng 15(4):721–737

Klove B, Ala-aho P, Bertrand G, Boukalova Z, Erturk A, Goldscheider N et al (2011) Groundwater dependent ecosystems. Part I: hydroecological status and trends. Environ Sci Policy 14(7):770–781

Kritikakis G, Kokinou E, Economou N, Andronikidis N, Brintakis J, Daliakopoulos IN et al (2022) Estimating soil clay content using an agro geophysical and agro geological approach: a case study in Chania Plain, Greece. Water 14(17):2625

Kumar L, Mutanga O (2018) Google Earth engine applications since inception: usage, trends, and potential. Remote Sens 10(10):1509

Kurniawan TA, Goh KC, Goh HH, Zhang D, Jiang M, Dai W, Ali I, Othman MHD, Aziz F, Anouzla A, Haddout S, Meidiana C (2024) Strengthening climate resilience: urban water technologies for heat-resilient physical infrastructure in Southeast Asia cities amidst extreme temperature events and El Nino challenges. ACS EST Water 4:3713–3724

Kurniawan TA, Othman MHD, Singh D, Avtar R, Hwang GH, Setiadi T et al (2022) Technological solutions for long-term storage of partially used nuclear waste: a critical review. Ann Nucl Energy 166:108736

Kyriakopoulos GL, Arabatzis G (2016) Electrical energy storage systems in electricity generation: energy policies, innovative technologies, and regulatory regimes. Renew Sustain Energy Rev 56:1044–1067

Lagesse RH, Hambling J, Gill JC, Dobbs M, Lim C, Ingvorsen P (2022) The role of engineering geology in delivering the United Nations Sustainable Development Goals. Quart J Eng Geol Hydrogeol 55

Lanzinger M, Garlandini A (2019) Local development and sustainable development goals: a museum experience. Museum Int 71(3–4):46–57

Lary DJ, Alavi AH, Gandomi AH, Walker AL (2016) Machine learning in geosciences and remote sensing. Geosci Front 7(1):3–10

Lear CH, Anand P, Blenkinsop T, Foster GL, Gagen M, Hoogakker B et al (2021). Geological Society of London scientific statement: what the geological record tells us about our present and future climate. J Geol Soc 178(1)

Lee YC, Liu YF (2023) Co-benefits of preserving urban farmland as climate change adaptation strategy: an energy approach. Ecol Indic 154:110722

Lehmler S, Forster M, Frick A (2023) Modelling green volume using Sentinel-1, -2, PALSAR-2 satellite data and machine learning for urban and semi-urban areas in Germany. Environ Manag 72:657–670

Leppanen PK, Kinnunen A, Hautamaki R, Jarvi L, Havu M, Junnila S, Tahvonen O (2024) Impact of changing urban typologies on residential vegetation and its climate-effects—a case study from Helsinki, Finland. Urban Urban Green 96:128343

Levin N (2023) Using night lights from space to assess areas impacted by the 2023 Turkey earthquake. Remote Sens 15(8):2120

Li J, Hong D, Gao L, Yao J, Zheng K, Zhang B, Chanussot J (2022) Deep learning in multimodal remote sensing data fusion: a comprehensive review. arXiv:2205.01380v1 (cs.CV)

Li X, Stringer LC, Dallimer M (2022b) The impacts of urbanisation and climate change on the urban thermal environment in Africa. Climate 10:164

Liao K, Deng S, Tan PY (2017) Blue-green infrastructure: new frontier for sustainable urban stormwater management. In: Tan PY, Jim CY (eds) Greening cities: forms and functions. Springer, pp 203–226

Liao Q-L, Evans LJ, Gu X, Fan D-F, Jin Y, Wang H (2007) A regional geochemical survey of soils in Jiangsu province, China: preliminary assessment of soil fertility and soil contamination. Geoderma 142(1):18–28

Liu J, Zuo J, Sun Z, Zillante G, Chen X (2013) Sustainability in hydropower development—a case study. Renew Sustain Energy Rev 19:230–237

Lomax A et al (2012) Earthquake hazard assessment using microzonation techniques. Geophys J Int 190(3):1369–1381

Losada-Iglesias R, García AM, Díaz-Varela E, Miranda D (2024) Proposed spatial decision support system for delineating ecological corridors in green infrastructure planning constrained by lack of data: a case study in Galicia, Spain. Landsc Ecol Eng 20:331–347

Manning DAC, Theodoro SH (2020) Enabling food security through use of local rocks and minerals. Extractive Indus Soc 7(2):480–487

Mansour MM, Ibrahim MG, Fujii M, Nasr M (2022) Sustainable development goals (SDGs) associated with flash flood hazard mapping and management measures through morphometric evaluation. Geocarto Int 37:11116–11133

Martin DA (2019) Linking fire and the United Nations Sustainable Development Goals. Sci Total Environ 662:547–558

Martinez-Frias J, Mederos EM, Lunar R (2017) The scientific and educational significance of geoparks as planetary analogues: the example of Lanzarote and Chinijo islands UNESCO Global Geopark. Episodes J Int Geosci 40(4):343–347

Marvin Ursula B (2002) Geology: from an Earth to a planetary science in the twentieth century. Geol Soc Lond 192(1):17–57

Mata-Perello JM, Mata-Lleonart R, Vintro-Sanchez C, Restrepo Martinez C (2012) Social geology: a new perspective on geology. DYNA 79:158–166

McElwee P, Calvin K, Campbell D, Cherubini F, Grassi G, Korotkov V et al (2020) The impact of interventions in the global land and agri-food sectors on nature's contributions to people and the UN Sustainable Development Goals. Glob Change Biol 26(9):4691–4721

Water MDPI (2023) Mapping groundwater potential zones in arid regions: a geographic information system and remote sensing approach for sustainable resource management in Khamis Mushait, Saudi Arabia. Water 17(6):782

Mead C, Buxner S, Bruce G, Taylor W, Semken S, Anbar AD (2019) Immersive, interactive virtual field trips promote science learning. J Geosci Educ 67(2):131–142

Mendoza A et al (2020) Geophysical characterization of informal urban settlements impacted by industrial pollution in Latin America. Environ Geophys 25(4):281–295

Meusburger K, Porto P, Mabit L, La Spada C, Arata L, Alewell C (2018) Excess lead-210 and plutonium-239+240: two suitable radiogenic soil erosion tracers for mountain grassland sites. Environ Res 160:195–202

Min J, Kc B, Kim S, Lee J (2020) The impact of disasters on a heritage tourist destination: a case study of Nepal earthquakes. Sustainability 12(15):6115

Mostafa MT, Nady HE, Gomaa RM, Abdelgawad HF, Farhat HI, Khalifa IH et al (2023) Geochemical baseline and pre-mining environmental assessment of heavy metals at iron exploration area, Northeastern Aswan, Egypt. Water Air Soil Pollut 234(7):456

Mudd GM (2021) Sustainable/Responsible mining and ethical issues related to the sustainable development goals. Geol Soc Lond 508(1):187–199

Mudd GM, Diesendorf M (2008) Sustainability of uranium mining and milling: toward quantifying resources and eco efficiency. Environ Sci Technol 42(7):2624–2630

Nab A, Gasselt SV (2014) A framework for planetary geologic mapping, cartography from pole to pole. Springer, pp 261–270

Nakrem HA, Lindemann F-J, Hurum JH, Hammer O (2023) Fossils from the Arctic in the collections of the natural history museum in Oslo, Norway. Collections 19(3):420–441

Noorollahi Y, Fadai D, Shirazi MA, Ghodsipour SH (2007) Geothermal energy development in Iran: a review of the current status. Renew Sustain Energy Rev 11(3):438–445

Nussbaumer SU, Hoelzle M, Husler F, Huggel C, Salzmann N, Zemp M (2017) Glacier monitoring and capacity building: important ingredients for sustainable mountain development. Mt Res Dev 37(1):141–152

Palsdottir A, Johannsdottir L (2021) Signs of the United Nations SDGs in university curriculum: the case of the University of Iceland. Sustainability 13(16):8958

Panagos P, Katsoyiannis A (2019) Soil erosion modelling: the new challenges as the result of policy developments in Europe. Elsevier, pp 470–474

Paulikas D, Katona S, Ilves E, Ali SH (2022) Deep-sea nodules versus land ores: a comparative systems analysis of mining and processing wastes for battery-metal supply chains. J Ind Ecol 26:2154–2177

Paxton AB, Riley TN, Steenrod CL, Smith CS, Zhang YS, Gittman RK, Silliman BR, Buckel CA, Viehman TS, Puckett BJ et al (2023) What evidence exists on the performance of nature-based solutions interventions for coastal protection in biogenic, shallow ecosystems? A systematic map protocols. Environ Evid 12:11

Peppoloni S, Di Capua G (2025) Geoethics and sustainability: addressing challenges in environmental education for achieving the SDGs. Sustainability 17(2):574. https://doi.org/10.3390/su17020574

Peppoloni S, Di Capua G, Bobrowsky P (2023) GOAL project: geoethics education for the next generation. Geosciences 13(10):302. https://doi.org/10.3390/geosciences13100302

Petterson MG (2019) Interconnected geoscience for international development. Int Union Geol Sci 42(3):225–233

Pimentel D, Kalyanaraman S (2021) Virtual climate scientist: a VR learning experience about Paleoclimatology for underrepresented Students. Interact Learn Environ 31:4426–4439

Pirasteh S, Rashidi P, Rastiveis H, Huang S, Zhu Q, Liu G et al (2019) Developing an algorithm for buildings extraction and determining changes from airborne LiDAR, and comparing with R-CNN method from Drone images. Remote Sens 11(11):1272

Planavsky N, Hood A, Tarhan L, Shen S, Johnson K (2020) Store and share ancient rocks. Nature 581(7807):137–139

Plant J, Smith D, Smith B, Williams L (2001) Environmental geochemistry at the global scale. Appl Geochem 16(11):1291–1308

Pratt C, Kingston K, Laycock B, Levett I, Pratt S (2020) Geo agriculture: reviewing opportunities through which the geosphere can help address emerging crop production challenges. Agronomy 10(7):971

Pugsley JH, Howell JA, Hartley A, Buckley SJ, Brackenridge R, Schofield N et al (2022) Virtual field trips utilizing virtual outcrop: construction, delivery and implications for the future. Geosci. Commun. 5(3):227–249

Ramboll (ed) (2016) Strengthening blue-green infrastructure in our cities: enhancing blue-green infrastructure & social performance in high density urban environments. In: Dreiseitl H, Wanschura B (eds)

Reynolds JM (2011) An introduction to applied and environmental geophysics (2nd ed.). Wiley-Blackwell

Rodriguez M, Fu G, Butler D, Yuan Z, Cook L (2024) The effect of green infrastructure on resilience performance in combined sewer systems under climate change. J Environ Manag 353:120229

Ruidas D, Pal SC, Biswas T, Saha A, Towfiqul Islam AR (2024) Extreme exposure of fluoride and arsenic contamination in shallow coastal aquifers of the Ganges delta, transboundary of the Indo-Bangladesh region. Geosci Front 15(1):101725

Ryabinin V, Barbiere J, Haugan P, Kullenberg G, Smith N, McLean C et al (2019) The UN decade of ocean science for sustainable development. Front Mar Sci 6:470

Rybach L (2003) Geothermal energy: sustainability and the environment. Geothermics 32(4–6):463–470

Salminen R, Gregorauskien V (2000) Considerations regarding the definition of a geochemical baseline of elements in the surficial materials in areas differing in basic geology. Appl Geochem 15(5):647–653

Saunders WS, Kelly S, Paisley S, Clarke LB (2020) Progress toward implementing the Sendai framework, the Paris agreement, and the sustainable development goals: policy from Aotearoa New Zealand. Int J Disaster Risk Sci 11(2):190–205

Saunders W, Becker J (2015) A discussion of resilience and sustainability: land use planning recovery from the Canterbury earthquake sequence, New Zealand. Int J Disaster Risk Reduc 14:73–81

Schipper CA, Dekker GG, de Visser B, Bolman B, Lodder Q (2021) Characterization of SDGs towards coastal management: sustainability performance and cross-linking consequences. Sustainability 13(3):1560

Schneider AE, Neuhuber T, Zawadzki W (2024) Understanding citizens willingness to contribute to urban greening programs. Urban Urban Green 2024(95):128293

Schultz RA, Hauber E, Kattenhorn SA, Okubo CH, Watters TR (2010) Interpretation and analysis of planetary structures. J Struct Geol 32(6):855–875

ScienceDirect (2024) Multi-criteria decision-making for groundwater potentiality zonation in a groundwater scarce region in central India using methods of compensatory aggregating functions. Groundwater Sustain Develop 25:101101

Scown MW (2020) The sustainable development goals need geoscience. Nat Geosci 13(11):714–715

Senger K (2024) Sustainable development goals and the geosciences: a review. Earth Sci Syst Soc 4:10124. https://doi.org/10.3389/esss.2024.10124

Senger K, Betlem P, Birchall T, Buckley SJ, Coakley B, Eide CH et al (2021) Using digital outcrops to make the high arctic more accessible through the Svalbox Database. J Geosci Educ 69(2):123–137

Sharma S, Baidya M, Poudel P, Panthi SR, Pote-Shrestha RR, Ghimire A et al (2021) Drinking water status in Nepal: an overview in the context of climate change. J Water Sanitation Hygiene Develop 11(6):859–866

Sharma R, Malaviya P (2021) Management of stormwater pollution using green infrastructure: the role of rain gardens. Wiley Interdiscip Rev Water 8:e1507

Sheffield J, Wood EF, Pan M, Beck H, Coccia G, Serrat-Capdevila A et al (2018) Satellite remote sensing for water resources management: potential for supporting sustainable development in data-poor regions. Water Resour Res 54(12):9724–9758

Shilston DT (2023) The 21st Glossop lecture: engineering geology and the geoscience time machine. Quart J Eng Geol Hydrogeol 56(1)

Shugar D, Jacquemart M, Shean D, Bhushan S, Upadhyay K, Sattar A et al (2021) A massive rock and ice avalanche caused the 2021 disaster at Chamoli, Indian Himalaya. Science 373(6552):300–306

Siehr SA, Sun M, Aranda Nucamendi JL (2022) Blue-green infrastructure for climate resilience and urban multifunctionality in Chinese cities. Wires Energy Environ 11(5):e447

Simandl G (2014) Geology and market-dependent significance of rare earth element resources. Miner Deposita 49(8):889–904

Smith DB, de Caritat P, Reimann C (2023) Introduction to the thematic collection: a review of continental-scale geochemical mapping projects in Australia, Europe, and the conterminous United States. Geochemistry Explor Environ Anal 23(1)

Smith P, Keesstra SD, Silver WL, Adhya TK, De Deyn GB, Carvalheiro LG et al (2021) Soil-derived nature's contributions to people and their contribution to the UN sustainable development goals. Philos Trans R Soc B Biol Sci 376(1834):20200185

Smith J et al (2020) Machine learning applications in hyperspectral remote sensing for mineral exploration. Ore Geol Rev 122:103456

Stemland HM, Johansen TA, Ruud BO, Romeyn R (2021) Active versus passive seismic monitoring of near-surface arctic thawing. In: NSG2021 27th European meeting of environmental and engineering geophysics, pp 1–5

Stern JC, McAdam AC, TenKate IL, Bish DL, Blake DF, Morris RV et al (2013) Isotopic and geochemical investigation of two distinct Mars analog environments using evolved gas techniques in Svalbard, Norway. Icarus 224(2):297–308

Stewart IS, Hurth V (2021) Selling planet Earth: re-purposing geoscience communications. Geol Soc Lond 508(1):265–283

Stewart IS, Gill JC (2017a) Social geology – Integrating sustainability concepts into Earth science education and practice. Proc Geol Assoc 128(2):165–172

Stewart IS, Gill JC (2017) Modernising the geoscience curriculum in higher education. Earth Sci Syst Soci 1(1)

Straume EO, Nummelin A, Gaina C, Nisancioglu KH (2022) Climate transition at the Eocene-Oligocene influenced by bathymetric changes to the Atlantic-Arctic oceanic gateways. Proc Natl Acad Sci 119(17):e2115346119

Sudmanns M, Augustin H, Killough B, Giuliani G, Tiede D, Leith A et al (2023) Think global, cube local: an Earth observation data cube's contribution to the digital Earth vision. Big Earth Data 7(3):831–859

Sussams LW, Sheate WR, Eales RP (2015) Green infrastructure as a climate change adaptation policy intervention: muddying the waters or clearing a path to a more secure future? J Environ Manag 2015(147):184–193

Takagi J, Wada A (2019) Recent earthquakes and the need for a new philosophy for earthquake-resistant design. Soil Dyn Earthq Eng 119:499–507

Tamiminia H, Salehi B, Mahdianpari M, Quackenbush L, Adeli S, Brisco B (2020) Google Earth engine for geo-big data applications: a meta-analysis and systematic review. ISPRS J Photogramm Remote Sens 164:152–170

Tarkowski R (2019) Underground hydrogen storage: characteristics and prospects. Renew Sustain Energy Rev 105:86–94

Taylor RG, Scanlon B, Doll P, Rodell M, van Beek R, Wada Y, Longuevergne L, Leblanc M, Famiglietti JS, Edmunds M, Konikow L, Green TR, Chen J, Taniguchi M, Bierkens MFP, MacDonald A, Fan Y, Maxwell RM, Yechieli Y, Gurdak JJ, Allen DM, Shamsudduha M, Hiscock K, Yeh PJ-F, Holman I, Treidel H (2013) Ground water and climate change. Nat Climate Change 12:322–329

The Guardian (2025) Global weirding': climate whiplash hitting world's biggest cities, study reveals. The Guardian. https://www.theguardian.com/environment/2025/mar/12/global-weirding-climate-whiplash-hitting-worlds-biggest-cities-study-reveals

Thompson L (2020) Challenges in AI applications for remote sensing in mineral exploration. Earth Sci Rev 209:103321

Tiangang W, Kumul C, Yuhao Z, Mosusu N, Bimin Z, Pei N et al (2024) National-scale geochemical baseline of 69 elements in Papua New Guinea stream sediments. J Geochem Explor 256:107355

Tonneijck F, Jansen B, Nierop K, Verstraten J, Sevink J, De Lange L (2010) Towards understanding of carbon stocks and stabilization in volcanic ash soils in natural Andean ecosystems of Northern Ecuador European. J Soil Sci 61(3):392–405

Trouet V, Van Oldenborgh GJ (2013) KNMI climate explorer: a web-based research tool for high-resolution Paleoclimatology. Tree-Ring Res 69(1):3–13

UNESCO (2022) The United Nations world water development report 2022: groundwater: making the invisible visible. UNESCO, Paris

UNDRR (United Nations Office for Disaster Risk Reduction) (2022) Global Assessment Report on Disaster Risk Reduction 2022. UNDRR, Geneva

United Nations (2022) Sustainable development goals. Available at https sdgs un org goals

United Nations (2015) Transforming our world: the 2030 agenda for sustainable development. https://sdgs.un.org/2030agenda

Untereiner E, Toboso-Chavero S, Farinas AV, Madrid-Lopez C, Villalba G, Durany XG (2024) Predicting willingness to pay and implement different rooftop strategies to characterize social perception of climate change mitigation and adaptation. Environ Res Commun 2024(6):015004

Van Aalst MK (2006) The impacts of climate change on the risk of natural disasters. Disasters 30(1):5–18

van der Zwaan B, Dalla Longa F (2019) Integrated assessment projections for global geothermal energy use. Geothermics 82:203–211

Van Straaten P (2006) Farming with rocks and minerals: challenges and opportunities. Anais Da Academia Brasileira De Ciencias 78(4):731–747

Varotsos C, Mazei Y, Novenko E, Tsyganov AN, Olchev A, Pampura T et al (2020) A new climate nowcasting tool based on Paleoclimatic data. Sustainability 12(14):5546
Velis M, Conti KI, Biermann F (2017) Groundwater and human development: synergies and trade-offs within the context of the sustainable development goals. Sustain Sci 12(6):1007–1017
Verheyen E, Abila R, Akoll P, Albertson C, Antunes D, Banda T et al (2016) Oil extraction imperils Africa's Great Lakes. Science 354(6312):561–562
Villacorta, J., et al. (2024). Reimagining geoscience education for sustainability. Earth Sci Syst Soc 1(2). https://doi.org/10.3389/esss.2024.10116
Wang J, Banzhaf E (2018) Towards a better understanding of Green Infrastructure: a critical review. Ecol Ind 85:758–772
von Schuckmann K, Holland E, Haugan P, Thomson P (2020) Ocean science, data, and services for the UN 2030 sustainable development goals. Mar Policy 121:104154. https://doi.org/10.1016/j.marpol.2020.104154
Wagaba B, Sayan C, Haemmerli H, Gill J, Parker A (2023) Understanding small NGOs access to and use of geological data and expertise in delivering SDG 6 in Eastern Africa. J Water Sanitation Hygiene Develop 13(5):312–321
Wang W, Wang X, Zhang B, Chi Q, Liu H, Liu X et al (2022) Concentrations and spatial distribution of chlorine in the pedosphere in China: based on the China geochemical baselines project. J Geochem Explor 242:107089
Wang C, Hazen RM, Cheng Q, Stephenson MH, Zhou C, Fox P et al (2021). The deep-time digital Earth program: data-driven discovery in geosciences. Nat Sci Rev 8(9):nwab027
Wei M, Zhang Z, Long T, He G, Wang G (2021) Monitoring landsat based burned area as an indicator of sustainable development goals. Earth's Future 9(6):e2020EF001960
Werner K, Muller J, Husum K, Spielhagen RF, Kandiano ES, Polyak L (2016) Holocene sea subsurface and surface water masses in the Fram Strait—comparisons of temperature and sea-ice reconstructions. Quatern Sci Rev 147:194–209
Wessel GR, Greenberg JK, Morgan FD (2015) Geoscience for the public good and global development. GSA Today 25(10):4–10
Whitmeyer SJ, Atchison C, Collins TD (2020) Using mobile technologies to enhance accessibility and inclusion in field-based learning. GSA Today 30:4–10
Whitmeyer SJ, Dordevic M (2020) Creating virtual geologic mapping exercises in a changing world. Geosphere 17(1):226–243
Wiley Online Library (2024) Blue-green infrastructure for climate resilience and urban sustainability. Wiley Interdisciplinary Reviews: Energy and Environment
Wilson AR, Vasile M (2023) The space sustainability paradox. J Clean Prod 423:138869
Witze A (2016) Iconic Antarctic geology lab gets the boot. Nature 534(7608):448
y Leon SB, Lindberg JC (2022) The humanitarian atom: the role of nuclear power in addressing the United Nations Sustainable Development Goals. Nuclear Law Global Debate 271–298
Zapata-Paulini J, Cabanillas-Carbonell M, Iparraguirre-Villanueva O, Sierra-Linan F, Baltozar-Clemente S, Alvarez-Risco A et al (2023) Augmented reality for innovation: education and analysis of the glacial retreat of the Peruvian Andean snow-capped mountains. J Open Innov Technol Market Complexity 9(3):100106
Zhang B, MacKenzie A (2024) Trade-offs and synergies in urban green infrastructure: a systematic review. Urban for. Urban Green. 94:128262
Zhao T, Wang S, Ouyang C, Chen M, Liu C, Zhang J, Yu L, Wang F, Xie Y, Li J, Wang F, Grunwald S, Wong BM, Zhang F, Qian Z, Xu Y, Yu C, Han W, Sun T, Shao Z, Qian T, Chen Z, Zeng J, Zhang H, Letu H, Zhang B, Wang L, Luo L, Shi C, Su H, Zhang H, Yin S, Huang N, Zhao W, Li N, Zheng C, Zhou Y, Huang C, Feng D, Xu Q, Wu Y, Hong D, Wang Z, Lin Y, Zhang T, Kumar P, Plaza A, Chanussot J, Zhang J, Shi J, Wang L (2024) Artificial intelligence for geoscience: progress, challenges, and perspectives. Innovation 5(5):100691. CellPress Partner Journal
Zheng Q, Lyu H-M, Zhou A, Shen S-L (2021) Risk assessment of geohazards along Cheng-Kun railway using fuzzy AHP incorporated into GIS. Geomat Nat Haz Risk 12(1):1508–1531

Chapter 11
Geophysics for Risk-Resilient and Critical Infrastructure

Safeguarding Vital Systems Through Proactive Subsurface Insights

Abstract This chapter focuses on how geophysical methods help protect critical infrastructure by assessing risks and enhancing resilience against natural hazards, especially earthquakes and climate impacts. It reviews current global resilience frameworks and explores how geophysical data can improve decision-making in planning and maintenance of utilities, transportation, and energy networks. The chapter highlights advanced techniques for early detection of subsurface changes and predictive maintenance. Emerging technologies such as AI, IoT, and digital twins are discussed as tools to strengthen infrastructure resilience. Several case studies from around the world illustrate practical applications, including metro systems, airports, and urban underground networks. The chapter concludes with recommendations for integrating geophysics into infrastructure policy, planning, and disaster risk reduction, emphasizing the vital role of proactive subsurface insights for safeguarding vital systems.

Modern critical infrastructure—comprising transportation networks, energy systems, utilities, and public services—forms the foundation of societal function and economic stability. Ensuring the resilience of these systems requires more than reactive repair; it demands proactive understanding of subsurface conditions that can lead to failure if left undetected. Engineering geophysics offers a suite of non-invasive tools that deliver crucial insights into ground conditions, enabling early risk detection and informed planning. This chapter explores how geophysical technologies can be leveraged to enhance the robustness, reliability, and longevity of critical infrastructure.

11.1 Introduction

Critical infrastructure such as transportation systems, energy grids, water supply networks, and communication facilities plays an indispensable role in maintaining societal function, economic activity, and public safety. The uninterrupted operation

D. C. Naskar, *Advances in Engineering Geophysics: Concepts and Applications*, Earth Systems Data and Models 9, https://doi.org/10.1007/978-3-032-28000-8_11

of these systems is vital, particularly as urban populations expand and environmental pressures intensify. In this context, infrastructure resilience—the ability of systems to absorb, adapt, and rapidly recover from disruptive events—has emerged as a key priority for planners, engineers, and policymakers.

Resilience is closely tied to the concept of vulnerability, which refers to the degree to which infrastructure is exposed to risk due to factors like design limitations, aging materials, ground instability, or exposure to natural hazards. Increasingly, subsurface conditions such as soil instability, hidden voids, and water infiltration are recognized as major contributors to infrastructure failure. Despite their importance, these subsurface factors often remain poorly understood or inadequately monitored until visible damage occurs.

The relational equation that connects hazards, vulnerability, and risk is often expressed as:

$$\text{Risk} = \text{Hazard} \times \text{Vulnerability} \tag{11.1}$$

In some models, capacity is also included to reflect the ability to cope with or mitigate the effects of hazards:

$$\text{Risk} = \frac{\text{Hazard} \times \text{Vulnerability}}{\text{Capacity}} \tag{11.2}$$

To understand how infrastructure systems, behave under disruptive conditions, it is useful to consider performance-based models of resilience. These models illustrate how a system's functionality evolves before, during, and after a disturbance, highlighting key components such as robustness, redundancy, resourcefulness, and rapidity. A graphical representation of these performance-based resilience components is shown in Fig. 11.1, which provides a conceptual foundation for subsequent discussions on infrastructure assessment and geophysical applications.

Engineering geophysics offers a suite of non-invasive, high-resolution techniques capable of imaging and monitoring the subsurface in real time. Methods such as Ground Penetrating Radar (GPR), Electrical Resistivity Tomography (ERT), and seismic surveys provide critical data on soil moisture, buried objects, structural voids, and geotechnical hazards. These tools enable early detection of anomalies, guide maintenance decisions, and reduce the risk of catastrophic failure—thereby serving as a cornerstone of proactive risk management in infrastructure systems (Gunn et al. 2023).

This chapter explores the integration of geophysical methods into infrastructure resilience planning and asset management. It examines how geophysics supports the safeguarding of utilities, transportation corridors, and energy networks, with a focus on pre-failure detection and preventive maintenance strategies. By presenting case studies, methodological frameworks, and future outlooks, the chapter aims to demonstrate how subsurface insights can enhance the safety, reliability, and sustainability of vital infrastructure systems.

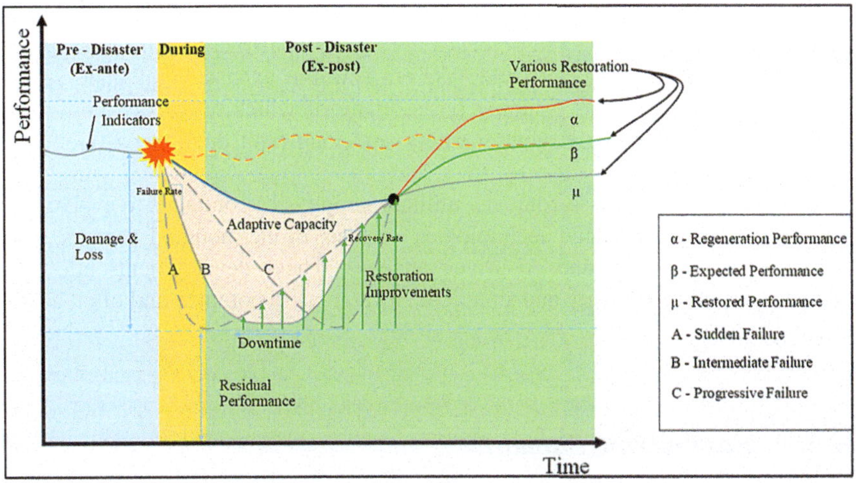

Fig. 11.1 Performance-based resilience components of a Critical Infrastructure System (CIS). The framework emphasizes system functionality over time, integrating aspects of robustness, redundancy, resourcefulness, and rapidity (*Source* Sathurshan et al. 2022—Open Access)

11.1.1 Assessing Earthquake Resilience in Critical Infrastructure

In recent years, a growing body of research has focused on developing and refining measurement frameworks to assess the resilience of critical infrastructure systems (CIS) (Osei-Kyei et al. 2021; Kim et al. 2019). These systems—such as transportation, power, telecommunications, water supply, and healthcare—form the backbone of societal function, especially during and after disasters. Ensuring their resilience is vital for reducing vulnerability and enhancing recovery capabilities. Resilience frameworks are therefore essential tools for identifying risks, evaluating performance, and supporting resilience-focused investments across disaster management phases.

Frameworks for resilience assessment have emerged in response to an increasing frequency and intensity of both natural and man-made hazards—including earthquakes, tsunamis, floods, landslides, technological failures, and acts of sabotage (Bach et al. 2013; Bertocchi et al. 2016; Satheeskumar 2016). Global institutions such as the United Nations Office for Disaster Risk Reduction (UNDRR), the Federal Emergency Management Agency (FEMA), and the International Organization for Standardization (ISO) have also contributed significantly by developing high-level guidelines and best practices.

To ensure methodological rigor, many studies—including Sathurshan et al. (2022)—adopt the PRISMA (Preferred Reporting Items for Systematic Reviews and Meta-Analyses) approach for identifying, screening, and selecting relevant literature.

This systematic method is illustrated in Fig. 11.2, which outlines the key stages in compiling and evaluating sources related to infrastructure resilience frameworks.

A comprehensive review by Sathurshan et al. (2022) focused on 24 critical infrastructure resilience (CIR) frameworks developed between 2015 and 2021—following the adoption of the Sendai Framework for Disaster Risk Reduction (SFDRR). This study targeted frameworks specifically designed to assess earthquake resilience. The frameworks were analyzed based on country of origin, methodological approach, disaster management phase addressed (pre-, during-, or post-disaster), and sectoral focus.

An analysis of the year-wise distribution of these studies (shown in Fig. 11.3) illustrates a rising trend in CIR research, especially after the introduction of the SFDRR in 2015, indicating increased global interest in resilience-based planning. The development approaches of these frameworks also vary, encompassing qualitative models, quantitative risk-based tools, hybrid systems, and simulation-based methods.

Figure 11.4 summarizes these diverse development methodologies, providing insight into the underlying philosophies and technical orientations guiding resilience framework design. The frameworks were also categorized by their geographical application. As shown in Fig. 11.5, most studies focused on either national or local contexts, with relatively few offerings regionally or globally transferable models.

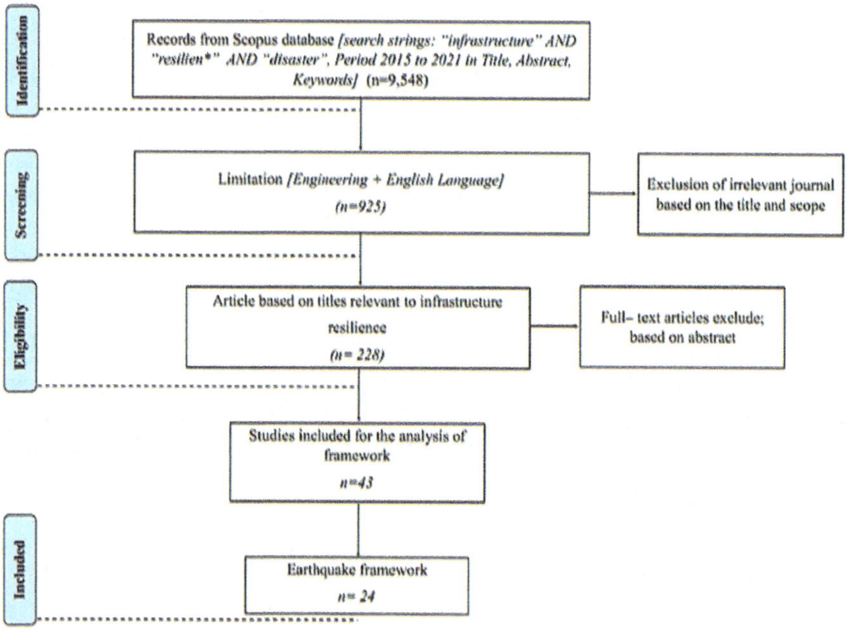

Fig. 11.2 The Preferred Reporting Items for Systematic Reviews and Meta-Analyses (PRISMA) flowchart used to identify and select studies relevant to infrastructure resilience frameworks (*Source* Sathurshan et al. 2022—Open Access)

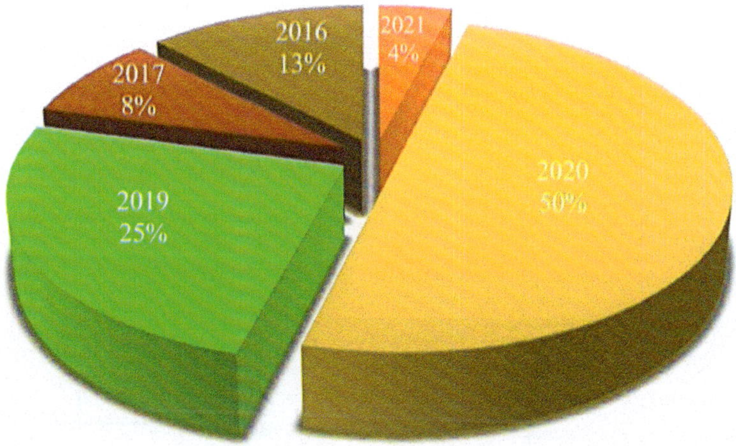

Fig. 11.3 The summary of studies evaluated by the year of publication (*Source* Sathurshan et al. 2022—Open Access)

This finding reinforces the need for more adaptable frameworks that can operate across a range of geographic and socio-economic environments.

Figure 11.6 shows the distribution of these frameworks by infrastructure type. The analysis revealed that the majority focused on housing and buildings (32%), followed by transportation systems (26%), electricity and power networks (23%), water infrastructure (11%), and telecommunications (8%). This distribution reflects a prioritization of sectors most vulnerable to seismic disruptions.

Despite growing efforts, the review uncovered wide-ranging disparities in how CIR is conceptualized and assessed. Many frameworks are context-specific—geographically or community-based—and lack adaptability across diverse settings. As a result, they fall short of providing a universal method for seismic resilience assessment. The absence of a consistent, adaptable approach impedes the ability of governments and stakeholders to prioritize and justify resilience-focused infrastructure investments.

Key attributes identified as essential for a comprehensive CIR framework include serviceability, functionality, and the development of standardized resilience performance indicators. These indicators should be applicable across all disaster phases to evaluate infrastructure performance consistently and to support evidence-based decision-making. Most existing frameworks emphasized: *Ex-ante*—Risk and reliability assessments, *During disaster*—Resourcefulness and emergency management strategies, and *Ex-post*—Restoration planning and long-term recovery. However, socioeconomic considerations and operational continuity during the immediate aftermath of a disaster were often underrepresented. In contrast, the framework proposed by Sathurshan et al. (2022) places a stronger emphasis on emergency protocols and socio-economic recovery mechanisms, particularly in the "during" and "post-disaster" phases.

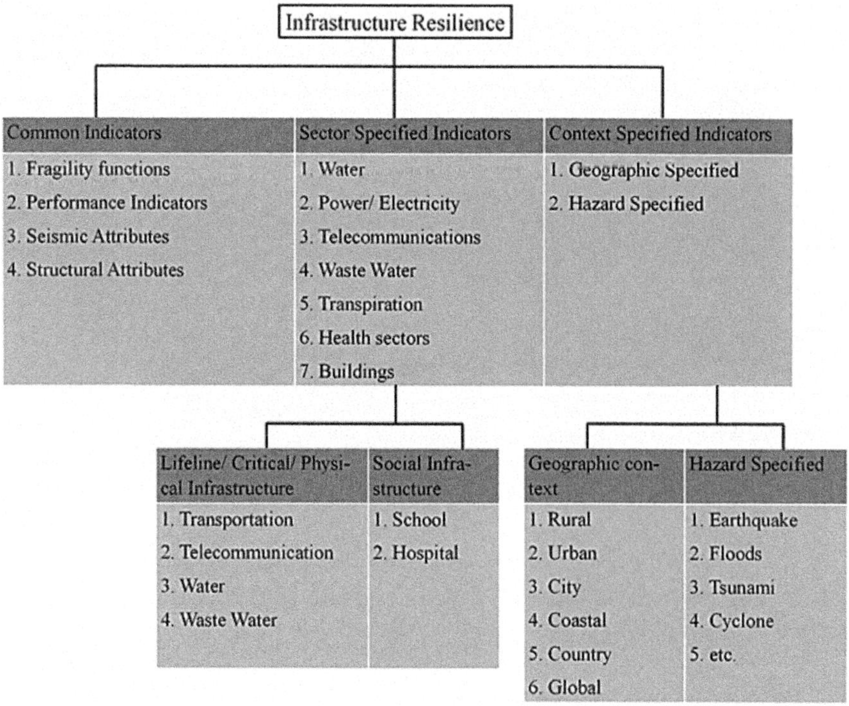

Fig. 11.4 Summary of infrastructure resilience framework development approaches (*Source* Sathurshan et al. 2022—Open Access)

Figure 11.7 provides a synthesis of attributes from damage-analysis-based frameworks, categorizing them into social, functional, and physical metrics. It highlights the range of indicators and methodologies used in past studies to assess infrastructure damage and recovery. Emerging trends in resilience assessment highlight the integration of geophysical techniques such as Ground Penetrating Radar (GPR) and Electrical Resistivity Tomography (ERT). These methods provide real-time, subsurface data critical to infrastructure health monitoring. Their application allows for the early detection of hidden structural vulnerabilities, thereby preventing catastrophic failures and informing maintenance strategies (Thomas et al. 2023a, b).

Additionally, the adoption of multi-scenario analysis is gaining traction. This approach simulates various hazard scenarios to evaluate the performance of infrastructure systems under stress, prioritize mitigation strategies, and identify weak points. Coupling multi-scenario planning with geophysical data strengthens the predictive capacity of resilience frameworks, offering a more comprehensive basis for infrastructure design and disaster preparedness.

Based on the systematic review, Sathurshan et al. (2022) propose an integrated and adaptable resilience assessment framework. It incorporates lessons from the 24 reviewed models and emphasizes: Applicability across multiple geographic contexts,

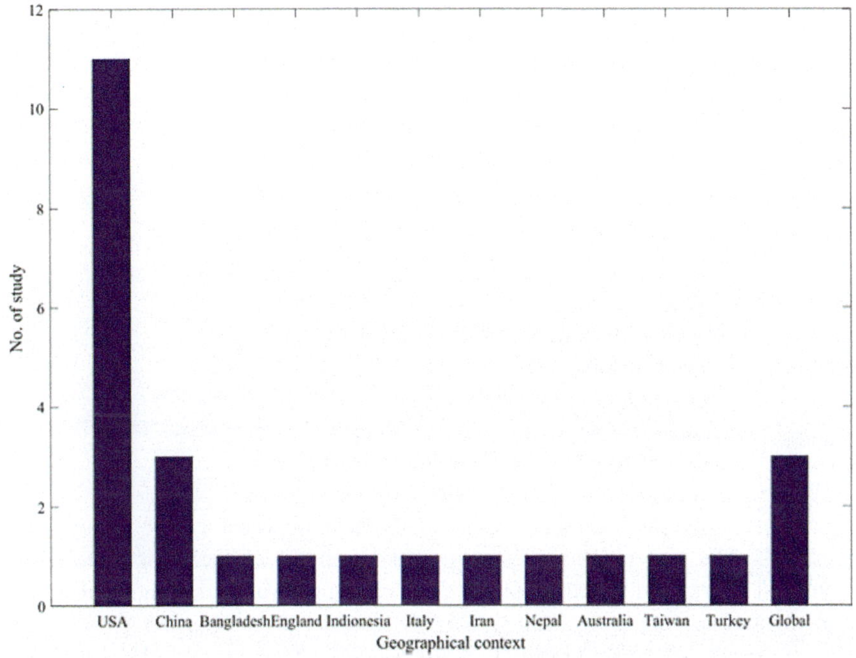

Fig. 11.5 The geographic scope of the infrastructure resilience frameworks evaluated (*Source* Sathurshan et al. 2022—Open Access)

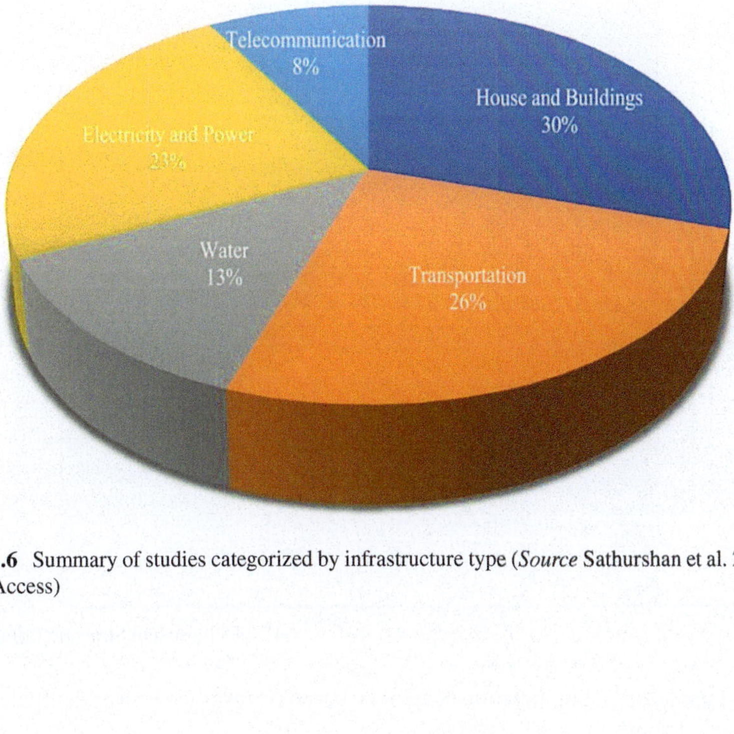

Fig. 11.6 Summary of studies categorized by infrastructure type (*Source* Sathurshan et al. 2022—Open Access)

Damage Modeling/ Analysis Framework

(Mazumder et al., Hadigheh et al., Kameshwar et al.)

Social & Serviceability measures	Functional impacts/ measures	Physical/ System infrastructure	Hazard specifications	Indicators	Methods adaptions
1.Time of occurrence of event 2. Preparedness 3. Resources 4. Societal response 5. Immediate occupancy 6. Life safety 7. Collapse prevention 8. Corrosion deterioration of pipelines	1. Recovery function/time (linear) 2. Retrofitting methods 3. Resilience values 4. Damage-economic loss relations 5. Damage-restoration time 6. Damage functionality relations 7. Repair rate 8. Component functionality 9. System functionality 10. Interdependent functionality 11. Renewal strategy of pipelines	1. Electricity/ Power 2. Buried water pipe-lines 3. Rehabilitated buildings (hospital) 4. Buildings 5. Transportation 6. Water	1. Earthquake/ Seismic 2. Coastal hazards 3. Tsunami inundation	1. Building codes 2. Retrofitting methods 3. Time of occurrence of event 4. Specification of the buildings 5. Seismic response 6. Design loads 7. Discrete variables 8. Hazard specifications 9. Guidelines 10. Pipelines network 11. Corrosion 12. Seismic repair rate 13. Characteristics of an earthquake	1. Performance-based design approach 2. Bayesian Network 3. Numerical modeling 4. Fragility functions

Fig. 11.7 Attributes of the damage-analysis-based framework from past studies in terms of social, functional, and physical metrics, considering the indicators and methodology used (*Source* Sathurshan et al. 2022—Open Access)

Use of consistent performance indicators, and Consideration of cascading failures and interdependencies among infrastructure systems. While the proposed framework marks a significant advancement, it remains preliminary and requires empirical validation. It should be tested across different disaster settings and infrastructure types to ensure scalability and relevance. Future work should also expand on the framework's treatment of interconnectivity, system redundancies, and secondary hazards.

In conclusion, this evolving approach supports a holistic and multidisciplinary understanding of infrastructure resilience—one that combines engineering, data science, and geophysics. It provides a forward-looking strategy to better predict, prepare for, and mitigate the impacts of disasters on critical infrastructure systems.

Metro Systems: Earthquake Resilience in Construction and Operation

This section provides an overview of metro systems by comparing findings from various international studies on metro construction, operation, and their impact on cities (Lin et al. 2024; Cardu and Seccatore 2016; Niu et al. 2022; Cong et al. 2022; Yang et al. 2022; Shi et al. 2020; Sonnenschein et al. 2022; Jothimani and Yamamura 1995). It highlights how metro systems, as vital urban infrastructure, require strong design and planning to ensure resilience, particularly in the face of natural hazards like earthquakes.

First, the development of metro construction methods has been reviewed. Key areas include construction techniques, safety risks on metro construction sites, cost and scheduling challenges, and the impacts of construction on nearby structures and environments. Although metro construction safety has been widely studied, there is still no comprehensive accident database available. Creating such a database would help researchers and engineers learn from past accidents and improve future safety practices, particularly in regions prone to seismic activity.

Second, several aspects related to the safe operation of metro systems have been examined. These include fire safety control, evacuation strategies, aerodynamic effects, and indoor air quality in metro tunnels and stations. As metro systems become more advanced—with deeper tunnels, complex transfer stations, and faster trains—the risks during emergencies, including earthquakes, increase. Traditional safety approaches may not be effective in these newer, more complex settings. Therefore, future research should explore safety strategies suited to modern metro environments, especially under extreme conditions like seismic events.

Third, the broader urban impacts of metro development have been analyzed. These include changes in land use, underground space utilization, property values, urban economy, travel behavior, air quality, and greenhouse gas emissions. Many studies confirm the long-term benefits of metro systems. However, the overall cost–benefit analysis of metro development—particularly how it alters underground water, soil, and rock conditions—has not been fully explored. In seismically active areas, such changes may affect ground stability and increase infrastructure vulnerability during earthquakes. Future research should therefore assess these factors more thoroughly to support resilient metro development. In conclusion, while metro systems play a vital role in urban transportation, their design, construction, and operation must increasingly account for natural hazards, especially earthquakes. Building a more comprehensive understanding of these risks will be essential to making metro infrastructure safer and more resilient in the decades to come.

11.2 Infrastructure Resilience Planning

11.2.1 Understanding Risk and Resilience

Infrastructure resilience is commonly defined as the capacity of a system to prepare for, absorb, recover from, and adapt to adverse events, including both sudden hazards like earthquakes and gradual threats such as subsurface degradation. Leading frameworks such as those from the United Nations Office for Disaster Risk Reduction (UNDRR), Federal Emergency Management Agency (FEMA), and International Organization for Standardization (ISO) 55,000 emphasize a systemic approach—assessing not only the robustness of physical structures but also their adaptive capacity and interdependence. Risk, in this context, is a function of hazard exposure, vulnerability, and the consequences of failure. Geophysical methods contribute uniquely to this equation by enabling the continuous assessment of subsurface conditions that directly impact infrastructure performance, such as soil stability, moisture variability, and hidden anomalies. By integrating geophysical data into risk modeling and decision-making, infrastructure managers can move from reactive to predictive maintenance strategies. Recent research highlights how real-time geophysical monitoring—particularly time-lapse electrical resistivity and passive seismic techniques—has been successfully incorporated into resilience assessments, improving both early-warning capabilities and infrastructure adaptation plans (Uhlemann et al. 2023).

11.2.2 Integration of Geophysical Techniques in Planning

Integrating geophysical techniques into infrastructure planning allows engineers and decision-makers to better characterize site conditions, assess potential hazards, and optimize design strategies before construction begins. Traditional approaches often rely on limited borehole data and visual inspections, which may miss critical subsurface variability. In contrast, non-invasive geophysical methods such as Ground Penetrating Radar (GPR), Electrical Resistivity Tomography (ERT), Multi-channel Analysis of Surface Waves (MASW), and microgravity surveys provide spatially continuous data over large areas, making them ideal for preliminary investigations, route selection, and foundation assessment. When combined with Geographic Information Systems (GIS) and Building Information Modeling (BIM), geophysical datasets enhance multi-criteria decision-making and risk analysis at both the planning and design stages. Recent advancements in sensor integration and automated interpretation have further improved the speed and reliability of these methods. As shown in a study by Bievre et al. (2023), early-stage geophysical mapping has been successfully used in infrastructure projects across Europe to detect unstable zones and guide the placement of critical utilities, demonstrating a shift toward data-informed resilience planning.

11.2.3 Decision-Making Tools and Data Integration

Modern infrastructure resilience planning increasingly relies on the integration of geophysical data with advanced digital tools such as Geographic Information Systems (GIS), numerical modeling platforms, and artificial intelligence (AI). GIS enables the spatial visualization and analysis of subsurface geophysical data—such as resistivity profiles or seismic velocity models—within the broader geographic and infrastructural context. This fusion enhances risk mapping, facilitates multi-parameter site assessments, and supports the identification of zones vulnerable to geotechnical failure. Coupled with simulation tools, these datasets allow engineers to model stress distributions, ground deformation, and the impact of subsurface anomalies on infrastructure over time. Furthermore, AI and machine learning algorithms are now being used to automate the interpretation of large geophysical datasets, detect patterns, and predict failure probabilities based on historical and real-time monitoring inputs. According to Zhang et al. (2023a, b), such integrative approaches have proven effective in urban planning scenarios, where complex subsurface conditions require fast, reliable, and scalable analysis to inform resilient infrastructure development.

11.2.4 Resilience Assessment Frameworks of Critical Infrastructures: State-of-the-Art Review

Global and Regional Frameworks (E.G., UNDRR, ISO 55000, NIST, ASCE)

Globally recognized organizations and national standard-setting bodies have contributed significantly to the development of resilience assessment frameworks. The United Nations Office for Disaster Risk Reduction (UNDRR) has promoted the Sendai Framework, which emphasizes disaster risk reduction and resilience across all infrastructure sectors. ISO 55000 provides guidance on asset management, focusing on infrastructure sustainability and lifecycle considerations. In the United States, the National Institute of Standards and Technology (NIST) offers scientifically grounded tools for measuring infrastructure resilience, particularly in urban environments. Similarly, the American Society of Civil Engineers (ASCE) has established hazard-specific standards and performance criteria for civil infrastructure systems.

These frameworks represent structured and systematic efforts to guide governments, engineers, and stakeholders in resilience planning. However, their implementation often depends on local context, and many of them lack the flexibility to address cascading hazards or region-specific vulnerabilities. Additionally, while global in outlook, their application remains uneven across countries and infrastructure types, often constrained by resource limitations, data availability, and governance capacity.

Integration of Geophysics in Existing Frameworks

A significant advancement in CI resilience assessment is the growing integration of geophysical techniques within existing frameworks. These methods—such as Ground Penetrating Radar (GPR), Electrical Resistivity Tomography (ERT), and seismic imaging—enhance the monitoring and evaluation of infrastructure by providing real-time, subsurface data. This capability is particularly valuable for detecting latent vulnerabilities and tracking infrastructure health before, during, and after hazardous events.

Geophysical integration supports early warning systems, real-time damage detection, and multi-scenario modeling, thereby enabling more informed and proactive decision-making. It also enhances risk mapping and resilience scoring by improving the accuracy of hazard exposure models. As frameworks evolve to incorporate such techniques, resilience assessments are becoming increasingly data-driven, predictive, and adaptable to complex, multi-hazard environments. Despite these benefits, the application of geophysics remains limited in many regions due to technical, financial, or institutional barriers, highlighting an opportunity for broader integration.

Gaps, Opportunities, and Future Trends

A detailed analysis of 24 resilience assessment frameworks published between 2011 and 2021 reveals common dimensions and indicators that define current practices in the field. The most frequently used dimensions are technical, organizational, social, and economic. Technical indicators include robustness, redundancy, recoverability, monitoring systems, and emergency preparedness. Organizational indicators encompass governance, adaptability, legislation, and leadership. However, social and economic dimensions remain underrepresented, with fewer indicators such as community awareness and crisis response budgeting.

This imbalance suggests the need to develop more holistic frameworks that capture the full spectrum of resilience—from engineering robustness to social and economic adaptability. Moreover, environmental considerations are often missing or underdeveloped, despite their growing relevance in climate-sensitive infrastructure planning.

Several opportunities exist to advance the field. First, more empirical studies are needed to assess how well existing frameworks are implemented in real-world contexts. Second, indicator sets should be revisited and refined to ensure they remain relevant, measurable, and actionable. Third, new frameworks should integrate environmental and interconnectivity factors, reflecting the interconnected nature of modern infrastructure systems. Lastly, developing a standardized rating system or index to evaluate resilience across different infrastructure types and geographic settings would provide a valuable benchmarking tool for practitioners and policymakers.

While the reviewed frameworks provide a strong foundation, they are still evolving. The transition from concept to implementation remains a challenge, requiring cross-sector collaboration, investment in technology, and the alignment of resilience planning with national development priorities. By incorporating multi-dimensional indicators and leveraging technological advances such as geophysics, future frameworks can become more effective tools for safeguarding critical infrastructure against complex and evolving risks.

11.2.5 Assessing Infrastructure Resilience: Theoretical Models, Scenario-Based Approaches, and Case Applications

Theoretical Frameworks for Resilience Assessment

As natural disasters become more frequent and severe, the need to protect critical infrastructure (CI) has become increasingly important. In response, researchers have developed several theoretical frameworks that help assess the resilience of such infrastructure systems (Wang et al. 2022). One of the recent contributions is the New Infrastructure Resilience Assessment Curve (NIRAC). Unlike earlier models that typically analyzed single-scenario cases, NIRAC is designed to handle multiple infrastructure scenarios (Irankunda et al. 2024; Davis 2021; Bruneau et al. 2003). It provides a broader understanding of how different systems perform under similar hazard conditions, with a focus on environmental sustainability. This theoretical model aims to help policymakers and engineers make more informed decisions about infrastructure design and recovery by considering environmental impacts and resilience together.

Another theoretical framework introduced is the Road Infrastructure Resilience Assessment (RIRA) model. RIRA is built around factors specifically relevant to road infrastructure and includes sustainability as a key component. To ensure its reliability, a structured survey was conducted among disaster response and road reconstruction professionals. The responses showed high levels of agreement on the importance of the identified factors, proving their relevance for assessing road resilience (Vugrin et al. 2011; Singhal et al. 2021). These theoretical models serve as essential tools

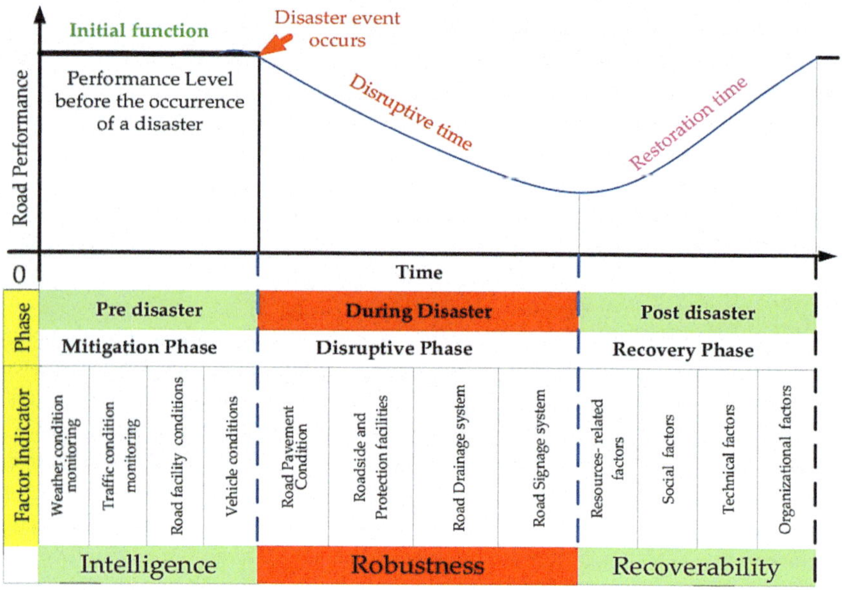

Fig. 11.8 Classification of resilience factors along with road performance curve and disaster phase (*Source* Irankunda et al. 2024—Open Access)

in improving our understanding of how infrastructure can adapt and recover from natural disasters while minimizing environmental harm. To visually support this discussion, Fig. 11.8 illustrates the classification of resilience factors along with the road performance curve and disaster phase. This figure helps conceptualize how various resilience factors interact across different disaster stages, providing a dynamic view of performance degradation and recovery over time.

Multi-scenario Modeling Techniques

Traditional infrastructure resilience studies often focus on a single disaster or scenario. However, infrastructure in real-world settings is exposed to many possible hazards, and therefore, a multi-scenario approach is more realistic and effective. The NIRAC model introduces this concept by evaluating how infrastructure performs across different types of disaster scenarios. This makes it possible to analyze the interactions between different infrastructure systems, such as roads, power lines, and communication networks, under shared stress conditions. The model emphasizes the need for sustainable and resilient planning, considering that different hazards can have overlapping effects on infrastructure.

The multi-scenario technique helps simulate and predict how infrastructure will respond not just to one event but to multiple, potentially simultaneous threats. It allows decision-makers to visualize system performance curves over time and under varied stress conditions, offering a clearer picture of recovery and functionality. Mapping these resilience factors over time can support better planning for pre-disaster readiness, emergency response, and post-disaster recovery. This modeling technique improves the practicality and usefulness of resilience assessments by acknowledging the complexity of real-world challenges.

Case Study Examples (Urban Transport, Power Grid, Etc.)

To test and validate the proposed resilience assessment frameworks, a real-world case study was conducted using National Road No. 9 (NR9) in Burundi, which sustained significant damage due to flooding from heavy rainfall (Irankunda et al. 2024). This study served as the application site for the Road Infrastructure Resilience Assessment (RIRA) framework, designed to evaluate road infrastructure resilience with a focus on environmental sustainability and disaster recovery.

The RIRA framework was applied to assess how well the damaged sections of the road could be restored using resilience-based and sustainable principles. Figure 11.9 shows the affected parts of NR9 at MP15 and MP28, offering a visual representation of the physical damage and the framework's applicability in a real-world post-disaster context.

The results of this application confirmed the framework's usefulness in identifying critical resilience indicators, guiding the prioritization of recovery efforts, and supporting decisions that were not only efficient but also environmentally responsible. To further support recovery operations, a detailed roadmap was developed to organize and guide all stages of intervention—from initial response to long-term restoration. This roadmap is presented in Fig. 11.10, which outlines the strategic planning process undertaken during the recovery of NR9.

Fig. 11.9 Affected part of road infrastructure at mileage points MP15 (**a**) and MP28 (**b**). The visible damage in these segments highlights the importance of sustainable recovery efforts that not only restore functionality but also account for long-term resilience against similar events (*Source* Irankunda et al. 2024—Open Access)

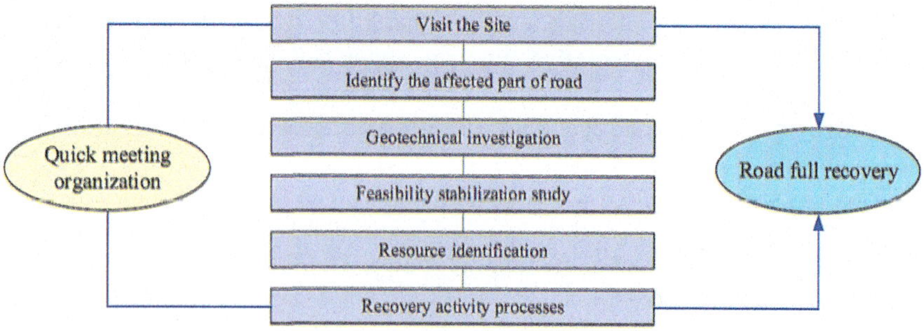

Fig. 11.10 NR9 recovery management roadmap (*Source* Irankunda et al. 2024—Open Access)

To demonstrate the practical outcomes of the applied resilience approach, Fig. 11.11 presents images of NR9 after recovery at MP15 (a) and MP28 (b), respectively. These post-recovery visuals emphasize the successful implementation of resilience-based strategies and provide tangible evidence of the road's restoration in alignment with the RIRA framework's goals.

Although this case focused on a rural road system, the insights gained are relevant to other types of infrastructure, including urban transportation networks, power grids, and telecommunication systems. The case highlights the importance of context-sensitive resilience planning that takes local environmental, technical, and social conditions into account.

However, it is also important to note certain limitations. The frameworks explored—NIRAC and RIRA—are still largely theoretical in nature and need to be tested on more complex, smart infrastructure systems, such as intelligent urban transport networks and digitalized utility grids. Future research should aim to apply these models across various infrastructure sectors and geographic regions to confirm their broader applicability, refine their components, and better support disaster preparedness and sustainable development goals.

Synthesis and Implications for Geophysical Integration

The findings of this study suggest that combining theoretical frameworks with multi-scenario modeling can significantly improve the assessment of infrastructure resilience. When these methods are applied in practice, especially with case-based evidence, they provide a more holistic understanding of system vulnerabilities and strengths. However, to further enhance these approaches, geophysical techniques such as Ground Penetrating Radar (GPR) and Electrical Resistivity Tomography (ERT) could be integrated. These methods allow for real-time monitoring and detection of hidden weaknesses in infrastructure before they lead to failure, supporting early intervention and proactive planning.

Fig. 11.11 NR9 after recovery at MP15 (a) and MP28 (b), respectively (*Source* Irankunda et al. 2024—Open Access)

Involving communities, local authorities, environmental experts, and infrastructure managers ensures that all relevant concerns are addressed in resilience planning. This inclusive approach improves public trust and increases the effectiveness of resilience strategies. Looking ahead, researchers are encouraged to expand the existing frameworks by applying them to new types of infrastructure and incorporating evolving environmental challenges. Enhanced Environmental Impact Assessments (EIAs) and stakeholder collaboration will be critical to balancing resilience, sustainability, and economic development goals.

11.3 Applications in Utilities, Transportation, and Energy Networks

11.3.1 Utilities (Water, Gas, Sewerage)

The integrity of buried utility networks—such as water pipes, gas lines, and sewer systems—is crucial for urban resilience and public safety. Traditional excavation-based inspection methods are costly, disruptive, and limited in scope. Geophysical techniques offer a non-invasive alternative for locating, mapping, and assessing these subsurface assets. Ground Penetrating Radar (GPR) is widely used to detect shallow utilities and identify anomalies such as pipe misalignments or deteriorated joints. Electrical Resistivity Tomography (ERT) is particularly effective in detecting moisture anomalies associated with leaks in pressurized water or sewer lines. Seismic methods, including surface wave analysis and seismic refraction, can be applied to detect voids or subsidence zones beneath critical infrastructure that may lead to pipe collapse or roadway failure. Integrating these techniques allows for comprehensive subsurface diagnostics that support both emergency response and proactive maintenance. A recent study by Abdalla et al. (2023) demonstrated the successful application of combined GPR and ERT techniques to detect leakage zones and underlying voids in aging urban utility networks, significantly reducing repair costs and minimizing service disruptions.

11.3.2 Transportation Systems (Roads, Railways, Airports)

Transportation infrastructure—such as roads, railways, and airports—relies heavily on subsurface stability for long-term performance and safety. Subsurface anomalies like voids, weakened zones, or water-saturated soils can lead to pavement deformation, track misalignment, or even sudden failures such as sinkholes. Geophysical methods offer powerful, non-destructive tools to evaluate the integrity of these systems. Surface wave methods (e.g., MASW), seismic refraction, and GPR are widely used to assess the stiffness and uniformity of pavement layers, identify zones of delamination, and locate buried anomalies beneath runways and railway beds. Electrical resistivity techniques are particularly valuable for identifying moisture-induced degradation, while time-lapse monitoring can track evolving conditions under high-traffic corridors. For example, Asakura et al. (2023) demonstrated the use of GPR combined with seismic testing for identifying early-stage structural damage in airport runways, enabling preventive maintenance with minimal disruption. Similarly, Wang et al. (2023) applied ERT and MASW for detecting sinkholes beneath a railway line in a karst-prone region (karst is a topography formed from the dissolution of soluble carbonate rocks such as limestone and dolomite), showing how geophysics can guide targeted interventions to avoid major service failures.

Global Study on Disaster Resilience of Airports

Airports are critical infrastructure. They connect cities and countries, making it possible for people and goods to move quickly across the world. In the past, airports were usually built far from city centers. Today, many airports have become central hubs, surrounded by growing real estate and commercial development. Larger airports act as both transport and economic centers, while smaller regional airports support local development and often lead to increased land value in nearby areas.

In many urban regions, land is scarce. As a result, some airports are built in areas that are more vulnerable to natural hazards—such as land below sea level, near coastlines, in floodplains, or on reclaimed land. This makes them more exposed to disasters like floods, storms, earthquakes, and rising sea levels. Since airports are long-term investments, they must be designed not only to handle today's risks but also to adapt to future challenges, including those caused by climate change (Link et al. 2014).

Airports also play a vital role during emergencies. After disasters, both large and small airports are key points for relief efforts. They allow search and rescue teams, medical staff, and supplies to reach affected areas quickly. During the COVID-19 pandemic, airports helped control the spread of disease and provided essential services. To do this effectively, airports need to be flexible. Their buildings and systems must adapt quickly to new threats and support emergency operations.

When airports are damaged by disasters, the impacts are not limited to infrastructure repair. Such events cause major economic losses and disrupt the lives and livelihoods of millions of people. Recognizing this, the Coalition for Disaster Resilient Infrastructure (CDRI) is launching a Global Study on Disaster Resilience of Airports. This study will explore how airports can be made safer and more resilient against natural and human-made hazards. The study will focus on several key areas, including *structural safety* such as runways, terminals, and control towers; *non-structural systems* like equipment, signage, and furnishings; and *operational and maintenance practices* that support the overall functionality and resilience of airport infrastructure.

It will include both airports that have suffered serious disaster impacts and those that have effectively managed extreme events. While some countries have conducted climate risk assessments focused on airline operations, few have looked at disaster resilience of airport infrastructure itself. Also, many emergency response plans do not take a full-systems approach to resilience.

To address this gap, the CDRI's study follows the Sendai Framework for Disaster Risk Reduction. It will view airport resilience from a systems perspective, which means looking at how all parts of the airport—including buildings, people, procedures, and technology—work together to handle risks. The study will also involve stakeholders such as airport authorities, government agencies, and infrastructure planners. The Global Study on Disaster Resilience of Airports will provide several key outputs. These include an inventory of natural and man-made hazards that airports commonly face, and an Airport Risk Index that measures the level of exposure, vulnerability, and adaptive capacity of airports. The study will also offer a set of policy recommendations based on lessons learned from airports that have experienced disasters. Additionally, it will provide insights into how disaster risks can be integrated into financial planning and investment decisions for airport development and operations.

11.3.3 Energy Networks (Pipelines, Substations, Renewables)

Energy infrastructure—including underground pipelines, electrical substations, and renewable energy installations—requires continuous monitoring and precise site evaluation to ensure operational safety and long-term stability. Geophysical techniques have become essential tools for assessing pipeline integrity, particularly in detecting corrosion, leakage, or ground movement that can threaten structural reliability. Methods such as Electrical Resistivity Tomography (ERT), Magnetic Flux Leakage (MFL), and Ground Penetrating Radar (GPR) are widely used to locate faults, map coating defects, and monitor moisture conditions around buried pipelines. For renewable energy projects like wind farms and solar parks, geophysical surveys support optimal siting by evaluating subsurface lithology, bedrock depth, and groundwater dynamics. Seismic refraction and MASW methods are particularly useful for foundation design and ground stiffness assessments, especially in areas prone to soil instability or seismic activity. Recent studies, such as that by Kumar et al. (2023), demonstrate the effectiveness of ERT in assessing pipeline corridors in shifting clay environments, while Lopez et al. (2023) highlight the role of GPR and seismic methods in siting offshore wind infrastructure with minimal environmental impact.

11.3.4 Building Disaster-Resilient Seaports: Exploring Adaptation in Practice

India is among the countries most vulnerable to climate change, especially to sea level rise and extreme weather events. Despite initiatives like the Maritime India Vision

2030, there is limited focus on protecting port infrastructure from climate-related risks. Seaports are critical for Small Island Developing States (SIDS), where trade, food, energy, and tourism depend entirely on maritime transport. These countries are highly exposed to hazards, and with low capacity to adapt, their risks are growing. According to the IPCC, risk is shaped by three factors: hazard, exposure, and vulnerability. For SIDS and coastal regions, all three are increasing. When ports are disrupted by storms or sea level rise, the effects extend far beyond the port itself—they impact cities, industries, and global supply chains (Source: www.cdri.world).

A study by Brazil's National Waterway Transport Agency (ANTAQ) assessed climate risks for 21 coastal ports. It developed a climate risk index for strong winds, thunderstorms, and sea level rise, and proposed 55 adaptation measures—including structural solutions like raised barriers and non-structural ones like better planning. In the Caribbean, the UNCTAD project created a Climate Risk and Vulnerability Assessment Framework for coastal transport infrastructure. This framework helps prioritize risks and plan effective adaptation. It can also be adapted for use in another SIDS worldwide. Several international best practices show how ports are adapting—The Port of Rotterdam has raised embankments and improved flood risk management, Port of Santiago built living shorelines with oysters and used eco-friendly concrete tide pools, and in South Africa, the eThekwini municipality developed a flood simulation and early warning tool.

Risk assessments are essential for effective planning. Port expansion can also affect coastal ecosystems, so a broader view is necessary. Planning should consider local conditions, infrastructure lifespan, and operational limits during extreme weather. Port strategies must be updated regularly with the latest climate data. Importantly, the adaptation measures themselves should not worsen climate change. Investments should be low-emission and climate-sensitive.

For SIDS, access to climate finance is critical. Grant-based funding is preferred over loans to avoid increasing debt. Insurance options for ports are limited and costly; affordable and inclusive insurance schemes are needed, especially those that reward risk-reducing actions.

Building climate-resilient ports is expensive, but not acting can result in far higher economic losses. More private sector investment should be encouraged, especially in sustainable infrastructure.

Collaboration is key. Port authorities, cities, and private stakeholders must coordinate and integrate climate resilience into all departments and business models. Strong governance structures are necessary to manage climate risks across large territories. Looking ahead, seaports must innovate. Ecosystem-based adaptation and nature-based solutions should be part of future strategies. Legal frameworks and policies must support resilience and be reflected in national disaster risk reduction plans and climate recovery programs. Finally, global trends such as the changing geography of trade, post-COVID recovery, and the shift toward renewable energy will reshape the maritime sector. Energy efficiency and decarbonization will support both mitigation and adaptation, helping ports become safer and more sustainable.

11.4 Pre-failure Detection and Maintenance Strategies

11.4.1 Subsurface Change Detection and Monitoring

Early identification of subsurface changes is critical to preventing infrastructure failure, particularly in aging systems and geologically sensitive areas. Time-lapse geophysics—also known as 4D geophysical monitoring—enables the tracking of changes in subsurface properties over time by repeatedly surveying the same area. Techniques such as Electrical Resistivity Tomography (ERT), seismic noise interferometry, and passive microseismic monitoring are increasingly used to detect stress accumulation, material fatigue, moisture migration, and void formation before visible damage occurs. These methods are particularly valuable in embankments, dams, tunnels, and transportation corridors, where gradual ground deformation or internal erosion may precede structural failure. Passive techniques, such as ambient seismic noise analysis, offer continuous, non-intrusive monitoring that can reveal evolving patterns of subsurface stress without the need for active sources. A recent study by Becker et al. (2023) demonstrated the effectiveness of time-lapse ERT and passive seismic monitoring in detecting stress build-up in railway embankments, leading to early intervention and risk mitigation before structural compromise.

11.4.2 Predictive Maintenance Using Geophysics

Predictive maintenance, supported by geophysical methods, represents a shift from reactive repair toward data-driven risk prevention in infrastructure management. By establishing geophysical baselines—such as resistivity profiles, seismic velocities, or natural frequency signatures—engineers can monitor for deviations that signal early-stage degradation, subsurface movement, or material fatigue. Thresholds for intervention can be defined using historical performance data and time-lapse measurements, allowing infrastructure managers to schedule targeted inspections or maintenance before visible signs of failure occur. This approach enhances safety while minimizing unplanned downtime and repair costs. Moreover, the economic benefits of predictive geophysics have been increasingly recognized; proactive strategies reduce life-cycle costs and extend asset longevity by addressing issues in their incipient phase. For example, Zhang et al. (2023a, b) demonstrated that combining geophysical monitoring with machine learning models in a highway network led to a 30% reduction in unplanned repairs over a two-year period, validating the cost-effectiveness of this approach.

11.4.3 Case Studies and Best Practices

Real-world case studies provide compelling evidence of the effectiveness of geophysical methods in preventing infrastructure failures. In one notable example, ERT and GPR were used in a metropolitan sewer rehabilitation project in Italy to identify hidden voids and saturated soils beneath aging pipelines. Early detection allowed for targeted grouting and reinforcement before any surface collapse occurred, saving both time and public disruption (Mariani et al. 2023). Similarly, a time-lapse seismic monitoring program implemented on a highway embankment in Japan successfully tracked progressive weakening due to increased traffic load and seasonal moisture variation. This led to a timely structural upgrade, avoiding what would have been a catastrophic failure during a heavy rain event (Tanaka et al. 2023). These case studies highlight several best practices: integrating multiple geophysical techniques, maintaining consistent monitoring schedules, and collaborating with civil and geotechnical teams from early project stages. Lessons learned include the importance of establishing data interpretation protocols and the value of automating anomaly detection to enhance responsiveness. Together, these insights reinforce the practical role of geophysics in extending the service life of critical infrastructure.

11.5 Challenges and Opportunities

11.5.1 Technical and Environmental Constraints

Despite the proven value of geophysical methods in infrastructure resilience, several technical and environmental constraints can limit their effectiveness. Technically, resolution and depth limitations vary between methods—while Ground Penetrating Radar (GPR) offers high-resolution imaging, its performance is hindered in clay-rich or water-saturated soils. Similarly, Electrical Resistivity Tomography (ERT) and seismic techniques can be affected by surface noise, access restrictions, and variable subsurface conditions. Data interpretation also remains a challenge, particularly when multiple anomalies overlap or when site heterogeneity reduces signal clarity. From an environmental perspective, urban settings often pose difficulties such as electromagnetic interference, buried debris, and limited physical space for sensor deployment. Additionally, natural environments like wetlands or mountainous terrain can restrict equipment mobility and long-term monitoring efforts. These constraints underscore the need for multi-method approaches, site-specific calibration, and advancements in sensor design. According to Kargl et al. (2023), addressing these challenges requires tighter integration between geophysical and civil engineering practices, along with standardized protocols to improve data reliability and repeatability in complex environments.

11.5.2 Standardization and Quality Control Needs

As geophysical methods become more widely adopted in infrastructure monitoring and planning, the need for standardization and rigorous quality control becomes increasingly urgent. The absence of universally accepted protocols for data acquisition, processing, and interpretation often leads to variability in results, limiting comparability across projects and reducing confidence among stakeholders. This is particularly problematic in infrastructure applications where decisions based on geophysical data can affect public safety and long-term investment. Establishing standardized workflows—including instrument calibration procedures, data quality benchmarks, and reporting formats—would enhance reliability and facilitate integration with engineering and regulatory frameworks. Quality assurance also extends to training and certification of personnel to ensure consistent application of methods in the field. Recent initiatives, such as those outlined by Thomas et al. (2023a, b), advocate for harmonized guidelines under international bodies like ISO and American Society for Testing and Materials (ASTM), aiming to bridge the gap between geophysics and civil infrastructure practices through better-defined technical standards and validation protocols.

11.5.3 Opportunities from AI, IOT, and Smart Infrastructure

The integration of Artificial Intelligence (AI) and the Internet of Things (IoT) is unlocking transformative opportunities in smart infrastructure development and maintenance. AI algorithms, when combined with IoT sensor networks, enable real-time monitoring, anomaly detection, and predictive maintenance across a wide range of infrastructure assets. These technologies allow engineers to automate diagnostics and forecast failures based on environmental and structural parameters collected from the field. For instance, embedded IoT sensors in bridges, tunnels, and utility pipelines can stream continuous geophysical and structural data to AI-based platforms, which analyze the data for early signs of stress, corrosion, or material fatigue. This approach enhances operational efficiency, reduces downtime, and supports sustainable asset management. Edge computing further optimizes this system by processing data locally for immediate decision-making, particularly in high-speed environments such as traffic systems or energy grids. According to Kumar et al. (2024), AI-driven predictive models have reduced false alarms by 40% and improved infrastructure health assessment accuracy in pilot smart city projects. Similarly, Al-Hassan and Becker (2025) emphasize that the integration of AI and IoT with geophysical data in critical infrastructure not only improves resilience but also enables real-time adaptation to changing subsurface conditions.

The Use of Artificial Intelligence to Bridge Multiple Narratives in Disaster Response

Artificial Intelligence (AI) is becoming an essential tool in disaster response, not only for analyzing data but also for connecting different perspectives and information sources during emergencies. In a disaster, governments, emergency teams, infrastructure managers, local communities, and media all have different viewpoints, priorities, and types of information. This makes coordination difficult and slows down response efforts. AI helps by bringing these different narratives together and creating one clear understanding of the situation (Amarel 2025).

AI systems can handle massive amounts of data, including satellite images, social media posts, sensor readings, and official reports. They use tools like natural language processing (NLP) to scan public messages and news reports in real-time, helping detect urgent needs or public concerns as they develop (Kopanov et al. 2024). At the same time, AI models can simulate disaster impacts on infrastructure—like water, energy, and transport systems—supporting better planning and faster response (Gong et al. 2025).

This ability to merge technical data with local feedback allows AI to create shared situational awareness. For example, if social media posts from local people report flooding, and sensor data confirms rising water levels, AI can match these with infrastructure maps to identify high-risk zones. Emergency teams can then act more quickly and confidently, knowing their response is based on both local input and scientific evidence.

AI also helps bridge the gap between experts and non-experts. Complex engineering or geophysical models can be translated into visuals and plain language that local officials and communities can understand, leading to better cooperation. This improves trust, accountability, and the ability to make decisions that include everyone affected (Veluru 2024). However, using AI in disasters comes with risks. The systems depend on good-quality data and must be designed to avoid bias. Ethical issues like privacy, fairness, and access also need to be addressed so that AI helps everyone, not just those with digital tools or data skills. In conclusion, AI is more than a technology for automation—it's a bridge that connects different viewpoints during disasters. It brings technical knowledge, community experiences, and real-time data together to support faster and fairer disaster management. Investing in AI systems that are transparent, ethical, and inclusive can help build more resilient infrastructure and communities worldwide.

11.6 Future Directions

11.6.1 Digital Twins and Geophysical Integration

Digital twins—virtual replicas of physical infrastructure systems—are rapidly emerging as a transformative tool in engineering geophysics. By integrating geophysical data into digital twin platforms, engineers and asset managers can simulate real-time responses of infrastructure to changing subsurface and environmental conditions. Geophysical inputs such as seismic velocity models, resistivity changes, and ground deformation data can be continuously updated from field sensors and remote monitoring systems. This integration allows for predictive scenario modeling, early warning capabilities, and dynamic maintenance scheduling. For example, time-lapse ERT or ambient seismic noise data can reveal early indicators of soil saturation or stress build-up, which are then visualized and tested within the digital twin environment. The use of AI within digital twins further enhances interpretation accuracy and decision-making under uncertainty. As noted by Fernandez et al. (2024), the fusion of geophysical insights with digital twin platforms offers unprecedented opportunities for resilient, adaptive infrastructure planning, especially in complex urban and geohazard-prone environments.

11.6.2 Role of Geophysics in Climate-Resilient Infrastructure Planning

As climate change intensifies the frequency and severity of environmental hazards—such as floods, landslides, droughts, and ground subsidence—the need for climate-resilient infrastructure has become a strategic priority worldwide. Geophysics plays a vital role in this context by enabling the continuous assessment of subsurface conditions that directly affect infrastructure vulnerability under shifting climate regimes. Techniques such as Electrical Resistivity Tomography (ERT), seismic imaging, and Ground Penetrating Radar (GPR) are particularly effective in detecting soil saturation, erosion potential, and dynamic groundwater movement—factors that are increasingly influenced by changing rainfall patterns and extreme weather events. These insights support informed site selection, foundation design, and early-warning systems for at-risk infrastructure assets. Moreover, time-lapse geophysical monitoring allows for real-time tracking of climate-induced changes in ground conditions, improving both adaptive response and maintenance planning. As highlighted by Oliveira et al. (2024), integrating geophysical data into climate resilience planning frameworks significantly enhances the predictive capability of infrastructure risk assessments, especially in regions prone to hydrological and geotechnical hazards.

11.6.3 Policy Recommendations and Funding Avenues

To fully leverage geophysics in infrastructure resilience and risk reduction, supportive policies and sustained funding mechanisms are essential. Policymakers should prioritize the inclusion of geophysical investigations in infrastructure planning regulations, disaster risk management frameworks, and environmental permitting processes. Standardizing geophysical data requirements within public infrastructure projects would ensure consistency, transparency, and accountability in subsurface assessment. Additionally, governments and funding agencies should allocate dedicated resources for long-term geophysical monitoring, especially in hazard-prone and climate-sensitive regions. Funding models could include public–private partnerships (PPPs), resilience bonds, or integration into national infrastructure investment plans. Cross-sector collaboration—linking geophysicists, engineers, urban planners, and policymakers—is also crucial for effective knowledge transfer and practical implementation. As emphasized by Delgado et al. (2024), embedding geophysical insights into urban resilience policies and unlocking funding for proactive monitoring can significantly reduce future repair

11.7 Case Studies

11.7.1 Case Study: Climate Disaster Vulnerability and Resilience of Railway Infrastructure in Northeast India

Climate disasters are becoming more frequent in India and have had a severe impact on the country's railway infrastructure. Every year, events like floods, landslides, cyclones, and earthquakes damage railway tracks, bridges, and signaling systems. This case study focuses on Northeast India, particularly the states of Assam, Mizoram, and Manipur (Roy et al. 2023), where railway systems face serious challenges due to both natural hazards and the complex geophysical structure of the region (Figs. 11.12 and 11.13).

This study analyzed three specific cases of climate disasters to assess their effects on railway infrastructure. The findings revealed that while physical damages were common, the impact of climate events went beyond immediate destruction. Disruption of services, delays in repairs, and safety risks were closely linked to the geological and geographical conditions of the region. The unique terrain of Northeast India makes it especially prone to landslides, floods, and earthquakes, which, in turn, increase the vulnerability of the railway network. The case study found that the vulnerability of railway infrastructure is strongly influenced by the local geophysical structure, such as unstable soil, steep slopes, and frequent seismic activity (Fig. 11.14). These natural features act as underlying risk factors that worsen the

Fig. 11.12 Major hydro-meteorological hazards and their impact on railway infrastructure in Assam, India, 2022 (*Source* Roy et al. 2023—Open Access)

Fig. 11.13 Railway tunnels (**a, b**) and bridges (**c, d**) along the railway network in Manipur (*Source* Roy et al. 2023—Open Access)

effects of climate events. As a result, railways in this region often experience long disruptions, affecting not only passenger travel but also the transport of goods and economic activity.

To address these challenges, Indian Railways has started implementing advanced technologies to track and predict natural disasters. Tools such as satellite imaging,

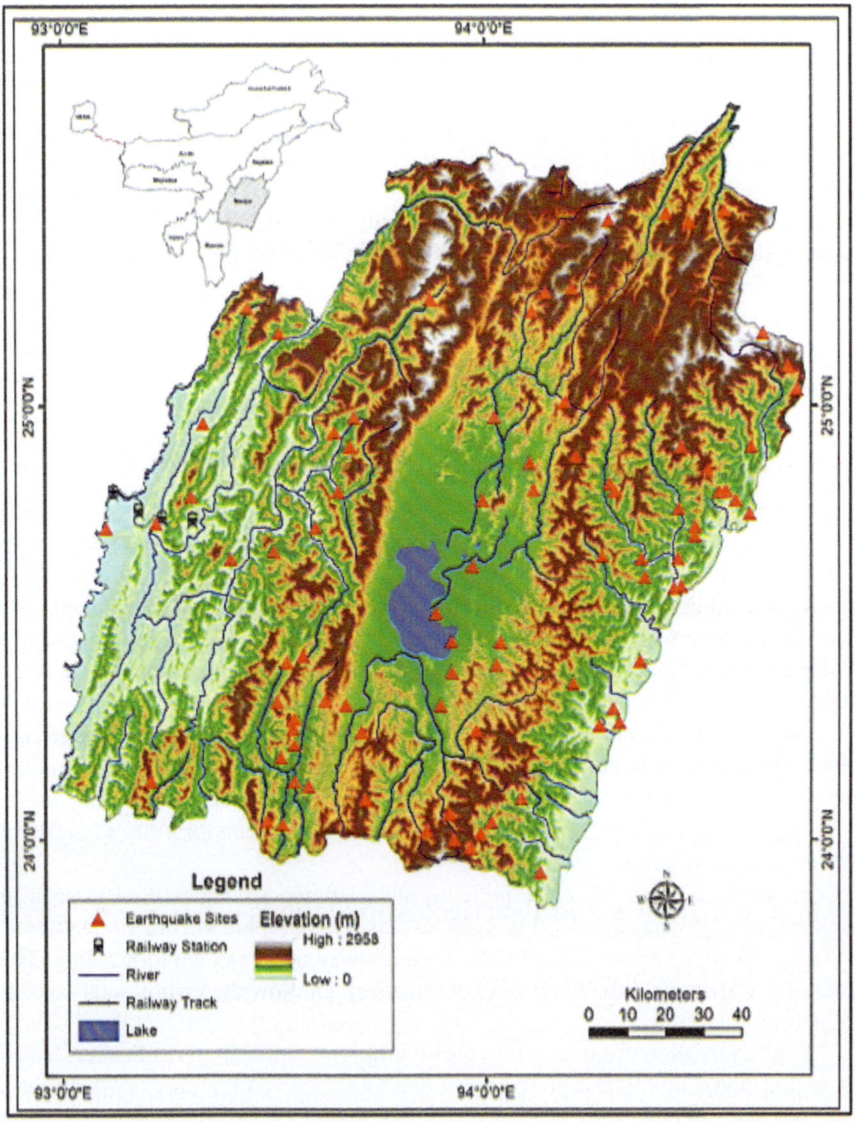

Fig. 11.14 Railway infrastructure and earthquake sites in Manipur, Northeast India (*Source* Roy et al. 2023—Open Access)

remote sensing, GIS (Geographic Information Systems), and data analytics are being used to assess risk-prone areas. This allows early action and helps minimize damage. However, technology alone is not enough. There is a need for continued investment in stronger, long-lasting infrastructure, including bridges, embankments, and safety systems that can withstand harsh weather conditions.

Policy interventions also play a key role. While environmental determinism (the idea that nature dictates development) is a significant factor in Northeast India, effective policies and proactive planning can help reduce disaster risk. The Indian Ministry of Railways, in collaboration with local governments, needs to focus on disaster-resistant design, improved maintenance systems, and community-based disaster awareness programs. In conclusion, climate disasters have a significant and multi-dimensional impact on railway infrastructure in Northeast India. While natural conditions create high risks, a combination of modern technology, strong infrastructure, and strategic policy planning can reduce vulnerability. Indian Railways must continue to strengthen its resilience strategy to ensure safe, efficient, and reliable transport in the face of growing climate threats.

11.7.2 Lessons from Case Studies of Flood Resilience: Institutions and Built Systems

Climate change and natural disasters have made it essential for cities to become more resilient, especially to flooding (Alphen et al. 2011; Aerts et al. 2014). This case study examines how three cities—New York in the United States, Tokyo in Japan, and Rotterdam in the Netherlands—have developed strategies to manage flood risks (Fig. 11.15). The focus is on how their institutional approaches affect the resilience of transportation expressways, which are critical during emergencies (Singh et al. 2021; North 1991; Amekudzi-Kennedy et al. 2018; Hofstede 2001).

Transportation systems are key during disasters because they help people evacuate, allow emergency vehicles to reach affected areas, and keep essential services running. If these systems are damaged by floods, it can lead to major disruptions. Therefore, ensuring that expressways are resilient to flooding is important for a city's long-term performance and safety.

This study looks at the laws, policies, and planning strategies used in each of the three cities to manage flood risks. It also uses geographic information systems, or GIS, to analyze how vulnerable their expressway networks are to flooding. The goal is to understand how flood risks are handled and how the institutional context influences outcomes.

The analysis shows that each city uses a different approach to resilience, shaped by its location, culture, and government structure. New York focuses on long-term planning and emergency response. Tokyo relies on heavy investment in infrastructure and has strict flood control regulations. Rotterdam combines national planning with strong involvement at the local level.

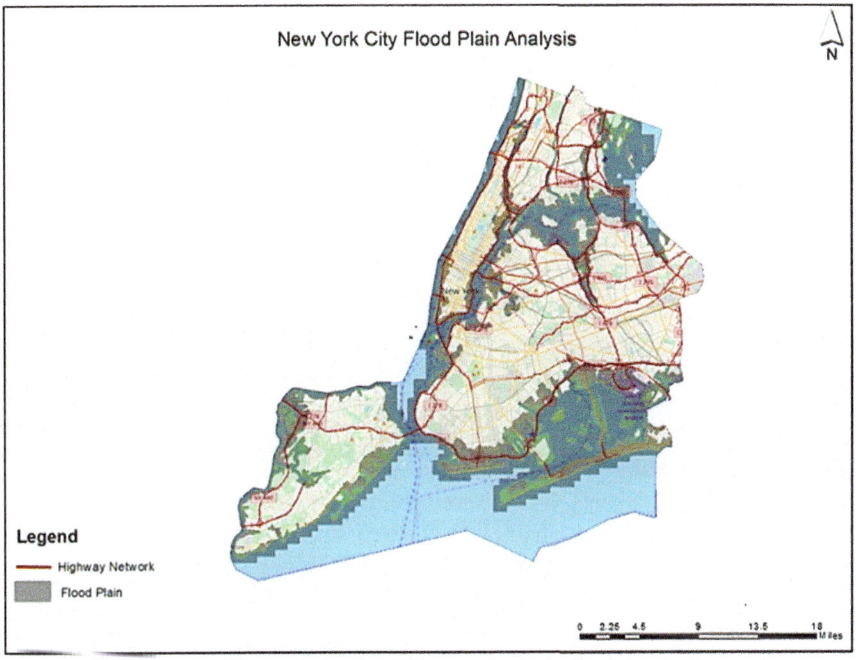

Fig. 11.15 Flood risk exposure of major highways in New York city: overlay with 100 year floodplain map (*Source* Singh et al. 2021—Open Access)

These different approaches have a clear impact on how well transportation systems function during floods. Cities with stronger institutional support and planning systems tend to have more resilient infrastructure. The study finds that system resilience is closely connected to the development of institutions—such as laws, policies, and government agencies—that support risk management. To illustrate the key differences between these cities, Table 11.1 presents a comparison of their flood management policies and the cultural or institutional factors that may explain these differences.

Table 11.1 shows how each city's flood resilience efforts reflect its governance structure and cultural context. New York's approach is shaped by a need to balance state and city responsibilities, Tokyo operates within a highly centralized national system, and Rotterdam demonstrates a collaborative model rooted in its long-standing relationship with water. The research identifies two main kinds of strategies. Defensive strategies are often led by central governments and focus on preventing damage, such as building flood barriers. Adaptive strategies are usually managed at the local level and aim to help communities adjust and recover after flooding, such as flexible urban planning and local emergency response systems.

Overall, the case study emphasizes that both physical systems and institutional systems must work together to achieve effective resilience. It shows that cities need a mix of short-term measures, long-term planning, and the ability to adapt to changing

Table 11.1 Observed differences in flood management policies and suggested contributing institutional and cultural factors (after Singh et al. 2021)

	Subject city country	New York City USA (Shorris 2015)	Rotterdam The Netherlands (Higgins 2012)	Tokyo Japan (Constitutional Law-Unitary and Federal Systems 2019)
Flood management policies	Federal/central	Defensive, responsive	Defensive, adaptive	Defensive, responsive, adaptive
	Local	Responsive	Adaptive	Adaptive
	State of resilience (adaptive) strategies	Nascent	Mature	Mature
Institutional factors	Power distribution in flood policymaking	More power to regional/state government	Distributed power with more central control	Centrally controlled
	State of Government (country level)	Federal	Federal	Unitary
Cultural dimensions score (scale of 100) (country level)	Power distance	40	38	54
	Uncertainty Avoidance	46	53	92
	Long-Term Orientation	26	67	88

conditions. The combination of technical solutions and strong, flexible institutions is key to preparing cities for future flood risks. These findings offer valuable insights for engineers, urban planners, and policymakers. As flood risks increase in many parts of the world, this study shows the importance of building not only stronger infrastructure but also stronger institutions to guide resilience efforts.

11.7.3 Social Sensing and Community Response to Flood Impacts: A Case Study of the 2018 Kerala Floods, India

Flooding is one of the most serious natural hazards, causing widespread damage and risks to people's health and safety (Kundzewicz et al. 2014; Field et al. 2012). In 2018, the Indian state of Kerala experienced one of the worst floods in its history (Figs. 11.16 and 11.17). This disaster killed 433 people and forced more than one million people to leave their homes (Young et al. 2022). Timely and accurate information can help reduce the impact of floods. For example, early warnings and real-time updates can improve emergency response and help coordinate relief. However, reliable data on flood impacts is often hard to get. Government reports may take time to prepare and are often costly. Insurance data is usually not available for many of the most vulnerable people.

This case study explores the use of *social sensing*—a method of collecting and analyzing data from social media—to track and understand flood impacts in Kerala. The goal is to find out whether social media can provide useful information about what happened, who was affected, and where help was needed. Data was gathered from several sources: Twitter, Telegram, a citizen-built platform called Kerala Rescue, and the official government database Rebuild Kerala. After sorting the posts to focus only on flood-related content, the data was analyzed to identify different types of impact, such as calls for help or reports of medical problems. The data was also used to create maps showing where flood damage was reported.

The findings show that the maps and information from social media platforms closely matched the data from Kerala Rescue and Rebuild Kerala. People shared useful, real-time information on Twitter and Telegram, including locations, requests, and on-the-ground reports. This shows that social media can act as a fast and reliable tool to understand what is happening during a flood. The study also highlights the value of using *multiple platforms*. Twitter and Telegram can provide early warnings and help track what is happening during the flood. Kerala Rescue, which was made by volunteers, worked as a hub to coordinate community help and relief. Government sources like Rebuild Kerala give more structured data, which can be used for longer-term planning and rebuilding.

This combination of real-time and official data helps create a full picture of the disaster, from immediate response to recovery. The results suggest that social sensing can improve how we plan for and respond to floods, especially in places where traditional data is slow or limited.

Fig. 11.16 Roads in Kerala, India destroyed by landslides following a flood in August 2018 (*Source* World Bank)

Fig. 11.17 Damaged roads and infrastructure in the Pamba Region of Kerala, India following a series of deadly floods and landslides in August 2018 (*Source* World Bank)

Importantly, this method is not limited to Kerala. As more people around the world use social media and mobile internet, this approach could be used in other flood-prone areas. It may also be useful for other types of natural disasters, like cyclones or landslides. In conclusion, this case study shows that social sensing using platforms like Twitter, Telegram, and Kerala Rescue can help communities and governments better understand and respond to floods. It can support faster decision-making, improve disaster management, and contribute to building long-term flood resilience.

11.7.4 Resilience of Urban Underground Infrastructure Under Multi-hazard Impacts: From Structural Integrity to Network-Level Functionality

Underground infrastructure, such as metro tunnels and road tunnels, plays a vital role in the development and operation of modern cities. As cities continue to expand and modernize, these underground systems have grown rapidly, especially in high-density urban regions. However, this fast development also brings increased risks. Underground infrastructure can be highly vulnerable to multiple types of hazards, including earthquakes, floods, and disruptions caused by human activities such as construction or accidents. These risks are becoming even more serious due to the effects of global climate change.

This case study presents a general framework to assess the resilience of underground infrastructure systems when they are exposed to multiple hazards. The framework addresses resilience on two levels. At the structural level, it focuses on how individual tunnels perform physically when they are impacted by hazards like earthquakes. At the network level, it evaluates how well the overall tunnel system maintains function and connectivity under disruptive events, and how quickly it can return to normal operation.

The types of hazards studied include both natural and man-made events, with special focus on earthquakes, urban flooding, and human-induced disturbances. The resilience of the infrastructure is evaluated based on its ability to absorb shock, continue functioning, and recover after being affected. The study uses real examples in Eastern China to show how this resilience framework can be applied in practice.

In the first example, the structural resilience of a shield tunnel in Shanghai is assessed under different levels of earthquake intensity. Using the proposed model, the tunnel's behavior is evaluated through structural indicators such as deformation and energy absorption. The results show that the tunnel, built in soft clay soil, performs well and maintains high resilience under design-level earthquake conditions.

The second example focuses on the network-level resilience of the road tunnel system in Suzhou under flood conditions (Figs. 11.18 and 11.19). As water levels rise due to extreme weather events, the performance of the tunnel system can drop significantly. To understand and quantify this, the concept of *network efficiency* is introduced. Mathematically, network efficiency reflects how easily and quickly people or services can move through a network. It is calculated using the following basic formula:

$$E_f = \frac{1}{N(N-1)} \sum_{i \neq j} \frac{1}{d_{ij}} \tag{11.3}$$

In this equation, E_f is the global network efficiency (Latora and Marchiori 2001), N is the number of nodes (or points of connection, such as tunnel entrances or intersections), and d_{ij} is the shortest distance or travel time between nodes i and j. A higher value of E means that the network is functioning well, allowing easy movement between different parts of the city.

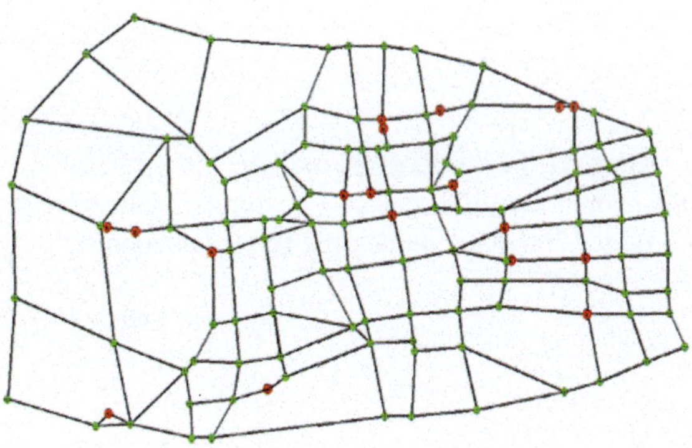

Fig. 11.18 Topographical map of the road tunnel network in Suzhou: Showing the spatial layout and elevation context of key underground routes (*Source* Huang et al. 2022—Open Access)

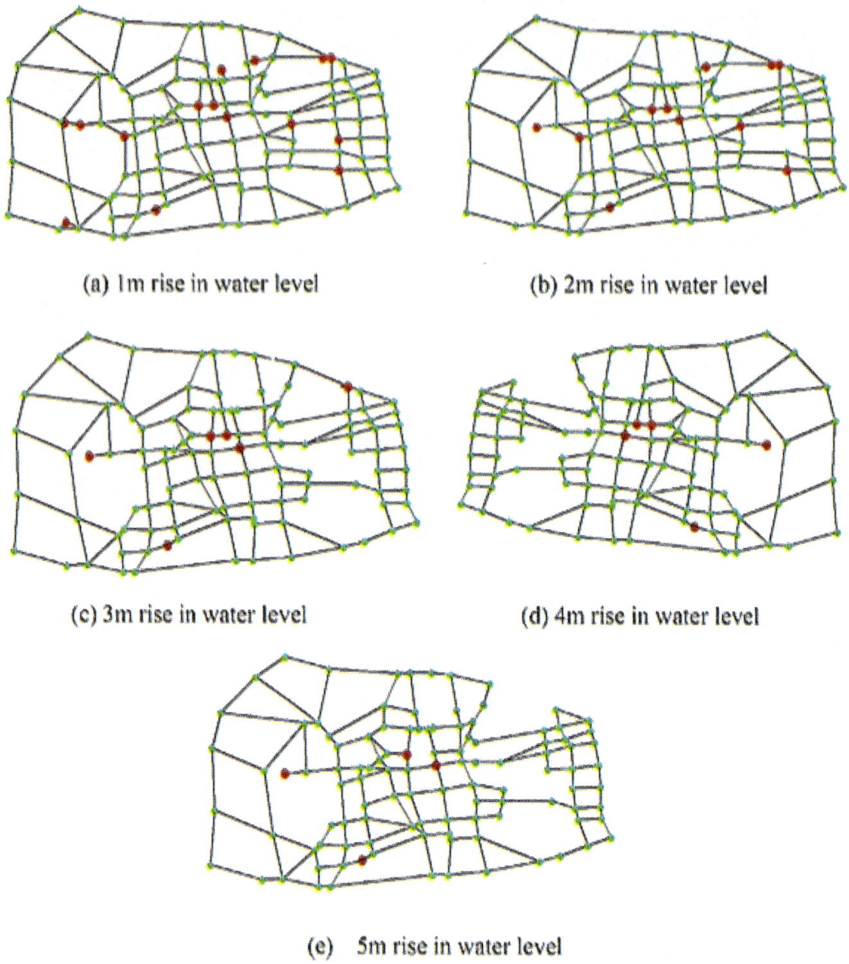

Fig. 11.19 Network disruption in the Suzhou road tunnel system under varying flood levels: Illustrating the impact of rising water on tunnel connectivity and efficiency (*Source* Huang et al. 2022—Open Access)

To make the model more realistic for urban tunnel systems, weights can be added to account for differences in passenger volume or physical tunnel capacity. In the *weighted efficiency* model, the formula becomes:

$$E_w = \frac{1}{W} \sum_{i \neq j} \frac{w_{ij}}{d_{ij}} \tag{11.4}$$

Here, w_{ij} represents the weight given to the connection between node i and node j, such as traffic flow or tunnel usage, and W is the total sum of all these weights in the network. This approach helps reflect the real importance of different tunnel routes based on how heavily they are used.

The influence of failure events on network performance is evaluated by relative network efficiency. Hence, network performance Q of underground infrastructure (UI) could be defined as the ratio of post-fault network efficiency $\left(E'_f\right)$ to initial network efficiency $\left(E_f^0\right)$:

$$Q(t) = \frac{E'_f(t)}{E_f^0} \tag{11.5}$$

Using this network efficiency model, the Suzhou road tunnel system was analyzed under different flood levels. As expected, the efficiency of the network decreased as water levels rose. Some tunnels became impassable, and the overall connectivity of the system dropped. To help deal with these challenges, the study proposes an *adaptive design strategy* based on real-time monitoring. Sensors and observational tools can provide real-time data about flood levels and tunnel conditions, allowing for dynamic responses such as rerouting traffic or closing affected sections in advance. This adaptive method was shown, through mathematical modeling, to significantly improve the resilience of the tunnel network.

In this framework, resilience is defined as the system's ability to maintain or quickly recover performance after being disrupted by a hazard. The system's performance over time is measured during three key phases: the initial disruption, the evolution or adjustment period, and the recovery phase. By comparing these performance phases with the normal state, the total resilience can be estimated. This gives city planners and engineers a way to compare different designs and responses.

The study concludes that both the structural behavior of individual tunnels and the system-level functioning of the entire network are important when assessing resilience. The Shanghai tunnel example highlights the importance of earthquake-resistant design, while the Suzhou network analysis shows how flooding can weaken overall system efficiency and how adaptive strategies can help maintain performance. The mathematical model also makes it possible to simulate the impact of hazards and test strategies for response and recovery.

While this study has looked at natural and human-made hazards separately, it does not yet include the combined or cascading effects of multiple hazards happening together. These interactions could increase risks and should be explored in future research to make resilience planning more practical and complete. In summary, this case study shows that strong design, real-time monitoring, and system-level planning can help cities improve the resilience of their underground infrastructure. This is increasingly important as climate change and urbanization continue to increase the exposure and vulnerability of such critical systems.

11.7.5 Risk Assessment and Management of Underground Metro Construction (Bengaluru Scenario)

This case study analyzes the risk assessment and management strategies for an underground metro construction project in Bengaluru, India (Faruq et al. 2022). Both qualitative and quantitative methods were used to evaluate the potential impact of various risk factors on the project's schedule and cost. A total of 87 risks were initially identified for underground metro projects. Out of these, 75 risks were applicable to the Bengaluru project. From this subset, 17 risks were categorized as major and high-priority risks, accounting for 34.5% of the total project risk. Among these, construction and geotechnical risks emerged as the most critical, followed by design and environmental risks.

The qualitative analysis prioritized risks based on their potential impact. For the project schedule, the high-priority risks included construction, design, environmental, and geotechnical issues. Financial, legal, management, political and policy, and safety risks were considered medium-priority, while market and social risks were categorized as low-priority. Regarding the cost impact, the high-priority risks were design, environmental, geotechnical, and financial. Meanwhile, construction, legal, management, market, political and policy, social, and safety risks were placed under medium-priority.

For the quantitative analysis, a pre-mitigation schedule risk analysis was performed using triangular distribution and Monte Carlo simulations with 1000 and 5000 iterations. The results showed a mean project duration of 1378 days in both cases, with maximum completion times of 1519 days (1000 iterations) and 1517 days (5000 iterations). The 95% probability of project completion stood at 1467 and 1464 days, respectively. Importantly, the minimum completion time in both simulations exceeded the estimated project duration, indicating a 0% probability of completing the project on time without mitigation measures.

Following this, a post-mitigation analysis was conducted by applying planned risk response strategies and re-running the simulations. The results were promising, with the mean duration reducing to 1311 days in both 1000 and 5000 iterations. The maximum project duration was reduced to 1418 and 1426 days, and the 95% probability of completion was 1364 and 1365 days. These results showed less variation and indicated a significant improvement in the project schedule reliability. The deterministic risk of delay dropped to less than 1%. The difference between pre- and post-mitigation results showed that 59 days were saved at the 50th percentile, 88 days at the 80th percentile, and the overall mean duration improved by 67 days. The detailed results for the 1000 iteration simulation after mitigation are summarized below in Table 11.2. This demonstrates that proper risk mitigation strategies can substantially reduce project delays.

The study emphasizes that risks are an integral part of any construction project and must be actively managed to avoid cost and time overruns. Identifying risks before project initiation and applying risk management techniques, such as qualitative analysis, quantitative modeling, and risk response planning, leads to more accurate and

Table 11.2 Post-mitigated schedule risk analysis results for iteration 1000 (after Faruq et al. 2022)

Summary statistics of post mitigated analysis for the total duration (iteration 1000)			
Percentile	Date of completion	Percentile (%)	Duration (dys)
0%	29/08/2022	0	1271
5%	06/09/2022	5	1279
10%	08/09/2022	10	1281
15%	14/09/2022	15	1287
20%	16/09/2022	20	1289
25%	20/09/2022	25	1293
30%	22/09/2022	30	1295
35%	23/09/2022	35	1296
40%	27/09/2022	40	1300
45%	29/09/2022	45	1302
50%	03/10/2022	50	1306
55%	05/10/2022	55	1308
60%	07/10/2022	60	1310
65%	12/10/2022	65	1315
70%	14/10/2022	70	1317
75%	19/10/2022	75	1322
80%	25/10/2022	80	1328
85%	01/11/2022	85	1335
90%	11/111/2022	90	1345
95%	30/11/2022	95	1364
100%	23/01/2023	100	1418

resilient project planning. Risk management also involves allocating risks appropriately among project stakeholders through contractual clauses and other mechanisms. Furthermore, the simulation approach provided more realistic and reliable schedule estimates compared to traditional deterministic scheduling methods, which tend to overlook risk events.

In addition to threats, the analysis also considered opportunities—positive risks that could benefit the project. To improve model accuracy, it is recommended to incorporate open-ended distributions, such as lognormal distributions, instead of the triangular distribution. The model could also be extended to include cost variables, enhancing its capability to deliver more comprehensive, project-specific results. Overall, this case study illustrates the importance of integrating geophysical and risk assessment tools in infrastructure projects to improve resilience and reduce uncertainty.

11.7.6 Case Study: Improving Climate Resilience at Kansai International Airport, Osaka Bay, Japan

Kansai International Airport, located on reclaimed low-lying land in Osaka Bay, Japan, is highly exposed to flooding and climate-related risks (see Figs. 11.20 and 11.21 for the location of Japan on the world map). In 2018, the airport experienced severe disruption due to a major typhoon, which caused a complete shutdown of operations for two weeks. This event revealed the airport's vulnerability and the urgent need for climate resilience planning. In response, the airport authorities launched a project to improve their ability to handle future disasters. The main objectives were to anticipate the impact of climate hazards on airport systems, protect critical infrastructure and personnel, and reduce operational downtime through cost-effective adaptation measures (Maeda 1991).

The project combined expertise in airport operations and flood resilience to create a practical solution tailored to Kansai's needs. A key part of the project involved using two digital tools: The Circle Tool and the Flood Simulation Tool developed by Royal HaskoningDHV's. The simulation tool was calibrated with real data from the 2018 typhoon to model potential future flood scenarios. The Circle Tool mapped out critical infrastructure and demonstrated how disruptions in one system, such as power to water pumps, could lead to cascading failures across the airport, including flooding of terminals and communication breakdowns.

With these insights, the project team worked with airport stakeholders to identify high-risk assets and test various mitigation strategies. Two final solutions were selected based on their effectiveness, feasibility, and ability to be implemented with minimal disruption to airport operations. The preferred solution was estimated to reduce disaster-related downtime from fourteen days to just two days. The project

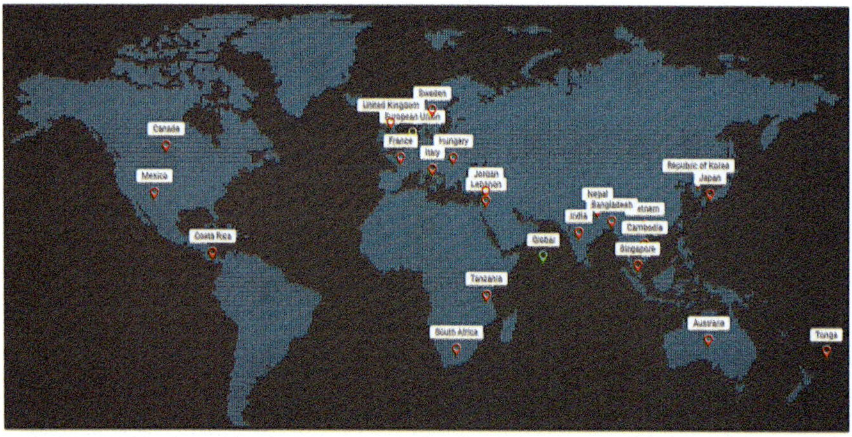

Fig. 11.20 Location of Japan on the world map—context for the Kansai International Airport case study (*Source* Coalition for Disaster Resilient Infrastructure)

Fig. 11.21 Osaka maps and orientation: Osaka, Kansai, Japan (*Source* World Guides)

also developed a step-by-step plan to implement the chosen solution without affecting ongoing airport activities.

The results were significant. The project improved the protection of passengers, staff, and essential systems, while also enhancing coordination among stakeholders. By introducing backup systems, movable equipment, and detailed emergency response plans, the airport was better prepared to continue functioning during and after flood events. This not only strengthened business continuity but also boosted confidence among airlines, passengers, and investors in the airport's ability to manage emergencies.

Furthermore, the project encouraged collaboration and knowledge sharing, offering a valuable example for other airports facing similar risks. It demonstrated how using digital platforms to visualize infrastructure interconnections can support informed decision-making. The approach taken at Kansai International Airport provides a practical model for airports in coastal and low-lying regions that are vulnerable to typhoons, sea-level rise, and heavy rainfall. By planning for long-term resilience, airports can protect lives, maintain operations, and improve passenger experience, even during extreme climate events.

11.7.7 An Integrated Infrastructure Resilience Approach: From the Geotechnical Asset to the Transport Network (Case Study: Greece)

This case study explores a comprehensive resilience framework for transport infrastructure in Greece, emphasizing how failures in geotechnical assets can disrupt entire road and highway networks. The methodology integrates the Societal, Ecological, and Technological (SET) framework to assess resilience at both the *asset scale* and the *network scale*, acknowledging the complex, interdependent nature of modern transport systems (Belokas et al. 2024). Figure 11.22 frames infrastructure systems as being embedded within SET environments, where their functionality and vulnerability are influenced by social needs, ecological risks, and technological performance. This systems-based approach provides a foundation for understanding how infrastructure resilience must account for cross-disciplinary and cross-scale dynamics (Markolf et al. 2018).

The SET framework is particularly valuable for analyzing real-world disruptions such as the Tsakona landslide, which occurred in Greece (Belokas and Dounias 2016). As shown in Fig. 11.23, the ecological interactions in the SET environment—specifically slope instability—led to progressive ground movement in the Tsakona area. Subfigures (a) and (b) illustrate the condition of the site before and after major slope activation, highlighting the temporal evolution of the hazard.

The impact of the Tsakona landslide is analyzed both at the network and asset scales. At the asset scale, the landslide caused the collapse of a critical highway embankment. At the network scale, this led to substantial rerouting and traffic delays (Helm et al. 2024). These perspectives are visualized in Fig. 11.24, which shows: (a) the network-wide disruptions caused by the landslide, and (b) the local failure point at the asset level. Such dual-scale analysis is crucial for understanding cascading effects and system-wide vulnerabilities.

Another significant geohazard event is the Tempi rockfall, which also occurred in Greece. The rockfall blocked a primary transport corridor, triggering extended closures. This event is similarly assessed at both scales to evaluate the resilience of the corridor and the infrastructure components involved (Saroglou 2019). As illustrated in Fig. 11.25, part (a) presents the rockfall's impact on the regional network, while part (b) zooms into the specific asset (a rock face adjacent to the roadway) whose failure initiated the disruption.

These events underline the need to quantify resilience using performance metrics that reflect instantaneous losses and long-term recovery trajectories. Resilience indicators applied to these two cases demonstrate how the failure of geotechnical assets can propagate across the entire network, reducing redundancy and adaptability. The study further shows that traditional risk analysis—while important—is insufficient for comprehensive resilience planning. It does not consider dynamic system behavior during post-disruption recovery, nor does it capture complex interdependencies. Therefore, an integrated approach, as framed through the SET model, is necessary for critical infrastructure planning in high-risk environments.

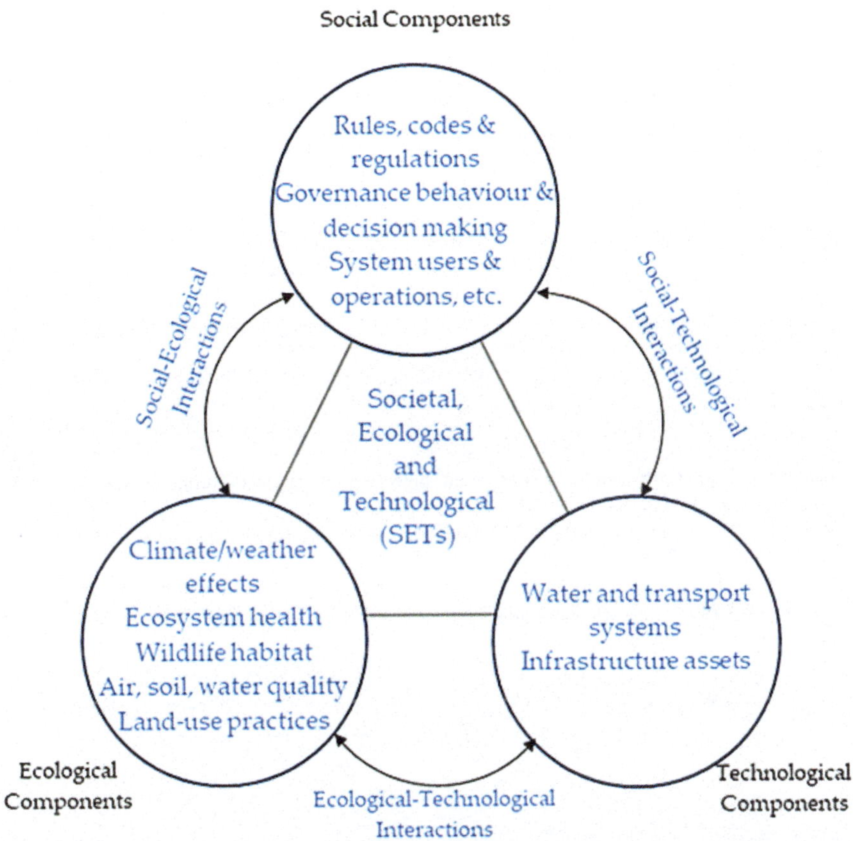

Fig. 11.22 Framing infrastructure as social, ecological, and technological (SET) systems (*Source* Belokas et al. 2024—Open Access)

Although the methodology has been validated using case studies from Greece, it is generalizable to other sectors and geographic contexts. Future research should aim to refine the resilience quantification framework, incorporate real-time operational data, and apply the methodology dynamically using advanced simulation tools. By combining multi-disciplinary knowledge, real-world case validation, and systems-level thinking, this case study offers a robust approach for enhancing infrastructure resilience. Importantly, it illustrates that the continuity and reliability of transport networks depend not just on asset performance but also on understanding and managing system-wide interconnections under stress.

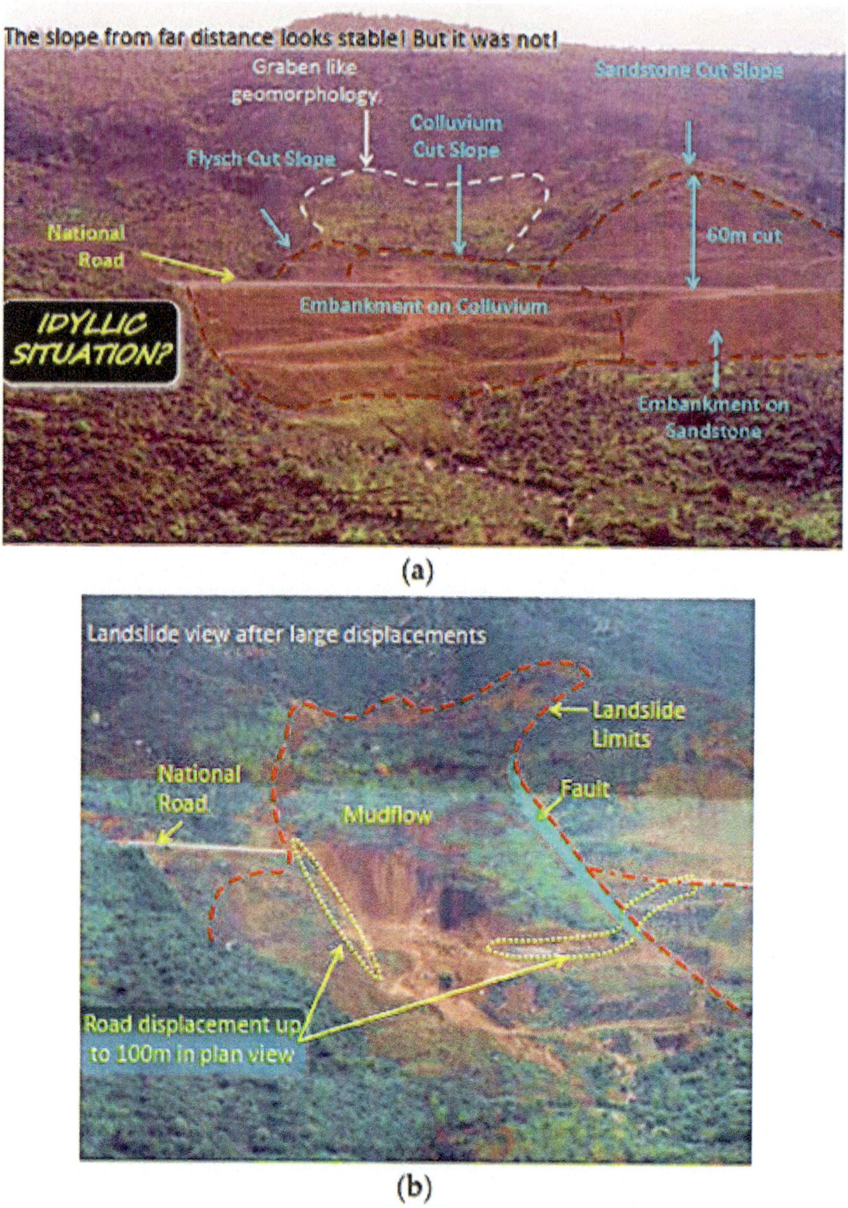

Fig. 11.23 SET environment interaction due to ecological component; case of Tsakona area (Greece) landslide: **a** prior to onset of major activation, **b** After major activation (*Source* Belokas et al. 2024—Open Access)

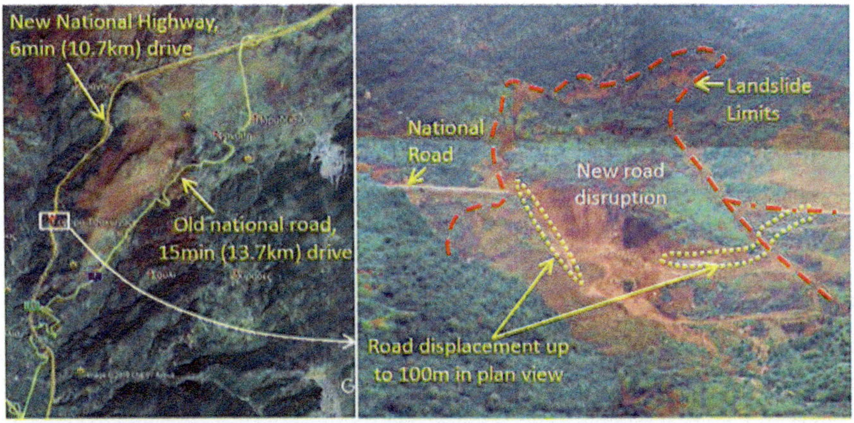

(a) (b)

Fig. 11.24 The Tsakona landslide: **a** the network scale—showing disruption across regional connectivity, **b** the asset scale—localized failure of the highway section (*Source* Belokas et al. 2024—Open Access)

(a) (b)

Fig. 11.25 The Tempi rockfall (Greece): **a** the network scale—regional transport impact, **b** the asset scale—rockfall site adjacent to roadway infrastructure (*Source* Belokas et al. 2024—Open Access)

11.8 Conclusions and Recommendations

In an era of intensifying natural and anthropogenic hazards, the proactive use of geophysical methods to enhance infrastructure resilience is no longer optional—it is essential. This chapter has explored how subsurface intelligence contributes to the lifecycle of critical infrastructure, from planning and construction to monitoring and adaptive response. This chapter underscores the value of geophysics in reducing systemic risks, informing decision-making, and integrating with broader resilience frameworks and digital tools.

Summary of Key Points

- Assessing Earthquake Resilience in Infrastructure (Sect. 11.1):

Subsurface evaluations are vital for earthquake-resistant design in metros and other urban systems. Geophysical tools inform foundational stability, material performance, and real-time risk management.

- Infrastructure Resilience Planning (Sect. 11.2):

Risk-informed infrastructure planning benefits from geophysical integration, particularly in multi-scenario modeling, data fusion, and the application of global resilience standards (e.g., ISO 55000, UNDRR). Current frameworks are evolving, but gaps in applying geophysics systematically remain.

- Application Across Utility, Transport, and Energy Sectors (Sect. 11.3):

Geophysical techniques support resilience in water and gas pipelines, road and rail networks, airports, and seaports. Case studies from Japan, India, and Greece reveal adaptive strategies and the importance of regional context.

- Pre-Failure Detection and Predictive Maintenance (Sect. 11.4):

Advanced geophysical monitoring detects early signs of subsurface changes, enabling predictive maintenance and reduced downtime. Real-time sensing enhances safety and cost efficiency.

- Challenges and Technological Opportunities (Sect. 11.5):

Barriers include data inconsistency, environmental variability, and lack of standardization. However, AI, IoT, and smart infrastructure open new pathways for automated analysis, integrated monitoring, and improved decision support.

- Future Directions in Resilience Innovation (Sect. 11.6):

Geophysics is crucial in developing digital twins and planning for climate-resilient infrastructure. Strategic funding, policy alignment, and interdisciplinary research are needed to maximize impact.

- Insights from Case Studies (Sect. 11.7):

Real-world cases—such as urban flood response, metro construction in Bengaluru, and resilience upgrades at Kansai Airport—highlight the benefits of combining geophysical analysis with social, institutional, and engineering strategies.

Strategic Recommendations for Future Research and Deployment

1. Institutionalize Geophysical Inputs in Resilience Frameworks:

Promote formal integration of geophysics into infrastructure design and risk management frameworks at national and international levels.

2. Develop Geophysical Decision-Support Tools:

Create modular, user-friendly platforms combining geophysical data, AI analytics, and simulation tools to support planners and engineers.

3. Strengthen Multi-hazard Assessment Capabilities:

Expand multi-scenario modeling using geophysical parameters for flood, earthquake, subsidence, and climate-related risks in complex infrastructure systems.

4. Foster Cross-Sector Collaboration and Data Sharing:

Encourage collaboration among utility operators, transport authorities, emergency agencies, and geophysics experts to standardize data and methodologies.

5. Invest in Smart Infrastructure Technologies:

Leverage AI, IoT, and digital twins to automate geophysical monitoring and enable adaptive infrastructure responses under dynamic conditions.

6. Promote Community-Based and Social Sensing Integration:

Combine geophysical insights with social data (e.g., citizen reports, mobile sensing) for more inclusive and responsive disaster resilience strategies.

7. Build Capacity and Interdisciplinary Training Programs:

Develop specialized curricula and training for engineers, planners, and geophysicists focused on resilient infrastructure and emergency preparedness.

The Path Forward for Resilient Infrastructure

The convergence of geophysical science with smart engineering, policy innovation, and community engagement offers a transformative pathway to resilience. As infrastructure systems face increasing pressures from urbanization and climate change, subsurface intelligence will remain a cornerstone of robust, adaptive, and future-ready infrastructure networks.

References

Abdalla A, El Naggar H, El-Baz F (2023) Integrated geophysical approach for detecting buried utilities and leak-induced voids in urban environments. J Appl Geophys 212:104821

Aerts JCJH, Botzen WJW, Emanuel K, Lin N, de Moel H, Michel-Kerjan EO (2014) Evaluating flood resilience strategies for coastal megacities. Science 344(6183):473

Al-Hassan M, Becker M (2025) Smart geoinfrastructure: leveraging AI and IoT for subsurface monitoring and adaptive resilience. Smart Cities Infrastruct J 6(2):112–126

Alphen JV, Bourger L, Elliott C, Fujita KI, Riedstra D, Rooke D, Tachi K (2011) Flood risk management approaches: as being practiced in Japan, Netherlands, United Kingdom and United States. IWR report 2011-R-08, published on behalf of US ACE, Rijkswaterstaat, MLIT and Environment Agency

Amarel T (2025) AI to the rescue: How Artificial Intelligence is transforming humanitarian and disaster response. Retrieved from https://tayloramarel.com/2025/04/ai-to-the-rescue-how-artificial-intelligence-is-transforming-humanitarian-and-disaster-response/

Amekudzi-Kennedy A, Woodall B, Karner A (2018) Institutional arrangements, transportation system investments and socio-economic outcomes: 2 affecting the development of shared regional prosperity

Asakura Y, Tanaka S, Fujita T (2023) Integrated geophysical survey for evaluating airport runway subgrade integrity. Transport Geotechn 44:100987

Bach C, Gupta AK, Nair SS, Birkmann J (2013) Training module, critical infrastructures and disaster risk reduction (in the context of natural hazards). National Institute of Disaster Management and Deutsche Gesellschaft fur internationale Zusammenarbeit GmbH (GIZ), New Delhi, India

Becker M, Uhlemann S, Chambers JE, Wilkinson PB (2023) Time-lapse geophysics for early warning: monitoring stress and deformation in critical infrastructure. Near Surf Geophys 21(2):105–120

Belokas G, Saroglou C, Moschovou TP, Vlahogianni EI (2024) An integrated infrastructure resilience approach, from the geotechnical asset to the transport network. Eng 5:819–833

Belokas G, Dounias G (2016) The Tsakona Landslide in Peloponnese—Greece. Riv Ital Geotech 50:34–50

Bertocchi G, Bologna S, Carducci G, Carrozzi L, Cavallini S, Lazari A, Oliva G, Traballesi A (2016) Guidelines for critical infrastructures resilience evaluation. AIIC, Rome, Italy

Bievre G, Doyen E, Jongmans D, Schmutz M (2023) Geophysical methods in infrastructure site selection: From hazard identification to design optimization. Eng Geol 322:107191

Bruneau M, Chang SE, Eguchi RT, Lee GC, O'Rourke TD, Reinhorn AM, Von Winterfeldt D (2003) A framework to quantitatively assess and enhance the seismic resilience of communities. Earthq Spectra 19:733–752

Cardu M, Seccatore J (2016) Quantifying the difficulty of tunnelling by drilling and blasting. Tunn Undergr Space Technol 60:178–182

Cong W, Shi L, Shi ZC, Peng M, Yang H, Cheng XD (2022) Numerical study on the ceiling gas temperature in a subway train with different fire locations. Build Simulation 15:549–560

Constitutional Law—Unitary and Federal Systems (2019) Encyclopedia Britannica, 7June 2019

Davis CA (2021) Understanding functionality and operability for infrastructure system resilience. Nat Hazards Rev 22:06020005

Delgado F, Nguyen T, Rhodes M (2024) Bridging policy and practice: Funding geophysical monitoring in national infrastructure resilience plans. Int J Disaster Risk Reduc 101:103238

Faruq SU, Archan YV, Reddy ML (2022) Risk assessment and management of underground metro construction (Bengaluru Scenario). Int J Innov Technol Explor Eng (IJITEE) 11(8). ISSN: 2278–3075 (Online)

Fernandez J, Müller C, Zhao H (2024) Digital twins and geophysical data integration for predictive infrastructure management. Comput Geosci 183:105228

Field CB, Barros V, Stocker TF, Dahe (eds) (2012) Managing the risks of extreme events and disasters to advance climate change adaptation: special report of the intergovernmental panel on climate change. Cambridge University Press, Cambridge

Gong X, McCarthy PX, Tian L, Rizoiu M-A (2025) Signals from the floods: AI-driven disaster analysis through multi-source data fusion. arXiv

Gunn DA, Chambers JE, Wilkinson PB, Uhlemann S, Meldrum PI (2023) Geophysical monitoring for infrastructure resilience: innovations, challenges and future prospects. Eng Geol 315:107027

Helm PR, Svalova A, Morsy AM, Rouainia M, Smith A, El-Hamalawi A, Wilkinson DJ, Postill H, Glendinning S (2024) Emulating long-term weather-driven transportation earthworks deterioration models to support asset management. Transp Geotech 44:101155

Higgins A (2012) Netherlands sets model of flood prevention. The New York Times

Hofstede G (2001) Culture's consequences: comparing values, behaviors, institutions and organizations across nations. SAGE Publications

Huang H, Zhang D, Huang Z (2022) Resilience of city underground infrastructure under multi-hazards impact: from structural level to network level. Resilient Cities Struct 1(2022):76–86

Irankunda G, Zhang W, Fernand M, Zhang J (2024) Assessing the resilience of critical infrastructure facilities toward a holistic and theoretical approach: a multi scenario evidence and case study. Sustainability 16:8735

Jothimani P, Yamamura E (1995) Spatial changes in urban landuse and ecosystems arising due to subway network—a case study in Sapporo, Japan using Geographical Information Systems. Stud Regional Sci 26:247–255

Kargl SG, Nocerino J, El-Hady MA (2023) Overcoming geophysical limitations in complex terrains: toward standardized urban infrastructure monitoring. J Environ Eng Geophys 28(2):167–181

Kim Y, Chester MV, Eisenberg DA, Redman CL (2019) The infrastructure trolley problem: positioning safe-to-fail infrastructure for climate change adaptation. Earth's Futur 7:704–717

Kopanov K, Varbanov V, Atanasova T (2024) Leveraging social media data and artificial intelligence for improving earthquake response efforts. arXiv.

Kumar V, Sharma P, Lin C (2024) AI-enhanced IoT frameworks for predictive maintenance in smart city infrastructure. J Infrastruct Syst 30(1):04023045

Kumar R, Singh H, Mehta M (2023) Geoelectrical imaging for integrity assessment of oil and gas pipelines in expansive soil zones. J Petrol Sci Eng 226:111635

Kundzewicz ZW, Kanae S, Seneviratne SI, Handmer J, Nicholls N, Peduzzi P, Mechler R, Bouwer LM, Arnell N, Mach K, Muir-Wood R, Brakenridge GR, Kron W, Benito G, Honda Y, Takahashi K, Sherstyukov B (2014) Flood risk and climate change: global and regional perspectives. Hydrol Sci J 59(1):1–28

Latora V, Marchiori M (2001) Efficient behavior of small-world networks. Phys Rev Lett 87(19):198701

Lin D, Zhou Z, Wong M, Broere W, Cui J (2024) Metro systems: construction, operation and impacts. Tunn Undergr Space Technol 143:105373

Link JFD, Smith, Hellingrath B, Babun Jr. TA (2014) Promoting port resiliency. In: Proceedings of the 5th International global disaster and risk conference IDRC, Davos, pp 1–4

Lopez D, Martínez A, Ruiz J (2023) Geophysical site characterization for offshore renewable energy projects: integrating GPR and seismic surveys. Renewable Energy 210:1234–1245

Maeda M (1991) The Kansai International Airport project and environmental impact assessment. Mar Pollut Bull 23:349–353

Mariani S, Colombo D, Rizzo E (2023) Geophysical diagnostics in urban utility corridors: a case study on sewer failure prevention. J Appl Geophys 214:104832

Markolf SA, Chester MV, Eisenberg DA, Iwaniec DM, Davidson CI, Zimmerman R, Miller TR, Ruddell BL, Chang H (2018) Interdependent infrastructure as linked social, ecological, and technological systems (SETSs) to address lock-in and enhance resilience. Earth's Future 6:1638–1659

Niu W, Feng XT, Yao Z, Bi X, Yang C, Hu L, Zhang W (2022) Types and occurrence time of rock bursts in tunnel affected by geological conditions and drilling & blasting procedures. Eng Geol 303:106671

North DC (1991) Institutions. J Econ Perspect 5(1):97–112

Oliveira R, Santos JA, Lin Y (2024) Geophysical tools for climate-resilient infrastructure: a proactive approach to subsurface variability. Clim Risk Manag 43:100522

Osei-Kyei R, Tam V, Ma M, Mashiri F (2021) Critical review of the threats affecting the building of critical infrastructure resilience. Int J Disaster Risk Reduct 60:102316

Roy S, Debnath P, Mitra S (2023) Impact of climate disasters on railway infrastructure: case study of Northeast India. Acadlore Trans. Geosci. 2(1):33–45

Saroglou C (2019) GIS-based rockfall susceptibility zoning in Greece. Geosciences 9:163

Satheeskumar N (2016) Wind load sharing and vertical load transfer from roof to wall in a timber-framed house. James Cook University, Townsville, Australia

Sathurshan M, Saja A, Thamboo J, Haraguchi M, Navaratnam S (2022) Resilience of critical infrastructure systems: a systematic literature review of measurement frameworks. Infrastructures 7:67

Shi C, Zhong M, Chen C, Jiao W, Li J, Zhang Y, Zhang L, Li Y, He L (2020) Metro train carriage combustion behaviors—full-scale experiment study. Tunn Undergr Space Technol 104:103544

Shorris, A., 2015. The plan for a strong and just city. 354.

Singh P, Amekudzi-Kennedy A, Woodallc B, Joshi S (2021) Lessons from case studies of flood resilience: institutions and built systems. Transport Res Interdiscip Perspect 9:100297

Singhal TK, Kwon OS, Bentz EC, Christopoulos C (2021) Development of a civil infrastructure resilience assessment framework and its application to a nuclear power plant. Struct Infrastruct Eng 18:1–14

Sonnenschein TS, Scheider S, Zheng S (2022) The rebirth of urban subcenters: how subway expansion impacts the spatial structure and mix of amenities in European cities. Environ Plan B Urban Anal City Sci 49:1266–1282

Tanaka S, Fujii T, Nakamura K (2023) Monitoring road embankment stability using time-lapse seismic techniques: a case from Japan. Eng Geol 318:107149

Thomas ME, Jenkins B, Arslan A (2023a) Toward standardization in geophysical applications for infrastructure: bridging gaps in practice and policy. Eng Geophys J 12(1):45–58

Thomas D, Miller J, Zhang S (2023b) Resilience measurement frameworks for critical infrastructure: a systematic review and integration of geophysical data. J Infrastruct Risk Manage 45(2):233–250

Uhlemann S, Chambers JE, Gunn DA, Wilkinson PB, Meldrum PI (2023) Real-time geoelectrical monitoring to support infrastructure resilience: developments and future directions. Near Surface Geophysics 21(1):29–42

Veluru CS (2024) Responsible artificial intelligence on large scale data to prevent misuse, unethical challenges and security breaches. J Artif Intell Cloud Comput. https://doi.org/10.47363/JAICC/2024(3)331

Vugrin ED, Warren DE, Ehlen MA (2011) A resilience assessment framework for infrastructure and economic systems: quantitative and qualitative resilience analysis of petrochemical supply chains to a hurricane. Process Saf Prog 30:280–290

Wang L, Zhang X, Li Y (2023) Application of MASW and ERT in sinkhole detection beneath railway infrastructure in karst terrains. Eng Geol 320:107165

Wang X, Mazumder RK, Salarieh B, Salman AM, Shafieezadeh A, Li Y (2022) Machine learning for risk and resilience assessment in structural engineering: progress and future trends. J Struct Eng 148:03122003

Yang X-X, Jiang H-L, Kang Y-L, Yang Y, Li Y-X, Yu C (2022) Passenger management strategy and evacuation in subway station under Covid-19. Chin Phys B 31:078901

Young JC, Arthur R, Spruce M, Williams HTP (2022) Social sensing of flood impacts in India: a case study of Kerala 2018. Int J Disaster Risk Reduc 74:102908

Zhang Y, Liu H, Wang Y, Zhou T (2023a) AI-driven integration of geophysical and geospatial data for infrastructure risk assessment and planning. Comput Environ Urban Syst 97:101967

Zhang L, He Q, Sun Y (2023b) Integrating geophysical monitoring and machine learning for predictive maintenance of road infrastructure. Autom Constr 151:104929

Chapter 12
Defense and Strategic Installations

Abstract This chapter explores the critical role of geophysical methods in securing national defense and strategic assets through subsurface intelligence. It examines techniques for detecting underground threats such as tunnels, bunkers, and concealed infrastructure, highlighting operational challenges and real-world applications. The chapter discusses military engineering geophysics, focusing on rapid assessment tools, engineering support, and multidisciplinary collaboration for tactical advantages. It also addresses the dual-use nature of geophysical technologies in homeland security and critical infrastructure protection, alongside emerging policy frameworks. Environmental and ethical considerations in defense-related geophysical surveys are reviewed, emphasizing compliance with international standards. The chapter further discusses interoperability, data-sharing challenges, and the limitations of current geophysical methods in defense contexts. It concludes with recommendations for advancing geophysical capabilities to enhance national security while balancing environmental and ethical responsibilities.

12.1 Securing National Assets with Subsurface Intelligence

This introductory section sets the stage for understanding how subsurface geophysical intelligence plays a pivotal role in safeguarding defense and strategic installations. It highlights the growing need for advanced geophysical tools to detect, monitor, and mitigate subsurface risks associated with critical national infrastructure. From underground military facilities and storage depots to communication hubs and surveillance networks, the security of such assets depends not only on surface measures but also on a comprehensive understanding of the underlying geology and potential hidden threats.

The section introduces core concepts such as the integration of geophysical data into risk assessments, site characterization for secure construction, and the role of non-invasive techniques in ongoing monitoring. It emphasizes how subsurface intelligence supports resilience against natural hazards, sabotage, tunneling, and other

D. C. Naskar, *Advances in Engineering Geophysics: Concepts and Applications*, Earth Systems Data and Models 9, https://doi.org/10.1007/978-3-032-28000-8_12

security challenges—forming the foundation for detailed applications discussed in the subsequent sections of the chapter.

12.2 Detection of Subsurface Threats: Tunnels, Bunkers, and Concealed Infrastructure

12.2.1 Nature of Subsurface Threats

Subsurface threats pose a persistent and evolving challenge to national defense and internal security frameworks. These threats typically manifest as covert underground structures—such as smuggling tunnels, fortified bunkers, and subterranean weapons storage facilities—engineered to bypass conventional detection and countermeasures. Their clandestine nature and strategic placement can compromise border integrity, undermine military installations, and facilitate transnational criminal or terrorist operations.

Historically, underground warfare has been a recurrent military tactic, from the siege tunnels of ancient warfare to extensive trench and tunnel systems in World War I and the Vietnam War. In modern contexts, non-state actors and insurgent groups continue to exploit the underground domain for tactical advantage. For instance, the extensive network of tunnels in Gaza (densely populated Palestinian territory located on the eastern coast of the Mediterranean Sea. It borders—Israel to the north and east, Egypt to the southwest, and The Mediterranean Sea to the west) has been used for the transport of weapons and personnel, while similar structures along the US-Mexico border have been designed for drug trafficking and human smuggling operations (Barak 2017; DHS 2020).

The strategic risk lies not only in the physical presence of these structures but in their potential to destabilize secure zones, disrupt critical infrastructure, and enable asymmetric warfare. Due to the increasing sophistication in tunnel construction and concealment, the detection and monitoring of such threats require the integration of advanced, non-invasive geophysical techniques.

12.2.2 Geophysical Detection Techniques

The detection of concealed subsurface threats—such as illicit tunnels (unauthorized underground passages constructed secretly, often for illegal purposes such as smuggling, human trafficking, arms transport, or militant infiltration across borders or into secured areas), fortified bunkers, and void spaces—relies heavily on the deployment of advanced geophysical techniques. These non-invasive methods provide critical spatial intelligence without disrupting the terrain or compromising operational secrecy. Each method offers unique sensitivity to different subsurface properties, enabling tailored approaches based on geological context, depth of interest, and the nature of the threat.

Ground Penetrating Radar (GPR)

GPR utilizes high-frequency electromagnetic waves to detect changes in the dielectric (dielectric, insulating material or a very poor conductor of electric current) properties of the subsurface. It is especially effective for shallow investigations, typically up to 5–10 m depth, depending on soil composition. The method offers high-resolution imaging suitable for identifying voids, disturbed ground, or artificial structures in dry or sandy conditions. However, signal attenuation in conductive soils (e.g., clay-rich or water-saturated zones) may limit its effectiveness (Daniels 2004).

Electrical Resistivity Tomography (ERT)

ERT measures variations in subsurface electrical resistivity by injecting low-voltage current through a series of electrodes. This technique is well-suited for detecting anomalies such as voids, buried structures, and altered soil zones by mapping resistivity contrasts. ERT is particularly valuable in complex geological settings where other methods may be limited, and can achieve greater penetration depths than GPR (Loke 2015).

Seismic Methods

Seismic refraction and reflection techniques detect changes in seismic wave velocities caused by discontinuities such as air-filled tunnels, fractured zones, or loose fill. These methods are effective for identifying voids and characterizing the geomechanical properties of the subsurface. Seismic approaches can be deployed in both active and passive modes, making them adaptable for a variety of field conditions, including urban or high-security environments (Telford et al. 1990).

Magnetometry and Gravimetry

Magnetometry detects variations in the Earth's magnetic field caused by ferrous materials or structural discontinuities. It is particularly useful for locating reinforced bunkers or steel-lined tunnels. Gravimetry, on the other hand, identifies local changes in gravitational acceleration associated with mass deficits (e.g., air-filled voids or tunnels). While both methods are sensitive to deeper or larger anomalies, they often require integration with other datasets for reliable interpretation (Nabighian et al. 2005).

In practice, the integration of multiple geophysical methods—often referred to as data fusion—enhances detection accuracy and reduces ambiguity. These techniques form a critical component of modern subsurface intelligence strategies for defense and national security applications.

12.2.3 Operational Challenges

While geophysical methods offer significant promise for the detection of subsurface threats, their practical deployment in defense and security contexts is often constrained by a range of operational challenges. These challenges arise from environmental conditions, technical limitations, and the complexity of interpreting subsurface anomalies in real-world scenarios.

Challenging Geological and Urban Conditions

Geological heterogeneity presents a significant obstacle to reliable detection. High-conductivity soils, such as clays and saline sediments, often attenuate electromagnetic signals, reducing the effectiveness of Ground Penetrating Radar (GPR) (Daniels 2004). Similarly, variations in moisture content, porosity, and layering can obscure or distort electrical resistivity measurements (Loke 2015). Seismic methods, while less susceptible to conductivity, may still face issues with signal scattering in fractured or unconsolidated formations (Telford et al. 1990).

In urban environments, the presence of subsurface infrastructure—including pipes, cables, reinforced concrete, and debris—introduces electromagnetic and mechanical noise that can obscure target signals. Furthermore, anthropogenic features may produce signatures that mimic those of tunnels or voids, complicating interpretation. Access limitations due to security protocols or built-up terrain further constrain the layout and density of survey grids, limiting resolution and spatial coverage (Nabighian et al. 2005).

Data Interpretation and False Positives

One of the most critical limitations in operational settings is the risk of false positives and misinterpretation. Anomalies that resemble threats—such as buried foundations, culverts, or geologic voids—can result in misclassification, leading to costly and time-consuming false alarms (Olhoeft 1998). Conversely, poorly defined or subtle anomalies may be dismissed as noise, allowing genuine threats to go undetected.

The interpretation of geophysical data is inherently complex, requiring expertise in both the physical principles of the method and the geological context of the site. Signal interference, low signal-to-noise ratios, and overlapping anomalies often make it difficult to distinguish between natural features and anthropogenic structures. These issues are particularly acute when decisions must be made rapidly in operational defense environments.

To mitigate these challenges, integrated geophysical approaches—combining GPR, ERT, seismic, and magnetic methods—are increasingly employed, allowing cross-verification of anomalies (Slob et al. 2010). Additionally, advances in data processing, including AI-assisted interpretation and machine learning classifiers, are beginning to reduce human bias and enhance detection reliability, although these technologies are still maturing and must be carefully validated in high-risk contexts.

12.2.4 Case Studies and Applications

The application of geophysical methods for detecting subsurface threats is no longer confined to theoretical exploration but has been demonstrated in multiple real-world defense and security contexts. These case studies highlight the operational significance of geophysical intelligence in safeguarding national borders and critical infrastructure.

Border Security Applications

One of the most widely recognized applications of geophysical techniques is in securing international borders from subterranean breaches (unauthorized or covert underground intrusions, such as tunnels or shafts, typically used to bypass surface-level security measures—for smuggling, infiltration, or sabotage). Along the US-Mexico border, the discovery and neutralization of smuggling tunnels have become a strategic priority for federal agencies. Since the early 1990s, hundreds of illicit tunnels—some extending over 800 m in length—have been uncovered. Ground Penetrating Radar (GPR), electrical resistivity, and microgravity surveys have been used in combination to detect these covert structures, which often exploit geotechnical vulnerabilities in sandy or alluvial terrains (DHS 2020; Hearty et al. 2014).

In the Gaza-Israel border region, tunnel detection has been elevated to a critical component of national defense due to the tactical use of subterranean networks by non-state actors. The Israeli Defense Forces (IDF) have developed specialized geophysical units that deploy seismic and acoustic sensors capable of identifying excavation activities in real time. These efforts are complemented by electromagnetic imaging and borehole-based ground truthing, often under operational constraints and hostile conditions (Barak 2017).

Protection of Critical Urban Infrastructure

Geophysical technologies have also been applied in urban environments to assess and protect critical infrastructure such as government buildings, embassies, transportation tunnels, and underground utilities. In many cases, urban infrastructure is vulnerable to both intentional threats—such as clandestine tunneling or sabotage—and unintentional hazards, including ground subsidence or undetected voids near foundations.

For example, in cities with aging infrastructure, GPR and seismic surveys have been used to monitor subgrade conditions beneath subways and utility corridors. These assessments help identify potential collapse zones or unauthorized excavations before they compromise structural integrity or public safety. Furthermore, urban security planning increasingly incorporates routine geophysical monitoring, especially in high-risk zones adjacent to secure facilities.

These case studies demonstrate the operational value of engineering geophysics as a defense tool, not only for threat detection but also for risk mitigation and strategic planning. They underscore the need for continuous innovation in sensor technology, data integration, and interpretation techniques to address evolving security challenges.

12.3 Military Engineering Geophysics: Field Applications and Tactical Advantage

12.3.1 Introduction to Military Geophysics

Military geophysics refers to the application of geophysical principles and techniques in support of defense operations, combat engineering, and national security infrastructure. This specialized field combines geoscientific knowledge with tactical and strategic military objectives, providing essential intelligence for terrain analysis, subsurface threat detection, structural assessment, and the construction or protection of defensive installations.

The relevance of military geophysics lies in its ability to offer rapid, non-invasive insights into subsurface conditions, which are critical for mission planning, logistical operations, and infrastructure resilience. Whether assessing the load-bearing capacity of terrain for armoured (military vehicles or units) movement, identifying natural or man-made obstacles, or locating underground threats such as tunnels or mines, geophysical tools are invaluable to modern combat engineering units.

Historically, geophysics has played a pivotal role in military operations. During World War II, tunneling techniques were employed both offensively (e.g., to undermine enemy fortifications) and defensively (e.g., to construct air-raid shelters and hidden command posts). The Cold War era saw the expansion of underground bunkers, missile silos, and listening posts, many of which relied on geophysical surveys for site selection, structural safety, and long-term monitoring (Goodman et al. 2007).

In contemporary defense contexts, military geophysics continues to evolve with the integration of advanced technologies such as drone-mounted sensors, real-time data transmission, and AI-driven analysis. These developments allow for enhanced operational agility and situational awareness in complex and often hostile environments.

12.3.2 Engineering Support

Geophysical methods serve as a vital toolset in military engineering, offering strategic support in evaluating terrain, subsurface stability, and environmental conditions essential for both temporary operations and long-term installations. The integration of geophysical data into engineering workflows enhances decision-making related to mobility, construction, and survivability in a variety of conflict environments.

Terrain Analysis for Troop and Equipment Movement

The safe and efficient movement of troops and military vehicles—particularly heavy, armoured units—requires detailed knowledge of ground conditions. Geophysical surveys such as seismic refraction and electrical resistivity imaging are used to assess soil strength, moisture content, and potential subsurface obstacles. These parameters are critical in identifying load-bearing zones, avoiding soft ground or sinkholes, and planning routes that minimize mechanical failure or ambush vulnerability (Telford et al. 1990).

In operations involving rapid deployment or manoeuvring in unfamiliar terrain, geophysical reconnaissance enables tactical mobility while reducing the risk of vehicle immobilization or terrain-induced delays.

Siting of Bases, Airstrips, and Underground Shelters

The selection of suitable locations for forward operating bases, temporary airstrips, underground bunkers, and other defensive infrastructure relies heavily on subsurface evaluation. Geophysical techniques help determine subsurface lithology, groundwater conditions, and geotechnical stability. Ground Penetrating Radar (GPR), seismic surveys, and electrical resistivity methods are often employed to map shallow subsurface layers and detect conditions such as high-water tables or unconsolidated sediments that could jeopardize structural integrity. In airstrip construction, for example, uniform compaction and drainage are essential. Geophysics aids in detecting subgrade variations or hidden features (e.g., old utilities, rubble) that could lead to differential settlement or surface failure under repeated aircraft loading.

The military base camp primarily fulfils an administrative and support role, with core mission capabilities centered around communication, coordination, and logistical support. Although the base camp itself is not a direct combat unit, it serves as a critical hub for sustaining operational readiness across multiple commands. Additional essential facilities associated within the base camp include: Medical clinic, Fire station, Barracks housing, Port operations, Operational staff commands, Maintenance facilities, Training command, Galley, and Morale, Welfare, and Recreation (MWR) services. The deployable assets of the base consist mainly of the ships docked at the pier.

Each operational stakeholder within the base contributes different mission capabilities. These are summarized in Table 12.1, which lists the specific functions and capacities of each stakeholder unit. Mission scenarios the base may be required to support can involve any combination of these capabilities. These scenarios often reflect simultaneous or overlapping priorities among tenant commands. For example: One ship may be undergoing scheduled maintenance, another ship might be engaged in simulator-based training, and a third vessel may be conducting sea trials prior to deployment. Such coordination requires strategic resource planning and operational flexibility to meet varied mission demands.

Table 12.1 Mission capabilities of operational stakeholders at the military base camp (after Mallery et al. 2022)

Command	Mission capabilities
Base command	Administration and communications
Fire station	Emergency response
Medical clinic	Routine healthcare and emergency response
Barracks housing	Hotel services
Movie theatre	Entertainment (MWR)
Hotel	Hotel services
Gymnasium	Exercise (MWR)
Fast food restaurant	Food options (MWR)
Galley	Provide basic sustenance
Training command	Fleet readiness: inspections, and classroom training
Port operations	Pilot, tug, security operations
Security	Emergency response, gate security
Operational staff command	Administration and communications
Maintenance facility	Equipment maintenance, overhaul, scaffolding, crane operations
Ships	Pier Work, Sea and Anchor Detail, Crane Operations (on/off load), Refuelling, Training, Maintenance, Deployment

To support resilient and effective infrastructure planning, Fig. 12.1 presents a fictitious military base map that illustrates how geophysical data informs strategic decision-making.

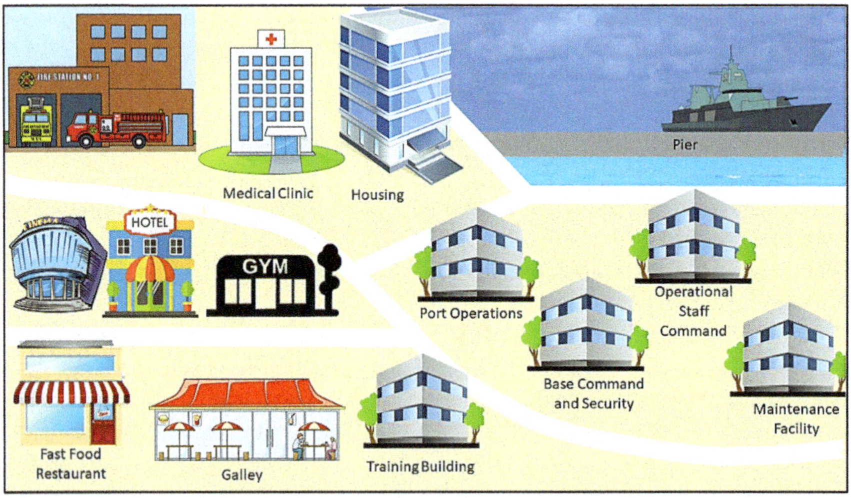

Fig. 12.1 Fictitious military base map (*Source* Mallery et al. 2022)

This figure highlights: Potential mission combinations, key terrain zones, and facility siting decisions influenced by subsurface geophysical conditions. The spatial integration of geophysical intelligence with operational requirements enhances infrastructure resilience and supports tactically advantageous planning (Bruneau et al. 2003).

Evaluating Ground Conditions for Blast Resistance and Survivability

Military installations must also be assessed for their ability to withstand explosive loads, both direct (e.g., from ordnance) and indirect (e.g., from shockwaves transmitted through the ground). Geophysical methods allow engineers to characterize soil and rock properties relevant to blast resistance, including shear strength, stiffness, and wave propagation velocities. Seismic velocity profiling and crosshole tomography, for instance, provide valuable input into models of blast impact and structural survivability (Andrus and Stokoe 2000).

Such evaluations support the design of protective berms (raised earthen barriers or embankments), revetments (protective structures), and buried shelters that can absorb or deflect energy from explosive events, thereby improving the safety of personnel and mission-critical infrastructure.

12.3.3 Rapid Assessment and Deployment Tools

Modern military and security operations demand rapid, adaptive, and mobile solutions for subsurface intelligence. To meet these requirements, a suite of deployable geophysical tools has emerged that enables real-time or near-real-time assessment of terrain and infrastructure with minimal logistical burden. These technologies are especially valuable in dynamic combat zones, forward deployment areas, and post-blast or post-conflict assessments.

Use of Drones, Mobile Platforms, and Handheld Instruments

Unmanned aerial vehicles (UAVs), commonly referred to as drones, have become a transformative asset in military geophysics. When equipped with ground-coupled radar, magnetometers, or LiDAR systems, drones can rapidly map surface and shallow subsurface features over large or hazardous areas. Their ability to operate in contested or inaccessible zones significantly reduces risk to personnel while improving the speed of data acquisition (Colomina and Molina 2014).

Mobile geophysical platforms, such as vehicle-mounted GPR systems or towed array resistivity and magnetometry rigs, allow for fast and systematic surveys along roads, borders, or suspected tunnel corridors. These platforms are often integrated with GPS and onboard processing units to provide immediate feedback to field operators. Handheld instruments—including portable GPR antennas, magnetic gradiometers, and compact resistivity meters—are essential for detailed site inspections, rapid verification of anomalies, and use in confined or built-up environments. Their portability and ease of use make them ideal for engineering reconnaissance, explosive ordnance detection, and entry-point assessments.

Integration with Satellite and Remote Sensing Data

To maximize spatial context and situational awareness, ground-based geophysical measurements are increasingly fused with satellite and remote sensing data. High-resolution optical imagery, synthetic aperture radar (SAR), thermal infrared imaging, and interferometric SAR (InSAR) are used to detect surface deformations, heat signatures, or terrain anomalies indicative of underground activity or structural instability (Massonnet and Feigl 1998).

Such integration allows for multi-scale analysis, where satellite data guide the planning of ground surveys and help prioritize zones of interest. In intelligence and security operations, this fusion of datasets enables more efficient resource deployment and increases the likelihood of detecting subtle or well-concealed subsurface threats.

12.3.4 Training and Multidisciplinary Teams

The effective application of military engineering geophysics depends not only on technology but also on the expertise and coordination of specialized personnel. The complexity of subsurface threat detection and terrain evaluation requires the integration of geophysical knowledge with military engineering and operational tactics. This necessity has led to the development of targeted training programs and the formation of multidisciplinary teams embedded within defense units.

Geophysicists Embedded with Military Engineering Units

Embedding geophysicists directly within military engineering units facilitates real-time interpretation of geophysical data, rapid problem-solving, and adaptation to dynamic field conditions. These specialists provide on-site expertise that bridges the gap between raw data acquisition and actionable intelligence, enhancing decision-making during critical operations. Training programs for such personnel emphasize not only technical proficiency in geophysical methods but also military protocols, communication skills, and situational awareness to operate effectively within combat environments (Hagrey 2007).

Joint Operations Involving Geoscientists, Engineers, and Defense Personnel

Successful subsurface threat mitigation and infrastructure protection increasingly rely on joint operations that combine the skills of geoscientists, civil and military engineers, and defense personnel. This collaborative approach ensures comprehensive assessment and response, leveraging the strengths of each discipline—from geological interpretation and geophysical surveying to structural design and tactical implementation.

Interdisciplinary coordination enhances operational efficiency and risk management, enabling integrated strategies that address the technical, environmental, and security challenges inherent in defense geophysics. Training exercises, simulations, and cross-disciplinary workshops are key to building trust, shared understanding, and seamless teamwork in the field (Boyer and Allard 2012).

12.4 National Security and Dual-Use Technologies

12.4.1 Dual-Use Potential of Geophysical Technologies

Geophysical technologies possess significant dual-use potential, serving critical roles in both military defense and civilian safety applications. This versatility enhances their strategic value while also raising important considerations regarding regulation and ethical deployment.

Tools Applicable in Both Military Defense and Civilian Safety

Many geophysical methods originally developed or refined for defense purposes—such as ground penetrating radar (GPR), electrical resistivity tomography (ERT), and seismic imaging—have found vital applications in civilian contexts. For instance, tunnel detection technology used to identify illicit underground passages for border security and military intelligence is also instrumental in disaster response, such as locating survivors trapped under rubble following earthquakes or building collapses (Katz et al. 2013).

Similarly, magnetometry and gravimetry, used to detect underground bunkers or weapon caches, support geotechnical investigations for critical infrastructure safety and environmental hazard assessment. This crossover enhances investment efficiency and fosters innovation, as advances in one sector often translate into improved capabilities in the other.

Benefits and Regulatory Considerations

The dual-use nature of geophysical technologies presents notable benefits, including resource optimization, enhanced interoperability between defense and civil agencies, and accelerated technological advancement through shared research and development. However, this duality also necessitates careful regulatory oversight to balance national security interests with privacy rights, public safety, and ethical considerations. Governments and international bodies often impose restrictions on the export, dissemination, and use of sensitive geophysical equipment to prevent misuse or proliferation of capabilities that could undermine security or civilian welfare (Meier et al. 2018).

Effective governance frameworks ensure that dual-use technologies contribute positively to both defense readiness and societal resilience without compromising ethical standards or escalating geopolitical tensions.

12.4.2 *Homeland and Critical Infrastructure Security*

The safeguarding of homeland security and critical infrastructure represents a paramount priority in modern national defense strategies. Geophysical technologies play an essential role in detecting and mitigating risks associated with sabotage, unauthorized subsurface intrusion, and structural vulnerabilities that threaten key facilities and urban centers.

Protection of Nuclear Facilities, Government Buildings, and Underground Transport

Nuclear power plants, government headquarters, and underground transportation networks (such as subways and utility tunnels) constitute high-value assets with profound strategic and societal importance. These facilities are inherently vulnerable to covert attacks, including tunneling operations aimed at breaching perimeters or compromising structural integrity.

Advanced geophysical techniques—including seismic tomography, electrical resistivity imaging, and ground penetrating radar—are employed to conduct regular subsurface surveys around and beneath such installations. These methods facilitate early detection of anomalies indicative of excavation activities, voids, or clandestine constructions that could signal sabotage or terrorist attempts (Ferguson and Sterling 2015). Figure 12.2 shows the system resilience curve, illustrating the ability of a critical infrastructure system to maintain or rapidly recover its operational functionality following a disruptive event. The figure highlights the importance of minimizing the duration and severity of system downtime—a key consideration when evaluating infrastructure security and emergency response capacity.

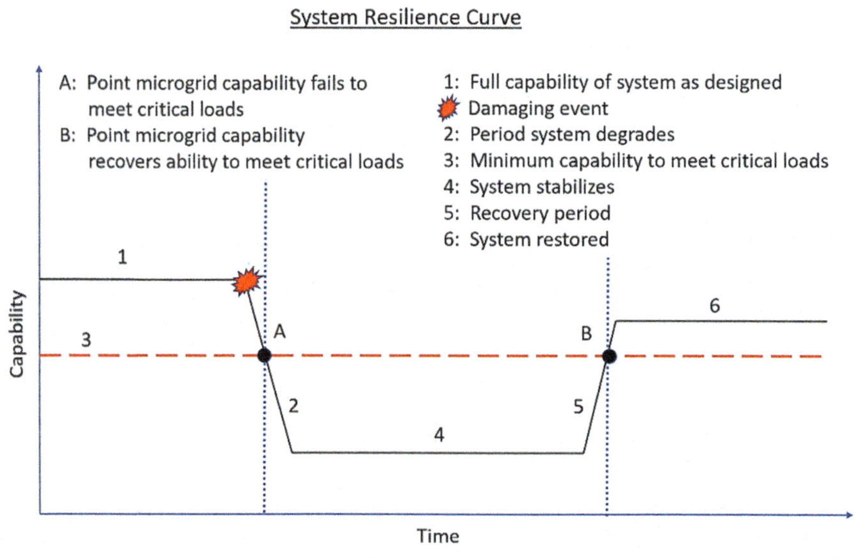

Fig. 12.2 System resilience curve (*Source* Mallery et al. 2022)

Monitoring for Sabotage, Subsurface Intrusions, and Structural Vulnerabilities

Continuous monitoring frameworks integrate geophysical data with sensor networks and surveillance systems to provide real-time situational awareness. For example, passive seismic sensors can detect vibrations caused by digging or tunneling, while magnetometry may reveal unauthorized metallic objects buried underground.

Beyond threat detection, geophysical assessments contribute to the structural health monitoring of critical infrastructure, identifying degradation, void formation, or material weaknesses that compromise safety and operational continuity. Figure 12.3 shows recommended mission priorities for critical load determination, categorizing infrastructure loads into vital, semi-vital, less vital, and non-vital. This prioritization aids in resource allocation during crises and supports infrastructure resilience planning.

Energy Resilience for Mission Continuity

In addition to protective monitoring, energy resilience is a critical component of homeland security. Energy supply systems must remain operational during emergencies and resist disruption due to cyber, physical, or environmental threats. Geophysics plays a key role in planning secure energy layouts, especially in mission-critical installations like military bases or emergency coordination centers. Figure 12.4 shows

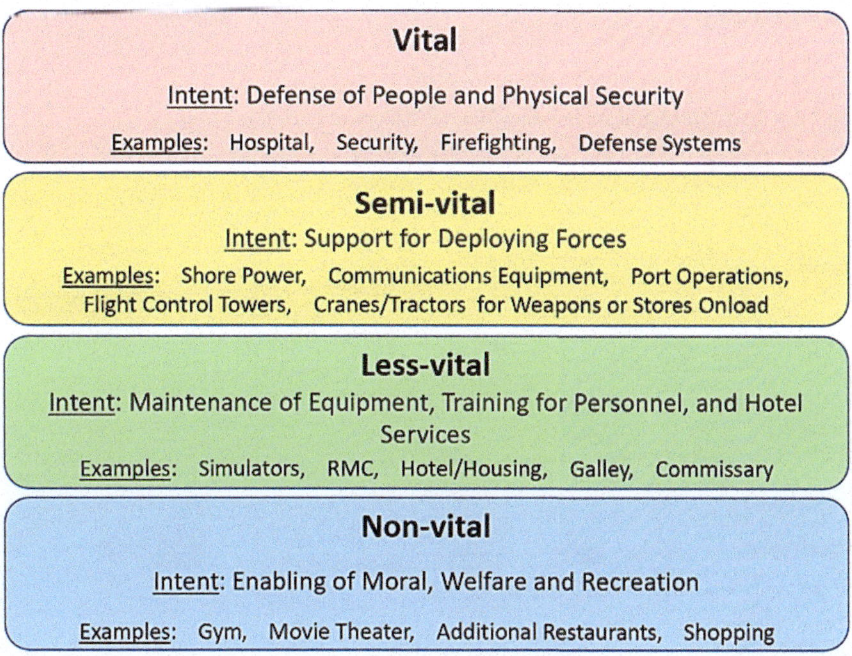

Fig. 12.3 Recommended mission priorities for critical load determination (*Source* Mallery et al. 2022)

the initial microgrid composition developed based on installed energy equipment and estimated demand.

This initial configuration reflects baseline infrastructure capability, serving as a starting point for designing robust and adaptable energy systems. Following scenario modeling and integration of resilience-enhancing technologies, Fig. 12.5 shows the final microgrid composition, which includes a diversified mix of diesel generators, solar photovoltaic (PV) arrays, and battery energy storage systems (BESS). This optimized setup exemplifies how energy resilience planning incorporates both geophysical data and strategic load forecasting to maintain operational continuity under duress.

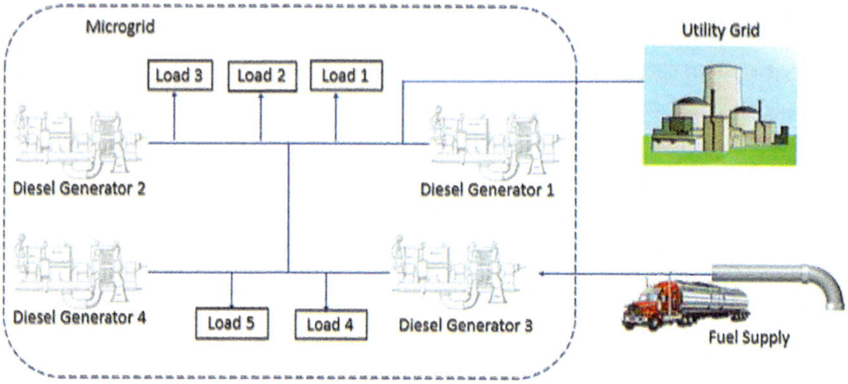

Fig. 12.4 Initial microgrid composition (*Source* Mallery et al. 2022)

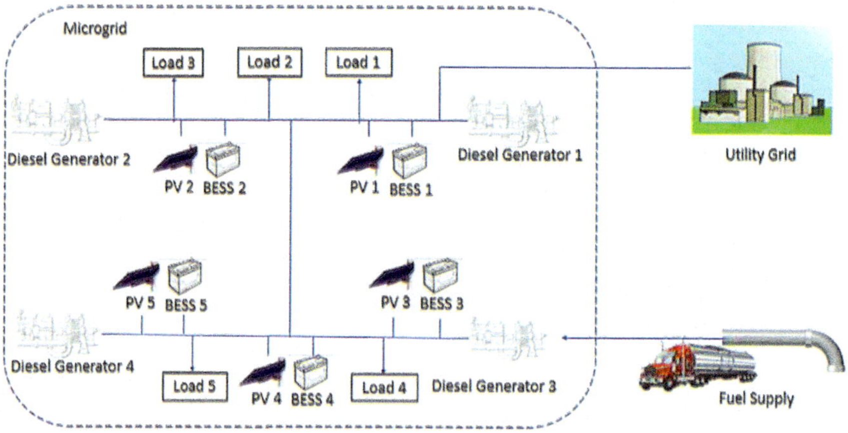

Fig. 12.5 Final microgrid composition (*Source* Mallery et al. 2022)

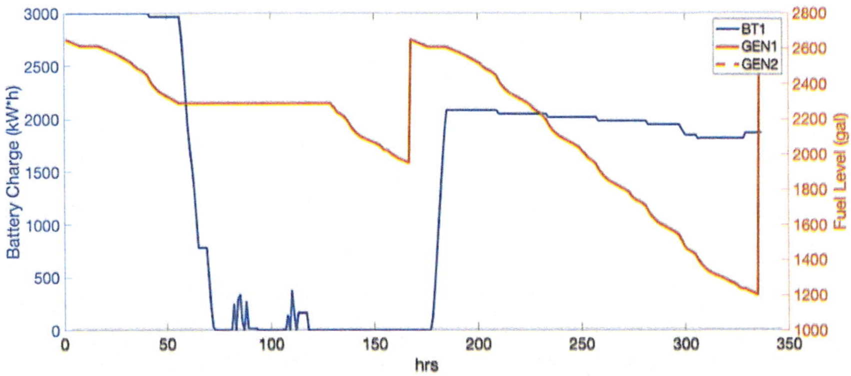

Fig. 12.6 Energy storage system (ESS) and generator fuel storage in 72-H generator failure scenario (*Source* Peterson et al. 2021—Open access)

Figure 12.6 depicts energy storage system (ESS) and generator fuel levels over a 72-h generator failure scenario, demonstrating how slaved generators supply identical output while energy storage systems maintain critical loads. This analysis is vital for designing resilient power systems in forward operating bases or critical command centers. Moreover, Fig. 12.7 shows ESS storage performance during a 72-h loss of the PV array, emphasizing the importance of distribution network flexibility. The illustration includes the operational link between BUS1 and BUS2 feeders, showcasing how energy routing and reserve capacity are managed to support infrastructure continuity in the absence of renewable inputs (Peterson et al. 2021).

In sum, the integration of geophysical monitoring, energy system planning, and real-time diagnostic tools into homeland security frameworks significantly enhances national resilience. These tools not only detect physical threats but also enable critical infrastructure to function through and recover from disruptive events, safeguarding the institutions and systems that sustain civil society.

Analyzing Mission Impact of Military Installations Microgrid for Resilience

In military installations, the assurance of continuous power supply is a fundamental requirement to maintain operations such as command and control, surveillance, communication, and life-support systems, particularly during contingency events (Peterson et al. 2021). Microgrids, supported by geophysical and engineering assessments, offer a resilient energy infrastructure capable of operating independently from the main grid. This section evaluates the mission-level impact of microgrid performance under critical failure scenarios, specifically addressing conditions involving the loss of primary generation sources such as diesel generators and photovoltaic (PV) arrays (Rickerson et al. 2019).

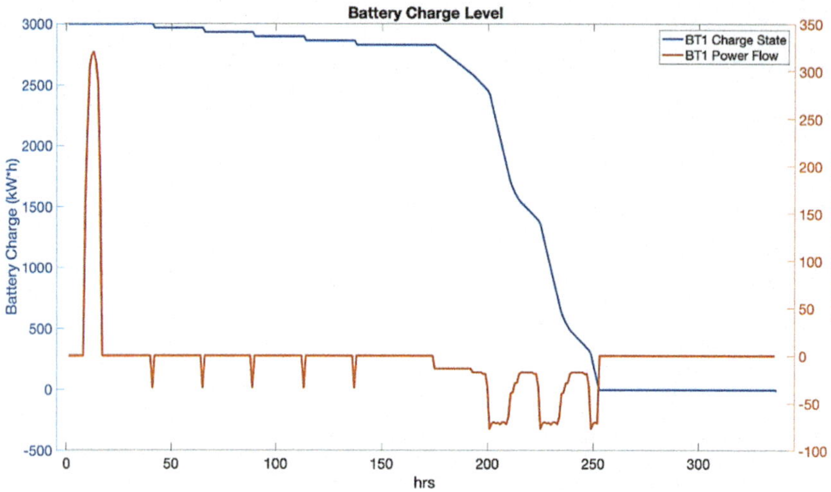

Fig. 12.7 ESS storage during 72-H loss of the PV array and the connection between the BUS1-BUS2 feeders. BUS1 and BUS2 typically refer to redundant or separate power distribution lines (*Source* Peterson et al. 2021—Open access)

To assess microgrid resilience, scenario-based modeling is conducted to simulate prolonged outages, including a 72-h loss of generator functionality or PV output. These simulations provide insight into the ability of the Energy Storage System (ESS) and supporting infrastructure to sustain mission-critical loads, redistribute energy through alternative feeders, and maintain operational continuity without reliance on external grid support (Giraldez et al. 2012). Under normal conditions, the combined contribution of generators, PV arrays, and ESS ensures a stable supply to both critical and non-critical loads. However, in degraded conditions, particularly during high-risk operations in hostile or disaster-prone environments, the reliance on ESS and reconfigurable feeder systems becomes central to energy security (Rusco and Lepore 2016).

Operational continuity during such failures depends significantly on the prioritization of loads. Mission areas such as intelligence processing, perimeter defense, field medical operations, and secure communications each exhibit varying degrees of criticality. In the context of a 72-h outage, it becomes essential to understand which subsystems can remain functional, which degrade in performance, and which must be temporarily shed or reduced. This informs not only the sizing of ESS and fuel reserves but also the operational planning and tactical response capabilities of the installation.

Defense Installation Energy Resilience for Changing Operational Requirements

Modern military operations are increasingly dynamic, requiring installations to support a diverse and rapidly evolving set of operational profiles. From traditional base defense to cyber operations, intelligence processing, humanitarian response, and unmanned systems control, the energy requirements of defense facilities have expanded in both complexity and criticality. This evolution necessitates a shift from static power systems to highly flexible and resilient energy architectures capable of adapting to mission-specific needs (Mallery et al. 2022).

Energy resilience in this context is defined not merely as the ability to maintain power during an outage but as the capacity to anticipate, absorb, adapt to, and rapidly recover from a wide range of disruptions while continuing to support changing operational objectives. Defense installations must therefore be equipped with scalable, reconfigurable microgrids that integrate distributed energy resources, advanced controls, and robust energy storage systems.

The fluctuating nature of operational demand means that a microgrid designed for one mission set may be insufficient or inefficient for another. For example, during active combat operations, the energy demand may be focused on perimeter defense, secure communications, and surveillance. In contrast, during disaster relief missions, medical facilities, water purification, and logistics hubs may become the primary energy consumers. This variability requires real-time prioritization and redistribution of available power resources to ensure that critical mission functions are maintained under all conditions.

Moreover, the threat landscape itself is changing. Military installations now face hybrid threats that combine physical attacks with cyber disruptions targeting energy infrastructure. The growing integration of digital control systems into power grids, while improving efficiency and automation, also creates new vulnerabilities. Energy resilience strategies must therefore include cyber-hardening measures, segmented control architectures, and operational continuity planning that accounts for cyber-physical interdependencies.

From an engineering perspective, energy systems must be both modular and mobile. This includes the use of containerized power units, portable battery banks, and deployable solar arrays that can be relocated or scaled up depending on mission scope and geographic setting. Such solutions are especially important for expeditionary forces and forward operating bases, where logistical constraints and hostile conditions demand both agility and self-sufficiency.

Operational adaptability also depends on situational awareness. Integrated monitoring systems, informed by geophysical assessments, load forecasting, and fault detection algorithms, are essential for dynamic decision-making. These tools enable operators to anticipate system stress, initiate preventive load shedding, and reconfigure microgrid components before a fault leads to mission degradation.

Policy frameworks and interagency collaboration further support energy resilience by standardizing approaches across installations and aligning technological capabilities with national defense objectives. This includes the development of shared resilience metrics, joint training programs, and unified procurement pathways that encourage interoperability and rapid deployment of resilient technologies.

In conclusion, defense installations must evolve from static energy consumers to intelligent, adaptive energy systems that support a broad spectrum of operational requirements. Resilience is no longer a secondary feature but a core mission enabler. By investing in modular infrastructure, cyber-secure controls, predictive analytics,

Table 12.2 Mission impact assessment of power loss scenarios

Mission area	Power criticality	Generator failure (72 h)	PV array failure (72 h)	Mitigation strategy
Command and control center	Vital	Operational (via ESS)	Operational	Prioritized load via BUS1
Surveillance systems	Semi-vital	Partially degraded	Fully operational	Load sharing via BUS2
Medical/field hospital	Vital	Operational (ESS + Gen)	Degraded	Backup ESS with BMS
Communications hub	Vital	Operational	Operational	Automatic switchover
Housing and non-essential ops	Non-vital	Offline	Offline	Load shedding
Perimeter defense systems	Vital	Partially degraded	Operational	Rerouted through ESS reserve

and integrated energy management, military facilities can ensure they remain capable, agile, and mission-ready in the face of uncertain and shifting demands.

To support this, Table 12.2 summarizes the results of simulated failure scenarios, comparing the impact of a 72-h loss of generators versus a 72-h loss of PV arrays. It outlines which mission areas remain operational, partially degraded, or completely disrupted based on energy availability from ESS and feeder reconfiguration (Deru et al. 2011).

The outcomes of these simulated conditions inform strategic planning across multiple dimensions, including the configuration of electrical buses, the specification of ESS capacity, the definition of load transfer sequences, and the protocols for automatic switchover in failure conditions. In parallel, geophysical data is used to ensure that ESS units, generation equipment, and distribution infrastructure are sited in locations that minimize risk from subsurface instability, water intrusion, or deliberate sabotage.

Through this type of integrated mission analysis, planners gain an enhanced understanding of both the vulnerabilities and adaptive capacities of military energy infrastructure. Ultimately, such evaluations contribute to the design and operation of resilient, mission-ready installations that can function effectively under conditions of stress, disruption, or attack.

12.4.3 Technology Integration and Policy Frameworks

The rapid advancement of artificial intelligence (AI), machine learning (ML), and data fusion technologies is revolutionizing the field of engineering geophysics, especially in the context of national defense and critical infrastructure security. These innovations enhance detection capabilities, streamline data interpretation, and facilitate decision-making, while simultaneously raising important ethical and legal considerations.

Role of AI, Machine Learning, and Data Fusion in Enhancing Detection Accuracy

AI and ML algorithms enable automated processing of vast and complex geophysical datasets, improving the identification and characterization of subsurface anomalies with greater speed and accuracy than traditional methods. By learning from labeled data, machine learning models can differentiate between benign geological features and potential threats such as tunnels, bunkers, or buried objects, reducing false positives and increasing operational efficiency (Zhang et al. 2020).

Data fusion techniques integrate multiple geophysical data types—such as seismic, electromagnetic, and gravimetric measurements—into unified analytical frameworks. This multi-modal approach leverages the complementary strengths of different methods, providing a more robust and comprehensive subsurface image critical for informed tactical and strategic decisions (Liu and Wang 2021).

Balancing National Security with Ethical and Legal Constraints

While the deployment of advanced technologies enhances national security, it also necessitates careful navigation of ethical and legal frameworks. The collection and analysis of subsurface data can implicate privacy concerns, particularly in urban or sensitive civilian areas. Ensuring transparency, data protection, and compliance with international laws governing surveillance and dual-use technologies is paramount to maintain public trust and avoid geopolitical friction (Roberts and Singh 2019).

Policymakers must establish clear guidelines that regulate the use, sharing, and export of advanced geophysical tools and AI systems. This includes protocols for data security, accountability in automated decision-making, and mechanisms for oversight to prevent misuse or unintended consequences. The integration of technology with policy frameworks is essential to harness the full potential of modern geophysics in defense applications while upholding democratic values and international norms.

12.4.4 Future Directions

As engineering geophysics continues to evolve, emerging technologies and shifting geopolitical landscapes will shape the future of subsurface detection and national security applications. Anticipating these trends is essential for maintaining strategic advantage and addressing new challenges effectively.

Emerging Technologies: Quantum Sensing and Real-Time 3D Geophysical Mapping

Quantum sensing represents a ground-breaking advancement in geophysical instrumentation, offering unprecedented sensitivity to magnetic and gravitational fields. These sensors exploit quantum phenomena to detect minute subsurface variations that conventional tools may overlook, enabling enhanced detection of tunnels, voids, and concealed infrastructure with higher resolution and accuracy (Budker and Romalis 2007).

Simultaneously, advances in real-time 3D geophysical mapping—combining rapid data acquisition, AI-driven processing, and immersive visualization—promise to revolutionize field operations. Operators will be able to visualize subsurface structures dynamically, facilitating immediate tactical decisions and improved situational awareness in complex environments (Smith et al. 2023).

International Collaboration and Export Control Issues

The global nature of security threats necessitates increased international collaboration in research, technology development, and operational deployment of geophysical defense tools. Sharing expertise and harmonizing standards can accelerate innovation and improve collective resilience against emerging risks.

However, this collaboration must be balanced with stringent export control policies to prevent proliferation of sensitive dual-use technologies. Nations face the challenge of fostering cooperation while safeguarding critical capabilities from misuse by hostile actors or unauthorized entities (Jackson and Patel 2020). Establishing transparent frameworks that facilitate responsible technology transfer and joint development initiatives will be vital to harness the benefits of cutting-edge geophysical methods while maintaining global security.

12.5 Environmental and Ethical Considerations in Defense Geophysics

As geophysical technologies are increasingly integrated into defense and security applications, there is a growing need to evaluate their broader environmental and ethical implications. While the primary aim of such technologies is to enhance situational awareness, detect subsurface threats, and support operational planning, their use often intersects with civilian, ecological, and legal domains. In areas like the Gaza-Israel border, the Line of Control in Kashmir, or urban centers with critical infrastructure, the deployment of geophysical surveys must be conducted with sensitivity to both the immediate tactical objectives and the potential long-term consequences to people and ecosystems.

12.5.1 Environmental Impact of Subsurface Surveys

Geophysical methods vary in their environmental footprint. Techniques such as ground penetrating radar (GPR) and magnetometry are generally non-invasive, but their deployment on wheeled platforms or by foot can disturb surface vegetation or compact soil, particularly in ecologically fragile environments. More intrusive methods—like seismic refraction or electrical resistivity tomography (ERT)—may require the insertion of electrodes, the generation of artificial seismic energy, or repeated passes over sensitive ground. In agricultural regions, protected habitats, or archaeological zones, such activities can disrupt local ecology, damage crops, or interfere with heritage preservation. Responsible geophysical practice must therefore include preliminary environmental assessments and the use of minimal-impact procedures whenever possible (Baird and Jennings 2018).

12.5.2 Ethical Issues in Military Geophysical Applications

Military geophysics often involves the scanning of subterranean environments in areas where civilians live, work, or travel. This raises ethical questions about consent, privacy, and proportionality. In contexts where urban underground infrastructure—such as tunnels, sewers, or shelters—is being surveyed, there is a risk of infringing on civilian spaces or collecting unintended intelligence on non-combatants. Moreover, the dual-use nature of geophysical technology complicates ethical oversight: a system used for post-earthquake search-and-rescue could also be repurposed for covert surveillance or target acquisition. These scenarios demand clear rules of engagement, transparent communication with civil authorities, and strict data protection protocols to ensure ethical compliance (Roberts and Singh 2019).

12.5.3 Compliance with International Norms and Treaties

The legal use of geophysical technologies in defense operations is governed by international humanitarian law, arms control treaties, and environmental protection agreements. The Geneva Conventions prohibit military activities that cause disproportionate harm to civilian property or infrastructure, while United Nations frameworks call for the safeguarding of the natural environment during conflict. In addition, the export and operational deployment of certain advanced geophysical sensors—such as quantum magnetometers or high-resolution synthetic aperture radar—are subject to international arms trade controls and national security regulations. Adhering to these frameworks is not only a matter of legality but also of maintaining legitimacy and accountability in both domestic and multinational defense operations (ICRC 2020; Jackson and Patel 2020).

12.5.4 Toward Responsible Defense Geophysics

A sustainable and ethically responsible approach to defense geophysics requires the integration of interdisciplinary oversight mechanisms at all stages—from research and development to deployment. Environmental engineers, legal advisors, geophysicists, and defense strategists must collaborate to design systems and protocols that minimize harm while maximizing effectiveness. Field personnel should receive ethics and compliance training alongside technical instruction. Additionally, the establishment of institutional review boards or mission-specific environmental and ethical audit teams can help ensure that military geophysics contributes not only to national security, but also to global standards of humanitarian responsibility and environmental stewardship.

12.6 Interoperability, Standards, and Data Sharing

In modern defense geophysics, where operations often span national borders, involve multiple agencies, and integrate diverse sensor technologies, interoperability and standardization have become critical. The ability to effectively share, interpret, and act on geophysical data is essential not only for enhancing operational efficiency but also for fostering collaboration among allied forces and civilian agencies. Without standardized protocols and interoperable systems, the value of geophysical intelligence can be significantly diminished, particularly in joint operations or rapid-response scenarios.

12.6.1 Geophysical Data Standards for Defense Applications

The effective use of geophysical data in defense requires a common framework for data acquisition, processing, interpretation, and exchange. Standardized formats, such as SEG-Y for seismic data and GPR-specific formats like RADAN or GSSI, enable compatibility across different software platforms and hardware systems. These standards facilitate efficient integration of datasets from various sources—ground-based, aerial, and satellite—into centralized command-and-control or geospatial information systems.

Defense applications also demand metadata standards that include contextual information such as geolocation, sensor calibration, and operational conditions. This ensures the reliability and reproducibility of results, especially when data must be verified under legal or security scrutiny. Agencies such as NATO and IEEE have published preliminary guidelines on geospatial and sensor data standards, but there is still a need for defense-specific extensions that address security classification, encryption, and tamper resistance (NATO STANAG 4609 2020; IEEE Geoscience and Remote Sensing Standards 2021).

12.6.2 Interagency and International Collaboration on Data Sharing

Many modern defense operations involve cooperation between military, intelligence, civil defense, and scientific agencies. Examples include tunnel detection on international borders, rapid urban infrastructure assessment in conflict zones, or cross-border disaster response where subsurface integrity is critical. Successful data sharing in such contexts depends on establishing common protocols, shared repositories, and secure communication channels.

International collaborations—such as those among NATO allies, the Five Eyes intelligence network, and joint UN peacekeeping missions—demonstrate the strategic value of harmonized geophysical frameworks. These collaborations often rely on shared geospatial platforms, encrypted data exchange formats, and joint field protocols to ensure seamless coordination. The inclusion of civilian geoscientists and contractors further underscores the need for data systems that are both secure and broadly accessible under controlled conditions (Garrity and Stone 2019).

12.6.3 Challenges and Solutions in Multinational Operations

Despite increasing interoperability efforts, multinational geophysical operations face several persistent challenges. These include inconsistent technical standards, incompatible hardware systems, and differing classification policies. For instance, one nation's seismic imaging data may be subject to export restrictions that prevent its use in allied analysis platforms, or hardware from different countries may produce data that require extensive reformatting before integration.

To address these issues, multinational defense consortia have begun developing joint interoperability frameworks and simulation environments to test cross-platform compatibility before field deployment. Establishing data "translation hubs" and using cloud-based collaborative platforms with tiered access control can also facilitate more efficient data sharing while maintaining national security requirements. Moreover, field training programs and joint technical workshops help harmonize practices among partner agencies and improve mutual understanding of system capabilities and limitations (Cunningham et al. 2022).

12.7 Limitations and Challenges of Current Geophysical Methods in Defense

Despite their growing role in defense and strategic security operations, geophysical methods are subject to a range of limitations that can constrain their effectiveness in complex or high-stakes environments. These limitations stem from both technological constraints and operational realities. Understanding these challenges is crucial not only for managing expectations but also for guiding future research, procurement, and deployment strategies in defense geophysics.

12.7.1 Technical and Environmental Limitations

Geophysical sensors and methods are inherently influenced by the geological and environmental context in which they are applied. For example, the effectiveness of ground-penetrating radar (GPR) diminishes significantly in high-conductivity soils such as clays or saturated terrains, limiting its utility in coastal or alluvial border regions. Seismic techniques, while powerful for detecting deep structures, can be affected by surface noise, anthropogenic vibrations, and heterogeneous subsurface conditions, which complicate signal interpretation. Similarly, electrical resistivity tomography (ERT) may yield distorted images in fractured or highly resistive environments.

In addition, physical limitations such as sensor depth penetration, resolution, and spatial coverage restrict the application of a single technique across diverse operational settings. This often necessitates the integration of multiple methods, which increases complexity, resource demand, and processing time (Butler and Lankston 2017). Environmental noise, both natural and artificial, also continues to be a major obstacle, particularly in urban warfare environments or along busy infrastructure corridors.

12.7.2 False Positives and Ambiguities in Data Interpretation

A major operational challenge in defense geophysics is the potential for false positives or ambiguous targets. Subsurface anomalies detected through magnetometry or seismic imaging may resemble illicit tunnels or voids but turn out to be naturally occurring geological features, abandoned infrastructure, or buried debris. These uncertainties can lead to unnecessary excavations, misallocated resources, or strategic miscalculations.

The complexity of signal interpretation requires highly skilled operators and analysts, often with site-specific experience and a multidisciplinary understanding of geology, engineering, and sensor behavior. However, in time-sensitive or remote military operations, such expertise may be limited or unavailable. Automated interpretation tools, while promising, still struggle with reliability in noisy, heterogeneous environments and often require manual validation (Williams et al. 2018).

12.7.3 Resource and Deployment Constraints

Operational deployment of geophysical systems in military contexts frequently encounters logistical constraints. Equipment must be rugged, mobile, and rapidly deployable—requirements that not all commercial systems are designed to meet. Transporting sensitive instruments into rough terrain, hostile environments, or denied areas presents additional risks and may require modifications to standard configurations.

Furthermore, many defense scenarios demand near-real-time results, yet most geophysical techniques still require substantial data processing and post-acquisition analysis. The lack of field-ready, automated interpretation systems limits the responsiveness of geophysical units. Personnel limitations also present a challenge; few defense forces maintain adequately trained geophysicists, and reliance on contractors or external agencies can delay operations or raise security concerns (Rao and Ellingson 2020).

12.8 Conclusions and Recommendations

The strategic use of geophysical technologies in defense and national security operations has become increasingly essential in safeguarding national assets, enhancing operational readiness, and responding to modern threat environments. This chapter has explored a comprehensive range of applications, from detecting subsurface threats to ensuring infrastructure resilience, with special attention to military base camps, field operations, and critical installations.

Summary of Key Points

- **Detection of Subsurface Threats** (Sect. 12.1): Subsurface threats such as tunnels, bunkers, and hidden infrastructure pose significant risks to military and civilian installations. Geophysical methods—such as ground-penetrating radar, seismic surveys, and electromagnetic sensing—play a vital role in identifying and mitigating these threats despite operational and environmental challenges.
- **Military Engineering Geophysics** (Sect. 12.2): Geophysics supports engineering missions in both construction and combat scenarios. Rapid assessment tools and interdisciplinary training enhance deployment speed and decision-making in field environments, ensuring tactical advantages.

- **Dual-Use Technologies and National Security** (Sect. 12.3): Geophysical systems often serve both military and civilian roles, contributing to homeland security, energy resilience (e.g., microgrids), and emergency preparedness. Successful integration requires robust policy frameworks and interagency coordination.
- **Environmental and Ethical Considerations** (Sect. 12.4): Responsible use of geophysical surveys must account for environmental protection and ethical standards. Compliance with international treaties and norms ensures legitimacy and minimizes negative impacts.
- **Standards, Interoperability, and Collaboration** (Sect. 12.5): Standardizing geophysical data across defense agencies and allied nations enables interoperability, smoother joint operations, and effective data-sharing. However, legal, political, and technical barriers remain.
- **Limitations and Challenges** (Sect. 12.6): Current geophysical tools face challenges such as resolution limits, data ambiguity, and deployment constraints in hostile or remote environments. These limitations must be addressed through ongoing innovation and adaptation.

Strategic Recommendations for Future Research and Deployment

1. **Invest in Next-Generation Sensing Technologies**: Research into more sensitive, compact, and AI-integrated systems can reduce false positives and enhance real-time threat detection.
2. **Strengthen Multidisciplinary Training Programs**: Develop specialized training modules that blend geophysics, tactical planning, and environmental awareness for military engineers and operators.
3. **Enhance Infrastructure Resilience Planning**: Integrate geophysical intelligence into early-stage planning for military base camps and critical infrastructure, with emphasis on climate adaptability and energy resilience.
4. **Foster Interagency and International Cooperation**: Expand collaborative frameworks for shared data standards, joint training exercises, and coordinated field operations to strengthen global defense capabilities.
5. **Implement Ethical Oversight and Environmental Safeguards**: Ensure that all applications of geophysical technologies in defense are guided by environmental stewardship and adherence to international humanitarian law.

The Role of Emerging Technologies and International Cooperation

Emerging technologies such as machine learning, autonomous sensor networks, and quantum geophysics are poised to redefine the landscape of defense-related subsurface exploration. However, their success will depend on deliberate international cooperation, transparent policies, and a commitment to ethical innovation. As threats evolve and infrastructure becomes more complex, geophysical intelligence must remain an integral, forward-looking component of national defense strategy—linking science, policy, and field operations into a cohesive and adaptive security framework.

References

Andrus RD, Stokoe KH (2000) Liquefaction resistance of soils from shear-wave velocity. J Geotech Geoenviron Eng 126(11):1015–1025

Baird TD, Jennings SA (2018) Responsible geophysical surveying in sensitive landscapes: ethics, law, and best practices. J Environ Geophys 23(2):85–94

Barak B (2017) Underground warfare and the changing face of military tunnels. Mil Strateg Aff 9(1):23–35

Boyer S, Allard M (2012) Interdisciplinary collaboration in military engineering projects: a review of best practices. Int J Defense Technol 8(1):45–54

Bruneau M, Chang SE, Eguchi RT, Lee GC, O'Rourke TD, Reinhorn AM, Shinozuka M, Tierney K, Wallace WA, von Winterfeldt D (2003) A framework to quantitatively assess and enhance the seismic resilience of communities. Earthq Spectra 19:733–752

Budker D, Romalis M (2007) Optical magnetometry. Nat Phys 3(4):227–234

Butler DK, Lankston RW (2017) Challenges in interpreting shallow geophysical data in complex environments. J Environ Eng Geophys 22(1):33–44

Colomina I, Molina P (2014) Unmanned aerial systems for photogrammetry and remote sensing: a review. ISPRS J Photogramm Remote Sens 92:79–97

Cunningham A, Zhao L, Okafor M (2022) Collaborative frameworks for multinational geophysical operations. J Defense Geotechnol 9(1):45–62

Daniels DJ (2004) Ground penetrating radar, 2nd edn. IET

Deru M, Field K, Studer D, Benne K, Griffith B, Torcellini P, Liu B, Halverson M, Winiarski D, Rosenberg M et al (2011) U.S. department of energy commercial reference building models of the national building stock. Technical report; National Renewable Energy Laboratory. https://doi.org/10.2172/1009264

DHS (Department of Homeland Security) (2020) Border tunnel threat assessment. Office of intelligence and analysis

Ferguson J, Sterling A (2015) Geophysical methods in critical infrastructure protection: applications and challenges. Int J Secur Stud 10(4):234–251

Garrity C, Stone R (2019) Enhancing operational geophysics through interagency data integration. Defense Sci J 14(3):112–121

Giraldez J, Booth S, Anderson K, Massey K (2012) Valuing energy security: customer damage function methodology and case studies at DoD installations. Technical report; U.S. Department of Energy Office of Scientific and Technical Information

Goodman D, Piro S, Nishimura Y (2007) Geophysical methods for subsurface mapping in military applications. J Environ Eng Geophys 12(3):223–234

Hagrey SA (2007) Geophysical methods in subsurface characterisation: their role in archaeological and forensic investigations. J Archaeol Sci 34(8):1266–1272

Hearty PJ, Martinez A, Delaney C (2014) Geophysical detection of illicit tunnels across the US-Mexico border. Lead Edge 33(12):1390–1396

ICRC (2020) Guidelines on the protection of the natural environment in armed conflict. International committee of the red cross

IEEE Geoscience and Remote Sensing Society (2021) Standards for data interoperability in geophysical systems

Jackson R, Patel M (2020) Balancing innovation and security: export control challenges for dual-use geophysical technologies. Defense Technol Rev 15(3):112–120

Katz O et al (2013) Applications of ground penetrating radar for search and rescue operations. J Appl Geophys 89:1–10

Liu Y, Wang J (2021) Data fusion techniques for integrated geophysical analysis in subsurface detection. J Appl Geophys 185:104289

Loke MH (2015) Tutorial: 2D and 3D electrical imaging surveys. Geotomo Softw

Mallery J, Van Bossuyt DL, Pollman A (2022) Defense installation energy resilience for changing operational requirements. Designs 6:28

Massonnet D, Feigl KL (1998) Radar interferometry and its application to changes in the earth's surface. Rev Geophys 36(4):441–500

Meier T et al (2018) Managing dual-use technologies in geosciences: policies and challenges. Geosci Policy Rev 5(2):122–135

Nabighian MN, Grauch VJS, Hansen RO et al (2005) The historical development of the magnetic method in exploration. Geophysics 70(6):33ND-61ND

NATO STANAG 4609 (2020) NATO standardization agreement for digital motion imagery

Olhoeft GR (1998) Electrical, magnetic, and geometric properties that determine ground penetrating radar performance. Proceedings of the seventh international conference on ground penetrating radar, pp 177–182

Peterson CJ, Van Bossuyt DL, Giachetti RE, Oriti G (2021) Analyzing mission impact of military installations microgrid for resilience. Systems 9:69

Rao N, Ellingson WA (2020) Logistical considerations in deploying geophysical sensors for field military operations. Mil Eng Rev 28(4):45–58

Rickerson W, Gillis J, Bulkele M (2019) The value of resilience for distributed energy resources: an overview of current analytical practices. Technical Report; National Association of Regulatory Utility Commissioners

Roberts T, Singh A (2019) Ethical considerations in the deployment of AI for national security. Int J Ethics Technol 14(2):87–103

Rusco F, Lepore BJ (2016) DoD renewable energy projects: improved guidance needed for analyzing and documenting costs and benefits. Technical Report; HT Media Ltd

Slob EC, Sato M, Olhoeft GR (2010) Surface and borehole ground-penetrating-radar developments. Geophysics 75(5):75A103-75A120

Smith L et al (2023) Real-time 3D mapping for subsurface exploration: a review of current capabilities and future prospects. J Geophys Eng 20(1):45–62

Telford WM, Geldart LP, Sheriff RE (1990) Applied geophysics, 2nd edn. Cambridge University Press

Williams J, Tanaka H, Alshamsi A (2018) Reducing false positives in tunnel detection: an integrated geophysical approach. Defense Appl Geophys 12(2):67–80

Zhang H et al (2020) Machine learning applications in geophysical data interpretation: advances and challenges. Geophys Res Lett 47(9):e2020GL087123

Chapter 13
Energy Resilience in Defense Installations: Geophysical and Engineering Perspectives

Subsurface Intelligence for Sustainable and Secure Power Infrastructure

Abstract This chapter highlights the crucial role of geophysical and engineering approaches in enhancing energy resilience within defense installations. It emphasizes the strategic importance of reliable power infrastructure for national security and examines subsurface investigations vital for designing and protecting energy systems. The chapter addresses natural hazard risk mitigation for critical energy assets and explores the integration of renewable energy sources with subsurface considerations. Special attention is given to the design of hardened energy facilities and the operational independence provided by microgrids and remote energy solutions. Finally, the chapter discusses policy frameworks and collaborative strategies to support sustainable, secure, and resilient energy infrastructure in defense contexts, concluding with recommendations for future advancements.

This subtitle frames the chapter's focus on how subsurface geophysical insights directly support the resilience, sustainability, and security of defense-related energy systems. It positions the content as a critical intersection of geophysical engineering and strategic infrastructure planning, especially for installations facing operational strain, climate change, or energy independence requirements.

13.1 Introduction: Strategic Importance of Energy Resilience

Modern defense installations face an evolving operational landscape shaped by rapid technological advancements and heightened security demands. Central to this transformation is the critical need for energy resilience—ensuring uninterrupted power supply amid complex and often contested environments. Energy resilience underpins operational effectiveness, enabling military bases to sustain electrification, autonomous systems, and cyber-physical operations essential for mission success.

The evolution of operational needs has brought electrification and autonomy to the forefront of military capabilities. Increased reliance on energy-intensive systems

D. C. Naskar, *Advances in Engineering Geophysics: Concepts and Applications*, Earth Systems Data and Models 9, https://doi.org/10.1007/978-3-032-28000-8_13

such as advanced communication networks, electric vehicles, and autonomous drones amplifies the demand for a reliable power supply (Hollinger et al. 2020). Additionally, the integration of cyber-physical systems—where digital control and physical processes are intertwined—necessitates continuous energy availability to prevent operational failures and safeguard against cyber-attacks that may target energy infrastructure (Lemieux and Anderson 2021).

Energy continuity is not merely a logistical concern but a cornerstone of mission readiness. Interruptions in power can degrade situational awareness, disrupt command and control, and compromise critical life-support systems, thereby jeopardizing personnel safety and operational integrity (U.S. Department of Defense [DoD] 2022). Studies highlight that even short-duration outages can result in significant operational setbacks, underscoring the imperative for resilient energy architectures (Smith and Zhao 2023).

Conventional grid-connected military bases are particularly vulnerable due to their dependence on centralized electrical grids that may be susceptible to natural disasters, physical attacks, and cyber intrusions (Wilson et al. 2019). The interconnected nature of traditional grids makes them prone to cascading failures, which can severely impair base functionality (Department of Energy [DOE] 2021). Furthermore, adversaries increasingly target critical infrastructure as a means of asymmetric warfare, elevating the risks associated with grid dependence (Government Accountability Office [GAO] 2020).

Consequently, there is a strategic shift toward distributed, secure, and hardened energy systems. Distributed energy resources (DERs), including microgrids, renewable generation, and energy storage, offer enhanced reliability by localizing power generation and reducing dependency on the external grid (Perera et al. 2022). Hardened energy infrastructures incorporate physical security measures and cyber protections to withstand and rapidly recover from disruptions (Federal Emergency Management Agency [FEMA] 2021). This paradigm not only improves energy resilience but also aligns with sustainability objectives and operational flexibility requirements. In summary, energy resilience in defense installations is a multifaceted challenge necessitating a holistic approach encompassing technological innovation, infrastructure hardening, and integrated management. The convergence of geophysical and engineering perspectives is essential for developing robust energy systems capable of supporting the critical missions of modern military operations.

As illustrated in Fig. 13.1, the concept of Installation Energy Resilience emerges from the intersection of energy resilience and installation resilience, emphasizing mission assurance as the central goal of energy planning in defense contexts.

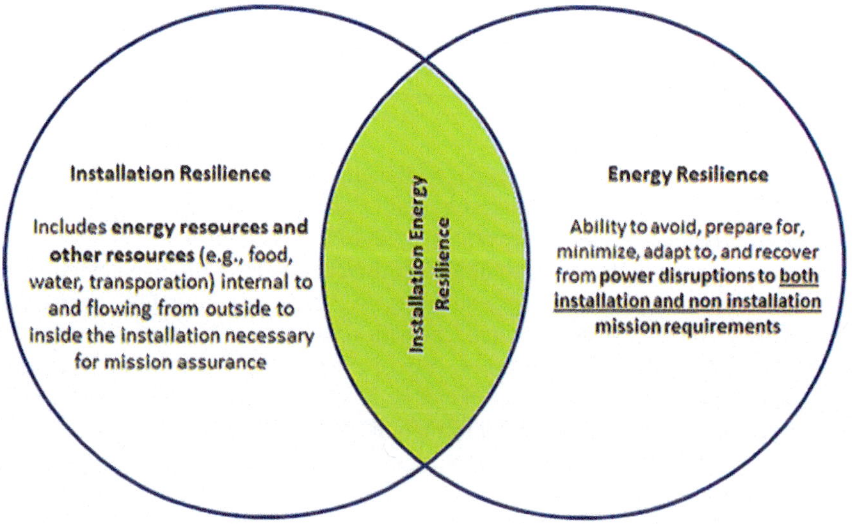

Fig. 13.1 Resilience Venn diagram. Defines installation energy resilience by showing the overlap between energy resilience and installation resilience, focusing on mission-critical energy use (*Source* Varley et al. 2022—Open Access)

13.2 Subsurface Investigations for Energy Infrastructure

The deployment of resilient energy infrastructure in defense installations hinges not only on robust surface systems but also on a deep understanding of the subsurface. Geophysical investigations are essential in siting and maintaining buried infrastructure such as power lines, fuel tanks, and distributed microgrid components. These subsurface systems are critical for secure, uninterrupted energy delivery and must be carefully placed to avoid geohazards, ensure longevity, and facilitate rapid maintenance or upgrades.

Geophysics in Energy Infrastructure Siting

Subsurface conditions dictate the viability and safety of buried infrastructure. Geophysical techniques provide non-invasive, spatially extensive tools to assess soil and rock properties, detect anomalies, and delineate subsurface structures prior to excavation or installation. For instance, siting underground power lines requires detailed knowledge of subsurface moisture content, voids, and anthropogenic obstructions—all of which can be identified using geophysical methods (Cardarelli and Di Filippo 2021). Similarly, the safe placement of underground fuel storage tanks near critical installations depends on identifying stable geological zones and avoiding corrosive soil environments, seepage paths, or seismic faults (Giao et al. 2022).

In the context of defense energy systems, microgrids—which often include underground cabling, conduit systems, and thermal storage—must be integrated without disrupting existing utilities or compromising subsurface security. Geophysics allows for precision mapping and minimizes the risk of interference with pre-existing buried infrastructure (Ahmed et al. 2023).

Methods for Subsurface Mapping

Among the most widely adopted geophysical methods are Ground Penetrating Radar (GPR), Electrical Resistivity Tomography (ERT), and seismic-based approaches such as seismic refraction and Multichannel Analysis of Surface Waves (MASW). GPR is especially effective in detecting shallow subsurface features with high spatial resolution, making it ideal for locating conduits, tanks, voids, and utility lines in resistive soils (Jol 2019). ERT, on the other hand, provides detailed resistivity profiles that are crucial for identifying moisture zones, corrosive soil conditions, or buried objects with distinct electrical properties (Loke et al. 2020). Seismic methods like MASW and refraction techniques offer vital information on subsurface stiffness, soil layering, and potential weak zones—particularly important when designing foundations for heavy energy equipment or infrastructure exposed to dynamic loads or seismic risk (Zhu et al. 2022). Integrating these techniques ensures a multi-dimensional understanding of subsurface conditions, supporting informed engineering decisions and reducing the likelihood of costly post-installation failures.

Geotechnical Risks and Energy Systems

Energy infrastructure is increasingly vulnerable to geotechnical risks such as landslides, sinkholes, liquefaction, and expansive soils—especially as climate change intensifies hydrological extremes. For military bases, where mission continuity is paramount, these risks can compromise energy delivery systems and demand pre-emptive mitigation.

For example, GPR and ERT have been deployed effectively in identifying shallow karst features that precede sinkhole formation beneath cable corridors and access roads (Fidelibus et al. 2021). In slope-prone regions, seismic methods have mapped weak layers susceptible to landslides, allowing rerouting of power infrastructure or the installation of slope stabilization measures (Tannant and Regmi 2020). These proactive investigations help fortify energy installations against natural and anthropogenic hazards.

Case Studies in Defense Energy Applications

Geothermal Installations: At a U.S. Army base in California, geophysical surveys—primarily ERT and seismic refraction—were used to characterize the shallow crust before drilling geothermal boreholes. The surveys helped identify thermal anomalies, optimize borehole placement, and reduce drilling costs by avoiding fractured zones (Nguyen et al. 2022).

Underground Cabling Projects: In a NATO-supported microgrid deployment in Eastern Europe, a combination of GPR and ERT was employed to design and install secure underground power lines. The geophysical surveys helped detect legacy infrastructure (e.g., buried metallic objects, old foundations) and mitigate risks associated with unexploded ordnance (UXO) zones (Paredes and Molnar 2023). These insights were critical in accelerating deployment while maintaining safety.

13.3 Natural Hazard Risk Mitigation for Critical Energy Systems

As military energy infrastructure becomes more complex and interconnected, the need for proactive mitigation of natural hazards grows increasingly urgent. Defense energy systems—ranging from high-voltage transformers and underground power lines to surface-mounted solar arrays and wind turbines—are often deployed in environments susceptible to seismic activity, flooding, soil instability, and extreme weather events. Engineering geophysics provides indispensable tools for evaluating and mitigating these risks, thereby supporting the resilience and operational continuity of mission-critical installations.

One of the primary natural threats to energy infrastructure is seismic activity. Earthquakes pose significant risk to energy assets such as substations, transformers, and fuel pipelines, which may suffer structural damage or operational failure when subjected to ground shaking. For example, transformer banks are highly sensitive to lateral and vertical accelerations, and failures in anchorage or insulation can lead to cascading outages (Mazzoni et al. 2021). Subsurface pipelines—particularly those traversing fault zones or liquefaction-prone soils—require rigorous geophysical site evaluations to model expected ground deformation and stress transmission (Ayoobi et al. 2023). Seismic microzonation using MASW, ambient noise methods, and borehole logging has been increasingly applied to identify risk-prone zones and inform seismic-resilient design practices in defense energy systems.

Flooding is another critical hazard that affects both underground and surface-mounted energy components. Subterranean cable networks, electrical vaults, and fuel storage tanks are highly vulnerable to inundation, which can lead to short circuits, corrosion, and system-wide outages. Surface-mounted assets such as transformers and power control units are also at risk if placed in low-lying or poorly drained areas. Geophysical methods like electrical resistivity tomography (ERT), combined with hydrogeological modeling, are widely used to delineate seasonal water table variations and predict flood infiltration pathways (Rogiers et al. 2022). Additionally, high-resolution digital elevation models (DEMs) generated from remote sensing and LiDAR data enable flood modeling at installation scale, allowing for the strategic siting or elevation of energy assets.

Wind loading and soil instability also pose significant threats, particularly for renewable energy infrastructure. Solar photovoltaic (PV) arrays and wind turbines must be installed with a deep understanding of local geotechnical conditions to avoid foundation failure under high wind or cyclic loading. Soil liquefaction, expansive clays, and frost heave are common causes of instability, which can compromise support structures and lead to misalignment, cracking, or total collapse. Geophysical surveys such as cone penetration testing with seismic sensors (SCPT), resistivity imaging, and surface wave analysis are increasingly employed to evaluate soil stiffness, bearing capacity, and drainage potential during the planning phase (Elkasabgy et al. 2021). In regions exposed to extreme weather, reinforced foundations, pile systems, and real-time monitoring using geotechnical sensors have become part of an integrated resilience strategy.

A cornerstone of hazard mitigation is the use of geohazard mapping, which integrates remote sensing and in situ data to produce spatial risk models. Remote sensing platforms—such as synthetic aperture radar (SAR), multispectral imagery, and UAV-based photogrammetry—offer broad-scale insights into terrain deformation, erosion trends, and hydrological changes. When combined with ground-based observations, including ERT, GPR, and seismic surveys, these data enable comprehensive hazard assessments tailored to defense energy systems (Fletcher and Wang 2022). The use of machine learning to integrate multisource geospatial data has further advanced geohazard prediction, particularly in rapidly changing or remote environments where traditional monitoring is limited (Liu et al. 2023).

Incorporating geophysical and remote sensing techniques into energy planning allows defense engineers to anticipate and mitigate the impacts of earthquakes, floods, high winds, and soil instability. This proactive approach not only protects physical infrastructure but also ensures sustained energy delivery in the face of evolving natural hazards—an essential component of energy resilience and national security.

13.4 Renewable Integration and Subsurface Considerations

As defense installations transition toward cleaner and more autonomous energy systems, integrating renewable technologies into operational landscapes demands careful evaluation of the subsurface environment. The strategic use of geophysics is central to supporting the reliable deployment of geothermal, solar, wind, and energy storage systems on military land. Subsurface conditions directly affect not only the placement and performance of renewable energy infrastructure but also the safety, resilience, and efficiency of such systems under operational stress and natural hazard exposure.

Geothermal energy is among the most site-sensitive renewable technologies, requiring detailed subsurface investigation to determine thermal gradient, lithology, porosity, and fluid content. Shallow and deep geothermal systems alike benefit from geophysical assessments such as electrical resistivity tomography (ERT), seismic refraction, and temperature-conductivity logging, which together reveal the spatial variability of thermal properties and help optimize borehole placement (Gholami et al. 2021). In low-enthalpy systems commonly used in military buildings and shelters, accurate knowledge of shallow thermal conductivity is vital to enhance heat exchange efficiency and avoid over-drilling. Military bases located in tectonically active or sediment-rich basins—such as those in the western United States or Southeast Europe—are particularly well-suited for geothermal integration when supported by robust geophysical mapping.

Site evaluation for solar farms and wind turbines also requires a firm understanding of subsurface geotechnical and geological conditions. Large-scale solar arrays exert minimal dynamic loads but demand uniform, stable terrain with minimal frost heave or expansive soil movement. In contrast, wind turbines impose significant dynamic and cyclic loading on foundations, requiring high-bearing capacity soils and minimal seismic amplification. Geophysical techniques such as multichannel analysis of surface waves (MASW), microtremor array methods, and ground penetrating radar (GPR) are increasingly employed during site selection to assess soil stiffness, detect voids, and map subsurface layering (Ali and Khan 2022). On defense lands, where previously disturbed areas or unexploded ordnance may exist, geophysics also plays a key role in hazard avoidance prior to foundation installation.

Energy storage systems—particularly buried battery arrays, underground hydrogen storage tanks, and embedded fuel cell systems—demand rigorous site vetting due to their sensitivity to thermal conditions, moisture ingress, and ground movement. Subsurface storage offers both physical protection and reduced surface clutter, but it introduces risks that must be mitigated through detailed geophysical and hydrogeological studies. For example, buried lithium-ion batteries must be isolated from zones of high salinity or corrosive soils, while underground hydrogen tanks require sealing and monitoring in stable, low-porosity formations (Zhu et al. 2023). ERT and borehole geophysical logging are used to assess soil saturation, resistivity, and void presence—critical variables for long-term storage resilience. Fuel cell modules embedded within protective enclosures also benefit from thermal mapping and subsurface drainage analysis to prevent overheating or water accumulation.

An often-overlooked aspect of renewable integration is the influence of subsurface thermal properties on the energy efficiency of shelters, command centers, and storage facilities. Subsurface thermal inertia can significantly enhance or impair passive thermal regulation. For example, in cold climates, shelters with high subsurface heat loss may require disproportionately high energy inputs for heating. Conversely, in hot regions, buildings embedded in thermally stable subsurface environments may benefit from passive cooling. Thermal conductivity and heat capacity of soil and rock materials can be mapped using thermal response tests (TRT), resistivity measurements, and time-domain reflectometry (Chen et al. 2022). This information feeds into the architectural and mechanical design of energy-efficient, hardened structures.

In conclusion, the successful integration of renewable energy technologies on defense installations depends heavily on subsurface characterization. Whether deploying geothermal wells, solar or wind systems, or energy storage units, geophysics provides the foundation for risk reduction, optimization, and long-term resilience. As defense energy systems grow more distributed and autonomous, subsurface understanding will remain a cornerstone of their reliability and sustainability.

13.5 Design and Protection of Hardened Energy Systems

Ensuring energy resilience in defense installations extends beyond generation and delivery—it demands deliberate and sophisticated hardening of infrastructure against physical, electromagnetic, and cyber threats. The integration of geophysics, engineering design, and advanced materials science forms the foundation for safeguarding energy systems that support mission-critical operations in hostile or unpredictable environments.

A primary consideration in hardened energy system design is blast-resistant siting. Energy assets such as backup generators, fuel depots, switchgear, and microgrid control units must be located in zones where the effects of explosive impacts, overpressure, and flying debris are minimized. Geophysical techniques, particularly seismic and resistivity surveys, are used to identify competent subsurface zones suitable for anchoring protective infrastructure. Shallow seismic refraction, for example, can help locate zones of low seismic amplification and delineate competent rock layers for embedded structures (Xu et al. 2023). When cables and microgrids are routed underground for additional protection, subsurface characterization ensures that routing avoids unstable soils, corrosive environments, or underground water intrusion, which could compromise performance over time (Lee and Tan 2021).

In parallel with physical protection, energy infrastructure must also withstand electromagnetic threats, including electromagnetic pulses (EMP) and high-powered microwave (HPM) attacks. Such phenomena, whether from solar storms or hostile action, can permanently disable unshielded electrical systems. Modern military energy nodes increasingly rely on EMP-hardened enclosures, bunkered substation modules, and shielded cable conduits installed within engineered trenches. Geophysical methods support the design and placement of these subsystems by mapping grounding quality, assessing subsurface conductivity, and determining electromagnetic field attenuation characteristics of local geologic materials (Cao et al. 2022). Subsurface routing of vulnerable assets not only offers physical shielding but also allows integration with natural Faraday shielding from conductive soil or bedrock, enhancing survivability.

A growing trend in defense energy planning is the implementation of dual-purpose infrastructure—energy systems designed for both civilian and military application. These nodes often support base operations while simultaneously powering surrounding communities during grid outages or disasters. Microgrids, energy storage facilities, and hybrid renewable systems (solar-geothermal-diesel) are being co-developed with resilience and redundancy in mind. This dual-function model requires highly adaptable and secure designs, incorporating geophysical assessments to ensure site suitability for continuous-use and emergency operations alike. Integrated geospatial databases, updated through repeated geophysical surveys, help manage overlapping utility zones, risk factors, and critical load priorities (Roehrig and Simmons 2021).

An emerging frontier in hardened system design is the convergence of material science and geophysics, enabling predictive monitoring and fault detection in energy infrastructure. Smart materials—such as piezoelectric coatings, self-healing composites, and fiber-optic-embedded conduits—can be embedded in critical subsurface elements to detect early-stage stress, corrosion, or ground movement. These sensing technologies interface with geophysical instruments, including distributed acoustic sensing (DAS) and resistivity monitoring, creating a dynamic diagnostic layer over buried systems (Wang et al. 2023). The coupling of geophysical data with advanced material responses allows for condition-based maintenance, greatly enhancing reliability while reducing lifecycle costs. For instance, early detection of cable sheath degradation or subgrade void formation through real-time geophysical monitoring can prevent catastrophic failures in microgrid networks.

In summary, the protection of hardened energy systems in defense installations requires a multidisciplinary approach—where geophysics informs structure placement, electromagnetic threats shape material selection, and sensing systems offer continuous diagnostic insight. As energy becomes increasingly weaponized in asymmetric conflict, such integrative design principles will be key to ensuring energy security, operational readiness, and infrastructure longevity.

13.6 Operational Independence: Microgrids and Remote Energy Solutions

Energy resilience in defense operations increasingly hinges on operational independence—the capacity to sustain mission-critical functions in isolation from centralized energy infrastructure. Whether operating in forward-deployed outposts, disaster zones, or austere training environments, modern military bases are adopting microgrids and hybrid remote energy systems to achieve self-sufficiency and rapid adaptability. In such environments, geophysical techniques—particularly mobile, modular applications—play a pivotal role in planning, siting, and protecting decentralized energy assets.

Mobile geophysics enables real-time site characterization in rapidly changing or undeveloped terrains, providing essential information for siting power components such as buried cables, ground-mounted solar arrays, thermal storage systems, and small-scale generators. Tools like portable ground penetrating radar (GPR), resistivity meters, and passive seismic sensors can be deployed quickly to assess subsurface hazards, identify stable installation zones, and locate hidden infrastructure or voids. These assessments are especially vital in remote or contested areas, where historical site data are unavailable and engineering tolerances are low (Kelley et al. 2023). For instance, in temporary or expeditionary base development, mobile geophysics has been used to delineate permafrost depth in Arctic zones, detect sabkha soils in desert environments, and assess soil saturation levels in tropical and littoral zones.

Mobile microgrids offer rapid deployment and shortened recovery times compared to conventional installation-based systems. Figure 13.2 provides a resilience curve

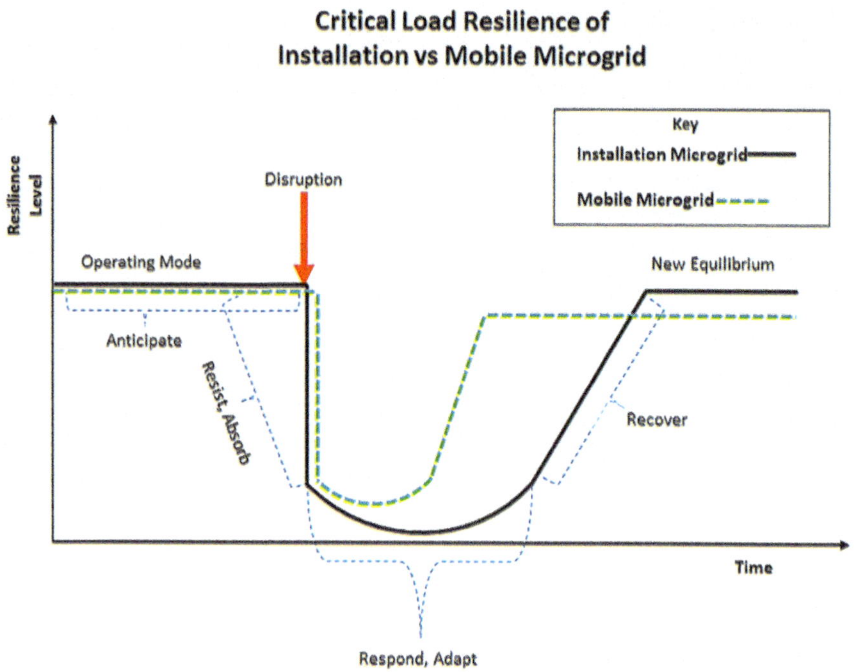

Fig. 13.2 Standard resilience curve—installation versus mobile microgrid recovery. A resilience performance comparison curve between an installation microgrid and a mobile microgrid post-disruption (*Source* Varley et al. 2022—Open Access)

comparing the two approaches, demonstrating the mobile system's advantage in the "Respond, Adapt, and Recover" phases.

The adoption of hybrid energy systems—typically composed of solar PV arrays, diesel generators, and energy storage modules managed through microgrid controllers—has become a standard solution for achieving energy autonomy in remote defense installations. From a logistical standpoint, these systems reduce fuel convoys and resupply missions, which are often vulnerable to attack or environmental disruption. However, hybrid systems require geotechnical and geophysical validation to optimize layout, ensure adequate load-bearing capacity, and avoid subsurface conflicts. Seismic methods (e.g., MASW) and ERT are used to evaluate foundation stability for heavy battery banks and generator pads, while thermal conductivity measurements help optimize battery placement for efficiency and thermal stability (Andersson and Huotari 2022). In addition, geophysics aids in routing low-voltage underground distribution safely around UXO, buried utilities, or natural obstacles.

The feasibility of mobile microgrid systems depends on efficient logistics and rapid energy deployment protocols. As shown in Fig. 13.3, a typical CONOPS scenario involves airlifting and activating a low-voltage mobile microgrid system to sustain operational headquarters in contingency environments.

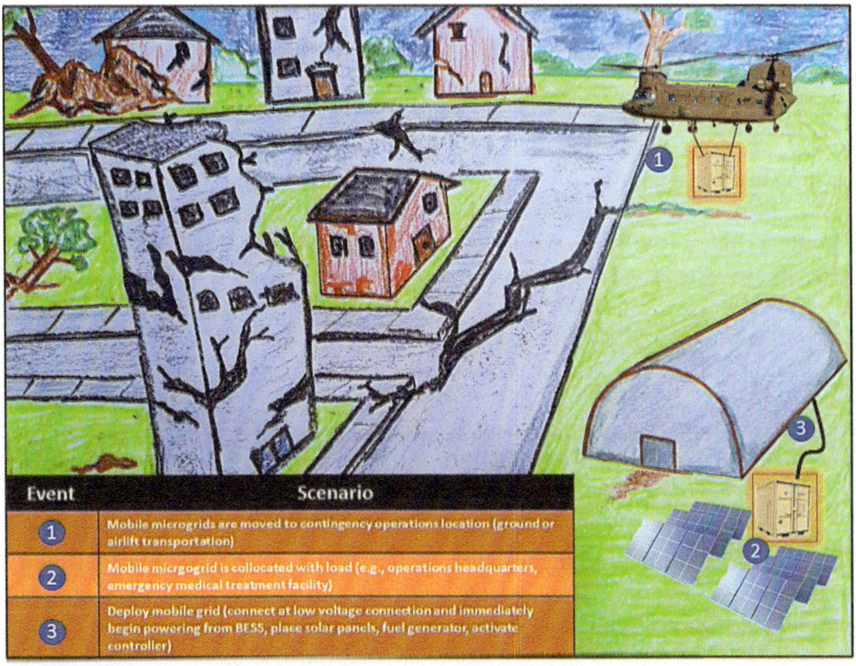

Fig. 13.3 CONOPS diagram—mobile microgrid deployment to contingency operation. Concept of operations (CONOPS) showing the logistical deployment and setup sequence of a mobile microgrid in a contingency environment (*Source* Varley et al. 2022—Open Access)

For forward-operating bases and islanded microgrids, resilience depends not only on generation but also on the robustness of site conditions under extreme climatic and operational stress. In field-deployable systems, particularly in regions with minimal infrastructure, geophysical assessments provide critical insights into terrain behavior, water table fluctuation, and thermal stability. In maritime and archipelagic environments, geophysics supports the identification of corrosion-prone soils and subsurface salinity gradients that can degrade cabling or storage systems (Li et al. 2023). Similarly, in Arctic deployments, mobile resistivity profiling and electromagnetic sensors have been used to map permafrost dynamics, ensuring that energy components are not installed over thermally unstable ground, which can thaw and shift during seasonal changes. These methods allow energy planners to design support systems that remain operational under conditions of ground movement, freeze–thaw cycling, or seasonal isolation.

Historical deployments in Arctic, desert, and maritime environments have highlighted several core lessons. In Arctic settings, such as northern Alaska and Greenland training ranges, the successful deployment of modular microgrids required not only redundant energy sources but also predictive ground monitoring to detect frost heave and active-layer thickening. In desert regions like CENTCOM (United States Central Command) training zones, airborne GPR and resistivity mapping were employed to locate sabkha zones—salt-crusted flats with low bearing strength that compromise solar farm stability (Schmidt et al. 2022). In maritime and coastal bases, subsurface radar and conductivity mapping have helped prevent corrosion failures by identifying groundwater salinity encroachment and stratification. These cases reinforce the importance of integrating mobile geophysics into the planning and operation of islanded energy systems.

In conclusion, geophysical methods—especially when mobile and modular—are essential enablers of operational energy independence. As military operations grow increasingly expeditionary and reliant on distributed power systems, the fusion of geophysics, energy engineering, and adaptive logistics will be fundamental to the success and survivability of remote defense energy solutions.

13.7 Policy, Collaboration, and Future Pathways

The future of energy resilience in defense infrastructure does not lie solely in technological advances but in the intersection of policy, scientific collaboration, and digital integration. As defense energy systems become more distributed, intelligent, and geophysically sensitive, strategic frameworks and interagency partnerships will play a decisive role in shaping secure, adaptable, and mission-aligned energy futures. Geophysics, long essential in tactical planning and infrastructure protection, is now emerging as a central tool in a broader network of energy decision-making.

The successful deployment of resilient energy systems increasingly depends on collaboration between defense agencies, national laboratories, geoscience institutions, and private industry. Programs like the Department of Defense's (DoD) Environmental Security Technology Certification Program (ESTCP) and the Department of Energy's (DOE) National Renewable Energy Laboratory (NREL) are key drivers of innovation in energy geoscience, supporting pilot projects and field validation of microgrid resilience, geothermal integration, and geophysical monitoring (Department of Defense [DoD] ESTCP 2022). These partnerships allow the military to rapidly prototype energy solutions under real-world conditions, while enabling academia and industry to refine tools for subsurface imaging, environmental sensing, and hazard forecasting in defense environments. The continued evolution of dual-use energy technologies—those benefiting both civilian infrastructure and military readiness—depends on these cooperative ecosystems.

From a regulatory perspective, policy frameworks are evolving to embed resilience, climate adaptation, and operational independence into defense energy mandates. The 2021 Executive Order 14057 on "Catalyzing Clean Energy Industries and Jobs Through Federal Sustainability" mandates that federal agencies—including the DoD—achieve net-zero emissions and adopt climate-resilient infrastructure practices (White House 2021). As part of implementation, geophysical assessments are increasingly mandated during site selection and system design to ensure compliance with floodplain avoidance, seismic zoning, and soil suitability requirements. At the base planning level, resilience-based codes are being coupled with climate-informed risk models to anticipate long-term environmental shifts—drought, thaw, coastal retreat—that could affect buried assets or generation sites.

Data integration is the cornerstone of modern energy resilience and one of the most transformative trends shaping future geophysical applications. With the rise of remote sensing platforms, IoT-enabled sensors, and AI-driven analytics, defense installations are moving toward continuous situational awareness of energy systems, including real-time monitoring of soil conditions, thermal properties, vibration anomalies, and electrical faults. Integrating remote sensing (e.g., synthetic aperture radar, hyperspectral imaging), ground-based geophysical data, and in-situ sensor arrays allows for multilayered mapping of terrain dynamics and system vulnerabilities (Xue et al. 2023). Meanwhile, AI models—trained on geophysical time-series data—are being used to forecast faults, detect subsurface movement, and optimize microgrid load balancing, turning passive monitoring into predictive resilience. As defense energy strategies continue to evolve, the integration of geophysical tools across multiple domains becomes central to resilience planning, site optimization, and long-term infrastructure adaptability. Table 13.1 outlines the primary geophysical methods currently used and emerging in the field, along with their energy-specific applications.

Looking ahead, several emerging technologies stand to redefine the energy-geoscience landscape. Among these are quantum batteries, which offer ultra-fast charging and robust energy density for mobile and isolated operations. Though early in development, such technologies could revolutionize power logistics in forward operating bases and rapid-deployment scenarios (Ferraro et al. 2021). At the same time, terrain-aware energy AI is under active research, leveraging high-resolution digital elevation models, soil geotechnical data, and dynamic environmental inputs to autonomously adapt energy routing, load prioritization, and threat response protocols. Additionally, the vision of resilient, self-healing energy grids—integrating smart materials, edge computing, and distributed geophysical intelligence—offers a model for infrastructure capable of anticipating and recovering from both environmental and adversarial disruptions. These modeling insights reinforce the importance of planning for performance under variable and extreme conditions. Figure 13.4 offers an example of such a simulation, where increased daytime loads interact with PV charging and battery storage. These capabilities, coupled with advances in sensing, AI, and policy frameworks, position geophysics as a foundational element in the future of defense energy systems.

In conclusion, the path forward for defense energy resilience lies at the convergence of advanced geophysics, smart data ecosystems, and policy-backed innovation. As climate instability, cyber-physical threats, and operational complexity continue to rise, the integration of geoscientific insight with digital foresight and interagency coordination will define the next generation of secure, adaptive energy infrastructure.

Table 13.1 Geophysical tools and applications for defense energy resilience

Geophysical tool	Primary function	Energy resilience application	Deployment context
Ground penetrating radar (GPR)	Detects shallow subsurface features, voids, and buried objects	Mapping existing utilities, siting buried cables and batteries	Urban bases, forward-deployed sites
Electrical resistivity tomography (ERT)	Maps subsurface resistivity profiles	Identifying corrosion-prone zones, groundwater intrusion, geothermal suitability	Coastal bases, deserts, storage areas
Seismic refraction/ reflection	Determines subsurface layering and mechanical properties	Site stability for microgrids, generator pads, wind turbines	Blast-resistant sites, seismic regions
MASW (Multichannel analysis of surface waves)	Evaluates shear wave velocity and dynamic soil behavior	Foundation design for wind turbines and solar farms	Arid zones, permafrost regions
Remote sensing (SAR, LIDAR, hyperspectral)	Monitors terrain change, surface movement, and flood exposure	Geohazard mapping, terrain-aware energy planning	Mountainous areas, flood-prone zones
Magnetometry and EM induction	Locates metallic objects, detects salinity and conductivity variations	Avoiding UXO during energy infrastructure placement, corrosion risk	Former conflict zones, coastal/marine settings
Thermal response testing (TRT)	Measures thermal conductivity and heat capacity of soils	Designing efficient geothermal and thermal storage systems	Shelters, underground storage nodes
IoT sensors and fiber optics	Continuous in-situ monitoring of stress, temperature, and moisture	Fault detection in buried cables, battery thermal regulation	Embedded in energy microgrids and storage fields
Satellite geodesy (InSAR)	Detects ground deformation over time	Early warning of subsidence or frost heave affecting energy assets	Arctic, coastal, or unstable geological zones

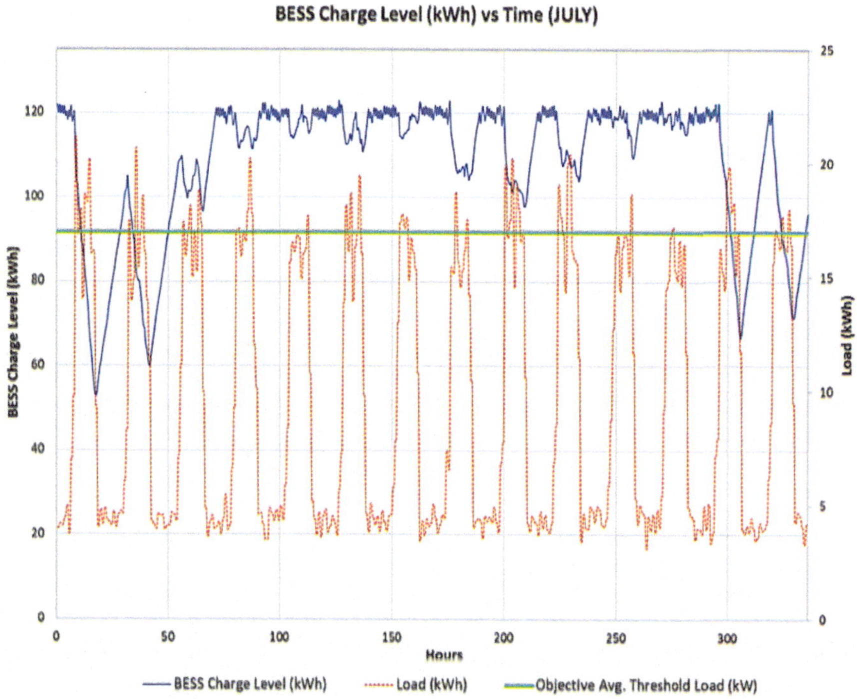

Fig. 13.4 System behavior snapshot during stress test conditions. Operational simulation of a mobile microgrid during a 14-day period under high load and solar input, illustrating system performance under stress (*Source* Varley et al. 2022—Open Access)

13.8 Conclusions and Recommendations

The increasing reliance on resilient, secure, and sustainable energy systems across defense installations has elevated the importance of geophysical and engineering perspectives in national security. This chapter has examined the strategic integration of subsurface intelligence into energy resilience planning, emphasizing the role of geotechnical insights in safeguarding power infrastructure, especially in remote and high-risk military environments.

Summary of Key Points

- Strategic Importance of Energy Resilience (Sect. 13.1):

Energy resilience is foundational to defense readiness. Interruptions in power supply can critically impair command, communication, and combat capabilities. Proactive energy planning ensures operational continuity under diverse threat conditions.

- Subsurface Investigations for Energy Infrastructure (Sect. 13.2):

Geophysical surveys are essential in selecting optimal sites for energy infrastructure, identifying geohazards, and informing design. Accurate subsurface data supports structural stability, efficient energy transmission, and long-term system integrity.

- Natural Hazard Risk Mitigation for Critical Energy Systems (Sect. 13.3):

Earthquakes, floods, and landslides pose substantial risks to energy installations. Integrated geophysical and geotechnical assessments enable hazard forecasting and protective design strategies, reducing vulnerability to natural disasters.

- Renewable Integration and Subsurface Considerations (Sect. 13.4):

Renewable energy systems such as solar, wind, and geothermal require thorough understanding of subsurface conditions. Geophysical inputs optimize siting, system longevity, and environmental impact, ensuring sustainable power generation.

- Design and Protection of Hardened Energy Systems (Sect. 13.5):

Critical energy nodes must be protected from both natural and hostile threats. Subsurface engineering supports the development of fortified infrastructure, including buried lines, shielded substations, and underground energy storage.

- Operational Independence: Microgrids and Remote Energy Solutions (Sect. 13.6):

Geophysics aids in deploying microgrids and self-sufficient energy systems in austere environments. Site-specific subsurface analysis facilitates efficient layout, reduces logistical burdens, and enhances energy autonomy.

- Policy, Collaboration, and Future Pathways (Sect. 13.7):

Achieving energy resilience requires cross-sector policy development, defense-private partnerships, and international collaboration. Unified data frameworks and innovation roadmaps are vital to long-term strategic success.

Strategic Recommendations for Future Research and Deployment

1. Advance Integrated Geophysical-Energy Modeling:

Develop tools that combine geophysical data with energy performance models to optimize site selection, risk assessment, and long-term planning.

2. Expand Subsurface Risk Mapping for Energy Security:

Create comprehensive geospatial databases that assess natural and anthropogenic risks to energy systems, supporting resilience planning across defense zones.

3. Support Renewable-Energy Resilience Research:

Invest in understanding the interplay between renewable technologies and subsurface environments to maximize performance in diverse military settings.

4. Design Hardened and Adaptive Infrastructure:

Incorporate geotechnical intelligence into the design of infrastructure capable of withstanding extreme events and attacks while maintaining energy continuity.

5. Accelerate Deployment of Remote Energy Systems:

Utilize geophysical support to streamline the development of off-grid microgrids and hybrid power systems in expeditionary and remote defense operations.

6. Foster Interagency and International Energy Resilience Cooperation:

Establish shared standards and collaborative protocols between military, civil, and allied institutions for integrated energy resilience strategies.

7. Ensure Sustainable and Ethical Energy Practices:

Prioritize environmental protection and local stakeholder engagement in all energy resilience projects, aligning defense operations with global sustainability goals.

The Future of Energy Resilience in Defense

As defense installations evolve to meet emerging security and climate challenges, the convergence of geophysics and energy engineering will become increasingly critical. Innovations such as autonomous subsurface mapping, AI-driven energy diagnostics, and adaptive infrastructure design must be matched by coherent policies and international cooperation. Ultimately, energy resilience is not only a matter of engineering—it is a strategic imperative shaping the future of global defense readiness.

References

Ahmed MN, Yusuf HA, Mukherjee A (2023) Integrated geophysical modeling for microgrid cable routing in urban and military environments. Geophys Prospect 71(1):34–49

Ali S, Khan MA (2022) Geophysical approaches for site evaluation of renewable energy infrastructure: case study from wind farm development. Renew Energy 193:44–58

Andersson L, Huotari J (2022) Geophysical and thermal assessments for hybrid microgrid systems in cold-climate military operations. Renew Energy 195:1234–1246

Ayoobi A, Adalier K, Youd TL (2023) Seismic vulnerability assessment of buried pipelines using advanced geophysical techniques. Soil Dyn Earthq Eng 168:107535

Cao M, Chen J, Zhang T (2022) Electromagnetic shielding analysis of underground energy infrastructure using geophysical conductivity profiling. IEEE Trans Electromagn Compat 64(1):102–112

Cardarelli E, Di Filippo G (2021) Application of GPR and ERT for shallow energy infrastructure mapping. Eng Geol 293:106308

Chen D, Zeng W, Luo J (2022) Subsurface thermal characterization for energy-efficient shelter design in arid and semi-arid regions. Energy Build 273:112379

Department of Defense (DoD) ESTCP (2022) Environmental security technology certification program annual report

Department of Energy (DOE) (2021) Grid resilience and modernization. U.S. DOE office of electricity

Elkasabgy M, Aly A, Sobaih M (2021) Geotechnical-geophysical hybrid methods for wind turbine foundation assessment. Renew Energy 178:768–779

Federal Emergency Management Agency (FEMA) (2021) Critical infrastructure protection and resilience. FEMA Publications

Ferraro D, Campaioli F, Binder FC (2021) Quantum batteries: physics, models, and future directions. Nat Rev Phys 3:374–386

Fidelibus C, Parise M, Fiore A (2021) Geophysical characterization of karst terrains for civil and energy engineering. J Appl Geophys 189:104304

Fletcher T, Wang J (2022) Remote sensing-based geohazard monitoring for military energy installations. Geosci Front 13(3):101208

Gholami R, Soltani M, Rezaee R (2021) Geophysical evaluation for shallow geothermal energy potential in military-grade installations. Geothermics 94:102102

Giao PH, Tran DQ, Le TH (2022) Subsurface characterization for underground fuel tank siting using integrated ERT and borehole logging. Environ Geotech 9(5):432–444

Government Accountability Office (GAO) (2020) Critical infrastructure protection: Actions needed to address cybersecurity risks. GAO-20-507

Hollinger M, Cheng H, Nguyen T (2020) Electrification and energy demand in military operations. J Defense Energy 15(3):110–126

Jol HM (2019) Ground penetrating radar: theory and applications, 2nd edn. Elsevier

Kelley B, Rogers D, Hasan M (2023) Portable geophysical applications in expeditionary energy site development. J Mil Eng Resilience 9(1):59–71

Lee SH, Tan KH (2021) Seismic and geoelectric screening for buried microgrid protection zones. Eng Geol 290:106176

Lemieux J, Anderson S (2021) Cyber-physical systems security in defense applications. IEEE Trans Aerosp Electron Syst 57(4):2560–2571

Li C, Ahmad R, Zhao F (2023) Geophysical risk profiling of coastal and island energy microgrids. J Coast Geotech 11(2):101–114

Liu Y, Chen Z, Zhao Q (2023) Machine learning integration of remote sensing and geophysics for landslide susceptibility mapping. ISPRS J Photogramm Remote Sens 200:12–28

Loke MH, Wilkinson PB, Chambers JE (2020) Advances in ERT for near-surface geophysical mapping. Near Surf Geophys 18(4):273–284

Mazzoni S, Pujol S, Gomez R (2021) Seismic fragility analysis of transformers and substation components. Eng Struct 243:112700

Nguyen A, Carter L, Bui V (2022) Geophysical site assessment for low-enthalpy geothermal energy in military installations. Renew Energy 187:1230–1242

Paredes R, Molnar T (2023) Subsurface geophysical reconnaissance for NATO microgrid installations. J Defense Infrastruct Eng 4(1):89–104

Perera A, Johnson M, Lee D (2022) Distributed energy resources in military microgrids: enhancing resilience and sustainability. Renew Energy Rev 135:110186

Roehrig P, Simmons D (2021) Resilient energy infrastructure in dual-use microgrid systems: planning and geotechnical integration. Defense Energy J 18(2):75–88

Rogiers B, Hermans T, Nguyen F (2022) Integrated ERT and hydrological modeling for flood risk analysis of underground infrastructure. Hydrogeol J 30(4):911–925

Schmidt JP, Al-Shehri A, Tanaka K (2022) Geotechnical mapping of desert sabkha soils for military solar PV array siting. Eng Geol 307:106812

Smith R, Zhao L (2023) Impact of power outages on military readiness: a case study. Defense Stud Quart 28(1):45–60

Tannant D, Regmi NR (2020) Geophysical detection of landslide-prone zones for energy infrastructure planning. Landslides 17(11):2631–2645

U.S. Department of Defense (DoD) (2022) Energy resilience for mission assurance. DoD Directive 3020:49

Varley DW, Van Bossuyt DL, Pollman A (2022) Feasibility analysis of a mobile microgrid design to support DoD energy resilience goals. Systems 10:74

Wang Z, Liu H, Chen Y (2023) Coupling geophysics with embedded sensing materials for subsurface energy fault detection. J Smart Infrastruct Mater 12(4):245–260
White House (2021) Executive order 14057: catalyzing clean energy industries and jobs through federal sustainability
Wilson T, Gupta R, Thomas K (2019) Vulnerabilities in conventional military energy systems. J Mil Eng 35(2):78–89
Xu J, Tang Y, Elbaz K (2023) Blast-resistant foundation siting using geophysical and dynamic response modeling. Soil Dyn Earthq Eng 165:107511
Xue L, Fan R, Chen M (2023) AI-assisted geophysical data fusion for terrain risk prediction and energy network resilience. Remote Sens Environ 293:113654
Zhu Q, Lin H, Zhang J (2022) Seismic imaging of soft-soil foundations for energy facilities using MASW and HVSR. Soil Dyn Earthq Eng 157:107204
Zhu H, Bai X, Deng W (2023) Geotechnical and geophysical considerations in underground hydrogen storage. Int J Hydrogen Energy 48(5):2137–2151

Part IV
Synthesis, Future Directions, and Interdisciplinarity

Chapter 14
Archaeological, Heritage, and Hidden Manmade Structures

Uncovering the Past and Protecting the Unseen

Abstract This chapter explores the application of geophysical methods in uncovering and preserving archaeological and heritage sites, as well as detecting hidden manmade structures. It provides a detailed overview of key subsurface exploration techniques such as ground penetrating radar, electrical resistivity tomography, magnetometry, seismic methods, and electromagnetic induction, emphasizing their complementary use for non-invasive investigations. Through a series of case studies—including ancient burial sites in Egypt, Pompeii's funerary areas, and heritage landmarks like the Leaning Tower of Pisa and the Sun Temple in India—the chapter demonstrates practical approaches to mapping, monitoring, and conserving cultural heritage. Additionally, it addresses the detection of clandestine tunnels and urban underground structures, highlighting the importance of advanced data processing and geospatial technologies. Ethical, legal, and conservation challenges are discussed, with a forward-looking view on emerging innovations and their role in safeguarding our shared past.

The subsurface holds remnants of civilizations long gone, secrets of ancient technologies, buried infrastructure, and clues to humanity's evolution. In the quest to explore and preserve these invaluable resources, engineering geophysics has emerged as a powerful, non-invasive toolkit. The subtitle "*Uncovering the Past and Protecting the Unseen*" encapsulates the dual mission of this discipline: to reveal what lies hidden below the earth's surface and to ensure its protection through informed and minimally invasive interventions.

Modern societies are increasingly faced with the challenge of balancing development and heritage conservation. Urban expansion, infrastructure projects, and land-use changes often occur on or near historically significant locations. Without the appropriate tools, these changes risk irreparable damage to cultural and historical artifacts. Traditional excavation methods, while effective in revealing historical details, are inherently destructive and labor-intensive. Geophysical methods, on the other hand, offer a scientific and non-destructive means of detection, mapping, and monitoring. Geophysics plays a critical role in: (i) Locating buried archaeological structures such as walls, foundations, roads, and tombs, (ii) Assessing the integrity

D. C. Naskar, *Advances in Engineering Geophysics: Concepts and Applications*, Earth Systems Data and Models 9, https://doi.org/10.1007/978-3-032-28000-8_14

of heritage buildings and underground historical features (iii) Discovering clandestine or undocumented structures like tunnels, bunkers, or escape routes, and (iv) Supporting conservation efforts through regular monitoring and risk assessment.

The use of Ground Penetrating Radar (GPR), Electrical Resistivity Tomography (ERT), magnetometry, and seismic methods has transformed archaeological prospection. These tools allow researchers and engineers to create high-resolution maps of subsurface features, enabling targeted excavation, structural assessments, and data-driven conservation strategies. Combined with modern technologies such as 3D visualization, machine learning, and remote sensing, the potential of engineering geophysics is expanding rapidly.

This chapter presents methodologies, case studies, and future directions for geophysical approaches applied in archaeological and heritage contexts. It aims to guide researchers, engineers, and policymakers toward a deeper understanding of how we can explore human history beneath the ground without disturbing the very context that gives it meaning. Ultimately, by uncovering the past and protecting the unseen, engineering geophysics serves not only science and history, but also the cultural identity and legacy of future generations.

14.1 Introduction

The exploration and preservation of subsurface cultural heritage have undergone a transformation with the advancement of non-invasive geophysical techniques. Traditional archaeological excavation methods, while valuable, are inherently destructive and often limited in scope. In contrast, geophysical methods enable broad, non-destructive investigation of buried archaeological features, offering an efficient, cost-effective, and environmentally sensitive alternative. These techniques provide critical insights into site stratigraphy, structural integrity, and material composition without disturbing the subsurface context (Neubauer 2021; Campana and Piro 2022).

Non-invasive geophysics has become indispensable in archaeological prospection, particularly in areas where excavation is either not permissible due to legal protection or is impractical due to urbanization. Techniques such as Ground Penetrating Radar (GPR), Electrical Resistivity Tomography (ERT), magnetometry, and seismic surveys allow for the detection of buried structures such as walls, roads, tombs, and entire settlement layouts with a high degree of spatial accuracy. Recent advancements in data processing, 3D modeling, and machine learning have further enhanced the interpretative power of these methods, making them critical to archaeological research and heritage management (Giocoli et al. 2020; Leucci 2023).

Engineering geophysics plays a central role in identifying, documenting, and preserving buried cultural assets. It supports heritage conservation by enabling: (i) Pre-excavation mapping of potential archaeological targets, (ii) Structural assessment of historical buildings and ruins, and (iii) Long-term monitoring of cultural heritage sites exposed to environmental or anthropogenic risks. Moreover, these techniques are not limited to academic research but have also been integrated into regulatory and development frameworks. For example, geophysical surveys are now routinely used in cultural resource management (CRM) and Environmental Impact Assessments (EIAs) across Europe and parts of Asia and the Americas (Sarris et al. 2022).

This chapter provides a comprehensive overview of how engineering geophysics contributes to the non-invasive exploration and preservation of archaeological, heritage, and hidden manmade structures. The main objectives are: (i) To introduce key geophysical techniques used in archaeological and heritage investigations, (ii) To present case studies that demonstrate their practical applications and effectiveness, and (iii) To discuss challenges, limitations, and the future outlook of non-destructive subsurface investigation methods. Through real-world examples and technical discussions, the chapter aims to inform researchers, practitioners, and policymakers about the capabilities and best practices of geophysical approaches in cultural heritage contexts.

14.2 Principles of Geophysical Methods for Subsurface Exploration

The application of geophysical methods in archaeology and heritage conservation has revolutionized how we understand, investigate, and protect buried cultural resources. These methods are fundamentally non-invasive and allow archaeologists and engineers to "see" beneath the surface without disrupting the stratigraphic or historical integrity of the site. They are particularly valuable in areas where excavation is restricted due to conservation policies, legal frameworks, or site sensitivity. This section introduces five core geophysical techniques commonly employed in subsurface exploration of archaeological and heritage sites. Each method is discussed in terms of its operating principles, advantages, limitations, and specific suitability for different types of buried features.

14.2.1 Ground Penetrating Radar (GPR)

Ground Penetrating Radar (GPR) is a high-resolution, radar-based technique that uses electromagnetic waves to detect and map subsurface features. It operates by emitting short pulses of high-frequency radio waves into the ground. When these waves encounter a material with contrasting dielectric properties—such as a buried wall, foundation, or void—some of the energy is reflected back to the surface and recorded by the receiver.

GPR excels in environments with dry, sandy, or homogenous soils, offering detailed imagery down to depths of 2 to 5 m depending on the antenna frequency and ground conditions. Its rapid data acquisition and high spatial resolution make it particularly well-suited for mapping complex archaeological structures such as floor plans, graves, and pathways. However, its performance declines in wet or clay-rich soils due to signal attenuation (Goodman et al. 2021).

14.2.2 Electrical Resistivity Tomography (ERT)

Electrical Resistivity Tomography (ERT) measures the electrical resistance of soils and buried materials by injecting electrical current through pairs of electrodes placed in the ground and recording potential differences. The resulting resistivity values provide insights into the subsurface composition, revealing features such as buried walls, ditches (a long, narrow excavation in the ground), cavities, and stratigraphic layers. ERT is especially advantageous for deep investigations (up to 20 m or more) and is highly effective in differentiating between materials such as stone, brick, and soil. It can be used to explore multilayered archaeological sites and to assess the condition of buried structures over time. Unlike GPR, ERT performs well in moist or clay-rich soils, making it a complementary technique (Drahor and Berge 2020).

14.2.3 Magnetometry

Magnetometry involves measuring variations in the Earth's magnetic field caused by subsurface features. Human activities such as burning, construction, or the presence of iron-containing materials can alter the natural magnetic properties of the soil. Magnetometers detect these variations, allowing archaeologists to identify features such as hearths (a floor-based structure used for fire), kilns (a chamber or structure built to reach high temperatures), furnaces, walls, and ditches. This method is particularly valuable for large-area surveys due to its rapid data collection capability. It is effective at detecting features at shallow depths (typically up to 2–3 m), especially in rural or minimally disturbed contexts. However, in urban or industrial areas, magnetic noise from modern infrastructure may reduce the clarity of results (Fassbinder 2022).

14.2.4 Seismic Methods

Seismic methods include a suite of techniques that analyze the propagation of seismic waves through the ground. These include seismic refraction, seismic reflection, and surface wave methods such as Multichannel Analysis of Surface Waves (MASW). Seismic velocities vary depending on the stiffness, density, and type of material encountered. In archaeological and heritage applications, seismic methods are used both to locate subsurface voids (such as tombs or tunnels) and to assess the condition of standing historical structures. Surface wave methods are particularly useful for shallow subsurface investigations and have been employed to evaluate the stability of monuments and detect subsurface weaknesses (Alani et al. 2021).

14.2.5 Electromagnetic (EM) Induction

Electromagnetic induction techniques rely on the generation and measurement of secondary electromagnetic fields in the subsurface, which result from variations in soil conductivity and magnetic susceptibility. EM methods are often deployed using frequency-domain or time-domain instruments and do not require ground contact, making them efficient for rapid surveys. These methods are suitable for mapping metal artifacts, buried utilities, changes in soil moisture, and anthropogenic backfill. EM surveys are frequently used for reconnaissance purposes in both rural and urban contexts and are particularly effective in delineating site boundaries or detecting clandestine features such as buried cables or pipelines (Kogut and Panissod 2023).

14.2.6 Comparative Suitability and Integration of Methods

The selection of a suitable geophysical method depends on multiple factors, including the expected depth of the target, the physical properties of the materials involved, the environmental conditions, and logistical constraints. Table 14.1 summarizes the comparative strengths and limitations of each technique.

Table 14.1 Comparative summary of geophysical methods for archaeological applications

Method	Best uses	Typical depth	Strengths	Limitations
GPR	Walls, pavements, graves, floor plans	0–5 m	High resolution, rapid, 2D/3D imaging	Attenuation in clay or wet soils
ERT	Deep stratigraphy, foundations	0–20 + m	Works in moist soils, 3D modeling	Slower setup and acquisition
Magnetometry	Burned structures, ditches, industrial sites	0–3 m	Fast over large areas	Sensitive to modern metal interference
Seismic	Voids, chambers, structural diagnostics	Variable	Evaluates material properties	Requires active sources and coupling
EM induction	Site mapping, metal detection	0–6 m	Non-contact, fast survey	Less effective in fine resolution mapping

In many archaeological investigations, a multi-method approach is employed. By integrating data from GPR, ERT, magnetometry, and other techniques, researchers can cross-validate findings, mitigate uncertainties, and generate more accurate and comprehensive site models. These combined methodologies are becoming increasingly important as heritage sites grow more complex and demand more nuanced investigation strategies.

14.3 Non-invasive Archaeological Mapping

The need to explore archaeological sites without physical disturbance has led to the increasing adoption of non-invasive geophysical mapping techniques. These methods enable archaeologists and geoscientists to detect, visualize, and interpret buried cultural features while preserving the site's integrity. Non-invasive archaeological mapping not only minimizes damage but also offers rapid, large-scale insights into subsurface configurations that traditional excavation would struggle to provide.

14.3.1 Strategies for Detecting Ancient Structures Without Excavation

Non-invasive archaeological prospection relies on identifying contrasts in the physical properties of subsurface materials, which often reflect human activity. The following strategic principles guide successful non-invasive detection:

(i) Preliminary Historical and Topographic Analysis: Review of historical maps, aerial photos, and LiDAR to identify anomalies or patterns on the surface (Ostrowski et al. 2021).
(ii) Grid-Based Geophysical Surveying: Setting up systematic grids for consistent data collection and accurate spatial referencing.
(iii) Selection of Appropriate Techniques: Based on site conditions, expected features (e.g., walls, roads, ditches), and soil properties.
(iv) Progressive Data Acquisition: Employing initial rapid-scan techniques (e.g., magnetometry) followed by targeted, higher-resolution methods (e.g., GPR or ERT).
(v) Non-linear Interpretation: Using iterative analysis, combining geophysical, archaeological, and environmental data to refine interpretations.

These strategies allow researchers to map urban layouts, religious sites, burial grounds, and agricultural features while preserving the archaeological record.

14.3.2 GPR and Magnetometry in Site Prospection

Ground Penetrating Radar (GPR) and magnetometry remain the leading techniques for subsurface archaeological prospection. GPR transmits high-frequency radar pulses into the ground and records the reflected signals from subsurface structures. It is especially effective for: (i) Detecting buried masonry, graves, floor layers, and voids (ii) Producing high-resolution depth slices and (iii) Constructing 3D models of structural layouts. GPR performs best in dry, resistive soils, and recent improvements in ultra-wideband antennas and real-time imaging have enhanced its precision (Ciminale et al. 2022).

Magnetometry measures subtle changes in the Earth's magnetic field caused by human activity—especially fire, construction, and soil disturbance. It is ideal for locating: (i) Kilns, furnaces, hearths, and burnt layers (ii) Ditches, postholes, and pits and (iii) Metal objects or magnetic inclusions. Fast and contactless, magnetometry is suited for large-scale surveys, although its sensitivity to modern metallic interference remains a challenge (Pozza et al. 2023).

Both methods are highly complementary—GPR provides vertical resolution, while magnetometry is excellent for wide horizontal coverage.

14.3.3 Multimethod Surveys and Data Integration

Increasingly, archaeological prospection employs multimethod surveys to achieve more robust and accurate interpretations of buried landscapes. Integrating multiple geophysical datasets compensates for individual method limitations and maximizes interpretive confidence.

Examples of effective multimethod combinations include: (i) Magnetometry + GPR: Offers both plan-view and depth-resolved imaging (ii) ERT + Seismic: Combines electrical resistivity with mechanical property mapping to detect buried voids, foundations, or stratigraphy and (iii) GPR + EM Induction: Useful in urban settings to balance detail with rapid data acquisition. These integrated datasets are often processed and visualized within Geographic Information Systems (GIS) and specialized software such as Geoplot, ReflexW, or GPR-Slice. Data fusion enhances the clarity of interpretations, especially in complex or multi-period sites (Lasaponara et al. 2021).

In conclusion, non-invasive archaeological mapping stands at the intersection of geophysical science, heritage conservation, and archaeological interpretation. Techniques such as GPR and magnetometry, when applied strategically and in combination, provide powerful insights into past human activity beneath the ground surface. As technology continues to evolve, these methods will play an increasingly central role in the ethical, efficient, and informed exploration of cultural heritage sites.

14.3.4 Case Study: Mapping an Ancient Burial Site—Qubbet el-Hawa (Egypt)

The Qubbet el-Hawa necropolis, located on the western bank of the Nile opposite Aswan, is one of Egypt's most significant burial sites for officials, governors, and priests dating back to the Old and Middle Kingdoms. Carved into a steep sandstone hillside, many of its tombs remain hidden beneath centuries of sediment, rockfall, and erosion. Due to the site's cultural value and fragile physical condition, archaeologists and geophysicists collaborated to investigate its subsurface features using entirely non-invasive methods (Stalder et al. 2023).

The primary goal of the survey was to detect hidden tombs, shafts, and chambers without disturbing the surface. Traditional excavation would have been risky, both in terms of damaging historical material and causing collapses in unstable zones. Instead, a multimethod geophysical approach was adopted to create a three-dimensional understanding of the buried tomb architecture. This allowed researchers to identify promising areas for future excavation while ensuring the long-term preservation of the site.

Ground Penetrating Radar (GPR) played a central role in this study (GPR with 400 MHz and 900 MHz). Using radar antennas of varying frequencies, researchers sent high-frequency electromagnetic waves into the ground and recorded the returning signals. Lower-frequency antennas penetrated deeper, while higher-frequency systems captured finer detail closer to the surface. The GPR data revealed the presence of tomb entrances, voids, and what appeared to be collapsed or sediment-filled chambers beneath the cliff terrace. Several strong reflections suggested the existence of multi-level burial structures, carved vertically into the rock.

Alongside GPR, magnetometry surveys were carried out to detect magnetic anomalies caused by human activity. Even though the site lacked modern metallic interference, subtle changes in the Earth's magnetic field could be linked to burned offerings, clay structures, or buried walls. This method was particularly useful in mapping features surrounding the tomb entrances, such as stairways, enclosures, or offering tables that had been covered by sand and debris over time.

Electrical Resistivity Tomography (ERT) was also employed to complement the radar and magnetic results. ERT involves introducing an electrical current into the ground and measuring how strongly the soil and rock resist the flow of electricity. Since empty tomb shafts and chambers have much lower resistivity than solid rock, the technique allowed the team to identify deeper features that GPR could not fully resolve. This method proved especially valuable on the sloped and uneven terrain of the site, where radar coverage was limited.

By integrating the results of all three methods—GPR, magnetometry, and ERT—the team created a comprehensive map of the subsurface burial landscape. They discovered several previously unknown entrances and shafts that had gone undetected by earlier surveys. The findings confirmed that many of the tombs were arranged in vertical tiers, sometimes with new burial chambers carved directly above or beside older ones. Importantly, they also identified unstable or dangerous areas that could collapse if excavated, helping to ensure the safety of future archaeological teams.

The Qubbet el-Hawa project stands as a successful example of how modern geophysics can support archaeological exploration without disturbing sensitive cultural heritage. By combining multiple non-invasive techniques, the researchers were able to uncover complex architectural details, improve excavation planning, and avoid potentially destructive interventions. This approach is now being adopted at other necropolises across Egypt and the Mediterranean, highlighting the growing role of engineering geophysics in heritage protection.

The results of this study not only expanded our understanding of the burial practices and architectural ingenuity of ancient Egypt but also demonstrated the practical benefits of a multimethod, non-destructive approach to archaeological investigation. As heritage sites face increasing pressure from urban development, tourism, and climate change, such techniques are becoming essential tools for responsible and informed site management.

14.4 Heritage Conservation with Geophysical Tools

As global concern grows for the protection of cultural heritage, geophysical methods have emerged as powerful tools for the non-destructive assessment, monitoring, and conservation of historic structures. Heritage sites—ranging from ancient temples and cathedrals to buried monuments and archaeological ruins—face threats from natural aging, environmental degradation, seismic activity, climate change, and human intervention. To mitigate these risks without altering the site physically, geophysics provides a scientific means of evaluating structural integrity, identifying hidden weaknesses, and monitoring subsurface and structural changes over time.

Assessing the Integrity of Built Heritage

Geophysical surveys are now routinely used to evaluate the condition of historic buildings, monuments, and foundations. Techniques such as Ground Penetrating Radar (GPR), Electrical Resistivity Tomography (ERT), Ultrasonic Tomography, and Infrared Thermography allow conservation scientists to detect cracks, voids, moisture ingress, material degradation, and other structural anomalies—without removing even a single stone.

For example, GPR is highly effective in mapping the internal stratigraphy of walls, floors, and domes, detecting detached plaster layers, or identifying past repair phases. Ultrasonic methods can assess the elasticity and cohesion of stone or brick, offering key insights into structural health (Binda et al. 2021). ERT has been applied to evaluate subsurface conditions beneath heritage sites, such as buried foundations and support systems, especially where surface symptoms suggest deeper instability (Leucci and Masini 2022). This non-invasive capacity is crucial when working with delicate, irreplaceable heritage assets where physical sampling or dismantling is not an option. Geophysical results can directly inform decisions about restoration, reinforcement, or preservation strategies.

Early Detection of Structural Vulnerabilities

One of the primary advantages of geophysical techniques in heritage conservation is their ability to identify latent structural problems before they become visible or critical. Cracks, voids, rising damp, salt crystallization, and internal erosion often begin well before signs emerge at the surface. By deploying high-resolution methods like microseismic monitoring or radar scanning, engineers can map these vulnerabilities in advance and track their development.

In particular, seismic refraction tomography and GPR can detect subsurface voids or sinkholes that threaten structural foundations. Similarly, techniques such as Infrared Thermography help identify moisture intrusion, which can weaken walls, frescoes (Frescoes are mural paintings made on freshly applied lime plaster), or buried vaults through slow but progressive damage (Abate et al. 2023). Early warnings enable preventive action, which is both cost-effective and conservation-sensitive, minimizing future repair needs.

Monitoring Changes Over Time Using Time-Lapse Geophysics

Long-term monitoring is essential for heritage sites exposed to seasonal variation, earthquakes, subsidence, or tourism-related stress. In such contexts, time-lapse (or 4D) geophysical methods offer a way to track changes in a structure or subsurface environment over time. These repeated measurements are compared to detect evolving patterns—such as shifts in water content, crack propagation, or subtle ground movement—which may compromise site stability.

For instance, time-lapse Electrical Resistivity Tomography has been used to monitor moisture dynamics beneath ancient pavements and saltwater intrusion in coastal ruins (Taviani et al. 2021). Acoustic emission sensors and passive seismic monitoring have been installed in heritage towers and walls to assess real-time responses to seismic or anthropogenic vibrations. The integration of such monitoring with Building Information Modeling (BIM) and digital twins is a growing trend in the smart management of historic infrastructure.

14.4.1 Case Study: Geophysical Monitoring of a UNESCO World Heritage Site—The Leaning Tower of Pisa (Italy)

A notable example of heritage monitoring through geophysical tools is the long-term study of the Leaning Tower of Pisa, one of the world's most recognized UNESCO World Heritage sites. Over the past decades, the tower has undergone an extensive conservation campaign to address concerns over its stability, due largely to foundation subsidence and soft ground conditions. To monitor the structure without causing disturbance, geophysicists employed a suite of techniques.

GPR was used to scan the floors and the internal masonry structure, identifying hidden voids and past restorations. Electrical Resistivity Tomography provided insights into groundwater changes and subsoil variations, which could influence future tilt. In addition, microtremor measurements and ambient vibration analyses were used to assess the dynamic behavior of the tower in response to natural and human-induced vibrations (Giorgio et al. 2022).

These geophysical assessments contributed significantly to engineering decisions, allowing restoration work to be precisely targeted. As a result of these monitoring efforts and controlled soil extraction, the tilt of the tower was reduced by 40 cm and stabilized for the long term—without compromising the historical appearance or structural integrity.

In conclusion, the application of geophysical tools in heritage conservation represents a fusion of modern science with ancient legacy. These non-invasive methods offer unique diagnostic and monitoring capabilities, enabling conservation professionals to understand, protect, and preserve cultural landmarks with minimal risk. By identifying hidden vulnerabilities, assessing existing damage, and enabling ongoing monitoring, geophysical techniques support sustainable preservation efforts and informed decision-making at local, national, and global levels. As climate change, urban development, and mass tourism place increasing stress on cultural heritage, the role of engineering geophysics will become even more central. When integrated with digital modeling, remote sensing, and conservation planning, geophysics provides a foundation for the next generation of resilient and respectful heritage stewardship.

14.4.2 Case Study: Foundation Features Investigation of the Sun Temple, Konark (India), Using Coupled Non-invasive and Invasive Techniques

The investigation of foundation systems in heritage structures presents unique challenges due to the need for minimal disturbance. This case study focuses on the Sun Temple at Konark, a UNESCO World Heritage Site located in Odisha, India (Dwivedi et al. 2022). Estimated to be over 700 years old, the temple is a massive stone masonry structure requiring careful and comprehensive evaluation of its subsurface features. The objective of this study was to determine the foundation type, depth, and lateral extent, as well as to characterize the subsurface soil stratigraphy beneath and around the temple using a combination of non-invasive and invasive geotechnical methods.

Various researchers have reported the application of GPR in analyzing the different features and details of the heritage structures (Sheets et al. 1985; Chauhan and Dwivedi 2013; Dwivedi and Chauhan 2013), buried mounds (Annan et al. 1990; Goodman and Nishimura 1993; Goodman 1994), mapping of the archaeological site and artifacts (Vaughan 1986; Conyers and Cameron 1998; Conyers 2006; Leucci and Negri 2006; Agarwal et al. 2018).

A large number of different applications of GPR, i.e., layer detection of clay liner, layering system, and stratigraphy, distress mapping in road embankment, sedimentary deposit analysis, water content determination, are also reported in the literature (Topp et al. 1980; Worsfold et al. 1986; Davis and Annan 1989; Jol and Smith 1991; Lafleche et al. 1991; Fisher et al. 1995; Overmeeren 1998; Guyonnet et al. 2003; Neal 2004; Shukla et al. 2008; Gomez-ortiz et al. 2010; Fortier et al. 2011; Slowik 2012; Dara et al. 2018). Wallace et al. (2018) presented a comprehensive review of GPR applications for different civil engineering problems.

Ground Penetrating Radar (GPR) was employed as the primary non-invasive technique. Both shielded and unshielded antennas with central frequencies of 100 MHz, 250 MHz, and 500 MHz were utilized to map subsurface features. The GPR investigation targeted the top of the platform, the plinth level, and surrounding areas. The GPR radargrams indicated a three-layer system beneath the temple platform: Top layer—lateritic stone, middle layer—clay-silt mix, and bottom layer—dense sand. Below the plinth, a two-layer system was identified: Upper layer—lateritic stone, and lower layer—sandy stratum. Beyond the plinth, in the adjacent lawn area, only natural soil strata were observed, with no extension of the foundation, suggesting a block-type foundation limited to the temple base.

Complementary invasive testing was conducted through borehole drilling at key locations: Drilling and soil sampling were performed at the platform and plinth levels, soil characterization confirmed a dense sandy stratum extending to a depth of 22 m from the existing ground level (EGL), the foundation depth was confirmed to be approximately 10.5 m from the platform top (or 5.5 m from EGL), and evidence of previous cement grouting was found at the plinth level—indicating past stabilization efforts. The non-invasive GPR results closely aligned with findings from borehole logs and soil analysis. Both techniques identified a consistent and uniform stratigraphy with dense sand supporting the foundation system. GPR proved effective in detecting foundation boundaries and subsurface layers without disturbing the structure, while borehole drilling provided confirmatory data.

This case study demonstrates the successful integration of non-invasive (GPR) and invasive (borehole) investigations to study the foundation system of the Sun Temple at Konark. The results revealed a block-type foundation resting on a dense sandy stratum, with a depth of 10.5 m from the temple platform. The application of GPR as a reconnaissance tool, validated by targeted invasive tests, illustrates the potential of such combined methodologies in the conservation of heritage structures. The findings emphasize GPR's value in archaeological mapping and structural assessment in heritage conservation.

14.4.3 Archaeological Excavation, Protection, and Display Engineering Design Practice: A Case Study in the Ruins of the Imperial City of the Minyue Kingdom

The Han Dynasty Ruins in Chengcun village, located in Wuyishan City, Fujian Province, China, are also known as the Ruins of the Imperial City of the Minyue Kingdom. Situated on a hilly slope southwest of Chengcun village in Xingtian Town, the site represents one of the most significant imperial ruins in East Asia. In 1999, it was inscribed as a world heritage site by the Wuyi Mountain World Cultural and Natural Heritage Committee, becoming the only Han Dynasty imperial city site in China to hold such recognition. It also stands as the most well-preserved large-scale imperial city site from the Middle Ages on the Pacific Rim (Figs. 14.1 and 14.2). This case study explores the integration of archaeological excavation, site protection, and display engineering design practices, with a focus on responding to the challenges posed by natural and anthropogenic impacts (Zhou et al. 2024).

A detailed geological survey was conducted to understand the terrain, soil composition, and foundation stability. The site lies on a hilly plateau characterized by stable

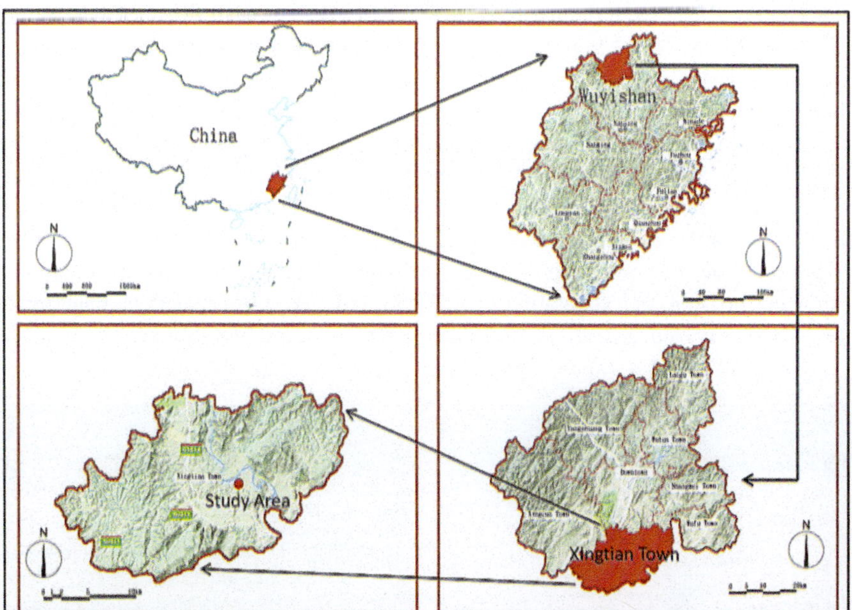

Fig. 14.1 Location analysis of the ruins of the imperial city of the Minyue Kingdom in Chengcun village, Wuyishan City, Fujian Province, China; the objective is to assess the site's geographical setting for archaeological excavation, structural preservation, and display planning (*Source* Zhou et al. 2024—Open Access)

Fig. 14.2 Typical erosion of the soil remains (image source: provided by the excavation team for the Ruins of the Imperial City of the Minyue Kingdom) (*Source* Zhou et al. 2024—Open Access)

and uniform soil layers, making it suitable for large palace foundations and modern protective engineering (Lombardo et al. 2023; Depauw and Dzierzbicka 2019; Hardesty and Little 2009; Jokilehto 2017). The preservation status of two key architectural features—the East Gate and the East City Wall—was investigated: (i) East Gate Site—Only the base components remain, consisting of rammed earth platforms and partial towers. The gate structure, 4.5 m wide, includes the remains of north and south guardrooms. Numerous collapsed tiles and charcoal burning traces from the Han Dynasty have been documented here. (ii) East City Wall Site—This rammed earth wall stands 6.3 m high and comprises 47 discernible layers. Severe damage has occurred due to erosion, partial collapse, and biological intrusion, including the growth of trees and shrubs. These issues are visually demonstrated in Fig. 14.2 (typical erosion of the soil remains), Fig. 14.3 (typical collapse of soil remains), and Fig. 14.4 (typical growth of trees and shrubs in the soil remains). An aerial photo gives a clear perspective on the extent and structural layout of the wall (Fig. 14.5, aerial photo of the remains of the East City Wall).

Multiple factors contribute to the degradation of the site: (i) Natural impacts—Rainwater erosion and vegetation growth weaken the masonry. (ii) Human activities—Unauthorized access, environmental stressors, and insufficient early-stage

Fig. 14.3 Typical collapse of soil remains (image source: provided by the excavation team for the Ruins of the Imperial City of the Minyue Kingdom) (*Source* Zhou et al. 2024—Open Access)

Fig. 14.4 Typical growth of trees and shrubs in the soil remains (image source: provided by the excavation team for the Ruins of the imperial city of the Minyue Kingdom) (*Source* Zhou et al. 2024—Open Access)

Fig. 14.5 Aerial photo of the remains of the east city wall (image source: provided by the excavation team for the Ruins of the imperial city of the Minyue Kingdom) (*Source* Zhou et al. 2024—Open Access)

protection have also accelerated deterioration. In response, a multi-pronged conservation strategy was developed:

1. Treating collapsed and eroded areas through stabilization and reinforcement.
2. Clearing invasive vegetation, while introducing native species with low root impact to stabilize soil.
3. Enhancing drainage to manage surface runoff and prevent further erosion.

These interventions were designed based on extensive archaeological excavation work, site condition analysis, and geotechnical surveys.

A comprehensive site display project was conceived to balance protection with educational accessibility, particularly at the East Gate. Key components include: (i) A public square and hydrophilic wooden plank road to guide visitor movement. (ii) A wooden observation platform and steel-structured protective shelter using modern materials that mimic Han Dynasty architectural styles. (iii) Pile foundation treatments and UV shielding for enhanced long-term durability and authenticity. This engineering design approach is visually presented in Fig. 14.6 (Rendering of the design scheme for the East City Gate Relic Exhibition Project).

This project demonstrates the critical role of multidisciplinary cooperation across archaeology, geophysics, geology, biology, architecture, and engineering. From geophysical and geological surveying to archaeological data recording, the study highlights the importance of integrating various domains to inform design and

Fig. 14.6 Rendering of the design scheme for the east city gate relic exhibition project (image source: provided by the excavation team for the Ruins of the imperial city of the Minyue Kingdom) (*Source* Zhou et al. 2024—Open Access)

conservation decisions. Moreover, the work at the Ruins of the Imperial City of the Minyue Kingdom provides valuable technical references for archaeological projects conducted in complex terrains and sensitive geological environments, particularly those associated with ancient capitals. It emphasizes the importance of breaking down disciplinary silos to generate a holistic understanding of site formation, historical evolution, and cultural significance.

This study not only contributes to domestic site preservation strategies but also offers a practical model for international heritage projects, encouraging innovation in the excavation, protection, and presentation of ancient cities worldwide.

14.4.4 Archaeological Analysis of the Newly Discovered Tomb with a Relief of a Couple at the Funerary Area of Porta Sarno in Pompeii

In July 2024, the "Investigating the Archaeology of Death in Pompeii Research Project" carried out a scientific and methodical excavation of the areas outside two of the gates to the city of Pompeii (Alapont et al. 2025). One of them is the funerary area of Porta Nola (next to the tomb of Obellio Firmo) and the other is outside Porta Sarno area (east of the tomb of Marcus Venerius Secundius). The investigated funerary area to the east of Porta Sarno corresponds with the area excavated in 1998 for the construction of the double Circumvesuviana rails. The 1998 excavations recorded the presence of more than 50 cremation burial sites, marked by stelae (columelle) and a monument with an arch, which are delineated by a boundary wall (Figs. 14.7 and 14.8).

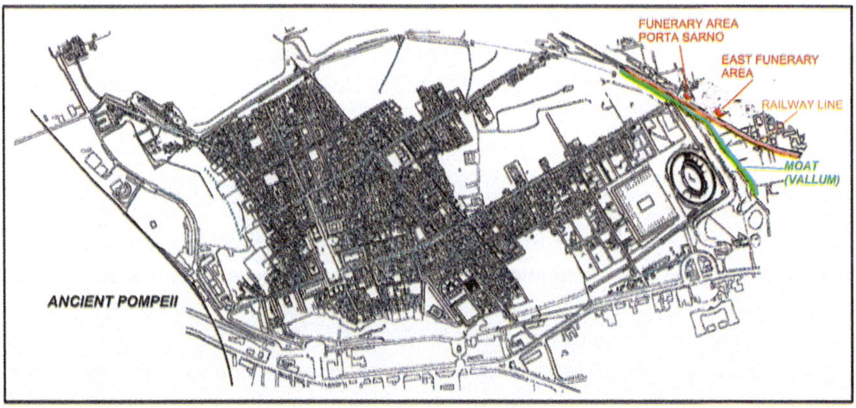

Fig. 14.7 Map of Pompeii. Location of the funerary areas outside Porta Sarno (map: Pompeii archaeological park) (*Source* Alapont et al. 2025—Open Access)

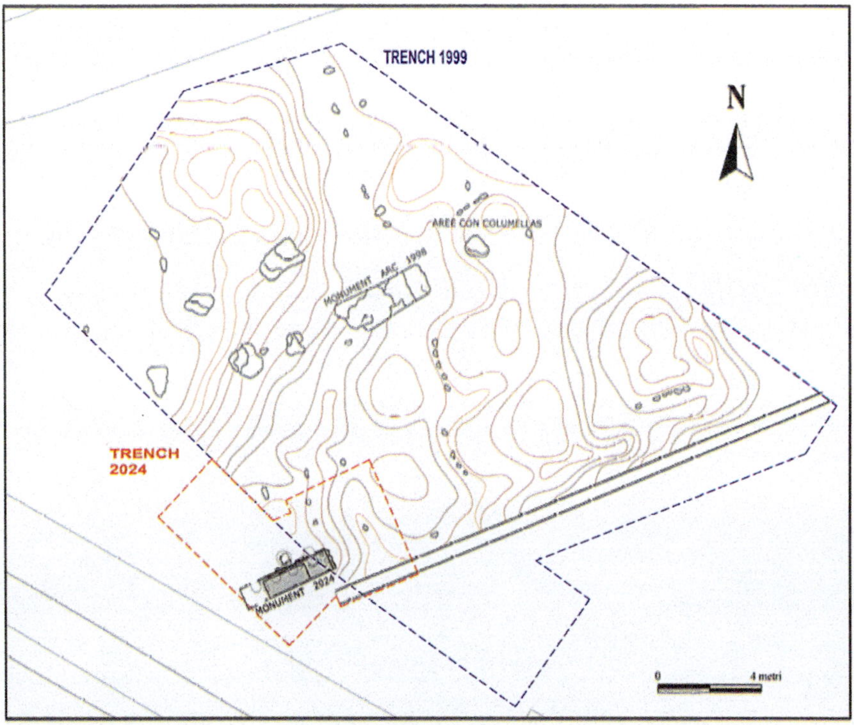

Fig. 14.8 General plan of the excavated trenches (map: Pilar Mas and Joaquin Alfonso) (*Source* Alapont et al. 2025—Open Access)

The tombs were initially dated to the Late Republican period. In order to carry out comprehensive studies of the funerary area uncovered in 1998, a four metre by four metre trenches was stratigraphically excavated. This investigation allowed mapping of the area and the carrying out of archaeological analysis and bioarchaeological studies in order to answer the questions that guided our archaeological research, such as whether the funerary area was abandoned and, if so, when? What was the chronological succession, monumentality, and prestige of this funerary space? Was it a single family and private funerary enclosure, or was it an open public space? How were this funerary area and the spaces destined to preserve the memory of the deceased managed? How were the funerary and mortuary rituals and gestures articulated and what did they consist of?

Methodical excavation discovered a monumental tomb which allows us to answer many of the questions raised by the research. This extraordinary monument consists of a wide wall with several niches containing the cremated remains of the deceased built into its structure and which is crowned by a relief of a young couple (Fig. 14.9). The symbolism of the carved accessories of the wife may identify her as a priestess of Ceres. Additionally, the quality of the carving in the sculptures and their archaic characteristics suggest a Republic period dating, which is uncommon in southern Italy.

In recent years, excavations around Porta Sarno have given its funerary area previously unexpected value. These investigations are bringing to light funerary enclosures

Fig. 14.9 Funerary monument located at a distance of 0.94 m from the boundary wall situated behind, suggesting that the monument is within a bounded enclosure (photo: Joaquin Alfonso) (*Source* Alapont et al. 2025—Open Access)

and monumental sepulchres, such as the one discussed in this paper. Our discoveries indicate that the funerary area may date back to the late-Republican period and may have been active until the eruption in AD 79, but with various changes, transformations, and even abandonments. The relevance of this area is probably due to the location of the funerary space along the access route to Pompeii, across the Sarno River and through the Via of the Abundance, as well as its age. The data provided by our study confirm that the funerary area was active during the years prior to the eruption that buried the city.

The tomb presented in this paper reveals interesting and diverse aspects of funerary customs and rituals, but above all, the presence of two particularly detailed reliefs. These sculptures belong to a large class of funerary reliefs made between the first century BC and the first century AD. Nevertheless, these types of sculptures are very rare in southern Italy. It is even more unusual to find reliefs of priestesses holding their religious objects. Even though other sculptures of priestesses are known, it is unusual that they show the iconography of their position. Although we can relate our relief to some known sculptural models, we preserve only a few images of a generic nature and these often exhibit unclear iconography.

There is no doubt that the laurel leaves held in this woman's hand indicate that she had a religious role, which identifies her as a priestess (Fig. 14.10). We understand that it is more controversial to ascribe her to the service of the goddess Ceres by means of other symbolic elements that she displays, in particular by the lunula on her necklace. However, the lunula is an apotropaic amulet of childhood that must be given to the lares before marriage, and therefore, in this case, it is devoid of traditional meaning and it is necessary to find another motif for it, which could well be its link with the goddess Ceres. The absence of a funerary inscription prevents us from knowing the names of this couple and who they were, and, therefore, from knowing with certainty whether the sculpted couple was a married couple. However, the parallels in known funerary models suggest this. The lack of a sepulchral tituli also prevents us from knowing with certainty whether this priestess served Ceres, although the epigraphy and the honorific and funerary architecture of Pompeii suggest this.

Roman women held a relatively limited but well-defined repertoire of priesthoods, in most cases doing so on an individual basis. The female relief therefore brings attention to the active role of women acting as priestesses in the religious life of their communities. Furthermore, our sculpture could be further proof that Ceres had a clear place in the officially sanctioned religion in Pompeii, having a dedicated priestess. Through the inscriptions presented in this article, it is clear that there were priestesses of Ceres in Pompeii, but this statue provides new evidence of the importance of the cult in the city. In addition, the cult of Ceres has been linked to the popular classes. However, the ostentation of the female relief may suggest that the status of priestess was still reserved for women belonging to a relatively high social standing (Fig. 14.11).

A sarcophagus with battle scenes from the graveyard in the Via Portuense, in Rome, uncovered in the Viale Gianicolense (1226), is provided here as a comparative reference for funerary sculpture styles (Fig. 14.12).

Fig. 14.10 The aspergillum of laurel leaves preserves elements of green paint (photo: Alfio Giannotti) (*Source* Alapont et al. 2025—Open Access)

14.5 Identification of Buried and Clandestine Structures

The ability to detect buried and concealed manmade structures is of vital importance not only in archaeology and heritage studies but also in civil engineering, urban development, and forensic investigations. Geophysical methods provide a non-invasive means to identify hidden tunnels, bunkers, buried foundations, clandestine chambers, and unauthorized underground constructions. Whether the objective is to recover historical knowledge, ensure public safety, or enforce regulatory compliance,

Fig. 14.11 Ring on the proximal phalanx of the ring finger (photo: Alfio Giannotti) (*Source* Alapont et al. 2025—Open Access)

these techniques enable professionals to "see" beneath the surface without excavation or demolition.

Detecting Hidden Tunnels, Bunkers, Foundations, and Illegal Constructions

Subsurface anomalies such as abandoned bunkers, historical war tunnels, forgotten foundations, or illicit constructions (like smuggling tunnels or unauthorized basements) often leave measurable geophysical signatures. Ground Penetrating Radar (GPR), Electrical Resistivity Tomography (ERT), and Seismic Refraction methods

Fig. 14.12 Sarcophagus with battle scenes from the graveyard in the via Portuense, in Rome, uncovered in the Viale Gianicolense (1226) (photo: Llorenç Alapont) (*Source* Alapont et al. 2025—Open Access)

are particularly effective in locating these structures, depending on their size, depth, construction materials, and surrounding soil conditions.

GPR can identify shallow buried voids, reinforced concrete elements, and tunnel-like structures based on strong reflection contrasts (Leucci 2022). ERT is useful in detecting voids and backfilled areas due to the significant resistivity differences between compacted soil and air- or water-filled cavities. Magnetometry and electromagnetic methods can also detect metallic reinforcements or construction debris.

These techniques are often deployed together to improve accuracy, especially when subsurface targets are complex or partially collapsed.

Forensic Geophysics in Cultural and Civil Investigations

Beyond archaeological applications, the identification of clandestine structures has important implications for forensic geophysics, which applies geophysical methods to civil, criminal, and legal investigations. This includes locating unmarked graves (a burial site), illicit burial (unlawful, interment or unauthorized burial) sites, concealed rooms, or hidden compartments within older buildings.

In heritage and cultural property contexts, forensic geophysics can be used to detect hidden chambers in religious buildings, smugglers' tunnels near border zones, or illegally modified cellars beneath protected monuments. These findings can assist law enforcement, urban planning authorities, and heritage conservators in managing risk and enforcing conservation regulations. Recent advancements in 3D imaging, drone-mounted sensors, and artificial intelligence-based anomaly detection have significantly improved the precision of forensic geophysical interpretation, even in cluttered or reinforced environments (Trinks et al. 2023).

Urban Applications and Challenges in Complex Environments

In urban areas, identifying clandestine structures presents unique challenges due to the presence of utilities, modern infrastructure, and electromagnetic interference. High-density construction materials, such as reinforced concrete and metal piping, can obscure or distort geophysical signals. Moreover, access may be restricted by property ownership or surface obstructions.

Despite these obstacles, specialized urban geophysical approaches—such as high-frequency GPR, cross-borehole ERT, and microgravity surveys—have proven effective in detecting anomalies under streets, pavements, and buildings (Simmonds and Colwell 2022). Proper calibration, shielding techniques, and the use of historical maps or engineering records are critical in guiding and validating interpretations. In cities with a long history of unrecorded construction, such as Rome, Istanbul, or Cairo, these technologies are indispensable for urban archaeology, infrastructure safety assessments, and construction planning.

14.5.1 Case Study: Discovery of Clandestine Tunnels Beneath Urban Infrastructure—Mexico–U.S. Border Zone

One of the most compelling examples of geophysical detection of clandestine structures involves the identification of illegal cross-border tunnels along the Mexico-United States boundary. These tunnels—often reinforced and equipped with electricity or ventilation—are designed to evade detection and are used for smuggling activities.

In recent years, geophysicists working with U.S. federal agencies have used a combination of microgravity surveys, seismic refraction, and 3D GPR scanning to identify anomalous voids beneath roads and warehouses. These tunnels often lie 5 to 15 m below the surface and can be extremely narrow or irregular in shape, complicating their detection (Sandoval et al. 2022).

A specific case in 2022 involved the detection of a tunnel nearly 400 m long beneath a commercial zone near San Diego, California. The investigation began with surface microgravity measurements that revealed subtle gravitational anomalies consistent with voids (Osella and De Donno 2021). Follow-up GPR surveys confirmed linear features consistent with reinforced tunnel walls. ERT imaging further verified the presence of an air-filled void that matched the tunnel's expected depth and trajectory.

This case not only highlights the technical feasibility of locating deeply buried and intentionally hidden structures but also showcases the importance of multidisciplinary collaboration between geophysicists, engineers, law enforcement, and urban authorities. The successful identification and closure of such tunnels protect not only national security but also nearby infrastructure and property from potential collapse or damage.

The identification of buried and clandestine manmade structures using geophysical methods is a rapidly expanding field at the intersection of engineering, archaeology, urban planning, and security. Whether uncovering hidden heritage features or combating illegal construction, these non-invasive techniques are indispensable tools for professionals working in sensitive or restricted environments. As subsurface detection becomes more refined and adaptable to urban constraints, geophysical technologies will continue to enhance our ability to explore, regulate, and preserve the built and buried environments of both the past and the present.

14.6 Data Processing and Interpretation Techniques

The success of geophysical investigations in archaeology and heritage conservation depends not only on the choice of field methods but also—critically—on how the collected data are processed, interpreted, and visualized. Subsurface features associated with ancient structures, buried artifacts, or degraded foundations often produce weak, complex, or ambiguous signals, particularly in urban or environmentally noisy settings. Therefore, the application of advanced data processing techniques, inversion algorithms, and visualization tools is essential for extracting meaningful and accurate insights.

Signal Enhancement and Noise Reduction in Heritage Surveys

Raw geophysical data—especially from Ground Penetrating Radar (GPR), magnetometry, and electrical methods—frequently contain high levels of ambient and cultural noise, including interference from modern infrastructure, vegetation, or soil variability. To address this, signal processing techniques such as bandpass filtering, background removal, dewow filtering, and gain adjustments are routinely applied to emphasize anomalies of archaeological interest.

For example, in GPR surveys conducted in historical cities or under building floors, clutter from rebar and pipes may obscure faint reflections from older features like burial shafts or foundations. Using time-slicing and migration algorithms can significantly enhance resolution and delineate linear or rectilinear features associated with manmade structures (Leckebusch 2021). Likewise, magnetometry data can benefit from high-pass filtering and destriping routines to eliminate sensor drift or terrain effects. Emerging approaches, such as machine learning-assisted denoising and automated anomaly classification, are also being introduced to improve the efficiency and objectivity of heritage-related interpretations (Li et al. 2023).

Inversion Techniques and 3D Visualization for Archaeogeophysics

Many geophysical methods—such as Electrical Resistivity Tomography (ERT), seismic tomography, and electromagnetic induction—require the use of inversion algorithms to convert raw measurements into models of the subsurface. These models estimate the distribution of physical properties (e.g., resistivity, velocity, conductivity) and are vital for interpreting archaeological targets, such as walls, voids, or gravel pits.

Modern inversion techniques employ finite-element modeling, constrained inversion, and robust optimization algorithms to increase accuracy, especially in layered or heterogeneous soils. 2D and 3D inversion outputs allow archaeologists and conservators to understand not just the presence of anomalies, but their depth, geometry, and internal structure.

The resulting models can be visualized using 3D rendering tools and virtual reconstruction software. When integrated with architectural plans or excavation data, these visualizations support interdisciplinary analysis, making them accessible to both technical experts and heritage stakeholders. Recent innovations also include augmented reality (AR) applications that project geophysical models directly onto site landscapes, enhancing decision-making during excavation or conservation work (Gaffney et al. 2022).

GIS and Remote Sensing Integration with Geophysical Data

To place geophysical results in broader spatial and historical context, Geographic Information Systems (GIS) and remote sensing platforms are increasingly used. GIS facilitates the spatial integration of geophysical datasets—such as GPR grids, magnetometry maps, and ERT profiles—with other spatial layers, including satellite imagery, historical maps, drone orthophotos, and LiDAR scans. This integration is particularly powerful in large-scale archaeological surveys, where identifying patterns across entire landscapes is crucial. For instance, GIS can be used to correlate anomalies detected in geophysical grids with known settlement boundaries or land-use zones. In heritage management, GIS enables risk mapping, helping authorities to prioritize preservation efforts based on structural vulnerability or exposure to urban development. Remote sensing tools, including hyperspectral imagery and thermal imaging, also support pre-survey planning and post-survey interpretation. When combined with in-situ geophysical results, these datasets create a multiscale, multitemporal perspective of buried heritage, extending the reach and depth of traditional fieldwork (Tapete and Cigna 2021).

In conclusion, advanced data processing and interpretation techniques are foundational to modern Archaeogeophysics. They not only improve the clarity and resolution of raw measurements but also facilitate deeper understanding through 3D modeling, spatial analysis, and contextual integration. As technologies evolve, the ability to combine these tools into coherent, actionable models of the past will continue to transform both research methodologies and conservation practices. The fusion of geophysics with digital technologies ensures that buried and vulnerable cultural heritage can be studied, interpreted, and protected in ever more sophisticated and sensitive ways.

14.6.1 Geospatial Technologies for Heritage: GIS and Remote Sensing in the DBAR-Heritage Mission

In recent decades, the application of Geographic Information Systems (GIS) and Remote Sensing (RS) in archaeological and cultural heritage (ACH) management has significantly expanded in both scope and depth. During the initial implementation phase (2016–2021) of the Digital Belt and Road Heritage (DBAR-Heritage) working group, numerous GIS-RS-based applications emerged, supporting conservation efforts across diverse cultural landscapes. To illustrate the transformative potential of these technologies, this section presents an overview of GIS- and RS-based applications in archaeological and cultural heritage (ACH), particularly focusing on the Silk Road region. These applications have proven essential for the identification of archaeological sites, risk assessment and monitoring of cultural heritage, and the digital management and visualization of cultural datasets (Yao et al. 2023).

Five illustrative case studies from the DBAR-Heritage initiative demonstrate how the integration of GIS and RS can enhance heritage recognition, interpretation, and preservation. These include the use of multi-source geospatial data for detecting linear archaeological features, fine-scale monitoring of the Angkor Wat heritage site, and reconstruction of ancient Silk Road routes. One such example is the Myra Valley Corridor in Tunisia, where satellite-based analysis—including Landsat 8 OLI (Operational Land Imager) and ASTER (Advanced Spaceborne Thermal Emission and Reflection Radiometer) GDEM (Global Digital Elevation Model) data—has been used to map and interpret terrain features and settlement patterns (Fig. 14.13).

These case studies form the foundation of emerging trends in GIS and RS applications in ACH along the Silk Road. They highlight how interdisciplinary, data-driven methods enable the development of intelligent tools for managing archaeological and cultural heritage in the era of Big Earth Data. The discussion also highlights both the advantages and limitations of GIS and RS in heritage contexts. Despite their potential, challenges such as data availability, processing complexity, and contextual

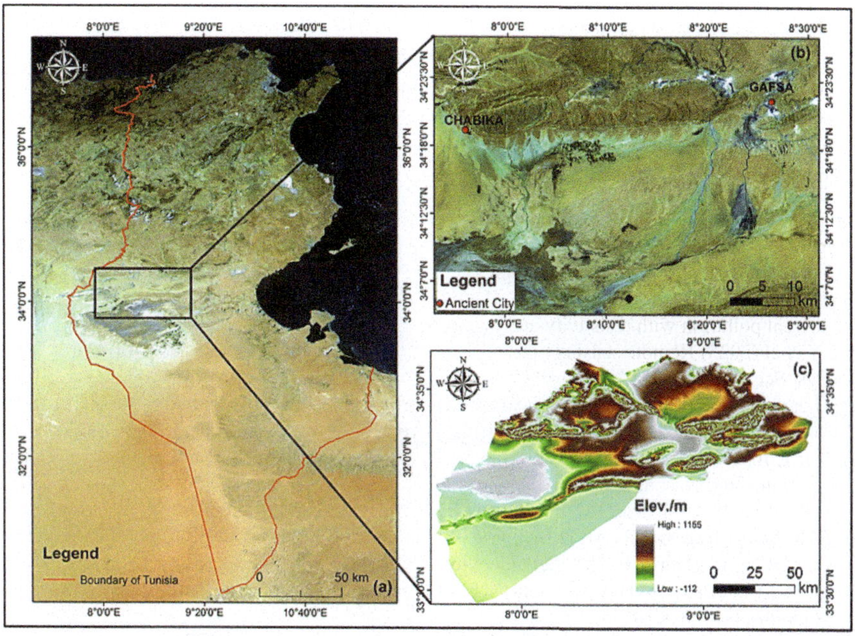

Fig. 14.13 Satellite imagery of the Myra Valley. **a** Location and scope of the Myra Valley Corridor within Tunisia, **b** Landsat 8 OLI data of the corridor. (RGB: bands 7, 4, and 3), and **c** Elevation of the corridor derived from ASTER GDEM data (*Source* Yao et al. 2023—Open Access)

interpretation remain. Nevertheless, the second implementation phase of DBAR-Heritage outlines ambitious plans to further integrate these technologies, promoting broader adoption and refinement of GIS-RS practices for ACH conservation along the Silk Road.

14.6.2 GPR Data Processing and Visualization for Cultural Heritage

Ground-Penetrating Radar (GPR) is among the core methodologies within the spectrum of near-surface geophysical techniques, widely applied in archaeological prospection and cultural heritage management (Martinho and Dionisio 2014; Viberg et al. 2011). Its non-invasive nature, high-resolution capability, and adaptability to various site conditions have made GPR a versatile tool in both field archaeology and the preservation of historical monuments.

In archaeological contexts, GPR has been extensively used to locate and map buried features such as walls, floors, tombs, and artifacts, as well as to assess the state of conservation of built heritage, including buildings, statues, frescoes, and mosaics (Trinks et al. 2010; Bornik et al. 2018). A significant advancement in the field has been the development of array-based multi-channel GPR systems, coupled with 3D data processing and visualization techniques. These innovations have not only accelerated data acquisition but also enabled full wave-field recording, resulting in more realistic and interpretable images of subsurface features (Trinks et al. 2010; Bornik et al. 2018).

The integration of 3D imaging has proven particularly transformative. It allows for the volumetric reconstruction of archaeological sites, supporting interpretations of their original configurations and facilitating virtual restoration. Notable applications include the investigation of prehistoric settlements, megalithic structures such as those at Stonehenge (Locker et al. 2013), and ancient necropolises (Lazar et al. 2011; Leucci et al. 2016), where GPR data have yielded both structural details and spatial relationships between buried elements (Bornik and Neubauer 2022).

In the domain of cultural heritage, GPR plays a pivotal role in preventive diagnostics and preservation planning. It has been applied to detect structural deterioration in heritage buildings, such as fractures, delamination, detachment, and moisture infiltration in walls and facades of churches and cathedrals (Perez-Gracia et al. 2013; Catapano et al. 2018), as well as to evaluate the condition of timber elements in ceilings and floors (Fontul et al. 2018). Additionally, GPR has facilitated the discovery of hidden burial chambers, sarcophagi, and crypts in various monumental structures (Novo et al. 2010; Leucci et al. 2021), and assessed the internal integrity of columns and decorative components, such as identifying fractures and iron fasteners in the rose window of the Troia Cathedral in southern Italy (Masini et al. 2007).

Other successful applications of GPR include the monitoring of fractures and voids in tombs (Santos-Assuncao et al. 2016), statues (Kadioglu 2013; Dimitriadis 2015), and ancient masonry arch bridges, where it has revealed construction details such as hidden arches, thickness variations, and material zoning (Solla et al. 2014; Alani et al. 2020; Lombardi et al. 2021). Though less common, GPR has also shown promise in evaluating the conservation status of artworks—detecting moisture, delamination, and adhesion issues in frescoes and paintings (Danese et al. 2018), and examining stratigraphy and moisture distribution in mosaics (Calia et al. 2013; Caldeira et al. 2019).

Moreover, GPR has begun to find application in palaeontology, particularly for the detection of vertebrate fossils and dinosaur remains (Carrozzo et al. 2003; Tinelli et al. 2012; Main and Hammon 2003). However, the use of advanced processing techniques, such as machine learning and deep learning algorithms, remains relatively underexplored in this domain, with only a limited number of studies currently available (Manataki et al. 2021; Kucukdemirci and Sarris 2020) (Fig. 14.14).

Large-scale archaeological surveys often integrate GPR with other complementary geophysical and remote sensing methods. Electrical Resistivity Tomography (ERT) is valuable for detecting voids, cavities, tunnels, and buried architectural remains, while magnetometry is effective for locating magnetic and conductive anomalies such as kilns, hearths, and metal artifacts. In heritage building diagnostics, GPR is frequently used in conjunction with non-destructive techniques (NDTs) including Infrared Thermography (IRT) for moisture analysis, ultrasonic methods (pulse-echo and impact-echo) for identifying cracks, voids, and delaminations, as

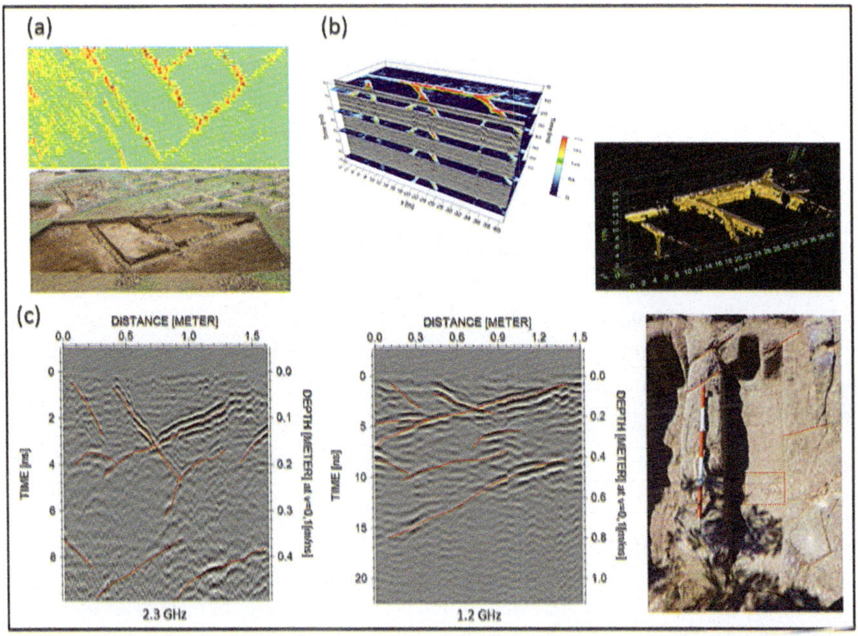

Fig. 14.14 Presents examples of GPR applications in archaeological and heritage investigations: **a** A 3D overlaid time-slice produced at the Roman site of Aquis Querquennis (Bande, Spain), revealing part of a buried barrack structure; **b** A 500 MHz 3D cube and volumetric reconstruction of an ancient underground tunnel with a central corridor and lateral galleries; **c** High-frequency radargrams (2.3 GHz and 1.2 GHz) from an anthropomorphic grave in the rock necropolis of San Vítor de Barxacova (Ourense, Spain), illustrating sub-vertical fractures marked in red (*Source* Lombardi et al. 2022—Open Access)

well as acoustic tomography and seismic reflection for evaluating elastic properties and wave velocities in structural elements such as walls and columns (Perez-Gracia et al. 2016).

14.7 Ethical, Legal, and Conservation Considerations

As geophysical methods become increasingly central to archaeological and heritage investigations, it is essential that their use be guided not only by scientific rigor but also by ethical responsibility, legal compliance, and conservation sensitivity. Unlike typical engineering surveys, geophysical work in cultural heritage contexts often interacts with sacred, fragile, or politically sensitive sites, where the implications of disturbance—intentional or unintentional—can be profound. Therefore, practitioners must be informed not just by technology, but also by an understanding of cultural values, stakeholder expectations, and best conservation practices.

Ethical Implications of Subsurface Investigations in Heritage Contexts

The ethics of archaeogeophysical investigations center around respect for cultural significance, community involvement, and minimal intervention. While non-invasive by nature, geophysical methods can still lead to decisions—such as excavation or site alteration—that impact historically or spiritually important locations. Ethically responsible practice includes obtaining informed consent from landowners, local communities, or descendant groups, especially where burial sites or sacred structures may be present. Furthermore, results from such investigations should be shared transparently and interpreted collaboratively with stakeholders, rather than imposed as purely scientific conclusions (Lafrenz Samuels 2021).

Practitioners should also be mindful of data stewardship—preserving digital models, maps, and interpretations for long-term use and access—ensuring that the knowledge generated benefits both the scholarly community and the broader public.

Legal Frameworks for Geophysical Investigations in Archaeological Sites

Legal considerations vary widely by country and jurisdiction but generally require permits and oversight for any subsurface investigations conducted at designated heritage or archaeological sites. In many regions, such as the European Union, United States, and parts of Asia and Africa, national heritage agencies or ministries of culture regulate the deployment of geophysical surveys in protected zones.

Geophysical investigations are typically categorized under non-destructive testing, but they may still fall under archaeological regulations, especially if data are intended to guide excavation, restoration, or development. In some countries, including Italy, Egypt, and Greece, the use of certain geophysical tools (e.g., GPR or magnetometers) on heritage-listed sites requires formal licensing and academic collaboration.

Failure to adhere to legal protocols can lead to data invalidation, legal sanctions, or reputational harm, both to individual researchers and institutions. As such, it is essential to conduct a legal due diligence review before fieldwork begins, particularly in multinational or UNESCO World Heritage contexts (ICOMOS 2022).

Conservation Best Practices and Minimal-Impact Policies

One of the primary benefits of geophysical techniques in heritage studies is their capacity for non-invasive diagnostics, which aligns strongly with international conservation principles. According to the Venice Charter (1964), Burra Charter (1979, revised 2013), and other conservation protocols, investigative methods should always seek to minimize physical intervention and preserve the site's authenticity.

To ensure this, geophysical surveys should be carefully planned in collaboration with archaeologists, conservators, architects, and site managers. Equipment should be selected to reduce site disturbance—e.g., using lightweight antennas on fragile surfaces or avoiding electrodes near deteriorated stonework. In post-survey reporting, emphasis should be placed on conservation recommendations, highlighting zones of vulnerability and advising against intrusive actions unless absolutely necessary.

Additionally, when data collection reveals sensitive or potentially controversial findings (e.g., unmarked graves, military remains, or religious relics), response plans should be in place to manage communications and engage affected communities with sensitivity and care.

In conclusion, ethical, legal, and conservation considerations are not peripheral concerns—they are foundational to the responsible application of geophysics in heritage contexts. As technology enables deeper and more detailed subsurface insights, it is the responsibility of professionals to ensure that their work honors the cultural significance, legal protections, and conservation needs of the sites they study. By adopting a holistic, interdisciplinary approach that integrates technical skill with ethical awareness and policy compliance, engineering geophysics can continue to make meaningful contributions to the discovery, understanding, and safeguarding of the world's cultural heritage.

14.7.1 Heritage Conservation in India: Challenges and New Paradigms

Conserving heritage structures in India is an interdisciplinary task that blends traditional knowledge of building materials and techniques with modern engineering tools and practices (Menon 2014). Around the world, it is widely accepted that structural safety must never be compromised in conservation efforts. However, in India, formal systems that treat conservation as a structured engineering discipline—particularly one where structural safety is central—are still largely missing.

India has one of the largest collections of heritage structures globally. Yet, there is a significant shortage of trained professionals to protect these monuments from ageing, natural hazards, and environmental degradation. The lack of both quality and quantity of skilled manpower remains a major hurdle. In addition to technical challenges, India's conservation efforts must also consider the country's strong spiritual and philosophical traditions. Ancient Indian architecture emphasized the *spirit* of the structure and accepted the *non-permanence* of materials. This is in contrast to Western conservation principles, which often view heritage structures as permanent vessels of historical memory. Therefore, India may need a different set of conservation guidelines, especially for living monuments that are still actively used by communities.

Capacity building is a pressing need. Conservation in India should evolve into a structural safety-focused discipline. This requires identifying existing but scattered expertise and bringing it together under a single, coordinated framework. A national knowledge pool should be created through focused research, education, and outreach in the field of heritage structure safety. One significant step in this direction is the establishment of the National Centre for Safety of Heritage Structures (NCSHS) at IIT Madras. The center aims to become a national platform for research, training, and collaboration. It will engage key stakeholder groups such as the Archaeological Survey of India, State Archaeology Departments, and academic institutions in engineering and architecture.

India's challenges in heritage conservation are both practical and philosophical. While there is a vast wealth of built heritage, formal education and research in conservation remain underdeveloped. Addressing this requires new paradigms that combine structural safety, traditional knowledge, and cultural sensitivity. Reviving ancient Indian wisdom through education reforms, fundamental research, and the promotion of traditional arts and crafts—also known as intangible heritage ("living" part of culture)—can strengthen the conservation movement. In the long term, this will also support the economic sustainability of heritage by creating skilled employment and renewing cultural practices.

14.7.2 Case Study: Archaeological Site Management and Tourism at Bibracte, France

Bibracte is described as a well-managed property with significant heritage value. Even as the site's managing body and its partners continue to deliver encouraging and engaging projects, there are barriers to success. Thus, the primary purpose of this study was to identify and analyze the possible future challenges that the current management may encounter. This study used a fieldwork methodology, comprising interviews, observations, guided visits, discussions, discourse analysis, and the review of an important corpus of material, from which conclusions were drawn (Ababneh 2021). The location of the Bibracte archaeological site and its rural setting play a critical role in understanding its cultural and environmental context (Fig. 14.15). Significant problems arising from various causes were identified. The results reveal that despite appropriate heritage management, interpretation, and tourism development practices, the heritage and tourism potential remain not fully captured (Moore et al. 2020; Para and Vourc'h 2012).

The management of the site has established its notoriety as an important research center for experimental and educational archaeology, rather than as a major tourist attraction. Thus, the site has failed to become a popular tourist destination. These difficulties have been shaped by site-specific limitations and reinforced by the obligations connected to its designation under the Reseau des Grands Sites de France. This research aims to highlight such underexplored management challenges, contributing valuable insights for both site staff and the visiting public. While Bibracte's heritage management is recognized as professionally and intellectually advanced, the study reveals persistent challenges in interpretation and tourism development. One key issue is that substantial effort has focused on excavation, conservation, and museum interpretation—using combined techniques such as tactile screens, 3D reconstructions, dioramas, and scale models (Fig. 14.16)—but often without fully integrating the broader context of the site's living and rural environment.

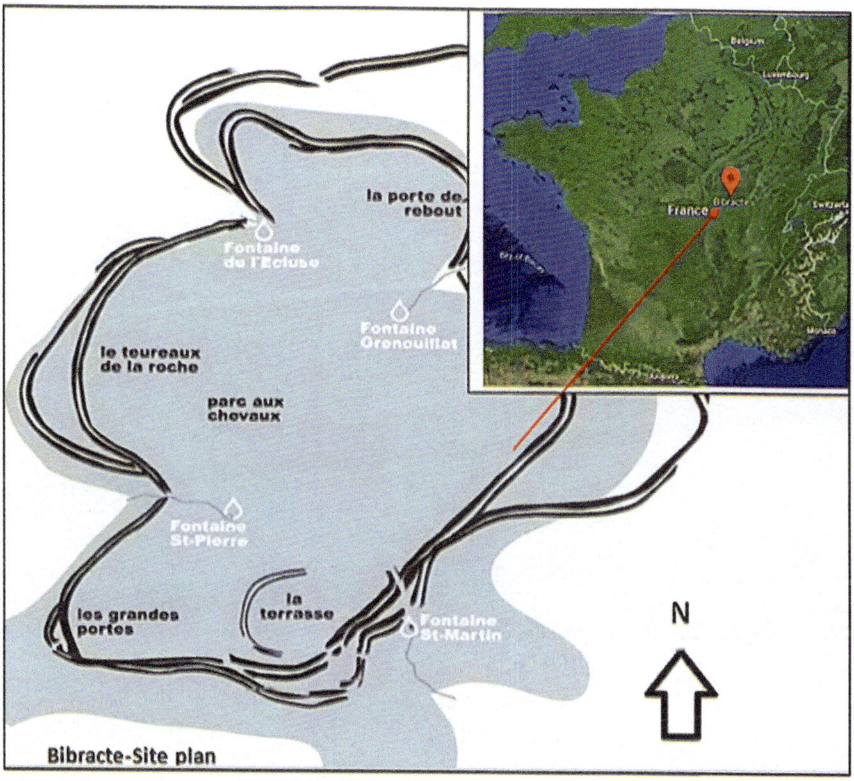

Fig. 14.15 Location of the Bibracte archaeological site on Mont Beuvray, Burgundy, France. This map illustrates the site's strategic setting within a rural landscape, highlighting its cultural and environmental context essential for heritage management planning (*Source* Ababneh 2021—Open Access)

These findings align with other studies of heritage sites globally, where issues of tourism planning, stakeholder coordination, and sustainability are influenced by complex social and political factors. At Bibracte, stakeholder cooperation and a lack of landscape-level planning appear as key bottlenecks in enhancing the site's public engagement and sustainable tourism potential. This study found that the most important challenge facing the site's current management is the limited integration of all heritage elements—cultural and natural, site-based and environmental—needed to manage, interpret, and develop the site's tourism effectively (Makuvaza 2017). The most pressing challenge is achieving a balance between the site's management and interpretation goals on one hand and the preservation of its surrounding living landscape on the other (Fig. 14.17).

Ultimately, the site holds significant untapped potential and offers a wide range of opportunities for development. To move forward, rural heritage sites like Bibracte must incorporate a landscape-wide perspective, reflecting the diverse values of the surrounding region. Therefore, this study recommends targeted investigations to

Fig. 14.16 Combined interpretive techniques including panels, tactile screens, 3D reconstruction, dioramas, and maquette (*Source* Ababneh, 2021—Open Access)

identify natural and cultural assets, assess visitor expectations, and revaluate site management strategies in relation to local community needs and tourism capacity (Alvarez 2014; Rotherham and Ardron 2006; Hughes and Carlsen 2010).

14.8 Future Trends and Innovations

The field of archaeological and heritage geophysics is entering a transformative phase, driven by rapid advancements in sensor technology, artificial intelligence, and aerial geophysical platforms. These innovations promise not only to improve the resolution and efficiency of subsurface investigations but also to make heritage diagnostics more accessible, mobile, and responsive to the needs of conservation professionals, archaeologists, and policy-makers. The next generation of geophysical tools will be increasingly integrated, autonomous, and capable of producing real-time insights with minimal environmental and cultural disruption.

Advances in Sensor Technology and Miniaturization

One of the most significant developments in recent years has been the miniaturization of geophysical sensors, which allows for portable, lightweight, and low-power

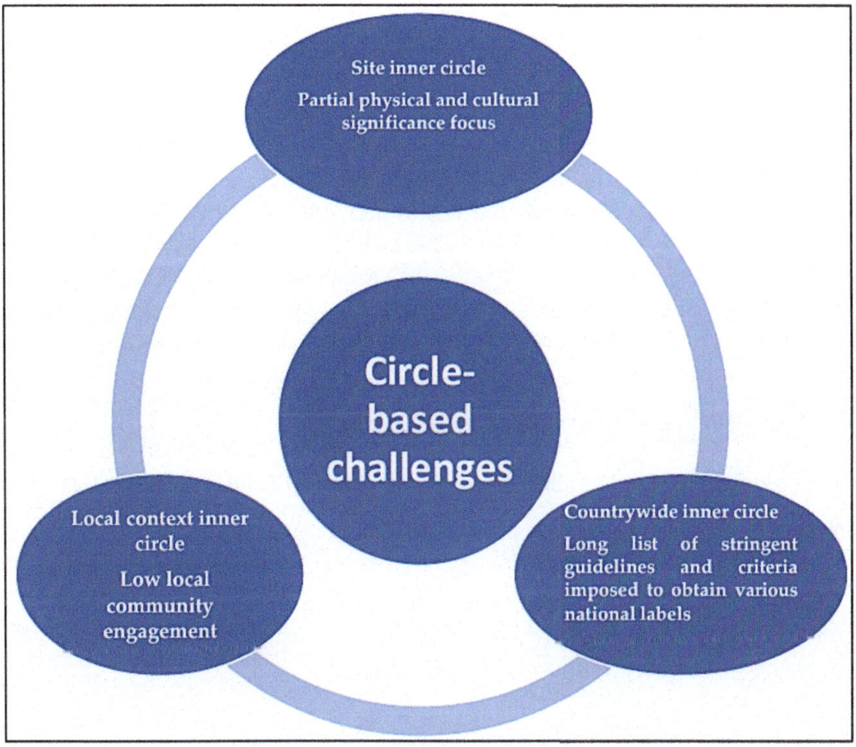

Fig. 14.17 Management problems faced by Bibracte using three circles and levels (*Source* Ababneh 2021—Open Access)

systems suitable for delicate or remote heritage contexts. Advances in materials science and microelectronics have led to compact versions of Ground Penetrating Radar (GPR), magnetometers, and multi-frequency electromagnetic (EM) sensors that can be mounted on handheld rigs, robotic platforms, or even drones.

For example, ultra-wideband GPR systems now offer sub-centimeter resolution and can be operated in tight indoor or archaeological excavation settings. MEMS (Micro-Electro-Mechanical Systems)-based accelerometers and magnetometers are being embedded in mobile survey platforms to monitor dynamic behavior or detect subtle subsurface variations in fragile structures (Xie et al. 2023). These portable sensors support rapid deployment, even in constrained heritage zones such as narrow crypts, cloisters (architecturally, a covered walkway in a monastery), or historic interiors, expanding the operational scope of non-invasive diagnostics.

AI and Machine Learning in Archaeological Data Interpretation

Artificial intelligence (AI), particularly machine learning (ML) and deep learning algorithms, is revolutionizing how geophysical data are interpreted. Traditionally, interpretation has relied heavily on expert knowledge and manual pattern recognition, but the growing volume and complexity of datasets now necessitate automated approaches.

Machine learning models are being trained to detect patterns in GPR and magnetometry data, classify subsurface anomalies, and even predict the likelihood of cultural features based on combined geophysical, spatial, and historical inputs (Chadha et al. 2022). Convolutional neural networks (CNNs) have been applied to GPR time slices to identify buried walls, ditches, and graves with impressive accuracy, reducing subjectivity and increasing reproducibility in data interpretation.

In the near future, AI-driven platforms may allow real-time processing and visualization of subsurface data on-site, giving archaeologists immediate insights that can guide excavation or conservation decisions. Coupled with cloud-based collaboration tools, such systems will enable distributed and interdisciplinary teams to work together on heritage projects globally and in real time.

Drone-Mounted Geophysics and Rapid Site Assessment Tools

The integration of unmanned aerial vehicles (UAVs), commonly referred to as drones, with geophysical instrumentation is emerging as a powerful method for rapid, non-contact surveys over large or inaccessible heritage areas. While drones have long been used for aerial photography and photogrammetry, recent innovations have adapted them to carry magnetometers, EM sensors, LiDAR units, and hyperspectral cameras.

Drone-mounted magnetometers, for instance, have proven effective in identifying subsurface structures such as buried walls or field boundaries in agricultural or forested areas without disturbing the surface (Powlesland et al. 2021). Similarly, drone-based EM induction systems can map shallow conductivity contrasts, useful for identifying graves, ditches, or moisture infiltration zones beneath historic landscapes.

These platforms enable high-resolution mapping across difficult terrain, such as marshes, deserts, and battlefields, where ground-based surveys would be logistically challenging or culturally intrusive. Paired with AI-assisted post-processing, UAV-based geophysics offers a scalable and efficient approach for prioritizing heritage sites, planning excavations, or monitoring environmental threats.

In conclusion, the future of archaeogeophysics lies in the convergence of automation, miniaturization, and intelligent interpretation. As sensors become smaller and smarter, and as AI enhances analytical capabilities, geophysicists will be equipped with powerful new tools for understanding the hidden layers of human history with unprecedented speed and precision.

These technologies not only support faster and more accurate heritage investigations but also align with the ethical imperatives of non-invasiveness, sustainability, and inclusivity. By embracing these innovations, the discipline will continue to evolve into a more collaborative, technologically integrated field—capable of safeguarding cultural heritage for generations to come.

14.9 Conclusions and Recommendations

Geophysical methods have become indispensable in archaeological discovery, heritage preservation, and the detection of buried or hidden anthropogenic structures. This chapter has demonstrated how interdisciplinary collaboration and non-invasive technologies can uncover, document, and protect cultural assets with precision and sensitivity. From ancient tombs and religious monuments to concealed tunnels and site-wide conservation efforts, geophysics now forms a core component of cultural heritage management.

Summary of Key Findings and Methodologies

- Non-invasive technologies such as Ground Penetrating Radar (GPR), Electrical Resistivity Tomography (ERT), and magnetometry have enhanced archaeological prospection and mapping without disturbing the physical integrity of sites (Sects. 14.2, 14.3).
- Multimethod integration is critical for accuracy and data reliability. Combining geophysical tools with GIS and remote sensing yields robust spatial and subsurface insights (Sects. 14.3.3, 14.6).
- Heritage engineering interventions informed by geophysical surveys have proven successful in preserving endangered monuments such as the Leaning Tower of Pisa and the Sun Temple at Konark (Sect. 14.4).
- Case studies, including Porta Sarno (Pompeii) and Bibracte (France), illustrate how geophysical data supports archaeological interpretation, tourism planning, and site management (Sects. 14.3.5, 14.7.2).
- GIS-RS applications under DBAR-Heritage reveal the growing global emphasis on satellite-based monitoring, big data analytics, and cross-regional cooperation for heritage protection along routes like the Silk Road (Sect. 14.6).
- Ethical and contextual considerations—including land use pressures, legal frameworks, and community involvement—are essential to responsible site interpretation and sustainable tourism (Sect. 14.7).

Strategic Recommendations

1. Standardize Geophysical Protocols for Heritage Sites

Develop international guidelines for using geophysics in archaeological and conservation practices, particularly at UNESCO World Heritage sites.

2. Invest in Integrated Geospatial Platforms

Build GIS-based databases that consolidate heritage data, real-time monitoring, conservation history, and visitor interaction interfaces.

3. Promote Multidisciplinary Training and Capacity Building

Support educational programs that bridge archaeology, engineering, digital mapping, and heritage management for future practitioners.

4. Balance Conservation with Public Engagement

Use geophysical findings to design protective yet accessible site layouts, enhancing public understanding while mitigating physical risk to the remains.

5. Advance Climate-Responsive Conservation

Utilize geophysical tools to assess threats from climate change, such as erosion, hydrological shifts, and vegetation encroachment on fragile structures.

6. Foster Global and Regional Collaborations

Encourage participation in transnational initiatives like DBAR-Heritage to share data, expertise, and innovations across borders.

7. Incorporate Local Knowledge and Participation

Engage local communities in heritage surveys and monitoring, blending geophysical insight with traditional custodianship and cultural memory.

Final Remarks

The evolving role of geophysics in archaeology and cultural heritage is marked by innovation, responsibility, and collaboration. As geophysical tools become more accessible and integrated with digital heritage practices, they provide a bridge between past and present—preserving what lies beneath, while informing sustainable futures above ground. Continued investment in interdisciplinary research, technology, and ethical stewardship is essential for safeguarding global heritage in the face of modern challenges.

References

Ababneh A (2021) Archaeological sites management, interpretation, and tourism development—a success story and future challenges: the case of Bibracte, France. Heritage 4:2261–2277

Abate D, Galli G, D'Ambrosio A (2023) Infrared thermography for historic building conservation: diagnosis and environmental interactions. J Cult Herit 59:168–178

Agarwal S, Majumder M, Bisht RS, Prashant A (2018) Archaeological studies at Dholavira using GPR. Curr Sci 114(4):879–897

Alani AM, Tosti F, Bianchini Ciampoli L, Gagliardi V, Benedetto A (2020) An integrated investigative approach in health monitoring of masonry arch bridges using GPR and InSAR technologies. NDT E Int 115:102288

Alani AM, Aboutalebi M, Kilic G (2021) Seismic applications in archaeological and cultural heritage sites. NDT E Int 121:102429

Alapont L, Cava R, Llorens JA, Lopez JJR, González AM, Hurtuna PM, Mullor TH, Revilla V, Palerm AP, Vila SA et al (2025) Archaeological analysis of the newly discovered tomb with a relief of a couple at the funerary area of Porta Sarno in Pompeii. Heritage 8:174

Alvarez MP (2014) Conservation realities and challenges at archaeological and historical sites in Colombia: seeking success by considering the context. In: Davis S (ed) Objects specialty group postprints, vol 21. The American Institute for Conservation of Historic and Artistic Works: Washington, DC, USA, pp 323–344

Annan AP, Scaife JE, Giamou P (1990) Mapping buried barrels with magnetics and ground penetrating radar. In: 60th annual international Mtg. social exploration geophysics, pp 422–423

Binda L, Saisi A, Anzani A (2021) Non-destructive testing techniques for masonry heritage structures: advances and applications. Constr Build Mater 302:124198

Bornik A, Wallner M, Hinterleitner A, Verhoeven GJJ, Neubauer W (2018) Integrated volume visualisation of archaeological ground-penetrating radar data. In: Proceedings of the 16th Eurographics work on graphics and cultural heritage, Vienna, Austria, 12–15 November 2018

Bornik A, Neubauer W (2022) 3D visualization techniques for analysis and archaeological interpretation of GPR data. Remote Sens 14:1709

Caldeira B, Oliveira RJ, Teixido T, Borges JF, Henriques R, Carneiro A, Pena JA (2019) Studying the construction of floor mosaics in the Roman villa of Pisoes (Portugal) using non-invasive methods: high-resolution 3D GPR and photogrammetry. Remote Sens 11:1882

Calia A, Lettieri M, Leucci G, Matera L, Persico R, Sileo M (2013) The mosaic of the crypt of St. Nicholas in Bari (Italy): integrated GPR and laboratory diagnostic study. J Archaeol Sci 40:4162–4169

Campana S, Piro S (2022) Seeing the unseen: geophysics and landscape archaeology. CRC Press

Carrozzo MT, Leucci G, Negri S, Pierri C, Varola A (2003) Ground-penetrating radar: preliminary results to locate vertebrate fossils. In: Proceedings of the SAGEEP 2003, environ and engineering geophysical society, San Antonio, TX, USA, 6–10 April 2003, pp 1017–1103

Catapano I, Ludeno G, Soldovieri F, Tosti F, Padeletti G (2018) Structural assessment via ground-penetrating radar at the Consoli palace of Gubbio (Italy). Remote Sens 10:45

Chadha A, Ricci M, Zanzi L (2022) Machine learning algorithms for GPR data classification in archaeological prospection. J Appl Geophys 202:104780

Chauhan PKS, Dwivedi A (2013) Investigation for foundation of a mosque minar using GPR. In: Proceedings of Indian geotechnical conference (IGC). IIT Roorkee, Roorkee, India (Paper ID: 1Th-21, 1–4. ISBN: 978-81-925548-1-5)

Ciminale M, Masini N, Leucci G (2022) Integrated GPR techniques for archaeological investigations: recent advancements. Remote Sens 14(3):611

Conyers LB (2006) Ground-penetrating radar techniques to discover and map historic graves. Hist Archaeol 40(3):64–73

Conyers LB, Cameron CM (1998) Ground-penetrating radar techniques and three- dimensional computer mapping in the American Southwest. J Field Archaeol 25(4):417–430

Danese M, Sileo M, Masini N (2018) Geophysical methods and spatial information for the analysis of decaying frescoes. Surv Geophys 2018(39):1149–1166

Dara R, Kettridge N, Rivett MO, Krause S, Ortiz GD (2018) Identification of floodplain and riverbed sediment heterogeneity in a meandering UK lowland stream by ground penetrating radar. J Appl Geophys 171

Davis JL, Annan AP (1989) Ground-penetrating radar for high-resolution mapping of soil and rock stratigraphy. Geophys Prospect 37(5):531–551

Depauw M, Dzierzbicka D (2019) Modern scholars at work in a digital and multidisciplinary setting. In: A companion to Greco-Roman and Late Antique Egypt; Wiley Inc.: Hoboken, NJ, USA, pp 15–31

Dimitriadis K (2015) GPR in the preservation of cultural heritage. In: COST action TU1208, proceedings of the TU1208 third general meeting, London, UK, 4–6 March 2015; Aracne, Ariccia: Rome, Italy; ISBN 978-88-548-8486-1

Drahor MG, Berge MA (2020) 2D and 3D electrical resistivity imaging in archaeological prospection. J Cult Herit 46:245–255

Dwivedi A, Samanta M, Mittal AK (2022) Foundation features investigation of a heritage structure in India through non-invasive and invasive techniques. J Appl Geophys 196(2022):104491

Dwivedi A, Chauhan PKS (2013) Delineation of distress area in mosque using GPR. In: Proceedings of Indian geotechnical conference (IGC-2013). IIT Roorkee, Roorkee, India (Paper ID: 1, Th-12, 1–4, ISBN: 978-81-925548-1-5)

Fassbinder JWE (2022) Magnetometry for archaeological prospection: state-of-the-art and future perspectives. Geosciences 12(2):93

Fisher TG, Jol HM, Smith DG (1995) Ground-penetrating radar used to assess aggregate in catastrophic flood deposits, northeast Alberta, Canada. Can Geotech J 32(5):871–879

Fontul S, Solla M, Cruz H, Machado JS, Pajewski L (2018) Ground-penetrating radar investigations in the Noble Hall of the Sao Carlos theatre in Lisbon, Portugal. Surv Geophys 39:1125–1147

Fortier R, LeBlanc A-M, Yu W (2011) Impacts of permafrost degradation on a road embankment at Umiujaq in Nunavik (Quebec). Canada Can Geotech J 48(5):720–740

Gaffney C, Neubauer W, Baldwin E (2022) Immersive visualization of 3D geophysical models for archaeological interpretation: an augmented reality approach. Digit Appl Archaeol Cult Heritage 25:e00240

Giocoli A, Marano M, Stabile TA (2020) Recent developments in archaeological geophysics: from 2D mapping to 3D and 4D imaging. J Appl Geophys 180:104103

Giorgio M, Milani G, Tralli A (2022) Monitoring and modeling the dynamic behavior of heritage towers: the case of the leaning tower of Pisa. Eng Struct 259:114150

Gomez-ortiz D, Martin-Crespo T, Martin-Velazquez S, Martinez-Pagan P, Higueras H, Manzano M (2010) Application of ground penetrating radar (GPR) to delineate clay layers in wetlands. A case study in the Soto Grande and Soto Chico watercourses, Donana (SW Spain). J Appl Geophys 72(2):107–113

Goodman D (1994) Ground penetrating radar simulation in engineering and archaeology. Geophysics 59(2):224–232

Goodman D, Nishimura Y (1993) Ground radar view of Japanese burial mounds. Antiquity 67(255):349–354

Goodman D, Piro S, Nishimura Y (2021) GPR remote sensing in archaeology, 2nd edn. Springer

Guyonnet D, Gourry JC, Bertrand L, Amraoui N (2003) Heterogeneity detection in an experimental clay liner. Can Geotech J 40(1):149–160

Hardesty DL, Little BJ (2009) Assessing site significance: a guide for archaeologists and historians. Altamira, Spain, Rowman

Hughes M, Carlsen J (2010) The business of cultural heritage tourism: critical success factors. J Herit Tour 5:17–32

ICOMOS (2022) Principles for the application of scientific techniques to archaeological heritage conservation. International council on monuments and sites

Jokilehto J (2017) A history of architectural conservation. Routledge, London, UK

Jol HM, Smith DG (1991) Ground penetrating radar of northern lacustrine deltas. Can J Earth Sci 28(12):1939–1947

Kadioglu S (2013) Transparent 2d/3d half bird's-eye view of ground-penetrating radar data set in archaeology and cultural heritage. In: Imaging and radioanalytical techniques in interdisciplinary research—fundamentals and cutting-edge applications. IntechOpen, London, UK

Kogut J, Panissod C (2023) EM induction methods for shallow archaeological mapping: an evolving geophysical frontier. Archaeol Prospect 30(1):23–39

Kucukdemirci M, Sarris A (2020) Deep learning based automated analysis of archaeo-geophysical images. Archaeol Prospect 27:107–118

Lafleche PT, Todoeschuck JP, Jensen OG, Judge AS (1991) Analysis of ground-probing radar data: predictive deconvolution. Can Geotech J 28(1):134–139

Lafrenz Samuels K (2021) Ethics in heritage practice: community engagement and respectful research. Int J Herit Stud 27(10):923–938

Lasaponara R, Masini N, Tapete D (2021) Emerging trends in archaeological remote sensing. Surv Geophys 42(3):563–593

Lazar C, Ene D, Parnic V, Popovici DN, Florea M (2011) Ground-penetrating radar prospections in Romania. M ariuta-la movil a necropolis, a case study. Mediterr Archaeol Archaeom 11:79–89

Leckebusch J (2021) Processing and interpretation of GPR data in archaeology: tools for resolution enhancement and feature recognition. Archaeol Prospect 28(3):195–208

Leucci G (2022) Ground penetrating radar and electrical resistivity tomography for the detection of buried archaeological and modern structures. Remote Sens 14(19):4837

Leucci G (2023) Geophysical techniques for archaeological and forensic investigations. Springer

Leucci G, Masini N (2022) Integrated geophysical methods for the conservation of cultural heritage: a critical review. Remote Sens 14(1):121

Leucci G, Negri S (2006) Use of ground-penetrating radar to map subsurface archaeological features in an urban area. J Archaeol Sci 33(4):502–512

Leucci G, De Giorgi L, Di Giacomo G, Ditaranto I, Miccoli I, Scardozzi G (2016) 3D GPR survey for the archaeological characterization of the ancient Messapian necropolis in Lecce, South Italy. J Archaeol Sci Rep 7:290–302

Leucci G, De Giorgi L, Ditaranto I, Miccoli I, Scardozzi G (2021) Ground-penetrating radar prospections in Lecce Cathedral: new data about the crypt and the structures under the Church. Remote Sens 13:1692

Li Z, Zhang X, Wang Y (2023) Machine learning approaches to GPR signal processing for archaeological target identification. Sensors 23(4):1123

Locker K, Baldwin E, Neubauer W, Gaffney W, Gaffney C, Hinterleitner A, Garwood PJ, Trinks I, Wallner M (2013) The Stonehenge hidden landscape project—data acquisition, processing, interpretation. In: Proceedings of the 10th international conference archaeological prospection, Vienna, Austria, 20 May–2 June 2013

Lombardi F, Lualdi M, Garavaglia E (2021) Masonry texture reconstruction for building seismic assessment: practical evaluation and potentials of ground-penetrating radar methodology. Constr Build Mater 299:124189

Lombardi F, Podd F, Solla M (2022) From its core to the niche: insights from GPR applications. Remote Sens 14:3033

Lombardo V, Karatas T, Gulmini M, Guidorzi L, Angelici D (2023) Transdisciplinary approach to archaeological investigations in a semantic web perspective. Semant Web 14:361–383

Main DJ, Hammon WS (2003) The application of ground-penetrating radar as a mapping technique at vertebrate fossil excavations in the Cretaceous of Texas. Cretaceous Res 24:335–345

Makuvaza S (ed) (2017) Aspects of management planning for cultural world heritage sites: principles, approaches and practices. Springer, Berlin/Heidelberg, Germany

Manataki M, Vafidis A, Sarris A (2021) GPR data interpretation approaches in archaeological prospection. Appl Sci 11:7531

Martinho E, Dionisio A (2014) Main geophysical techniques used for non-destructive evaluation in cultural built heritage: a review. J Geophys Eng 11:053001

Masini N, Nuzzo L, Rizzo E (2007) GPR investigations for the study and the restoration of the rose window of Troia Cathedral (Southern Italy). Near Surf Geophys 5:287–300

Menon A (2014) Heritage conservation in India: challenges and new paradigms. In: Peña F, Chavez M (eds) SAHC2014—9th international conference on structural analysis of historical constructions. Mexico City, Mexico, 14–17 October 2014, p 17

Moore T, Guichard V, Alvarez Sanchs J (2020) The place of archaeology in integrated cultural landscape management. A case study comparing landscapes with iron age oppida in England, France and Spain. J Eur Landsc 1:9–28

Neal A (2004) Ground-penetrating radar and its use in sedimentology: principles, problems, and progress. Earth Sci Rev 66(3–4):261–330

Neubauer W (2021) Non-invasive archaeology: Advances in remote sensing and geophysical prospection. Annu Rev Earth Planet Sci 49:345–371

Novo A, Lorenzo H, Rial FI, Solla M (2010) Three-dimensional ground-penetrating radar strategies over an indoor archaeological site: convent of Santo Domingo (Lugo, Spain). Archaeol Prospect 17:213–222

Osella A, De Donno G (2021) High-resolution microgravity surveys in urban areas: Identifying subsurface voids and anomalies. Geophys Prospect 69(6):1451–1464

Ostrowski W, Mleczek P, Lercari N (2021) Integrating remote sensing, GIS, and historical data in archaeological site modeling. Digit Appl Archaeol Cult Heritage 22:e00192

Overmeeren RA (1998) Radar facies of unconsolidated sediments in the Netherlands: a radar stratigraphy interpretation method for hydrogeology. J Appl Geophys 40:1–18

Para L, Vourc'h A (2012) Heritage sites and sustainable tourism as leverage for local development: the grand site de France. P3T Public Policies Territ vol 1, p 6

Perez-Gracia V, Caselles JO, Clapes J, Martinez G, Osorio R (2013) Non-destructive analysis in cultural heritage buildings: evaluating the Mallorca cathedral supporting structures. NDT E Int 59:40–47

Perez-Gracia V, Fontul S, Santos-Assuncao S, Marecos V (2016) Geophysics: fundamentals and applications in structures and infrastructures. In: Riveiro B, Solla M (eds) NDT for the evaluation of structures and infrastructure. CRC Press/Balkema: Leiden, The Netherlands

Powlesland D, Linford N, Gaffney C (2021) UAV-mounted magnetometry for landscape archaeology: a case study in Yorkshire, UK. Archaeol Prospection 28(4):345–359

Pozza A, Verdonck L, Van Meirvenne M (2023) High-resolution magnetometry for archaeological prospection: performance analysis using real site data. J Archaeol Sci Rep 48:103909

Rotherham ID, Ardron PA (2006) The archaeology of woodland landscapes: issues for managers based on the case-study of Sheffield, England and four thousand years of human impact. Arboric J 29:229–243

Sandoval J et al (2022) Geophysical discovery of an underground tunnel near the U.S.–Mexico border using GPR and microgravity. Eng Geol 303:106610

Santos-Assuncao S, Dimitriadis K, Konstantakis Y, Perez-Gracia V, Anagnostopoulou E, Gonzalez-Drigo R (2016) Ground penetrating radar evaluation of the ancient Mycenaean monument Tholos Acharnon tomb. Near Surf Geophys 14:197–205

Sarris A, Trinks I, Gaffney C (2022) Geophysical survey in cultural heritage management: applications and global perspectives. Geosciences 12(1):45

Sheets PD, Loker WM, Spetzler HAW, Ware RW (1985) Geophysical exploration for ancient Maya housing at Ceren, El Salvador. Natl Geogr Res Rep 20:645–656

Shukla SB, Patidar AK, Bhatt N (2008) Application of GPR in the study of shallow subsurface sedimentary architecture of Modwa spit, Gulf of Kachchh. J Earth Syst Sci 117:33–40

Simmonds J, Colwell M (2022) Detection of concealed underground constructions using multi-method geophysics: applications at border security sites. Near Surf Geophys 20(1):75–89

Slowik M (2012) Influence of measurement conditions on depth range and resolution of GPR images: the example of lowland valley alluvial fill (the Obra River, Poland). J Appl Geophys 85:1–14

Solla M, Riveiro B, Lorenzo H, Armesto J (2014) Ancient stone bridge surveying by ground-penetrating radar and numerical modeling methods. J Bridge Eng 19:110–119

Stalder Y, Amrani M, Abdelwahab A (2023) Multimethod geophysical exploration of the Qubbet el-Hawa necropolis: a model for non-invasive burial mapping. Near Surf Geophys 21(4):289–304

Tapete D, Cigna F (2021) Integrating remote sensing and GIS with non-invasive geophysical surveys: monitoring change and assessing threats to cultural heritage. J Cult Herit 50:74–85

Taviani S, Cosentino P, Di Filippo M (2021) Time-lapse geoelectrical monitoring of buried archaeological structures under environmental stress. Near Surf Geophys 19(3):273–283

Tinelli C, Ribolini A, Bianucci G, Bini M, Landini W (2012) Ground-penetrating radar and palaeontology: the detection of sirenian fossil bones under a sunflower field in Tuscany (Italy). C.R. Palevol 11:445–454

Topp GC, Davis JL, Annan AP (1980) Electromagnetic determination of soil water content: measurements in coaxial transmission lines. Water Resour Res 16(3):574–582

Trinks I, Johansson B, Gustafsson J, Emilsson J, Friborg J, Gustafsson C, Nissen J, Hinterleitner A (2010) Efficient, large-scale archaeological prospection using a true three-dimensional ground-penetrating radar array system. Archaeol Prospect 17:175–186

Trinks I, Neubauer W, Eder-Hinterleitner A (2023) Urban geophysics for subsurface mapping: addressing the challenges of complexity and noise. J Appl Geophys 210:104789

Vaughan CJ (1986) Ground-penetrating radar surveys used in archaeological investigations. Geophysics 51(3):595–604

Viberg A, Gustafsson J, Triantaphyllou M (2011) Geophysical methods in archaeology: Applications of ground-penetrating radar and other near-surface techniques for cultural heritage studies. Journal of Archaeological Science 38(10):2572–2582. https://doi.org/10.1016/j.jas.2011.06.014

Wallace WL, Xavier D, Annan P (2018) A review of ground penetrating radar application in civil engineering: a 30-year journey from locating and testing to imaging and diagnosis. NDT E Int 96:58–78

Worsfold RD, Parashar SK, Perrott T (1986) Depth profiling of peat deposits with impulse radar. Can Geotech J 23(2):142–145

Xie M, Wang J, Wang Z (2023) Miniaturized geophysical sensors for heritage diagnostics: trends and technological challenges. Sens Actuators, A 352:114090

Yao Y, Wang X, Luo L, Wan H, Ren H (2023) An overview of GIS-RS applications for archaeological and cultural heritage under the DBAR-heritage mission. Remote Sens 15:5766

Zhou S, Zhang L, Chen Y, Zheng L, Lei N, Zhang J (2024) Archaeological excavation, protection, and display engineering design practice: a case study in the ruins of the imperial city of the Minyue Kingdom. Coatings 14:1220

Chapter 15
Interdisciplinary Approaches in Engineering Geophysics

Abstract This chapter highlights the vital role of interdisciplinary collaboration in advancing geophysical research and applications. By integrating geophysics with geology, hydrology, remote sensing, environmental science, and engineering, complex subsurface and environmental challenges are addressed more effectively. The chapter presents detailed case studies ranging from groundwater contamination and carbon sequestration to flood risk assessment and infrastructure projects, showcasing how cross-disciplinary methods enhance problem-solving capabilities. Special attention is given to the synergy between geophysical techniques and emerging technologies such as machine learning, remote sensing, and geospatial analysis, emphasizing data fusion and collaborative approaches. Challenges in data integration, technological advancement, and education are discussed, along with future pathways to strengthen interdisciplinary efforts for sustainable and resilient environmental and engineering solutions.

15.1 Integrating Geophysics with Geology, Hydrology, and Remote Sensing

Section 9.1 explores the importance of interdisciplinary approaches in advancing geophysical research and applications. It focuses on the integration of geophysics with complementary fields such as geology, hydrology, and remote sensing. These collaborations enhance the understanding of subsurface conditions, fluid dynamics, and environmental processes. By combining geophysical methods with geological and hydrological data, and leveraging remote sensing technologies, more accurate and comprehensive models can be developed, leading to better solutions for complex environmental and engineering problems. This interdisciplinary approach is crucial for addressing challenges in resource management, environmental monitoring, and hazard assessment.

D. C. Naskar, *Advances in Engineering Geophysics: Concepts and Applications*, Earth Systems Data and Models 9, https://doi.org/10.1007/978-3-032-28000-8_15

15.1.1 *The Role of Interdisciplinary Collaborations in Solving Complex Geophysical Problems*

In the rapidly evolving field of engineering geophysics, the complexity of real-world geophysical problems often requires expertise from multiple scientific disciplines. Interdisciplinary collaboration is essential in providing a holistic approach to understanding and solving the challenges presented by the Earth's subsurface. Combining geophysical methods with the knowledge and tools from other fields such as geology, hydrology, remote sensing, and environmental science enables a comprehensive understanding of subsurface properties, processes, and behaviors (Telford et al. 1990; Schlumberger 2010).

Geophysics alone can provide valuable data regarding the subsurface, including the structure, material composition, and physical properties of the ground. However, this information often needs to be interpreted in the context of larger geological, hydrological, and environmental systems. Through collaboration, geophysicists can benefit from the expertise of geologists who understand the processes that shape the Earth's surface and subsurface. Hydrologists provide insights into fluid dynamics and water movement within porous media, while remote sensing specialists contribute surface-level data that can complement and extend geophysical observations (Reeves 2014; Auken et al. 2014; Li et al. 2022).

Geophysics and Geology

Geologists play a key role in interpreting geophysical data by providing the necessary context for understanding subsurface structures, such as faults, fractures, and stratigraphy. For instance, seismic methods that detect changes in rock density or velocity may be better understood when geologists provide knowledge of regional tectonics or sedimentary layers. Integrated geophysical and geological approaches can improve the identification of mineral deposits, groundwater aquifers, and contamination zones, offering critical information for mining, water resource management, and environmental clean-up projects (Jones et al. 2013).

One example of interdisciplinary collaboration between geophysics and geology is the use of geophysical inversion techniques. Geophysical inversion converts observed geophysical data, such as seismic waves or electrical resistivity, into meaningful geological parameters. The inversion process benefits from geological knowledge of the subsurfaces physical properties to constrain the model, improving the accuracy of the results and providing a more realistic representation of the Earth's subsurface (Oldenburg and Li 2017).

Geophysics and Hydrology

Geophysics plays an important role in hydrology, particularly in the study of groundwater systems, aquifers, and contaminant plumes. Geophysical methods such as electrical resistivity imaging (ERI) and ground-penetrating radar (GPR) are used to map the distribution of water, identify aquifer boundaries, and monitor changes in water content within subsurface materials (Binley et al. 2015; Telford et al. 1990). Hydrologists, on the other hand, offer an understanding of groundwater flow, the behavior of contaminants, and the physical properties of soil and rock in relation to water movement (Freeze and Cherry 1979).

By integrating geophysics with hydrology, more accurate models of groundwater flow can be created. For example, electrical resistivity surveys can map areas of high or low water content, while hydrological models can simulate how groundwater moves and how pollutants spread (Leroy and Auken 2017). The combination of these data allows for better management of water resources and more effective monitoring of contamination, which is critical in areas affected by industrial pollution, agricultural runoff, or waste disposal sites (Berglund et al. 2012).

Geophysics and Remote Sensing

Remote sensing technologies, such as satellite imagery, LiDAR, and aerial photography, provide valuable surface-level data that can enhance geophysical studies. Remote sensing enables large-scale monitoring of environmental changes, such as land degradation, vegetation loss, or urbanization, which may have subsurface implications. When combined with geophysical surveys, remote sensing offers a comprehensive view of the environment, allowing for the detection of subsurface anomalies, mapping of surface deformation, and identification of potential areas for further geophysical investigation (Foster and Smith 2016; Pettigrew et al. 2009).

The integration of geophysical data with remote sensing can be particularly beneficial for large-scale environmental monitoring. For example, combining remote sensing data on vegetation and surface water with geophysical data on subsurface water flow allows researchers to assess the relationship between surface features and groundwater systems. This interdisciplinary approach is crucial in monitoring environmental conditions such as drought, flooding, and the effects of climate change (Melesse and Shrestha 2013; Weng and Hu 2008).

Mathematics of Interdisciplinary Integration

The integration of geophysics with other disciplines is often facilitated by advanced mathematical and computational methods. These methods are essential for combining data from multiple sources and for creating models that accurately represent the subsurface environment.

1. **Inversion Techniques**: Geophysical inversion methods, such as those used in seismic tomography or electrical resistivity imaging, rely heavily on mathematical techniques to estimate subsurface properties from measured data. These techniques often involve solving systems of nonlinear equations and employing iterative methods to find the best-fit model that describes the subsurface. For example, in electrical resistivity tomography (ERT), the forward problem (predicting resistivity measurements) and the inverse problem (estimating resistivity distribution from data) are both modeled mathematically:

$$\mathrm{G} \cdot \mathrm{m} = \mathrm{d} \tag{15.1}$$

   where G is the forward model matrix, m is the vector of model parameters (e.g., subsurface resistivity), and d is the vector of observed data.
2. **Geophysics in Ecosystem and Biodiversity Monitoring**: With growing concerns over biodiversity loss and ecosystem degradation, geophysics is increasingly being integrated with ecological data to monitor changes in soil composition, water content, and land use. Geophysical tools, when combined with remote sensing data, can provide valuable insights into ecosystem health, helping to track habitat loss, desertification, and other environmental issues that impact biodiversity. The integration of geophysics with environmental science allows for a better understanding of how subsurface processes affect surface ecosystems (Yuan et al. 2022).

   When analyzing soil moisture content or subsurface water movement in an ecosystem, geophysical methods like electrical resistivity can be combined with hydrological models to predict water flow using Darcy's Law. The equation for groundwater flow is:

$$\mathrm{Q} = -\mathrm{K} \cdot \mathrm{A} \cdot \frac{\Delta \mathrm{h}}{\Delta \mathrm{x}} \tag{15.2}$$

   where Q is the volumetric flow rate of water, K is the hydraulic conductivity of the material, A is the cross-sectional area of flow, and $\frac{\Delta h}{\Delta x}$ is the hydraulic gradient.
3. **Artificial Intelligence and Machine Learning in Geophysical Data Integration**: AI and machine learning (ML) have rapidly become indispensable in geophysical applications. ML techniques are being used to automate data interpretation, enhance inversion algorithms, and improve predictions of subsurface properties. Machine learning algorithms, such as neural networks and support vector machines, have shown great potential in processing large datasets and extracting patterns from geophysical, geological, and hydrological data (Zhu et al. 2021). These techniques are particularly useful in environmental monitoring, such as detecting contaminant plumes or identifying subsurface hazards from complex datasets (Auken et al. 2014). ML algorithms such as neural networks often rely on optimization techniques to minimize the error between predicted

and observed values. A common loss function used in regression tasks, such as predicting groundwater contamination levels, is the mean squared error (MSE):

$$MSE = \frac{1}{n}\sum_{i=1}^{n}(y_i - \widehat{y_i})^2 \tag{15.3}$$

where y_i is the observed value, $\widehat{y_i}$ is the predicted value, and n is the number of data points.

4. **Big Data and Data Fusion for Subsurface Modeling**: When combining remote sensing data with geophysical data, data fusion techniques such as Principal Component Analysis (PCA) or Multivariate Statistical Analysis (MSA) are used to extract relevant features from each data set. The mathematical formulations for data fusion typically involve transforming original data matrices into new variables that reveal the most significant patterns across multiple data sources:

$$Z = X \cdot P \tag{15.4}$$

where Z is the transformed data matrix, X is the original remote sensing or geophysical data matrix, and P is the matrix of eigenvectors or transformation coefficients. These mathematical tools enable the seamless integration of diverse data sources and provide a framework for making informed decisions in environmental monitoring, resource management, and hazard mitigation.

5. **Climate Change and Environmental Monitoring with Geophysics**: Climate change impacts such as sea-level rise, glacier melt, and land subsidence can be monitored using a combination of geophysical and remote sensing data. This interdisciplinary approach allows for the large-scale monitoring of environmental conditions and subsurface changes over time. Geophysical methods like seismic tomography, electrical resistivity imaging (ERI), and ground-penetrating radar (GPR) can provide real-time data on environmental changes such as flooding, subsidence, or the formation of new faults, all of which are essential for understanding the effects of climate change on subsurface systems (Mohamed et al. 2021).
6. **Subsurface Carbon Storage and Geophysical Monitoring**: In light of the global push for carbon sequestration, geophysical methods are being used to monitor CO_2 storage sites, ensuring that stored CO_2 does not leak into the atmosphere. Seismic surveys, combined with geological and hydrological data, can help monitor the integrity of carbon storage reservoirs. The interdisciplinary collaboration between geophysicists, hydrologists, and geologists is crucial for assessing the long-term stability of CO_2 storage sites and ensuring the safety of these solutions in the fight against climate change (Simmons et al. 2020).

The monitoring of CO_2 injection and storage often involves the use of time-lapse seismic data to track changes in seismic velocity, which can indicate the presence of CO_2 in the subsurface. The change in seismic velocity can be modeled using the velocity–pressure relationship in porous media:

$$\Delta V_p = \alpha \cdot \Delta P \tag{15.5}$$

where ΔV_p is the change in seismic velocity, α is the velocity–pressure coefficient, and ΔP is the change in pressure due to CO_2 injection.

7. **Autonomous Geophysical Surveys Using Drones and Robotics**: Autonomous drones and robots are increasingly used for remote geophysical data collection, particularly in hazardous or hard-to-reach locations. These platforms can collect high-resolution geophysical and remote sensing data in real-time, enabling faster and more efficient surveys. AI-powered data processing algorithms are then applied to interpret this data and identify subsurface features or environmental hazards (He et al. 2021). This interdisciplinary approach allows for more cost-effective and comprehensive geophysical surveys, particularly in the context of environmental monitoring, disaster management, and infrastructure inspection.

Interdisciplinary collaborations between geophysics, geology, hydrology, and remote sensing are crucial for solving complex problems related to understanding the Earth's subsurface. By integrating these disciplines, researchers and practitioners can develop more accurate models, improve predictions, and create more sustainable solutions for both environmental and engineering challenges. The fusion of data from multiple scientific fields enhances our ability to tackle pressing issues such as groundwater contamination, resource management, and environmental monitoring, thereby contributing to a more comprehensive understanding of the Earth's systems (Auken et al. 2014; Berglund et al. 2012).

These recent advancements underscore the increasing importance of interdisciplinary cooperation in geophysical research. Innovations in machine learning, big data integration, and autonomous survey technologies are transforming environmental monitoring, resource exploration, and climate change assessments. The combination of geophysics with other scientific disciplines not only enables more efficient and accurate data collection but also offers more robust solutions to complex environmental challenges. By continuing to bridge these disciplines, we are better equipped to address the environmental challenges of the future.

Figure 15.1 emphasizes the importance of these interdisciplinary collaborations and highlights how combining diverse scientific techniques leads to improved outcomes in environmental and engineering applications. This flowchart serves as a visual guide to understanding the interconnectedness of different scientific approaches and their collective role in addressing complex geophysical challenges.

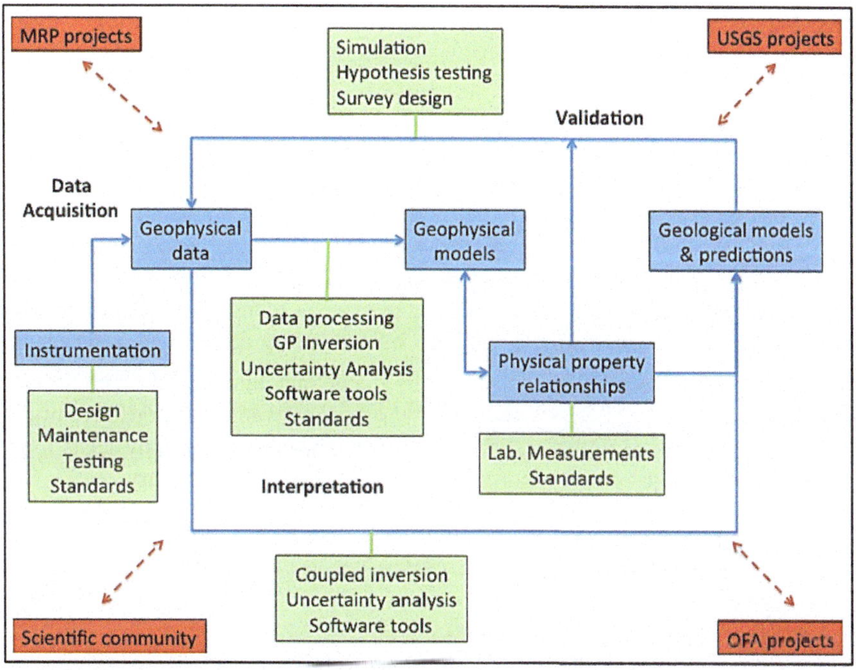

Fig. 15.1 Geophysical methods and applications project flowchart: interdisciplinary approaches in environmental and engineering challenges

15.1.2 Interdisciplinary Approaches to Geophysical Challenges: Case Studies

Case Study 1: Groundwater Contamination Monitoring Using Geophysics and Remote Sensing

In an ongoing project at a former industrial site in the Midwest United States, interdisciplinary collaboration has been employed to monitor and remediate groundwater contamination. Geophysicists used electrical resistivity imaging (ERI) to detect contaminated zones in the subsurface, while hydsrologists analyzed groundwater flow and contaminant transport using tracer studies. Remote sensing was used to monitor surface water and soil contamination levels. By integrating geophysical data with hydrological models, the team was able to pinpoint areas of the subsurface most affected by contamination, helping to design more effective remediation strategies.

Outcome: The collaborative approach improved the accuracy of contamination mapping, guided the placement of monitoring wells, and facilitated the optimization of the remediation efforts. This interdisciplinary effort highlighted the importance of combining geophysical, hydrological, and remote sensing techniques in groundwater management and pollution control (Hubbard et al. 2016).

Case Study 2: Subsurface Characterization for Carbon Sequestration at the Sleipner Project

The Sleipner Carbon Capture and Storage (CCS) project in the North Sea is a well-known example of interdisciplinary collaboration in geophysics, geology, and hydrology. The project aims to store CO_2 in a deep saline aquifer beneath the seabed. Geophysical methods such as time-lapse seismic monitoring (also known as monitoring by "4D seismic") were integrated with geological and hydrological data to monitor CO_2 injection and its movement within the storage reservoir.

Outcome: The collaboration between geophysicists, geologists, and hydrologists led to the development of highly accurate subsurface models that allowed for the detection of any potential leakage and ensured the safe long-term storage of CO_2. This case study illustrates the critical role of interdisciplinary approaches in the monitoring and safety of CCS projects (Arts et al. 2004).

Case Study 3: Remote Sensing and Geophysics for Flood Risk Assessment in Jakarta

In Jakarta, Indonesia, urbanization and climate change have exacerbated the risk of flooding. A study integrating remote sensing, geophysics, and hydrology was conducted to assess flood risk in the city. Remote sensing data from satellites provided high-resolution images of the urban landscape, while geophysical methods such as ground-penetrating radar (GPR) and electrical resistivity tomography (ERT) were employed to assess subsurface conditions, including soil composition and groundwater levels. Hydrological models were used to predict how water would flow during heavy rainfall events.

Outcome: This interdisciplinary collaboration resulted in improved flood risk maps and provided insights into how urban infrastructure and the natural environment interact during flooding events. The integration of these data allowed urban planners to implement more effective flood mitigation strategies (Kusumastuti et al. 2018).

Case Study 4: Landslide Hazard Assessment Using Geophysics, Geology, and Hydrology in the Italian Alps

In the Italian Alps, an interdisciplinary approach combining geophysics, geology, and hydrology was used to assess landslide hazards in mountainous regions prone to seismic activity and heavy rainfall. Geophysical methods such as seismic refraction and ground-penetrating radar (GPR) were used to map subsurface layers and detect any anomalies indicative of landslide risks. Geologists studied rock formations and past landslide occurrences, while hydrologists modeled how rainfall affected groundwater flow and slope stability.

Outcome: By integrating these disciplines, the team developed a comprehensive risk assessment model that allowed local authorities to implement effective landslide mitigation strategies and early warning systems, significantly reducing the risk to communities living in these areas (Lanzano et al. 2016).

Case Study 5: Coastal Erosion Monitoring Using Remote Sensing, Geophysics, and Geology in California

In California, coastal erosion has become a major concern due to rising sea levels and increased storm activity. An interdisciplinary team combined remote sensing data from satellite imagery and LiDAR, geophysical surveys (such as seismic and GPR), and geological analysis to monitor coastal erosion and the impact of human activity on shoreline stability.

Outcome: The combination of geophysical and remote sensing data helped create detailed models of coastal erosion dynamics. These models informed the development of coastal management strategies to protect vulnerable areas and prevent further erosion. The study also helped identify key areas where human intervention (such as construction or land reclamation) was exacerbating the problem, leading to more sustainable practices (Sallenger et al. 2012).

These case studies demonstrate the power of interdisciplinary collaboration in geophysics, highlighting the importance of integrating multiple scientific disciplines to address complex environmental and engineering challenges. By combining data and techniques from geophysics, geology, hydrology, and remote sensing, scientists and engineers can develop more accurate models, make better predictions, and implement more effective and sustainable solutions to pressing issues such as groundwater contamination, resource management, carbon sequestration, and environmental hazards like flooding and coastal erosion.

Case Study 6: Coastal Erosion Monitoring Using Remote Sensing and Sea Surface Temperature Trends in Gujarat, India

This study aimed to monitor coastline changes in 16 coastal districts of Gujarat, India (Fig. 15.2), using an integrated approach of Remote Sensing and GIS (Fig. 15.3) over a 42-year period from 1978 to 2020 (*Source* Patel et al. 2022).

The analysis revealed that approximately 723.6 km of the total 1600 km coastline, or 45.9%, experienced erosion during this time. The study examined the spatial and temporal variations in erosion across the coastal districts, providing critical data for integrated coastal zone management (ICZM) in Gujarat. Additionally, Sea Surface Temperature (SST) data from the Gulf of Kachchh, Khambhat, and Saurashtra coastlines were analyzed for trends from 1860 to 2020 (Fig. 15.4). The results indicated a gradual increase in SST across all three regions, with the Saurashtra coast showing the highest SST values and the Gulf of Khambhat recording the lowest. The Gulf of Khambhat, however, showed the most significant increase in SST by 1.50 °C, followed by the Saurashtra coast with an increase of 1.0 °C, and the Gulf of Kachchh with a rise of 0.75 °C.

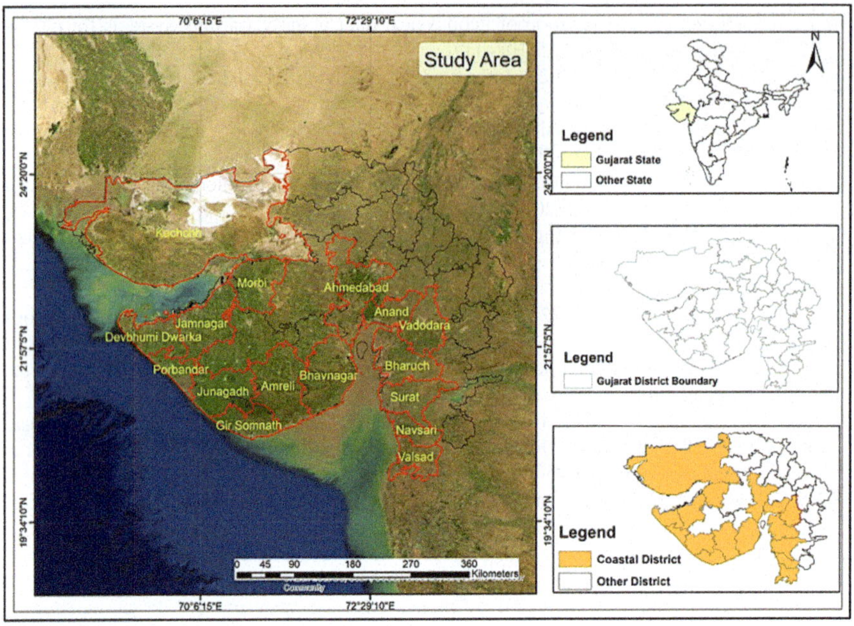

Fig. 15.2 Location map of coastal districts in Gujarat State, India (*Source* Patel 2022—Open Access)

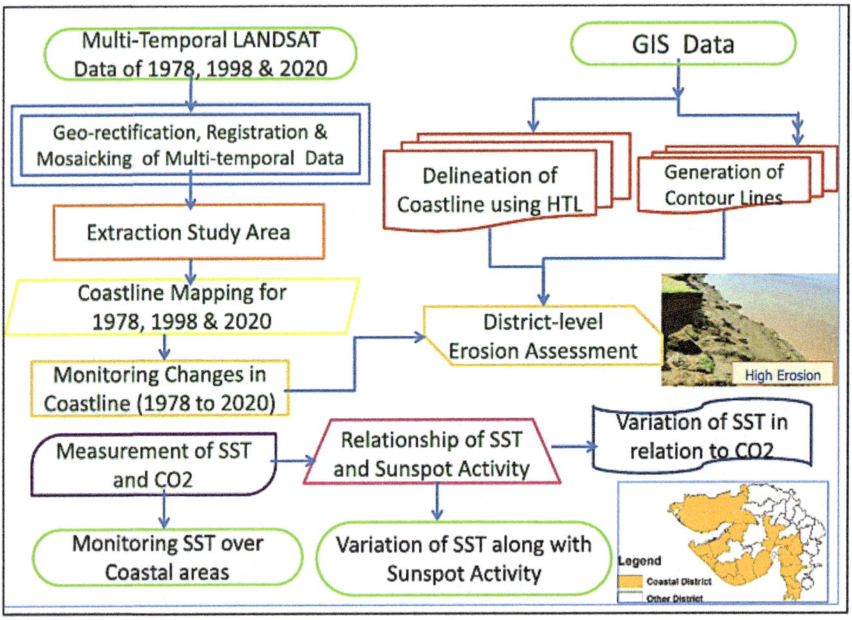

Fig. 15.3 Methodology flow-chart of Shoreline and erosion monitoring in Gujarat state (*Source* Patel 2022—Open Access)

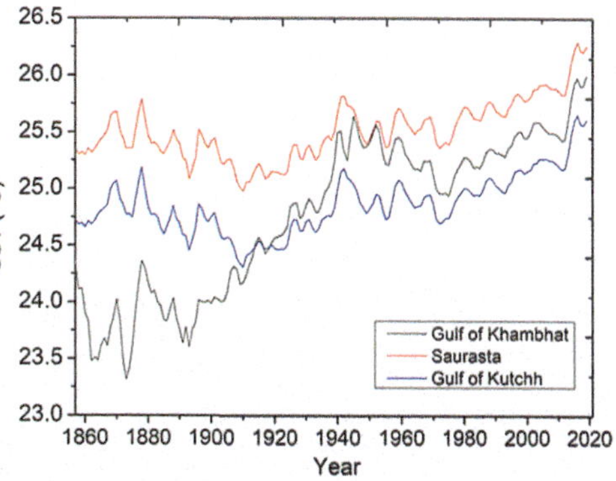

Fig. 15.4 Normalized sea surface temperature (SST) data covering Gujarat coastline from 1860 to 2020: analysis of Gulf of Kachchh, Khambhat, and Saurashtra coast (*Source* Patel 2022—Open Access)

The study also explored the relationship between SST variations and sunspot activity, finding that SST tended to increase in tandem with sunspot cycles (Fig. 15.5). However, the recent rise in SST, despite a decline in sunspot numbers, was linked to the increase in CO_2 emissions, likely due to industrial growth and higher galactic cosmic ray flux. The analysis of coastline changes at two-decade intervals (1978–1998 and 1998–2020) classified the extent of erosion into three categories: high, moderate, and low (Fig. 15.6). These classifications were based on the observed rate of shoreline erosion in different coastal districts.

The findings underscore the significant role of multi-temporal Remote Sensing satellite data in monitoring coastline changes and inform integrated coastal management strategies. The results highlight the need for the Government of Gujarat to adopt sustainable development policies that address coastal erosion. By utilizing the data on coastline erosion, SST trends, and the impacts of CO_2 emissions (Fig. 15.7), effective land-use planning and Integrated Coastal Zone Management (ICZM) policies can be developed to protect coastal villages from further erosion and promote the long-term sustainability of coastal ecosystems.

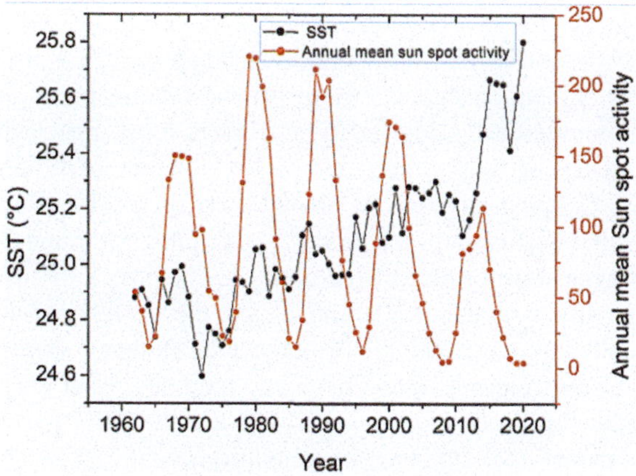

Fig. 15.5 Sea surface temperature (SST) and annual mean sunspot activity data from 1960 to 2020: analysis of Gujarat coastline, including Gulf of Kachchh, Khambhat, and Saurashtra coast (*Source* Patel 2022—Open Access)

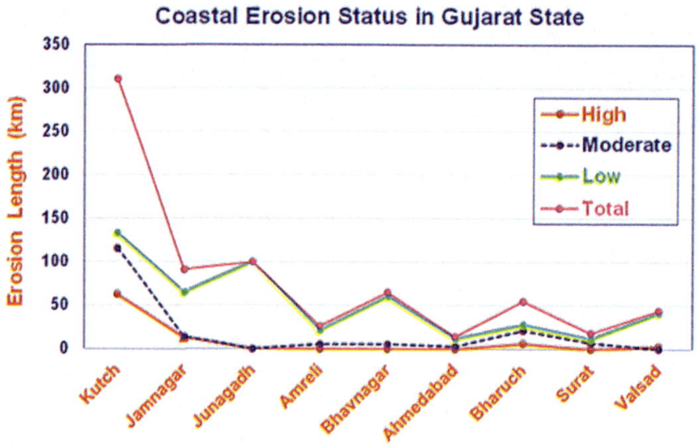

Fig. 15.6 Estimated coastal erosion under different categories during the last 42 years (1978–2020) in Gujarat state (*Source* Patel 2022—Open Access)

This case study emphasizes the importance of monitoring coastal erosion and climate change using advanced technologies like Remote Sensing and GIS, contributing to informed decision-making for the sustainable management of coastal areas in Gujarat.

15.2 Integration of Geophysics with Other Scientific Fields to Address Environmental and Engineering Challenges

The world today faces an array of environmental and engineering challenges, including climate change, natural disasters, resource depletion, and infrastructure decay. These problems are often interconnected and require interdisciplinary solutions for effective mitigation and management. For instance, rising sea levels, induced by climate change, contribute to coastal erosion, flooding, and the displacement of communities. Additionally, engineers must deal with infrastructure failure due to increasing seismic activity, soil liquefaction, or erosion.

Geophysics plays a pivotal role in tackling these issues by providing critical insights into subsurface conditions without invasive methods. By studying physical properties of the earth, geophysics aids in understanding and mitigating environmental hazards, enhancing the design and maintenance of infrastructure, and improving natural resource management. Geophysical techniques, including seismic, electrical, and magnetic methods, offer crucial information about ground stability, groundwater availability, and potential geohazards, which are essential for decision-making in engineering projects.

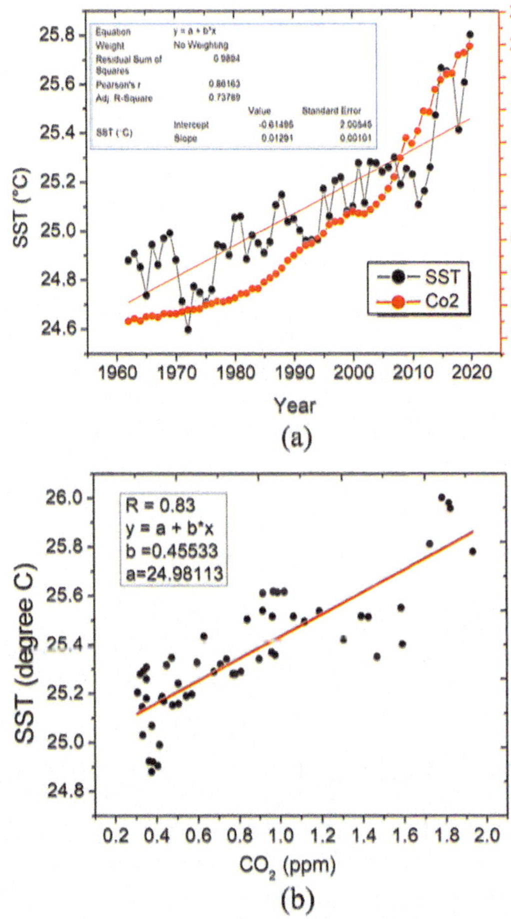

Fig. 15.7 **a**: Relationship between sea surface temperature (SST) and CO_2 emissions: analysis of Gujarat coastline, including Gulf of Kachchh, Khambhat, and Saurashtra coast (*Source* Patel 2022—Open Access). **b** Regression analysis of sea surface temperature (SST) and CO_2 emissions: analysis of Gujarat coastline, including Gulf of Kachchh, Khambhat, and Saurashtra coast (*Source* Patel, 2022—Open Access)

15.2.1 Geophysics and Environmental Science Integration

Environmental Monitoring and Hazard Prediction

Environmental monitoring often involves detecting, mapping, and analyzing changes in the earth's surface, especially concerning natural hazards such as earthquakes, landslides, and floods. When geophysics integrates with environmental sciences, it enhances the accuracy and predictive power of environmental hazard monitoring.

One of the primary applications is the use of Ground Penetrating Radar (GPR) in landfill sites. GPR is used to assess potential leakage from waste disposal areas and track subsurface water movement that could result in contamination of surrounding ecosystems (Smith et al. 2023a, b). This technology, when combined with environmental science expertise, allows for accurate, non-invasive monitoring of waste site integrity, helping to prevent future environmental damage.

Recent Application: Seismic data integration has proven valuable in predicting soil liquefaction in earthquake-prone regions. By combining seismic wave propagation data with soil composition and historical earthquake data, engineers and environmental scientists can assess which areas are most vulnerable to liquefaction, thus enabling better infrastructure planning and disaster response strategies (Jones et al. 2024).

Geophysical Methods in Climate Change Studies

Geophysics offers valuable tools to track and assess the environmental consequences of climate change, particularly in vulnerable regions such as the Arctic, where permafrost is retreating rapidly. Electrical Resistivity Tomography (ERT) is a key tool in studying the degradation of permafrost due to rising temperatures. ERT allows scientists to map subsurface water content and ice distribution, which are crucial indicators of permafrost stability. Such studies are critical for understanding how the thawing of permafrost could release large amounts of methane, exacerbating climate change (Nguyen et al. 2022). Remote sensing and geophysical methods are increasingly integrated to monitor forest cover, soil erosion, and glacial retreat. Satellite data combined with ground-based geophysical surveys help track environmental changes on a larger scale and provide more accurate models for climate impact predictions.

15.2.2 Integration of Geophysics with Geology and Hydrology

Geophysical Tools for Groundwater Exploration and Management

The management of groundwater resources is crucial in regions facing water scarcity. Geophysical techniques like seismic refraction, electrical resistivity, and magnetometry are used in groundwater exploration to determine the location and quality of aquifers. In arid regions, where groundwater is often the only source of drinking water, magnetotellurics (MT) is an effective method for mapping deep groundwater resources that are not detectable with conventional methods. MT surveys allow scientists to map deep geological formations and identify water-bearing zones that can be tapped for sustainable water management (Levy et al. 2023). Combining these geophysical methods with hydrological models allows for better management of water resources, ensuring both environmental sustainability and human needs. Hydrologists working with geophysicists can accurately estimate groundwater recharge rates, assess contamination risks, and design better water extraction strategies in fragile ecosystems (Gomez 2022).

Geophysics for Geohazard Assessment and Mitigation

Geohazards such as landslides, sinkholes, and faults pose significant risks to both the environment and human infrastructure. Geophysics, when integrated with geological surveys, offers critical insights into subsurface structures and behaviors that contribute to these hazards. The use of high-resolution seismic surveys to monitor volcanic activity has enabled real-time tracking of magma movement, gas emissions, and potential eruptions. By combining seismic imaging with geological data, scientists can create dynamic models that predict volcanic eruptions with greater precision (Williams 2023). By integrating geophysical methods like Ground Penetrating Radar (GPR) with geological data, engineers can assess slope stability and understand the role of subsurface water in triggering landslides, particularly in mountainous regions. This interdisciplinary approach has been instrumental in designing more resilient infrastructure in landslide-prone areas.

Soil and Sediment Analysis for Engineering Projects

The properties of soil and sediment are fundamental to the stability of engineering projects. Geophysical surveys can map sediment types, bedrock depth, and groundwater flow, all of which inform decisions regarding the design and location of infrastructure such as dams, tunnels, and bridges. Using integrated geophysical techniques to study sediment accumulation in dam reservoirs helps engineers evaluate the impact of sedimentation on the dam's storage capacity over time. This information is critical for planning maintenance and making informed decisions about dam operations (Kumar et al. 2021).

15.2.3 Geophysics and Remote Sensing: Synergistic Collaboration

Improved Spatial Resolution Through Remote Sensing and Geophysical Integration

The combination of remote sensing and geophysical surveys enhances spatial resolution and accuracy. Remote sensing techniques such as LiDAR, UAV surveys, and satellite imagery provide broad spatial coverage, while geophysics provides high-resolution, in-situ data. LiDAR data combined with ground-based seismic imaging can be used to monitor changes in forest cover, land use, and underground geological hazards. LiDAR allows for the precise mapping of terrain, which, when integrated with geophysical data, helps predict areas prone to soil erosion or landslides (Harper 2024).

Geophysical Remote Sensing for Environmental Monitoring

Remote sensing via satellites or UAVs, when combined with geophysical data, can monitor a wide range of environmental parameters, such as soil moisture, land surface temperature, vegetation health, and atmospheric conditions. These tools are indispensable for assessing the impact of climate change, pollution, and deforestation on ecosystems. By combining passive seismic monitoring with satellite imagery, researchers can assess deforestation and track the potential mudslide risks in mountainous regions. The integration of data helps in early warning systems and real-time monitoring (Tan 2023).

Cross-Disciplinary Methods for Real-Time Disaster Monitoring

Geophysics and remote sensing are key components in real-time monitoring of disasters. Data from seismic instruments, GPS networks, and satellite sensors can be combined to provide early warning systems for earthquakes, tsunamis, and floods. In coastal areas, an integrated approach that uses remote sensing and geophysics enables scientists to monitor coastal erosion, storm surges, and flooding due to severe weather events. By using a combination of seismic, geophysical, and remote sensing techniques, researchers can develop real-time prediction systems for communities at risk of disasters (Fernandez 2023).

15.2.4 Case Studies and Applications in Engineering Projects

Infrastructure Development and Maintenance

Geophysics is crucial in planning and maintaining infrastructure projects. By using non-invasive methods, engineers can assess the conditions of soil, bedrock, and groundwater to ensure the stability and safety of large structures. Seismic site investigations are critical in areas prone to earthquakes, where understanding soil composition and bedrock structure helps engineers design safer foundations. Geophysical surveys provide essential data for decision-making, helping to avoid costly infrastructure failures in seismically active regions (Davis 2024).

Case Study 1: Pir Panjal Railway Tunnel, Jammu and Kashmir

The Pir Panjal Railway Tunnel is a vital infrastructure project that forms part of the Udhampur-Srinagar-Baramulla Rail Link (USBRL) project in Jammu and Kashmir. Stretching over 11.215 km, this tunnel is the longest railway tunnel in India and one of the longest in Asia. It plays a crucial role in connecting the Kashmir Valley with the rest of India's railway network, providing essential connectivity that is particularly important for the region's economic development and accessibility (Gupta 2014).

The project encountered several geological challenges due to the complex topography of the Pir Panjal mountain range. The area is characterized by seismic activity, unpredictable weather, and difficult rock formations, all of which created challenges such as unstable ground, water ingress, and the risk of tunnel collapse, which could compromise the safety of workers and the tunnel itself (Kumar and Sharma 2015). The high seismicity of the region further complicated construction efforts, as the tunnel needed to be designed to withstand potential earthquakes while maintaining structural integrity (Sharma et al. 2016).

In response to these challenges, the engineers chose the New Austrian Tunnelling Method (NATM) for its flexibility and ability to adapt to changing ground conditions. NATM is a continuous, monitoring-based technique where the tunnel's behavior is observed in real-time and adjustments are made accordingly to ensure its stability. The method includes the use of shotcrete, steel ribs, and mesh for immediate support during excavation, making it ideal for the unstable rock formations encountered. Geotechnical instruments such as extensometers and load cells were used to continuously monitor the stability of the tunnel, ensuring any necessary interventions could be made promptly (Singh and Sood 2013). Additionally, careful consideration of environmental impacts was considered, with controlled blasting techniques and minimal disruption to the surrounding ecosystem (Chand and Verma 2017).

The completion of the Pir Panjal Railway Tunnel has been a milestone for the region. It has significantly improved connectivity in the Kashmir Valley, which was previously isolated by the mountainous terrain. The tunnel has facilitated trade, tourism, and economic activities, helping to integrate the region with the broader Indian economy and improving the quality of life for the local population (Thakur and Joshi 2018).

Case Study 2: Rana Pratap Sagar Dam Project, Rajasthan

The Rana Pratap Sagar Dam is located on the Chambal River in Rajasthan, an arid region in northwestern India. The dam serves a multi-purpose role, primarily providing water for irrigation, drinking, and industrial use, as well as generating hydroelectric power. The dam is essential for the region's water management, especially given the area's semi-arid climate, where water scarcity can severely limit agricultural and industrial growth (Sharma and Joshi 2019). It also plays a crucial role in flood control during the monsoon season, protecting the surrounding areas from potential damage.

Building the dam in Rajasthan presented unique geological challenges, including rocky terrain, fault lines, and seasonal flooding. The geological conditions required thorough testing to assess the strength and permeability of the underlying rock, as any weaknesses could jeopardize the dam's stability (Kumar et al. 2016). Additionally, the region is prone to heavy rains during the monsoon season, which posed risks to the structural integrity of the dam and reservoir, requiring careful consideration of flood management systems (Rathi and Shah 2017).

To address these challenges, the project used advanced engineering techniques such as grouting to stabilize the foundation (Jain and Sharma 2015). This technique involved injecting cement-based mixtures into the bedrock to fill voids and improve its strength. This process ensured that the foundation could bear the immense pressure of the stored water and helped prevent any erosion or shifting of the rock beneath the dam. Reinforced concrete was also used throughout the structure, providing the necessary tensile strength to withstand the pressure exerted by the water and any seismic activity in the region (Bhatnagar and Mehta 2018). The dam's design incorporated a sophisticated spillway system, which was critical for managing excess water during the monsoon season. The spillway allows controlled release of water, preventing flooding while maintaining the safety of the dam and surrounding areas (Gupta 2014).

The impact of the Rana Pratap Sagar Dam has been profound. It has significantly boosted agricultural productivity by ensuring a consistent water supply for irrigation, which is crucial in a region that experiences long periods of drought. The dam's hydroelectric power generation capacity has also helped meet the energy demands of Rajasthan, contributing to the state's economic growth (Singh and Patel 2017). Additionally, the flood control measures have protected communities and infrastructure from seasonal flooding, enhancing the resilience of the local population (Sharma et al. 2016). The project exemplifies how careful planning and advanced engineering can mitigate environmental challenges and improve the quality of life in arid regions.

Key Findings and Takeaways

Both projects, the Pir Panjal Railway Tunnel and the Rana Pratap Sagar Dam, illustrate the role of innovative engineering in overcoming geographical and geological challenges in India. Through the application of advanced techniques like NATM for tunneling and grouting for dam foundation stabilization, these projects demonstrate how infrastructure development can improve transportation, water management, and economic opportunities for local communities.

15.2.5 Environmental Remediation Projects

Geophysics is used to monitor and evaluate the progress of environmental remediation projects, such as those involving contaminated land or polluted water sources.

Induced polarization (IP) methods are used to monitor the clean-up of contaminated industrial sites. IP surveys allow engineers to track the movement of pollutants over time, aiding in the design of more effective clean-up strategies (Peterson 2022).

Results of Applying Resistivity/IP Methods to Environmental Issues

In the last two decades, numerous studies have explored the use of resistivity and Induced Polarization (IP) methods for addressing environmental concerns. These methods have been applied in four primary areas: (i) landfills and industrial wastes, (ii) seawater intrusion, (iii) oil products contamination, and (iv) contamination from agricultural activities.

Landfills and Industrial Wastes

A significant portion of research on the use of resistivity/IP methods focuses on landfills and industrial wastes, with 81% of studies addressing landfills and 19% related to industrial waste (Boudreault et al. 2010; Gazoty et al. 2012). Most landfills or industrial waste sites are located in areas with unconsolidated sediments like alluvium, sands, silts, and clay (De Carlo et al. 2013). These permeable formations account for 60% of investigations aimed at identifying and mapping leachate contamination plumes, monitoring their movement, and assessing groundwater contamination risks (Chambers et al. 2006). Geological, hydrogeological, and geochemical data, along with complementary geophysical methods, are often employed to reduce ambiguity in the interpretation of results (Bernstone et al. 2000).

In these studies, the resistivity/IP methods are commonly integrated with borehole logs, physico-chemical analyses, and sometimes soil analysis (Casado et al. 2015a, b). In recent research, additional data such as methane concentrations and petrological data were used to validate resistivity/IP results (Frohlich et al. 2008). The contamination plumes from landfills are typically conductive due to the high ionic concentration of the leachate, and the correlation of resistivity/IP data with geological and hydrogeological information is crucial to differentiate between contamination and natural clayey formations (Orozco et al. 2020).

Key hydrogeological and geochemical parameters like the water table depth, hydraulic conductivity, groundwater flow direction, electrical conductivity (EC), pH, and temperature are measured to assess contamination levels (Maurya et al. 2017). For urban landfills with heterogeneous waste, additional parameters such as metal concentrations and the presence of non-aqueous phase liquids (NAPLs) are also analyzed (Martinho and Almeida 2006). In many studies, resistivity/IP methods helped identify areas of low resistivity associated with leachate, and variations in ionic strength were used to gauge contamination levels (Bording et al. 2019).

Electromagnetic (EM) methods are frequently combined with resistivity/IP methods in the reconnaissance phase to map apparent conductivity. These methods, like EM31 and EM34, are useful for quickly detecting contamination distribution

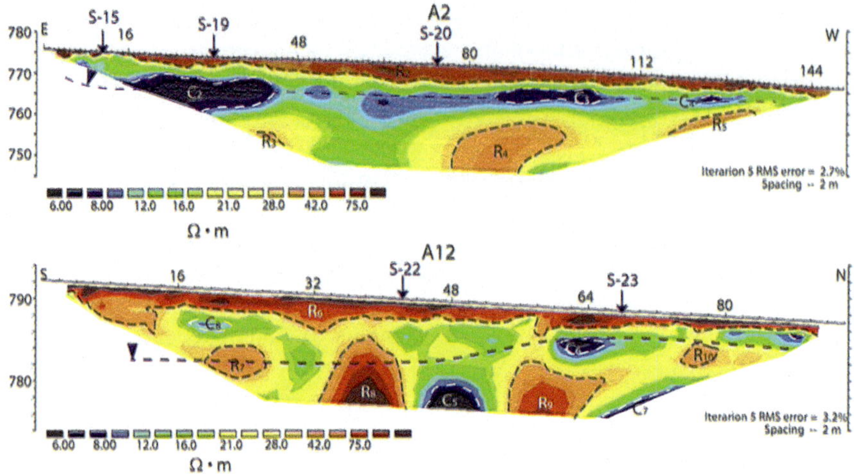

Fig. 15.8 Resistivity profiles illustrating leachate migration in uncontrolled landfill sites in Portugal (*Source* Martinho 2023—Open Access, after Casado et al. 2015a, b)

but do not offer detailed depth information (Abdullahi et al. 2010). Complementary techniques like Ground Penetrating Radar (GPR) and Time-Domain Electromagnetic (TEM) methods were also applied in some studies (Buselli and Lu 2001). However, in some cases, the interpretation of induced polarization (IP) results was challenging, particularly when strong IP effects from clay formations overlapped with contamination signatures (Vaudelet et al. 2011).

A review of 98% of the papers shows that resistivity methods were successful in identifying contamination zones, mapping leachate flow, and assessing groundwater pollution risks (Maurya et al. 2017; Park et al. 2016). While IP methods provided additional insights, their interpretation in some studies was less clear (Jacome et al. 2021). Research also investigated the internal structure and moisture content of landfills, including studies on leachate recirculation and bioreactor technologies (Dumont et al. 2016). In these investigations, 2D Electrical Resistivity Tomography (ERT) was the most commonly used method, with some studies incorporating 3D ERT for more detailed imaging (Ling et al. 2013).

Figure 15.8 illustrates how resistivity and complementary geophysical methods were used to identify contamination and monitor leachate migration within landfill sites (Casado et al. 2015a, b).

Seawater Intrusion Investigation Using Resistivity Methods

The application of resistivity techniques for investigating seawater intrusion has gained significant attention in recent years due to its relevance in understanding the processes of saline intrusion, its relationship with climate change, and the challenges it presents for freshwater management. Numerous studies have been conducted on coastal aquifer systems to address two main aspects: (i) assessing the distribution of salinity in groundwater (both vertically and laterally) and the factors controlling this distribution, and (ii) defining the geometry of aquifer systems.

Seawater intrusion is a widespread issue affecting coastal regions worldwide, as illustrated by various studies. These investigations predominantly focus on aquifers located in Quaternary sediments (sand, gravel, silt, clay, sandy clay, sandy silty, and clayey silty), with 93% of the papers in this field covering these types of aquifers. In contrast, only 7% of the papers investigate bedrock aquifers. The depths of investigation in these studies range from 4 m (Greggio et al. 2018) to 280 m (Goebel et al. 2017), with most studies concentrating on depths of less than 60 m.

To calibrate resistivity data, researchers often use physicochemical properties of groundwater and/or hydrogeological and lithological data, with 80% of the studies incorporating these data. Groundwater salinity, a complex phenomenon that may result from seawater intrusion, water–rock interaction, or anthropogenic pollution, is often investigated using integrated methodologies.

The following physical and chemical parameters are commonly measured:

1. Electrical conductivity (EC): An indicator of the salinity level of groundwater.
2. Total dissolved solids (TDS): A measure of the total mineral content in the water, typically expressed in mg/L.
3. Concentrations of major ions: Such as Na^+, K^+, Ca^{2+}, Mg^{2+}, Cl^-, SO_4^{2-}, HCO_3^-, and NO_3^-, which help identify the origin of salinity (seawater, water–rock interaction, or anthropogenic pollution).
4. pH: The acidity or alkalinity of the groundwater.

Additionally, some authors have measured salinity and temperature to further understand aquifer-ocean interactions (Folch et al. 2020). For instance, EC values are typically used to represent the degree of salinity in groundwater, with freshwater having EC values below 3 mS/cm, brackish water ranging from 3 to 5 mS/cm, and seawater or saline water exhibiting values above 5 mS/cm. TDS values can vary significantly across regions, with values exceeding 550 mg/L commonly associated with seawater intrusion (Wilson et al. 2006). Chloride concentrations, a major ion in seawater, are often used as indicators of seawater intrusion, with concentrations exceeding 500 mg/L marking the presence of saline water (Edet and Okereke 2001).

While many of the coastal aquifers studied contain clayey formations, the Induced Polarization (IP) method was only employed in 5 studies (Adeyemo et al. 2017; Ammar et al. 2021; Attwa et al. 2011; Kumar et al. 2021; Martinho et al. 2004). The IP method proved valuable for defining the sedimentary structure and identifying the presence of saline water, with high chargeability values helping to delineate clayey materials.

In addition to resistivity, complementary geophysical methods such as refraction and reflection seismic, electromagnetic (EM) surveys, gamma logs, and ground-penetrating radar (GPR) were used in only 9 studies (20%). These methods aided in identifying lithological units, material types, thicknesses, and depths, and were used to calibrate resistivity data.

Resistivity techniques, particularly 2D Electrical Resistivity Tomography (ERT) and cross-well 3D ERT, have proven effective in studying coastal aquifer systems, groundwater salinity distribution, and the factors influencing this distribution. Groundwater physico-chemical properties and lithological data were critical for calibrating resistivity results, particularly due to the complex geological formations found in most coastal aquifers.

For example, in a study by Oghenero et al. (2017), the groundwater geochemistry suggested that poor water quality was not due to saline intrusion but rather the presence of iron, which could not be detected by resistivity methods. Furthermore, 2D and 3D ERT methods have been applied in studies monitoring the migration of saline tracers (such as sodium chloride). Perri et al. (2012) used a saline tracer test to compare 2D ERT and borehole ERT techniques for hydraulic characterization, showing that 2D ERT, despite some limitations, provided excellent imaging of saline tracer tests. Similarly, Singha and Gorelick (2005) and Cassiani et al. (2006) successfully used cross-well 3D ERT and 2D ERT, respectively, to monitor saline tracer migration, revealing aquifer heterogeneities and hydraulic connections between aquifers.

Among the various data acquisition techniques, 2D ERT was the most commonly used (employed in 75% of studies), while 3D ERT was less frequently applied (only in two studies). The time-lapse methodology was used in two studies (5%) to track changes over time using 2D ERT. The application of resistivity methods in seawater intrusion studies has proven invaluable in understanding the dynamics of groundwater salinization, identifying intrusion mechanisms, and assisting in freshwater management efforts in coastal regions (Fig. 15.9).

Hydrocarbon Contamination

The context of hydrocarbon contamination studies highlights a division between field and laboratory work, with field studies slightly more predominant (55%). Field investigations commonly employ 2D ERT or 2D ERT/IP techniques, while laboratory experiments focus on spectral induced polarization (SIP) for pore space characterization and understanding fluid-mineral interactions involved in biophysical and chemical processes in soils or rock/organic mixtures.

Key objectives of these investigations include: (i) detecting hydrocarbons and delineating contamination plumes; (ii) monitoring bioremediation processes; and (iii) evaluating ERT and SIP resolution for these goals. Research from the 1990s showed varying resistivity outcomes: fresh hydrocarbons resulted in high resistivity in controlled environments, while field observations often indicated lower resistivities in contaminated zones, particularly in aged plumes due to factors like mineral surface dissolution and biogeochemical activities (Atekwana et al. 2004; Sauck 2000).

Field studies typically focus on contamination detection in unconsolidated sediments (e.g., sand, clay), including LNAPLs and DNAPLs, and on monitoring bioremediation processes using resistivity/IP techniques, which are promising for mapping contamination and microbial activity (Atekwana et al. 2000; Orozco et al. 2021). The

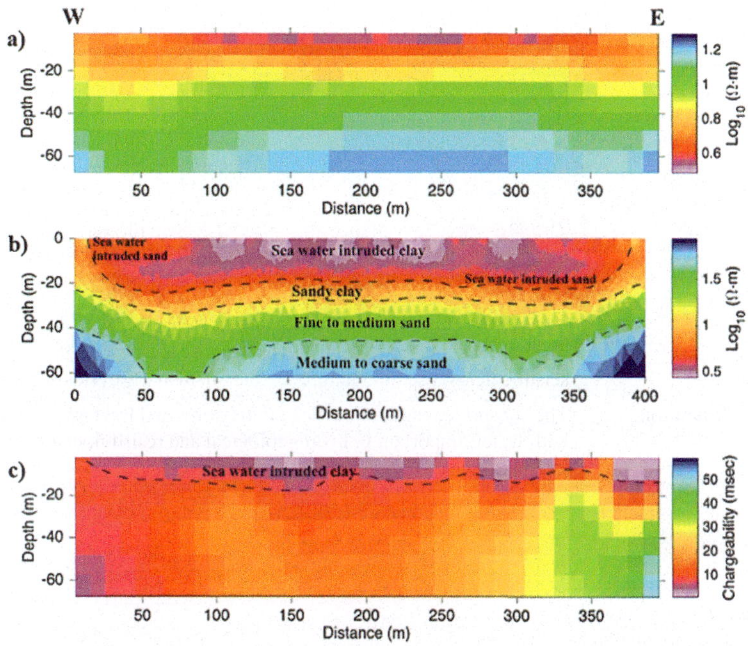

Fig. 15.9 Interpretation of 2D electrical resistivity tomography (ERT) and induced polarization (IP) inversion models for Profile 1, obtained in the Digha area (West Bengal, India). The aim of the study was to delineate possible zones of shallow aquifers contaminated by seawater. The resistivity values (ρ) are indicated as 10^x, where x represents the values corresponding to the color scale shown in the figure (for a to c) (*Source* Martinho 2023—Open Access, after Kumar et al. 2021)

use of IP methods, particularly 2D IP time-domain, has helped in the detection of contaminants and correlation with known concentration levels (Sogade et al. 2006). Complex resistivity measurements, such as crosshole IP imaging, have provided valuable data on hydrocarbon plumes (Kemna et al. 2004), although results can be inconsistent (Fig. 15.10). For instance, low phase values were recorded in some contaminated regions, contradicting expected high-phase responses (Orozco et al. 2021).

In laboratory settings, SIP has been utilized for monitoring biodegradation, microbial growth, and biofilm formation, which influence electrical properties like resistivity and chargeability (Williams et al. 2005; Abdel Aal et al. 2006). These processes, especially during hydrocarbon degradation, alter pore fluid conductivity, ion concentration, and microbial activity, contributing to changes in soil resistivity (Martinho et al. 2010). Some studies highlight how microbial processes such as sulphate reduction and bioremediation affect the electrical response of contaminated environments, enhancing IP signals (Sparrenbom et al. 2017).

Furthermore, SIP and resistivity measurements in different soils (e.g., sand, clay) under varying oil-wet and water-wet conditions have shown complex patterns in conductivity and phase magnitude (Schmutz et al. 2010; Deng et al. 2018). These

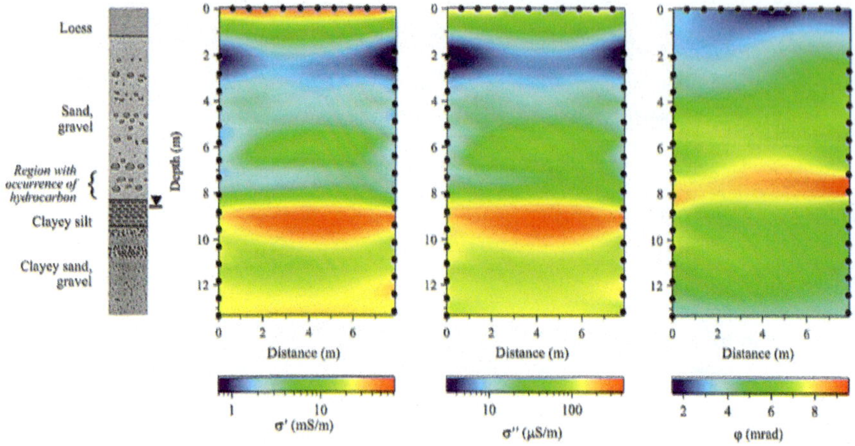

Fig. 15.10 Crosshole imaging results from a hydrocarbon-contaminated site in Strasbourg-Entzheim, France. **a** Real component, **b** Imaginary component, and **c** Phase of complex conductivity (*Source* Martinho 2023—Open Access, after Kemna et al. 2004)

findings, while insightful, suggest that further research is required to fully understand the influence of oil types and microbial processes on electrical signatures.

Field studies often include microbial analysis and chemical characterization of soil and groundwater, particularly when assessing the impact of contaminants (e.g., LNAPLs and DNAPLs) on electrical properties and bioremediation progress (Batayneh 2005; Halihan et al. 2005). The primary parameters measured in these studies are groundwater electrical conductivity, microbial population, and the concentration of major ions and hydrocarbons, which provide essential data for understanding contamination and remediation dynamics.

Overall, both field and laboratory studies underscore the importance of resistivity/IP techniques in monitoring hydrocarbon contamination, mapping plumes, and understanding microbial processes involved in bioremediation (Abdel Aal and Atekwana 2014). However, inconsistencies in results across studies highlight the need for further investigation, especially in relation to soil properties and microbial activities that influence the electrical behavior of contaminated sites.

Contamination from Agricultural Activities

Few studies have specifically applied the resistivity method to investigate contamination originating from agricultural activities. As a result, no table will be presented in this context. Casas et al. (2008) utilized 2D Electrical Resistivity Tomography (ERT) to evaluate the vulnerability of shallow aquifers in the Emporda basin (NE Spain), a region heavily impacted by diffuse nitrate pollution. The study area consists of a vast alluvial plain. The main objective was to assess the thickness and lateral extent of protective layers above the aquifer. Significant variability in electrical resistivity was observed, which was correlated with soil parameters like clay content and hydraulic conductivity. The authors concluded that continuous areas of low resistivity (clay formations) acted as protective barriers against contamination, while areas with high resistivity (coarse sediments) provided rapid infiltration paths for pollutants.

Sainato et al. (2010, 2012) employed 2D and 3D ERT to detect contamination caused by animal waste at cattle farms. Conductive anomalies identified through ERT were linked to increases in soil and groundwater contaminants, such as nitrates, sulphates, bicarbonates, and phosphates in soils, and chlorides in groundwater. The ERT findings were confirmed by analyzing the physicochemical properties of both the soil and groundwater.

Soil Parameters:

1. Electrical conductivity (EC)
2. pH
3. Concentration of major ions: Ca^{2+}, K^{+}, Na^{+}, Mg^{2+}, Cl^{-}, NO_3^{-}, SO_4^{2-}
4. Cation exchange capacity (CEC)
5. Total organic carbon
6. Soil texture

Groundwater Parameters:

1. Electrical conductivity (EC)
2. pH
3. Concentration of major ions: Ca^{2+}, K^{+}, Na^{+}, Mg^{2+}, Cl^{-}, NO_3^{-}, SO_4^{2-}, HCO_3^{-}, Total Dissolved Solids (TDS)

The concentration of ions significantly influences conductivity, and these ions are closely associated with animal manure. Soil texture is another crucial factor, as it determines the proportion of sand, silt, and clay fractions. Texture can act as an indicator of soil permeability, which affects the spread of contamination, given its strong relationship with clay content.

15.2.6 Geophysical Monitoring for Disaster Mitigation

Integrated monitoring systems using geophysical data help mitigate disasters by providing early warnings and long-term monitoring of potential hazards such as landslides and earthquakes. Landslide prediction using integrated seismic, geoelectrical, and remote sensing data allows for real-time monitoring of unstable slopes, alerting authorities and communities to imminent risk (Tao 2023).

15.2.7 Challenges and Opportunities

Data Integration and Interpretation

One of the primary challenges of integrating geophysics with other scientific fields lies in the complexity of data integration. Geophysical data often come in large volumes and varied formats, requiring specialized methods to combine them with data from fields like geology, hydrology, and remote sensing. Artificial intelligence (AI) and machine learning algorithms are expected to play a pivotal role in improving the integration and interpretation of large geophysical datasets, providing insights that were previously difficult to obtain.

Technological and Methodological Advancements

Advances in geophysical sensors, such as higher-resolution seismic arrays and more sensitive electrical resistivity tools, along with improvements in computing power and spatial modeling, have opened up new opportunities for interdisciplinary collaboration. As geophysical instrumentation continues to evolve, real-time, high-resolution data collection will be more feasible, improving disaster prediction, environmental monitoring, and infrastructure design (Miller et al. 2024).

Interdisciplinary Collaboration and Education

Addressing environmental and engineering challenges requires the collaboration of experts across multiple disciplines. Geophysicists, hydrologists, engineers, environmental scientists, and remote sensing specialists must work together to provide solutions. Developing interdisciplinary programs and fostering collaborative research efforts can break down barriers between scientific fields and enable more effective problem-solving. Collaborative education programs will ensure that professionals are well-versed in multiple disciplines, promoting more integrated approaches to real-world challenges.

15.3 The Synergy Between Geophysical Methods and Other Technologies for Enhanced Solutions

15.3.1 Introduction to Synergy Between Geophysics and Other Technologies

In the rapidly evolving field of engineering geophysics, the integration of geophysical methods with other technological advancements has become crucial in addressing complex subsurface challenges. This synergy is particularly important in providing more accurate, efficient, and scalable solutions to problems that span across various domains such as environmental monitoring, resource exploration, urban planning, and natural hazard assessment. The merging of geophysics with technologies such as remote sensing, machine learning, geographic information systems (GIS), and environmental monitoring tools enhances the precision and scope of geophysical investigations, leading to more robust results and informed decision-making processes.

One of the key drivers of this interdisciplinary approach is the need for better spatial and temporal resolution of subsurface and surface data. Geophysics, with its non-invasive techniques, offers a wealth of data on subsurface properties, while other technologies complement these methods by enhancing data acquisition, processing, and interpretation (Smith et al. 2023a, b). This section explores how these synergies have been applied in practice and the transformative impact they have had on the field of engineering geophysics.

15.3.2 Geophysics and Remote Sensing Technologies

Remote sensing technologies, such as satellite imagery, unmanned aerial vehicles (UAVs), and Light Detection and Ranging (LiDAR), are increasingly being used alongside geophysical techniques to enhance the depth and spatial resolution of geophysical surveys. Remote sensing allows for the collection of large-scale, high-resolution data that provides a comprehensive view of the surface and subsurface features, especially in remote or difficult-to-access areas. These technologies offer valuable supplemental data, which, when integrated with geophysical methods, leads to more comprehensive models and interpretations.

For instance, the integration of ground penetrating radar (GPR) with UAV-based LiDAR data allows for detailed subsurface mapping, which is invaluable in fields such as archaeology, civil engineering, and environmental monitoring. In environmental applications, combining electromagnetic methods with satellite remote sensing enables the monitoring of groundwater contamination and vegetation health over large areas (Baker et al. 2022). This integrated approach helps to bridge the gap between surface-level observations and subsurface conditions, offering more accurate insights into complex geological settings.

15.3.3 Geophysics and Geospatial Technologies

Geospatial technologies, including Geographic Information Systems (GIS) and Global Positioning Systems (GPS), have revolutionized the way geophysical data is mapped, analyzed, and interpreted. Geophysical surveys typically produce large amounts of spatially distributed data that can be difficult to visualize or interpret without appropriate tools. GIS provides a platform for integrating and analyzing geophysical data alongside other spatial datasets, such as topography, soil composition, and land use. This integration enhances the ability to interpret geophysical results in a spatial context, allowing for the creation of detailed, georeferenced models.

The synergy between geophysics and GIS is particularly evident in urban planning and resource management. For example, in seismic hazard assessments, geophysical methods such as seismic reflection and refraction can be integrated with GIS to model subsurface conditions and assess earthquake risk in urban areas. Furthermore, GPS data can be combined with geophysical measurements to improve the accuracy of surface displacement measurements, which is vital for monitoring land subsidence and tectonic activity (Feng and Liu 2021). The ability to merge geophysical data with geospatial technologies allows for better decision-making in infrastructure development, environmental management, and disaster response.

15.3.4 Geophysics and Machine Learning/Artificial Intelligence

The application of machine learning (ML) and artificial intelligence (AI) in geophysics is an emerging area of research that holds significant promise for improving data interpretation and predictive modeling. Machine learning algorithms can process large and complex geophysical datasets, identify patterns, and automate the interpretation process. These technologies can be particularly useful in cases where traditional methods struggle to provide clear answers, such as in the interpretation of noisy or sparse data.

In mineral exploration, for example, ML algorithms can be trained to recognize geological features from geophysical data, enhancing the accuracy and speed of mineral resource assessment. AI techniques, such as neural networks, can be applied to predict subsurface conditions based on surface geophysical measurements, improving the effectiveness of geophysical surveys in areas like oil and gas exploration or groundwater detection. Additionally, AI-driven tools can automate data analysis, reducing human error and allowing for the faster processing of large datasets (Jones and Patel 2024). By leveraging machine learning, geophysics becomes more efficient, reliable, and capable of tackling more complex subsurface challenges.

15.3.5 Geophysics and Environmental Monitoring Technologies

The integration of geophysics with environmental monitoring technologies is a powerful approach to address pressing environmental challenges. Technologies such as sensor networks, drones, and autonomous vehicles provide real-time, continuous data that can be integrated with geophysical surveys for dynamic environmental monitoring. This is particularly valuable in areas such as groundwater contamination, soil health, and geohazard monitoring.

For instance, in environmental monitoring, the combination of electromagnetic geophysical methods with sensor networks allows for the detection of contaminants in groundwater and soil at various depths. Real-time geophysical data, collected using mobile sensor platforms or drones, can be transmitted to central systems for immediate analysis and decision-making. Such capabilities are increasingly being employed in natural disaster management, where geophysical data integrated with real-time environmental monitoring helps provide early warning systems for hazards such as earthquakes, floods, or landslides (Nakamura et al. 2023). This integration enhances the efficiency of monitoring systems and ensures a more proactive response to environmental risks.

15.3.6 Geophysics and Biotechnology

An exciting and emerging area of interdisciplinary collaboration is between geophysics and biotechnology. Geophysical methods are increasingly being applied to monitor and support biotechnological processes, particularly in the areas of soil bioremediation and environmental restoration. For example, geophysical techniques can be used to track microbial activity in soils, monitor the movement of bioremediation agents, and assess the effectiveness of environmental clean-up efforts.

Geophysical tools, such as electrical resistivity tomography (ERT) and ground-penetrating radar (GPR), are particularly useful in detecting changes in the subsurface environment that occur during bioremediation processes. These techniques can provide valuable data on the progress of soil and groundwater remediation projects, allowing for more targeted interventions and more efficient use of resources (Wu and Chen 2023). The synergy between geophysics and biotechnology is creating new opportunities for the monitoring and enhancement of ecological restoration efforts, particularly in areas affected by industrial pollution or contamination.

15.3.7 Geophysics and Civil Engineering Technologies

Geophysics plays a critical role in civil engineering, particularly in site investigation, foundation design, and infrastructure monitoring. The integration of geophysical data with civil engineering technologies, such as building information modeling (BIM), helps improve the safety, durability, and cost-effectiveness of construction projects. Geophysical methods can be used to assess the subsurface conditions of a site, allowing engineers to design more appropriate foundations and avoid costly and time-consuming construction issues later on.

In large infrastructure projects such as tunnels, dams, and bridges, geophysical surveys can provide real-time data on subsurface conditions, which can then be integrated with other civil engineering technologies to monitor the integrity of the structure throughout its lifecycle. For instance, seismic and geoelectrical data can be integrated with structural health monitoring systems to detect shifts in the subsurface that may affect the stability of the infrastructure (Jin and Zhang 2022). By integrating geophysical methods into civil engineering projects, professionals can improve risk assessment, optimize design, and ensure the longevity of structures.

15.3.8 The Role of Data Fusion and Multidisciplinary Collaboration

Data fusion plays a critical role in geophysics by integrating multiple, often disparate, data sources into a single, coherent model to aid decision-making. This process combines geophysical data with non-geophysical data such as geological, hydrological, and remote sensing information to enhance the understanding of subsurface and surface conditions. For example, seismic survey data can be integrated with ground-penetrating radar (GPR) results to produce a more comprehensive image of subsurface features, crucial for applications such as infrastructure development and environmental monitoring. Advanced algorithms and machine learning techniques are increasingly being used to facilitate the fusion of large and complex datasets, which improves model accuracy and efficiency. Real-time data fusion, especially when integrating satellite remote sensing and continuous geophysical monitoring, enhances dynamic monitoring capabilities, allowing for more accurate predictions and assessments of environmental and geological changes (Yang and Zheng 2024).

Collaboration among geophysicists and professionals from other fields such as engineering, environmental science, and geology is essential for producing enhanced geophysical solutions. Geophysicists bring specialized knowledge of subsurface conditions, while engineers and environmental scientists contribute their expertise in surface processes, infrastructure, and ecological considerations. This multidisciplinary approach leads to more precise hazard assessments, better resource management, and more effective environmental monitoring. For instance, collaboration between hydrologists and geophysicists significantly improves groundwater flow modeling, which is vital for water resource management and contamination studies. By working together, these diverse experts ensure that geophysical models are not only technically robust but also practical and relevant to real-world applications (Feng and Liu 2021).

Despite the clear advantages, data fusion also presents several challenges, including issues with data compatibility, quality, and the need for standardization. Geophysical and non-geophysical datasets often come from different sources and may have varying formats, units, or spatial resolutions, making integration difficult. For instance, remote sensing data might have a different resolution compared to geophysical survey data, complicating the fusion process. Additionally, data quality issues such as noise or missing data can affect the accuracy of the final model, and this must be addressed before fusion. Furthermore, there is a pressing need for standardization in data collection methods and processing protocols. Without common standards, it becomes challenging to ensure that data from different sources can be seamlessly integrated, particularly in large-scale projects involving multiple teams and disciplines. Establishing standard protocols for data formats, quality control, and interpretation is essential to ensure the success of data fusion in interdisciplinary collaborations (Jones and Patel 2024).

15.3.9 Future Directions in Geophysics and Technology Synergy

The future of geophysics is increasingly intertwined with emerging technologies such as quantum computing, 5G, and blockchain, all of which hold the potential to revolutionize geophysical methods and their applications. Quantum computing, for instance, could significantly enhance computational capabilities, enabling the processing of vast datasets much faster than current systems, which could improve real-time monitoring and more accurate modeling of subsurface conditions (Zhang et al. 2023). 5G technology, with its high-speed data transfer and low latency, could enable seamless integration of geophysical data collected from remote sensors and field-based instruments, providing more precise and timely information for decision-making (Wang and Li 2022). Additionally, blockchain could play a crucial role in ensuring data security, transparency, and traceability in large-scale geophysical projects, particularly in collaborative international initiatives (Sharma and Khan 2023). The integration of these advanced technologies with traditional geophysical methods is poised to create innovative solutions for complex engineering challenges, such as infrastructure development, resource exploration, and environmental conservation. Looking ahead, the next generation of interdisciplinary collaboration will be key in addressing global challenges like climate change, disaster management, and resource scarcity. By combining the strengths of geophysics with cutting-edge technologies, experts from diverse fields can work together to develop solutions that are not only scientifically sound but also practically applicable to pressing global issues, driving sustainability and resilience in the face of future uncertainties.

15.4 Conclusions

The synergy between geophysics and other technologies represents a powerful tool for tackling complex engineering and environmental challenges (Wang and Li 2024). As the integration of geophysics with remote sensing, machine learning, GIS, biotechnology, and civil engineering technologies continues to evolve, it opens up new possibilities for more efficient, accurate, and sustainable solutions. These interdisciplinary collaborations will undoubtedly play a central role in the future of engineering geophysics, allowing professionals to address the increasingly complex problems of the twenty-first century (Huang and Liu 2025).

The future promises even more innovations, as emerging technologies such as quantum computing, blockchain, and 5G are likely to further revolutionize the way geophysical data is collected, processed, and applied. Continued collaboration across disciplines will be key to unlocking the full potential of geophysical methods and ensuring their continued relevance in addressing global challenges such as climate change, urbanization, and resource management.

References

Abdel Aal GZ, Atekwana EA (2014) Spectral induced polarization (SIP) response of biodegraded oil in porous media. Geophys J Int 196:804–817

Abdel Aal GZ, Slater LD, Atekwana EA (2006) Induced-polarization measurements on unconsolidated sediments from a site of active hydrocarbon biodegradation. Geophysics 71:H13–H24

Abdullahi NK, Osazuwa IB, Onugba A (2010) Detecting municipal solid waste leachate plumes through electrical resistivity survey and physio-chemical analysis of ground water samples. J Am Sci 6:540–548

Adeyemo IA, Omosuy GO, Adelusi AO (2017) Geoelectric soundings for delineation of saline water intrusion into aquifers in part of Eastern Dahomey Basin, Nigeria. J Geosci Environ Protec Tion 5:213–232

Ammar AI, Gomaa M, Kamal KA (2021) Applying of SP, DC-Resistivity, DC-TDIP and TDEM soundings in high saline coastal aquifer. Heliyon 7:07617

Arts R, Chadwick A, Gibson M (2004) The Sleipner CO_2 storage project: monitoring and verification. Geol Soc, Lond, Spec Publ 233:49–63

Atekwana EA, Atekwana EA, Rowe RS, Werkema DD Jr, Legall FD (2004) The relationship of total dissolved solids measurements to bulk electrical conductivity in an aquifer contaminated with hydrocarbon. J Appl Geophys 56:281–294

Atekwana EA, Sauck WA, Werkema DD Jr (2000) Investigations of geoelectrical signatures at a hydrocar bon contaminated site. J Appl Geophys 44:167–180

Attwa M, Gunther T, Grinat M, Binot F (2011) Evaluation of DC, FDEM and IP resistivity methods for imaging perched saltwater and a shallow channel within coastal tidal flat sediments. J Appl Geophys 75:656–670

Auken E, Christiansen AV, Sørensen KI (2014) Geophysical methods for environmental and engineering applications. Springer
Baker L et al (2022) Remote sensing and ground penetrating radar integration for archaeological surveys. Remote Sens 14(6):1502
Batayneh AT (2005) 2D electrical imaging of an LNAPL contamination, Al Amiriyya fuel station. Jordan J Appl Sci 5:52–59
Berglund EA, Telford WM, Yule D (2012) Groundwater contamination: Geophysical techniques for investigating subsurface fluid movement. Geophys J Int 181(1):325–332
Bernstone C, Dahlin T, Ohlsson T, Hogland W (2000) DC-resistivity mapping of internal landfill structures: two pre-excavation surveys. Environ Geol 39:360–371
Bhatnagar V, Mehta R (2018) Engineering solutions for large Dams in Seismic zones: a case study of the Rana Pratap Sagar project. Indian J Civil Eng 44(3):128–139
Binley A, Glaser B, Dafflon B (2015) Geophysical methods for groundwater and contamination monitoring. CRC Press
Bording TS, Fiandaca G, Maurya PK, Auken E, Christiansen AV, Tuxen N, Klint KES, Larsen TH (2019) Cross-borehole tomography with full-decay spectral time-domain induced polarization for mapping of potential contaminant flow-paths. J Contam Hydrol 226:103523
Boudreault JP, Dube JS, Chouteau M, Winiarski T, Hardy E (2010) Geophysical characterization of contaminated urban fills. Eng Geol 116:196–206
Buselli G, Lu K (2001) Groundwater contamination monitoring with multichannel electrical and electromagnetic methods. J Appl Geophys 48:11–23
Casado C, Mahjoub M, Lovera P, Fernandez T, Casas A (2015a) 3D behaviour of contamination in landfill sites using 2D resistivity/IP imaging: case studies in Portugal. Environ Earth Sci 74(3):2271–2283
Casado I, Mahjoub H, Lovera R, Fernández J, Casas A (2015b) Use of electrical tomography methods to determinate the extension and main migration routes of uncontrolled landfill leachates in fractured areas. Sci Total Environ 506–507:546–553
Casas A, Himi M, Diaz Y, Pinto V, Font X, Tapias JC (2008) Assessing aquifer vulnerability to pollutants by electrical resistivity tomography (ERT) at a nitrate vulnerable zone in NE Spain. Environ Geol 54:515–520
Cassiani G, Bruno V, Villa A, Fusi N, Binley AM (2006) A saline trace test monitored via time-lapse sur face electrical resistivity tomography. J Appl Geophys 59:244–259
Chambers JC, Kuras O, Meldrum PI, Ogilvy RD, Hollands J (2006) Case history: electrical resistivity tomography applied to geologic, hydrogeologic, and engineering investigations at a former waste-disposal site. Geophysics 71:B231–B239
Chand R, Verma S (2017) Environmental management in Tunnel construction: The Pir Panjal case study. J Environ Eng 52(6):560–573
Davis F et al (2024) Geophysical site investigations for infrastructure in seismic zones. Eng Geol
De Carlo L, Perri MT, Caputo MC, Deiana R, Cassiani G (2013) Characterization of a dismissed landfill via electrical resistivity tomography and mise-a-la masse method. J Appl Geophys 98:1–10
Deng Y, Shi X, Revil A, Wu J, Ghorbani A (2018) Complex conductivity of oil-contaminated clayey soils. J Hydrol 561:930–942
Dumont G, Pilawski T, Dzaomuho-Lenieregue P, Hiligsmann S, Delvigne F, Thonart P, Robert T, Nguyen F, Hermans T (2016) Gravimetric water distribution assessment from geoelectrical methods (ERT and EMI) in municipal solid waste landfill. Waste Manage 55:129–140
Edet AE, Okereke CS (2001) A regional study of salt water intrusion in southeastern Nigeria based on the analysis of geoelectrical and hydrochemical data. Environ Geol 40:1278–1289
Feng X, Liu Z (2021) GIS integration with geophysical data for urban earthquake risk assessment. Geospatial Eng 36(4):201–217
Fernandez G et al (2023) Integrated monitoring systems for coastal erosion. Environ Monit Assess
Folch A, del Val L, Luquot L, Martínez-Perez L, Bell munt F, Le Lay H, Rodellas V, Ferrer N, Palacios A, Fernandez S, Marazuela MA, Diego-Feliu M, Pool M, Goyetche T, Ledo J., Pezard

P, Bour O, Queralt P, Marcuello A et al (2020) Combining fiber optic DTS, cross-hole ERT and time-lapse induction logging to characterize and monitor a coastal aquifer. J Hydrol 588:125050.
Foster R, Smith M (2016) Remote sensing for geophysical applications. Wiley
Freeze RA, Cherry JA (1979) Groundwater. Prentice-Hall
Frohlich RK, Barosh PJ, Boving T (2008) Investigating changes of electrical characteristics of the saturated zone affected by hazardous organic waste. J Appl Geophys 64:25–36
Gazoty A, Fiandaca G, Pedersen J, Auken E, Christiansen AV (2012) Mapping of landfills using time domain spectral induced polarization data: the Eskelund case study. Near Surf Geophys 10:575–586
Goebel M, Pidlisecky A, Knight R (2017) Resistivity imaging reveals complex pattern of saltwater intrusion along Monterey coast. J Hydrol 551:746755
Gomez P et al (2022) Hydrogeophysical approaches for sustainable water resource management. J Hydrol
Greggio N, Giambastiani BMS, Balugani E, Amaini C, Antonellini M (2018) High-resolution electrical resistivity tomography (ERT) to characterize the spatial extension of freshwater lenses in a salinized coastal aquifer. Water 10:1067
Gupta P (2014) Seismic activity and tunneling techniques in the Himalayas: challenges and solutions. Geotech Eng Rev 48(5):25–36
Halihan T, Paxton S, Graham I, Fenstemaker T, Riley M (2005) Post-remediation evaluation of a LNAPL site using electrical resistivity imaging. J Environ Mental Monit 7:283–287
Harper L et al (2024) Combining LiDAR and seismic imaging for environmental monitoring. Remote Sens Environ
He Q, Zhang X, Wu M (2021) Autonomous robotic systems for geophysical surveys. J Geophys Res 126(7):2345–2356
Huang C, Liu J (2025) The future of geophysical technologies: a vision for the next decade. Geophys Rev 39(1):1–15
Hubbard S, Oldham C, Knight R (2016) Geophysics for groundwater remediation: integrating hydrology and geophysics for groundwater contamination. Environ Sci Technol 50(12):6163–6173
Jain S, Sharma A (2015) Grouting techniques for dam foundations: a study of the Rana Pratap Sagar project. Int J Geotech Eng 13(7):217–225
Jacome M, Costanzo-Alvarez V, Aldana M, Patraskovic P, Drielsma C, Galatro D, Amon C (2021) A methodology to characterize a sanitary landfill combining, through a numerical approach, a geoelectrical survey with methane point-source concentrations. Environ Technol Innov 21:101225
Jin L, Zhang H (2022) Application of geophysical methods in site investigation for large civil engineering projects. J Civil Eng Geophys 12(1):31–44
Jones D, Patel A (2024) Machine learning-enhanced geophysical data interpretation for mineral exploration. J Appl Geophys 56(3):87–101
Jones B et al (2024) Seismic data integration in soil liquefaction prediction. J Seismolog Earthq Eng
Jones DR, Telford WM, Haskins R (2013) Geophysics for resource exploration and environmental monitoring. Cambridge University Press
Kemna A, Binley A, Slater L (2004) Crosshole IP imaging for engineering and environmental applications. Geophysics 69:97–107
Kumar N, Singh A, Rao YS (2016) Integration of geophysical, geological and hydrological data for subsurface characterization in complex terrains. Environ Earth Sci 75:1284. https://doi.org/10.1007/s12665-016-6083-9
Kumar P, Tiwari P, Singh A, Biswas A, Acharya T (2021) Electrical resistivity and induced polarization signatures to delineate the near-surface aquifers contaminated with seawater invasion in Digha, West-Bengal. India. CATENA 207:105596
Kumar V, Sharma S (2015) Geotechnical considerations in Tunnel Design: the Pir Panjal railway tunnel. J Rock Mech Geotech Eng 7(3):145–154

Kusumastuti D, Suwignyo S, Nurfiani D (2018) Flood risk assessment using remote sensing and geophysical methods in Jakarta, Indonesia. Nat Hazards 90:243–262

Lanzano G, Pizzi C, Sinopoli C (2016) Landslide hazard mapping using interdisciplinary geophysical, geological, and hydrological data in the Italian Alps. Eng Geol 204:121–132

Leroy P, Auken E (2017) Geophysical methods for hydrological studies. Elsevier

Levy R et al (2023) Magnetotellurics for groundwater mapping in arid regions. Hydrol Geophys

Li X, Zhang C, Wang Y (2022) Big data integration and geophysical data fusion: recent advancements. Geophys J Int 218(3):1124–1135

Ling C, Zhou Q, Xue Y, Zhang Y, Li R, Liu J (2013) Application of electrical resistivity tomography to evaluate the variation in moisture content of waste during 2 months of degradation. Environ Earth Sci 68:57–67

Martinho E (2023) Electrical resistivity and induced polarization methods for environmental investigations: an overview. Water Air Soil Pollut 234:215

Martinho E, Abreu MM, Pampulha ME, Alegria F, Oliveira A, Almeida F (2010) An experimental study of the diesel biodegradation effects on soil bio geophysical parameters. Water Air Soil Pollut 206:139–154

Martinho E, Almeida F (2006) 3D behavior of contamination in landfill sites using 2D resistivity/IP imaging: case studies in Portugal. Environ Geol 49:1071–1078

Martinho E, Almeida F, Senos Matias MJ (2004) Time domain induced polarization in determination of the salt/freshwater interface (Aveiro–Portugal). Groundwater Saline Intrusion, 18th salt water intrusion meeting, Cart Agena, 2004, Spain, pp 385–393

Maurya PK, Ronde VK, Fiandaca G, Balbarini N, Auken E, Bjerg PL, Christiansen AV (2017) Detailed landfill leachate plume mapping using 2D and 3D electrical resistivity tomography—with correlation to ionic strength measured in screens. J Appl Geophys 138:1–8

Melesse AM, Shrestha R (2013) Remote sensing and geophysical applications for monitoring environmental changes. Springer

Miller C et al (2024) AI-based analysis for geophysical data integration. Comput Geosci

Mohamed A, Neff B, Davies D (2021) Geophysical monitoring of climate change impacts on the Earth's surface. Earth-Sci Rev 215:103657

Nakamura H et al (2023) Real-time geophysical monitoring systems for environmental hazard Management. Environ Geophys 42(2):110–123

Nguyen M et al (2022) Electrical resistivity tomography for permafrost degradation studies. J Geophys Res: Earth Surf

Oghenero OA, Felix CI, Emmanuel EO (2017) Salt water intrusion appraisal of shallow aquifer in Burutu area of the Western Niger Delta with 2D electrical resistivity tomography. J Appl Sci Environ Manage 21:372–377

Oldenburg DW, Li Y (2017) Geophysical inversion and data analysis techniques. Oxford University Press

Orozco AF, Ciampi P, Katona T, Censini M, Papini MP, Deidda GP, Cassiani G (2021) Delineation of hydrocarbon contaminants with multi-frequency com plex conductivity imaging. Sci Total Environ 768:144997

Orozco AF, Gallistl J, Steiner M, Brandstatter C, Fellner J (2020) Mapping biogeochemically active zones in landfills with induced polarization imaging: the Heferlbach landfill. Waste Manage 107:121–132

Park S, Yi MJ, Kim JH, Shin SW (2016) Electrical resistivity imaging (ERI) monitoring for groundwater contamination in an uncontrolled landfill, South Korea. J Appl Geophys 135:1–7

Patel V (2022) Advances in multidisciplinary geophysical imaging techniques for subsurface investigation. Earth Science Informatics 15:789–804. https://doi.org/10.1007/s12145-022-007 45-2

Patel V, Sharma S, Singh R (2022) Integrated geophysical and remote sensing approaches for subsurface characterization and environmental applications. J Appl Geophys 200:104623. https://doi.org/10.1016/j.jappgeo.2022.104623

Perri MT, Cassiani G, Gervasio I, Deiana R, Binley A (2012) A saline tracer test monitored via both surface and cross-borehole electrical resistivity tomography: comparison of time-lapse results. J Appl Geophys 79:6–16
Peterson M et al (2022) Time-lapse geophysical surveys for remediation monitoring. J Environ Eng
Pettigrew J, McGrath R, Fraser D (2009) Geophysical exploration and remote sensing: a methodological overview. J Environ Eng 31(2):75–81
Rathi S, Shah M (2017) Flood control and management for Dams in Semi-Arid regions. J Water Resour Eng 39(2):89–98
Reeves P (2014) Geophysics and remote sensing for environmental monitoring. Springer
Sainato CM, Losinno BN, Malleville HJ (2010) Electrical resistivity tomography applied to detect contamination on a dairy farm in the Pampean region, Argentina. Near Surf Geophys 8:163–172
Sainato CM, Losinno BN, Malleville HJ (2012) Assessment of contamination by intensive cattle activity through electrical resistivity tomography. J Appl Geophys 76:82–91
Sallenger A, Dorman L, Howd P (2012) Coastal erosion monitoring using remote sensing and geophysical techniques. J Coastal Res 28(5):1224–1234
Sauck WA (2000) A model for the resistivity structure of LNAPL plumes and their environs in sandy sediments. J Appl Geophys 44:151–165
Schlumberger Limited (2010) Oilfield review: Integrated geoscience and engineering approaches for subsurface characterization. Schlumberger. https://www.slb.com
Schmutz M, Revil A, Vaudelet P, Batzle M, Femenía Vinao P, Werkema DD (2010) Influence of oil saturation upon spectral induced polarization of oil-bearing sands. Geophys J Int 183:211–224
Sharma R, Khan A (2023) Blockchain for geophysical data management and security. Int J Geophys 34(2):123–135
Sharma P, Joshi A (2019) Hydroelectric power and irrigation: the role of the Rana Pratap Sagar dam. J Renew Energy 42(5):263–275
Sharma R, Joshi R, Singh M (2016) Geotechnical monitoring during tunnel excavation: a study of the Pir Panjal railway tunnel. J Eng Geol 59(4):322–331
Simmons M, Nelson A, Hughes M (2020) Geophysics in carbon capture and storage monitoring. Geophys Prospect 68(2):303–319
Singh R, Patel S (2017) Hydroelectric power generation in Rajasthan: a case study of the Rana Pratap Sagar Dam. Energy Power J 12(4):191–202
Singh D, Sood A (2013) Application of the New Austrian tunnelling method in challenging geotechnical conditions: the Pir Panjal tunnel. Tunnel Eng J 18(2):71–79
Singha K, Gorelick SM (2005) Saline tracer visualized with three-dimensional electrical resistivity tomography: field-scale spatial moment analysis. Water Resour Res 41:W05023
Smith R et al (2023a) Integration of geophysics with machine learning in environmental studies. J Eng Geophys 28(2):145–158
Smith A et al (2023b) Use of ground penetrating radar for landfill monitoring. Environ Geophys J
Sogade JA, Scira-Scappuzzo F, Vichabian Y, Shi W, Rodi W, Lesmes DP, Morgan FD (2006) Induced-polarization detection and mapping of contaminant plumes. Geophysics 71:B75–B84
Sparrenbom CJ, Akesson S, Johansson S, Hagerberg D, Dahlin T (2017) Investigation of chlorinated solvent pollution with resistivity and induced polarization. Sci Total Environ 575:767–778
Tan Y et al (2023) Remote sensing and seismic data for mudslide risk analysis. Nat Hazards Earth Syst Sci
Tao J et al (2023) Integrated geophysical approaches for landslide early warning. Nat Hazards
Telford WM, Geldart LP, Sheriff RE (1990) Applied geophysics. Cambridge University Press
Thakur P, Joshi S (2018) The impact of the Pir Panjal tunnel on economic and social development in Jammu and Kashmir. J Infrastruct Dev 26(3):212–225
Vaudelet P, Schmutz M, Pessel M, Franceschi M, Guerin R, Atteia O, Blondel A, Ngomseu C, Galaup S, Rejiba F, Begassat P (2011) Mapping of contaminant plumes with geoelectrical methods. a case study in urban context. J Appl Geophys 75:738–751
Wang P, Li S (2024) Synergistic approaches in modern geophysics: a path to innovation and solutions. Geophys J 12(3):200–215

Wang H, Li J (2022) The role of 5G in geophysical monitoring: opportunities and challenges. J Remote Sens 15(4):99–112

Weng Q, Hu C (2008) Remote sensing applications for environmental monitoring. CRC Press

Williams J et al (2023) Seismic monitoring for volcanic hazard prediction. Geophys J Int

Williams KH, Ntarlagiannis D, Slater LD, Dohnalkova A, Hubbard SS, Banfield JF (2005) Geophysical imaging of stimulated microbial biomineralization. Environ Sci Technol 39:7592–7600

Wilson SR, Ingham M, McConchie JA (2006) The applicability of earth resistivity methods for saline inter face definition. J Hydrol 316:3012–3312

Wu X, Chen Y (2023) Geophysical applications in biotechnology: monitoring microbial remediation of contaminated soil. Environ Biotechnol 20(4):48–64

Yang Z, Zheng M (2024) Data fusion approaches in engineering geophysics: a review of multidisciplinary integration. J Geophys Eng 18(2):97–112

Yuan S, Zhang Z, Wu L (2022) Geophysical and ecological monitoring for biodiversity conservation. Ecol Ind 125:107417

Zhang X, Liu Y, Zhao J (2023) Quantum computing applications in geophysical data processing. Geophys J Int 202(1):45–59

Zhu Z, Liu Y, Wang Y (2021) Machine learning for geophysical data interpretation. J Appl Geophys 179:103676

Chapter 16
The Next Frontier: Emerging Trends and Technologies

Abstract This chapter explores the transformative technologies redefining the future of geophysics, from quantum sensing and robotics to artificial intelligence and digital infrastructure. It begins with an introduction to quantum sensors and their superior sensitivity in geophysical exploration, followed by advancements in autonomous platforms such as UAVs, ROVs, and swarm robotics for surveying remote and hazardous environments. Artificial intelligence and machine learning are examined as critical tools for data interpretation, anomaly detection, and predictive modeling. The chapter also emphasizes geophysics' role in climate adaptation, smart city development, and next-generation subsurface exploration for geothermal energy, carbon capture, and hydrogen storage. Key innovations such as real-time sensor fusion, digital twins, and integration with BIM and GIS are detailed. Ethical challenges, including data privacy and equitable access, are critically reviewed. The chapter concludes by identifying barriers and opportunities in translating emerging technologies into scalable, field-ready solutions, while highlighting the need for interdisciplinary collaboration, skilled talent, and supportive policy frameworks.

16.1 Introduction

Engineering geophysics stands at the cusp of a transformative era characterized by rapid technological advancements and shifting global demands. The field is experiencing paradigm shifts driven by the convergence of innovative sensing technologies, sustainable development imperatives, and the increasing need for precise subsurface characterization. These changes are redefining how geophysical investigations are conducted, interpreted, and applied across diverse sectors including infrastructure development, environmental monitoring, disaster risk management, and resource exploration.

Central to this evolution is the interplay between innovation, sustainability, and precision. Cutting-edge technologies such as quantum sensing, autonomous robotic platforms, and artificial intelligence are enhancing the accuracy and efficiency of geophysical surveys. At the same time, the growing emphasis on sustainability

D. C. Naskar, *Advances in Engineering Geophysics: Concepts and Applications*, Earth Systems Data and Models 9, https://doi.org/10.1007/978-3-032-28000-8_16

compels the geophysics community to develop non-invasive, cost-effective methods that minimize environmental impact while maximizing data quality. The integration of digital tools, including Internet of Things (IoT) frameworks and digital twins, further empowers stakeholders to make informed, real-time decisions in complex urban and natural environments (Bongs et al. 2023; Tatham 2024; FT 2024; Udhyog Mantra 2024; Tobias and Zhang 2023).

Emerging societal and environmental demands—from climate adaptation and smart city infrastructure to resilient defense installations and sustainable resource management—are shaping the trajectory of engineering geophysics. These drivers are not only expanding the scope of geophysical applications but are also prompting interdisciplinary collaborations that bridge geosciences with computer science, robotics, and environmental engineering. This chapter explores these frontiers, highlighting key technologies and trends that will define the future of engineering geophysics on a global scale (Technmagazine 2024a; Wired UK 2024a).

16.2 Quantum Sensing in Geophysics

Quantum sensing represents one of the most promising frontiers in applied geophysics. By exploiting quantum phenomena such as superposition, entanglement, and matter-wave interferometry, quantum sensors offer sensitivity and precision far beyond conventional instruments. This technology is increasingly being deployed in gravity surveying, infrastructure monitoring, volcanology, and environmental assessments.

16.2.1 Principles of Quantum Sensing

Quantum sensors rely on the behavior of atoms at extremely low temperatures. Atom interferometers, for example, use laser-cooled atoms (such as rubidium or cesium) dropped in a vacuum chamber. Their wave-like behavior under gravity allows precise measurements of the local gravitational field. Unlike classical gravimeters that rely on mechanical components, quantum gravimeters use the interference pattern of matter waves to determine gravitational acceleration. This eliminates drift, reduces noise, and allows long-term monitoring with high repeatability. According to a 2024 paper published in Nature Physics, quantum sensors have demonstrated accuracy improvements up to tenfold over classical instruments in field conditions (Bongs et al. 2023).

16.2.2 Real-World Applications

A. **Volcanology**

One of the landmark deployments of quantum gravimetry was carried out on Mount Etna by the Italian National Institute of Geophysics and Volcanology (INGV). The Absolute Quantum Gravimeter (AQG), developed by iXblue (now Exail), successfully operated under harsh environmental conditions including seismic noise and temperature swings. This deployment, reported by Exail in 2023, marked the first operational use of a quantum gravimeter for active volcano monitoring (Exail 2023).

B. **Urban Infrastructure and Civil Engineering**

In urban environments, quantum gravimeters are increasingly being used for detecting voids, sinkholes, and underground infrastructure. Their non-invasive nature and mobility make them ideal for pre-construction surveys and risk mitigation. A 2024 report by M2 Lasers highlighted successful trials in the UK using quantum gravimetry to detect buried tunnels beneath roadways and rail lines with sub-meter accuracy (M2 Lasers 2024).

C. **Environmental and Hydrological Studies**

Quantum sensors are also being explored for monitoring groundwater recharge and subsurface fluid migration. Their sensitivity allows detection of minor mass redistributions, providing insights into water table dynamics and drought impacts. A recent study published on arXiv emphasizes the use of portable quantum sensors for real-time groundwater monitoring, especially in regions vulnerable to climate stress.

16.2.3 Advantages Over Classical Instruments

Quantum sensors offer a range of compelling advantages over traditional gravimetric instruments, positioning them as the next-generation tools in geophysical investigations. One of the most notable benefits is their exceptional sensitivity. Quantum gravimeters, which rely on matter-wave interferometry, are capable of detecting changes in gravitational acceleration on the order of 10^{-9} g, far surpassing the resolution of classical spring-based devices. This level of precision enables the identification of subtle mass redistributions in the subsurface, such as small voids, aquifer movements, or geological faults, which might otherwise go undetected using conventional methods.

Another major advantage lies in their long-term stability. Traditional mechanical gravimeters often suffer from drift due to the degradation of components or environmental influences such as temperature and pressure changes. In contrast, quantum gravimeters use atomic references that are inherently stable over time, allowing for more consistent measurements without frequent recalibration. This stability makes them especially well-suited for long-term monitoring projects, such as volcano surveillance or groundwater level tracking.

Portability has also been a focus of recent technological development. Many modern quantum gravimeters are now engineered for field deployment, with ruggedized designs that withstand variable environmental conditions. This mobility allows for rapid surveys in both urban and remote locations, expanding the range of geophysical applications.

Importantly, quantum sensing is entirely non-invasive. These instruments can detect underground features without any need for physical penetration or excavation. As a result, they are particularly valuable in sensitive environments—such as heritage sites, urban infrastructure, or ecological reserves—where intrusive methods like drilling or trenching would be undesirable or prohibited. Taken together, these advantages underscore the transformative potential of quantum sensing in engineering geophysics.

16.2.4 Challenges and Future Directions

Despite the clear advantages and growing interest in quantum sensing technologies, several challenges must be addressed before their widespread adoption in engineering geophysics can be realized. One of the most pressing issues is the high cost and complexity of these systems. Quantum gravimeters, for example, require ultra-stable laser systems, vacuum chambers, and vibration isolation platforms, all of which contribute to both financial and operational barriers. In addition, operating and maintaining such instruments often demands specialized training in quantum optics and instrumentation, limiting their accessibility to institutions with advanced technical capabilities (Sorrentino et al. 2024).

Another significant challenge is the sensitivity of quantum sensors to environmental disturbances. Although the technology has matured considerably in terms of field robustness, instruments remain vulnerable to noise sources such as temperature fluctuations, electromagnetic interference, and ground vibrations. These factors can reduce the reliability of measurements, particularly in noisy urban or industrial environments where geophysical surveys are frequently conducted.

Furthermore, the integration of quantum sensors into existing geophysical workflows presents both technical and organizational difficulties. At present, there is a lack of standardized protocols for data acquisition, processing, and interpretation when using quantum systems alongside classical instruments. Bridging this gap will require the development of unified data formats, calibration procedures, and interdisciplinary training programs that align with the needs of both geophysicists and quantum technologists.

Looking ahead, the future of quantum sensing in geophysics will likely hinge on continued improvements in system portability, cost reduction through manufacturing scale-up, and the seamless integration of quantum-derived data with traditional survey methods and artificial intelligence-driven interpretation platforms. As these barriers are progressively overcome, quantum sensors are expected to become a core component of advanced, non-invasive geophysical diagnostics in both research and applied contexts.

16.3 Robotics and Autonomous Geophysical Platforms

The advent of robotics and autonomous platforms has revolutionized geophysical surveying by enabling safer, faster, and more precise data acquisition in environments that are often challenging or hazardous for human operators. Unmanned Aerial Vehicles (UAVs), Remotely Operated Vehicles (ROVs), and autonomous land rovers now play critical roles in expanding the reach and resolution of geophysical investigations (*Source* Maritime Robotics).

16.3.1 Development of UAVs, ROVs, and Autonomous Land Rovers

Recent advances have seen rapid development of UAVs, ROVs, and autonomous rovers specifically designed for geophysical applications. UAVs equipped with magnetometers, radiometric sensors, and LiDAR systems offer efficient aerial mapping and monitoring over large and difficult terrains, such as volcanic regions or dense urban environments. ROVs extend these capabilities underwater, conducting detailed seismic and magnetic surveys in deep and hazardous marine areas. Autonomous land rovers, meanwhile, navigate rough terrestrial terrains, carrying instruments like ground-penetrating radar (GPR) and electromagnetic sensors for subsurface exploration. These platforms reduce human exposure to dangerous environments and improve spatial data density (Aeromag Online 2023).

16.3.2 Integration of Geophysical Instruments with Mobile Platforms

The integration of advanced geophysical instruments with these robotic platforms has enhanced data collection capabilities. Miniaturized and ruggedized sensors are now designed to be compatible with UAVs, ROVs, and rovers, allowing simultaneous collection of multiple geophysical parameters such as magnetic fields, gravity variations, and ground conductivity. Data can be transmitted in real time to control stations, enabling immediate analysis and adaptive survey strategies. This integration has improved the precision and versatility of geophysical investigations, facilitating complex multi-parameter surveys that were previously impractical (*Source* ULC Technologies).

16.3.3 Remote and Hazardous Site Surveying

Robotic and autonomous platforms have unlocked the ability to conduct geophysical surveys in remote and hazardous locations, including active volcanic zones, polar ice sheets, underwater environments, and contaminated sites. These platforms operate autonomously or semi-autonomously under extreme conditions, collecting high-quality data without risking human safety. For example, autonomous underwater vehicles have been deployed beneath Arctic ice to perform seismic and magnetometric surveys, while ground rovers have been utilized for mapping subsurface features in contaminated industrial sites (Aeromag Online 2023).

In hazardous or inaccessible environments—such as conflict zones, disaster sites, or contaminated areas—the deployment of autonomous geophysical platforms offers transformative advantages in safety, efficiency, and data quality. Ground Penetrating Radar (GPR) is increasingly being integrated into unmanned aerial vehicles (UAVs), remotely operated vehicles (ROVs), and autonomous ground rovers for detailed subsurface imaging.

Figure 16.1 illustrates key examples where GPR systems, when combined with autonomous deployment, have enabled successful surveys in challenging forensic and demining contexts (Barone and Di Maggio 2019; Olhoeft and Modroo 2006). These include the detection of shallow graves, subsurface cavities, and landmines—all without exposing personnel to direct risk. The integration of AI-based object recognition and real-time data streaming into such platforms further enhances their operational potential in both civil and military geoscience missions.

16.3.4 Swarm Robotics and Synchronized Sensor Arrays

The concept of swarm robotics is emerging as a transformative approach in geophysical surveys. Multiple autonomous agents equipped with synchronized sensor arrays can operate collaboratively, covering extensive areas more efficiently than single-platform surveys. This approach allows dynamic reconfiguration of the swarm based on real-time data, improving spatial resolution and reducing survey time. Swarm systems can adapt to environmental challenges, optimizing data acquisition strategies in complex terrains and enhancing the robustness of geophysical datasets (*Source* Maritime Robotics).

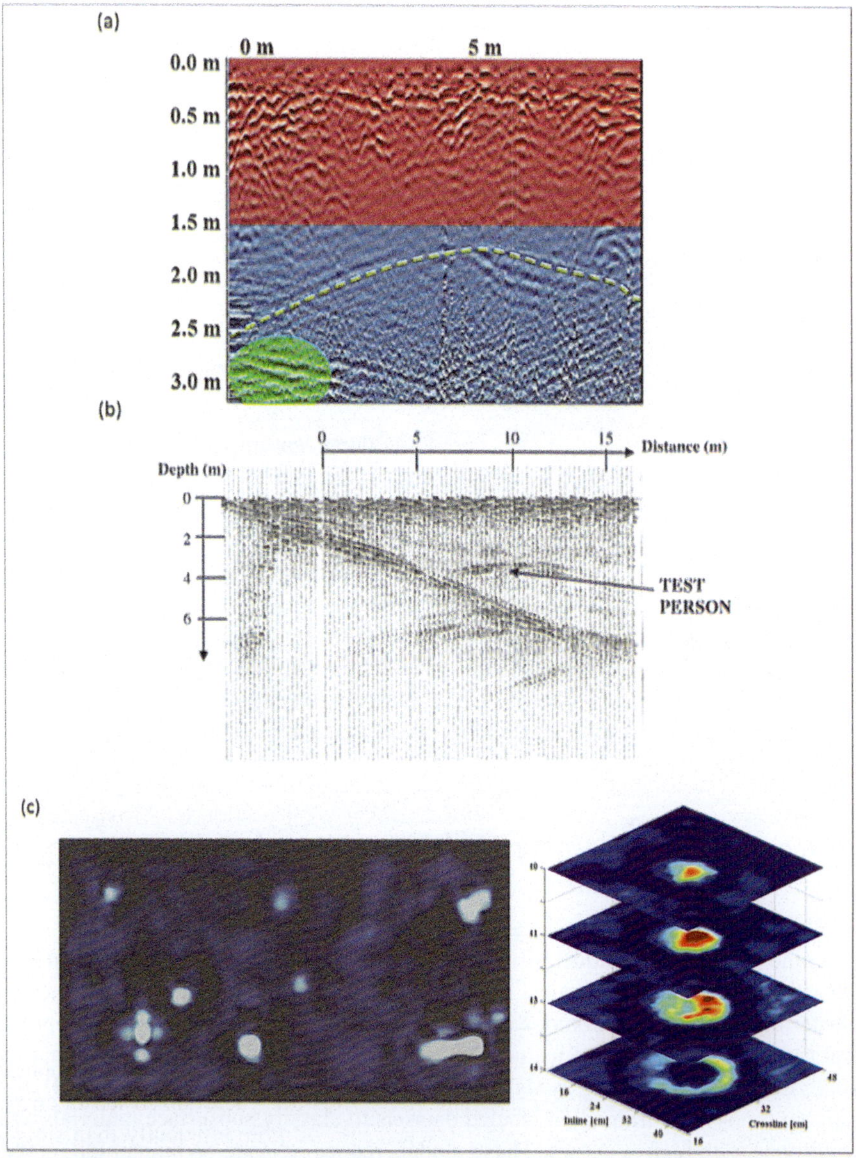

Fig. 16.1 Ground penetrating radar (GPR) applications in forensic and demining scenarios, demonstrating the integration of GPR with autonomous platforms for hazardous site investigations: **a** a 500 MHz radargram revealing a potential shallow grave (green shadow), **b** a 900 MHz radargram detecting a subsurface cavity with a test subject, **c** a 2 GHz 3D GPR dataset mapping inert landmines and mortar shells in a controlled minefield, and **d** high-resolution radar slices showing the internal structure of a simulated antipersonnel landmine. These examples highlight the capability of GPR-equipped unmanned systems to conduct high-risk surveys safely and effectively. *Source* Lombardi et al. (2022)—Open Access

16.4 Artificial Intelligence and Machine Learning

The rapid evolution of Artificial Intelligence (AI) and Machine Learning (ML) technologies has profoundly impacted the field of engineering geophysics. By enabling automated data processing, enhanced pattern recognition, and advanced predictive modeling, these technologies are transforming how geophysical data is interpreted and applied. The following subsections explore the key roles AI and ML play in modern geophysical workflows.

16.4.1 AI for Data Processing and Pattern Recognition

AI algorithms, particularly those based on deep learning, have revolutionized geophysical data processing by automating the extraction of meaningful features from complex datasets. Traditional manual processing methods are often time-consuming and prone to subjective bias. AI models, however, excel at recognizing subtle patterns within seismic, electromagnetic, and other geophysical signals, thereby enhancing signal-to-noise ratio and data quality. This capability is critical for identifying geological features such as faults, stratigraphic boundaries, and fluid contacts. Additionally, AI-driven pattern recognition facilitates real-time monitoring applications, enabling rapid detection of changes in subsurface conditions that could indicate hazards or resource movements (Technmagazine 2024b).

16.4.2 Machine Learning Models for Anomaly Detection and Classification

Machine learning techniques provide powerful tools for anomaly detection and classification within geophysical datasets. Unsupervised learning methods, such as clustering and principal component analysis (PCA), can isolate unusual geophysical signatures indicative of features like mineral deposits, voids, or contamination zones. Supervised learning models, including support vector machines (SVM) and random forests, are trained on labeled datasets to classify subsurface materials and conditions with high accuracy. These models reduce reliance on human interpretation and increase consistency in identifying anomalies, making them invaluable for applications in mineral exploration, environmental monitoring, and infrastructure assessment.

16.4.3 Predictive Modeling and Inversion Optimization

AI and ML have also enhanced the process of predictive modeling and inversion optimization in geophysics. Predictive models use historical and current data to forecast geological phenomena such as seismic events, groundwater flow, or contaminant migration. Inversion techniques, which involve estimating subsurface properties from surface measurements, are computationally intensive and often non-unique. Machine learning accelerates inversion by approximating complex physical models and optimizing parameter searches, resulting in faster and more reliable subsurface imaging. These advancements support more informed decision-making in engineering design, risk assessment, and resource management (Technmagazine 2024c).

16.4.4 Case Studies in Smart Geophysical Interpretation

Several case studies highlight the successful integration of AI and ML in geophysical interpretation. For example, AI-powered seismic interpretation has been used to delineate hydrocarbon reservoirs with improved precision, reducing exploration risk and costs. In urban environments, ML models have been applied to GPR data to map underground utilities and detect anomalies related to structural integrity. Additionally, AI-driven inversion methods have enhanced the characterization of contaminant plumes in groundwater studies. These practical applications demonstrate the transformative potential of AI and ML in delivering faster, more accurate, and cost-effective geophysical solutions.

16.5 Geophysics in Climate Adaptation and Resilience

As climate change intensifies, geophysical methods have become indispensable tools for monitoring, modeling, and mitigating climate-related hazards. Through subsurface characterization and continuous environmental monitoring, geophysics supports adaptive strategies aimed at enhancing resilience in vulnerable regions.

16.5.1 Subsurface Monitoring for Flood, Drought, and Permafrost Change

Geophysical techniques such as electrical resistivity tomography (ERT), seismic surveys, and ground-penetrating radar (GPR) provide critical insights into subsurface hydrological changes linked to climate phenomena like flooding, drought, and permafrost thaw. For instance, ERT has been widely employed to monitor soil moisture dynamics and groundwater levels, helping to predict drought severity and manage water resources. Similarly, seismic and resistivity methods have been applied to assess permafrost degradation in Arctic regions, where thawing permafrost threatens infrastructure stability and releases greenhouse gases. Continuous geophysical monitoring enables early detection of subsurface changes, informing climate adaptation policies and infrastructure planning (Smith et al. 2023).

Among these tools, GPR is increasingly being deployed in environmental monitoring for its ability to produce high-resolution, non-invasive images of the subsurface. It is particularly valuable in delineating aquifers, mapping glacier beds, detecting sinkholes, and identifying geological discontinuities. Its adaptability across various terrains and frequencies makes it well suited for dynamic, climate-impacted environments.

Figure 16.2 presents a range of environmental and climate monitoring applications using GPR, including aquifer structure imaging, glacier profiling, sinkhole detection, and subsurface anomaly identification (Irving et al. 2019; Shokri et al. 2022; Grab et al. 2021; Francke 2012). These examples underscore the importance of integrating GPR into broader monitoring systems that incorporate AI analytics, IoT sensors, and predictive climate models. Such integration can significantly improve decision-making in water management, disaster mitigation, and climate resilience strategies.

16.5.2 Role in Early Warning Systems and Risk Modeling

Geophysical data feeds directly into early warning systems for natural hazards exacerbated by climate change, such as floods, landslides, and subsidence. By integrating seismic, geodetic, and hydrological measurements, advanced models can forecast hazardous events with greater accuracy. Geophysical methods contribute to quantifying risk by characterizing unstable slopes, aquifer depletion, or soil liquefaction potential. This integration allows emergency management authorities to implement timely mitigation actions, reducing human and economic losses. Recent advances in AI-assisted geophysical data analysis further enhance the predictive capability of these systems (Jones and Kumar 2024).

16.5.3 Ground Stability Assessments for Climate-Impacted Regions

The assessment of ground stability is critical in regions facing increased climate variability. Geophysical methods evaluate soil and rock properties that influence slope stability, foundation integrity, and susceptibility to erosion. Techniques such as seismic refraction, microgravity, and electrical resistivity surveys have been applied to monitor changes in terrain and subsurface conditions in areas affected by

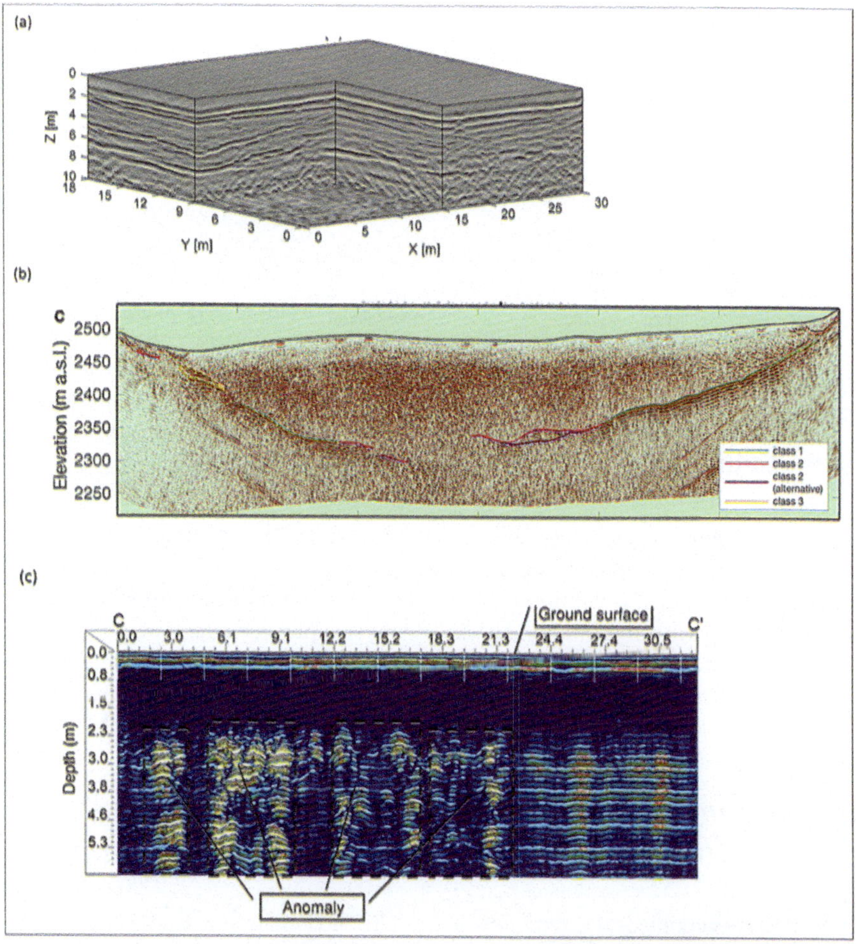

Fig. 16.2 GPR applications in environmental monitoring: **a** 100 MHz 3D data delineating alluvial aquifer geometry, **b** 25 MHz profile revealing glacier bed and reflection classifications, **c** 270 MHz data showing sinkhole anomalies, and **d** 30 MHz radar profile of a kimberlite pipe. These applications support climate resilience planning. *Source* Lombardi et al. (2022)—Open Access

increased rainfall, thawing permafrost, or drought-induced subsidence. These assessments inform the design of resilient infrastructure and land-use planning, helping communities adapt to changing environmental conditions (Lee et al. 2023).

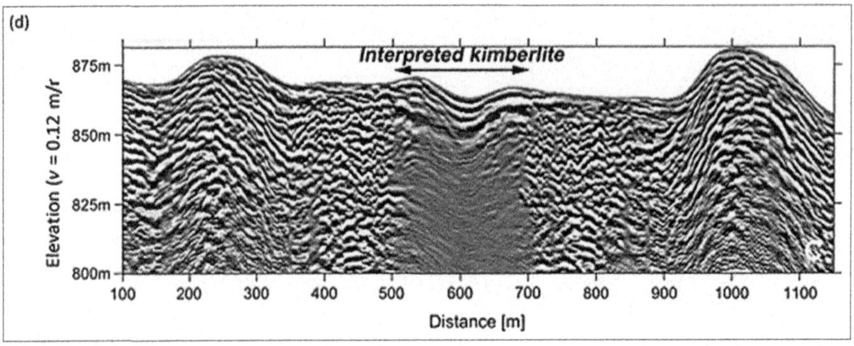

Fig. 16.2 (continued)

16.6 Smart Cities and Urban Geophysics

In the evolution toward smart cities, geophysics plays a critical yet often under-appreciated role. As urban environments become increasingly digitized and interconnected, the need for continuous, non-invasive monitoring of subsurface infrastructure has grown substantially. Urban geophysics provides the tools to visualize, monitor, and manage what lies beneath cities—improving resilience, efficiency, and sustainability in line with smart city frameworks.

16.6.1 Non-invasive Assessment of Underground Infrastructure

Non-invasive geophysical techniques such as Ground Penetrating Radar (GPR), electrical resistivity tomography (ERT), and seismic methods have become foundational tools in assessing underground infrastructure without the need for excavation. These methods allow for the inspection of buried utilities, foundation integrity, tunnel conditions, and aging sewer systems. For instance, GPR is routinely used to detect cracks, voids, or corrosion in subsurface pipes and concrete structures, helping municipal authorities to prioritize maintenance without disrupting surface activities. This is particularly valuable in densely populated cities where surface access is limited and infrastructure age is a growing concern (Zhao et al. 2023a).

Ground Penetrating Radar (GPR) remains a foundational tool in geophysical surveying. In smart cities, GPR can be integrated into intelligent monitoring systems to non-invasively assess aging or undocumented infrastructure beneath urban surfaces (Fig. 16.3), forming part of digital twins and predictive maintenance strategies (Prego et al. 2017; Solla and Fernandez 2022).

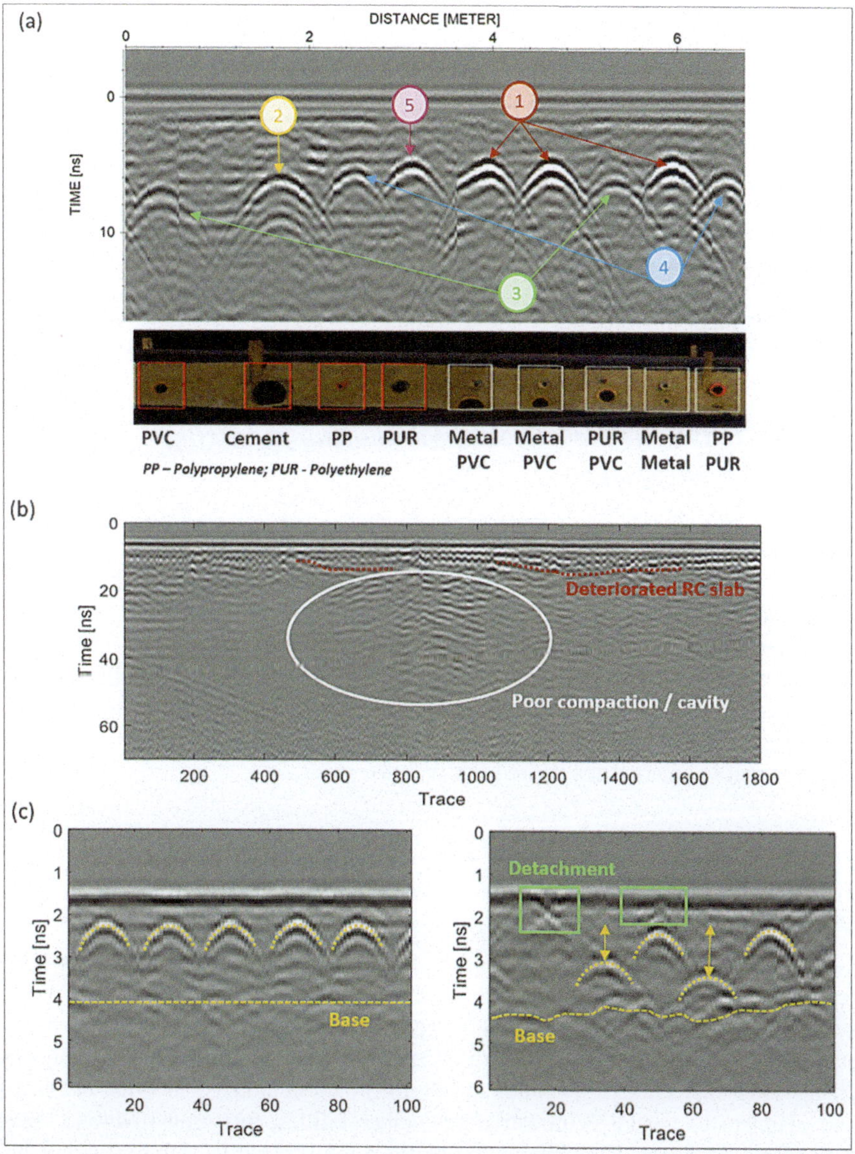

Fig. 16.3 Ground penetrating radar (GPR) applications for subsurface utility detection and infrastructure assessment in urban environments: **a** 1 GHz GPR data revealing underground pipes made of various materials—metal (1), cement (2), PVC (3), polypropylene (4), and polyethylene (5); **b** 500 MHz GPR scan of a rigid pavement structure; **c** 2.3 GHz data used for non-invasive inspection of a reinforced concrete (RC) building slab. These examples demonstrate the integration potential of GPR into smart city systems for underground asset management, infrastructure maintenance, and digital twin modeling. *Source* Lombardi et al. (2022)—Open Access

16.6.2 Utility Mapping and Sinkhole Prediction

Precise mapping of underground utilities is crucial for urban planning, construction safety, and emergency response. Geophysical methods enable the creation of accurate 3D models of subsurface utility networks, minimizing risks associated with accidental damage. Additionally, sinkhole prediction—an increasingly urgent concern in urban environments affected by aging drainage systems and climate-driven hydrological shifts—relies heavily on techniques like microgravity, resistivity surveys, and electromagnetic imaging. These methods can detect voids, cavities, and subsurface water movement that precede sinkhole formation (Lee and Santos 2024).

16.6.3 Real-Time Geophysical Monitoring in Intelligent Urban Systems

The integration of real-time geophysical sensors into urban infrastructure systems allows for continuous monitoring of subsurface dynamics. For example, seismic sensors embedded in roads and tunnels can detect vibrations associated with subsidence, vehicular loads, or underground faults. Similarly, fiber-optic Distributed Acoustic Sensing (DAS) networks are increasingly deployed to monitor temperature changes, fluid leaks, or stress accumulation in pipelines and foundations. These systems feed into urban control centers, enabling proactive response and asset management in smart cities (Wired UK 2024b).

16.6.4 Integrating Geophysics with IoT and Digital Twins

Geophysical data is now being integrated into broader Internet of Things (IoT) frameworks and digital twin models to create real-time, interactive simulations of urban systems. Digital twins of cities incorporate dynamic geophysical data to model subsurface conditions in response to construction, natural hazards, or environmental change. IoT-enabled sensors continuously update these models, allowing engineers and planners to visualize infrastructure vulnerabilities, test scenarios, and make data-driven decisions. This integration marks a transformative step toward predictive maintenance and resilience-based urban design (Technmagazine 2024d; Arasteh et al. 2024).

16.7 Deep Subsurface Mapping and Exploration

As global energy systems transition toward sustainability and resilience, the demand for accurate deep subsurface mapping has increased significantly. Engineering geophysics now plays a central role in addressing complex challenges in energy exploration, underground storage, and geological hazard mitigation. With the deployment of advanced seismic, electromagnetic (EM), and gravity-based technologies, engineers and geoscientists are able to construct detailed 3D and even 4D models of the subsurface. These developments support emerging energy strategies such as geothermal exploitation, carbon capture and storage (CCS), and hydrogen storage—all of which require precise and long-term subsurface characterization.

16.7.1 Next-Gen Seismic, EM, and Gravity Technologies

Recent advances in seismic, electromagnetic, and gravity methods have vastly improved the resolution and penetration depth of subsurface investigations. Full waveform inversion (FWI) in seismic imaging now enables fine-scale resolution of subsurface heterogeneities, essential for identifying fracture zones, fluid pathways, and subtle lithological changes. In EM methods, controlled-source EM (CSEM) and magnetotellurics (MT) are used for detecting resistivity anomalies linked to geothermal reservoirs or saline aquifers. Furthermore, new quantum-enhanced gravimetry instruments, offering sensitivity in the nanogal range, allow the detection of minor density contrasts deep within the Earth's crust (Technmagazine 2024e). These integrated methods enable multi-parameter imaging of subsurface systems with greater precision and efficiency than ever before.

16.7.2 High-Resolution 3D/4D Subsurface Modeling

The ability to model the subsurface in three and four dimensions (3D/4D) represents a leap forward in engineering geophysics. High-resolution 3D modeling provides static snapshots of subsurface architecture, while 4D modeling captures temporal changes—such as fluid migration, thermal evolution, or stress redistribution—over time. These models are particularly valuable for monitoring active processes in energy reservoirs, mining operations, and engineered storage systems. Software platforms integrating seismic, EM, resistivity, and borehole data now offer dynamic visualizations that guide decision-making in both exploration and long-term monitoring (Jones et al. 2023). The emergence of cloud-based platforms and AI-assisted modeling workflows further accelerates real-time interpretation.

16.7.3 Role in Geothermal Energy, Carbon Capture, and Hydrogen Storage

Geophysical exploration is foundational to several low-carbon energy strategies. In geothermal energy development, seismic and magnetotelluric (MT) surveys are used to map deep heat sources, fracture systems, and cap rocks. For carbon capture and storage (CCS), time-lapse seismic and gravity monitoring are employed to verify CO_2 injection and ensure long-term containment. Similarly, as hydrogen emerges as a clean energy carrier, the identification and characterization of suitable geological formations—including salt caverns and depleted gas reservoirs—relies on integrated geophysical surveys. These technologies not only support feasibility assessments but also meet regulatory demands for monitoring and risk assurance (Wired UK 2024c).

While traditional methods such as seismic and EM remain dominant, new applications of Ground Penetrating Radar (GPR) are expanding the toolkit for subsurface investigations. Though typically limited to shallow depths, GPR's use in extreme environments is opening pathways for its potential enhancement in deep geoscience.

Figure 16.4 highlights a ground-breaking example of GPR use beyond Earth, in the Chang'E-4 lunar mission. A 500 MHz radar mounted on a lunar rover successfully imaged subsurface stratigraphy on the Moon's far side, revealing layered regolith and internal structures. Such applications demonstrate the robustness of GPR in harsh, untested conditions, offering insights into how this technology might evolve for high-resolution imaging in deep geothermal reservoirs, hydrogen storage caverns, and CO_2 injection monitoring (Molron et al. 2020; Ribolini et al. 2021; Li et al. 2020; Ruffell and Parker 2021).

16.8 Digital Transformation of Infrastructure Geophysics

The digital revolution is profoundly reshaping infrastructure geophysics, transforming traditional survey and monitoring approaches into highly integrated, data-driven processes. This digital transformation enhances the precision, efficiency, and responsiveness of geophysical investigations, aligning them with modern infrastructure demands and smart city initiatives.

16.8.1 Sensor Fusion and Edge Computing in Field Systems

Modern geophysical surveys increasingly rely on sensor fusion—combining data from multiple sensor types such as seismic, electromagnetic, thermal, and inertial units—to achieve a comprehensive subsurface picture. Edge computing technology allows this data to be processed locally on-site in real time, reducing latency and bandwidth requirements for large data transfers. This capability enables rapid

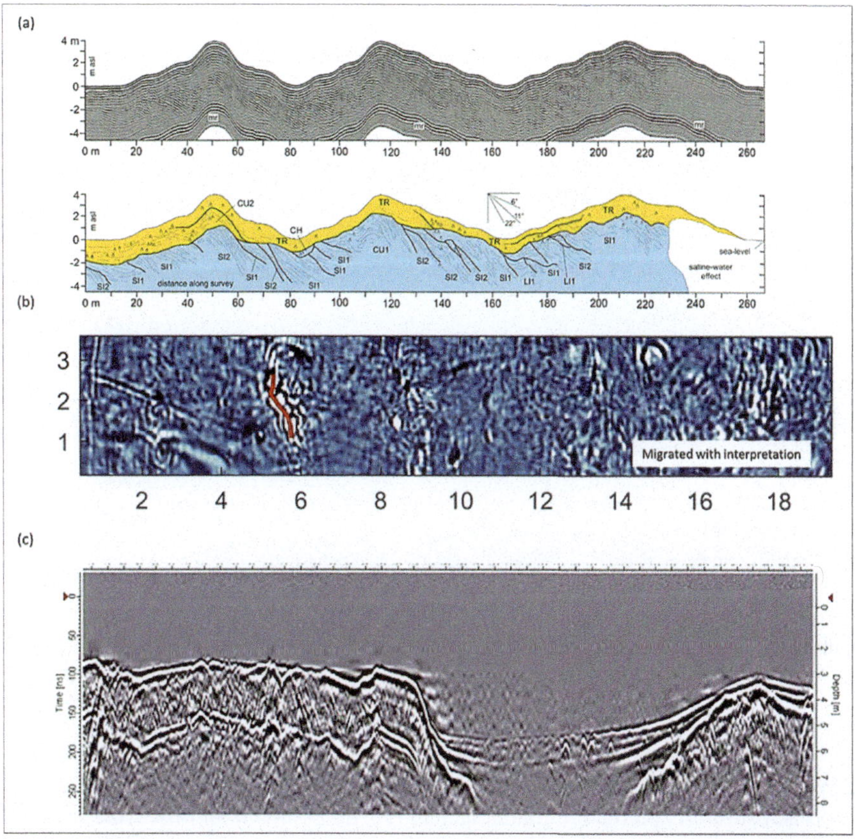

Fig. 16.4 Subsurface structural imaging from the Chang'E-4 mission on the far side of the Moon using a 500 MHz GPR system. Data reveal stratified regolith and underlying structures, showcasing how GPR extends beyond terrestrial geophysics and supports space-based exploration. *Source* Lombardi et al. (2022)—Open Access

decision-making during surveys, adaptive data acquisition, and early identification of anomalies or faults. Field-deployable edge devices equipped with AI algorithms enhance data quality and reliability even in remote or bandwidth-limited environments, streamlining the geophysical workflow (Technmagazine 2024f).

16.8.2 Digital Twins of Civil Structures Informed by Geophysical Data

Digital twins—virtual replicas of physical infrastructure—are increasingly being developed using geophysical data to reflect real-time subsurface conditions and

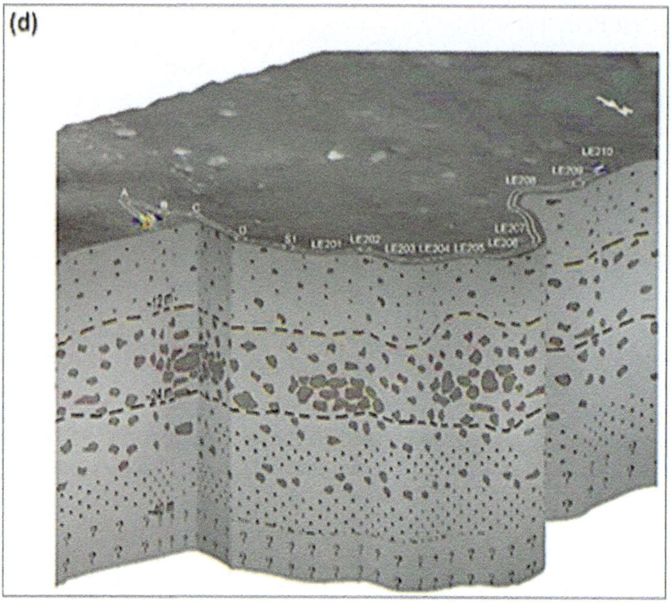

Fig. 16.4 (continued)

structural health. Integrating data from ground-penetrating radar, seismic sensors, and other geophysical sources enables the creation of dynamic models that simulate stress, deformation, and material integrity beneath and around civil structures. This approach supports predictive maintenance, risk assessment, and lifecycle management of infrastructure such as bridges, tunnels, and foundations, enhancing resilience and reducing operational costs (Wired UK 2024d).

16.8.3 Big Data Platforms and Real-Time Interpretation

The proliferation of high-resolution geophysical sensors generates massive volumes of data, necessitating the use of big data platforms capable of ingesting, storing, and analyzing these datasets efficiently. Cloud computing combined with machine learning algorithms allows for real-time interpretation of geophysical signals, anomaly detection, and automated reporting. These platforms facilitate collaboration among multidisciplinary teams and enable continuous monitoring of infrastructure, helping stakeholders respond promptly to emerging issues and optimize resource allocation (Zhao et al. 2023b).

16.8.4 Integration with BIM and GIS Technologies

Integrating geophysical data with Building Information Modeling (BIM) and Geographic Information Systems (GIS) enhances spatial visualization and contextual analysis of underground and aboveground infrastructure. This integration allows urban planners, engineers, and geophysicists to visualize subsurface conditions alongside architectural and environmental data, facilitating informed decision-making in construction, maintenance, and urban development. The seamless incorporation of geophysical insights into BIM/GIS workflows supports comprehensive asset management and risk mitigation in complex urban environments (Lee and Santos 2024).

16.9 Ethical and Policy Considerations

As engineering geophysics enters a new era marked by advanced sensing, AI integration, and digital transformation, ethical and policy challenges are becoming increasingly significant. Emerging technologies in geophysics are not only technical innovations—they also carry implications for data ownership, transparency, equity, and regulatory governance. Addressing these dimensions is critical to ensuring that geophysical advancements are responsibly developed and equitably applied.

16.9.1 Data Privacy and Proprietary Models

The integration of AI and sensor technologies in geophysics has led to the generation and collection of vast amounts of data, often encompassing sensitive information about subsurface structures and environmental conditions. This raises significant concerns regarding data privacy and the ownership of proprietary models. The opacity surrounding proprietary algorithms and data usage can lead to imbalances in access and control, potentially marginalizing communities without the resources to engage with or benefit from these technologies. To mitigate these issues, it is essential to establish clear data governance frameworks that prioritize transparency, consent, and equitable access, ensuring that the benefits of geophysical advancements are shared broadly and responsibly (Percy et al. 2021).

16.9.2 AI Transparency and Decision-Making Accountability

The deployment of AI in geophysical applications, such as automated interpretation of seismic data and predictive modeling, necessitates a commitment to transparency and accountability. AI systems, particularly those utilizing deep learning, often operate as "black boxes," making it challenging to understand how decisions are made. This lack of explainability can undermine trust and hinder the ability to audit and validate AI-driven outcomes. To address these challenges, it is crucial to implement Explainable AI (XAI) methodologies that provide insights into the decision-making processes of AI systems. Additionally, establishing regulatory standards and ethical guidelines, such as those proposed by UNESCO and the European Commission, can promote accountability and ensure that AI technologies are developed and applied in ways that are consistent with societal values and human rights (UNESCO 2024a; European Commission 2024a).

16.9.3 Equitable Access to Emerging Geophysical Technologies

The rapid advancement of geophysical technologies, including AI-driven sensors and modeling tools, has the potential to revolutionize resource management and environmental monitoring. However, disparities in access to these technologies can exacerbate existing inequalities, particularly in low-resource settings. To promote equitable access, it is essential to develop policies that facilitate the sharing of geophysical data and technologies, support capacity-building initiatives, and encourage international collaboration. By fostering an inclusive approach to technological development and deployment, the geophysical community can ensure that the benefits of innovation are accessible to all, contributing to sustainable and resilient infrastructure and environmental stewardship worldwide (Tiltack 2025).

16.10 The Road Ahead: Opportunities and Barriers

As engineering geophysics embraces an era of rapid innovation, the discipline is poised to play a central role in global efforts toward sustainable infrastructure, environmental resilience, and energy transition. However, the journey from conceptual innovation to real-world impact is not without its challenges. Bridging research and application, fostering cross-disciplinary expertise, and addressing institutional constraints are essential steps in shaping the future of the field.

16.10.1 Translating Research into Practical Field Applications

One of the central challenges facing engineering geophysics today is the translation of academic research into scalable, field-deployable solutions. While universities and research institutes continue to develop cutting-edge technologies—such as high-resolution imaging techniques, AI-driven anomaly detection, and quantum sensors—these innovations often face delays in deployment due to logistical, financial, or technical constraints. Bridging this gap requires closer collaboration between researchers, private industry, and public agencies to pilot, test, and adapt innovations in diverse field environments. Initiatives such as "Geo4Infra" by the European Commission have demonstrated how research consortia can align technological innovation with infrastructure development needs across member states (European Commission 2024b; Bonaventura et al. 2017).

16.10.2 Talent Development and Interdisciplinary Collaboration

As geophysical technologies become more integrated with artificial intelligence, data science, and environmental policy, the demand for cross-disciplinary expertise has grown substantially. The next generation of geophysicists must be equipped not only with traditional skills in subsurface imaging and signal processing but also with competencies in coding, machine learning, and systems integration. This requires the reform of academic curricula, new modes of professional development, and joint research programs that blend geoscience, engineering, and computational disciplines. Moreover, diverse and inclusive talent pipelines are critical to ensure a broad range of perspectives and solutions—particularly as geophysics becomes more integral to global sustainability and disaster resilience efforts (UNESCO 2024b).

16.10.3 Funding, Infrastructure, and Regulatory Needs

Despite the strategic relevance of geophysics to infrastructure resilience, environmental monitoring, and climate adaptation, the sector often suffers from underinvestment and fragmented policy support. Many innovative tools remain inaccessible due to high costs, limited deployment infrastructure, or lack of regulatory frameworks for their use. Public funding mechanisms must be expanded to support early-stage innovation, field trials, and large-scale deployment of geophysical systems. Additionally, there is a need for updated regulatory guidelines that accommodate emerging technologies, such as AI-based interpretation systems and remote sensing platforms.

Coordinated efforts between governments, research councils, and multilateral institutions will be essential in overcoming these structural barriers and unlocking the full potential of engineering geophysics (World Bank 2023; World Economic Forum 2018; PwC 2023).

16.11 Summary and Outlook

As this chapter has explored, the field of engineering geophysics is undergoing a profound transformation, shaped by emerging technologies, interdisciplinary integration, and the growing demands of a rapidly changing world. From quantum sensors and AI-powered analytics to autonomous survey platforms and digital infrastructure modeling, the innovations discussed throughout this chapter are not merely incremental—they are redefining the scope, accuracy, and impact of geophysical investigations.

The transformative potential of these trends lies in their ability to enhance decision-making, reduce uncertainty, and support sustainable development across sectors such as construction, energy, defense, environment, and urban planning. For example, the integration of geophysical data with digital twins and smart city infrastructure enables proactive maintenance and climate adaptation strategies, while advanced subsurface imaging contributes to safe and efficient site characterization for critical projects. These applications demonstrate how geophysics is shifting from a traditionally observational science toward a proactive, data-rich discipline at the core of modern engineering and policy design.

Looking ahead, the role of geophysics will be defined not only by technological capability but also by ethical stewardship, inclusive innovation, and international cooperation. Ensuring equitable access to emerging tools, building frameworks for responsible AI deployment, and addressing privacy and data sovereignty will be vital to maintaining public trust and maximizing societal benefit.

Equally important is the preparation of the next generation of engineering geophysicists. This entails reimagining education and training to embrace interdisciplinary learning, digital fluency, and systems thinking. Tomorrow's professionals must be equipped to navigate both physical and digital terrains—interpreting complex datasets, collaborating across sectors, and contributing to resilient infrastructure and climate-conscious design.

In essence, engineering geophysics stands at a pivotal juncture. The path forward will depend on our collective ability to harness innovation responsibly, bridge academic and practical domains, and empower a new cadre of geophysicists ready to meet the challenges—and opportunities—of the twenty-first century.

References

Aeromag Online (2023) Army-2023: Rubin demonstrates robotic system for seismic prospecting. https://www.aeromagonline.com/2023/08/22/army-2023-rubin-demonstrates-robotic-system-for-seismic-prospecting/

Arasteh S et al (2024) AI-based data fusion for enhanced urban subsurface mapping. arXiv preprint. https://arxiv.org/abs/2412.18994

Barone PM, Di Maggio RM (2019) Forensic geophysics: ground-penetrating radar (GPR) techniques and missing persons investigations. Forensic Sci Res 4:337–340

Bonaventura M, Latora V, Nicosia V, Panzarasa P (2017) The advantages of interdisciplinarity in modern science. arXiv. https://arxiv.org/abs/1712.07910

Bongs K, Smith J et al (2023) Field applications of quantum gravity sensors. Nat Phys

European Commission (2024a) Assessment list for trustworthy artificial intelligence (ALTAI). https://ec.europa.eu/digital-strategy/our-policies/ethics-guidelines

European Commission (2024b) Geo4Infra: advancing geotechnologies for resilient infrastructure. https://europa.eu/infrastructure-geo4infra

Exail (2023) iXblue and INGV present the first application of quantum technology to volcano monitoring on Mount Etna. https://www.ixblue.com/north-america/ixblue-and-the-national-institute-of-geophysics-and-volcanology-present-the-first-application-of-quantum-technology-to-volcano-monitoring-on-mount-etna

Francke J (2012) A review of selected ground-penetrating radar applications to mineral resource evaluations. J Appl Geophys 81:29–37

FT (Financial Times) (2024) UK leads quantum sensing push in infrastructure and security

Grab M, Mattea E, Bauder A, Huss M, Rabenstein L, Hodel E, Linsbauer A, Langhammer L, Schmid L, Church G et al (2021) Ice thickness distribution of all Swiss glaciers based on extended ground-penetrating radar data and glaciological modeling. J Glaciol 67:1074–1092

Irving J, Xu Z, Lindsay KM, Bradford J, Zhu P, Holliger K (2019) Determination of the correlation structure of an alluvial aquifer from multi-frequency 3D GPR reflection measurements. In: AGU fall meeting abstract 2019, H43F-2040

Jones P, Kumar S (2024) Integration of geophysical data in early warning systems for climate resilience. Clim Risk Manag J 15:102–117

Jones L, Ahmed T, Ferreira D (2023) Integrated 4D modeling for subsurface energy systems. J Geophys Eng 20(2):245–263

Lee H, Santos V (2024) Geophysical techniques for urban sinkhole risk management. Urban Geohazards J 29(1):32–47

Lee D, Patel R, Gomez A (2023) Ground stability and infrastructure resilience in climate-impacted regions. Eng Geol Adv 45(2):120–135

Li C, Su Y, Pettinelli E, Xing S, Ding C, Liu J, Ren X, Lauro SE, Soldovieri F, Zeng X et al (2020) The Moon's far side shallow subsurface structure unveiled by Chang'E-4 lunar penetrating radar. Sci Adv 6:eaay6898

Lombardi F, Podd F, Solla M (2022) From its core to the niche: insights from GPR applications. Remote Sens 14:3033

M2 Lasers (2024) Quantum gravimetry for civil and urban engineering. https://m2lasers.com/quantum-gravimeter.html

Molron J, Linde N, Baron L, Selroos J-O, Darcel C, Davy P (2020) Which fractures are imaged with ground-penetrating radar? Results from an experiment in the Aspo Hardrock Laboratory, Sweden. Eng Geol 273:105674

Olhoeft GR, Modroo JJ (2006) Locating and identifying avalanche victims with GPR. Lead Edge 25:306–308

Percy C, Dragicevic S, Sarkar S, d'Avila Garcez AS (2021) Accountability in AI: from principles to industry-specific accreditation. AI Soc. https://doi.org/10.3233/AIC-210080

Prego F, Solla M, Puente I, Arias P (2017) Efficient GPR data acquisition to detect underground pipes. NDT E Int 91:22–31

PwC (2023) Global infrastructure trends—developments in financing. https://www.pwc.com/gx/en/industries/capitalprojectsinfrastructurepublicationsinfrastructure
Ribolini A, Bertoni D, Bini M, Sarti G (2021) Ground-penetrating radar prospections to image the inner structure of coastal dunes at sites characterized by erosion and accretion (Northern Tuscany, Italy). Appl Sci 11:11260
Ruffell A, Parker R (2021) Water penetrating radar. J Hydrol 597:126300
Shokri M, Gao Y, Kibler KM, Wang D, Wightman MJ, Rice N (2022) Contaminant transport from stormwater management areas to a freshwater karst spring in Florida: results of near-surface geophysical investigations and tracer experiments. J Hydrol Reg Stud 40:101055
Smith R, Johnson L, Zhao M (2023) Geophysical monitoring of subsurface hydrological changes for climate adaptation. J Environ Geophys 58(4):345–360
Solla M, Fernandez N (2022) GPR analysis to detect subsidence: a case study on a loaded reinforced concrete pavement. Int J Pavement Eng 1–15
Sorrentino F et al (2024) Field deployment of quantum sensors for environmental monitoring. arXiv preprint. https://arxiv.org/abs/2405.10844
Tatham D (2024) Robotic platforms for seismic imaging in extreme environments. J Appl Geophys
Technmagazine (2024a) Emerging technologies in engineering geophysics. https://technmagazine.com/emerging-geophysics
Technmagazine (2024b) Digital transformation in geophysics: revolutionizing earth sciences. https://technmagazine.com/digital-transformation-geophysics
Technmagazine (2024c) Top tech innovations of 2024: AI and quantum computing. https://technmagazine.com/top-tech-innovation
Technmagazine (2024d) Digital twins and IoT in urban geophysics. https://technmagazine.com/digital-twins-iot-geophysics/
Technmagazine (2024e) Subsurface imaging revolution: quantum sensors and AI integration. https://technmagazine.com/emerging-geophysics
Technmagazine (2024f) Edge computing and sensor fusion for geophysical applications. https://technmagazine.com/edge-computing-geophysics
Tiltack T (2025) AEJIM: a real-time AI framework for crowdsourced, transparent, and ethical environmental hazard detection and reporting. https://arxiv.org/abs/2503.17401v2
Tobias L, Zhang H (2023) Digital twins and AI in smart urban infrastructure. J Infrastruct Syst
Udhyog Mantra (2024) Digital transformation trends in geophysics
UNESCO (2024a) Recommendation on the ethics of artificial intelligence. https://www.unesco.org/en/legal-affairs/recommendation-ethics-artificial-intelligence
UNESCO (2024b) STEM education and workforce development in earth sciences. United Nations Educational, Scientific and Cultural Organization, Paris
Wired UK (2024a) Innovation in geophysical sensing and data integration. https://www.wired.co.uk/article/geophysical-technology-innovation
Wired UK (2024b) Sensors beneath the streets: how cities are becoming smarter from the ground up. https://www.wired.co.uk/article/smart-cities-infrastructure-monitoring
Wired UK (2024c) Geophysical frontiers: mapping for energy transition. https://www.wired.co.uk/article/geophysical-technology-innovation
Wired UK (2024d) Digital twins and smart infrastructure monitoring. https://www.wired.co.uk/article/digital-twins-infrastructure
World Bank (2023) Financing infrastructure resilience: the role of geotechnologies. International Bank for Reconstruction and Development, Washington, DC. https://www.worldbank.org/en/topic/infrastructure/publication/geotechnology-financing
World Economic Forum (2018) We need to invest more in infrastructure technologies. Our future depends on it. https://www.weforum.org/agenda/2018/06/investing-in-the-future/
Zhao M, Clarke J, Singh P (2023a) Non-invasive subsurface diagnostics for urban infrastructure. J Appl Geophys 210:105065
Zhao M, Clarke J, Singh P (2023b) Big data and machine learning for real-time geophysical interpretation. J Appl Geophys 210:105065

Part V
Reference Materials

Chapter 17
Conclusions: Bridging Innovation and Practical Application in Geophysics

Abstract This final chapter synthesizes the key innovations and cross-sectoral impacts of geophysical science covered throughout the volume, emphasizing its expanding strategic role in addressing global challenges. It highlights the evolution of geophysics from traditional exploration toward integrated, data-driven, and sustainability-aligned applications in sectors such as energy, infrastructure, defense, climate resilience, heritage preservation, and urban development. The chapter critically reflects on existing limitations—technical, ethical, and operational—and offers actionable recommendations for policymakers, industry leaders, and researchers. It advocates for a robust integration of geophysics into education and workforce development, promoting interdisciplinary training and ethical awareness. Finally, it presents a forward-looking vision for geophysics as a transformative science at the heart of innovation, resilience, and responsible stewardship of the planet.

17.1 Summary of Advances

Throughout this volume, engineering geophysics has been presented as a rapidly evolving discipline that plays a pivotal role in addressing subsurface challenges across diverse fields. Technological advancements—such as full waveform inversion, machine learning in geophysical interpretation, and integrated survey techniques—have significantly enhanced the precision, efficiency, and applicability of geophysical investigations.

These developments have shifted the perception of geophysics from a predominantly research-driven field to a practical, solution-oriented discipline. Applications now span a wide spectrum, including site characterization, infrastructure assessment, environmental remediation, resource exploration, and urban planning. The ability to non-invasively visualize subsurface features continues to improve risk mitigation strategies and engineering designs.

D. C. Naskar, *Advances in Engineering Geophysics: Concepts and Applications*, Earth Systems Data and Models 9, https://doi.org/10.1007/978-3-032-28000-8_17

17.2 Strategic Importance for Global Challenges

As the world grapples with a range of pressing challenges—climate change, rapid urbanization, environmental degradation, natural hazards, and the need for sustainable infrastructure—engineering geophysics stands at the intersection of science, engineering, and policy. Its ability to deliver non-invasive, high-resolution insights into subsurface conditions makes it a critical enabler for informed decision-making.

In particular, the integration of geophysics with Sustainable Development Goals (SDGs) has emerged as a strategic imperative. Whether it's supporting clean water access through aquifer mapping, reducing disaster risks through early warning systems, or contributing to low-carbon energy transitions, engineering geophysics provides essential data for sustainability-driven planning. Furthermore, its applications in defense and critical infrastructure highlight its role in national security, energy resilience, and the protection of cultural heritage.

17.3 Reflections on Future Impact Across Sectors

Looking ahead, the influence of engineering geophysics will likely deepen and diversify. Integration with artificial intelligence, edge computing, and satellite technologies will further enhance data acquisition and interpretation capabilities. The field is also poised to benefit from increased collaboration with disciplines such as environmental science, structural engineering, and urban planning.

In civil engineering, smart cities will demand continuous monitoring of underground utilities and structures—tasks ideally suited to geophysical techniques. In climate adaptation, geophysics will help model groundwater behavior, assess coastal vulnerability, and monitor permafrost degradation. In post-conflict zones or disaster-hit areas, it will facilitate rapid and safe assessments of structural stability and buried hazards. To realize this future, it will be essential to invest in education, cross-sector partnerships, open data initiatives, and technology transfer mechanisms that democratize access to advanced geophysical tools.

17.4 Challenges and Limitations

Despite the transformative advancements in engineering geophysics, the field continues to face a number of practical and technical limitations. One of the most common challenges lies in the trade-off between resolution and depth of investigation. For instance, while seismic methods are effective for deeper exploration, they may lack the resolution needed for detailed imaging in shallow or complex subsurface environments. Similarly, techniques such as ground-penetrating radar offer high-resolution imaging but are limited to shallow depths and specific soil conditions.

Data interpretation remains another significant limitation. Geophysical data are often non-unique, meaning that multiple subsurface models can explain the same dataset. As a result, accurate interpretation typically requires integration with geological, geotechnical, or hydrological data—resources that may not always be available or accessible. Additionally, the high cost of advanced geophysical equipment and data processing software can hinder the use of these methods, particularly in low-resource or remote areas.

Operational challenges also include limitations in survey access due to environmental constraints, land ownership, or political sensitivities, which can restrict fieldwork in critical regions. Finally, the shortage of trained personnel with interdisciplinary expertise continues to affect the quality and reliability of geophysical assessments. A lack of understanding among non-specialists regarding the capabilities and limitations of geophysical tools can also lead to misapplication or underutilization in engineering contexts.

17.5 Recommendations for Practice and Policy

To fully leverage the potential of engineering geophysics, a number of strategic actions are recommended for both practitioners and policymakers. First, the development and adoption of standardized methodologies across different geophysical techniques will significantly enhance data consistency, interpretation, and integration across projects. Uniform guidelines can help minimize errors and improve collaboration between engineers, geophysicists, and regulatory authorities.

It is also crucial that geophysical surveys be more explicitly included in regulatory frameworks governing infrastructure development, environmental monitoring, and natural hazard assessment. Mandating geophysical investigations during the planning and design phases of critical infrastructure can lead to more informed decisions and reduced project risks.

Public–private partnerships should be encouraged to promote the development and dissemination of geophysical technologies, especially in emerging economies. Collaborative efforts between universities, research institutions, government agencies, and industry players can facilitate technology transfer and foster innovation.

Equally important is the promotion of open-access data repositories and geophysical information systems. Shared platforms can reduce duplication of effort, promote transparency, and support cross-disciplinary research. Financial incentives, grants, and technical support can further enhance the accessibility and application of geophysical tools in underserved regions and communities.

17.6 Role in Education and Workforce Development

The advancement of engineering geophysics depends not only on technology but also on a skilled and knowledgeable workforce. Currently, many engineering programs offer limited exposure to geophysical principles, resulting in a knowledge gap that hinders integration with engineering design and analysis. To address this, geophysical content should be incorporated into undergraduate and graduate curricula in civil, environmental, and mining engineering. This integration should include both theoretical instruction and practical fieldwork to develop applied skills.

Continuing education and professional development programs are equally important. Short-term certification courses, online modules, and hands-on workshops can provide practicing engineers and technicians with the necessary expertise to understand and apply geophysical methods effectively. Industry-led training initiatives can also help align academic learning with current market needs and technologies.

Academic institutions should collaborate more closely with industry to create internship opportunities, joint research projects, and case-based learning models. These partnerships can provide students and early-career professionals with exposure to real-world problems and foster interdisciplinary thinking. In addition, scholarships and funding support for students in geophysics can help attract talent and ensure the long-term growth of the field.

17.7 Ethical and Social Considerations

As geophysical technologies become more widespread and powerful, it is essential to address the ethical and social implications of their application. One critical consideration is the need for informed consent and transparency when conducting surveys in inhabited or privately-owned areas. Clear communication with stakeholders, including property owners and local communities, is vital to build trust and ensure the responsible use of geophysical methods.

Respect for cultural and archaeological heritage must also guide the use of geophysical surveys. Investigations near heritage sites or indigenous lands require collaboration with archaeologists, anthropologists, and local leaders to avoid unintended damage or misinterpretation. Ethical guidelines should be established to govern such activities, ensuring cultural sensitivity and respect for local values.

Environmental impacts, though generally minimal, should not be overlooked. In sensitive ecosystems, certain survey methods—such as those involving ground disturbance or the use of vehicles—can have localized effects. Proper environmental assessments and adherence to best practices can minimize these risks.

Equally important is the ethical responsibility in interpreting and presenting geophysical data. Overstating confidence in results or ignoring uncertainties can lead to poor decision-making, particularly in high-stakes projects like dam construction or hazardous waste monitoring. Practitioners must maintain scientific integrity, communicate limitations clearly, and advocate for cross-validation with other data sources when appropriate.

17.8 Vision for the Discipline

The future of engineering geophysics is characterized by increasing integration with digital, sustainable, and resilient engineering practices. As infrastructure systems become smarter and more interconnected, real-time geophysical monitoring will play a central role in detecting structural changes, identifying subsurface anomalies, and supporting predictive maintenance. Innovations in sensor technology, cloud computing, and edge analytics will enable continuous data collection and interpretation, bringing geophysics into the realm of proactive infrastructure management.

In the context of global climate adaptation, geophysical methods will be essential for monitoring permafrost degradation, assessing flood risks, managing groundwater resources, and analyzing coastal erosion. Their ability to provide rapid, non-invasive data will support sustainable land use and disaster preparedness in vulnerable regions. Interdisciplinary collaboration is expected to shape the next phase of geophysical innovation. The combination of geophysics with artificial intelligence, remote sensing, structural mechanics, and geospatial modeling will allow for more accurate and comprehensive understanding of subsurface conditions. This integration will enable engineers and planners to make more informed, data-driven decisions.

Ultimately, engineering geophysics is evolving into a core discipline that not only investigates the subsurface but also informs policy, safeguards infrastructure, protects the environment, and contributes to global resilience. With continued investment in research, education, and policy alignment, the discipline is poised to deliver lasting impact across sectors and societies.

17.9 Final Thoughts

Engineering geophysics now occupies a central position in modern engineering and environmental practices. As technological innovation continues to enhance its capabilities, the field is well-positioned to contribute meaningfully to resilient infrastructure, sustainable development, and strategic decision-making across sectors. By effectively bridging scientific research with real-world application, engineering geophysics will remain a cornerstone of efforts to understand, manage, and protect the subsurface environment for future generations.

Chapter 18
Technical Reference and Glossary: A Comprehensive Companion for Terminology, Symbols, and Equations

Abstract This chapter serves as a foundational technical companion to the preceding chapters, offering readers a structured and accessible reference for terminology, acronyms, symbols, and core mathematical formulations used throughout engineering geophysics. It begins with a comprehensive glossary and list of standard abbreviations, supporting clarity and consistency in multidisciplinary contexts. The chapter further consolidates key mathematical symbols and essential equations relevant to seismic, electrical, electromagnetic, and thermal geophysical methods. A practical reference for SI and CGS unit conversions, commonly used measurement scales, and physical constants ensures accurate interpretation and application of data. To extend learning and professional engagement, curated suggestions for textbooks, academic journals, datasets, and chapter-specific readings are included. This chapter is designed as both a quick-reference tool for practitioners and a foundational resource for students and researchers entering the field.

18.1 Glossary of Terms

This section provides standardized definitions of essential terminology used in engineering geophysics, civil engineering, environmental science, and remote sensing. Each entry is concise and consistent with international usage to help ensure clarity across disciplines.

Term	Definition
A	
Acoustic impedance	The product of a material's density and seismic wave velocity; determines the reflection and transmission of seismic waves at interfaces
Amplitude spectrum	A plot representing the distribution of amplitudes across different frequencies in a signal
Aquifer	A body of permeable rock or sediment that can contain or transmit groundwater

(continued)

D. C. Naskar, *Advances in Engineering Geophysics: Concepts and Applications*, Earth Systems Data and Models 9, https://doi.org/10.1007/978-3-032-28000-8_18

(continued)

Term	Definition
Attenuation coefficient	A parameter that quantifies the rate at which a wave's amplitude decreases with distance in a medium
Anisotropy	A property of a material where physical characteristics varies with direction. In geophysics, commonly seen in seismic velocity or electrical resistivity
Attenuation	Reduction in amplitude and energy of a wave as it propagates through a medium, due to scattering, absorption, or geometric spreading
Azimuth	The angle between a reference direction (usually true north) and the line connecting the observer to a point of interest, measured clockwise from the reference
B	
Borehole	A narrow, deep hole drilled into the ground to obtain subsurface information, such as soil samples or geophysical data
Bulk density	The mass of soil or rock per unit volume, including the air space; important for geotechnical and geophysical studies
C	
Capacitively coupled resistivity (CCR)	A resistivity technique using capacitive coupling rather than galvanic contact with the ground. Useful on paved surfaces
Capacitance	The ability of a system to store an electric charge; relevant in capacitive-based geophysical methods
Cohesion	The force that holds particles of a material together; important in soil mechanics and geotechnical engineering
Conductivity	A measure of a material's ability to conduct an electric current; the inverse of resistivity
Confining layer	A layer of material that restricts the movement of groundwater, often due to low permeability
Critical damping	The condition in which a system returns to equilibrium without oscillating; important in seismic analysis of structures
Curie point	The temperature above which certain magnetic materials lose their permanent magnetic properties; relevant in magnetic susceptibility studies
D	
Dielectric permittivity	A measure of a material's ability to store electrical energy in an electric field; critical in GPR interpretation
Damping	The reduction in amplitude of a wave or oscillation due to energy loss; significant in seismic wave propagation
Dipole–dipole method	A geophysical technique used in resistivity surveys, where two electrodes are used to inject current and measure potential difference
Dispersion curve	A graph showing the variation of wave velocity with frequency; used in surface wave analysis
Ductility	The ability of a material to undergo significant plastic deformation before rupture; important in structural engineering

(continued)

(continued)

Term	Definition
Dynamic range	The ratio between the largest and smallest values of a signal or measurement; relevant in geophysical data analysis
E	
Elastic modulus	A mechanical property describing material stiffness, often derived indirectly from seismic velocity data
Eddy current	A circulating flow of electric charge induced in a conductor by a changing magnetic field; utilized in electromagnetic induction methods
Elasticity	The ability of a material to return to its original shape after deformation; fundamental in seismic wave propagation studies
Electromagnetic induction	The process by which a changing magnetic field induces an electric current in a conductor; basis for many geophysical methods
Electrode	A conductor through which electricity enters or leaves a medium; used in various geophysical survey techniques
Ellipticity	A measure of the deviation of a wave's shape from a perfect circle; relevant in surface wave analysis
End-member	A pure, well-defined component of a mixture; used in geochemical and geophysical modeling
Energy density	The amount of energy stored in a given system or region of space per unit volume; important in wave propagation studies
F	
Faraday's Law	A fundamental law of electromagnetism stating that a change in magnetic flux through a circuit induces an electromotive force; foundational in EM methods
Fresnel Zone	A series of concentric ellipsoidal regions around a wave's path, within which the phase difference between rays is negligible; important in seismic wave analysis
G	
Galvanic	Describes geophysical techniques that require direct contact with the ground in order to pass current; the alternative is to induce currents in the earth
Geophone	A device that converts ground movement (velocity) into voltage, used in seismic surveys to detect and record seismic waves
Geoelectric section	A vertical profile showing the variation of electrical resistivity with depth; constructed from resistivity survey data
Geophysical logging	The practice of recording measurements of physical properties of the subsurface from a borehole; used in various geophysical investigations
Geotechnical engineering	A branch of civil engineering concerned with the behavior of earth materials and their application in construction
Gravimeter	An instrument used to measure the acceleration due to gravity at a specific location; used in gravity surveys
Ground penetrating radar (GPR)	A non-invasive geophysical method that uses radar pulses to image the subsurface; effective for shallow investigations

(continued)

(continued)

Term	Definition
Groundwater table	The upper surface of the zone of saturation in an aquifer; indicates the level below which the ground is saturated with water
H	
Half-space model	A simplified geophysical model representing the Earth as a uniform, infinite medium; used as a reference in various analyses
Hardness	A measure of a material's resistance to deformation, particularly permanent deformation, indentation, or scratching; relevant in material science
Harmonic analysis	The decomposition of a signal into its constituent sinusoids; used in the analysis of seismic and electromagnetic data
Hydraulic conductivity	A property of a porous material that describes the ease with which a fluid can move through pore spaces; crucial in hydrogeology
Hydraulic head	The potential energy available to drive the flow of groundwater; a combination of elevation and pressure head
Hydrogeology	The study of the distribution
I	
Inversion (geophysics)	The process of estimating subsurface properties by mathematically adjusting model parameters to fit measured data
Induced polarization (IP)	A geophysical effect whereby electrical charge is momentarily polarized within a material, usually a disseminated ore or a clay; this effect is the basis for the IP method
Impedance	The resistance of a medium to the flow of an alternating current; in geophysics, it relates to the opposition to seismic wave propagation
J	
Joint inversion	A geophysical technique that simultaneously inverts multiple datasets (e.g., seismic and resistivity) to obtain a more comprehensive subsurface model
K	
Kernels	Mathematical functions used in geophysical inversion to relate model parameters to observed data; essential in the construction of inversion algorithms
L	
Lithology	The physical characteristics of a rock or sediment, including composition, grain size, and texture; important in geophysical interpretation
Lateral resolution	The ability of a geophysical method to distinguish between two closely spaced features in the horizontal plane
M	
Magnetotellurics (MT)	A passive EM technique that uses natural variations in the Earth's magnetic and electric fields to image resistivity at depth

(continued)

(continued)

Term	Definition
Microseisms	A more or less continuous motion of the ground that is unrelated to an earthquake and that has a period of 1.0–9.0 s; caused by various natural and artificial phenomena
N	
Noise	Unwanted signals or disturbances that interfere with the measurement of geophysical data; can arise from various sources, including equipment and environmental factors
Normal fault	A type of fault where the hanging wall moves downward relative to the footwall; indicative of extensional tectonic environments
O	
Overburden	The layers of soil and rock that lie above a deposit of economic value; in geophysics, it refers to the material overlying a target zone
Offset	The horizontal displacement between two points on opposite sides of a fault; used in seismic interpretation to estimate fault movement
P	
P wave	Primary or compressional seismic wave; the fastest seismic wave and the first to be detected by seismographs
Paleoseismology	The study of prehistoric earthquakes, based on the geological record, to understand earthquake behavior and recurrence intervals
Q	
Quality factor (QF)	A dimensionless parameter that describes the damping of seismic waves; higher Q indicates less energy loss and better wave propagation
Quasi-static	A condition where changes occur slowly enough that dynamic effects can be neglected; used in certain geophysical modeling scenarios
R	
Resistivity	A measure of a material's resistance to the flow of electric current; used in various geophysical methods to infer subsurface properties
Rayleigh wave	A type of surface seismic wave that causes both vertical and horizontal ground motion; important in surface wave analysis
S	
Shear wave velocity (V_s)	The speed at which shear waves travel through a material; important for seismic site classification and dynamic analysis
Seismic refraction	A geophysical method that measures the bending of seismic waves as they pass through different subsurface layers, providing information about layer depths and velocities
T	
Tomography	A 2D or 3D image reconstruction method used to visualize subsurface properties using data from multiple paths or angles
Tidal wave	Often used interchangeably with tsunami; refers to a series of waves caused by the displacement of water, typically due to seismic activity
U	

(continued)

(continued)

Term	Definition
Ultrasonic testing	A non-destructive testing method that uses high-frequency sound waves to detect imperfections or changes in material properties. Commonly used in concrete inspection, weld integrity testing, and flaw detection in engineering structures
Up-hole survey	A seismic technique used to determine the velocity structure of near-surface layers, typically for optimizing the depth of explosive sources in reflection surveys
V	
Velocity model	A mathematical representation of how seismic wave velocities vary with depth or position in the subsurface; essential in seismic imaging and migration
Volumetric strain	A measure of the change in volume of a material under stress, important in geomechanics and soil behavior analysis
W	
Wavelet	A short-duration oscillation used in seismic data analysis to model seismic pulses. The shape and frequency of the wavelet influence resolution and interpretation
Well logging	The process of recording detailed geophysical data from boreholes using downhole tools; includes resistivity, gamma-ray, density, and sonic logs
X	
X-component (geophysics)	Refers to the horizontal component of a vector field (e.g., magnetic, seismic, or electric field) measured along the X-axis, typically in east-west orientation
X-ray fluorescence (XRF)	An analytical technique for determining the elemental composition of materials; used in geochemical and archaeological investigations
Y	
Young's modulus (E)	A measure of the stiffness of a solid material; defined as stress divided by strain. Commonly derived from seismic data in geophysical applications
Yield strength	The stress at which a material begins to deform plastically. In geotechnical applications, it's crucial for understanding slope stability and foundation performance
Z	
Z-component (geophysics)	The vertical component of a vector quantity, such as magnetic or seismic fields; used in interpreting data in 3D geophysical surveys
Zone of saturation	The subsurface region in which all the pores and fractures are filled with water; important in hydrogeological and resistivity interpretations

18.2 List of Acronyms and Abbreviations in Engineering Geophysics

Acronym	Full form	Domain
AC	Alternating Current	Electrical Engineering
ACF	Auto-Correlation Function	Signal Processing
AGC	Automatic Gain Control	Instrumentation
AGU	American Geophysical Union	Professional Organization
AMT	Audio-Magnetotellurics	Geophysics
API	American Petroleum Institute	Industry Standards
APS	Acoustic Logging Tool	Well Logging
AR	Augmented Reality	Visualization Technology
ASI	Acoustic Source Imaging	Imaging Techniques
ASTM	American Society for Testing and Materials	Standards Organization
ATEX	Atmospheres Explosible	Safety Standards
AWI	Alfred Wegener Institute	Research Institution
BEM	Boundary Element Method	Computational Modeling
BGR	Bundesanstalt für Geowissenschaften und Rohstoffe	Research Institute
BHT	Bottom Hole Temperature	Well Logging
BOM	Bill of Materials	Project Management
BPS	Borehole Positioning System	Surveying
CBM	Coal Bed Methane	Energy Resources
CDP	Common Depth Point	Seismic Surveying
CSEM	Controlled-Source Electromagnetic	Geophysical Method
CIG	Common Image Gather	Seismic Imaging
CIV	Civil Engineering	Engineering Discipline
CME	Coronal Mass Ejection	Space Weather
CMG	Coastal and Marine Geology	Geosciences
CMT	Centroid Moment Tensor	Seismology
CNS	Central Nervous System	Biology
CPT	Cone Penetration Test	Geotechnical Engineering
CRS	Coordinate Reference System	Geospatial Science
CTD	Conductivity, Temperature, Depth	Oceanography
CTI	Chesapeake Technology, Incorporated	Technology Company
CVS	Constant Voltage Source	Electronics
DAS	Distributed Acoustic Sensing	Fiber Optic Sensing
DB	Decibel	Measurement Unit
DBA	Designated Baseline Assessment	Environmental Management

(continued)

(continued)

Acronym	Full form	Domain
DCS	Distributed Control System	Automation
DGPS	Differential Global Positioning System	Navigation
DHI	Direct Hydrocarbon Indicator	Exploration Geophysics
DIP	Dipole–Dipole	Geophysical Surveying
DIT	Dual Induction Tool	Well Logging
DMT	Dilatometer Test	Geotechnical Testing
DNL	Day-Night Average Sound Level	Acoustics
DOAS	Differential Optical Absorption Spectroscopy	Atmospheric Science
DOI	Digital Object Identifier	Publishing
DOS	Disk Operating System	Computing
DSS	Decision Support System	Data Analysis
DTH	Down-The-Hole	Drilling Technology
DTS	Distributed Temperature Sensing	Fiber Optic Sensing
DTV	Digital Television	Broadcasting
DVR	Digital Video Recorder	Electronics
DWL	Downhole Logging	Well Logging
EM	Electromagnetic	Geophysical Surveying
EAS	Emergency Alert System	Communication
ECG	Electrocardiogram	Medical
ECGR	Environmentally Corrected Gamma Ray	Well Logging
EDTC	Environmental Data and Technology Center	Research Facility
EEM	Environmental Effects Monitoring	Environmental Science
EFT	Environmental Flow Test	Hydrology
EMI	Electromagnetic Interference	Electronics
EMR	Electromagnetic Radiation	Physics
EMS	Environmental Management System	Environmental Science
EMT	Electromagnetic Tomography	Imaging Technique
ENVI	Environment for Visualizing Images	Software
EPA	Environmental Protection Agency	Government Agency
EPS	Encapsulated PostScript	Graphics Format
ERD	Earth Resistivity Device	Geophysical Instrumentation
ESRI	Environmental Systems Research Institute	Software Company
ETL	Extract, Transform, Load	Data Processing
ETM	Enhanced Thematic Mapper	Remote Sensing
ETP	Environmental Testing Protocol	Environmental Science
FEM	Finite Element Method	Computational Modeling

(continued)

(continued)

Acronym	Full form	Domain
GPR	Ground Penetrating Radar	Near-Surface Geophysics
HEM	Helicopter Electromagnetic Survey	Airborne Geophysics
IP	Induced Polarization	Electrical Geophysics
JSG	Joint Seismic-Geoelectric Modeling	Multiphysics Modeling
KML	Keyhole Markup Language	GIS/Remote Sensing
LIDAR	Light Detection and Ranging	Remote Sensing/Topography
MT	Magnetotellurics	Deep Crustal Geophysics
NDT	Non-Destructive Testing	Structural/Geophysical Inspection
OEM	Original Equipment Manufacturer	Instrumentation/Hardware
P-Wave	Primary Wave	Seismology
Q-Value	Quality Factor	Seismic Attenuation Analysis
RMT	Radio Magnetotellurics	Environmental Geophysics
SAR	Synthetic Aperture Radar	Radar Remote Sensing
TDEM	Time-Domain Electromagnetic	Electrical/Electromagnetic
UAV	Unmanned Aerial Vehicle	Drone-Based Surveying
VSP	Vertical Seismic Profiling	Borehole Geophysics
WMS	Web Map Service	GIS and Data Integration
XCT	X-ray Computed Tomography	Rock Physics/Subsurface Imaging
YIG	Yttrium Iron Garnet (magnetometer core)	Magnetic Survey Instrumentation
ZEM	Zero-Electromagnetic Method	Experimental Geophysical Method

18.3 Mathematical Symbols and Notation

This section lists the most commonly used symbols in geophysical equations, along with definitions, units, and dimensional context. A standardized table aids in uniform understanding across applications.

Symbol	Meaning	Unit	Dimension
ρ	Electrical resistivity	Ohm-m (Ωm)	$[ML^3T^{-3}I^{-2}]$
σ	Electrical conductivity ($1/\rho$)	S/m	$[M^{-1}L^{-3}T^3I^2]$
v	Seismic wave velocity	m/s	$[LT^{-1}]$
α	Attenuation coefficient	dB/m or Np/m	
f	Frequency	Hz	$[T^{-1}]$
λ	Wavelength	m	[L]
T	Period	s	[T]

(continued)

(continued)

Symbol	Meaning	Unit	Dimension
k	Wavenumber	rad/m	$[L^{-1}]$
G	Gravitational constant	$m^3/kg\ s^2$	$[L^3M^{-1}T^{-2}]$
μ	Magnetic permeability	H/m	$[MLT^{-2}I^{-2}]$
ε	Electric permittivity	F/m	$[M^{-1}L^{-3}T^4I^2]$
∇	Nabla (vector differential)	–	$[L^{-1}]$ (operator)
∂	Partial derivative	–	– (operator)
φ	Porosity (fraction)	– or %	Dimensionless
τ	Relaxation time	s	[T]
Q	Quality factor (damping)	–	Dimensionless
Z	Impedance	Pa s/m or Ohm	$[ML^{-2}T^{-2}]$
θ	Angle of incidence/reflection/refraction	Degrees or radians	$[L^{-1}]$
Δ	Laplacian (∇^2 operator)	–	$[L^{-2}]$
ζ	Damping ratio	–	Dimensionless
P	Pressure	Pa	$[ML^{-1}T^{-2}]$
E	Elastic modulus	Pa	$[ML^{-1}T^{-2}]$
A	Amplitude	V (or arbitrary)	Depends on context
B	Magnetic flux density	Tesla (T)	$[MT^{-2}I^{-1}]$
C	Capacitance	Farad (F)	$[M^{-1}L^{-2}T^4I^2]$
D	Electric displacement	C/m^2	$[TIL^{-2}]$
E	Electric field strength	V/m	$[MLT^{-3}I^{-1}]$
F	Force	Newton (N)	$[MLT^{-2}]$
g	Acceleration due to gravity	m/s^2	$[LT^{-2}]$
h	Depth or thickness	m	[L]
I	Electric current	Ampere (A)	[I]
J	Current density	A/m^2	$[IL^{-2}]$
K	Hydraulic conductivity	m/s	$[LT^{-1}]$
κ (kappa)	Thermal conductivity	W/m K	$[MLT^{-3}\theta^{-1}]$
L	Length or offset	m	[L]
M	Mass	kg	[M]
n	Refractive index/porosity exponent	–	Dimensionless
p	Pore pressure	Pa	$[ML^{-1}T^{-2}]$
q	Specific discharge/flux	m/s	$[LT^{-1}]$
r	Radial distance	m	[L]
R	Resistance	Ohm (Ω)	$[ML^2T^{-3}I^{-2}]$
S	Storage coefficient/saturation	– or %	Dimensionless
t	Time	s	[T]
u	Displacement vector	m	[L]

(continued)

(continued)

Symbol	Meaning	Unit	Dimension
V	Voltage or electric potential	Volt (V)	$[ML^2T^{-3}I^{-1}]$
W	Work or energy	Joule (J)	$[ML^2T^{-2}]$
x, y, z	Cartesian coordinates	m	[L]
Units are in SI; Dimensions use the base units: M = Mass, L = Length, T = Time, I = Electric Current (Dimensionless quantities have no units); θ (theta): Used to denote temperature dimension; You can optionally add Greek letters (e.g., η for viscosity, ψ for electric potential, χ for magnetic susceptibility) based on the chapters where they appear			

18.4 Key Mathematical Equations and Models

Description	Equation
Chapter 3	
Wave equation	$\frac{\partial^2 u}{\partial t^2} = v^2 \nabla^2 u$ where: u = displacement (m), v = wave velocity (m/s), ∇^2 = Laplacian operator, t = time (s) Explanation: Describes how seismic (elastic) waves propagate through the subsurface. Fundamental in modeling body waves (P-waves and S-waves) and surface waves
Hooke's Law	$\sigma = E \cdot \varepsilon$ where: stress (or conductivity in other contexts), E = Young's modulus, ε = strain Explanation: Defines the stress–strain relationship in linear elastic materials—crucial for understanding how seismic waves interact with different subsurface media
Ohm's Law for subsurface resistivity	$\rho = \frac{RA}{L}$ where: ρ = resistivity (Ωm), R = resistance (Ω), A = cross-sectional area (m^2), L = length between electrodes (m) Explanation: Foundation of electrical resistivity methods used in geophysical exploration and site assessment

(continued)

(continued)

Description	Equation
Darcy's Law (for groundwater flow modeling)	$Q = -KA\frac{dh}{dl}$ where: Q = volumetric flow rate (m^3/s), K = hydraulic conductivity (m/s), A = cross-sectional area (m^2), $\frac{dh}{dl}$ = hydraulic gradient Explanation: Important for understanding fluid flow in porous media; relevant in hydrogeophysics and environmental geophysics
Chapter 4	
Travel time equation (for seismic refraction and GPR)	$t = \frac{d}{v}$ where: t = travel time (s), d = distance travelled by wave (m), v = velocity of wave in the material (m/s) Explanation: Used to estimate how long it takes for a seismic or radar wave to reach a receiver. Important for locating layer boundaries and detecting anomalies underground
Apparent resistivity (from electrical resistivity surveys) Equation (Wenner array)	$\rho_a = 2\pi aR$ where: ρ_a = apparent resistivity (Ωm), a = electrode spacing (m), R = measured resistance (Ω) Explanation: This basic equation helps interpret subsurface layering based on how easily current flows through the ground. Different materials (clay, sand, rock) have different resistivities
Magnetic anomaly intensity (simplified form) Equation	$\Delta T = T_{measured} - T_{expected}$ where: ΔT = magnetic anomaly (nT), $T_{measured}$ = measured total field (nT), $T_{expected}$ = regional/normal field (nT) Explanation: Used in magnetic surveys to highlight buried objects (e.g., tanks, utilities) or geological features like dikes or faults
Basic EM induction (time-domain electromagnetics) Equation	$E = -\frac{dB}{dt}$ where: E = induced electric field (V/m), B = magnetic field (T), t = time (s) Explanation: Simplified Faraday's Law—used in time-domain EM to detect subsurface conductivity changes, e.g., groundwater or contamination zones

(continued)

(continued)

Description	Equation
Depth of investigation (rule of thumb) Equation (for resistivity surveys)	Depth $\approx \frac{a}{2}$ where: a = electrode spacing (m) Explanation: A quick estimation of how deep the survey can "see" based on how far apart electrodes are placed. Larger spacing = deeper penetration
Chapter 5	
Snell's Law (seismic refraction)	$\frac{\sin\theta_1}{v_1} = \frac{\sin\theta_2}{v_2}$ where: θ_1 = angle of incidence, θ_2 = angle of refraction, v_1, v_2 = seismic wave velocities in layers 1 and 2 Explanation: Determines the bending of seismic waves as they pass between layers with different velocities—foundational in subsurface layer mapping
Reflection coefficient (seismic boundaries)	$R = \frac{Z_2-Z_1}{Z_2+Z_1}$ where: Z_1, Z_2 = seismic impedances of upper and lower layers, $Z = \rho \cdot v$, where ρ is density and v is velocity Explanation: Used to estimate how much seismic energy reflects off a boundary—important in interpreting reflection seismic data
Seismic impedance	$Z = \rho \cdot v$ where: ρ = density of the material, v = velocity of seismic wave in that material Explanation: Seismic impedance controls reflection strength—key for understanding contrasts between subsurface layers
Chapter 6	
Electrical conductivity	$\sigma = \frac{1}{\rho}$ where: σ = electrical conductivity, ρ = electrical resistivity Explanation: Used to identify zones of contamination, salinity, or water saturation—important in environmental assessments
GPR time-to-depth conversion	$d = \frac{vt}{2}$ where: d = depth to target, v = wave velocity in the material, t = two-way travel time Explanation: Converts radar signal time to actual depth for mapping buried utilities, structures, or contaminants
Chapter 7	

(continued)

(continued)

Description	Equation
Seismic moment	$M_0 = \mu AD$ where: M_0 = seismic moment, μ = shear modulus of the rock, A = area of fault rupture, D = displacement along the fault Explanation: Quantifies the energy released during an earthquake—key for assessing seismic hazard
Richter magnitude (simplified)	$M_L = \log_{10}(A) + \text{correction}$ where: A = amplitude of seismic wave, "correction" = function of distance and site conditions Explanation: Estimates earthquake magnitude from seismograph readings
Liquefaction potential index (conceptual)	$LPI = \int_0^z F(z) \cdot W(z)\,dz$ where: F(z) = factor indicating severity of liquefaction at depth z, W(z) = weighting function for depth Explanation: Used to evaluate potential for soil liquefaction during earthquakes—important in site safety
Chapter 8	
Elastic modulus from seismic velocity	$E = \rho V_p^2 \cdot \frac{(1+\upsilon)(1-2\upsilon)}{(1-\upsilon)}$ where: E = Young's modulus, ρ = density, V_p = P-wave velocity, υ = Poisson's ratio Explanation: Links dynamic geophysical properties to mechanical stiffness of soils/rocks—critical for foundation design
Poisson's ratio (from velocities)	$\upsilon = \frac{V_p^2 - 2V_s^2}{2\left(V_p^2 - V_s^2\right)}$ where: V_p = compressional wave velocity, V_s = shear wave velocity Explanation: Helps characterize elastic properties of materials—important for dynamic response modeling
Chapter 9	
Gravitational anomaly (simple form)	$\Delta g = G\frac{m}{r^2}$ where: Δg = gravity change due to mass, G = gravitational constant, m = buried mass, r = distance from mass Explanation: Used to detect ore bodies or cavities underground based on mass differences

(continued)

(continued)

Description	Equation
Magnetic susceptibility effect	$B = \mu H$ where: B = magnetic field, μ = permeability, H = applied magnetic field Explanation: Measures how subsurface materials respond to magnetic fields—useful in mineral exploration
Induced polarization (IP) decay curve	$V(t) = V_0 e^{-\frac{t}{\tau}}$ where: V(t) = decaying voltage, V_0 = initial voltage, t = time, τ = decay constant Explanation: Characterizes chargeability of materials—useful for detecting disseminated mineral deposits
Chapter 10	
Water table estimation	$z = \frac{vt}{2}$ (Same as GPR time-to-depth) Explanation: Used in groundwater mapping to locate the depth of the water table
Electrical conductivity for soil health	$EC = \frac{1}{\rho}$ Explanation: Indicates salinity and fertility of soil—essential in agricultural geophysics and land management
Chapter 11	
Shear modulus (G)	$G = \rho V_s^2$ where: G = shear modulus, ρ = density, V_s = shear wave velocity Explanation: Assesses how resistant soil/rock is to shearing—crucial for seismic site response analysis
Damping ratio	$\zeta = \frac{c}{2\sqrt{km}}$ where: ζ = damping ratio, c = damping coefficient, k = stiffness, m = mass Explanation: Indicates how quickly vibrations diminish—important in structural response under seismic loads
Chapter 12	
Anomaly detection (magnetic/EM)	$\Delta T = T_{measured} - T_{background}$ Explanation: Simple calculation to highlight buried objects or infrastructure based on magnetic differences

(continued)

(continued)

Description	Equation
Radar range equation (simplified)	$R = \sqrt{\frac{P_t G^2 \lambda^2 \sigma}{(4\pi)^3 P_r}}$ where: R = radar range, P_t = transmitted power, G = antenna gain, λ = wavelength, σ = radar cross-section, P_r = received power Explanation: Basic radar detection principle—useful for security-related subsurface imaging
Chapter 13	
Thermal resistivity	$R_t = \frac{L}{kA}$ where: R_t = thermal resistance, L = thickness, k = thermal conductivity, A = area Explanation: Evaluates insulation efficiency—relevant for buried cables or geothermal studies
Fourier's Law (heat flow)	$q = -k\frac{dT}{dx}$ where: q = heat flux, k = thermal conductivity, $\frac{dT}{dx}$ = temperature gradient Explanation: Used to model subsurface heat movement—essential in energy installations and geothermal engineering
Chapter 14	
GPR travel time (simplified)	$t = \frac{2d}{v}$ where: d = depth, v = wave speed, t = round-trip travel time Explanation: Helps locate buried walls, artifacts, or tombs using radar
Magnetic dipole anomaly	$B = \frac{\mu_0}{4\pi} \cdot \frac{m}{r^3}$ where: B = magnetic field intensity, μ_0 = magnetic constant, m = magnetic moment, r = distance from dipole Explanation: Used to detect ferrous buried artifacts or structural remains
Chapter 15	
EM for agriculture	$EC = \frac{1}{\rho}$ Explanation: Monitors moisture and salinity—essential for precision farming
Soil moisture (capacitance sensor model)	$\theta = \frac{C}{C_{sat}}$ where: θ = volumetric water content, C = measured capacitance, C_{sat} = capacitance when saturated Explanation: Assesses how much water is held in the soil—interdisciplinary use in hydrology and agriculture

(continued)

(continued)

Description	Equation
Chapter 16	
Model accuracy in machine learning	$\text{Accuracy} = \frac{\text{TP+TN}}{\text{TP+TN+FP+FN}}$ where: TP = true positives, TN = true negatives, FP = false positives, FN = false negatives Explanation: Helps validate AI-driven interpretations in geophysical data analysis
Inversion objective function (simplified)	$\Phi = \|d_{obs} - d_{calc}\|^2 + \lambda\|m\|$ where: d_{obs} = observed data, d_{calc} = calculated data, m = model parameters, λ = regularization factor Explanation: Used in geophysical inversion to balance data fitting and model smoothness

Notes Some symbols are used in multiple contexts (e.g., ρ, σ, m), so they should be defined per context in the main text

18.5 Data and Measurement Units Reference

18.5.1 SI and CGS Unit Conversions (Select Parameters)

Quantity	SI unit	CGS unit	Conversion
Length	meter (m)	centimeter (cm)	1 m = 100 cm
Mass	kilogram (kg)	gram (g)	1 kg = 1000 g
Time	second (s)	second (s)	Same
Velocity	m/s	cm/s	1 m/s = 100 cm/s
Acceleration	m/s^2	cm/s^2	1 m/s^2 = 100 cm/s^2
Force	Newton (N)	dyne	1 N = 10^5 dyne
Pressure/stress	Pascal (Pa)	Barye (Ba)	1 Pa = 10 Ba
Density	kg/m^3	g/cm^3	1 kg/m^3 = 0.001 g/cm^3
Energy	Joule (J)	erg	1 J = 10^7 erg
Power	Watt (W)	erg/s	1 W = 10^7 erg/s
Magnetic field (H)	A/m	Oersted (Oe)	1 A/m = 0.01257 Oe
Magnetic flux density	Tesla (T)	Gauss (G)	1 T = 10^4 G
Electrical resistivity	Ohm-m	Ohm-cm	1 Ohm-m = 100 Ohm-cm

18.5.2 Common Measurement Scales and Units

Frequency

- Unit: Hertz (Hz)
- Definition: 1 Hz = 1 cycle per second
- Typical Range in Geophysics:
  - Seismic: 1–100 Hz
  - GPR: 10 MHz–2 GHz.

Attenuation

- Unit: decibel (dB)
- Definition:

$$\text{Attenuation (dB)} = 10\log_{10}\left(\frac{P_{in}}{P_{out}}\right)$$

- Used for: Signal loss in seismic, EM, or GPR propagation.

Amplitude Ratios

- Often expressed in dB for comparisons
- Example: − 6 dB means power is reduced to ¼.

Electrical Conductivity

- Unit: Siemens per meter (S/m)
- Alternative: milliSiemens per centimeter (mS/cm) in water/soil analysis.

Magnetic Susceptibility

- SI Unit: dimensionless (volume susceptibility)
- Alternative CGS: emu/cm^3 (older texts).

18.5.3 Temperature and Heat Flow

Quantity	Unit	Notes
Temperature	°C, K	0 °C = 273.15 K
Thermal conductivity	W/m K	Used in Fourier's Law
Heat flow	W/m^2	Earth heat flux ranges: 40–100 mW/m^2

18.5.4 Seismic and Wave Units

Parameter	Unit	Comment
Seismic velocity	m/s	V_p, V_s
Acoustic impedance	kg/m^2 s	$Z = \rho \cdot v$
Wavelength	m	$\lambda = \frac{v}{f}$
Frequency	Hz	As above

18.6 Further Reading and Online Resources (Optional)

18.6.1 Core Textbooks on Engineering Geophysics

Title	Author(s)	Publisher	Year
Applied Geophysics	W. M. Telford et al.	Cambridge University Press	1990
Near-Surface Geophysics	D. K. Butler (Ed.)	SEG	2005
Geotechnical and Environmental Geophysics (Vols. I–III)	S. H. Ward	SEG	1990
Introduction to Applied Geophysics	H. Robert Burger et al.	W.W. Norton	2006
Environmental and Engineering Geophysics	P. Kearey et al.	Wiley-Blackwell	2002

18.6.2 Journals for Advanced Study and Research

Journal	Publisher	Focus
Geophysics	Society of Exploration Geophysicists (SEG)	Core applied geophysics research
Journal of Environmental and Engineering Geophysics (JEEG)	EEGS	Near-surface and site characterization
Engineering Geology	Elsevier	Geotechnical integration
Geophysical Prospecting	EAGE	Exploration and inversion modeling

(continued)

(continued)

Journal	Publisher	Focus
Near Surface Geophysics	EAGE	GPR, EM, shallow seismic, and more
Seismological Research Letters	SSA	Earthquake and hazard monitoring

18.6.3 Online Resources and Tools

Educational Platforms

- OpenGeophysics.org—Open-source tools and tutorials
- SEG Wiki—Methods, models, and case histories
- IRIS Education—Seismic learning tools and visualizations
- Geophysics for Practitioners (YouTube)—Video explainers and software demos.

Interactive Tools and Simulators

- GeoGebra Interactive Seismic Models—For Snell's Law, wavefronts, etc.
- EMGeo (EM modeling simulator)—Open-source electromagnetic inversion tool
- SeisLab—MATLAB Tools for Seismic Analysis
- Colab Geophysics Notebooks—Inversion, GPR, resistivity.

18.6.4 Open Datasets and Field Examples

Dataset	Host	Description
USGS Geophysical Data	USGS.gov	Magnetics, gravity, seismic
NGU Geophysical Maps	geo.ngu.no	Norway's national geophysical surveys
SEG Wiki Case Studies	wiki.seg.org	Real-world industry examples
IRIS Seismic Data Browser	iris.edu	Live earthquake data and stations
Geoscience Australia Datasets	ga.gov.au	Gravity, EM, magnetic, GPR

18.6.5 Suggested Reading by Chapter

Chapter	Suggested resource
Subsurface Exploration	Telford et al., Applied Geophysics
Environmental Monitoring	Butler, Near-Surface Geophysics
Disaster Risk and Hazards	IRIS/SSA Earthquake Education Resources
Civil Infrastructure	Engineering Geology Journal
Machine Learning	Colab Notebooks on Inversion and Classification

Book Synopsis: Advances in Engineering Geophysics: Concepts and Applications

Dr. Dulal Chandra Naskar

This book presents a comprehensive and interdisciplinary exploration of the role of geophysics in engineering applications, focusing on infrastructure development, disaster risk mitigation, environmental assessment, and emerging technologies. Authored by a seasoned expert with over 35 years of field, academic, and policy-level experience, this work bridges classical geophysical principles with cutting-edge innovations to address 21st-century challenges.

Structured across 17 detailed chapters, the book progresses from foundational methods to complex, real-world case studies and future-forward perspectives. It opens with a strong theoretical and methodological base, covering seismic, electrical, electromagnetic, and gravimetric techniques, before advancing to applications in geo-engineering, hydrology, critical infrastructure, energy security, defense, archaeology, and smart cities. Particular emphasis is placed on risk-resilient infrastructure planning, subsurface characterization, geohazard detection, and digital integration through technologies like AI, digital twins, and IoT-based monitoring.

Special chapters are devoted to geophysics in defense installations, resilience of urban and transportation networks, climate adaptation, and non-invasive heritage preservation—making this volume uniquely holistic. More than 60 regionally and globally significant case studies from India, Japan, France, Italy, Greece, and the U.S. reinforce practical application. The book culminates in discussions on quantum sensing, robotics, and autonomous platforms, offering a vision for the next frontier in geophysical sciences.

Dr. D. C. Naskar
M.Sc. (Physics), Ph.D. (Science), Former Director (Geophysics), Geological Survey of India (GSI), Kolkata, India

Chief Consultant in Geophysics, DMT Pvt. Solution Ltd., Kolkata, India

Senior Consultant, National Centre for Seismology (NCS), New Delhi, India

Ministry of Earth Sciences (MoES), Government of India, Kolkata, India

D. C. Naskar, *Advances in Engineering Geophysics: Concepts and Applications*, Earth Systems Data and Models 9, https://doi.org/10.1007/978-3-032-28000-8

With its seamless integration of theoretical depth, methodological rigor, and field-tested insights, this volume stands as a key reference for geoscientists, civil engineers, urban planners, disaster risk professionals, policy-makers, and graduate-level students. The work also addresses the ethical, regulatory, and collaborative dimensions of geophysics in today's evolving socio-technical landscape.

As climate change, urbanization, and infrastructural complexities escalate, *Advances in Engineering Geophysics* provides the critical knowledge needed to design, protect, and sustain systems that support modern civilization. The book not only consolidates past advances but also illuminates a strategic roadmap for the future of applied geophysics in global development and risk governance.

The manufacturer's authorised representative in the EU is Springer Nature Customer Service Centre GmbH, Europaplatz 3, 69115 Heidelberg, Germany. If you have any concerns regarding our products, please contact ProductSafety@springernature.com

Printed and bound by CPI Group (UK) Ltd, Croydon, CR0 4YY
15/06/2026
02138333-0002